Thirty Great Inventions of China

Jueming Hua · Lisheng Feng
Editors

Thirty Great Inventions of China

From Millet Agriculture to Artemisinin

大象出版社

Springer

Editors
Jueming Hua
Institute for the History of Natural
Science of Chinese Academy of Sciences
Beijing, China

Lisheng Feng
Tsinghua University
Beijing, China

经典中国国际出版工程
China Classics International

ISBN 978-981-15-6524-3 ISBN 978-981-15-6525-0 (eBook)
https://doi.org/10.1007/978-981-15-6525-0

Jointly published with Elephant Press Co., Ltd
The print edition is not for sale in China (Mainland). Customers from China (Mainland) please order the print book from: Elephant Press Co., Ltd.
ISBN of the Co-Publisher's edition: 978-7-5342-8625-9

This Springer imprint is published by the registered company Springer Nature Singapore Pte Ltd.
The registered company address is: 152 Beach Road, #21-01/04 Gateway East, Singapore 189721, Singapore

Preface

The book is written out of the pursuit of cultural self-confidence. Culture is the gene on which one nation or country depends for its survival, development, and maintenance of its own characteristics, so whether one nation or country holds the consciousness and idea of cultural self-confidence or not functions as the spiritual premise of the continuous cultural inheritance and revitalization of the nation or country itself. Similarly, any invention or creation has to hold the same consciousness and idea of cultural self-confidence, which is embodied in the following two aspects in terms of cultural independence and self-discipline:

First, on the basis of the researches of Chinese science and technology in the last 100 year, further exploration, evaluation, and sort out have to make so as to select the major inventions with originality and distinctive characteristics that have produced outstanding contributions to an extensive influence over the progress of civilization in China, and even the world. For each of these inventions, we have to discuss its occurrence, development, and relations with social humanities, economy, politics, and people's daily life, and compare it with those similar inventions made in other regions of the world, so as to improve our research and argumentation on major inventions. Just as what the article *Introduction of the Thirty Great Inventions of China* says, "Chinese people have the particular responsibility to make these inventions and creations clear to the world".

Second, the research and evaluation of inventions is a serious matter. The scientific spirit of seeking truth from facts is the criterion that we should strictly abide by in our research and evaluation. It is a manifestation of nihilism to ignore the achievements of the nation's own invention and creation, as well as to ignore the wisdom and innovative spirit of our forefathers and posterities. On the contrary, it is a chauvinistic attitude to exaggerate or even fabricate the nation's own invention and creation, as well as to ignore the creativity and ingenuity of other nations or countries. Neither is desirable. Respecting objective facts, abiding by academic norms, striving to be rigorous and accurate in research, sort out, evaluation and elucidation, all are the principles we should follow when dealing with the problems of invention and creation. Cultural self-discipline is a kind of self-restraint and also a guarantee that our research and exposition can stand any test.

This book includes 32 chapters, 31 of which talk about the 30 great inventions made in China, from ancient to contemporary times: millet agriculture, rice farming, sericulture and silk production, Chinese characters, algorism and rod calculus, bronze metallurgy, pig iron metallurgy, canal and navigation lock, plow, grain sowing implement, waterwheel, traditional technique of lacquer, papermaking, traditional Chinese medicine diagnosis and treatment (including acne vaccination), porcelain, the technique of Chinese wooden structure, Chinese cooking, carriage hitching technique and stirrup, printing, tea planting and making, the invention and use of celestial equatorial measurements, bulkhead, gunpowder, compass, the technique of ultra deep drilling, cultivation and eco-agriculture, abacus, ferment of distiller's yeast, fire arrow and projection firearm, artemisinin, hybrid rice, etc. The preface *Introduction of the Thirty Great Inventions of China* concerns the problems proposed, the definition, classification and evaluation of invention, the origin of the 30 great inventions, the enlightenments obtained, and the other 30 important inventions enumerated for follow-up researches to refer to.

The whole book includes more than half a million Chinese characters, as well as nearly 600 diagrams. The 30 great inventions described in the book relate to such disciplines as mathematics, physics, chemistry, astronomy, geology, biology, agriculture, medicine, philology, textile, mining and metallurgy, water conservancy, machinery, traditional technique of lacquer, papermaking, porcelain making, shipbuilding, architecture, carriage hitching technique and stirrup, printing, instrumentation, gunpowder and firearms, ultra deep drilling, microbial engineering, and so on. The book was written by 32 scholars in the above fields, among whom there are 24 professors or researchers, 6 associate professors or associate researchers, 1 lecturer or assistant researcher, and 1 postdoctoral fellow. 11 of them are gray-headed scholars, 14 middle-aged scholars, and 7 young scholars.

Major inventions have always been paid much attention by the academia, as well as the public at home and abroad, and are of especial practical significance nowadays when an innovation-oriented social construction is being advocated. This book is just the first attempt in this field, and we would like to ask the academic community and all the readers of the book to oblige us with their valuable comments. It is also expected that more experts will participate in the discussion so as to further improve this research.

Last but not least, we would like to thank the China Publication & Promotion Association, Elephant Press for their positive and fruitful work on the publication of this book.

Beijing, China
December 2016

Jueming Hua
Lisheng Feng

Contents

Editors and Contributors

About the Editors

Jueming Hua Born in Wuxi, Jiangsu, in 1933, graduated from the Department of Mechanical Engineering of Tsinghua University in 1958. In 1967, he got his masters degree in the Institute of Natural Science History of the Chinese Academy of Sciences. In 1986, he served as a researcher at the Institute of Natural Science History of the Chinese Academy of Sciences, served as deputy director from 1988 to 1993, and retired in the same year. From 1993 to 2003, he served as the director of the Institute of Science History and Ancient Literature, Tsinghua University.

Research Interests: History of technology; the philosophy of technology; traditional crafts.

Major Works: A Collection of Papers on History of Chinese Metallurgy, *Ancient Chinese Metal Technology: Civilization Made by Copper and Iron, Complete Works of Chinese Traditional Crafts, Chinese Handcrafts and Collections of Hua Jueming's Works.*

Awards: The first prize of scientific and technological achievements granted by the Ministry of Culture;

the first prize of the National excellent publications;

the Lifetime Achievement Award for the Foundry industry.

Lisheng Feng Born in Wuhai, Inner Mongolia, in 1962, graduated from Inner Mongolia Institute of Technology with a bachelor's degree in engineering in 1983; then he got a Master of Science degree in 1987, from the Institute of History of Science, Inner Mongolia Normal University. In 1999, he got his doctoral degree in science from the Department of Mathematics, Northwestern University.

He was a professor at Inner Mongolia Normal University and currently the director of the Institute of Science and Technology History and Ancient Literature of Tsinghua University, and professor and doctoral supervisor of the Institute of Science and Technology History of Inner Mongolia Normal University. He has been a visiting scholar and visiting professor at Japan's Tohoku University, Tokyo University, and Central University. In 2004, he was selected into the "New Century Outstanding Talent Support Program" by the Ministry of Education. He also serves as a member of the National Intangible Cultural Heritage Protection Committee, executive director of the Chinese Academy of Science and Technology History, president of the Chinese Traditional Arts and Crafts Research Association, and vice-chairman of the National Mathematical History Society.

Research Interests: History of mathematics; History of Technology; Scientific and technological literature; Protection of Science and Technology Cultural Heritage.

Major Works: History of Ancient Chinese metrology, History of China-Japan Mathematical Relations.

Awards: The First Prize of Scientific Research Achievements of the Inner Mongolia Normal University;

Co-editing Notes of Chou Ren Records won the first prize of the National Outstanding Ancient Books Award;

the *Encyclopedia of Color Map Science and Technology* that he participated in the compilation won the second prize of the 2009 National Science and Technology Progress Award;

History of Chinese Science and Technology: Mathematics won the first prize of the Fourth Guo Moruo History Award.

Contributors

Peijun An North China University of Technology, Beijing, China

Bei Chang Jiangsu Research Institute of Culture and History, Beijing, China

Wei Chen The Institute for History of Natural Sciences, Chinese Academy of Sciences, Beijing, China

Nianzu Dai The Institute for History of Natural Sciences, Chinese Academy of Sciences, Beijing, China

Lisheng Feng Institute for History of Science and Technology & Ancient Texts, Tsinghua University, Tsinghua, China

Xiaowu Guan The Institute for History of Natural Sciences, Chinese Academy of Sciences, Beijing, China

Daiheng Guo Tsinghua University, Beijing, China

Shuchun Guo The Institute for the History of Natural Sciences, Chinese Academy of Sciences, Beijing, China

Qi Han The Institute for History of Natural Sciences, Chinese Academy of Sciences, Beijing, China

Jueming Hua The Insititute for History of Natural Sciences, CAS, Beijing, China

Xing Huang The Institute for History of Natural Sciences, Chinese Academy of Sciences, Beijing, China

Hui Liu Institute for the Natural Science History, Chinese Academy of Sciences, Beijing, China

Benshan Lu Guangdong Beauty and Cosmetic Museum, Guangzhou, China

Zongdian Min China Agricultural Museum, Beijing, China

Yahua Niu Institute of Information on Traditional Chinese Medicine, China Academy of Chinese Medical Sciences, Beijing, China

Jixing Pan The Institute for History of Natural Sciences, Chinese Academy of Sciences, Beijing, China

Yu Peng School of History Culture and Ethnology, Southwest University, Chongqing, China

Pangtong Qiu Expert Guidance Committee of China Cuisine Association, Beijing, China

Yunli Shi University of Science and Technology of China, Hefei, China

Rongyu Su The Institute for History of Natural Sciences, Chinese Academy of Sciences, Beijing, China

Longfei Xi China Maritime History Studies Association, Beijing, China

Yeyun Xin China National Hybrid Rice R&D Center, Changsha, China

Yongshan Yang Tsinghua University, Beijing, China

Zhanhong You Institute for History of Science and Technology & Ancient Texts, Tsinghua University, Beijing, China

Xiongsheng Zeng The Institute of the History of Natural Sicences, Chinese Academy of Sciences, Beijing, China

Baichun Zhang The Institute for History of Natural Sciences, Chinese Academy of Sciences, Beijing, China

Xiumin Zhang The Institute for History of Natural Sciences, Chinese Academy of Sciences, Beijing, China

Weibing Zhang Institute of Water Resources and Hydropower Research, Beijing, China

Feng Zhao China National Silk Museum, Hangzhou, China

Jiahua Zhou The Institute for History of Natural Sciences, Chinese Academy of Sciences, Beijing, China

Kuiyi Zhou The Society for Research on Hydraulic Engineering History, Chinese Hydraulic Engineering Society, Beijing, China

About the Translators and the Division of Translation

Wei Chen: Ph.D., English professor of the School of Translation Studies in Jinan University, who independently translated the following sections and chapters of the whole book: Preface; Canal and Navigation Lock; Plows and Seed Plows; Water Wheel; Traditional Technique of Lacquer; Papermaking; Traditional Chinese Medicine Diagnosis and Treatment; Intensive Cultivation and Eco-agriculture; Inoculation of Traditional Chinese Medicine Diagnosis and Treatment; The Invention and Civilization (as Postscript).

Aiping Mo: Ph.D., English professor of the National Demonstration Center for Experimental Simultaneous Interpretation Education in Guangdong University of Foreign Studies (GDUFS), who was in charge of the translation of the rest of chapters. He would like to express his thanks to **WANG Guoju** (Lecturer at Guangzhou Institute of Technology), **DENG Xianyi** (Associate Professor at Guangxi Normal University for Nationalities), **ZHAO Hui** (Lecturer at Guangdong Vocational Colleges of Science and Trade), **LIU Xianghui** (Lecturer at Gannan Normal University), **CHEN Yulian** (Associate Professor at South China Business College, GDUFS), **DENG Laiying** (Lecturer at Guangzhou City Polytechnic), **MO Juan** (Assistant at Center for Network and Informatization, GDUFS), as well as some of his MA students (**ZHAO Shengnan**, **CHEN Zhuohui**, **PAN Xiaoyan**, **SHI Xiao'an**, and **ZHEN Peijue**) and MTI students (**MA Delun**, **ZHOUZhou, YU Yixuan, XIAO Chunhui, TANG Zijian, YU Xiaolin**) from the School of Interpreting and Translation Studies in GDUFS, for their initial translation of his chapters or support and help.

The Invention and Civilization (As Postscript)

Jueming Hua and Lisheng Feng

The role and position of inventions, especially great inventions in human history is the topic that has been paid wide attention to and needs to be further clarified. Related to this topic, such issues as the value orientation of inventors, the incentive mechanism of invention and creation, etc., remain to be discussed. We hereby attempt to make a brief discussion about them.

First, the Invention as the Driving Force for the Progress of Human Civilization

Throughout the evolutionary history of human civilization, people's daily life and customs, as well as the social economy, politics, military, culture, and arts were all related to the invention and discovery in science and technology. The practice of drilling wood to make fire is undoubtedly the first great invention ever made by mankind. From knowing how to use fire two or three million years ago to learn how to preserve fire about half a million years ago, then to inventing the technology of artificially making fire, ancient human beings had experienced hundreds of thousands of years of exploration, failure, and final success, the difficulties involved in which are obviously beyond the imagination of contemporary people. Drilling wood to make fire, that is, to generate heat and fire by friction, was a successful application of the principles of physics, and also the first domination of natural force by mankind. Fire brought warmth and light to the living people then, strengthened their ability of fighting against wild animals and natural disasters, and meanwhile created conditions for the agricultural revolution which was characterized by the "slash-and-burn" farming method. The cooking with fire provided people with cooked food so as to strengthen their physiques. With fire, it became easier for our ancestors to overcome the restrictions of climate and thus expand their existence. With the widespread use of fire, all kinds of inventions and creations were achieved largely, such as Emperor Shennong making pottery, Su Sha producing salt, Nvwa refining stone, Yellow Emperor making arrow, which finally made human beings transit from the uncivilized to the civilized. And, we certainly know very well the following events concerning the great inventions ever made: the invention of steam engine led mankind into the industrial society; the use of electricity brought us into the era of electricity; the invention of nuclear energy brought mankind into the nuclear age; and the invention of the internet introduced us into the information age. As a result, people's daily life and customs, as well as the social economy, politics, military, culture, and arts have experienced great changes: from the farming civilization to the industrialization society, and then to the informational age, from the theocracy to the monarchy and then to the democracy, from the theology to the metaphysics and then to the science. All these achievements depended on the driving force of the progress of human civilization: the invention. Then, why has the invention in science and technology become the driving force of the progress of human civilization? We argue that human beings, the wisest of all creatures, have to deal well with the relationship between human beings and that between man and nature, which are interdependent, and reinforce and neutralize each other. The invention in science and technology always aims to improve the relationship between man and nature. If the relationship between man and nature has changed, the relationship between human beings will accordingly change. If the relationship between man and nature is improved, the relationship between human beings may also be improved, so human civilization will tend to move forward. Otherwise, it will drawback.

The invention and creation as a kind of innovation include such fields of invention and creation in science and technology, in a social system such as the metrology, the civil service, and the democracy, in literature and art such as the rhythm of poetry, ink painting and oil painting and the orchestra, in sports entertainment such as football, acrobatics, poker, and mahjong, and so on. This book only deals with the field of invention and creation in science and technology.

Second, the Incentive Mechanism of Invention and Creation

In examining the above great inventions made in Chinese history, we can find that the creativity of Chinese people experienced a fluctuant performance in different historical periods.

In general, as shown in the figure below, there appeared 1 great invention in an average of 1,000 years during the middle and late Neolithic times; 1 great invention every 450 years in the Xia, Shang, and Zhou dynasties; 1 great invention every 45 years in the Han Dynasty, which was at the peak of the creativity; 1 great invention in more than 370 years in the Wei, Jin, Southern and Northern dynasties, which was at the bottom of the creativity; 1 great invention every 75 years in the Sui, Tang, and the Five dynasties; 1 great invention every 65 years in the Song and Yuan dynasties, which was at the peak of the science and technology of ancient China; 1 great invention in more than 270 years in the Ming Dynasty whose old and decadent political and economic system began to decline; and even no great invention in the reign of 268 years of the Qing Dynasty whose autocratic rule of imperial power was on the verge of collapse. In 1911 when the imperial system was overthrown and the Republic was founded, the ancient China started a new journey of cultural recovery and national rejuvenation. During the later one hundred years, two major inventions were made. This is the vertical comparison of Chinese creativity. If we make a horizontal comparison, we can find that during the following 500 years since 1609, when Galileo invented the twentyfold telescope, there were hundreds of major inventions made in the whole world, among which there were only two made in China, and they were made after the founding of the People's Republic of China. This is a historical fact that we Chinese people have to face up to, which deserves our serious consideration.

The examples of the major inventions made all over the world at all times tell us that the innovation always originates from the development of people's wisdom, the liberation of the mind, the pursuit of interest, and the encouragement of utilitarian gain. All these, however, must be premised on certain social conditions, that is, fair and clear politics, prosperous economy, sound and effective education, well-developed science, free thoughts, and independent academic activities. So long as the innovation is encouraged and the rights and interests of innovators are protected, the invention and creation can be achieved endlessly. The incentive mechanism of invention and creation is just composed of the combination and interaction of the above factors.

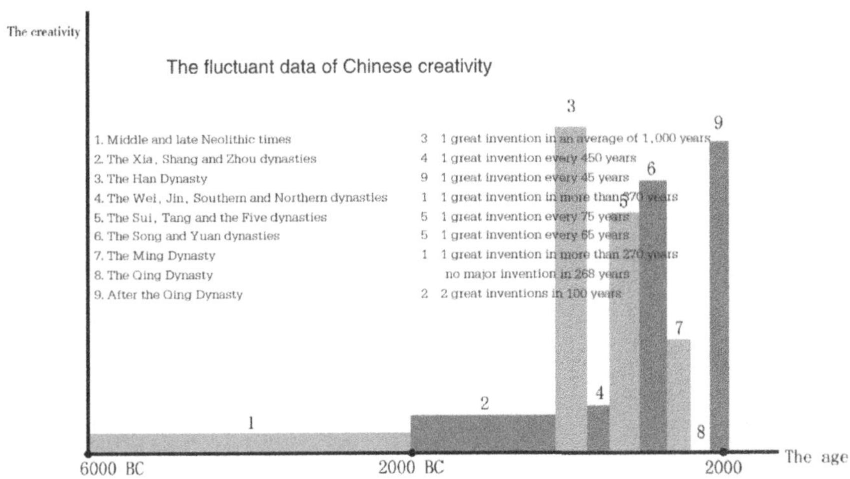

The fluctuant data of Chinese creativity

Third, the Invention and Creation as the Aspiration, Mission, and Undertaking of Generations of Chinese

The period between the late Qing Dynasty and the Republic of China was the transitional stage of China from the autocratic society to the democratic society, so the lack of major inventions at that time could be explained. The successes of Yuan Longping in planting hybrid rice stem from the urgent need for increasing grain production and Yuan's own concern for people's livelihood, innovative consciousness, and dedication, as well as a current advanced social system and good social environment. Of course, even in relatively backward developing countries, it is still possible to produce any great original inventions. In this regard, we should have full understanding and confidence.

In ancient mythology of China, there are such great inventors as Emperor Shennong, Cang Jie, Yellow Emperor, Lei Zu, and so on, and in historical records, there are such real famous inventors as Lu Ban, Bian Que, Ou Ye, Cai Lun, Bi Sheng, etc. In the vernacular movement of modern China, Liu Bannong created a character "她" (she), which was soon adopted by Chinese people as the creation was reasonable and applicable. It was actually a continuation of the work initiated by Cang Jie who created the Chinese writing system. Similarly, Wang Xuan can be said to be the heir of Bi Sheng who invented the movable-type printing in the 1040's, Yuan Longping and Tu Youyou to be the successors of Emperor Shennong who was the legendary inventor of agriculture and medicine in ancient China. Lu Ban, Bian Que, Ou Ye, and Cai Lun also have their own successors.

The innovation can lead to civilization. The invention and creation has been the aspiration, mission, and undertaking of generations of Chinese. With the development and progress of the society, it is expected that more inventions and creations will be produced in China, so as to make greater contributions to the process of human civilization.

Chapter 1
Introduction of the Thirty Great Inventions of China

Jueming Hua

1.1 Question: How Many Great Inventions Are There in Chinese History?

The Four Great Inventions of China, paper-making, printing technique, gunpowder, and compass are known to all. However, the public doesn't really understand the great inventions in Chinese history.

In 1550, Italian mathematician Jerome Cardan regarded the compass, printing technique, and gunpowder as three great inventions of China, commenting "the whole of antiquity has nothing equal to show" [1]. Later Jean Bodin also restated the status of these three [2].

In his *Novum Organum*, Francis Bacon put forward further in 1620, that "we should notice the force, effect, and consequences of inventions, which are nowhere more conspicuous than in those three unknown to the ancients, namely, printing, gunpowder, and the compass. For these three have changed the appearance and state of the whole world: first in literature, then in warfare, and lastly in navigation; and innumerable changes have thence been derived, so that no empire, sect, or star, appears to have exercised a greater power and influence on human affairs than these mechanical discoveries."

First published in 1585, and finally completed in 1638, Johannes Stradanus's *Nova Reperta* presents an illustration of the ranking of nine great discoveries and great inventions. They are the discovery of American Continent, compass, gunpowder, printing machine, lignum vitae, distillation, silk, and stirrup, among which China's Great Three rank the top place [2].

At the same time, Karl Marx pointed out in *The Application of Machinery, Natural Force and Science* "Gunpowder, the compass, and the printing press were the three great inventions which ushered in bourgeois society. Gunpowder blew up the knightly

---

J. Hua (✉)
The Insititute for History of Natural Sciences, CAS, Beijing, China
e-mail: huajueming@vip.sina.com

J. Hua and L. Feng (eds.), *Thirty Great Inventions of China*,
https://doi.org/10.1007/978-981-15-6525-0_1

class, the compass discovered the world market and found the colonies, and the printing press was the instrument of Protestantism and the regeneration of science in general; the most powerful lever for creating the intellectual prerequisites" [3].

Paper has a close relationship with the printing technique for the reason that the former is the carrier of the latter. Adding the fourth great invention seems to be a matter of course in this sense. The concept of "the Four Great Inventions" emerged across China, in the 1920s [4], at the latest and was widely accepted in the 1940s or 1950s [5].

In addition, in his speech in October, 1946, at the UNESCO conference at Paris, Joseph Needham mentioned "Gunpowder, the magnetic compass, and paper and printing were the three most important inventions invented in China". Although he followed the suit of his predecessors, the accession of paper-making technique shaped the new concept of "the four great inventions". At the same time, he believed "without gunpowder, paper, printing, and magnetic compass, the eradication of feudalism in Europe would be unimaginable" [5, 6].

Referring to the above evidences, the idea of China's three great inventions was formulated, in the first place, by Western scholars, reflecting those inventions' influences on human civilization, especially the modern Western society. And later it evolved into "the Four Great Inventions of China", which has both classical significance and specific meaning in that particular context. A failure to learn about the cause may lead to a misunderstanding that these four inventions are the most important, top-ranking inventions in Chinese history, which actually is not appropriate. And this misconception is not the original intention of scholars such as Francis Bacon and Joseph Needham. Over the years, the commonly suggested "the fifth great invention of China" was the exact result of that illusion. In fact, China's major inventions and their rankings in history have never been studied in research of scientific and technological history neither at home nor abroad, let alone a definitive conclusion.

Hence, a question may be raised—how many great inventions are there in Chinese history?

1.2 Definitions: What Can Be Counted as a Great Invention?

In this part, the evaluation criteria of inventions, especially great inventions, will be discussed.

The definition of "invention" in *Cihai*, an authoritative Chinese dictionary in China, reads "creating new things or to innovate new methods of production" [7], in which "new" is twofold. For example, the automobile is undoubtedly an invention, while manufacturing a brand new car also fits into the concept of "new" in the definition above. Hence, using "creating new things" to define "invention" could not be further from rigorousness, and it is ambiguous and distracting. The other definition "to innovate new methods of production" seems to be in accordance with what an invention is. Trying to examine the instance of telescope is not much of a method of production but an observational measure. So the definition made in *Cihai* is uncertain and incomplete, which is not sufficient enough to define all inventions.

Here are some definitions of "invention" by the author.

- All inventions are original, and those without originality shall not be called inventions.
- Inventing is accidental or unintentional, purposeful or scheduled. The common denominator of all inventions is the possession of the function of recognizing, adapting, and changing nature, society, and human beings.
- Inventions fall within the scope of means and methods. The invention can be presented in a materialized form, such as a telescope, or in a non-materialized form, such as characters or words.
- In the front of "means and methods', a qualifier "technical" should be added to distinguish it from the intellectual and institutional means and methods, and to limit the invention to the extent where people can understand and accept, conforming to the public's conventional perception of invention. Such a qualifier can help prevent the definition from being ambiguity. For instance, democracy and KeJu, an imperial examination system in ancient China, are both innovations instead of inventions in its general sense; employing operations research to predict horse racing can be treated as an intellectual means (or common wisdom) rather than an invention.
- On the grounds of various impacts on nature, society and human beings, all inventions can be classified into four categories, namely great invention, important invention, invention, and gizmo. Again, taking the automobile as an example, it is undisputed that the automobile is a great invention and the airbag of it can also be sorted as an important invention because of its protection on the drivers. While the wiper is a general one and the auto windows a gizmo. Similarly, paper-making is obviously a great invention and letter paper is an important invention. The general invention may be the improved paper curtain, whilst mechanization of paper drying can only be regarded as a gizmo.
- Some inventions have no conclusive evidence that they are non-original, but they do show creativity and significant influences. Here are some instances. The technology bronze smelting and casting in ancient China consists of techniques such as mining (featured by its unique wooden structure support), smelting (featured by the shaft furnaces and the roasting and refining of copper sulfide ore) and casting (featured by "Liu Qi" alloy preparation rules, mold method and waxing method and wax stripping method), powering the glorious in the Shang and Zhou dynasties, also known as the Bronze Civilization of around 2,000 years, and exerting far-reaching influences on later generations. Such inventions should be counted as great inventions.
- Why are paper-making, printing, gunpowder, and compasses recognized as the four great inventions in ancient China? One of the reasons is their international influences on the advancement of modern civilization. What if we date back to

middle ancient times or even earlier? The inventions back then pushed forward no civilization, so how can we grade them?

- The author holds that in the early phase, human civilizations were relatively discrete. With the help of increasing contacts, exchanges, and communications, a trend of globalization is gradually developed and this process is far from completion. Accordingly, the grading of the inventions in middle ancient times or even earlier should be on the reality basis, which means the significance of the inventions and their influences on the state and the surrounding areas should be taken into account. For example, the pig iron-based steel casting technique in ancient China and the sponge iron-based technology in ancient Europe were two different technical systems. In terms of the technical concept, the former is more advanced than the latter, and modern steel technology is consistent with the ancient Chinese steel technology in terms of casting process. China was able to smelt pig iron as early as in the late Western Zhou Dynasty. Taking this opportunity, Chinese invented techniques such as the softening of cast iron, the casting of iron mold, and the scrambling of iron, which were finally learned by Europeans some 1,400 or 1,500 years later. And then the irrigation method of steel was found, thus creating a steel smelting system with China's own characteristics, and building up a glorious steel civilization that has long been in the forefront of the world. Furthermore, this system demonstrated a vital impact on China's neighboring countries like Japan, North Korea, Vietnam, etc. Therefore, it is rational and tenable to champion the great invention status of China's pig iron-based steel casting technique. In this respect, millet cultivation, rice cultivation, and wooden construction technique all fall into the same category as the steel casting technique.
- The validation of a great invention is not related to the inventing time. Under different conditions, inventions can occur either earlier or later, or even happen in some areas simultaneously. There are a host of exhibits in the long course of scientific and technological history from ancient times to now. The validation of great invention should be differentiated from the time of invention and cannot be confused with each other.
- For a scientific definition of the invention, the following restrictions are indispensable:

  (1) Scientific discoveries and pure academic research results are different from inventions, and should not be defined as "invention".
  (2) During an engineering project, existing inventions may be utilized or new inventions will be made to meet the needs of the project. Nevertheless, the project per se doesn't belong to the class of invention.
  (3) Some major inventions, such as "drilling wood to make fire", are common to early human civilization and do not fall into the great invention category of a certain region or country.
  (4) Inventions with significant controversy should be further studied before conclusions are made.

- Lives resemble an evergreen tree and reality is more diverse than the theories. The definition of the invention is crucial, but when it comes to assessing and grading an invention, there will be some problems that need to be carefully measured and properly handled.

In one of his published articles, the author listed Chinese medicine as one of the twenty-four inventions in China. Some scholars do not agree with that from their own points of view. Simply taking Chinese medicine as an invention, or more specifically a great invention, is not justifiable. Rather, replacing it with the name "diagnosis and treatment technique of Chinese medicine" does not seem inappropriate. Acupuncture, pulse-taking, bone-setting, prescriptions, herb processing, and other TCM techniques have played an extremely important role in Chinese people's well-being, health preservation, survival, and reproduction for thousands of years, which is an objective existence and should be fully affirmed.

Another example is about Chinese cookery. Some scholars argue against categorizing the Chinese cookery as an important invention, and even advocate that it is not an invention at all. Such issues should also be further explored later. The author believes that among all Chinese cultural heritage, perhaps the most potential and most promising prospects will go to Chinese characters, Chinese medicine, and Chinese cookery. The ascending number of Chinese learners, the gradual acceptance of acupuncture and moxibustion in many countries, the honor awarded to artemisinin and the boom in Chinese restaurants in towns of Europe, the US, Japan, and Australia are all lively evidences. With the growth of China's national strength and international influence, this trend is expected to augment. This situation reminds us that we must pay great attention to those three inventions: Chinese characters, diagnosis, and treatment techniques of Chinese medicine, and Chinese cookery.

The assessment and grading of inventions, of course, is based on a scientific judgment. Whether an item is defined as an invention or not, it is necessary to understand it and make clear the reason. Such a process should not be underestimated. The Four Great Inventions has been talked about for nearly a hundred years, but it has not been explained clearly, so that many people mistakenly believe that these four inventions were the most important and top-ranking inventions in ancient China, misleading some scholars, including the author, to wrongly propose the fifth great invention (such as pig iron, Qu'nie fermentation, and hybrid rice). From the A to Z, Joseph Needham listed twenty-six inventions in ancient China; Robert Temple believed that there were one hundred scientific achievements in ancient China, and Jin Qiupeng wrote the book *One Hundred Chinese Inventions*. None of them have graded and sorted these inventions, and some have mistakenly placed scientific discoveries and engineering projects among the inventions. The study of Chinese science and technology history in the modern sense has a history of nearly one hundred years and the researches of each branch reach a considerable depth and are backed by a group of authoritative scholars. But what are the great inventions in China's history? Our countrymen and the researchers around the world deserve an answer. The answer is no way perfect and won't satisfy everybody. But it is better to try to answer rather than standing aloof in the face of the issue. Furthermore, we are still studying and discussing, gradually trekking from an imperfect answer to a relatively satisfactory one, Chinese themselves should figure out their own issues. And on the issue of invention and creation, China must have its own right to speak.

1.3 The Origin of the Thirty Great Inventions of China

The answers to the question "What are the great inventions in China's history and how many of them" always vary and there must be various voices and controversies.

In 2008, the author published an article "A Review on China's Four Great Inventions and China's Twenty-four Inventions" in *Science News*. Later in 2013, in another paper "What exactly are the great inventions of China" (*Research on the History of Natural Science*, Volume 32, Issue 4), the formulation of "the twenty-four inventions" is still retained. They are: millet cultivation and rice cultivation, silkworm woven silk, jade polishing, Chinese characters, smelting and casting technique of bronze copper, iron-based steel casting technology, canal excavation, plow, hoe, water wheel, lacquer decoration, paper-making, diagnosis and treatment technique of Chinese medicine, porcelain, Chinese wooden structure construction skills, Chinese cookery, printing, tea cultivation and production, gunpowder, deep well drilling technology, intensive farming and agronomy, compass, flaming arrows, abacus calculation, and Qu'nie fermentation.

In March, 2014, the Chinese Culture Promotion Society hosted a consultation meeting entitled "What Did the Chinese Invent?", attended by senior scholars in the research community of scientific and technological history, such as LONG Fei, YANG Yongshan, GUO Shuchun, LUO Jianjin, ZHOU Jiahua, WAN Fubin, WANG Yusheng, JIANG Zhenhuan, LI Ling, LIU Changhua, HU Huakai, GUAN Zengjian, SU Rongyu, ZHONG Shaoyi, QIAN Wei, HUA Jueming, and some young scholars as well, like WANG Li (representing Mr. ZHOU Kuiyi) and CHEN Xiaoshan. After repeated discussions at the meeting, we added into the list weights and measures, stern rudder, horse harnesses and stirrup, armillary sphere and abridged armillary, blunderbuss, hybrid rice, and so forth. After the meeting, the original millet cultivation and rice cultivation were divided into two items; since some skills have not yet been confirmed, the jade polishing technique was temporarily crossed out; human pox inoculation was added into the entry of diagnosis and treatment technique of Chinese medicine and the tofu making and application into the item of Chinese cookery; the blunderbuss and flaming arrows were combined. Afterward, some scholars argued that weights and measures are more of an institutional innovation so that they recommended to exclude it. In addition, they pointed out that the research on stern rudder had to be carried forward. We have adopted their opinions after careful consideration. And all those discussions constitute the thirty great inventions of China, in this book: millet cultivation, rice cultivation, silk weaving, Chinese characters, decimal place value notation and rod arithmetic, bronze smelting and casting technique, pig iron-based steel smelting technology, canal and navigation lock, plow and hoe, water wheel, lacquer decoration, paper-making, diagnosis and treatment technique of Chinese medicine (human pox inoculation included), porcelain, Chinese wooden structure construction technology, Chinese cookery, horse harnesses, and stirrup, printing, tea cultivation and production, armillary sphere and abridged armillary, watertight bulkhead, gunpowder, compass, deep well drilling technology, intensive farming and eco-agronomy, abacus calculation, Qu'nie fermentation, blunderbuss and fire arrow, artemisinin, and hybrid rice.

Among those thirty great inventions, rice cultivation, silk weaving, decimal place value notation, porcelain, paper-making, printing, tea, gunpowder, compass, fire arrows, artemisinin, and hybrid rice have made salient contributions to human civilization. Besides, Chinese characters, Chinese cookery, and diagnosis and treatment techniques of Chinese medicine have demonstrated great importance and huge potential. Syncing with the elevation of China's national strength and international position, their impacts are expected to increase and there will be much room for development.

There are more than 30 inventions, and there must be a sorting procedure. The ordering of those inventions in this book is determined by the age of inventing and maturing. There will be a process of defining the great invention, and the thirty inventions mentioned above are only practices of various definition schemes. In order to facilitate the study, the following thirty important inventions can be used for reference: jade polishing, bow and crossbow, clepsydra, wheelbarrow, dragon bone waterlift, seismograph, Chinese brush making, ink making, cartography, geodesy, Yellow River Control Project, species variation, glazed pottery in low temperature, master coin method, astronomical clock tower, hall flower art, the forging of gong and cymbals, double acting air blower, chain bridge, biological solid base technique in bridge construction, pattern dye, tie dye, wax die, Ming-dynasty-style furniture making, twelve-tone equal temperament, Zheng He's treasure ship, foil forging, Chinese garden designing, violet sand earthenware, paktong, zinc smelting in the crucible and vertical windmill.

References

1. Xiaohe, Cang. 1988. *A Compendium of the History of Natural Science: The Role of Science in History and the Influence of History on Science*, 267. Beijing Publishing House.
2. Needham, Joseph. 1999. Part 2: Mechanical Engineering, vol. 4: Physics and Physical Technology, *Science and Civilization in China*, 6. Science Press, Shanghai Classics Publishing House.
3. Marx, K., and Frederick Engels. 1979. *Central Compilation and Translation Bureau*, vol. 47, 427. People's Publishing House.
4. Evidences can be found in The Origin of Printing Technique, One of the Four Great Inventions, 55–72, Issue 5, *Middle School Student*. The author of the passage is Xiang Da, styling himself Jueming.
5. During this Period, the Concept of The Three Great Inventions Still Existed, Which Can be Looked up in The Three Great Inventions-Notes on Books of Scientific and Technological History by Lin Zhuai. *Science Times*, p. 15, Issue 6, 1946 and in The Three Great Inventions in Ancient China by Qian Weichang in *China Youth*, pp. 10–13, Issue 61, 1951.

6. Needham, Joseph. 1986. *China's Contributions to Science and Technology*, complied by Pan Jixin and translated by Chen Yangzheng, etc. *Collected Works of Joseph Needham- Papers and Speeches about the Scientific and Technological History of China (1944–1984)*, vol. 118, 123. Liaoning Science and Technology Publishing House.
7. Hai, Ci. 2010. Shanghai Lexicographical Publishing House: 452.

Jueming Hua Born in Wuxi, Jiangsu, in 1933, graduated from the Department of Mechanical Engineering of Tsinghua University in 1958. In 1967, he got the master degree in the Institute of Natural Science History of the Chinese Academy of Sciences. In 1986, he served as a researcher at the Institute of Natural Science History of the Chinese Academy of Sciences, served as deputy director from 1988 to 1993, and retired in the same year. From 1993 to 2003, he served as the director of the Institute of Science History and Ancient Literature, Tsinghua University.

Research interests: History of technology; the philosophy of technology; traditional crafts.

Major works: *A Collection of Papers on History of Chinese Metallurgy, Ancient Chinese Metal Technology: Civilization Made by Copper and Iron, Complete Works of Chinese Traditional Crafts, Chinese Handcrafts and Collections of Hua Jueming's Works.*

Awards: The first prize of scientific and technological achievements granted by the Ministry of Culture;

the first prize of the National excellent publications;

the Lifetime Achievement Award for the Foundry industry.

Chapter 2
Millet Agriculture

Xiongsheng Zeng

Millet agriculture, one of the earliest inventions of China, was invented by transforming wild plants into cultivated crops with a complete set of cultivation techniques. About 1,000 A.D., millet had fed the world's largest population and nurtured the brilliant Chinese civilization.

2.1 The Origin and Early Development of Millet Agriculture in China

Millet to be discussed here includes two kinds of crops: foxtail millet (Setariaitalica) and broomcorn millet (Panicummiliaceum). They are of different genera in terms of plant classification, yet the two are often classified as the same category in terms of origin, dissemination, cultivation, and distribution. Moreover, they have similar physiological characteristics and cultivation conditions. Therefore, foxtail millet and broomcorn millet are regarded as one kind of crop in some regions of Europe and Asia. Foxtail millet, also called *Ji* in Chinese by the elderly people, is an annual herbaceous plant with round and oval seeds. In North China, it is generally called *guzi*, which to a great extent refers to its seeds. After being peeled, its seeds become millet that can be cooked. What's more, compared with other cereal crops, such as rice and wheat, its seed is small in size, so it is also called as *xiaomi* in Chinese. Broomcorn millet is also a kind of annual herbaceous plant. Its seeds are called *shuzi* in Chinese, most of which are light yellow. After being ground and peeled, they become yellow particles with longer diameter than millet. Plants of foxtail millet and broomcorn millet are also known as seedlings. From the perspective of today's global staple crops, foxtail millet and broomcorn millet have played a less important

X. Zeng (✉)
The Institute of the History of Natural Sicences, Chinese Academy of Sciences, Beijing, China
e-mail: zeng@ihns.ac.cn

J. Hua and L. Feng (eds.), *Thirty Great Inventions of China*,
https://doi.org/10.1007/978-981-15-6525-0_2

role in grain supply and even fallen behind corn, but, in history, foxtail millet and broomcorn millet were the main food crops.

Then, where is the origin of millet? China (Eastern Asia), Europe, Southern Asia? The wild ancestors of foxtail millet are widely distributed around the world. Before agriculture appeared, wild millet had become people's food to pick from the nature. The appearance of agriculture might begin with trying planting such cereal plants when they began the domestication and cultivation of them. In theory, the place wherein the wild millet and early humans' activities were found may be the original area of millet. Nevertheless, more and more evidences reveal that China is one of the original areas of millet cultivation agriculture in the world.

The wild species of millet (green bristle grass) is a native plant in the semi-arid losses area of North China. Broomcorn millet has stronger cold tolerance than foxtail millet. From Northeast China, Inner Mongolia to Gansu, Xinjiang Uygur Autonomous Region, there is the distribution of wild broomcorn millet (called *lang* in Ancient China). The northerners in ancient China very early began to take advantage of such a plant. The analysis on the extracts from the surface residues of unearthed stone and pottery wares and the ancient starch remains in the cultural deposits in the Nanzhuangtou Site (earlier than 11,000 years ago), Xushui County, Hebei and the Donghulin Site (as early as 11,000–9,500 years ago), Mentougou District, Beijing showed that 11,000 years ago, starch granules with domestication characteristics appeared in ancient starch remains, proving that human beings had started to domesticate wild foxtail millet and broomcorn millet [1].

In 1976, archeological workers found lots of the remains of millet in Cishan Village in Wuan, Hebei. And then, archeological workers carried out archeological excavation of three stages in Cishan Site, finding a total of 476 ash pits, of which 88 caves had an accumulative millet ash layer, which is 0.2–2 m thick in general, and 10 caves had accumulative food layer over 2 m thick. The food reserves are estimated to be 5–6 tons [2]. The considerable amount and accumulative thickness are exceedingly rare in excavated Neolithic cultural remains. In the 1980s, broomcorn millet remains were found to be more than 7,000 years ago in Dadiwan Site, Qinan County in Gansu [3]. From 2001 to 2003, the earliest cultivated foxtail millet and broomcorn millet (Fig. 2.1) were found to be between 7,700 and 8,000 years ago in Xinglonggou Site, Chifeng in Inner Mongolia Autonomous Region. In 2009, archeologists discovered the best-preserved Xiajadian Lower Cultural Site in Erdaojingzi, Chifeng in Inner Mongolia. They find that there include charring broomcorn millet, millet granules, spiked charring food crops, and some furs and straw fabrics, adding new evidences to prove that millet cultivation was originated from China.

Up to the beginning of the twenty-first century, the number of foxtail millet sites found in China has amounted to nearly 70 and broomcorn millet sites 14 [4]. These sites spread over the provinces like Henan, Hebei, Shandong, Shanxi, Liaoning, Inner Mongolia, Heilongjiang, Shaanxi, Gansu, Qinghai, Xinjiang, Yunnan, Tibet, and Taiwan. The millet sites in China are the earliest and continuous in temporal distribution, and the most concentrated and extensive in spatial distribution, intensifying the proof of the origin of millet cultivation in China. Beyond people's expectation, the earliest crop cultivated in Tibet wherein the highland barley has been regarded

Fig. 2.1 Charring broomcorn millet unearthed in Xinglonggou site

as the staple food for a long time, is also millet. Crop remains found in Karuo Site (about 5,000 years ago) in Changdu, Tibet were only millet, which was gradually replaced by the highland barley and wheat.

The existing archeological discoveries traced the origin of millet cultivation in Eastern Asia to at least 10,000 years ago when human beings invented agriculture in response to the global warming. The wild ancestors of foxtail millet and broomcorn millet became the preferred crops of the ancient Chinese because of their characteristics such as extremely strong resistance, short growth period, easy to plant, and easy to store. Planting foxtail millet and broomcorn millet marks the beginning of the primitive agriculture in North China.

In Chinese ancient fables and tales, *Shennong* is the first ancestor of agriculture. After *Pangu* created the Heaven and Earth, *Suiren* and *Baoxi* invented the method of making fire, fruit-picking, and nets for hunting. These inventions facilitated the reproduction of human beings themselves. By the time of *Shennong*, there were so many people but few beasts and a shortage of food. Therefore, *Shennong* wanted to plant crops to feed the people. For this purpose, he tasted the seeds of all kinds of herbs, checking the sourness and bitterness *(New Language, Daoji).* Finally, he picked out several main species such as foxtail millet, broomcorn millet, rice, and so on among the numerous wild plants and made them the new source of food. Later, according to the climate change of the four seasons in a year, taking advantage of the terrain conditions, *Shennong* made plow and taught people how to cultivate land. And then agriculture became another means of living ensuing fruit-picking and hunting.

At the beginning of agriculture, foxtail millet and broomcorn millet might not stand out. They were just one of the abundant crops early humans tried planting and domesticating. Afterward, people gradually recognized such advantages of them as early maturity, drought and heat resistance, few damage by disease and insect, and easy cultivation. After defeating other rivals, they established their place in food supply. By 13C dating, which was used to measure the recipe of ancient people, we can know that millet food accounted for only 50% in the period of Yangshao Culture while 70% in the period of Longshan Culture, indicating that the millet agriculture gained further development [5].

Like broomcorn millet, foxtail millet didn't have an edge. Although foxtail millet became the representative crop of the dry farming in North China later, it enjoyed even lower priority than broomcorn millet in the early stage. Originally, the archeologists believed that foxtail millet excavated in Cishan Site, Wuan County in Hebei was found to be 7,800 years ago. Being remeasured by the phytolith method, it was found that between 10,000 and 8,700 years ago, broomcorn millet was the early crop preserved in Cishan Site, and during the period from 7,500 to 8,700 years ago, a small quantity of phytoliths of millet appeared [6]; the Xinglonggou Site unearthed in Chifeng, Inner Mongolia also showed that more broomcorn millet was found than foxtail millet. The unearthed broomcorn millet was up to about 1500 grains, while foxtail millet only about 10 grains. From 2000 B.C. to 1500 B.C., broomcorn millet took the majority among charring grains in the caves of Xiajiadian Lower Cultural Site, in Erdaojingzi of Chifeng in Inner Mongolia. This phenomenon continued until the Shang Dynasty.

Among the oracle inscriptions in the Shang Dynasty Ruins, there are 106 items of the words of broomcorn millet and 36 items of the words of foxtail millet [7]. In *the Book of Songs: National Wind*, the poetic verses "Fat rats, fat rats, don't eat my broomcorn millet" reflect the fact that broomcorn millet at that time was the staple food. Until the Warring States Period, people still ranked broomcorn millet first in ordering top nine, five or four cereal crops. In some alpine regions, broomcorn millet is the only cereal crop. Mencius once mentioned in *Talking to Zizhang of volume* 12 of *Mencius Annotations* that Racoon Dog, a Northern ethnic minority, only plant boomcorn millet because in the alpine region only broomcorn millet can grow and ripens early.

However, in the Zhou Dynasty, foxtail millet gradually took the place of broomcorn millet. The ancestor of the Zhou people is Houji who enjoyed playing games about planting in his childhood. After growing up, he liked farming and was promoted to be a farm official by Emperor Yao, taking charge of agricultural production. Ji is foxtail millet in old Chinese; Houji, in Chinese, meant that he was in charge of agricultural production and foxtail millet cultivation. The Zhou tribe therein Houji was born, was originally an agricultural tribe. The word "Zhou" in oracle inscriptions and inscriptions on ancient bronze object looks like the actions of sowing and fertilizing in the field. The Zhou people regarded this word as their tribe's name to show the importance they attached to agriculture. Afterward, under the domination of King Wu of Zhou, this tribe defeated the Shang Dynasty and established the Zhou Dynasty. Later, the young King Cheng of Zhou established the basic national policy of developing agriculture under the assistance of the Duke of Zhou, which made foxtail millet the national staple food. Its cultivation area even extended to the south of the Yangtze River, which was originally dominated by rice cultivation.

From the Zhou Dynasty to the Song Dynasty, foxtail millet had been the most important food crop of the northern Chinese, honored as the leading crops. The poor harvest of foxtail millet meant the coming of famine. *Spring and Autumn Annals*, compiled by Confucius, recorded that no other crops but the poor harvest of millet, indicating that millet had an important status in the Spring-and-Autumn Period.

2.2 The Improvement and Maturity of Millet Agriculture Technology

Since the ancient times, the efforts of Chinese people in improving the technology of millet cultivation have never stopped. Before the epoch, traditional millet cultivation technology in China gradually stepped into maturity. Selecting seeds was one of the earliest ways to improve its production. *The Book of Songs* already mentioned the concept of seeds selection. The big and plump seeds were the best choice. Adjusting to different needs of cultivation, seeds at that time are different in planting and maturing time: early planting but ripening late, and late planting but early ripening. In *The Book of Songs*, there are also descriptions of sharp farm tools such as *Si* used for soil preparation and *Bo* for weeding fields. The requirement of soil preparation is the degree of depth. As *The Spring and Autumn of Lv Buwei: Rendi* goes, the cultivation depth must reach the moist soil where there is no sight of weeds and vermin; in this way, we can reap the harvest of well-grown millet and wheat. *Tso Chuan: Yingong for Seven Years* pointed out that intertillage must eliminate weeds and weed out their roots in case that they revive. Clearing weeds can not only wipe out weeds and prevent them from competing with crops for water and nutrients, but also provide nutrients for crops with the decayed weeds.

There is a key link between soil preparation and intertillage—sowing seeds. In sowing seeds, *You,* a kind of farm tool like wood hammer, is necessary, and it is used to soil blocks and let them cover the seeds evenly. Farmers attached great importance to soil preparation, covering seeds, intertillage, and other farming links. *Chuang Tzu* illustrated the benefits of deep cultivation and careful planting with his own experience, saying that "when I was a farmer, I paid less attention to spring plowing and summer weeding; unfortunately, I had a poor harvest in autumn. The next year, by deep cultivation and careful planting, my millet grew very well and secured me enough food".

Farmers also utilized fertilizer to improve millet production. Fertilization mainly used base manure, in addition to green manure made by planting mung beans and small beans. The main source of fertilizer is the animal manure, silkworm excrement, horse bones, and so on. Besides the nurturing soil, these manures are directly used for seeds treatment and as seeds fertilizer. The method of seeds covered with manure (*Souzhong* Method in Chinese) is a method of seeds treatment, in which three kinds of materials, namely, snow water, bone marrow, and silkworm excrement and sheep excrement, monkshood are mixed up and then processed. The early record of this

Fig. 2.2 Sketch map of ridge tillage

method was seen in *The Book of Fan Shengzhi* of the Han Dynasty. The legend goes that such a method was invented and improved by *Shennong* and *Houyi*. Seeds, treated by this method, are covered with a layer of manure shell with silkworm and sheep excrement as main materials, which cannot only help alleviate the damage of pests, but also strengthen the drought assistance of seedlings, thus increasing the yield of millet. Therefore, it is the pioneer of today's coated seeds.

Due to their better drought resistance, foxtail millet and broomcorn millet became the first group of crops planted in North China. People tried to provide a sound environment for the growth of foxtail millet and broomcorn millet by soil cultivation and preparation. The invention of ridge tillage (Fig. 2.2) is just one of their efforts in this regard. Ridge tillage, composed of ditch and ridge, have relatively fixed orientation, generally to the southeast for the convenience of drainage. In *The Book of Songs*, there are some poetic lines describing this. As for the trimming of ditch and ridge, *The Spring and Autumn of Lv Buwei: Biandi* indicates that ridge should be broad and flat while ditch should be shallow and deep, and surface soil should be loose and fine while subsoil should be solid, providing good growing conditions for crops. In sowing, the level terrain and soil water content are crucial for deciding the sowing position. For example, seeds are sowed in the ditch if the field is in a higher terrain; seeds are sowed in the ridge if the field is in lower terrain.

Later, there appeared the method of regional planting and the method of switched field. As the legend has it, the method of regional planting was invented by Prime Minister *Yiyin* in the early years of the Shang Dynasty. He divided cultivated land into several regions wherein the fertilizer and water management was concentrated in order to tackle drought and increase the yield. The method of the regional planting was tested on a variety of crops, including foxtail millet and broomcorn millet. According to the description in *The Book of Fan Shengzhi*, the method of regional planting has high technological requirements. The switched field consisted of ditches and ridges alternately. Seeds are sowed in the ditches. When they sprout, the ridge soil was flatted in combination with intertillage to play the role of wind prevention, lodging resistance, increase of soil moisture, and drought resistance. The position of ditch and ridge will change after a year to let the land alternately rest and work (Fig. 2.3). These features, plus a raft of matched farm tools and scheduled promotion measures, helped the method of switched field achieve good results and increase the yields of millet. As *the Book of Han: Shi Huo Zhi* mentions, using less efforts would be helpful in gaining lots of millet. The legend has it that the method of switched

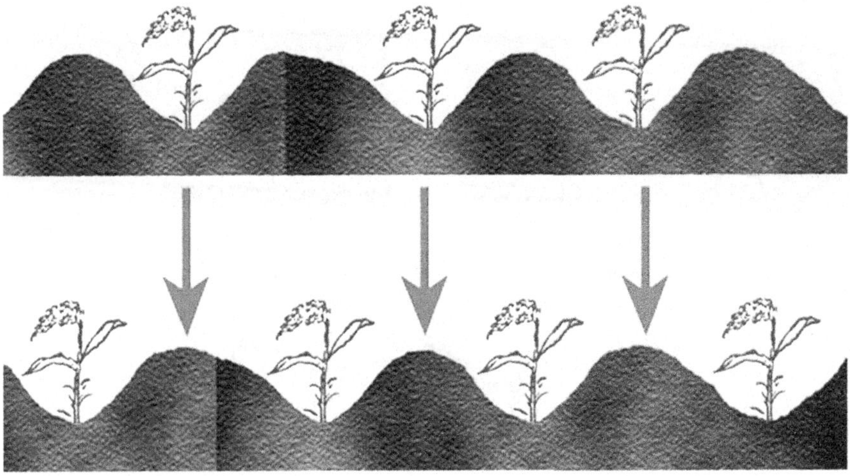

Fig. 2.3 Sketch map of the method of switched field

field was invented by Houji, the ancestor of the Zhou Dynasty, and in the last years of the reign of Emperor Wu in Han Dynasty, the method was spread by Zhao Guo, an official responsible for searching for foxtail millet the method. Therefore, the method of the switched field played an important role in recovering war wounds and developing social economy. It is an important invention in the history of the dryland agricultural technology in northern China.

When Zhao Guo advocated the method of the switched field, he also invented a kind of animal seeder. *Political Comment,* written by Cui Yan, mentioned its principle that three plows are led by a cow and the cow is pulled by a person. When sowing, one should hold the animal-drawn seed plow, which is available for him picking the seeds. In this way, a field of 100 *mu* can be sowed in a day. The so-called "three plows led by a cow" here is the prototype of the late three-foot seed plow. The animal-drawn seed plow consists of plow funnel and plow feet. Plow feet are straightly connected with plow funnel, where seeds are stored. When it is working, its front is led by the animal power and controlled by one person, and the seeds are sowed in the ground through plow feet. A farmer, by utilizing the animal-drawn seed plow, can finish the work of ditching, sowing seeds, and covering soil on his own. The invention of the animal-drawn seed plow can not only improve the planting seeds sowing but also ensure the even spacing, depth, and density of sowing, facilitating ventilation, light transmission and field management, and providing conditions for mechanical fertilization and harvest.

During the Han and Tang Dynasty, the traditional millet technology in China was already quite mature and the species of millet increased significantly. The number of the millet species recorded in *Qi Min Yao Shu* is 106. Among these species, some have high yields, some are delicious, precocious, late-maturing, drought-resistant, water-tolerant, wind-resistant, while others are free from insects, birds, and are

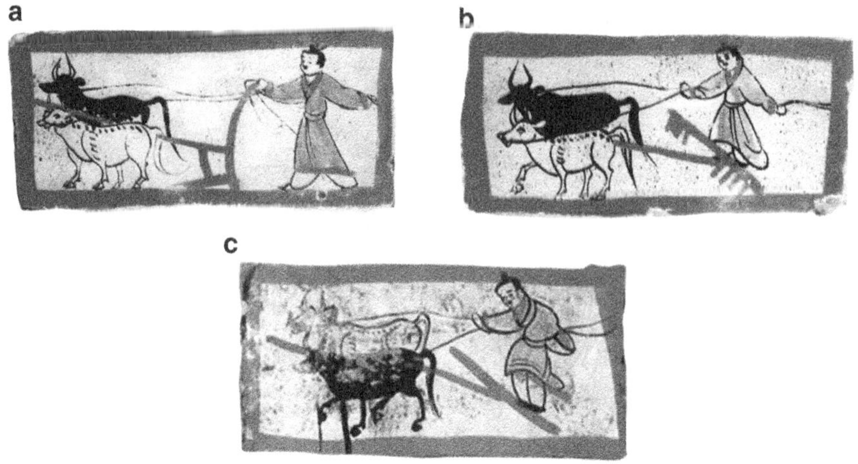

Fig. 2.4 Frescoes of Tombs in the Wei and Jin Dynasties in Jiayuguan, Gansu. **a** plowing the land **b** harrowing the land **c** leveling the land

easily mashed. These species open up the possibility for farmers to plant selectively according to different conditions. After the publication of *Qi Min Yao Shu*, the number of millet species continued to increase. Emperor Kangxi of the Qing Dynasty planted the white millet found in the holes of trees by the people of Ula (now Jilin Province) in Chengde Summer Resort. The result of this trial planting was high-yielding, precocious, and good-quality millet.

While attaching importance to selecting and cultivating seeds, people intensified their efforts to do soil preparation. In respect of the soil cultivation technology, the technology system of the soil preparation which combines plowing, harrowing and leveling, and the intertillage technology (Fig. 2.4), which was the beginning of the combined use of hoe, feng, and jiang. Plowing is to seek deep cultivation and increase soil moisture; harrowing and leveling are to cultivate intensively and preserve soil moisture; intertillage emphasizes more hoeing to increase the yields. Farmers also took other measures of intensive cultivation, including suppressing, thinning out the seedlings, earthing up the seedlings, irrigation, and frost and the dew prevention.

2.3 Millet Agriculture and the Development of Chinese Civilization

These farming technologies, through the summary of the farmers' experiences of all ages, constitute the cornerstone of Chinese civilization. The Chinese population in Han dynasty reached 60 million, while amounted to 80–90 million in Tang dynasty. To a great extent, these people mainly fed on millet rice. Millet later became synonymous with food. Moreover, millet was also the main source of the tax revenue for the ancient governments and a major symbol of social wealth. For thousands of years, China had been a country of agriculture where the tradition of worshiping millet god was handed down from generation to generation. In ancient times, the worship of the millet god enjoyed the same status as that of the god of the land. She Ji (millet) in the Chinese became the representative of the nation. Since the Zhou and Qin dynasties, the governments had been ranking the policy of valuing agriculture and millet as the first in their administration, which had been the basic national policy of the country. During the reign of Emperor Wen of the Western Han Dynasty, Chao Cuo, a well-known thinker in history, proposed the policy of valuing millet as the breakthrough of developing agriculture. In response to the worsening social contradictions, he believed that the primary task of the central government is to encourage farmers to work in agriculture and value millet production, which became a standard of reward and punishment for the citizens. Big millet producers might obtain titles by donating millet to the country, and criminals could also be exempt from punishment in this way.

Millet culture has also penetrated into fields such as calendar, arithmetic, and character, and then exerted an influence on all respects of people's life. The pattern of the Chinese character "nian" (year) in oracle bone inscriptions consists of He (millet) in Chinese at the top and Ren (person) in Chinese at the bottom, which means a person is bearing the ripe millet on the back. The original meaning of "year" is the harvest of millet. As millet is harvested once a year, "year" became the unit of time. The Chinese character "He" for millet plants has become the common side of many Chinese characters for cereal crops. In *Kangxi Dictionary*, the number of characters with "He" as a component increased to 448. The issue of millet rate resulting from the calculation of millet yield rate became the method for specifically discussing the conversion of different ratios. The basic unit of measurement is composed of foxtail millet and broomcorn millet as benchmarks.

In a sense, millet cultivation laid the foundation for the rise of Chinese civilization, and it also exerted influence on the world with the booming of Chinese civilization. Crops of foxtail millet genre and broomcorn millet genre in East Asia, Southeast Asia, Indian Peninsula in South Asia, Eurasian Steppe, Eastern Europe, and Central Europe were all imported from China, in the prehistorical and historical times [8]. Over 3,000 years ago, millet entered the Korean Peninsula from the Yellow River basin in North China, which was proved by the seeds of rice, wheat, and millet (1205 B.C. by 14C dating) unearthed from the pottery pots of Hunamni Site in Namhangang, Korean Peninsula in 1978. In Japan, millet was also the staple food from the late Jyomon period to the early Yayoi period. Regions (in Southeast Asian islands and Taiwan of China) were first planted with millet and taro before rice was introduced. Millet also reached Europe very early. Together with wheat, they became the important symbols of the prehistoric human food globalization. Such ancient agricultural books that mainly recorded millet cultivation and other dryland farming technologies as *Qi Min Yao Shu* and *Nong Sang JiYao* (Fig. 2.5), were introduced to Japan and Korea, very early. Completed in the sixth century, *Qi Min Yao Shu* entered Japan, in the ninth century in the form of handwritten copy, and until the twentieth century, the dryland farming technology in it has been researched by the Japanese

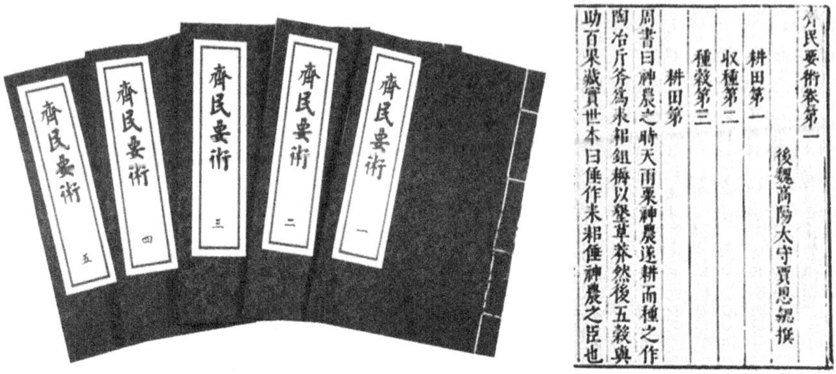

Fig. 2.5 *Qi Min Yao Shu*

scholars. *Straight Talk about Agriculture*, a Korean agricultural book published in 1430, mainly cited *Nong Sang Ji Yao* completed in China, in 1273. As Korea borders the North China, sharing a similar climate, agriculture in Korea has been developing under the influence of the agricultural technologies in North China, represented by millet [9].

The contribution of the Chinese millet to the world civilization is also reflected in language. In Sanskrit, millet is called Cinaka, which means China; in Hindi, millet is called Chena or Cheen, and in Bengali, called Cheena, which are all homonyms of "Qin". The Persian word Shu-shu is probably a homonym for the Chinese pronunciation of millet; the English word millet comes from the Medieval French, and it can be further traced to milium in Latin and mele in Indo-European, which both means crushing and grinding. Mele has evolved into mill, molar, millstone, and so on, which are words derived from the use of stone mill to shell and grind in primitive agriculture. This is similar to the situation of millet in China. Millet is easily associated with Chinese characters mi and mo (Fig. 2.6), as foxtail millet and broomcorn millet are both grains with seed coats and must be milled to become the edible food.

Fig. 2.6 The stone roller, quern unearthed in Piligang site, Xinzheng, Henan

For thousands of years, millet has nurtured people scattered around the world with its drought and infertility resistance. Today, rice and wheat have played absolute leading roles at the table. However, the position of millet in the origin of agriculture and its contribution to early human civilization will never be forgotten. As a poem of the Tang Dynasty goes, hoeing underneath the midday sun, sweat dripped onto the toiled soil; did you know on a plate of meal, every grain you eat come from such toil? For the treasure of millet, we should start with no waste of grain.

References

1. Xiaoyan, Yang, Wan Zhiwei, Linda Perry, et al. 2012. Early Millet Use in Northern China. *PNAS* 109 (10): 3726–3730.
2. Weiqing, Tong. 1984. Primitive Agricultural Remains of Cishan Site and Related Issues. *Agricultural Archaeology* 1: 197.
3. Anzhimin. 1984. On Early Neolithic Culture in North China. *Archaeology* (10): 939.
4. Xinglin, Liu. 2004. *Prehistoric Agricultural Research*. Huangshan Book Society: 64–68.
5. Bingqi, Su. 1991. Reconstructing the Ancient Age of Chinese Ancient History, Preface to Volume II of a General History of China. *Research on History of Historiography* 3(3): 5.
6. Houyuan, Lu, Zhang Jianping, Wu Naiqin, et al. Phytolith Analysis for Differentiating between Foxtail Millet (Setariaitalica) and Green Foxtail (Setariaviridis). https://dx.doi.org/10.1371/journal.pone.0019726.
7. Xingwu, Yu. 1957. Cereals in the Shang Dynasty. *Journal of Humanities of Northeast People's University* 1: 81–107.
8. Bingdi, He.1969. *The Origin of Loess and Chinese Agriculture*, 133. Chinese University of Hong Kong Press.
9. Yanhua, Pak. 2001. A Comparison Between Korea's *Straight Talk about Agriculture* and China's *Nong Sang Ji Yao*. *Journal of Yanbian University (Social Sciences Edition)* 3: 92–95.

Xiongsheng Zeng Born in Xingan, Jiangxi in 1962. Researcher at the Institute of Natural Science History, Chinese Academy of Sciences.

Research interests: History of agronomy; General history of science.

Major works: *Chinese Agricultural History; General History of Chinese Agriculture: Volumes of Song, Liao, Xia, Jin, Yuan Dynasties.*

Awards: *Chinese Agricultural History* was selected into the Second "Three One Hundred" Original Publishing Engineering Science and Technology Books by the State Administration of Press, Publication, Radio, Film and Television of the People's Republic of China, the Nomination Award of the 3rd China Outstanding Publication Books;

The History of Ancient Chinese Science and Technology Entering the Palace won the 3rd China Outstanding Publication Book Award.

Chapter 3
Rice Farming

Xiongsheng Zeng

The country faced unprecedented stress when the number of Chinese living on millet cultivation had reached between 80,000,000 and 90,000,000 in the Tang Dynasty. Millet cultivation couldn't meet the needs of the growing population. Li Shen, a poet of the Tang Dynasty, once wrote in a poem that: "spring sows one grain of millet, autumn reaps in ten thousand more; no fields fallow the Four Seas over, farmers, nonetheless, perish from hunger." As a result, the Chinese people had to search for another crop with high yields. When millet resigned from the leading post, rice first grown in southern China became the main crop in China. The transition period was in about 1000 B.C.; since then, rice has played a key role in the food constitution in China.

Though rice is grown in southern China and called *da mi* in Chinese, while millet is grown in North China and called *xiao mi* in Chinese, they have something in common. Rice was called *he* or *gu* in southern China, the same as millet was called in the North. The Chinese character *he* is a hieroglyph, referring to plants more often. The other character *gu* refers to the fruit of rice, which is also called *mi* in Chinese after being threshed. For the grain of rice is bigger than that of millet, rice is called *da mi*, millet is called *xiao mi*.

Another common ground they have is their husk, "pi" in Chinese. Ancient people called the husk "jia"—a hieroglyph, referring to the shell of plants and fruits. Written in small seal script, "jia" resembles the husk breaking when the seed is sprouting. Crops with "pi" as the protective covering can be kept in storage for a long time and bear the long-distance transport, which is particularly crucial for rice (especially one of its kind, japonica rice). Since the Tang and Song dynasties, million tons of rice had been transported from the south to the north of China, supporting the operation of the huge empire.

X. Zeng (✉)
The Institute of the History of Natural Sciences, Chinese Academy of Sciences, Beijing, China
e-mail: zeng@ihns.ac.cn

J. Hua and L. Feng (eds.), *Thirty Great Inventions of China*,
https://doi.org/10.1007/978-981-15-6525-0_3

3.1 The Origin of Rice Farming

Before rice was continuously transported from the south to the north of China and became the main food supply, it had occupied the dinning tables of half the Chinese population for a long time. According to the ancient myth of South China, since Pangu created the world, the people of Miao, Yao, Yi, Han, and Yi ethnic groups living in the South have had rice as their main food. The inventor of agriculture, Shennong, was also living in the South. If so, the cereal crops originally planted by him should include rice, which is the most common plant in southern China.

In terms of botany, there are 20–25 species of Oryza plants all over the world. Two of them are cultivated ones, Oryzasativa and Oryzaglaberrima, and the rest are all uncultivated ones. The ancestor species of Oryzasativa is known to be Oryzaperennis or Oryzarufipogon, as well as Oryzarufipogon Griff. Oryzaperennis mainly reproduces by perennial roots, and its blossom can proceed generative propagation.

In ancient Chinese books, those which can grow without cultivation are called "ni, lv or lu", including probably the ancestor of cultivated rice—Oryzarufipogon. There are more than 10 records about the self-generation of Oryzarufipogon in the course of Chinese history such as the story that Wusunquan had changed the name of Youquan County into "Hexing" when he knew the Oryzarufipogon could grow without cultivation in 231 A.D. The places where they tended to grow could be lined up as a curve, which first started at Quzhou (in Sichuan) in the upper reaches of the Yangtze River, Xiangyang and Jiangling in the mid-stream, the north of Zhejiang and the south of Suzhou both located in Taihu Lake area, as well as the lower reaches of the Yangtze River, the north and middle of the Suzhou and Huaibei in this Province, and at last, ended at Lucheng (Ganzhou today) in the Bohai Bay. Nowadays, the growing area of Oryzarufipogon within China, ranges south from Ya County, Hainan Island in Guangdong (N18°09′), north to Dongxiang in Jiangxi (N28°14′), west to Yingjiang in Yunnan (E97°56′), and east to Taoyuan in Taiwan (E121°15′).

The cultivated rice is grown from Oryzarufipogon through artificial domestication. Same as the millet's ancestor, the ancestor of rice—Oryzarufipogon ever spread not only in China, but also to the areas of Southeast Asia and South Asia. Theoretically, Northeastern India, northern Bangladesh, Myanmar, Thailand, Laos, Vietnam, and the areas bordering South China could be the earliest domestication center of the wild rice in the world. Part of this theory was supported by documents of linguistics, paleo meteorology, and ethnography. However, the claim that China is the origin of rice has been testified by more and more proofs from anthropology, linguistics, archeology, and genetics. In the Yangtze River Basin and areas to the south of it where Oryzarufipogon is mainly grown, many ethnic groups of people had lived since the Paleolithic age. They were called "Baiyue People" by the historians. The Baiyue people gathered Oryzarufipogon and then kept on its planting, which is actually the start of rice cultivation. In the southwestern regions where the Baiyue's descendents are living the old pronunciation of rice has been kept until now: "Khau" and "K'ao", which is the origin of "he" and "gu"—the pronunciation of rice in the dialect of South China. After the introduction of these dialects to North China, people got confused with rice and millet for millet was also called “he” or “gu” in the north. Therefore, the written version of rice in Chinese, “dao”, was coined to fix the confusion.

The most convincing proof of the claim that southern China is the origin of rice cultivation came from archeological excavation. In the 1970s, the Hemudu Site found in Zhejiang demonstrated that the beginning of China’s rice cultivation might date back to 7,000 years ago, which aroused people’s attention to its origin. In the early twenty-first century, the Tianluoshan Site found near the Hemudu Site was another new evidence. By the time of 2004, 182 sites of charring rice, paddy fields or remains of rice stems, spores, and phytoliths had been unearthed throughout the country [1]. There are 140 sites located in the Yangtze River Basin, accounting for 76.92% of all, including 56 in the lower reaches, accounting for 30.77%, 75 in the middle reaches, accounting for 41.2% and 9 in the upper reaches, accounting for 4.95%. Moreover, there are 7 sites located in the southeast coastal areas, accounting for 3.85% of all, 13 between the Yangtze River and the Huai River, accounting for 7.14%, and 22 between the Yellow River and the Huai River, accounting for 12.09%.

The unearthed sites of rice in Neolithic age kept increasing in amount, enlarging in space and dating earlier in time. In terms of time, the earliest site existed over 10,000 years ago and the latest one is connected with the historical records, i.e., the time span of about six thousand years from 10,000 to 4,000 years ago [2]. So far, there are three rice sites over 10,000 years old: the Yuchanyan Site in Dao County in Hunan, the Wannian Xianren Cave Site, and the Diaotonghuan Site in Jiangxi. By 14C dating, the Yuchanyan Site is 12320 +120 years old now. The rice phytoliths found in Xianrendong and Diaotonghuan sites appeared between 14,000 and 11,000 years ago. The sites about 10,000 years old also include the Niulandong Site in Yingde in Guangdong and the Shangshan Site in Pujiang County in Zhejiang. The rice biosilica found in the second and third cultural layers of the Niulandong Site is 11,000 to 8,000 years old. The Shangshan Site with more rice husk imprints, rice husk, and phytoliths excavated is 11,000 to 9,000 years old. Among the Neolithic rice sites about 8000–4000 years old, the typical ones are the Pengtoushan Bashidang Site (8500–7500 B.P.) in Li County in Hunan, the Zaoshi Xiaceng Site (7000 B.P.), the Hemudu Site (6950 + 130 B.P.) and the Tianluoshan Site in Yuyao, the Luojiajiao Site (7040 + 150 B.P.) in Tongxiang in Zhejiang, and the Caoxieshan Site (7000 B.P.) in Wu County, Jiangsu. In addition to the rice remains unearthed from these sites, bone rakes for plowing field and many remains of fields (Figs. 3.1 and 3.2) were found there.

The existing archeological findings show an obvious trend that the middle and lower reaches of the Yangtze River are the earliest rice growing areas discovered so far, and then the rice planting spread to the north and south of China: first, to Henan and Shaanxi in the Yellow River Basin through the middle reaches of the Yangtze River; second, to Shandong in the lower reaches of the Yellow River and the north of Jiangsu and Anhui in the lower reaches of the Huaihe River through the lower reaches of the Yangtze River; third, to the southeastern coastal areas, Taiwan and southwestern China.

The studies of plant genetics have also confirmed that rice originated in southern China. Genomics researchers traced the history of rice evolution through large-scale

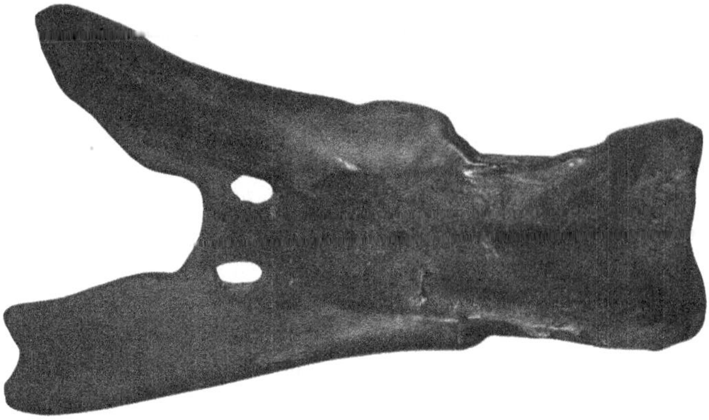

Fig. 3.1 Bone *si* (a spade-shaped farm tool) unearthed in the Tianluoshan Site (provided by Zhejiang Institute of Cultural Relics and Archaeology)

Fig. 3.2 The rice unearthed in the Hemudu Site (provided by Zhejiang Institute of Cultural Relics and Archaeology)

gene sequencing. The results of their research show that the earliest occurrence of rice domestication dates back to 9,000 years ago in the Yangtze River Basin of China [3]. There are two different subspecies of cultivated rice, namely, japonica rice and indica rice. The previous studies suggest that the japonica rice and indica rice have their own ancestors in Oryzarufipogon, so the theory of multiple origins of Asian cultivated rice has been put forward. The indica rice originated in Southeast Asia or South Asia, while the japonica rice originated in East Asia. It is also said that Oryzarufipogon was first domesticated into indica rice, and then gradually evolved into japonica rice when spreading to the regions of high latitudes and altitudes. Recently, the scientists constructed a fine map of the rice genome-wide genetic variation and found that human ancestors first domesticated the japonica rice in the Pearl River Basin by using local species of wild rice, which then gradually spread northward after a long period of artificial selection. Another branch of the cultivated rice entered Southeast Asia, wherein the indica rice was produced after hybridizing with the local wild rice and experiencing continuous selection. Although this result is different from the existing archeological evidence, it proves once again that southern China is the origin of rice cultivation.

3.2 Improvement and Development of Rice Farming Technology

Rice farming, though originating in southern China ten thousand years ago, was developed considerably only a thousand years ago. After the Tang and Song Dynasties, with the increase of population and the southward shift of economic center in the country, the relationship between man and land tended to be tense. To produce as much rice as possible, the Chinese people at that time continuously expanded the rice planting areas. On the one hand, they reformed the drylands into the paddy fields; on the other hand, they opened up fields from mountains and waters by poldering, terracing, and erecting fields. At the same time, they developed intensive cultivation technologies, mainly including paddy field soil preparation technology with plow, harrow and *chao* (leveling land with a harrow-like implement) as the main body (Figs. 3.3, 3.4 and 3.5), the seeding technology with seedling transplantation as the main body and field management technology with the field cultivation and drying fields as the main body. The high-yielding rice becomes an important economic basis for supporting the late development of the empires, exerting influence on China's development in the last thousand years, and making important contributions to East Asia and the world.

The development of rice agriculture in South China also began with the improvement of the farm tools. In Tang Dynasty, Jiangdong Plow appeared in the areas of Jiangdong (the lower reaches of the Yangtze River) as a new farm tool. According to the Leisi Book by Lu Guimeng of the Tang Dynasty, Jiangdong Plow, characterized by flexible operation, is mainly composed of eleven parts, including plow shovel and plow wall made of iron, and the wooden-made plow bottom, shovel, policy quota, plow arrow, plow shaft, plow tip, plow bar, plow key, and plow plate. The application of Jiangdong plow indicates that the structure of the plow in China had been basically finalized. The Jiangdong Curved-Shaft Plow is widely used in the rice growing areas of China and Southeast Asia. In the seventeenth century, in Java, Indonesia, and other places, the Dutch people saw the local Chinese immigrants using this plow, and soon introduced it into the Netherlands, which later had an important impact on the improvement of this farm tool in modern Europe. The new plowing system

Fig. 3.3 Plowing field in *Plowing and Weaving Pictures*

Fig. 3.4 Harrowing field in *Plowing and Weaving Pictures*

Fig. 3.5 *Chao*-ing in *Plowing and Weaving Pictures*

formed after absorbing the characteristics of the Chinese plow then ushered in the revolution of western agricultural technology.

The Jiangdong plow, together with the harrow, "Liuzhou" (stone roller for threshing grain), "lize", "chao" (leveling land with a harrow-like implement) and other farm tools, formed the foundation of paddy field soil preparation technology in the South after the Tang and Song Dynasties. Different from the dryland farming technology system formed in the north of the country during the period of the Wei, Jin, Southern, and Northern Dynasties, which centered on plow-harrow- "mo" (a farm tool for leveling land), paddy field soil preparation technology in southern China features the use of "chao", which is used to level the field to facilitate rice growth following plowing and harrowing. Planting rice requires an even level of irrigation water in the paddy field to ensure that the rice seedlings grow uniformly and neatly. This requires that the plowed soil should be loose and flat. The application of "chao" marks the formation of the paddy field soil preparation technology system in South China.

Like millet, rice cultivation also resorts to seeds selection and breeding to expand the rice planting area and increase the yields under different natural and social conditions (Fig. 3.6). In the Northern Song Dynasty, the Champa Rice, which originated in the Champa (now south-central Vietnam), was introduced from Fujian to the areas along the Yangtze River, the Huaihe River, and Zhejiang province. Because of its

Fig. 3.6 Soaking seeds

characteristics of early maturity, drought tolerance, and no choice of land, especially being suitable for growing in high-lying land, it promoted the development of terraces and the increase of the rice yields. The Huang Lu Rice (Yellow Rapid Ripening Rice), because of its early maturity and waterlogging resistance, can grow normally when the water level of the paddy field exceeds the actual needs and has made a great contribution to the development of the lowland. Even the emperors of the time joined farmers in seeds breeding and popularization. In Qing Dynasty, Kangxi successfully bred a new fine variety of rice named "Yudao" (royal rice) by using the single plant selection method to consciously select the mutant rice plants and popularized it all over the country. C.R. Darwin, a British biologist, highly commented on Kangxi's move, pointing out that this variety of rice was valuable as it was the only species that could grow to the north of the Great Wall. Through the efforts of generations of the Chinese people, the species of rice kept increasing. The number of the rice species included in *Shou Shi Tong Kao* published in 1742, during the reign of Emperor Qianlong in Qing Dynasty is as large as 3429 kinds.

Different from the millet cultivation using direct seeding, the rice cultivation was mainly based on seedling raising and transplanting after the Tang and Song Dynasties (Fig. 3.7). As rice cultivation is carried out in water, weeding in the paddy fields becomes difficult. First, people pull out the rice seedlings which grow to a certain height together with weeds, and then plant them back when weeds are removed. This method is also used to replenish the seedlings. Later, it was found that the centralized seedling cultivation in the seedling field could facilitate centralized management

Fig. 3.7 The seedling transplanting

at the seedling stage and field management after seedlings transplanting, leaving sufficient time for the growth of the previous crop in the field and paddy field soil preparation, which may create conditions for the development of the multi-cropping system of rice. After the Tang and Song Dynasties, the technology of raising and transplanting the seedlings was widely popularized, and the "seedling rice" became the synonym of rice cultivation.

Transplantation facilitates rice field management. In the stage of paddy field management, the field cultivation and drying fields in the sunshine are the most critical steps (Fig. 3.8). Field cultivation has the function of loosening the soil and earthing up the seedlings besides weeding. The traditional rice cultivation attaches great importance to the field cultivation, which should be carried out three or four times. There are two main ways of rice field cultivation: one is hand cultivation; the other is foot cultivation. In Song and Yuan Dynasties, a new tool and method of the field cultivation was invented, named *Yun Dang* in Chinese, based on the method of using a hoe on the dry land and the improvement of the hoe itself.

Cultivating fields is a hard job, especially by hand, which requires the farmers to use their eyes, body, and hands well. Therefore, as a saying goes to describe hand cultivation, "pain on the waist and hands, focus in the eyes". Moreover, due to the hot weather and the mosquito bites, this job became much more painful. In order to reduce labor intensity and better protect themselves, the ancients invented a series of

Fig. 3.8 The fields cultivation

Fig. 3.9 The dragon bone water lift

the auxiliary farm tools, such as "yunzhao", "haoma", "fukc", "bigou", "tongzanh", etc.

The Field Cultivation is often carried out in conjunction with irrigation. Rice is one of the crops with large water demand, especially the late rice species in autumn. Water shortage seriously affects the yields of rice. To solve the problem of rice irrigation, the ancients also invented a variety of irrigation tools, such as the dragon bone water lift, scoop waterwheel, irrigation buckets, shaduf, water shuttles, and so on (Fig. 3.9, 3.10, 3.11 and 3.12), while building water conservancy projects.

Irrigation can not only meet the needs of rice for water, but also is a means to regulate the temperature of the paddy fields. In the Western Han Dynasty, people creatively used the location arrangement of the inlet and outlet of water flow to regulate water temperature according to the requirement of the rice growth for water temperature. In spring, when the weather is still cold, the water in the field should be kept warm and absorb more sunshine, so that the inlet and outlet should be on the same straight line. In summer, in order to prevent the water temperature from rising too fast, the water inlet and outlet are interlaced so that water flow in the field is facilitated to reduce the water temperature.

Fig. 3.10 The water lift by cow

Fig. 3.11 The scoop waterwheel

The drying fields are also a measure to control water and temperature in the paddy field. During the Great Heat period in summer when the rice seedlings grow vigorously, the water in the paddy field is discharged to expose the soil to the sun so as to fix the roots of the seedlings, which is called "drying fields", and then water is re-introduced into the field, which is called "returning water". Drying the paddy field can improve its environment, prevent the seedlings from overgrowth and lodging, and improve drought resistance and yields of the rice. This technology was seen in *Qi Min Yao Shu*, matured in the Song and Yuan Dynasties, and got widely used in the paddy fields to the south of the Yangtze River in the Ming and Qing Dynasties.

The technology of the rice cultivation successfully supported the development of China's agriculture after the Song Dynasty. The reason why Chinese civilization originated in the Yellow River Basin, which was established by millet and maintained by wheat, did not follow the fate of the disappearance of the ancient Egyptian and Babylonian civilizations after thousands of years, lies in the fact that the Yangtze River Basin, backed up by rice with great potential for increasing production, relieved the burden of the Yellow River Basin and continued the mission of promoting the prosperity of Chinese culture. Around 1000 A.D., China's population surpassed 100 million for the first time, while more than half of these people depend on the rice to survive. According to the estimation of Song Yingxing in the late Ming Dynasty, rice accounted for about 70% of the country's food supply, while wheat only 30%. Since

Fig. 3.12 The irrigation buckets

the Southern Song Dynasty, there have been the folk saying "Harvest of rice in Suzhou and Huzhou ensures the full supply of food in the country" and "Harvest of rice in Hunan and Hubei ensures the full supply of food in the whole country". The middle and lower reaches of the Yangtze River wherein Suzhou, Huzhou, Hunan, and Hubei are located were just the main rice producing areas in China. In 1935, Hu Huanyong, a geographer, drew a horizontal line from Aihui Town, Heilongjiang in northeastern China to Tengchong, Yunnan. The southeast regions to this line accounted for 36% of China's territory, while the population there is 96%; the northwest regions to the line accounted for 64%, while the population there is only 4%. A significant number of people in the southeast regions fed on rice as the staple food.

3.3 The Significance of Invention and Development of Rice Farming in China

As a great invention of China, the rice cultivation is of great importance to the world. Today, rice has become the staple food of more than 30 countries, and more than half of the world's population lives on rice. In Asia, 2 billion people consume 60–70% calories and 20% protein from rice or rice products. China is the largest rice producer among all countries, with the rice yields accounting for 35% of all around the world. In India, about 65% of the total population lives on rice as the staple food. Japan, the ninth largest rice producer in the world, has about 2.3 million rice farmers, taking rice as its "national food". In South Korea, straw has become an integral part of the Korean culture as symbolized by the "straw roof". In 2004, the United Nations set up "the international rice year" for the first time, the theme of which is "Rice Is Life"; this is the first time in the history of the United Nations to make such an arrangement for a crop.

All these mentioned above can be originated from the invention of rice cultivation in China. More than 2,000 years ago, the people of the states of Wu and Yue, who lived in the middle and lower reaches of the Yangtze River of China, crossed the sea to today's Kyushu area of Japan for fleeing from the chaos caused by war, bringing rice cultivation techniques of rice there. This is the beginning of rice cultivation in Japan, where people who were engaged in it are called Yayoi people, and the culture caused by it is called Yayoi culture. The pronunciation of rice in Japanese "稲(いね)" is just the way the people of Wu and Yue called rice in ancient China. Before adopting rice cultivation, Japan had been in the fishing and hunting period, namely the "Jōmon Culture" period, and had been taking China as an example in rice growing for a long time before the Ming and Qing Dynasties. The rice named "Datang Rice" imported from China in the twelfthand thirteenthcenturies became an indispensable variety of crop for low-temperature cultivation in reclaiming land from the sea of Japan. Korea was also influenced by China's rice culture. The technique of raising seedlings on a dry field, which originated from the Song Dynasty of China, was introduced into Korea, and then "Dry Field Rice" appeared in the agricultural books of Korea in the seventeenthcentury. Two dry field rice species "Mouzu (rice)" and "Ruizu (rice)" of Korea were imported from China.

The Rice Cultivation in Southeast Asia also developed under the influence of China. In the Han Dynasty of about 2,000 years ago, although Cochin Rice, grew in a place of northern Vietnam, was heard of in the Chinese mainland, the agricultural level of the Southeast Asian countries still lagged behind that of China. There plowing by the cow was developed under the influence of China mainland. In the Eastern Han Dynasty, Ren Yan, the Chief of Jiuzhen (now northern Vietnam) first introduced the plowing techniques from China to the areas under his jurisdiction, which then spread to the neighboring areas, such as Cochin and Xianglin (now central Vietnam). After 600 years of development, these places have caught up with China in agricultural technology. The bronze drums widely found in the Southwestern provinces of China like Yunnan, Guizhou, Guangxi, and Guangdong and neighboring countries such as Vietnam, Laos, Myanmar, Thailand, Malaysia, and Indonesian can serve as an important evidence of cultural links between China and Southeast Asia.

After thousands of years of evolution, the rice cultivation of China has been continuously making contributions to the world. Dr. Te-Tzu Chang (1927–2006), a Chinese at the International Rice Research Institute in the Philippines, worked out the "Miracle Rice" by utilizing a kind of indigenous rice of Taiwan, which greatly improved the yields of rice in Southeast Asian countries. Yuan Longping, a Chinese scientist on hybrid rice technology, has made outstanding contributions to maintaining world food security, with the increased production of rice feeding 35 million people in the world every year. The paper "Sequence Map of the Working Framework of the Rice (Indica Rice) Genome", written by the Chinese scientists independently, was published in the world-renowned American journal *Science* on April 5, 2002, and appraised as "the most significant milestone work which has a global impact on human's health and survival in the new century and has changed our botanical research" and "a landmark contribution of China to science and humanity".

References

1. Anping, Pei. 2004. *Rice Culture in the Yangtze River Basin*, 36–46. Hubei Education Press.
2. Chi, Zhang. 2000. The Remains of Pottery and Rice Phytoliths in Jiangxi Province Ten Thousand Years Ago. In See *The Origin of Rice, Pottery and Cities*, Yan Wenming and Yasuda Heung Heon, editors-in-Chief, 43–49. Cultural Relics Publishing House.
3. J. Molina, M. Sikora, N. Garud, J.M. Flowers, S. Rubinstein, A. Reynolds, P. Huang, S. Jackson, B.A. Schaal, C.D. Bustamante, A.R. Boyko, and M.D. Purugganan. 2011. Molecular Evidence for a Single Evolutionary Origin of Domesticated Rice. *Proceedings of the National Academy of Science* 108 (20): 8351. https://doi.org/10.1073/pnas.1104686108.

Xiongsheng Zeng Born in Xingan, Jiangxi in 1962. Researcher at the Institute of Natural Science History, Chinese Academy of Sciences.

Research interests: History of agronomy; General history of science.

Major works: *Chinese Agricultural History; General History of Chinese Agriculture: Volumes of Song, Liao, Xia, Jin, Yuan Dynasties.*

Awards: *Chinese Agricultural History* was selected into the Second "Three One Hundred" Original Publishing Engineering Science and Technology Books by the State Administration of Press, Publication, Radio, Film and Television of the People's Republic of China, the Nomination Award of the 3rd China Outstanding Publication Books;

The History of Ancient Chinese Science and Technology Entering the Palace won the 3rd China Outstanding Publication Book Award.

Chapter 4
Sericulture and Silk Production

Feng Zhao and Hui Liu

China is the cradle of silk. Mulberry planting, silkworm breeding, silk reeling, and weaving are all great inventions of the ancient Chinese people. With its unique charm, gorgeous colors, smooth and soft texture, and strong cultural and artistic characteristics, Chinese silk has not only contributed to beautifying the lives of the people in the world for thousands of years, but also linked the continents of Europe and Asia through the Silk Road, spreading the spirits of peace, freedom, cooperation, and tolerance. As the bridge of economic, political, and cultural exchanges between China and other countries, it has exerted far-reaching influence in the world.

4.1 Mulberry Cultivation and Silkworm Breeding

4.1.1 Cultivation of Mulberry

According to the archeological data, there were abundant wild mulberry trees in the vast areas along the ancient Yellow River and Yangtze River, which are fundamental for the development of silk industry. The origin of the artificial cultivation of mulberry remains unknown, but judging from the records of the mulberry forest sacrifices in oracle bone inscriptions of the Shang Dynasty, we speculate there might have been a certain scale of mulberry planting [1].

About one thousand species of the cultivated mulberries constitute three main systems of mulberry: Lu mulberry, white mulberry, and Jing mulberry. Originated

F. Zhao (✉)
China National Silk Museum, Hangzhou, China
e-mail: 1007043621@qq.com

H. Liu
Institute for the Natural Science History, Chinese Academy of Sciences, Beijing, China

J. Hua and L. Feng (eds.), *Thirty Great Inventions of China*,
https://doi.org/10.1007/978-981-15-6525-0_4

in Shandong before the Han Dynasty, Lu mulberry is known as an excellent and widespread variety of mulberry. In Jia Sixie's *Qi Min Yao Shu* completed in the Northern Wei Dynasty, a farming proverb goes: "With hundreds of Lu mulberries, abundant silk comes", which shows the importance of Lu mulberry at that time. Nowadays, the most common variety of Hu mulberry in Jiangsu and Zhejiang belongs to the Lu mulberry system. White mulberry and Jing mulberry are also widely distributed mulberry species, but they are not so extensively used as Lu mulberry.

There are several methods of reproducing mulberry, including sowing, cuttage, layering, and grafting, of which sowing is the oldest method where mulberry seedlings are to be transplanted after sowing and sprouting. Grafting is an important achievement in cultivating mulberry. As an advanced asexual propagation method, it is of great significance to the rejuvenation and regeneration of the old mulberry trees, the preservation of good properties of them, the acceleration of reproduction of the mulberry seedlings and the cultivation of fine varieties. The technique of grafting has a long history in China, appearing in the late Warring States Period at the latest [2], and its principles and procedures are found in *Qi Min Yao Shu*, but its application in mulberry cultivation was first seen in Song Dynasty. Chen Fu recorded mulberry grafting in *Agriculture Book*. In Yuan Dynasty, the mulberry grafting technique was popularized in a wide range, as discussed in many agriculture books in detail. For example, four kinds of grafting methods were proposed in *Nong Sang Ji Yao*, including cuttage grafting, cleft grafting, dimple grafting, and batch grafting. Wang Zhen summarized six commonly used grafting methods in *Agriculture Book* and pointed out that the Jing mulberry could be grafted with the Lu mulberry to improve the performance of the mulberry trees.

Among the keys to mulberry cultivation, cultivation of the tree type is the most important. There are two tree types of mulberry, namely tree (tall) and mulberry (short); early in northern China, the former was mostly used, and later in the southern areas, the latter was more planted.

4.1.2 Domestication of Silkworm

Silkworm belongs to the Bombycidae of Lepidoptera. Its life will go through four stages: egg, larva (silkworm), pupa, and adult (moth). Hatched from eggs, the silkworms will slough their skin 3–4 times; about 30 days later, they mature, spin cocoons and become pupae; a week later, the moths come out of the cocoons to mate and lay eggs before they die (Fig. 4.1). Silk is extracted from cocoons to be the raw materials for weaving the silk products. It is by a long-term domestication that the wild silkworms became the domestic silkworms. The modern archeological discoveries directed proved when this domestication began. In 1926, half a silkworm cocoon (Fig. 4.2) was unearthed in the Yangshao Cultural Site (6,000–5,600 years ago) in Xiyin Village, Xia County, Shanxi [3]. In 1958, some textiles and belts like sheets, threads, and ribbons (Fig. 4.3), were excavated in the Qianshanyang Site (4,400–4,200 years ago) in Huzhou, Zhejiang. In 1984, silk yarns and ribbon fabrics

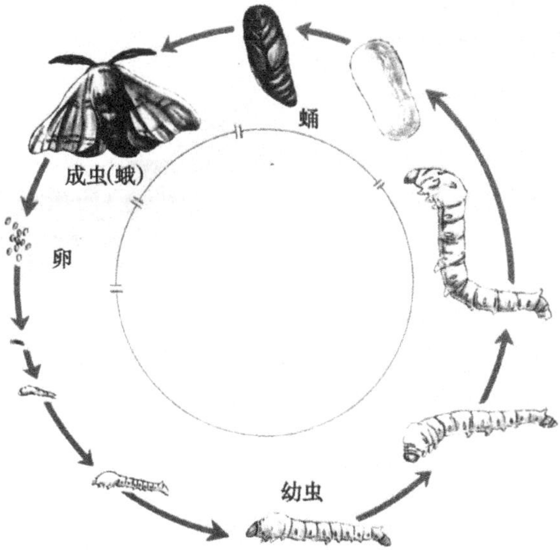

Fig. 4.1 Life of silkworm, Bombyx mori

Fig. 4.2 Half of silkworm cocoon (6,000–5,600 years ago, unearthed in Xiyin Village, Xia County, Shanxi)

Fig. 4.3 Silk sheet (4,400–4,200 years ago, excavated in Qianshan Yang Ruins, Huzhou, Zhejiang)

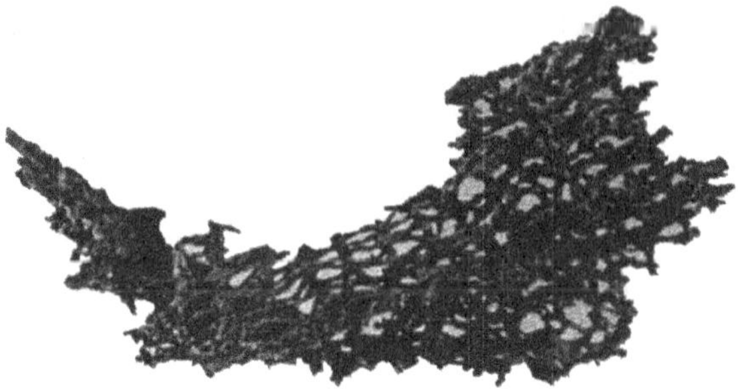

Fig. 4.4 Woven fabric (about 5,500 years ago, excavated in Qingtai Village of Xingyang, Henan)

(Fig. 4.4), were unearthed in the Yangshao Cultural Site (about 5,500 years ago) in Qingtai Village of Xingyang, Henan. The silk fabrics unearthed from the last two sites have been identified to be made of the silk of the mulberry silkworm, because the cross-section of the silk fibers used is triangular [4]. The cocoon found in the Xiyin Village remains uncertain in terms of which kind of silkworm it is made. It is said to be a product of the wild silkworm [5]; another saying goes that it is a cocoon of the domestic silkworm, though it is small due to its insufficient evolution [6]. In addition, there are still many carvings of silkworms or pupae of the Neolithic period. For example, four pairs of the worm-shaped images on a tooth carving unearthed in the Hemudu site (about 5000 B.C.) in Yuyao, Zhejiang were regarded as the silkworm pattern (Fig. 4.5), by many scholars [7]. These relics show that the silkworms had become the attention of our ancestors more than 7,000 years ago, while 5,000 years ago, China began the domestication of the silkworm and silk production.

During the Warring States Period, people had a better understanding of the physiology and ecology of silkworms. For example, in *Ode to Silkworm*, Xun Kuang summarized the habits of silkworms as "sleeping in winter and crawling in summer,

Fig. 4.5 Tooth carving in silkworm pattern (about 5,000 B.C., excavated in Hemudu site in Yuyao, Zhejiang)

eating mulberry and spinning silk, getting peace after turbulence, born in summer but hating heat, loving dampness but disliking rain, sleeping and waking up three times in a life, which were the species commonly raised during the Warring States Period and would start spinning cocoons in 21 or 22 days after its birth". At the same time, Xun Kuang also put forward the proposition that the silkworms can be of two kinds: males and females.

Though there is only one species of the domestic silkworm in animal taxonomy, its varieties are many, different in voltinism and moultinism. Voltinism refers to the times of silkworm being hatched in a year without the human factors, and moultinism the times of molting in the larval stage, of which people had a deep understanding during the Wei-Jin, Southern and Northern Dynasties period. In *Qi Min Yao Shu*, Jia Sixie made various classifications of silkworms from different perspectives, such as white-headed silkworms, Jieshi silkworms, Chu silkworms, black silkworms, and gray child silkworms in terms of the body color and markings, autumn mother silkworms, autumn middle silkworms, old autumn silkworms, and the late autumn silkworms in terms of the breeding time, the floss silkworms and co-cocoon silkworms in terms of the cocoon shape. In his *Record of Yongjia County*, Zheng Qizhi made a very important summary of the character, generation relationship, and low-temperature control of the hatching time of the silkworms.

In Song Dynasty, one-voltinism and three-moultinism silkworms, with stronger disease resistance and adaptability, were raised in the north of China, while one-voltinism or two-voltinism and four-moultinism silkworms, which spin more silk, were bred in the south with more suitable climate for the growth of silkworms. In addition, due to their damage to mulberry and poor quality of silkworm cocoons, two and multi-voltinism silkworms could hardly be seen in the north and a small number of them were raised in the south until the Song and Yuan Dynasties, and with the development of the silkworm raising technology, they were paid attention gradually in the north. In Ming Dynasty, the species of the silkworm gained further improvement. As recorded in *Tian Gong Kai Wu*, there are two kinds of silkworms: early silkworm and late silkworm. The late worms are hatched out five or six days earlier than the early ones, making cocoons first and laying eggs as moths when the latter are cocooning. The "early silkworm" and the "late silkworm" mentioned above refer to the one-voltinism silkworm and the two-voltinism silkworm, respectively.

Breeding new silkworm with hybrid is a great creation in the development of the silkworm raising technology. According to *Tian Gong Kai Wu*, better species of the silkworm was cultivated by mating the "early male" silkworm (one-voltinism male silkworm) with the "late female" silkworm (two-voltinism female silkworm), and hybridizing the white male silkworm with the yellow female silkworm would give birth to silkworm spinning brown cocoons. It can be seen that people in the Ming Dynasty have learned to improve the variety of the silkworm by utilizing the biological genetic characteristics and the method of mating hybridization. This technology came hundreds of years earlier than the discovery of Mendel's Law in the West, which fully reflects the wisdom of the working people in China, but unfortunately, it has not been further studied and applied.

4.2 Silk Reeling and Weaving

4.2.1 Silk Reeling

Silk reeling is to separate silk from cocoons to get filaments, which then go through the process of winding, blending, and twisting. According to the silk fabrics unearthed from the Qianshanyang Site in Huzhou, Zhejiang, this technology appeared in the late Neolithic Age and got quite mature in the Shang Dynasty. The silk fragments found on the bronze and jade wares of the Shang Dynasty are all filaments, which are smooth and even, proving that the silk reeling technology at that time had reached a certain high level.

There are two important technological parameters in silk reeling: water temperature and humidity. In the Song Dynasty, people paid great attention to the water temperature for soaking cocoons, which was controlled to be about 80 °C; as noted in *Silkworm Book* authored by Qin Guan, the boiling water used for soaking cocoons should be like the eyes of crabs. At the same time, the reeled filaments were heated over a fire to facilitate the ensuing processes and keep fresh of the silk color. Chen Fu wrote in *Agricultural Book* that frequent change of water when boiling and baking with fire may keep the silk bright and clean. In order not to affect the brightness of the silk, dry, and smokeless firewood should be used. This technique is recorded in Song Yingxing's *Tian Gong Kai Wu*.

According to some poems of the Tang Dynasty, the silk reeling machines were quite widespread at that time. In Song Dynasty, the design of the pedal reeling machine was basically fixed. As described in *Silkworm Book* (by Qin Guan) with many pages and depicted vividly in *Silkworm Weaving Pictures,* the pedal reeling machine in Song Dynasty has three parts: the frame part, the button and twisting sheath part, and the winding part. The latter two parts are more important. The button part includes Coin's Eye (button), Lock Star (drum wheel), etc. In *Silkworm Weaving Pictures*, the Coin's Eye of silk reeling machine can't be clearly seen, only a bamboo pole is erected and then put out horizontally, and two locking stars are installed on it.

The winding part is the most complex of the pedal reeling machine, including the winding device, *si yue* (frame for winding silk), and transmission structure. In *Silkworm Book*, Qin Guan accurately describes the reeling device of the reeling machine at that time. In function, the drum is equivalent to today's eccentric disk. The ladder is called the connecting rod, and the hook on the ladder is the twisting device. The transmission mechanism drives the drum to rotate, causing the winding rod to move horizontally, so that the filaments can be evenly wound on *si yue*. Although there is no mention of the pedal drive device in Qin Guan's *Silkworm Book*, it is clearly depicted in *Silkworm Weaving Pictures* (Fig. 4.6) in Empress Wu's annotated edition and the *Silkworm Weaving Pictures* (Fig. 4.7) in Liang Kai's annotated edition.

Developed on the basis of the hand reeling machine, the pedal reeling machine drives the rotation of *si yue* and eccentric disk through the up-and-down movement

Fig. 4.6 The silk reeling in *Silkworm Weaving Pictures* (in Empress Wu's annotated edition of the Southern Song Dynasty, Heilongjiang Provincial Museum)

of the pedal rod. In this way, the workers can free their hands to do the job of seeking floss and adding floss, thus greatly improving the productivity.

4.2.2 Silk Fabrics

During the Shang and Zhou Dynasties, the silk weaving technology in China developed and improved greatly. By the Warring States Period, there had been a complete system of silk varieties, including *juan* (thin, tough silk), *luo* (silk gauze), *qi* (the figured woven silk material), *jin* (brocade), *xiu* (embroidery), *zhui* (stitch), and *bian* (weaving). From the Qin Dynasty to the Song Dynasty, with the improvement of spinning, weaving, dyeing, and smoothing technology, rich varieties of the silk fabrics were developed. The "three elementary weaves" in modern textile histology—plain weave, twill weave, and satin weave—all appeared in Song Dynasty, and the transition from warp-weave to weft-weave was completed in Tang Dynasty. By the Ming

Fig. 4.7 The silk reeling in *Silkworm Weaving Pictures* (in Liang kai's annotated edition of the Southern Song Dynasty, Cleveland Museum, America)

Dynasty, there had been a clear method of classification of the silk fabrics and a complete system of the designation of silk fabrics. In Qing Dynasty, there were many more varieties of silk. In addition to brocade, damask silk, *luo* (silk gauze), satin, *juan* (thin, tough silk), and velvet, which were classified according to their fabric characteristics, there were many varieties with local features, such as *yun* brocade and *song* brocade.

In the development of the silk weaving technology, brocade is the most representative of the weave technology and weaving level. As said in an ancient book *Explanation of Names*, brocade, which is as expensive as gold, is hard to make due to its complicated technology and superb weaving technology. Brocade is a kind of gloss silk fabric and colorful fabric, showing a variety of patterns in various colors through the change of weave. The analysis of the fabrics at that time shows that most brocades are heavy fabrics. In Song and Yuan Dynasties, a large number of the gloss silk fabrics or heavy fabrics appeared, resulting in less use of the name of brocade; they were mostly called *luo* or satin.

In the literature, the word "brocade" first appeared in *Xiangbo of Siao Ya, The Book of Songs*. But the earliest brocade identified was the warp brocade which emerged in the Western Zhou Dynasty. It was woven by the rearrangement of warp thread, so it was named the warp brocade. In the tombs of the Spring and Autumn Period and the Warring States Period, the warp brocade became a common brocade species, the most famous of which is the dancer animal brocade (Fig. 4.8), from the Mashan Chu tomb in Jiangling County, Hubei. It uses deep red, dark yellow, and brown warp threads and brown weft threads, with the patterns of dancer, Chinese dragons, phoenix, kylin, and geometric patterns. With weft layout, it is 5.5 cm in warp and

Fig. 4.8 The brocade with dancer and animal pattern in the Warring States Period (excavated in Mashan, Jiangling County, Hubei)

49.1 cm in weft, showing that the warp brocade weaving at that time had adopted the multi-heddle jacquard loom [8]. Brocade was also very popular in Han Dynasty. Among the brocades unearthed from the Mawangdui Han Tomb in Changsha, Hunan, there are velvet brocade and convex brocade. They were made by weaving a layer of velvet pattern on the pattern of brocade. This kind of brocade is probably the "brocade embroidery" recorded in the literature of the Han Dynasty [9].

Weft brocade first appeared in the Wei, Jin, and Southern and Northern Dynasties. A large number of the plain weft brocades were unearthed from the tombs of the fifth century in Turpan, Yingpan, and Huahai of Xinjiang, and most of them are the simple animal cloud patterns, indicating that weft brocades began to be used in silk weaving. At the beginning of the Tang Dynasty, the twill weft brocade made its debut and became popular. But for weft brocade, we can still divide them into two types according to the details of weaving. The so-called Western type of the weft brocade can also be called the Persian brocade, Sogdian brocade, and *satalaqi*; with a strong Z-twist on its warp and patterns of a clear Western Regions' style (Fig. 4.9), it might originate in the Sogdian region of Central Asia. Another type is the Tang-style weft brocade (Fig. 4.10). It uses S-twist warp, with the main patterns of treasure flowers or flowers and birds, mostly produced in the Central Plains.

By the middle and late Tang Dynasty, this kind of weft brocade had changed greatly in basic structure and weaving technology, and a new kind of weft brocade, called Liao-style weft brocade, appeared. As this kind of brocade was discovered in the process of analyzing the silk fabrics of the Liao Dynasty and it meets the basic characteristics of the brocades in Liao Dynasty, so we call it the Liao-style weft brocade. The basic difference between Liao-style weft brocade and Tang brocade lies the different functions of their Ming warp. The Ming warp in Tang weft brocade is only consolidated on the surface and produces twill effect, while the Ming warp in Liao-style weft brocade appears on the surface and the reverse side of the fabric once, the rest lying between the upper and lower weft lines, the same position as the clamp warp (Figs. 4.11 and 4.12).

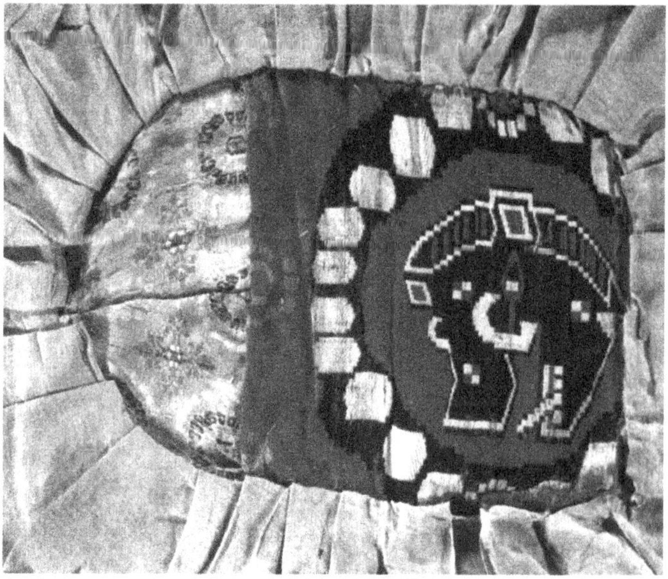

Fig. 4.9 The bead pig's head brocade of the Tang Dynasty (unearthed in Ashtar, Turpan, Xinjiang)

Fig. 4.10 The Xiaoke Baohua brocade (unearthed in Ashtar, Turpan, Xinjiang)

Fig. 4.11 Part of the Liao-style twill weft brocade

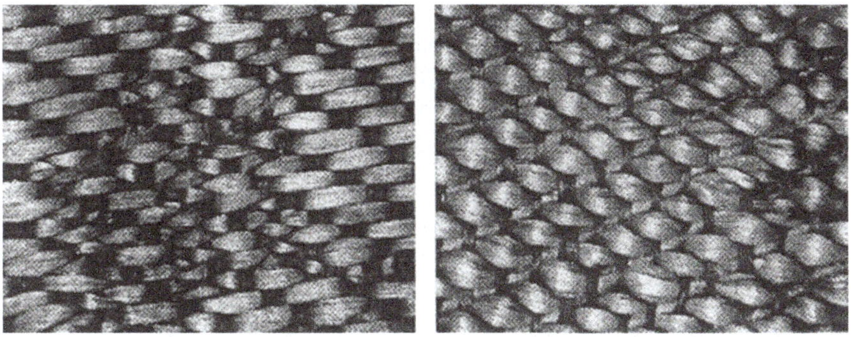

Fig. 4.12 Part of the Tang twill weft brocade

The Liao-style weft brocade falls into two main categories: the twill weft brocade and the satin weft brocade. The twill weft brocade can also be divided into the ordinary Liao-style twill weft brocade, the Liao-style floating twill weft brocade, the makeup gold twill weft brocade and the Liao-style diamond twill weft brocade. The Satin weft brocade refers to the double-weft heavy fabric with satin as the basic consolidation structure divided into the ordinary Liao-style satin weft brocade, the floating satin weft brocade, and the makeup gold weft brocade. The warp of Liao-style weft brocade is generally not twisted obviously, the Ming warp is always single and the clamp warp is usually two or even three a pair; the weft is of five to seven colors. The structure of such weft brocade is also widely applied in the brocades of the Song Dynasty. The clouding flower brocade of the Northern Song Dynasty unearthed from the Ruiguang Tower in Suzhou, the Five-Dynasties brocade unearthed from the underground palace of Leifeng Tower in Hangzhou, and the Diamond Sutra brocade of the Liang Dynasty in Liaoning Museum (Fig. 4.13), all adopted this structure, but changed in color matching, twisting, and other respects.

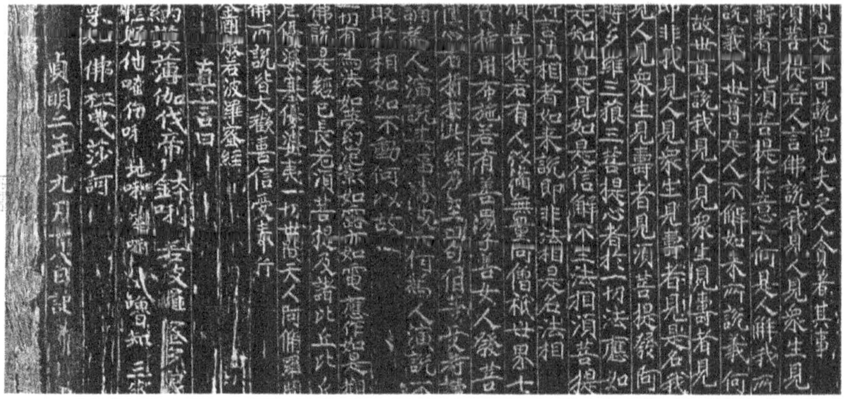

Fig. 4.13 The Late-Liang Dynasty Diamond Sutra Brocade (Liaoning Provincial Museum)

The brocade, with gold woven in it, is called the gold brocade. Nasich (Fig. 4.14), a Persian word for the gold brocade, is the most famous and characteristic gold brocade in the Yuan Dynasty [10]. At that time, most of the high-grade clothes of the government officials were made of nasich, and the government set up the dyeing and

Fig. 4.14 Griffin nasich in the Yuan Dynasty (private collection)

Fig. 4.15 Part of the satin brocade

weaving divisions in qualified areas throughout the country to concentrate weavers on weaving nasich as materials for making the daily-used curtains, mattresses, chair cushions and cushions for the brick beds. According to Marco Polo, the military tent was also made of this brocade.

In the Ming and Qing Dynasties, the brocade was mainly made of special-structure heavy weave, but the flower weave is of various kinds. The plain weave, twill weave, and satin weave were all used at random, and they are mostly with warp ground and weft flower. When the ground structure was satin structure, the brocade was called brocade satin. The brocade satin, a kind of popular brocade in Qing Dynasty, mostly uses six irregular satins as the ground and inserts one consolidation warp every three or six ground warps. The structure of the consolidation warp and the weave weft is of twill or plain weave kinds. Because the warp float of the satin brocade is long, the ground part of the satin brocade is delicate, but also relaxed, not as flat as the twill (Fig. 4.15). The brocade satin often uses the gold and silver thread in its weft, so it has gorgeous decoration and more floral patterns.

4.3 Loom

4.3.1 Pedal Loom

The pedal looms probably appeared during the Warring States Period, and it was commended by Dr. Joseph Needham as a great contribution of China to the development of the textile technology in the world. In the story *Jichang Learning Shooting*

Arrow recorded in the book *Lieh-tzu*, Jichang ever lied down under the loom of his wife to practice his gazing ability by observing the up-and-down movement of the pedals. On the stone reliefs of the Eastern Han Dynasty, there are many images of pedal looms, such as the looms found in Hongdaoyuan and Longyangdian in Teng County, Wuliang Temple in Jiaxiang County, Xiaotangshan Guoju Temple in Changqing County, Ciyun Temple in Jinyang Mountain in Jinning, Shandong, Liucheng in Pei County, Honglou in Tongshan District, Caozhuang in Sihong County, Jiangsu, and Zengjiabao in Chengdu, Sichuan. Seen from the stone reliefs unearthed from the Wuliang Temple, Honglou, Caozhuang and other places, the method of connecting the pedal and the heddle on the looms is very special. Under the warp surface and above the middle of the loom, there seem to be two mutually perpendicular short rods, which connect two pedals by flexible ropes or rigid wooden poles respectively (Fig. 4.16). It is inferred from the later pedal vertical looms that this kind of slant loom should have adopted a center shaft device, in which a pair of the right-angled short rods form two connecting rod structures with two pedals by crank or rope. This seems to be more clearly demonstrated in a glazed pottery model of the looms in the Eastern Han Dynasty collected by the French scholar Madame Ribb (Fig. 4.17). Referring to the evidences above, we restored a pedal slant loom of the Han Dynasty (Fig. 4.18). According to its principle, we can call it a middle axis pedal slant loom.

The pedal slant loom became gradually rare, and its remains can be seen in *Ziren's Legacy* of the Yuan Dynasty. It was then called the vertical loom, with a basically vertical frame. The earliest images of the vertical loom are found on the pictures in *Huayan Relief Sculpture* on the north wall of K98 in the Dunhuang Mogao Grottoes of Gansu (Fig. 4.19), but in the Dunhuang documents at the end of the Tang Dynasty, the cotton fabric named vertical loom appeared. Later, the image of the vertical loom could be seen on the frescoes of the Northern Song Dynasty in Kaihua Temple of Gaoping, Shanxi, and *Silkworm Palace Picture* of the Ming Dynasty collected in the National Museum. However, the most detailed records of the vertical loom are in

Fig. 4.16 The textile stone relief of the Eastern Han Dynasty (unearthed in Caozhuang, Sihong County, Jiangsu)

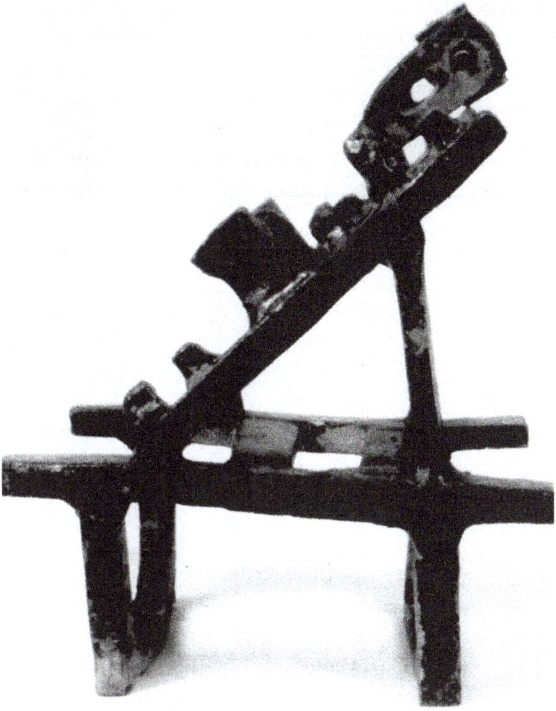

Fig. 4.17 The glazed pottery model of the loom of the Han Dynasty

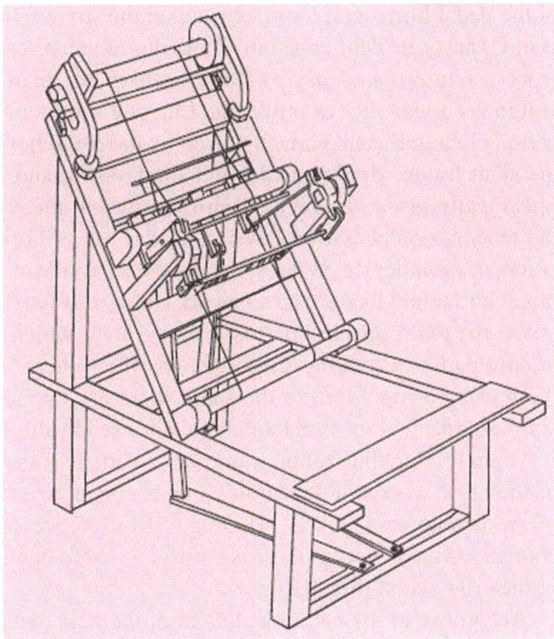

Fig. 4.18 The restored middle axis pedal slant loom of the Han Dynasty (Zhao Feng)

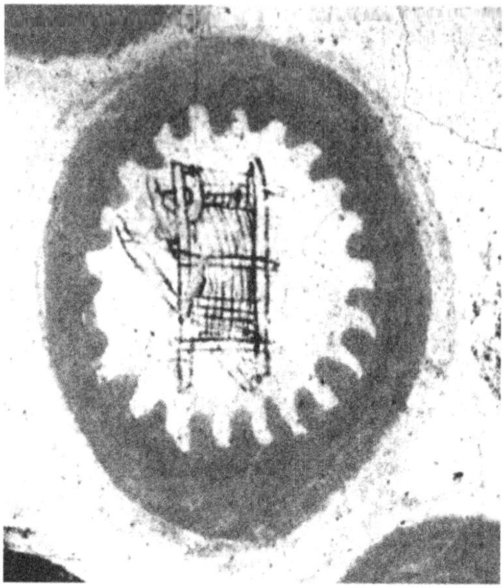

Fig. 4.19 The image of the Five-dynasties vertical loom (K98 north wall, Mogao Grottoes, Dunhuang, Gansu)

Ziren Legacy by Xue Jingshi in the Yuan Dynasty, and it can be restored according to the recorded sizes and graphic illustrations in the book.

There are many more varieties of the pedal looms. One of them is the pedal horizontal loom that controls the tension by the weaver's body. The earliest image of the pedal horizontal loom appears on the stone reliefs of the tomb of the Eastern Han Dynasty in Zengjiabao of Chengdu, Sichuan, and the most explicit records of it are also in *Ziren Legacy* by Xue Jingshi in the Yuan Dynasty. This kind of loom is still in use today. For example, the Liuyang summer cloth in Hunan and the Fufeng cotton cloth in Shaanxi are all woven by the pedal horizontal loom, characterized by the slant frame, single heddle, and single lever, and control of tension with waist. Specifically, it can be divided into two types: the direct horizontal loom without the tension compensation device and the lifting horizontal loom with the tension compensation device. What's being used in Yao Minority Area of Hunan is just the direct horizontal loom. Its framework consists of two horizontal frames and two foot poles; the main opening part is a lifting lever which is mounted on the frame; the middle part is a rotating shaft; behind which there is a short rod, connected with the foot by a rope. Before the shaft, there are two short rods to lift a heddle. The simplest lifting horizontal loom is the one used by the Tujia people in western Hunan. It also has a slant horizontal frame and a vertical frame in the middle. There is a pair of the crow woods on it, with two ends connecting the pedal rod and the opening of the heddle respectively. The biggest difference of this loom lies in its tension compensation device, as there is a warping rod to connect the pedal and the back end of the crow wood (Fig. 4.20).

According to the paintings left from the past dynasties, the pedal looms mostly adopted double heddles, using two levers to control the two heddles to weave fabrics

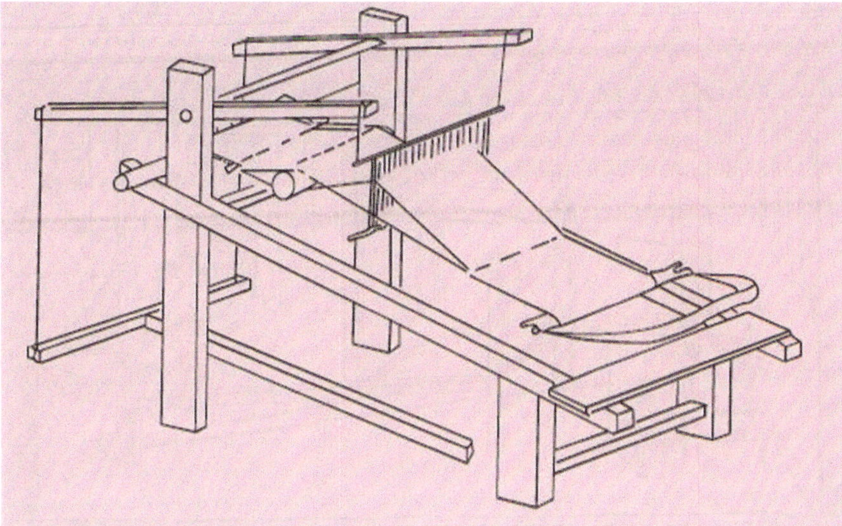

Fig. 4.20 The restored lifting horizontal loom

flat. The warp face is generally horizontal. The double-heddle pedal looms were depicted in *Silkworm Weaving Pictures* by Liang Kai in Southern Song Dynasty and *Plow and Weave Pictures* by Cheng Qi in the Yuan Dynasty. The looms in the two books are basically the same in shape, they have two pedals with the long one connected with a long crow wood to control one of the heddles, the short one connected with two short crow woods to control the other one. Two groups of the crow woods stand on the frame in the middle of the loom equivalent to the early Horse Head, but much larger than the horse's head. The warp face is no longer so slant as the slant loom of the Han Dynasty. The warp axis is slightly higher, with a warp wood in the middle to press warp yarns lower, which is also a tension compensation device (Fig. 4.21). Same as the looms depicted in *Pictures for Convenience of People* of the Ming Dynasty, this double-heddle loom is opened by lifting the heddles up through the crow wood with the pedals, while the two heddles are driven and lifted by the pedals independently. Therefore, this double-heddle pedal loom is also called the single-acting double-heddle loom, which is still in use today.

The existing silk tapestry loom is also of the same kind, but its crow wood is transversely placed. On top of the rack, there is an axis in the same direction with the warp. Two pieces of crow wood in the same direction with the weft are placed on the axis. Under the frame, there are two pedal rods in the same direction with the crow wood. The rods are connected with the crow wood by rope along the frame. This device is quite similar to the Fanzi Device on the jacquard loom in the Ming and Qing Dynasties.

At the turn of the Yuan and Ming Dynasties, the interactive double-lever and double-heddle loom (Fig. 4.22) emerged. This kind of loom is characterized by the

Fig. 4.21 The pedal double-heddle loom in *Silkworm Weaving Pictures* (Liang Kai, Southern Song Dynasty, Cleveland Museum)

use of the lower-pressure heddle, in which two pedals are connected to the lower end of the two heddles, while, on the top of the loom, the levers are used to connect the upper ends of the two heddles. So when the weaver steps down on a pedal, a heddle will press down a group of warp. At the same time, the upper part of the heddle pulls the lever on the top of the frame to lift another heddle so that a clearer shed is formed. To open another shed, the weaver will step down on another pedal. The shedding structure is very clear and became very popular in Europe in the twelfth and thirteenth Century. The single-acting looms shifted to interactive ones in China, probably owes a lot to the prosperity of cultural exchanges between the East and West in the tirteenthcentury.

4.3.2 Jacquard Loom

As the most complicated loom, the jacquard loom uses the most complicated weaving technology named jacquard technology, which is also a complex information storage technology. The complex jacquard information of the patterned silk fabrics must be stored with the jacquard devices installed on the loom so that the shed information in memory can be repeatedly used like today's compiled computer programs. According to the brocades unearthed from the Chu Tomb in Mashan of Jiangling County in Hubei, the jacquard silk weaving technology of China had already been matured during the Warring States Period.

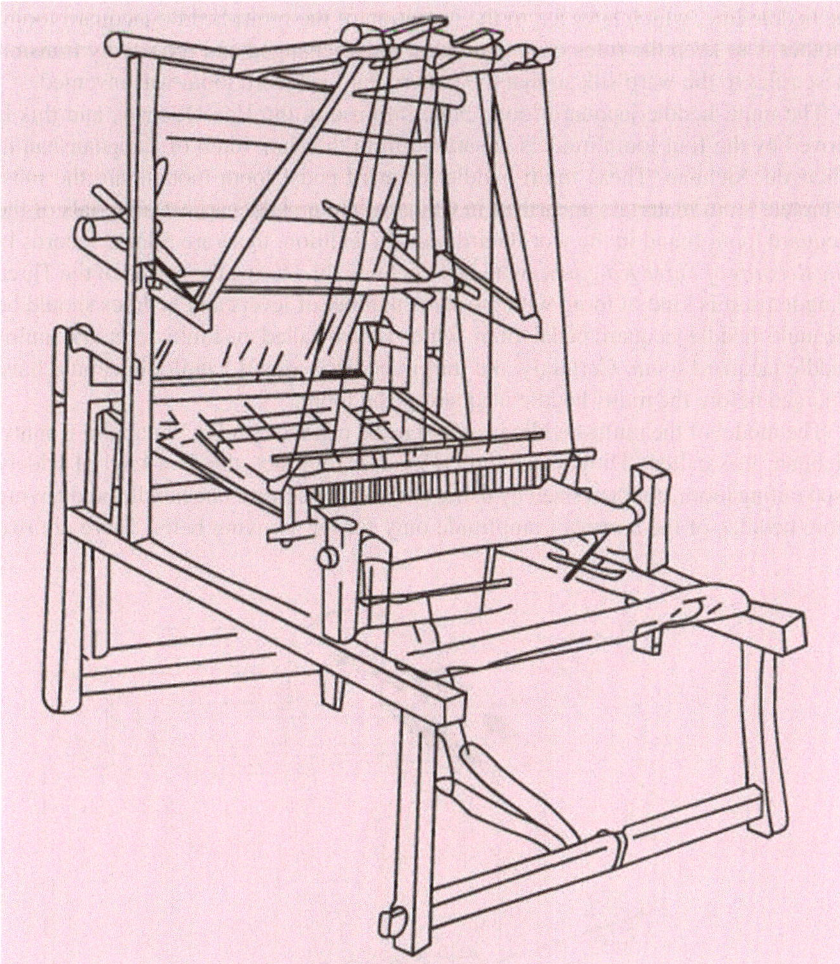

Fig. 4.22 The interactive double-lever and double-heddle loom

However, the development of the jacquard technology was not accomplished overnight, and it experienced a process from cross stitching to jacquard. All looms can be used to pick patterns on them by the stitching rod; especially the original waist looms, slant looms and horizontal looms were ever used to weave the fancy fabrics. There are two methods of cross-stitch: one is to stitch one weft and weave one weft, which requires high competence and confidence of the weavers; the other is to stitch one cycle and weave one cycle, which is more widely used. However, these methods still could not improve the efficiency, because the information of the cross-stitch could not be stored and reused for a long-term. To solve this problem, the ancients worked out another two approaches, which then inspired the jacquard technology. One is to "soften" the stitching rod, i.e., to replace the stitching rod with the heddle line, which later led to the invention of the multi-heddle jacquard loom; another is to keep the rules of the stitching rod unchanged and repeatedly transmit these rules to the warp silk so that the pattern-card jacquard loom was invented.

The multi-heddle jacquard loom came into use in the Han Dynasty, and this is proved by the four loom models unearthed from the Han Tomb of Laoguanshan in Chengdu, Sichuan. These multi-heddle jacquard pedal loom models are the most complete loom materials unearthed in China so far, and the earliest materials of the jacquard loom found in the world until now. In addition, there are related records in *The Records of Three Kingdoms* noted by Pei Songzhi. During the period of the Three Kingdoms, this kind of loom with the same number of levers and heddles should be the multi-heddle jacquard pedal loom, which is now called the multi-lever and multi-heddle jacquard loom. Certainly, the multi-heddle jacquard handloom should have emerged before the multi-heddle jacquard pedal loom.

The model of the multi-heddle jacquard pedal can be found in Shuangliu County, Sichuan. It is called "Dingqiao loom" (Fig. 4.23). In fact, this is a kind of widely used railing loom, characterized by using one lever to control one heddle, and having more heddles of the narrower amplitude only for the weaving belts. There are two

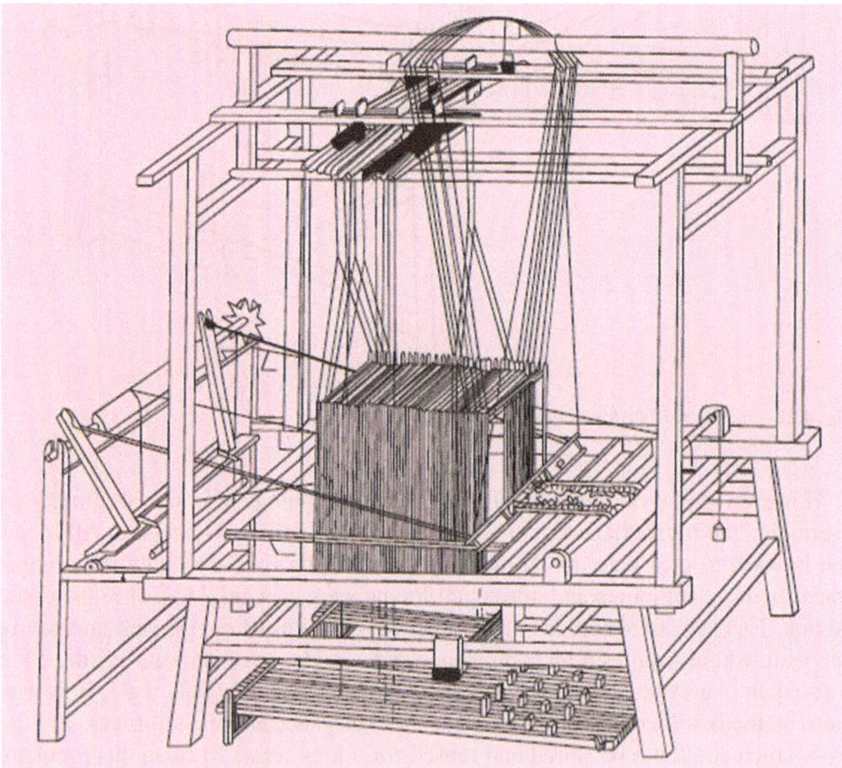

Fig. 4.23 The Dingqiao Loom (Hu Yuduan, etc.)

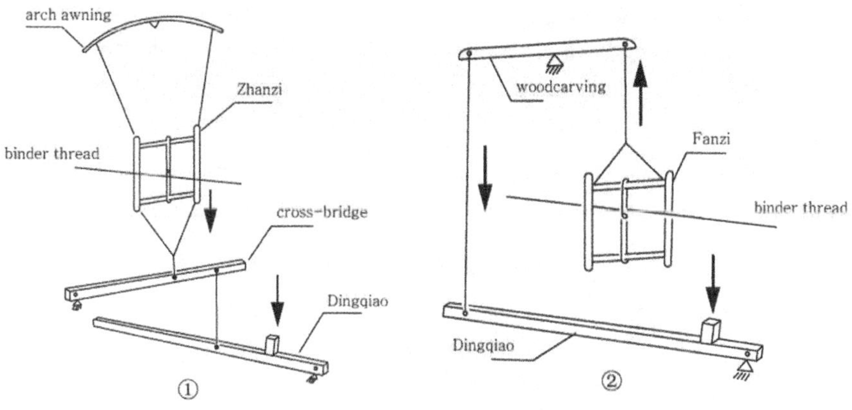

Fig. 4.24 The opening of the heddles on the Dingqiao loom

kinds of heddles used on Dingqiao loom: one is shaft raising, also named Fan Zi lifted by the pedal through the crow wood; the other is called Zhan Zi, whose lower border is pulled by the peal to press the heddle lower (Fig. 4.24). The Dingqiao loom is not one kind of the multi-heddle pedal looms of the Han Dynasty, because in Han Dynasty there were no shaft raising and the width of the fabrics at that time is much larger than those made on the Dingqiao loom. But their main principles should be the same.

Around the early Tang Dynasty, the bundle-heddle jacquard loom made its debut in history. Featured by threaded pattern-cards, the bundle-heddle jacquard loom was influenced by the 1-N harness system on the weft brocade loom in Central Asia. A brocade with a Chinese character "吉" and the paired sheep pattern, unearthed from Turpan in Xinjiang, may be a proof of the existence of the bundle-heddle jacquard loom at that time. In the early Tang Dynasty, a large number of the small pattern brocades emerged, and they are clearly the products of the bundle-heddle jacquard loom with 1-N harness device [11]. The jacquard described in *Poetry of Women Weavers* by Yuan Zhen in the Tang Dynasty is just this kind of jacquard, but its image was not seen until the Southern Song Dynasty. In *Silkworm Weaving Pictures* collected by Heilongjiang Provincial Museum, there is a bundle-heddle jacquard loom (Fig. 4.25) with a straight frame, a raised flower tower in the middle and a hanging pattern-card on the tower, and two ground heddles before the flower tower to lift the pedals by the crow woods. According to the geographical and historical background and the pictures in *Silkworm Weaving Pictures*, it is a typical damask silk loom [12], also the earliest and fairly complete image of the jacquard loom so far.

This kind of jacquard loom with the straight frame, commonly used in the areas south of the Yangtze River, is mainly suitable for weaving the light and thin fabrics, such as silks and satin. The Luo jacquard loom in the *Plow and Weave Pictures* (Fig. 4.26), collected by the National Museum of China, is also of this kind, and its

Fig. 4.25 The bundle-heddle jacquard loom in *Silkworm Weaving Pictures* (Southern Song Dynasty, collected by Heilongjiang Provincial Museum)

shape is similar to that of the damask silk loom described in the *Silkworm Weaving Pictures*, but it has a double warp axis and four heddles.

By the Ming Dynasty, the bundle-heddle jacquard technology had been quite complete and popular. The records of a flower-tower jacquard loom and its images in *Tian Gong Kai Wu* (Fig. 4.27), by Song Yingxing demonstrate the level of the bundle-heddle jacquard weaving technology at that time. The jacquard loom in Qing Dynasty is featured by the increased inclination of the frame.

The peak of the development of the bundle-heddle jacquard loom is the big flower-tower loom, such as the Nanjing copy satin loom and the makeup loom. Its records can be found in books such as *Can Sang Cui Bian* by Wei Jie, *Bin Feng Guang Yi* by Yang Shen and *Feng Lu Xiao Zhi* by Chen Zuolin in Qing Dynasty. The big flower-tower loom is characterized by large and circular pattern-card, and it can weave the fabrics for making the garments of emperors, with the patterns circulating over 10 m. The number of the ground heddles used by the jacquard loom is two, five or eight, depends on the kind of fabrics to be woven (Fig. 4.28).

Fig. 4.26 The luo jacquard loom in *Plow and Weave Pictures* (the Southern Song Dynasty, collected by National Museum of China)

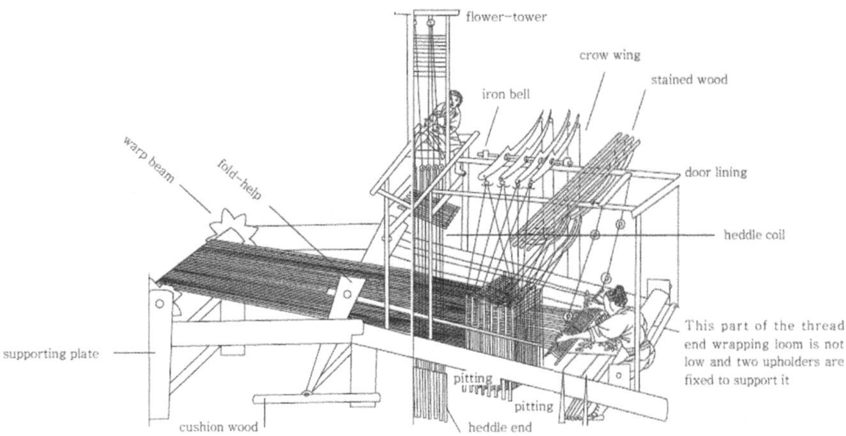

Fig. 4.27 The flower-tower loom in *Tian Gong Kai Wu* (the Ming Dynasty, Song Yingxing)

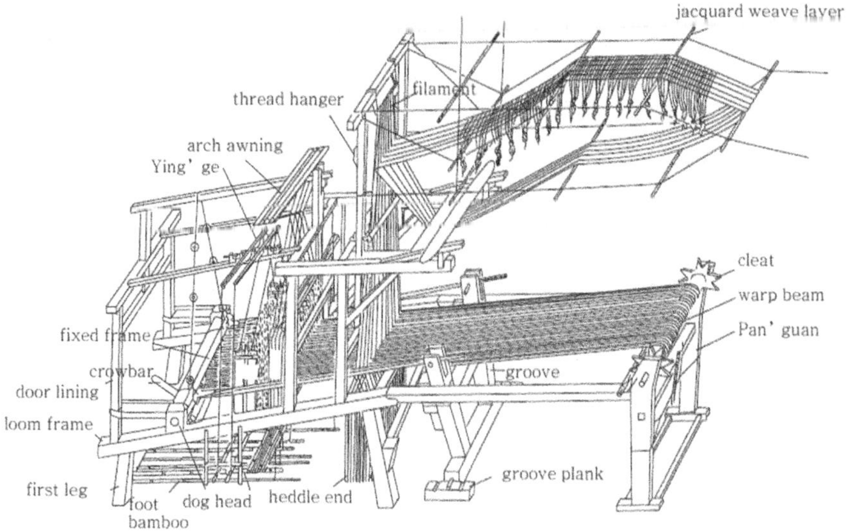

Fig. 4.28 The big flower-tower loom

The pattern-card is the most complex and peculiar part of the bundle-heddle jacquard loom. Song Yingxing made a classical explanation of the pattern-card in his book *Tian Gong Kai Wu* as follows:

All craftsmen who make the pattern-card have the most delicate skills. The painter first draws colors on paper, the pattern-card maker measures them with the silk lines and calculates the distance, and then the weaver picks up the shuttle to materialize the patterns.

Later, this thready pattern-card was developed into the pattern board used on the Jiaka jacquard loom, using the punched paper card and steel needles to control the loom in weaving different patterns. Later, the punched paper card inspired the principle of transmission of telegraph signals, so it is the prototype of the early computers. Therefore, the invention of jacquard loom in ancient China exerted great influences on the history of modern science and technology in the world.

References

1. Feng, Zhao. 2005. *General History of Chinese Silk*, 44. Suzhou University Press.
2. Weiji, Chen. 1984. *History of Chinese Textile Science and Technology* (Ancient Part), 120. Science and Technology Press.
3. Ji, Li. 1927. *Prehistoric Remains of the West Yin Village*, 22. Tsinghua Research Academy.
4. Hui, Xu., et al. 1981. Verification of Silk Fabrics Unearthed from Qianshan Yang. *Silk* 2: 43–45.
5. Shunro, Ume. 1979. *The Origin of Silkworm and Ancient Silk*, 165, Yuzankaku.
6. Feng, Zhao. 2005. *General History Of Chinese Silk*, 13. Suzhou University Press.

7. Hemudu Site Archaeological Team. 1980. The Main Gains of the Second Excavation of the Hemudu Site in Zhejiang Province, *Cultural Relics* (5): 1–15.
8. Hubei Jingzhou Regional Museum. 1985. *Chu Tomb No. 1, Ma Shan, Jiangling*, 41. Cultural Relics Publishing House.
9. Wang, Xu. 2001. The Discovery and Research of Silk Fabrics in the Han Dynasty. See *Wang Xu and Textile Archaeology* by Zhao Feng, 51–68. Art and Garments Press.
10. Rulin, Han. 1983. A New Probe into the Scalded-Animal Banquet in the Yuan Dynasty. *Qionglu Collection*, 251. Shanghai People's Publishing House.
11. Zhao Feng, Jin. 2004. Taquete and Samite Silks: The Evolution of Textiles along the Silk Road. *China:Dawn of a Golden Age (200–750AD)*, 67–77. The Metropolitan Museum of Art and Yale University Press.
12. Feng, Zhao. 1986. The Version of Silkworm Weaving Pictures and Silkworm Weaving Techniques in the Southern Song Dynasty. *Agricultural Archaeology* 1: 345–359.

Feng Zhao Born in Zhuji, Zhejiang Province in 1961, Ph.D. and researcher. He is currently the director of the China Silk Museum, the director of the China Textile Identification and Protection Center, the director of the key scientific research bases of the State Administration for Textile Cultural Relics Protection, the executive director of the International Ancient Textile Research Center and Member of the National Heritage Identification Committee.

Research interests: History of textile science and technology; textile cultural relics identification and protection; Silk Road and cultural exchange between China and Western countries.

Major works: He has published more than 100 papers, including: "Study on the Restoration of the Pedal Weaver in Han Dynasty"; "Study on Cotton Thread Brocade in Xinjiang"; *General History of Chinese Silk, Chinese Silk Art History, Complete Works of Dunhuang Silk Art.*

Awards: Chinese Silk History won the China Outstanding Publication Award; the English version of *Chinese Silk Art History* won the National Textile Association Book Award; *Study on Textile Weaving Technology of Eastern Zhou Dynasty* won the second prize of scientific and technological innovation in national cultural relics protection.

Hui Liu Born in Donga, Shandong in 1980, Ph.D., assistant researcher at the Institute of Natural Science History, Chinese Academy of Sciences.

Research interests: Textile history.

Major works: *Study on Weaving Technology of Double Layer in Han and Jin Dynasties; Identify the Type of Loom used in the Han and Jin Dynasties Based on Unearthed Objects in Xinjiang.*

Chapter 5
Chinese Characters

Yu Peng

There are enough reasons for Chinese character (also called Han character) to be listed as one of the most important inventions in China, for this well-deserved treasure is both the great creation of the Chinese nation and the symbol of Chinese civilization with the condensed Chinese culture, strong cultural influence, unique configuration system, and incomparable aesthetic connotation of calligraphy.

The Chinese characters are the recording symbols in Chinese. "Han" here cannot be simply understood as the Han Nationality or the Han Dynasty, whose historical record is much later than that of the Chinese characters. Chinese language and Chinese characters were originally the languages (dialects) and writing systems used by the people of all nationalities in today's Chinese territory, with the Han Nationality as the main body. For example, in the Spring and Autumn Period and the Warring States Period, the social discipline was chaotic in the civil war. Many countries and nations had their own writing systems then, e.g., Qi characters, Chu characters, etc. Today's ancient philologists can clearly point out the differences between them, but these characters are still within the scope of the study of Chinese characters from the perspective of the historical development of writing.

Chinese characters are divided into ancient and modern categories by a watershed set by the philologists. The characters before the Qin Dynasty are collectively called ancient characters, including oracle bone scripts, bronze inscriptions, Warring States scripts, Xiaozhuan scripts, and so on. After the Qin Dynasty, the shape of Chinese characters was basically stereotyped. Xiao Zhuan and Zao Li were for the official use and Li Writing was common among the civilians. The transformation from Xiao Zhuan to Li Writing is a milestone in the historical development of Chinese characters. Since then, the shape of the Chinese characters has not changed as much as that of the ancient times, the character structure and the ideographic function are more stable with changes only in the style of calligraphy.

Y. Peng (✉)
School of History Culture and Ethnology, Southwest University, Chongqing, China
e-mail: pengyu0305@126.com

J. Hua and L. Feng (eds.), *Thirty Great Inventions of China*,
https://doi.org/10.1007/978-981-15-6525-0_5

Fig. 5.1 The stone statue of Cang Jie on his General Tomb, Beizhai village, Yinan County, Shandong

There are many stories and legends about the origin of Chinese characters. According to the records of *Mosou* written in the Tang Dynasty, Baoxishi began to write Longshu to record the dragon story. Shennongshi created *Basui Book* to issue seasoning records.

The historiographer of Emperor Huang—Cang Jie imitated the bird traces as writing and made the Zhuan Script; Jintianshi created the book of Luanfeng, Emperor Yao (Tao Tangshi) the book of tortoise, and Xia Houshi the Bell Script. Among the above stories, Cang Jie's Book is the most long-lasted and popular versions.

According to *The Annals of Lü Buwei.Jun Shou*, Xi Zhong made a carriage, Cang Jie made a book, Houji made crops, Gaoyao set rules for punishment, Kunwu made pottery, and Xiagun built the city. These six people were real great inventors.

Since the Han Dynasty, the legend has been gradually added some touch of mystery. In the story of *The Huai-nan-tzu: Ben Jing*, when Cang Jie wrote the book, the ghosts cried in the heavy night rain.

The excavations of the cultural relics show the popularity of the legend of Cang Jie's Book in Han Dynasty, the stones carved with Cang Jie's portrait and inscriptions (Fig. 5.1), the bamboo slips of the Han Dynasty with sentences "Cang Jie wrote

Fig. 5.2 Cang Jie's inscription rubbings of bronze mirrors unearthed in Xi'an

books to teach the heirs", and bronze mirror of the Three Kingdoms' Period, the East Han Dynasty, with casted texts "Cang Jie Made Books and Suirenshi Made Fire" (Fig. 5.2).

All great inventions are rooted from the mass, yet carried forward by the prominent figures or official management. The same is true for Chinese characters. The Shang Dynasty oracle scripts, the earliest known writing systems, cannot be the original stage of writing, for it's quite mature in structure and quantity.

Archaeology is an important means to trace the origin and early evolution of Chinese characters from the oracle bone scripts. The eye picture on the inscribed tortoise shell unearthed from Peiligang Ruins in Jiahu Village, Wuyang County, Henan, is much like the Chinese character "目" in shape (Fig. 5.3), and the door shape much like "户", which cannot be explained by accident. The test made by the Radiation Laboratory of Institute of Archaeology, Chinese Academy of Social Sciences, shows that the Peiligang charcoal samples has a history of 7,000–8,000 years, over 1,000 years earlier than Emperor Huang and Cang Jie's period which could be the possible time to collect and regulate writing systems accordingly.

Among the collected ancient Chinese characters, the Shang Dynasty Oracle Scripts, dated back to more than 3,000 years (Figs. 5.4 and 5.5), are undoubtedly the earliest. Oracle Scripts, also called carved scripts (Qiwen), oracular inscriptions, tortoise shell scripts, or Shang oracle bone writing, are inscribed writing on tortoise shells or animal bones (Fig. 5.6), which can be dated back to the Shang Dynasty therein the rulers were very superstitious about ghosts and gods, and used tortoise shells and animal bones for divination when encountering great events. Then the content and the results of the divination were engraved on the oracle bones in a

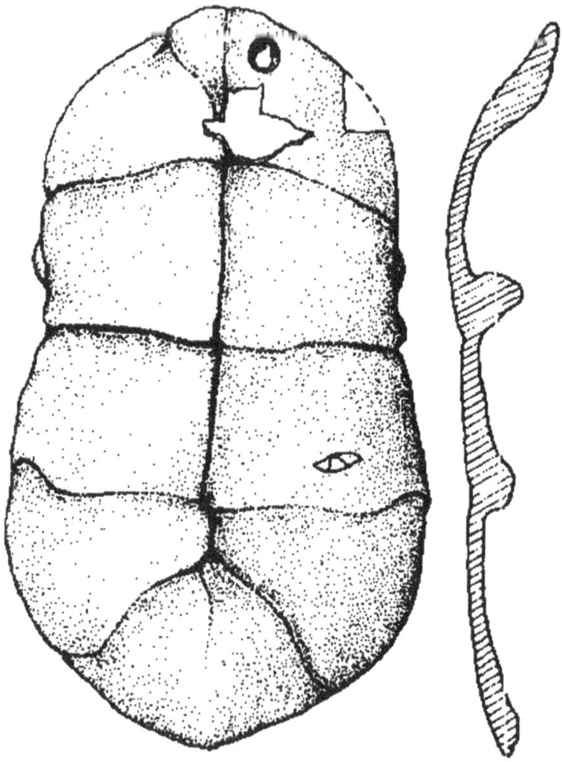

Fig. 5.3 The inscribed tortoise shell unearthed from Peiligang Ruins in Jiahu village, Wuyang County, Henan

stable format as today's official files. These documents dealt mostly with ceremonial sacrifice, hunting activities, weather, military campaigns, and diseases to illuminate the Shang Dynasty.

The oracle bone had been a prescription of traditional Chinese medicine for a long time. It was not until the end of nineteenth century that Wang Yirong, who was then ill, accidentally found scripts on the oracle bones and collected them at high prices. It is said that before Wang's discovery, the drug dealers had noticed the nicking, but they all regarded it ominous and scraped it off. In this way, Wang's discovery is really sheer luck.

Today, the oracle research has become an independent science. After Wang's death, Liu E, Luo Zhenyu, and others successively engaged in the collection and sorting of oracle bones (Fig. 5.7). Wang Guowei, Guo Moruo, Dong Zuobin, Hu Houxuan, Yu Sheng-wu, Qiu Xigui, Huang Tianshu, and other scholars have made great contributions to interpreting the oracle scripts. It is worth mentioning that Mr. Wang Guowei linked the records of the Shang Dynasty oracle scripts to *Annals of Yin (the Yin-Shang Dynasty),* thus making the List of Shang Kings a credible history.

So far, more than 4,000 Chinese characters have been accumulated in over 150,000 pieces of oracle bones mostly unearthed in Anyang, Henan, but less than one third is fully identified by scholars. Due to the robbing excavations or personal collecting,

Fig. 5.4 The front page of P.376, *Jiaguwen Heji (A Collection of Chinese Oracle Scripts)*

the oracle bones are scattered in many countries and cities, even exhibited in many local museums.

The early Shang Dynasty moved the capital frequently, for 13 times, according to historical records. After Pan Geng moved the capital to Yin, a vast area located across the Heng-River sides in the northwestern suburb of Anyang, Henan. Yin became the center for political movements, economics, and culture of the Shang Dynasty. In Yin Ruins, the archaeologists found abundant remains of the Shang Dynasty culture, including thousands of the oracle bones and a large number of bronze wares, many of which were casted with inscriptions.

The interpretation of the oracle scripts is different from that of the writing systems in other ancient civilizations such as the saint letters of Egypt, because the oracle scripts, unlike those long-lost ancient Egyptian letters, have been passed onto Chinese characters in late Qin and Han Dynasties. In addition, there are systemic literal books such as *the Collection of Ancient Chinese Characters* and *Shuowen Jiezi* (*Explaining Graphs and Analyzing Chinese Characters*), as well as epigraphy study flourishing in the Song Dynasty and philology peak in the Qing Dynasty, which make the profound oracle scripts interpreted and less mysterious.

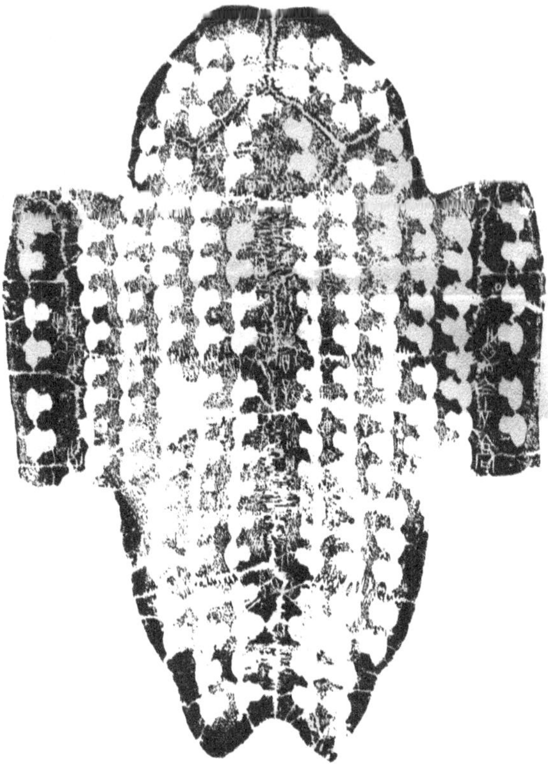

Fig. 5.5 The reverse page of P.376, *Jiaguwen Heji (A Collection of Chinese Oracle Scripts)*

It is worth mentioning that the archaeological data show that the oracle bone scripts are not exclusively used in the Shang Dynasty but also in the Zhou Dynasty (Fig. 5.8). In 1977, more than 10,000 shells and 300 bones were unearthed in the Xizhou Ruins in Chufeng Village, Qishan County, Shanxi, 289 shells inscribed with scripts (30 scripts to the most). In 1979, 22 oracle bones (six bones inscribed) were collected in the nearby Qi Village, Fufeng County. Qishan and Fufeng Counties were called Zhouyuan, the ancient capital city of the Zhou Dynasty. Thus the oracle bones unearthed there are called the Zhouyuan Oracle Bones, which is also echoed in the lines in the *Collection of Poetry (Shijing):* The land of Zhouyuan is very fertile and vegetables are very sweet. We began to plan and consult, and then carved tortoise shell to see the divination.

The Oracle scripts are a writing system with a large number of hieroglyphs, many of which are derived from copying and imitating natural things. For example,  (moon) looks like a piece of a crescent moon,  (tortoise) the side shape of a tortoise,  (horse) the main horse body with mane and tail,  (hare) a short-tailed hare,  (vehicle) a simple vehicle with wheels, axle, and carriage, 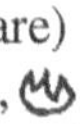 (fire) the fire flame, etc. Those pictographs were drawn as their objects, as explained in *SHUOWEN JIEZI*.

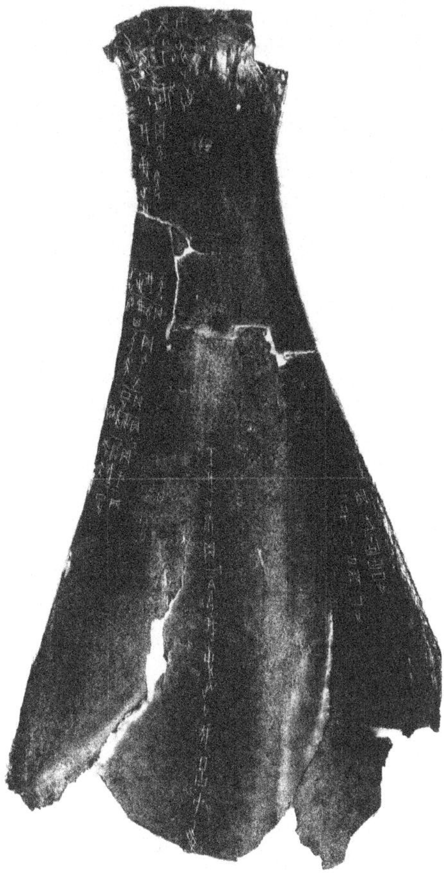

Fig. 5.6 The front page of P.1804, *Supplement Edition of Jiaguwen Heji (A Collection of Chinese Oracle Scripts)*

The pictographic characters, a most primitive method of word-formation or the most easily created and popularized writing system in the initial stage of human civilization, might be originated from and related to the early characterization of symbols with strong graphic character but a weak symbolic feature. In addition to the Shang Dynasty Oracle Scripts, other pictographic character systems were created and used in many ancient civilizations and nations, such as the ancient Egyptian hieroglyphs (Fig. 5.9), Cuneiform scripts (Fig. 5.10), Indus script used in Indian Harappan period (Fig. 5.11), Dongba scripts in Yunnan (Fig. 5.12), Shui Scripts (used in Shui ethnic group) (Fig. 5.13). However, pictographs have huge limitations, because some physical and abstract objects are hard to be drawn, big, small, more, less, thick and thin, etc. So the wise ancients invented some other ways to express different concepts. Scholars in Han Dynasty classified Chinese characters as pictographs (象形), ideographic (指事), compound ideographs (会意), phono-semantic compounds (形声), phonetic loan characters (假借), and derivative cognates(转注), collectively known as Liu Shu, or "Four Bodies and Two Usages" (phonetic loan characters

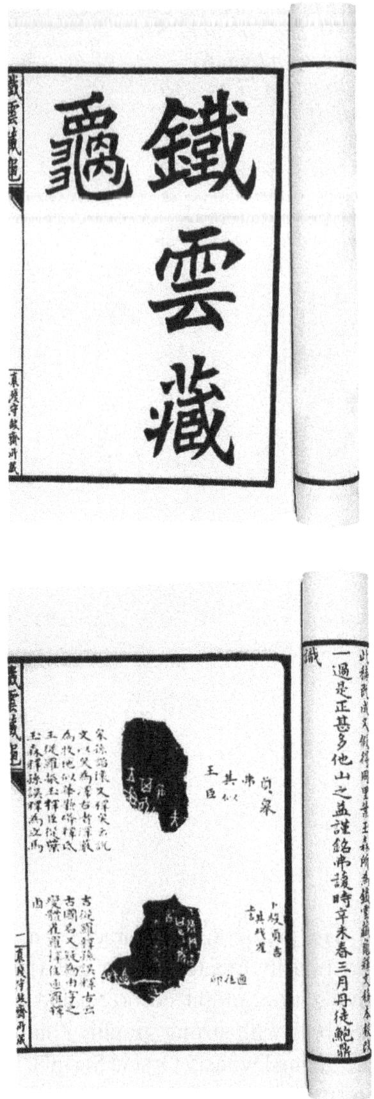

Fig. 5.7 *Initial Seal Collection by TIEYUN* (Liu E)

and derivative cognates are the two usages of characters, and the other four are the methods of making characters). These new writing formations still have to be based on the pictographics by adding a new character or a sound. The Four Bodies writings were found in the Shang Dynasty Oracle Scripts with a large number of the phono-semantic compounds, which shows that the oracle scripts are a mature writing system and the Four Bodies are the main character formation method in the following dynasties.

Fig. 5.8 The Zhouyuan Oracle bones

The oracle scripts are not only found on unearthed shells and bones, but on bronze wares (Figs. 5.14 and 5.15). Bronze wares were called Jijin (lucky metal, hard metal, or durable metal) in ancient times because they were used in sacrifices, one of the two considered great events in ancient China, the other being wars. Therefore the scripts on bronze wares were called the bronze inscriptions.

The oracle scripts were mainly found in the capital of the Shang Dynasty, but the bronze wares were distributed in a much wider range, from the south of the Yangtze river, to Gansu, Shandong, and Liaoning with the famous excavations in Panlong, Hubei, Ningxiang County in Hunan (Fig. 5.16), Dayangzhou town in Jiangxi (Fig. 5.17), Subu village in Shandong, and Shilou town, in Shanxi, etc.

It is generally recognized that the oracle script represents a divination culture and the bronze inscription a sacrificial culture in Shang Dynasty. Compared with the former, the latter is more elegant and solemn with huge investment on its reproduction and refinement. In Western Zhou Dynasty, there were great changes in shapes and decorations on bronzewares (Fig. 5.18), even with long inscriptions, such as the Mao Gong Ding with up to 497 inscriptions exhibited in Taipei Palace Museum (Fig. 5.18). The content of the Xizhou bronzeware inscriptions is rich, mainly on the brilliance of the times, not only limited to the sacrifice rites as the Shang Dynasty Oracle scripts.

In the past, the bronze inscription was often referred to as the "Ding (Bell) script." This was because the Shang and Zhou ceremonial vessels were represented by the

Fig. 5.9 Ancient Egyptian Hieroglyphs

bronze "Ding" and Musical Instruments by "Bell." The study of the bronze ware inscriptions on the huge body of the Ding (Bell) is known as epigraphy.

Initiated by the Song Dynasty, the study of epigraphy focused on the stone and metal inscription, with the bronze ware inscription as the mainstream. The Song scholars paid special attention to the collection and sorting of inscriptions on the bronze wares in their works, *Kao Gu Tu* compiled by Lv Dalin, *Bo Gu Tu* by Wang Fu, *Xiaotang Jigulu* by Wang Qiu, *Lidai Zhongding Yiqi Kuanshi Fatie (Model Calligraphy, Compendia of the Inscriptions on Bronze Wares)* by Xue Shanggong. In the Qing Dynasty, considerable research and collections on the bronze ware inscriptions were in process, *Inscriptions on Ritual Bronzewares collected by Jiguzhai* compiled by Ruanyuan, *Study on Inscriptions on Bronze Vessels from Zhuiyizhai* by Fang Ruiyi, to name just a few. *Remains of the Auspicious Bronze Inscriptions from the Three Dynasties* (Fig. 5.19) by Luo Zhenyu in 1937 collected more than 4,000 pieces of the bronze ware inscriptions in exquisite printing.

Nowadays, with the development of science and technology, and thanks to the efforts of scholars, reference books about the bronze scripts are increasing in number, making related research more convenient. For instance, *The Collection of Bronze Inscriptions in the Yin & Zhou Dynasties* (CBI) compiled by the Research Institute of Archaeology of the Chinese Academy of Social Sciences is a very comprehensive collection of inscriptions on bronze wares. It includes the details of more than ten thousand different wares from the important ancient Chinese historical periods

Fig. 5.10 Cuneiform scripts

Fig. 5.11 Indus script in Indian Harappan period

Fig. 5.12 Dongba scripts in Yunnan

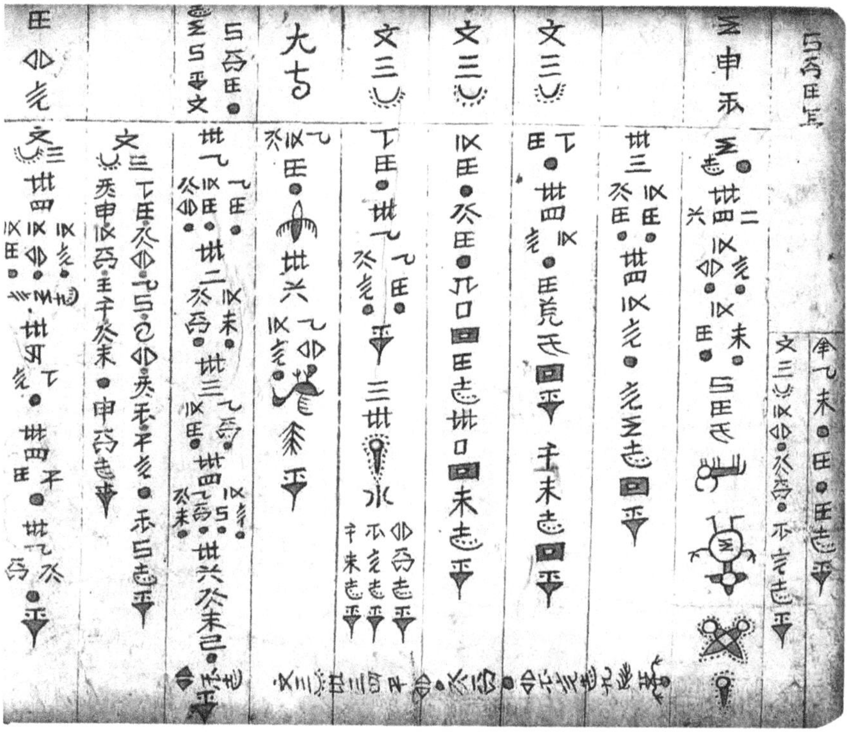

Fig. 5.13 Shui scripts used in Shui ethnic group

before the country was unified, including the Yin and Zhou Dynasties (1600–256 B.C.), the Western Zhou Dynasties (1046–776 B.C.), the Spring and Autumn Period (770–476 B.C.), and the Warring States Period (475–221 B.C.). It also catalogues various excavations recorded by different experts or newly unearthed in different locations from the Song Dynasty (420–479 A.D.), as well as collections of major museums both at home and abroad. Linking up to CBI, *The Newly Collected Bronze Inscriptions in the Yin and Zhou Dynasties* is another famous book on bronze scripts

Fig. 5.14 The Houmuwu Bronzeware

which starts at the date of the closing of each volume of the latter and ends at the end of May 1999. In addition, *The Collected Inscriptions and Images of Bronze Wares in the Shang and Zhou Dynasties* compiled by Wu Zhengfeng has a complete collection of the bronze wares. The whole book contains 16,704 bronze wares handed down or newly unearthed. It contains both rubbings and images of the inscriptions, and arranges the explanations and relevant background material together to provide scholars with directions, which is of great benefit to the related research.

The Bird-insect script, also commonly referred to as the bird-insect seal, is a kind of bronze inscription with special forms from the Zhou Dynasty (1046–256 B.C.) as is shown in Fig. 5.20. Bird-insect scripts, emerging roughly in the middle and late Spring and Autumn Period and prevailing in the Warring States Period, were mainly popular in the middle and lower reaches of the Yangtze River, spreading to the Central Plains. The inscriptions are usually found incised on bronzes of the Wu, Yue, Chu, Cai, Xu, and Song State, usually in the form of inlaid gold. The basic forms of some scripts are embellished with decorative forms such as birds by the curved lines, as is the case with the Bronze Sword of King Gou Jian[1] and the Bronze Sword of King Zhou Gou.[2] And some scripts are made up of winding strokes, like the shape of insects, as is the case with the Wang Ziwu[3] tripod from the late Spring and Autumn Period which is of changing strokes that are even difficult to identify.

The bird-insect scripts are trans figures of the decorative characters, not another writing system. Though they look luxuriant, the scope of use is extremely limited. The bird-insect scripts that we can see today are mainly cast on the bronze wares, especially on weapons, and a small number on vessels and seals. Since the Warring

Fig. 5.15 Inscription rubbings on the Houmuwu Bronzeware

Fig. 5.16 The Four-ram Zun unearthed in Ningxiang county, Hunan

Fig. 5.17 The bronzeware unearthed in Dayangzhou town, Jiangxi

Fig. 5.18 **a** Qin Gui (bronze ware to feed chicken) in the Western Zhou Dynasty and the inscription rubbings on it **b** The rubbings of Mao Gong Ding

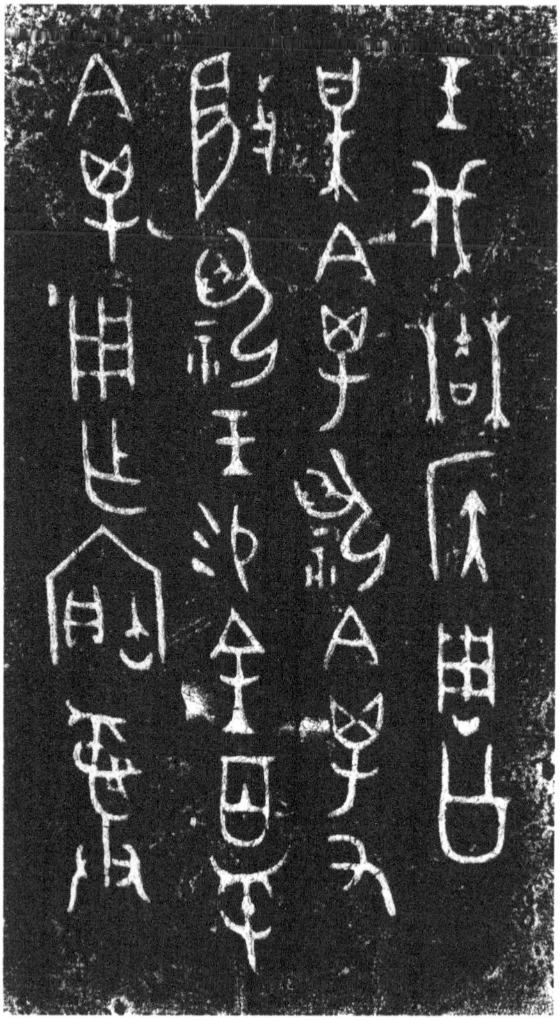

Fig. 5.18 (continued)

States Period, the bird-insect scripts have been used less and less, but they were still visible on ritual vessels and seals in Han Dynasty (206 B.C.–220 A.D.) and on stone tablets of the Tang Dynasty (618–907 A.D.).

Before the Spring and Autumn Period, the bronze scripts were almost always cast when the wares were produced, but since the middle of the Warring States Period, the inscriptions were often engraved with knives after the bronze was cast. For example, the inscriptions on weapons in the Qin State were basically carved with knives (Fig. 5.21). The inscriptions of the Spring and Autumn Period and the Warring States Period are not only engraved on bronze wares, but also on some other carriers, such as seals (Fig. 5.22), currency (Fig. 5.23), pottery (Fig. 5.24), lacquer (Figs. 5.25 and 5.26), jade (Figs. 5.27 and 5.28), bamboo, and silk (Fig. 5.29).

Fig. 5.19 The printed version of *San Dai Ji Jin Wen Cun*

Before the popularization of the plant fiber paper, bamboo slips and silk were the most commonly used writing materials of the ancients. According to literature, the use of bamboo slips seemed to have started at the beginning of the Shang Dynasty (1675–1029 B.C.), and the use of silk as a writing material might be a little bit later. As bamboo and silk are subject to damage and rotting, the early silk and bamboo slip scripts are hard to preserve. The earliest known were from the Warring States Period.

All the bamboo slip inscriptions found from the Chu State were written in ink with a brush, most of which are records of the funeral objects or horses and chariots. In addition, there are records of divination, judicial documents, books on time divination, as well as literature concerning ancient books like *Shang Shu, Yi Zhou Shu,* and *Shijing.*[4]

So far, only one silk book with a complete text has been found in the Warring States Period. The book was excavated from the Chu Tomb in Zhidanku of Changsha, Hunan, and has flowed into the United States before 1950s. With 900 characters in

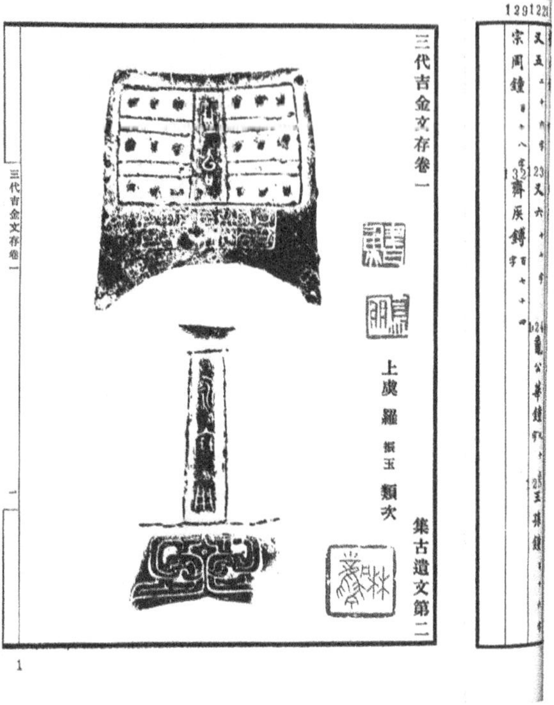

Fig. 5.20 The bird-insect scripts on the dagger-axe of Song Gongluan

Fig. 5.21 The rubbings of Qin Zi Ge

black ink and 12 weird idols around the characters, the silk book is mainly about myths and taboos circulating in the Chu State, especially those about the celestial phenomena, cataclysms, four seasons operation, and monthly order. At the four

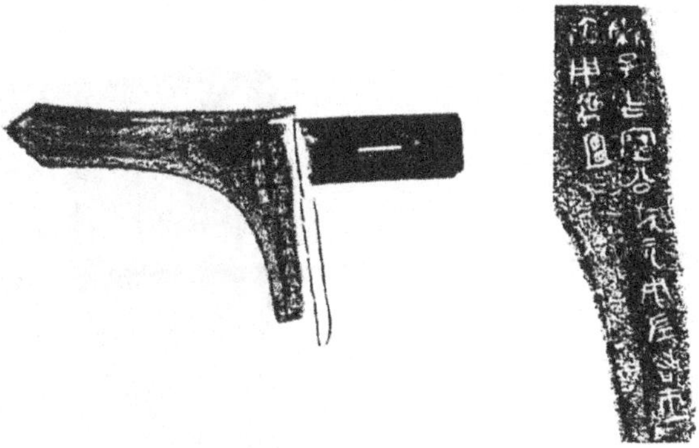

Fig. 5.22 The seal inscriptions

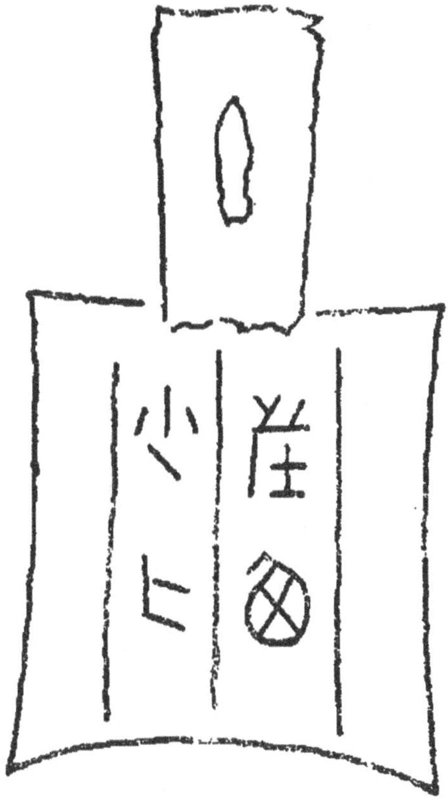

Fig. 5.23 The currency inscriptions

corners of the book, there are images of trees painted in green, red, white, and black (Fig. 5.30).

After the Qin State annexed the other six warring states (Chu, Qi, Yan, Zhao, Wei, Han) and unified China, the small seal script was introduced all over the country. The First Emperor[5] traveled around the country, engraving stone inscriptions as records

Fig. 5.24 The pottery inscriptions

of merit or victory in places like Yishan, Taishan, Langya, Jieshi, and Jijie. During the reign of the second emperor, imperial edicts were additionally inscribed on every carved stone to show that the First Emperor had engraved the stone inscriptions. These engraved stones are the best materials for studying the small seal scripts. Unfortunately, most of the original materials have been destroyed, only some remnants remain and some copies that have been handed down.

As the richest and most systematic written data of characters from the Qin Dynasty, *Shuo Wen Jie Zi* has collected more than 9000 small seal scripts. However, *Shuo Wen Jie Zi* was written in the middle of the Eastern Han Dynasty (25–220 A.D.); there were many errors in the small seal scripts written by people at that time. Moreover, scholars, including Xu Shen, inevitably had some misunderstandings of the glyph structure of the small seal scripts, which in turn led to the erroneous modification of the glyph of small seal scripts in the book. Since the publishing of *Shuo Wen Jie Zi*, it has been copied and carved through the ages and some mistakes have been made by the calligraphers, carvers, and some poor collators. Therefore, some of the small seal scripts in the dictionary are unreliable and need to be corrected by referring to small seal scripts inscribed on bronzes and stone tablets of the Qin and Han Dynasties.

Since the 1970s, a large number of the bamboo slip inscriptions and silk inscriptions have been unearthed, providing us with important data for studying the characters of the Qin Dynasty. Containing more than 1100 bamboo slips and focusing on the codes of the Qin Dynasty, *Rishu*,[6] the records of great events with the bamboo texts discovered in late 1975 in Shuhudi of Yunmeng, Hubei, is a famous collection. In 1986, more than 460 bamboo slips and four wooden planks (Fig. 5.31) with characters were excavated from the tomb of Fangmatan of Tianshui, Gansu. In 2002, more than 38,000 Qin bamboo slips (Fig. 5.32) were unearthed in the Ancient City Ruins of Liye,[7] Hunan, among which there were as many as 18,000 slips with inscriptions. This number greatly exceeds the total number of bamboo slips unearthed over the years. Since then, Yuelu Academy of Hunan University and Peking University

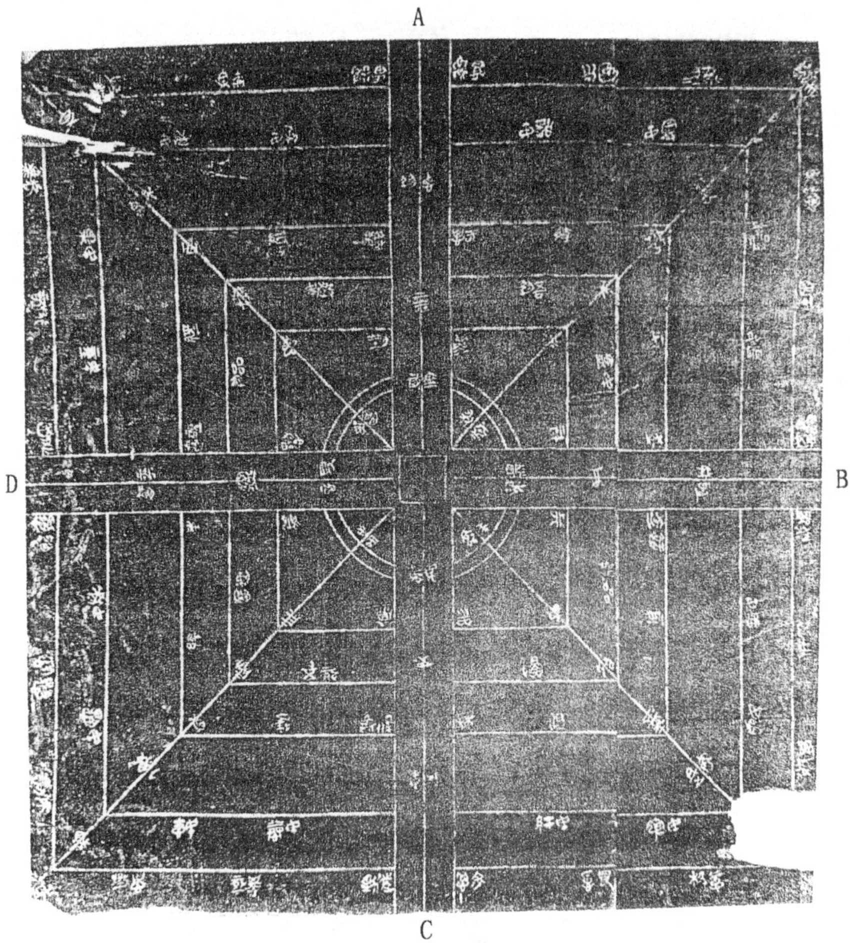

Fig. 5.25 The scripts on the lacquer chess unearthed in Chu tomb of Jinmen

have respectively purchased and collected a bamboo slip of the Qin Dynasty, the inscriptions on which are of vital importance.

According to the previous research, since small seal scripts have gradually evolved from the inscriptions of the Qin State in the Spring and Autumn and the Warring States Period, they are not completely different from each other. It is since the Han Dynasty that the names such as great seal scripts (大篆, read as/dà zhuàn/), Qin seal scripts (秦篆, read as/qín zhuàn/) and small seal scripts (小篆, read as/xiǎo zhuàn/) have been used. In Qin Dynasty, probably only "篆," the general name of this kind of script was used. "篆 (seals)" and "瑑" (engraved texts, read as/zhuàn/,) are homophones. *The Lüshi Chunqiu*[8] reads "meritorious masterpieces are carved on dishes; seal inscriptions are carved on pots." In people's minds at that time, the clerk

Fig. 5.26 The scripts on the lacquer suitcase discovered in the tomb of Zeng Houyi

Fig. 5.27 The jade inscriptions in Houma alliance treaty

scripts used by the clerks, and government officials were not professional enough, and only seal inscriptions were qualified to be engraved on wares and stones.

As early as the Spring and Autumn Period, the script styles in the Qin State and the other Warring states were strikingly divergent. In the Warring States Period, dramatic changes in the scripts of the eastern states made the distinctions more

Fig.5.28 The inscriptions on drum-shaped stone blocks

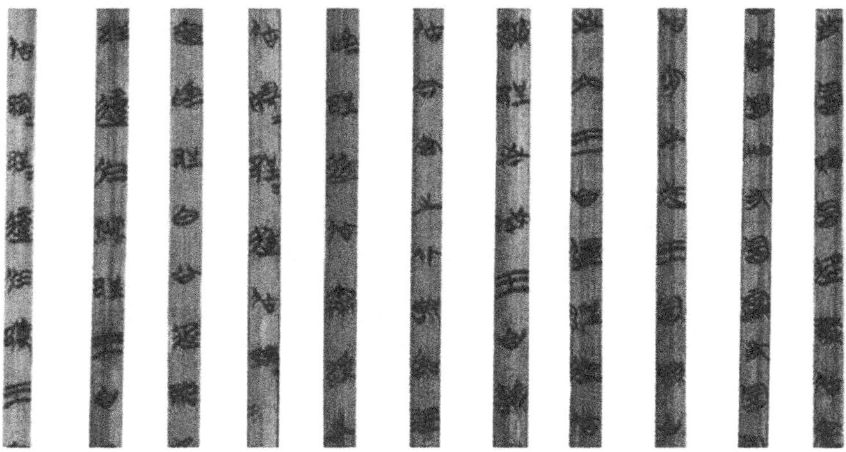

Fig. 5.29 The bamboo slips of Tsinghua

obvious, which greatly influenced the regional economic and cultural exchanges and was not favorable to the rule of the Qin Dynasty. Therefore, after the Qin State unified the other states, the policy of "Towards a Single Written Language" was rapidly implemented, and the inscriptions of the Qin State were used as the standard to unify the Chinese characters. In fact, prior to the unification, the Qin State had

Fig. 5.30 The silk book discovered in Zidanku, Changsha

gradually adopted this practice in the newly occupied areas, which can be verified by the relevant literary materials unearthed in various places.

As is well known, people's writing habits are formed by long-term practice. So it is imaginable that it would not be easy for the Qin State to change the writing habits of the people in the conquered areas. The written materials unearthed today reveal that although the Qin State rapidly unified written language through draconian laws and severe punishment, the influence of the other six states' scripts still lingered on. The bamboo slip scripts in the early Western Han Dynasty, for instance, still retained traces of the Chu State characters. Regardless of all this, this practice basically eliminated the simultaneous existence of different characters at that time and occupied a very important position in the development of Chinese characters.

Since then, in today's territory of China, both the official and the civilian have begun to use a unified writing system as a tool for recording language and social communication. Chinese characters play an indispensable role in strengthening the ethnic exchanges and enhancing the sense of identity of the Han people.

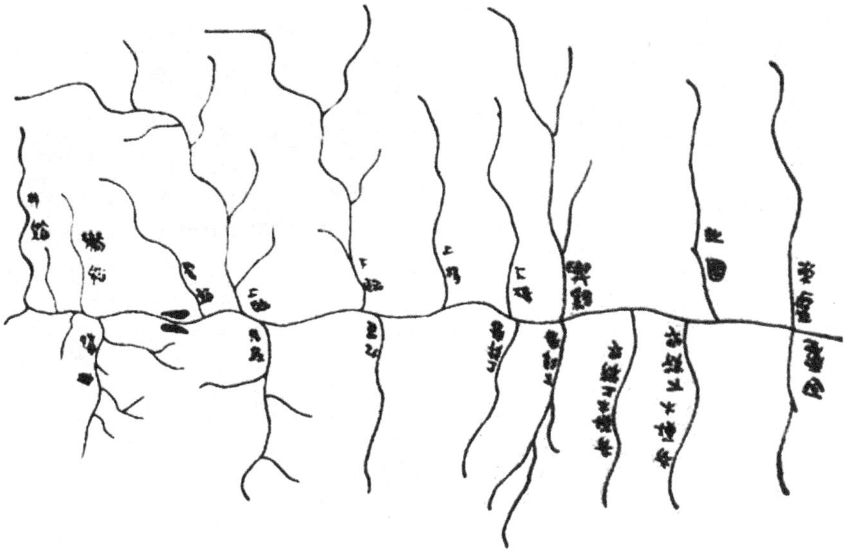

Fig. 5.31 The map unearthed in Qin tomb in Fangmatan, Tianshui county

Since the Han Dynasty when clerk scripts replaced small seal scripts as the official script, Chinese characters have developed into the stages of clerk scripts and regular scripts. *Han Shu Yi Wen Zhi*[9] says that clerk scripts were simplified characters created in the Qin Dynasty to cope with the busy official matters, but it is not the case.

The archaeological data show that the clerk scripts were created in the late Warring States Period. At that time, for the convenience of writing, people in the Qin State constantly transformed the shapes of the standardized characters; the non-standard style of the characters came into being, whose shape usually changed from long and round to flat and square, having a strong favor of the clerk scripts.

The coexistence of both the standardized characters and the non-standard styles have been witnessed in the inscriptions in the times of Qin Xiaogong,[10] while the characters on the bamboo slips in Shuihudi have proved that clerk scripts had already been basically formed in the era in which these bamboo slips were made. See Fig. 5.33. Though not yet fully mature as new auxiliary characters, clerk scripts developed rapidly as they are much easier to write than small seal scripts.

It is generally believed that the clerk scripts became mature during the reign of Emperor Wu of Han,[11] and because of this, the clerk scripts in early Qin Dynasty and early Western Han Dynasty were called the early clerk scripts. These scripts were not well-developed and the shape of many characters was obviously similar to that of the seal scripts, a situation which has lasted until the late Western Han Dynasty.

Development of the seal scripts into the clerk scripts is the most important change in the evolution of Chinese characters, which has brought profound changes to not only their appearance but also to their structure, such as abolishing the seal scripts,

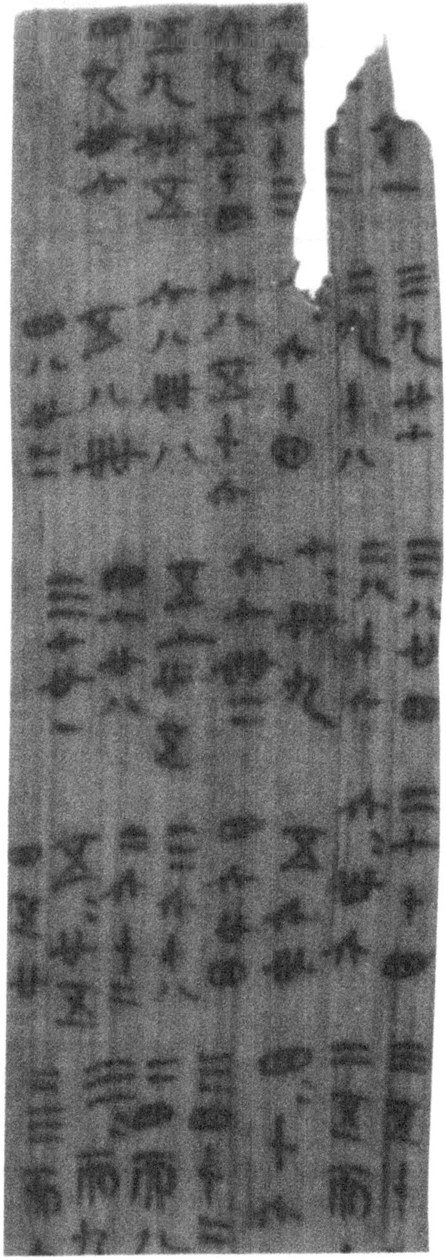

Fig.5.32 The bamboo slip inscriptions found in Liye count

Fig. 5.33 The scripts on Qin Bamboo slips unearthed in Shuihudi, Yunmeng

changing curves to straight lines, and merging or omitting strokes of the seal scripts. (Fig. 5.34).

Besides the clerk scripts, the cursive scripts as shown in Fig. 5.35 were also used in the Han Dynasty which has both broad and narrow meanings. In the broad sense, all scribbled characters in any era can be counted as cursive. In a narrow sense, the cursive scripts refer to a specific type of script formed in Han Dynasty which has been named as the "ordinary cursive scripts"(章草, read as/zhāng cǎo/) since the Eastern Jin Dynasty (317–420 A.D.). In order to distinguish it from the new cursive script at that time, it was called "present cursive scripts" (今草, read as "Jīn cǎo). "章" means order and organization and "章草" is probably named after its more organized structure.

Fig. 5.34 The clerk scripts on the Stele of Cao Quan

The cursive scripts, as simplified scripts supplementary to the clerk scripts, are not only used for drafting manuscripts and letters, but also favored by the calligraphers. It is said that several calligraphers at that time wrote *Ji Jiu Pian*[12] in cursive scripts, copies of which have come down to modern times. Nowadays, people can still have access to the cursive scripts written at that time in the unearthed bamboo slips of the Han Dynasty. However, since the cursive scripts are of very simple forms and easy to be confused with each other, they can not replace the clerk scripts as the main characters.

The running script is a new type of Chinese characters that sprouted in the late Eastern Han Dynasty (Fig. 5.36). The running scripts with which we are familiar today have both the features of the regular scripts and the present cursive scripts. Experts speculate that it has been named so because it can be written quickly.

Entering the Northern and Southern Dynasties (420–589 A.D.), the regular scripts have finally been taken as the main characters (Fig. 5.37). "楷" (read as/kǎi/) has the meaning of "a model," so the original meaning of "regular scripts" refers to characters that can be used as models, rather than the specialized name for a certain type of scripts. The form of the regular scripts has continuously changed; however, no obvious change has occurred in the style of calligraphy. Owing to the writers' distinct writing habits, numerous characters in a popular form and many variants

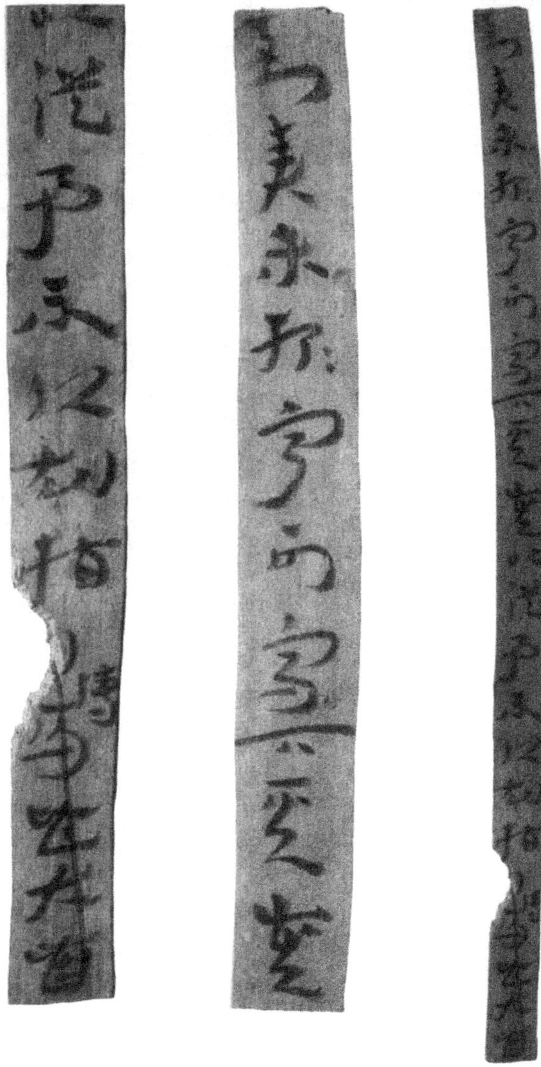

Fig. 5.35 The cursive scripts on Han bamboo slips discovered in Dunhuang

have arisen, which have been included in later ages in books on Chinese characters like *Yu Pian*[13] and *Long Kan Shou Jing*.[14]

Reviewing the evolutionary process of the Chinese characters, it is not difficult to figure out the development rules. Specifically, in form, the Chinese characters have gradually developed from graphics to strokes and from complexity to simplicity, and in the principle of word-formation, from pictographic and ideographic characters to phonographic characters. The total number of Chinese characters has been increasing, but only three or four thousand characters are commonly used. It is worth mentioning that, after the New Culture Movement (around the time of the May 4th Movement

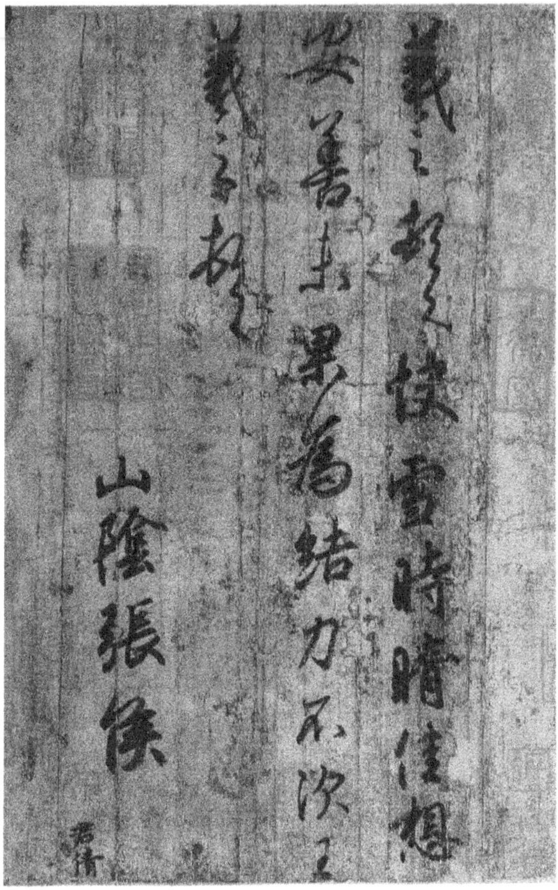

Fig. 5.36 The running scripts in famous works of Wang Xizhi

in 1919), there was a view that Chinese characters were outdated and should be eliminated. Lao She, a famous Chinese novelist, once said, "Chinese characters are hard to recognize and hard to write, and education will never develop without trying to change them." In recent years, many people still believe that alphabetic writing is superior to Chinese characters. Some Chinese characters have too many strokes to write and a lot of Chinese characters are polysemous, which is not helpful for the study of Chinese. This view may not be right for the following reasons.

On the one hand, it makes sense linguistically to record Chinese language with Chinese characters. Chinese differs from other languages. In alphabetic writing, one writing unit can represent a word, but in Chinese, there is no one-to-one relationship between a character and a word. Since there are a great many monosyllabic morphemes and homophonic morphemes, it is obviously appropriate to use the Chinese character system to record such a language. The combination of phonology, form, and meaning of the Chinese characters enables the users to distinguish homophones quickly, which is more evident in pictophonetic characters. Take "歪" (read

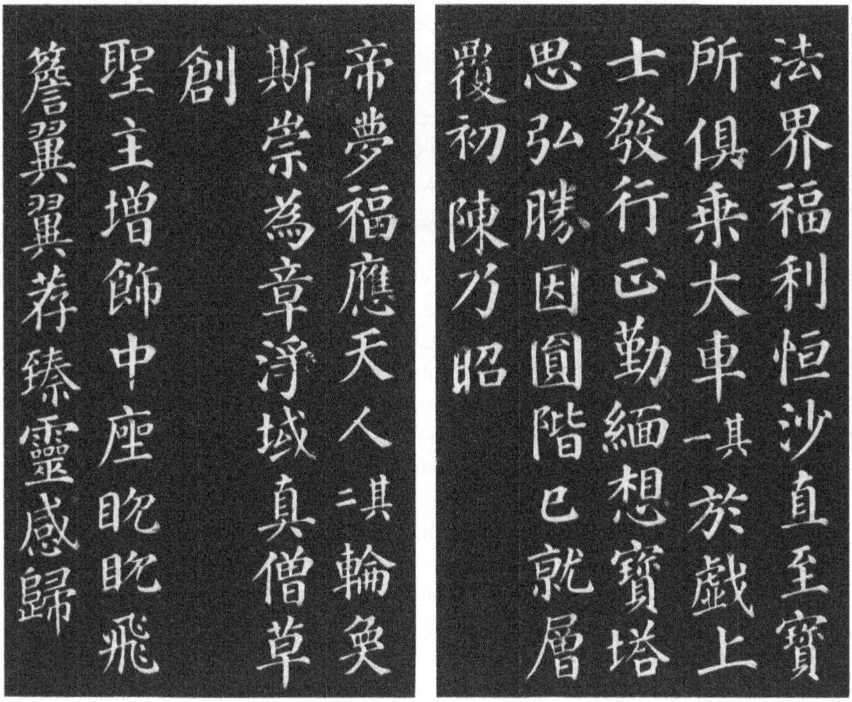

Fig. 5.37 The regular scripts in famous works of Yan Zhenqin

as/yú/) and "龢" (read as/yú/) for example. Although their pronunciation is the same, they belong to different semantic radicals, thus people can immediately understand that "盉" is related to vessels while the latter is related to the musical instruments.

On the other hand, every kind of writing system has to experience a long-term practice in recording language and needs to conform to people's language habits and modes of thinking. In thousands of years of development, Chinese characters and Chinese language interact with and infiltrate into each other. Just as language shapes thinking, Chinese characters also shape people's thinking. When people create characters, they often put into them their own life experience and ideas. So when the later generations study and use these characters, they not only use them to communicate, but also learn and inherit life experience and thoughts of their predecessors. For example, the word "协" (read as/xié/) used to be written as "㤺" (read as/xié/) which is formed by the radical "忄" (heart) and the character "劦" (joint efforts, read as/xié/) and well expresses the profound meaning that people will work together to achieve the same purpose. It is precisely because Chinese characters play a subtle role in molding the way people learn and use Chinese characters, they unconsciously accept the civilization and culture conveyed, which thus enhances their cultural and psychological identity. Therefore, the policy of "Towards a single written language" is of great significance in enhancing national identity and promoting national integration, contributing much to the formation and unity of the Chinese nation.

Another important feature of Chinese characters is their practicality and innovation. *Kangxi Dictionary*,[15] for instance, has included more than 47,000 Chinese characters; however, it is evident through careful reading that most characters once used in history are no longer used today, which indicates that Chinese characters have also experienced a process of keeping pace with the times and self-elimination. This is the main reason why Chinese characters have been able to outshine texts of other ancient civilizations when they were successively on the wane. Owing to their innovation and practicality, Chinese characters can well satisfy the needs of different groups of people in every dynasty. Moreover, because there have been a fixed number of Chinese characters in common use for thousands of years, Chinese characters themselves have a powerful force. The situation can be compared to the need to compete for a job—these common characters were undoubtedly the ones most needed and recognized by people at that time. Since they moved with the times, they exhibited a great vitality and appeal, and people were led to use them as a norm to standardize other characters in the contemporaneous character systems by separating and creating new characters, or absorbing and integrating them. Consequently, the standardized characters regained people's favor and survived; renewal and development of the whole language system were completed.

Interestingly, different from character systems of other countries, Chinese characters and Chinese are not completely historically equivalent. In addition to truthfully recording the language texts commonly used in communication, Chinese characters have the ability to produce new systematized characters, such as Mantra used by people in the Song Dynasty and Women's script in the Qing Dynasty (1644–1911 A.D.), which are related to Chinese characters but have their own systems of form, sound, and meaning. Although limited in the scope of use, they can meet the needs of the relevant groups. This is also a manifestation of the strong vitality of Chinese characters.

With the deepening of the research, linguists have found that the functions of Chinese characters are far more powerful than people think. In the period of the Republic of China (1912–1949 A.D.), the typewriters became more and more popular, but due to the technical constraints, only English input was used, and Chinese characters input was excluded from the new trend. However, with the innovation of technology in the past decade, people have now found that Chinese characters can be input into computers and into the typesetting and printing systems, and the present Chinese character input method is a much faster method of recording language than that of English. In addition, Chinese falls into the category of language with the smallest number of common characters—there is a fixed number of 2500 common characters; therefore in the future big data era when input is calculated through cloud computing, associative memory, and other methods, Chinese characters will show their strength.

Entering the information age, more and more Chinese people tend to use computers for daily writing, and the need to write with a pen is much less than in the past. One of the shortcomings of the Chinese character system is the fact that the characters are complicated and contain many stokes; while improvements have been made to make writing easier, many people begin to worry about whether the traditions of Chinese characters can be preserved, and whether children will still have a unique feeling for Chinese characters in the future. Furthermore, because of the increasingly closer interchanges between countries in the world in the information age, a large number of loanwords and new words including network parlance have been introduced into the Chinese character system. Whether the purity of Chinese characters can be maintained has also aroused people's concern.

In fact, the development of Chinese characters has revealed that Chinese characters in history have always been complicated. The greatest advantage of Chinese characters is to be able to absorb and keep pace with the times. If Chinese characters were not able to absorb and record the latest language, they would lose their functional advantage and be abandoned sooner or later. Judging from the characteristics of Chinese characters, it seems that there is no need to worry about this. Taking classical Chinese and vernacular Chinese as an example, the changes in people's expressing habits can be deemed as enormous, yet the Chinese characters used have not changed a lot. This shows that Chinese characters are very competent in recording Chinese language and can comply with great changes. Though network terms such as "囧" (embarrassed, read as/jiǒng/) and "槑" (stupid, read as/dāi/) have been endowed with new meanings, the scope of use nevertheless is limited, so it is difficult for them to be classified into the system of commonly used Chinese characters, not to mention dominating the Chinese character system. This is another aspect of the mightiness of Chinese characters.

As one of the oldest character systems in the world, Chinese characters are not only the most important means of knowledge dissemination, but also produce a new art form—calligraphy. For thousands of years, people have enjoyed it and the calligraphy masters have emerged in large numbers, which is also a cultural phenomenon worth exploring.

Additionally, Chinese is also one of the six official languages stipulated in the Charter of the United Nations, and the role of the simplified Chinese characters in international affairs has naturally attracted much attention. Chinese characters are well known for their simplicity, and it is said that the thinnest version of the UN official documents in the six designated languages must be the one written in Chinese characters.

Having gone through great historical evolution, Chinese characters still remain vibrant. They are not only one of the creations of the Chinese nation, but also a symbol of Chinese civilization. Their strong cultural influence, unique configuration system and unparalleled aesthetic connotation of calligraphy have condensed the basic characteristics of Chinese culture. Being a well-deserved treasure in the world cultural heritage, Chinese characters can be listed as one of the most important inventions in China.

Translators' Notes

1. Gou Jian, the king of the Yue State in the late Spring and Autumn Period (reigned 497–465 B.C.).
2. Zhou Gou, the great-great-grandchild of Gou Jian and the king of the Yue State (reigned 448–412 B.C.).
3. Wang Ziwu was also called Lingyin Zigeng, a combination of his title plus his courtesy name. Lingyin (令尹, read as/*Lìngyǐn*/), translated as prime minister or chancellor, was an official government position established in the Chu state during the Spring and Autumn Period (771–475 B.C.).
4. *Shijing* or *Shih-ching*, has been translated variously as *the Book of Songs*, *Book of Odes*, or is simply known as *the Odes or Poetry* (诗, read as/Shī/). It is the oldest existing collection of Chinese poetry, comprising 305 works dating from the eleventh to seventh centuries B.C.
5. The First Emperor (始皇帝) was born Ying Zheng (嬴政), a prince of the state of Qin. He became Zheng, the King of Qin (秦王政) when he was thirteen, then China's first emperor when he was 38 after the Qin State had conquered all of the other Warring States and unified all of China in 221 B.C.
6. *Rishu* (日书), is one of the divinatory books discovered in late Warring States Period which has great cultural significance in ancient and medieval China.
7. Liye (里耶镇, read as/lǐyē zhèn/) is a town of Longshan County, Hunan Province. It is an ancient town with a history of over 2,300 years. It was established by Chu State in the late Warring States Period (403–221 B.C.).
8. *The Lüshi Chunqiu* (《吕氏春秋》), also known in English as *Master Lü's Spring and Autumn Annals*, is an encyclopedic Chinese classic text compiled around 239 B.C. under the patronage of the Qin Dynasty Chancellor Lü Buwei.
9. *Han Shu Yi Wen Zi (*《汉书艺文志》), the earliest bibliographic literature in China.
10. Qin Xiaogong (秦孝公) was the ruler of the Qin State from 361 to 338 B.C. during the Warring States Period of Chinese history.
11. Emperor Wu of Han (西汉武帝) was the seventh emperor of the Han Dynasty, ruling from 141–87 B.C.
12. *Ji Jiu Pian* (《急救篇》), a textbook of children in Han Dynasty.
13. *Yu Pian* (《玉篇》), the first Chinese character dictionary categorized according to radicals.
14. *Long Kan Shou Jing* (《龙龛手镜》), a book of Chinese characters compiled according to radicals and tones.
15. *Kangxi Dictionary* (《康熙字典》) was the standard Chinese dictionary during the 18th and 19th centuries and it is named after the Emperor's era name.

Acknowledgements Many of the views and materials in this paper are quoted from Mr. Qiu Xigui's *Chinese Writing*. Mr. Hua Jueming has given much guidance and provided many suggestions for revision in the writing process. And Mr. Feng Lisheng of Tsinghua University has also offered a lot of help. The author would like to show sincere gratitude to all those who have helped in the completion of this paper.

Yu Peng Born in Guzhen, Anhui in 1982, Doctorate of the Unearthed Literature and Ancient Literature Research Center of Fudan University, former assistant researcher in the History Department of Tsinghua University and current lecturer at the School of History and Culture of Southwest University.

Research interests: Unearthed documents, ancient Chinese characters; Ancient Chinese bronze Ware.

Major works: He has participated in many major national social sciences programs and independently presided over 1 National Social Science Fund Youth Project. He has published more than 30 papers in publications such as Cultural Relics, Archaeology and *Cultural Relics, Jianghan Archaeology, Huaxia Archaeology, Chinese History and National Museum of China.*

Chapter 6
Algorism and Rod Calculus

Shuchun Guo

With a very different form from the ancient Greek mathematics, traditional Chinese mathematics was good at calculation and advanced in the world from the second or third century B.C. to the beginning of the fourteenth century. There are many reasons, of course. Among them, the most important one is China's earliest use of the most superior notation system in the world—the algorism, and the most convenient computing tool in the early civilization—counting rods. The algorism is still widely used at present, while the calculation with counting rods—rod calculus, evolved into the abacus arithmetic in the Southern Song Dynasty (1127–1279) at the latest, then was simultaneously used with and completely replaced by abacus arithmetic by the middle of the Ming Dynasty (1368–1644).

Evidently, most of the achievements of traditional Chinese mathematics give credit to the rod calculus.

6.1 The Development of Algorism and the Invention of Rod Calculus

6.1.1 The Development of Number Concept and Algorism

The understanding of number is a very long process. People initially completed the abstraction of the number concept when using a number, such as two, to represent two people, two apples, or other objects. According to *Book of Origins* (second century B.C.), an official of Emperor Yellow, "historian Lishou invented the numbers" [1] 5,000 years ago. Another record in *Book of Changes* shows that "in the highest

S. Guo (✉)
The Institute for the History of Natural Sciences, Chinese Academy of Sciences, Beijing, China
e-mail: scguo@ihns.ac.cn

J. Hua and L. Feng (eds.), *Thirty Great Inventions of China*,
https://doi.org/10.1007/978-981-15-6525-0_6

antiquity, government was carried on successfully by the use of knotted cords (to preserve the memory of things) [2]. In subsequent ages the sages substituted for these written characters and bonds." By the 1950s, knotted cords (Fig. 6.1) and carved woods for recording were still adopted by some minorities in Yunnan. On the other hand, the number symbols in oracle-bone inscriptions (Fig. 6.2) of the Shang Dynasty (sixteenth to eleventh century B.C.) are the earliest available data on the algorism.

The exact time of establishing the algorism remains unknown by now. Some texts in *The Book of Master Mo* written in about 330 B.C. show that the Mohist scholars had realized that the same digit could be placed with different values in different digits.

In Master Mo's interpretation, "One is less than two yet more than five. Explanation is given under 'establishing a position'" [3] (the last "one" is in a higher-than-unit position); and "There are ones in five and fives in one. The last 'one' is at the next

Fig. 6.1 Knotted cords

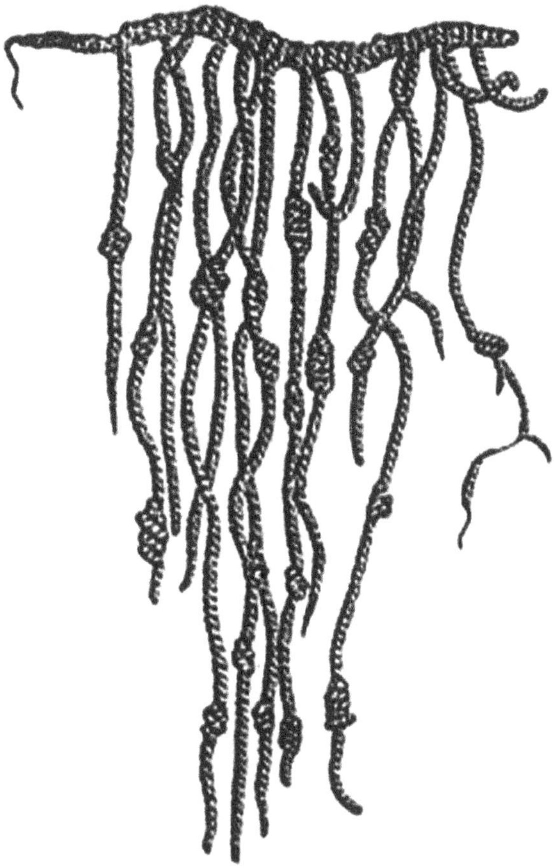

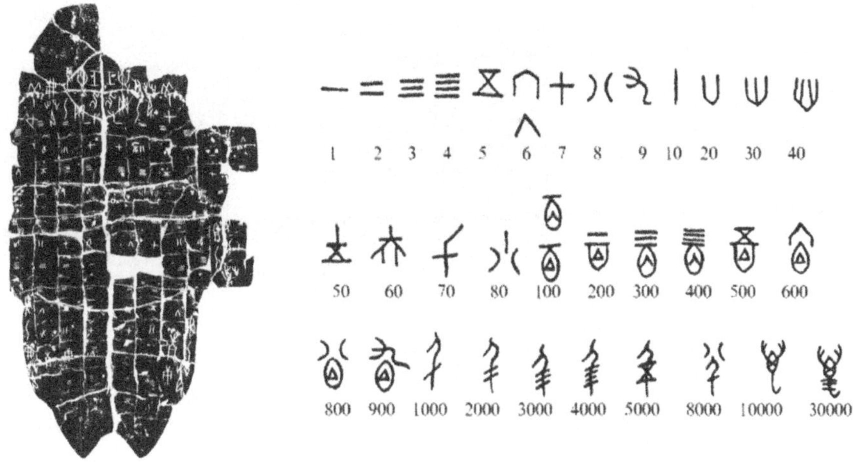

Fig. 6.2 The oracle-bone inscriptions

higher place and therefore contains two 'fives'." [4] As you can see, the algorism had been quite perfect in the Spring and Autumn Period (770–476 B.C.) at the latest.

6.1.2 Computing Tool—Counting Rod

6.1.2.1 Counting Rod

The counting rods, known as *Suan Chou* in Chinese, are made of bamboo, wood, ivory or bone. Nowadays, no one knows when it came into being. Much cited is the saying of Laozi in *Canon of the Dao and Its Virtue* at the end of the Spring and Autumn Period: "Good mathematicians can do without the counting rods." [5] A riddle recorded in 543 B.C. said that an old man's age was equal to a Chinese seal character "亥". [6] According to the rod numeral and rod configuration in rod calculus, the character "亥" could be divided into =T⊥T and represents 26,660 days, i.e., 73 years. All the above-mentioned data show that the counting rods had been the major computing tool before the end of the Spring and Autumn Period and surely it was invented much earlier. After its birth, the counting rod gradually changed from long to short, and from cylindrical to square in cross-section. According to "Memoir on the Calendar" in *History of the Han Dynasty,* [7] the counting rods were 0.23 cm in diameter and 13.86 cm long. Since the 1950s, a number of ancient counting rods have been unearthed in archeological excavations. The rods of the Western Han Dynasty unearthed in Qianyang County [8] and Xunyang County (Fig. 6.3) of Shanxi are true to the descriptions in *History of the Han Dynasty*. But the rods of the Eastern Han Dynasty unearthed in Shijiazhuang, Hebei are square and 7.8–8.9 cm long [9]. Furthermore, three red paint spots are found on the two rods unearthed from

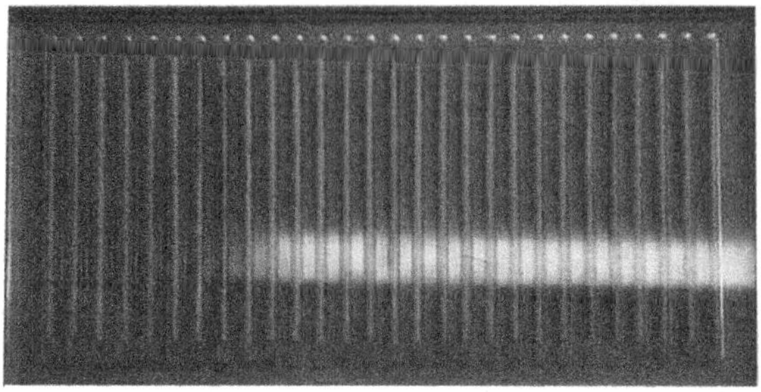

Fig. 6.3 The counting rods unearthed at Qianyang

the tomb of Emperor Wenjing in Western Han Dynasty. According to Liu Hui, if different colored rods were used, then the red ones represented *zheng* (positive) and the black ones represented *fu* (negative). Thus, those colored rods would have been used in the calculation of positive and negative numbers [10].

The first nine natural numbers formed from counting rods were as follows:

Numeral	1	2	3	4	5	6	7	8	9
Vertical	𝍠	𝍡	𝍢	𝍣	𝍤	𝍥	𝍦	𝍧	𝍨
Horizontal	𝍩	𝍪	𝍫	𝍬	𝍭	𝍮	𝍯	𝍰	𝍱

The rod numeral system adopted algorism. Today, we cannot find any primer on mathematics in ancient China. But the earliest records of notation could be seen in *Sun Zi Suanjing (The Mathematical Classic of Sun Zi)* [11] as follows: "When doing calculation (with rods), one should first determine the positions (of the rod numerals). The units are vertical and the tens horizontal, the hundreds stand and the thousands lie; thousands and tens look alike and so do ten thousands and hundreds"; and another description is almost the same to the text in *Xiahou Yang suanjing (Xiahou Yang's Mathematics Manual)* (Fig. 6.4) [12]: "For 6 and above, a rod on top to represent five. 6 is not expressed in 6 rods and 5 is not just only 1 rod." This means that for numerals 6 to 9, the single rod on top represented the quantity five; 6 was not represented by |||||| nor ䷀, and 5 on its own was not represented by a single rod. With different configurations and a blank space to represent zero, the counting rods could indicate any natural number. For example, 597031 in rod numeral notation appeared like this: 𝍭𝍨𝍯 𝍫𝍠.

6.1.2.2 "○" and Numeral

There is no conclusive data on the date of invention of "○" for zero. Misunderstanding may arise when rod numerals indicated zero with the blank space. In ancient China, a square symbol "□" was usually served as the omission of words and therefore stood for zero. Later the "□" gradually evolved in "○", which appeared first in *Records of the Ming Dynasty* edited in Jin Dynasty. In this book there were some numbers containing "0," like "四百○三" (403).

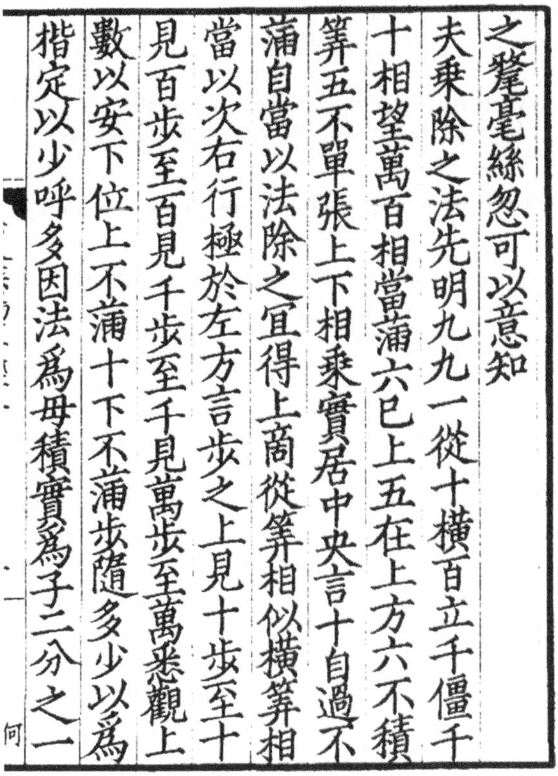
之氂毫絲忽可以意知
夫乘除之法先明九九一從十橫百立千僵
十相望萬百相當滿六已上五在上方六不積
筭五不單張上下相乘實居中央言十自過不
滿自當以法除之宜得上商從筭相似橫筭相
當以次右行極於左方言步之上見十步至十
見百步至百見千步至千見萬步至萬悉觀上
數以安下位上不滿十下不滿步隨多少以爲
指定以少呼多因法爲母積實爲子二分之一

Fig. 6.4 The photocopy of Xiahouyang Suanjing

In the middle of thirteenth century, "0" was repeatedly used in Qin Jiushao's and Li Ye's mathematics works. For instance, in Vol. 7, Probs. 2 of *Ceyuan hai jing (Sea Mirror of Circular Measurement)*, there was a number "一千四百五十万○○八百六十四" [13], in which the first "0" stood for "and" and the second for the blank in the thousands position. No one knows when "0" was introduced into calculation with rods. Anyways, "0" could be found in algorithms in Qin Jiushao's and Li Ye's mathematics classics.

After the middle of the Tang Dynasty, people began to employ the rod numeral system. The extant records of these numerals were first seen in *Li Cheng Suanjing (The Mathematical Classic of Li Cheng)*. For the convenience of writing, the ancient character "╳" was borrowed to represent five and "十" for ten. After the invention of "0," "ō" or "ȯ" (a horizontal or vertical line on the top of "0") was used for five while "╳" indicated four instead. For the same reason, nine appeared as "╳̄" or "╳̇" basing on "╳." For example, Qin Jiushao wrote the number "40642560000" as "×○⊥×‖ȯ⊥○○○○" and "16900" as "丨⊥╳̄○○" [14]. In his notation "╳" was four instead of five, while "ȯ" was five and "╳̄" was nine. In this way, a new numeral system was gradually established in Southern Song Dynasty as follows:

With the invention of the abacus, the rod numerals gradually developed and evolved in a series of unified symbols as follows:

These symbols, with the cursive form of five and nine, are known as Suzhou numerals, traditionally used by the shopkeepers to mark prices to the first half of the twentieth century [15].

6.1.2.3 Rod Notation of Common Fractions and Decimal Fractions

The rod numerals could be used to express common fractions, decimal fractions and negative numbers. A common fraction was displayed in two rows with the numerator in the upper and the denominator in the lower. For example, the common fraction $\frac{34}{99}$ was notated as . A mixed number was displayed in three rows with the integer in the upper, the numerator in the middle and the denominator in the lower. For example, the common fraction $56\frac{16}{65}$ was notated as .

In the history of mathematics, decimal fractions were produced much later than common fractions. The ancient Chinese mathematicians were the earliest to possess the notation of decimal fractions. In a text "37 men 5 *fen,*" the answer to Ch. 3, Probs. 2 of *Sun Zi Suanjing (The Mathematical Classic of Sun Zi),* the "5 *fen*" was just 0.5. The invention of the decimal fraction should attribute to the conversion of the non-decimal units. After the middle of the Tang Dynasty, there was an urgency to change those non-decimal units into decimal numerals due to the increasing need for massive and fast calculation and the inconvenience in converting *duan, chi* and *cun* or *jin* and *Liang*. In the answer to Ch. 3, Probs. 11 of *Xiahou Yang suanjing (Xiahou Yang's Mathematics Manual),* "1525 *pi* 3 *zhang* 7 *chi* 5 *cun*" was expressed in "152549375." The notation was quite similar to our present decimal fractions. In Song and Yuan dynasties, the decimal fractions were widely used.

There were four major expressions to depict decimal fractions:

A. The one used in *Sun Zi Suanjing (The Mathematical Classic of Sun Zi),* expressing the fractional parts as *fen* in words.
B. The one expressed as the place-value system and rod numerals, adding the character *fen* under the first place in fractional parts. For example, in Vol. 8, Probs. 5 of *Ceyuan hai jing (Sea Mirror of Circular Measurement),* the coefficient of first order with unknown x in a polynomial equation "" was just 5.5χ.
C. The one replacing integral part with a "O". For example, in Ch. 1 of *Yigu yanduan (New Steps in Computation),* the coefficient of the second order with unknown x in a quadratic equation "" was just 0.75χ^2.

D. The one adding the unit name under the integral part. For example, in the answer to a problem of capacity in Ch. Storage of grains, *Shu Shu jiuzhang (Mathematical Treatise in Nine Sections)* by Qin Jiushao in Southern Song Dynasty, the bucket with a "depth of 1 *chi* 5 *cun* 9 *fen* 2 *li*" was expressed in "1592", that is, 15.92 *cun*.

Some expressions of the decimal fractions in Li Ye's books were the same as that of Qin Jiushao. In 1585, Belgium mathematician Simon Stevin established the notation and laws of the decimal system of tenths. But his operations were rather complex.

The method of *jin* and *liang* was a formula in verse with the decimal fraction numeration, which was created in the middle of the Tang Dynasty and converted the unit *liang* into *jin* in the scale system. The song of *jin* and scale (*jin xia liu fa*) in *Suanxue qimeng (Introduction to mathematical studies)* was almost the same as that of the present. For example, converting one *liang* into the *jin* system could be expressed in "1 of 16 getting 625," i.e., 0.0625 *jin*.

6.2 The Fundamental Operations of Arithmetic

6.2.1 The Fundamental Operations of Integer

The law of the *fundamental operations* of integer *was* not recorded in the mathematic texts in the Qin and Han dynasties (third century B.C. to third century A.D.) as well as *Jiu zhang suanshu* (*The Nine Chapters on the Mathematical Art)* (- third century A.D.). But early in the Spring and Autumn Period, the times table was widely circulated, and people were able to skillfully use the law of multiplication and division of integer, which first appeared in Ch. 1 of *Sun Zi Suanjing (The Mathematical Classic of Sun Zi),* one of the extant ancient *arithmetic* books, and was elaborated in this book as below:

> When doing multiplication, we need to redeploy the counting rods in rows. A number is put in the top row, and another number in the bottom with the ones digit right under the first digit in the top from the left. In between is the blank row for operation of product. If the number in the top row is a tens digit, move the ones digit in the bottom row to the place right under the tens in the top row. If the top is a hundreds, move to the hundreds. If the top is a thousands, move to the thousands and so on.
>
> Now, the operation goes as follows.
>
> First, multiply the digits in the bottom row one by one from left to right by the first digit in the top row; put the first product in the middle row and add the subsequent products up to the obtained ones in turn. If the product is a tens digit, carry one more place on the left. If the product is a ones, leave it at the normal place. If there are over 5 counting-rods in a certain digit, reduce them with a specific posing of a rod for 5. Of course, 5 or less rods in a digit are allowed. Second, take away the first multiplier digit in the top row and move the whole number in the bottom row one place to the right. Third, multiply the number in the bottom

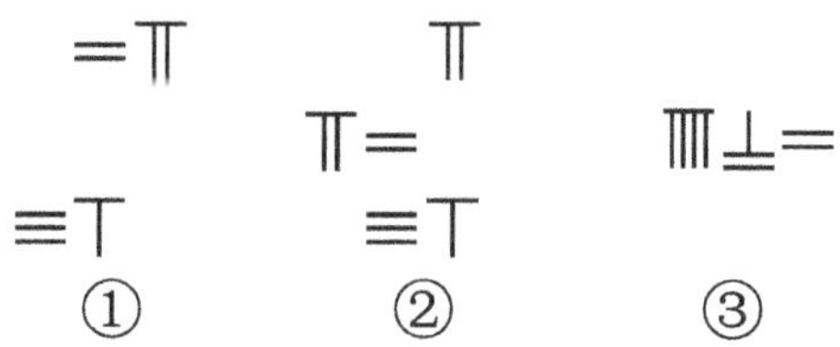

Fig. 6.5 Counting rods deployment of 36 × 27

row in this way by the second digit in the top row and keep going like this until all the digits in the top row are removed. In the end, only a sum is left in the middle row and you get the final product.

Let's illustrate the law with the operation of 36 × 27.

The rod numerals would be displayed as shown in Fig. 6.5①, with 27 in the upper position, 36 in the lower position, and the "6" of 36 directly below the "2" of 27. First, multiply the 3 in the first digit of the lower numeral by the 2 in the first digit of the upper numeral commencing from the left, and place the product of 600 in the middle row. Secondly, multiply the 6 in the second digit of the lower by 2 and add the product of 120 to the previous 600 to get a sum of 720. Now you can take away the 2 in the first digit of the upper and move the number 36 to the bottom row one digit to the right (Fig. 6.5②). Then you can multiply the digits in the lower numeral one by one from left to right by the 7 in the upper numeral and get the products of 210 and 42 in order. Following the operations as before, add the product of 210 to the previous 720 to get a sum of 930, and likewise, plus 42 to 930 to make another sum of 972. Finally, all the digits of the upper numeral are removed and the lower numeral is then removed leaving the result of 972 (Fig. 6.5③).

Beginning at the digit of the highest rank was a basic law of multiplication in ancient China. Since the 1980s, some short-cut counting methods were claimed to be invented with the idea of multiplying from the highest rank. In fact, it was just a reproduction of traditional methods before the late Qing Dynasty.

The operation of division in *Sun Zi Suanjing (The Mathematical Classic of Sun Zi)* was as follows:

> When doing division, it is reversed to multiplication. The quotient is placed in the upper row while the product in the middle row. Suppose 6 is the divisor and 100 is the dividend. When dividing 100 by 6, move the rods of 6 two places to the left directly below the hundreds. It is impossible to divide 1 by 6 since the divisor (6) is greater than the dividend (1). Then shift the rods of 6 to the right directly below the tens. Remove the dividend 10 by the divisor 6. It is possible since 40 remains when 60 subtracts from 100. If the divisor is less than the dividend right above its position, then keep it below the hundreds and do not move it.

> It follows that if the ones of the divisor are below the tens of the dividend, the place value of the digit of the quotient is tens; if they are below the hundreds, the place value of the digit of the quotient is hundreds. The rest of the rule is the same as multiplication. In the end, if there is a reminder of the dividend, assign it to the divisor with the divisor as the denominator and the dividend as the numerator.

Since mathematicians laid emphasis on practical uses in ancient China, the dividend was expressed in *shi* (object) involving objective things like grain output, length

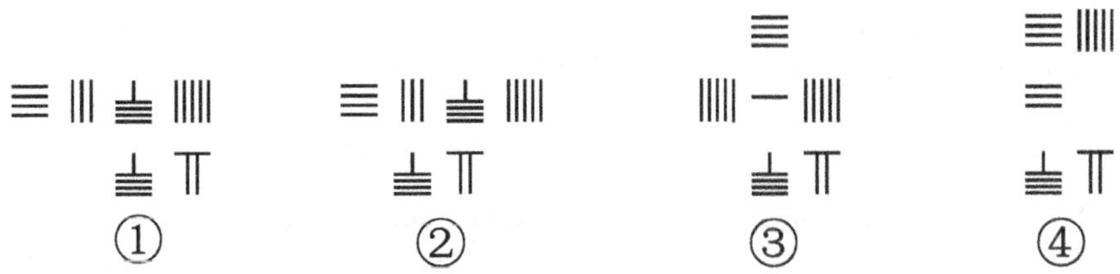

Fig. 6.6 The counting rods deployment of 4395÷97

of cloth, area, volume, etc. Afterwards, the constant term of an equation was called *shi* as well. The divisor was expressed in *fa* (law) with the original meaning of standard. Actually, the operation of division was to divide something by a certain standard, and then the quantity of standard became the divisor. Afterwards, the linear term of an equation was called *fa* as well.

Let's depict the Chinese method of division in rod numerals. The *shi* (dividend) is put in the middle row with the quotient in the upper and the *fa* (divisor) in the lower. First, place the first digit of divisor aligned with the first digit of the dividend. If that part of the dividend immediately above the divisor is numerically smaller than the divisor, division is not possible, and the divisor is shifted one place to the right to negotiate the first digit of the quotient. Secondly, the digit of the quotient multiplies each digit of the divisor commencing from the left. Each product is subtracted from the digits of the dividend occupying the same columns. Thirdly, when subtraction is completed, the second digit of the quotient multiplies each digit of the divisor from the left to right, and then those products are subtracted from the digits of the dividend. The same procedure is repeated with the remaining dividend until there is either no remainder or the remainder is less than the divisor. If the latter occurs, make a decimal fraction with the divisor as denominator and the remainder as numerator.

Take the operation of dividing 4395 by 97 as example. The dividend 4395 is initially displaced in the middle row and the divisor in the lower row (Fig. 6.6①). The divisor 6 is then shifted to the extreme left such that its first digit is aligned with the first digit of the dividend. Division is not possible since the number 43 in the first two digits of dividend is less than the number 97 in the first two digits of divisor. The divisor is shifted one place to the right such that its first digit 9 is aligned with the second digit 4 of the dividend (Fig. 6.6②). Figure out the first digit 4 of the quotient and place it above the tens of the dividend. Multiply the first digit 9 of the dividend by 4. The product of 3600 is subtracted from the dividend. 4395 is reduced to 795. Then multiply the second digit 7 of the dividend by 4. The product of 280 is subtracted from the previous reminder of dividend. 795 is reduced to 515. The divisor 6 is shifted to the right by one place (Fig. 6.6③). Figure out the second digit 5 of the quotient. Multiply the first digit 9 of the dividend by 5. The product of 450 is subtracted from the previous reminder of dividend. 515 is reduced to 65. Then multiply the second digit 7 of the dividend by 5. The product of 35 is subtracted from the previous reminder of dividend. 65 is reduced to 30 (Fig. 6.6④). By this step, division is not possible. Finally, make a decimal fraction and the quotient is $45\frac{50}{97}$.[16]

6.2.2 *The Fundamental Operations of Common Fraction*

The mathematical texts of Qin and Han Dynasties and *Jiu zhang suanshu* (*The Nine Chapters on the Mathematical Art)* all gave out the laws of *fundamental operations* of fractions, including reduction (*yue fe*), addition (*he fen*), subtraction (*jian fen*), comparison (*ke fen*), averaging (*ping fen*), multiplication (*cheng fen*) and division (*jing fen*). *The Nine Chapters on the Mathematical Art* depicted the addition (*he fen*) of fractions as follows:

> Multiply each numerator by the denominator in the other fraction, and take the sum of the two products as the dividend; multiply the two denominators, and take the product as the divisor. If the dividend is equal to the divisor, the answer is just one. If the dividend is less than the divisor, assign it to the divisor to make a fraction. If the denominators are the same, directly sum up the two numerators.

Given two fractions of $\frac{b}{a}$ and $\frac{d}{c}$, the law of addition could be $\frac{b}{a} + \frac{d}{c} = \frac{bc}{ac} + \frac{ad}{ac} + \frac{bc+ad}{ac}$. The mathematicians in ancient China all operated the problems on fractions in rod numerals and rod calculus, and expressed the laws of fundamental operations of fractions in rather abstract ways. Those laws were true to that at present except for the clear idea of the least common multiple in the reduction of fractions to a common denominator when adding and subtracting the fractions as well as the reciprocal multiplication in the division of fractions (even though the reciprocal multiplication was employed once in a problem of *Suan shu shu: Writings on Reckoning*). This is the earliest in the history of world mathematics.

Besides, according to the mathematical texts in Qin and Han Dynasties and *Jiu zhang suanshu* (*The Nine Chapters on the Mathematical Art)*, the rod numerals and rod calculus were used in the operations of surplus and deficit, proportion and proportional distribution. Qian Baocong, Joseph Needham and other scholars all believed that the operational law of surplus and deficit became the main method of solving problems in Europe after its introduction from ancient China.

6.3 Calculation of π Approximation

When it came to the area and volume of circle, the ratio of the circumference (*zhou*) of a circle to its diameter (*jing*) was taken to be 3 to 1 in *Zhou bi suanjing (The Mathematical Classic of the Zhou Gnomon), Jiu zhang suanshu (The Nine Chapters on the Mathematical Art)* and other mathematical texts in Qin and Han Dynasties. After proving the formula of area of a circle $S = \frac{1}{2}Lr$, Liu Hui claimed that the ratio of the circumference of a circle to its diameter should be an infinite number instead.

Thus, the correct method for calculating the exact approximation of π was stated. Beginning with a circle 2 *chi* in diameter, he progressively cut the circle by bisecting the inscribed hexagon as follows:

> Method of cutting an inscribed hexagon into an inscribed dodecagon. If the diameter of the circle is 2 *chi*, the length of one side of the hexagon is just half of it, that is, 1 *chi*. Take the radius of length 1 *chi* as the hypotenuse and a half of a side of the hexagon, 5 *cun*, as the *gou* (the shorter side of the right-angled triangle). Then determine the *gu* (the longer orthogonal side). Subtract 25 *cun*, which is the square of the orthogonal square, from the hypotenuse square to get a reminder of 75 *cun*. Extract the square root up to the lower place values of *fen*, *li*, *miao* and *hu*. Again, move the number in the divisor two places back to determine the other more minute numbers. Since these numbers are so minute that there is no name for them, convert them into a fraction with 10 as the denominator. The fraction is approximately $^2/_5$. Hence the *gu* is 8 *cun* 6 *fen* 6 *li* 2 *miao* $5^2/_5$*hu*. Subtract this from the radius to leave a remainder of 1 *cun* 3 *fen* 3 *li* 9 *hao* 7 *miao* $4^3/_5$*hu*. That is the smaller *gou* of a smaller right-angled triangle. Then half the side of the hexagon to get the smaller *gu* of the smaller right-angled triangle. Determine the hypotenuse. Then the square of the hypotenuse is 267949193445 square *hu*. Discard the remaining terms. In the end, extract the square root to find the length of one side of a dodecagon.

According to the Pythagorean theorem (*gou gu* theorem), the algorithms on extraction of a root and the equivalence between the side of inscribed hexagon and the circle radius, Liu Hui worked out the hexagon's apothem and then excess radius, and moved forward to the length of the dodecagon's side. Following the same procedure, Liu Hui continued repetitively to determine the apothem and excess radius of the dodecagon and the side's length, apothem and excess radius of the polygon of 48 sides, and recursively reached to the area of the polygon of 96 sides $S_4 = 313\frac{584}{625}$ and the polygon of 192 sides $S = 314\frac{64}{625}$. Because the integral parts of S5 and S4 + 2(S5−S4) both were 314 *cun*, Liu Hui proposed the value of 314 *cun* for the approximation of the area of the circle and took the circumference of the circle as 6 *chi* 2 *cun* 8 *fen* by substituting the approximation in the formula expressed in *Jiu zhang suanshu (Nine Chapters of the Mathematical Art)*. He continued to compare the circumference with the diameter, and the ratio of the circumference to the diameter was obtained as 157:50, that is $\pi = \frac{157}{50}$.[1]

Liu Hui stated that the circumference still had other smaller units in the π. Therefore, he persistently worked out the length of the sides of the polygon consisting of 1536 sides and the area of the polygon of 3072 sides, then subtracted the residual area and obtained the approximation of the circumference 6 *chi* 2 *cun* $8\frac{8}{25}$*fen*, and compared it with the diameter 2 *chi* to get 1250 for the diameter and 3927 for the circumference. The ratio of the circumference to the diameter was thus obtained as $\pi = \frac{3927}{1250}$.

Liu Hui's calculation of the ratio of the circumference to the diameter caught up with and then surpassed the Archimedean in ancient Greece. He laid the theoretical and methodological foundation for Chinese π calculation to lead the world in more than a millennium.

Based on Liu Hui's theory and method, Zu Chongzhi (429–500 A.D.) corrected the π to the seventh decimal place: $\pi = \frac{355}{113}$.

Liu's value for π was not surpassed by mathematician al-Kashi from Central Asia until 1247. As for Zu, he had to compute the square root of decimals in more digits to obtain the π with 8 significant figures from a fraction with the denominator less than 1660423. There was no way for him without using the algorism and the formula of square root in rod calculus.

6.4 Extraction of Root—Solution of Linear Equation

6.4.1 Extraction of Root in Ancient China

6.4.1.1 Extraction of Root in *Jiu Zhang Suanshu* (*The Nine Chapters on the Mathematical Art*)

The extraction of root at present generally refers to the process of solving the root of a form of quadratic equation, and the equality of the form is called equation. But both of them were expressed in extraction of root (*kai fang*) in ancient China. The cube root was depicted in *Jiu zhang suanshu (The Nine Chapters on the Mathematical Art)* as follows:

> Take the cube of a certain unknown number as the dividend and put it in the second row from the top of the counting board. Put one counting rod in the bottom row and move it three places from right to left. This counting rod in new place is called the rod borrowed. Negotiate the product of the first root digit under trial multiplied by the first rod borrowed. Then multiply the first rod borrowed twice by the selected first Figure of the root. Take the product as the divisor, put it in the third row from the top and divide the dividend. Put the first remainder in the second row from the top. When finishing the division, multiply the divisor by three to get the first fixed divisor. Move the product one place back in preparation for the next division operation. Multiply the product by three and put the first product in the middle row. Once again, borrow another rod and put it in the bottom row. Move the first product in the middle row two places and the second rod borrowed in the bottom row three places to the right. Then negotiate the second digit of the root. Multiply the first product in the middle row by the second digit of the root to get the second product in the middle row. Multiply the second rod borrowed twice by the second Figure of the root to get the product in second row from the bottom. Add the two products one by one to the first fixed divisor and put them in the third row from the top. Divide the first remainder by the second fixed divisor and put the second remainder in the second row from the top. After finishing the division, double the product in second row from the bottom. Then add the third product to the second product in the middle row. Again, add the sum to the second fixed divisor and get the subordinate fixed divisor. Move the subordinate fixed divisor one place back in preparation for the next division operation. Continue the process in the second row from the bottom in a similar way.

The text above had long been very obscure and the problems of root extraction were solved with rod calculus. The extraction of root in *Jiu Zhang Suanshu (The Nine Chapters on the Mathematical Art)* was revised by Liu Hui and later evolved in the extraction method by means of a chart initiated by Jia Xian in the Northern Song Dynasty.

6.4.1.2 Jia Xian's *Li Cheng Shi Suo Fa* (Unlocking the Coefficients by Means of a Chart) and Triangle

Exactly *shi suo fa* (unlocking the coefficients) was the extraction of root since extracting could be seen as unlocking. *Li cheng,* originally compiled by the mathematicians in Tang and Song Dynasties, was computing some charts with a list of constants. Therefore, *li cheng shi suo fa* was just an extraction method of unlocking the coefficients by means of a chart with a list of constants and the *li cheng* here referred to Jia Xian's Triangle (Pascal Triangle) [17] (Fig. 6.7).

Fig. 6.7 Jia Xian's triangle from *Yong Le da diun (Great encyclopedia of the Yong Le reign)*

Jia Xian's Triangle was originally named as *kai fang zuo fa ben yuan* (origin of extraction method), also known as *shi suo qiu lian ben yuan* (origin of unlocking the coefficients). In math textbooks for middle school and many popular science books, it was called Yang Hui's Triangle, incorrectly relayed from an erroneous message. In *Xiang jie jiuzhang suanfa (A Detailed Analysis of the Methods of Computations in Nine Chapters)* (1261) from *Yong Le da diun (Great Encyclopedia of the Yong Le Reign)*, a commentary stated that "Jia Xian used this technique in *Shi Suo Suan Shu (Mathematical Book on Unlocking the Coefficients)*". Obviously it should be called Jia Xian's Triangle since it was originated from Jia Xian.

Jia Xian's Triangle was an isosceles triangle chart with the array of the coefficients of the integral binomial expansions, listing from top to bottom. In Jia Xian's Triangle from *Yong Le da dian (Great Encyclopedia of the Yong Le Reign)*, the numerals were expressed in Chinese characters but in rod numerals instead in the "Diagram of the old method up to the eighth degree" from Zhu Shijie's work *Si yuan yu jian (Jade Mirror of Four Unknowns)* (1303, Fig. 6.8) [18]. The numbers of the triangle in Zhu's work were connected by two crosswise lines so that the diagram could be used for extracting of root and solving *duo ji shu* (the sum of arithmetic series of higher

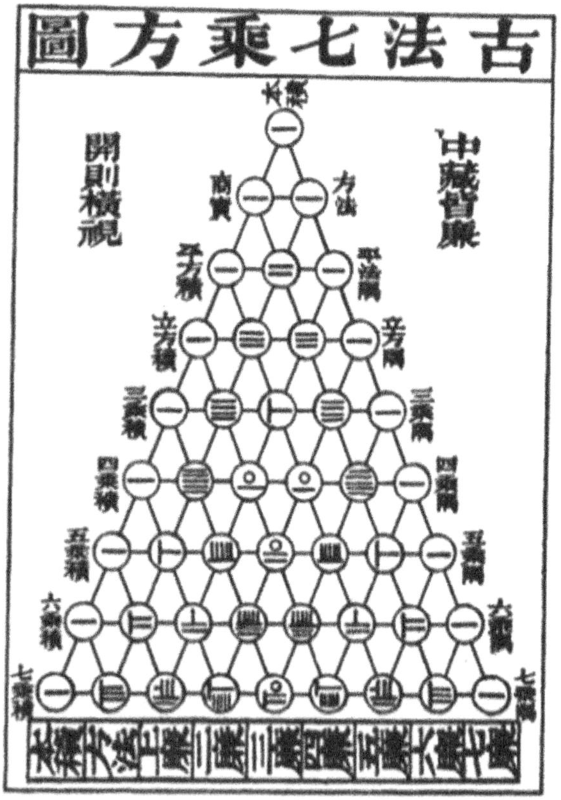

Fig. 6.8 Diagram of the old method up to the eighth degree in *Si yuan yu jian (Jade Mirror of Four Unknowns)*

order). Later, the similar triangle, called Pascal Triangle in Europe, appeared in Arab region. But it postdated Jia Xian by 500–600 years.

6.4.2 *Zeng Cheng Kai Fang Fa*

6.4.2.1 *Zeng Cheng Kai Fang Fa* by Jia Xian

Zeng cheng kai fang fa (extraction method of adding and multiplying) initiated by Jia Xian was the significant progress of root extraction in Song and Yuan Dynasties. His *Zeng cheng kai ping fang fa* (square root extraction method of adding and multiplying) and *Zeng cheng fang fa* (cube root extraction method of adding and multiplying) were found in *A Detailed Analysis of the Methods of Computations in Nine Chapters* by Yang Hui [19]. In *Great Encyclopedia of the Yong Le Reign*, we could also see Jia Xian's depiction of the extraction of the fourth root. For example:

Problem 6. Find the fourth root of 1,336,336 *chi*. Answer: 34 *chi*.

The computation of the extraction of the fourth root: When the number for the quotient in the upper row is determined, move up the divisor in the lower row to the cube and then subtract the product from the dividend. Take the given cube as the dividend and separately place one counting rod to be the divisor in the lower row. Move it from the end of the dividend three places to the left to make it directly below the dividend. Move one place in the square root extraction and three places in the fourth root extraction to make the divisor in the lower row directly below the ten thousandth place of the fixed dividend. So, the number for the quotient in the upper row is determined (30), and multiply the divisor in the lower row by this number to get the coefficient of x^3 (30). Next, multiply the coefficient of x^3 by the quotient in the upper row to get the coefficient of x^3 (900). Again, multiply the coefficient of x^2 by the quotient in the upper row to get the cube (27,000). Multiply this number by the quotient in the upper row and subtract the product from the dividend to get a remainder of 526,336.

The computation of the number in the second place for the quotient: multiply the divisor in the lower row by the quotient in the upper row and add the product to the coefficient of x^3. Get the sum of 60. Multiply the coefficient of x^3 (60) by the quotient in the upper row add the product to the coefficient of x^2. Get the sum of 2,700. Next, multiply the coefficient of x^2 (2,700) by the quotient in the upper row and add the product to the cube. Get the sum of 108,000. Again, multiply the divisor in the lower row by the quotient in the upper row and add the product to the coefficient of x^3 (60). Get the sum of 90. Multiply the coefficient of x^3 (90) by the quotient in the upper row and add the product to the coefficient of x^2 (2,700). Get the sum of 5,400. Again, multiply the divisor in the lower row by the quotient in the upper row and add the product to the coefficient of x^3 (90). Get the sum of 120. Move the cube one place to the right, the coefficient of x^2 two places, the coefficient of x^3 three places and the divisor in the lower row four places. Then the cube is 108,000, the coefficient of x^2 is 5,400, the coefficient of x^3 is 120 and the divisor in the lower row is 1. Furthermore, put down its adjacent number next to the quotient in the upper row. The number in the second place is obtained (4). Multiply the divisor in the lower row by this number and add the product to the coefficient of x^3. Multiply the coefficient of x^3 (124) by the second number in the quotient (4) and add the product to the coefficient of x^2. Get the sum of 5,896. Multiply to the coefficient of x^2 (124) by the second number in the quotient (4) and add the product to the cube. Get

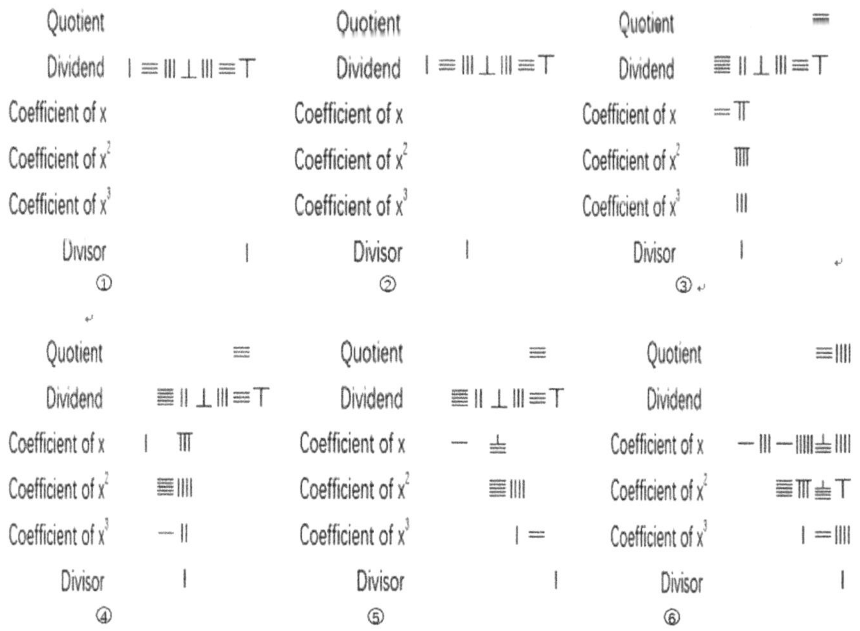

Fig. 6.9 The counting rods configuration for the positive root of $\chi^4 = 1336336$

> the sum of 131,584. Multiply this number by the quotient in the upper row and subtract the product from the dividend. No remainder left. Thus, the fourth root is determined. To compute the third digit of the cube root, continue the operation from the second digit when the remaining dividend is not zero.

The result was the positive root. According to the abstract principle and the detailed description, the counting rods could be arranged as Fig. 6.9.

The key point of *Zeng cheng kai fang fa* is to determine a number in the root. To continue the extraction, sequentially complete the multiplying and adding with the number in the place for the quotient from below and stop at every lower position to reduce the root equation. This method, similar to the synthesis division in middle school's mathematics textbooks at present, is more regulated, organized and mechanized compared to the *li cheng shi suo fa*. If we rearrange, place and shift the rods in the right way, the method could be applied to solving numerical equations of any degree.

The identical methods appeared in Arab region later and were developed independently in the West by Ruffini and Horner in the early nineteenth century.

6.4.2.2 Qin Jiushao's Extraction Method of Positive–Negative Rule

Listed in Area of Cropland Chapter, *Mathematical Treatise in Nine Sections*, the problem of area was an algorithm for obtaining a numerical solution of quartic equations. Qin Jiushao showed in detail the disposition on the counting board, "*zheng fu kai san cheng fang tu*" (diagram of the solution of a "positive–negative" equation of fourth degree). The display of the rod numerals in the different stages of calculations took up 21 panels in his book. We generalize them in eight steps (Fig. 6.10) with serial number ①–⑧ added by the author.

The method goes as follows: The sign of the quotient (*shang*) is positive; the sign of the constant term (*shi*) is negative; the sign of the following term is positive (*cong*); the sign of the first term is negative (*yi*).

In the end, Qin Jiushao pointed out that the diagrams above were the solution of an equation of fourth degree, and the method could be applied to all kinds of equations below. So, this was a universal approach.

Since the coefficients of Qin Jiushao's equation had no limitation in the real number range, the irrational coefficients could be rationalized. For the first time in the extant historical records of Chinese mathematics, Qin Jiushao proposed the method of estimating root, "*yi fang yue shi*" (dividing coefficient by constant term). When the constant terms changed from negative to positive or the absolute values went up, his method could be valid for all the exceptional cases.

Li Ye and Zhu Shijie both made contributions to the root extraction with rod numerals.

6.5 Methods of Solving Equations

6.5.1 Fang Cheng Method

Fang cheng method (method of setting up linear equations) is the most remarkable achievement in *Nine Chapters on the Mathematical Art*. Xu Shen, a famous scholar in Han Dynasty, cited in his book *Shuo Wen Jie Zi* that *fang* corresponds to paralleling boats; and *cheng*, initially acted as a metric unit, is extended to standardize, to measure and to examine things as shown in the saying "metric units like *'fa'*, *'fen'*, and *'cun'* were equaled by ten times successively from the first to the last". Thus, *fang cheng* means merging, juxtaposing the quantitative relationships among objects and assessing their metrics.

Ch. *fang cheng,* Probs. 1 in *Nine Chapters on the Mathematical Art* as follows:

> There are three bundles of top grade cereal, two bundles of medium grade cereal and one bundle of low grade cereal, which yield thirty-nine *dou* (of grains) as *shi*; two bundles of top grade cereal, three bundles of medium grade cereal and one bundle of low grade cereal yield thirty-four *dou* as *shi*; one bundle of top grade cereal, two bundles of medium grade cereal and three bundles of low grade cereal yield three hundred and twenty-six *dou* as *shi*. What is the measure of grains in each bundle of the top, medium and low grade cereal respectively?

After the answer was given in *Nine Chapters on the Mathematical Art*, *fang cheng methods* (equations) were then proposed:

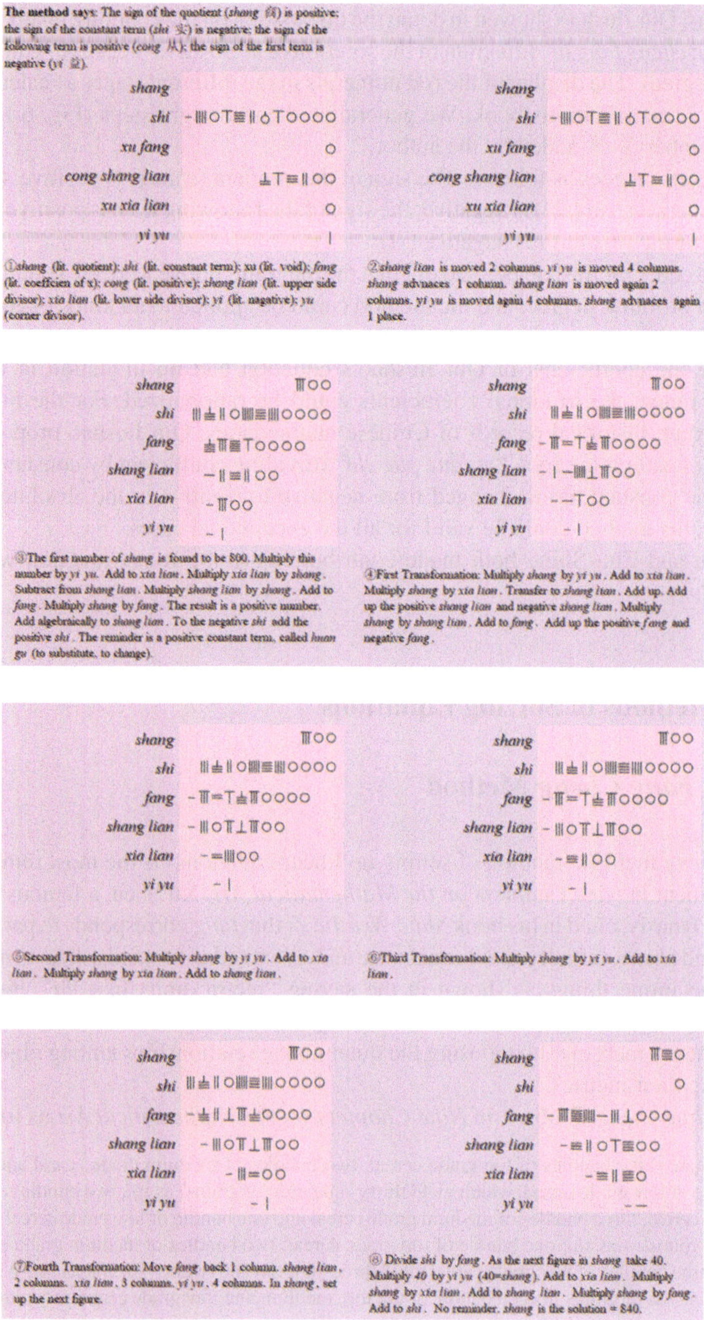

Fig. 6.10 Diagram of solution to a "positive-negative" equation of fourth degree

> Put down three bundles of top grade cereal, two bundles of medium grade cereal, one bundle of low grade cereal and thirty-nine *dou* as *shi* in a column on the right. Set up the columns in the centre and on the left in the same way as the column on the right. Take the (number representing) top grade cereal in the right column to multiply all nine (numbers) in the central column, and then use (the method) of direct subtractions (*zhi chu*). Once again multiply (the numbers) in the next column (i.e. the left column, by the number representing top grade cereal in the right column), and then use (the method of) direct subtractions (*zhi chu*). Next multiply all (the numbers in) the left column by the remaining (number representing) medium grade cereal in the central column, and then use (the method of) direct subtractions. The left column has the remaining number (representing) low grade cereal. The *fa* is above and the *shi* below; the *shi* here is the *shi* for low grade cereal. To find (the measure for) medium grade cereal, multiply the *shi* in the central column by the *fa* (of the left column), and subtract the *shi* for low grade cereal. The remainder is divided by the number of bundles of medium grade cereal (in the central column), yielding the *shi* for medium grade cereal. To find (the measure for) top grade cereal, multiply once again the *shi* in the right column by the *fa* (of the left column), and subtract (the respective) *shi* for low and medium grades. The remainder is divided by the number of bundles of top-grade cereal (in the right column), yielding the *shi* for top grade cereal. The *shi* for all (grades) are each divided by the *fa* to yield the measures (per bundle of the respective grades).

What is mentioned above was a general procedure to solve simultaneous linear equations. As Liu Hui said, "This is the way to solve the problem. It's hard to tell by empty words only, so it's the grade cereal of the special system that could resolve this problem." Based on this method, the solution through a process of elimination is depicted in Fig. 6.11 as follows:

Afterwards, Liu Hui invented a reciprocal multiplication and elimination method, which is the same as that of today's simultaneous linear equations.

6.5.2 *Sun Yi* Method *(Loss and Profit Method)*

Sun yi method (loss and profit method) was an important way to set up an equation in *Nine Chapters on the Mathematical Art.* The Ch. *fang cheng,* Probs. 2 put forward: "the loss is said to be the profit, and the profit is to be the loss," referring to increases and decreases of numbers. The term "the loss is said to be the profit" is to say that one end of the relationship loses a certain amount, equaling to the same increased amount

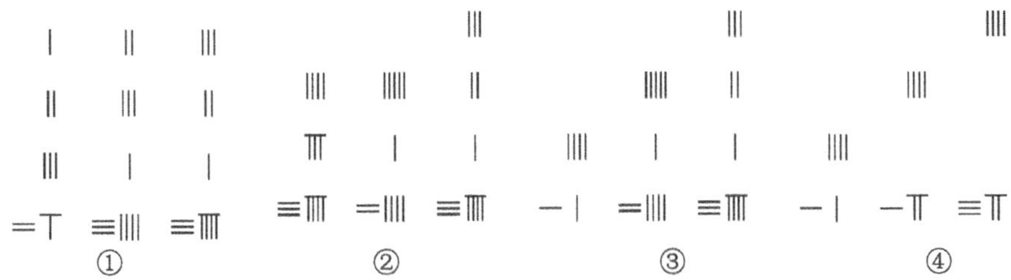

Fig. 6.11 The schematic diagram of the elimination method

at the other end; and vice versa. Loss and profit method had helped to establish the equation for Ch. *fang cheng*, Probs. 11 in *Nine Chapters on the Mathematical Art*. The perception of loss and profit, originally a dialectical thought of the Pre-Qin Philosophers, could be traced back to *Book of Changes* (He who makes a loss makes a gain) [20] and *Canon of the Dao and Its Virtue* (Everything will be increased if derogated and derogated if increased) [21].

It is generally believed that the term "algebra" derives from Arabic "*al-jabr*" in an algebraic work *Algorithm and Algebra* by Khwarizmi (~783–850) [22]. "*Al-jabr*," "restoration" or "transforming" in Arabic, refers to shifting the negative number from one end to the other and enable it to become a positive one when solving an equation. "*Al-muq Bala*," meaning "cancellation," refers to eliminating or merging the same items at both ends [23]. Obviously, the method of restoring and merging similar items in *Nine Chapters on the Mathematical Art* is about 1000 or earlier than that of Khwarizmi.

6.5.3 Zheng Fu Method *(Positive and Negative Rules)*

It is an important achievement of ancient Chinese mathematics to introduce negative numbers into equations in *Nine Chapters on the Mathematical Art* and propose the rules of the addition and subtraction of positive and negative numbers. The rules are stated as follows:

> When names are the same, perform *chu* (subtraction); when names are different, perform *yi* (addition). A positive number matches with nothing becomes negative; a negative number matches nothing becomes positive. When names are different, perform subtraction; when names are the same, perform addition. A positive number matched with nothing becomes positive and a negative number matched with nothing becomes negative.

The "same names" stated in the above rule denote that both numbers are positive or negative, while "different names" refer to the fact that one of them is positive, the other is negative, and vice versa. "*Chu*" is subtraction, and *Yi* is addition. Nothing refers to no match.

Here are the examples for the two rules illustrated in the following:

Rule 1: Subtraction of numerals

When names are the same and both are positive, subtract mutually.

$$(\pm a) - (\pm b) = \pm(a - b),\ a \geq b$$
$$(\pm a) - (\pm b) = \mp(b - a),\ a \leq b$$

When names are different, then positive minus nothing becomes negative, and negative minus nothing becomes positive.

$$0 - a = -a$$

$$(\pm a) - (\mp b) = \pm(a + b)$$

$$0 - (-a) = a$$

Rule 2: Addition of numerals

When names are different, subtract mutually.

$$(\pm a) + (\mp b) = \pm(a - b),\ a \geq b$$
$$(\pm a) + (\mp b) = \mp(b - a),\ a \leq b$$

When names are the same, positive plus nothing becomes positive, negative plus nothing becomes negative.

$$0 + a = a$$

$$(\pm a) + (\pm b) = \pm(a + b)$$

$$0 + (-a) = -a$$

There are a large number of positive and negative multiplication and division operations in *Nine Chapters of Arithmetic*. Later, Zhu Shijie, a mathematician in the Yuan Dynasty, first proposed the principle of multiplication and division of positive and negative numbers in his book *Suanxue qimeng*. Liu Hui defined positive and negative numbers in this way: "Red counting rods represent positive numbers and black rods represent negative numbers, or upright counting rods represent positive numbers and oblique rods represent negative numbers." A slash on the last digit of the negative number often appeared in the mathematical works of the Song and Yuan Dynasties. The concept of negative numbers and the addition and subtraction of positive and negative numbers were proposed in China much earlier, over several centuries or even 1000 years, than in other countries. In 628 A.D., Indian mathematician Brahmagupta used negative numbers to represent debt and used positive numbers to represent credit. He was the first scholar outside China to use the negative numbers. Later, even though the negative numbers were introduced into Europe, many scholars did not recognize them in the fifteenth to seventeenth century.

6.6 *Tian Yuan* Method and *Si Yuan* Method

6.6.1 Tian Yuan Method

Tian yuan method (methods of setting up quadratic equations), a way to construct equations of unknown numbers, initiates the semi-symbolic algebra in China. This method was explored in ancient books like Liu Nuxie's *Ruji shisuo* [24]. However, all these ancient books have been lost, and Li Ye's *Ceyuan hai jing (Sea Mirror of Circular Measurement)* is the earliest existing work on the use of *tian yuan* (celestial element), a mature method in his age. Li Ye's *tian yuan* method is very similar to the modern column equation. First, characters *tian* or *yuan* are used to represent an unknown number. Secondly, two equal algebraic equations are listed according to the conditions in the problem, and then subtracted to obtain an equation with one end at zero.

In *tian yuan* notations, a character *tian* is written next to a one degree term, or a *tai* next to a constant. The power of other items depends on the relative position of *tai* or *yuan*. In *Ceyuan hai jing*, the higher power is placed on the top and the low power at the bottom. A reversed representation is used in *Yi Gu Yan Duan* and its subsequent works, and becomes the standard form of *tian yuan* notations. *As* a polynomial of unknown numbers rather than *kaifang shi* (equation), *tian yuan* notation has been misunderstood since the middle of Qing Dynasty. For example, the *tian yuan* notations of Vol. 3, Probs. 5 in *Ceyuan hai jing* is , which represents $144x^2 + 5184x + 2488320$. Sometimes *tai* or *yuan* is omitted, as shown in the *tian yuan* notations in Problem 1 of *Yi Gu Yan Duan* 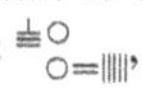, which represents $0.25x^2 + 80x + 1600$ [25] and in the notation of Problem 39 in *Yi Gu Yan Duan* , which represents $\chi^2 + 228\chi + 3780$.

6.6.2 Si Yuan Method

Si yuan method (four-unknown method) was the representation, establishment and solution of higher order equations of two variables, three variables and four variables. With the existence of *tian yuan* method, other related methods followed one after another, such as *er yuan* method, *san yuan* method and *si yuan* method. On the basis of Li Dezai's *er yuan* method and Liu Dajian's *san yuan* method, mathematician Zhu Shijie in the Yuan Dynasty invented *si yuan* method. Zhu's book, *Jade Mirror of the Four Unknowns*, proposed 36 problems of two unknowns, 13 problems of three unknowns and 7 problems of four unknowns. The four typical samples of solving equations illustrated his methods of solving unknowns, among which three problems involved representations of higher order equations for two unknowns, three unknowns and four unknowns, establishment of equations and eliminations of unknowns.

According to Fig. 6.12, the constant term "*tai*" is centered, and denotations x, y, z and w located at the lower, left, right and upper parts of "*tai*" represent the four elements of Heaven, Earth, Man and Matter, respectively. Its power is determined by the distance from "*tai*." The farther the distance is, the higher the power is. Depending on the position, the product of each adjacent unknown number and its power lies at the intersection of the corresponding row and column, with non-adjacent unknowns and their powers placed in the corresponding crevice, as displayed in the blank places in Fig. 6.12. The counting rod models (*choushi*) could find their modern equivalents, i.e., one counting rod model is similar to present one equation in one unknown number. Accordingly, there would be simultaneous equations in two unknowns for two counting rod models, and the same is true to three or even four counting rod models. This is a representation of separating coefficient, which is very convenient for establishing higher order simultaneous equations and eliminating variables.

The key to *si yuan* method relies on the elimination of four variables. According to the four typical samples of solving equations in *Jade Mirror of the Four Unknowns,* it could be divided into three steps—first, eliminate the equations of three or four variables into binary equations of higher order which is called the former and the posterior equations; secondly, reduce the binary equations of the first step to simultaneous equations of first order for a certain variable, which is called left form and right equations; lastly, eliminate the equation in the previous step to the higher order one of one variable finding its positive root by *Zeng cheng kai fang fa* (Extraction Method of Adding and Multiplying). Take *sancai yun yuan* (three talents) as an example demonstrated below:

> Question: Given the addition of *gou, gu* and *xian* divides the difference by *xian* subtracting *gu* equals to *zhi ji* (multiplication by *go* and *gou*), the addition of *gou, gu* and *xian* divides

Fig. 6.12 The representation of *si yuan* method

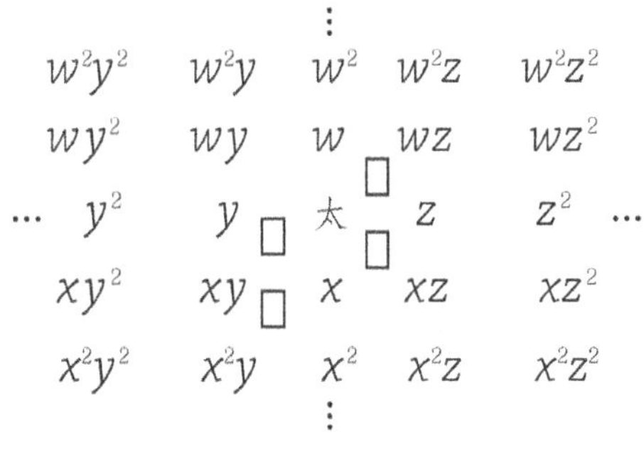

the difference by *xian* subtracting *go* equals to *go*, and the addition of both *gou* and *gu*'s square root equals to *xian*'s square root, then what is *xian*?

Answer: Set up *unitary tian* as the *gou* (base), *di* as *gu* (vertical) and *ren* as *xian* (hypothenus). By matching *sancai,* we obtain the following three equations

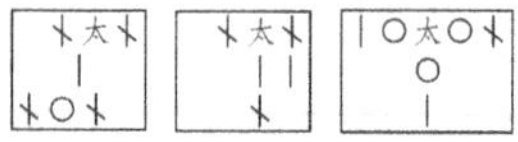

Eliminating of unknown between II and III by manipulation of exchange of variables, we obtain two new equations

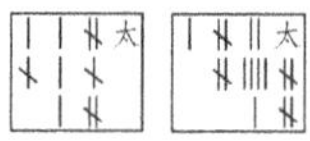

Eliminating of unknown between IV and V we obtain the following set of equations:

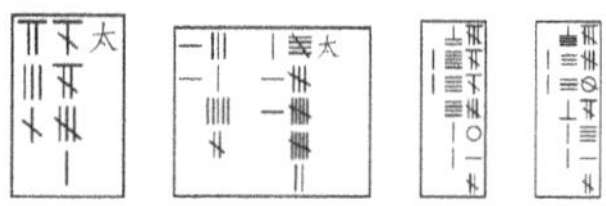

Successive elimination of unknowns to get 3rd order equation:

Solve this equation to obtain the answer that the *xian* (hypothenus) is 5 paces, which is the right one to the question [26].

It is not until 1775 that Bezout in Europe put forward the same solutions to the above-mentioned equations.

6.7 *Da Yan* Method

With the counting rods and rod calculus, ancient Chinese mathematics had led the world in mathematics for a long time. Their achievements ranged from *duo ji* method (the summation of higher arithmetic progression), *zhao cha* method (the 3-power interpolation of modern mathematics) and *da yan* method (linear congruence equations) to the solutions for area, volume and right-angled triangles (Pythagorean theorem). Here introduced only is *da yan* method.

Number games of "King Qin (Li Shimin, 598–649) Counting Soldiers," "Hanxin Counting Soldiers," "*Gui Gu* Arithmetic," "Calculation behind the Wall" and "Tube Cutting" were spread widely in ancient China and handed down from generation to generation. All of them in nature are problems of congruence equations which originated from the last volume of *Sun Zi Suanjing* (*The Mathematical Classic of Sun Zi*), i.e., "The Unknown Number of Things." It writes:

> Here is a problem of the unknown number of things. The remainder is 2 when the things are counted by threes; the remainder is 3 when they are counted by fives; the remainder is 2 when they are counted by sevens. Then we ask: what is the number of the things?

> The method to solve this problem: "The remainder is 2 when the things are counted by threes", then we write down 140 here; "the remainder is 3 when they are counted by fives", then we write down 63 here; "the remainder is 2 when they are counted by sevens", then we write down 30 here. Add the three numbers together to get the number 233 which subtracts 210 to get the number of the things. Every time when we count things by threes with the remainder being 1, we write down 70; when we count things by fives with the remainder being 1, we write down 21; when we count things by sevens with the remainder being 1, we write down 15. If the sum is over 106, then the result can be obtained by subtracting 105.

This actually, in the modern number theory, is a problem of congruence equations to find the minimum positive integer satisfying the condition of $N \equiv 2(\text{mod}3) \equiv 3(\text{mod}5) \equiv 2(\text{mod}7)$. The method here introduces the solution $N = 140 + 63 + 30 - 210 = 23$. This arithmetic problem proposed in *Sun Zi Suanjing* is the first one of congruence equations in mathematic works around the world.

Qin Jiushao (1208–1268) who made a further development came up with "*da yan* method" which in fact proposed a theorem: If $A_i(i = 1, 2, \ldots, n)$ are co-prime positive integers and $R_i < A_i$ while R_i is also a positive integer($i = 1, 2, \ldots, n$), and the positive N satisfies the condition of the congruence equations $N \equiv R_i(\text{mod } A_i)$with $i = 1, 2, \ldots, n$; if the positive integer K_i can be found, satisfying the condition $k_i(\prod_{j=1}^{n} A_j \div A_i \equiv 1(\text{mod}A_i)$ with $i = 1, 2, \ldots, n$, then $N = \sum_{i=1}^{n} [R_i k_i(\prod_{j=1}^{n} A_j \div A_i)](\text{mod} \prod_{j=1}^{n} A_j)$, hence providing with a general solution to the problem of linear congruence equations. Qin Jiushao named K_i as multiplying term (*cheng lu*), A_i as fixed parent (*ding shu*), $\prod_{j=1}^{n} A_j$ as extension parent (*yan mu*), and $\prod_{j=1}^{n} A_j \div A_i$ as extension number (*yan shu*). In this equation, A_i must be relatively prime positive integers in pairs, whereas in real problems A_i may not be co-primes, sometimes even not integers. Qin Jiushao presented different procedures of reduction accordingly to solve different problems. The famous German mathematician Carl Friedrich Gauss also wrote this theorem in his Latin book *Disquisitiones Arithemeticae* published in 1801.

The core of *da yan* method lies in the seeking-one method (*qiu-yi shu*). For convenience, we use G for $\prod_{j=1}^{n} A_j \div A_i$, A for A_i and k for k_i. If $G > A$, while $G = g(\text{mod } A)$ and $0 < g < A$, then $kg \equiv 1(\text{mod} A)$ and $kG \equiv 1(\text{mod} A)$ equate the same in value, which reflects the transitivity in modern congruence equation theory. Consequently, the problem turns out to be seeking k by satisfying $kg \equiv 1(\text{mod} A)$. At the same time, Qin Jiushao termed g as Surplus Number and wrote in his book *Shushu Jiu Zhang* (*Mathematical Treatise in Nine Sections*):

The *da yan* seeking-one method (*da yan qiu-yi shu*) goes like this: Put the Surplus Number on the upper right, the Fixed Parent on the lower right, and the important number 1 on the upper left. First, we use the upper right number to divide the lower right number, of which the quotient is used to multiply the upper left number 1, of which the product is put on the lower left (the two numbers in the upper row stay the same). Then we still use the smaller number to divide the bigger one on the right column, of which the quotient is used to multiply the unchanged number on the upper or lower left, of which the product together plus the former number in the position is put on the lower or upper left. The rest patterns can be done in the same way until the upper right number remains 1. Finally, we check the upper left number which is the Multiplying Term. It should be known that only when the remainder becomes 1, the Multiplying Term can be sought out. Next let's take the story "Inferring How Much Rice Stolen from How Much Remained" as an example to show how the *da yan* seeking-one method operates. The story goes: There was a rice store owner who complained that three baskets (*kuang*, a kind of round bamboo basket) of rice had been stolen yet with the number of how much being unknown. In the left basket, there left 1 *ge* (a unit of volume measurement in ancient times),in the middle one, 14 *ge*, and in the right one, 1 *ge*. Sometime later, three thieves called A, B and C were caught. A said he fumbled a *ma shao* (an ancient carrying tool whose shape is like a large ladle with a short handle) and used it to fetch the rice from the left basket to his cloth bag. Then B confessed that he kicked a wooden shoe and used it to fetch the rice from the middle basket to his own cloth bag. At last C admitted that he groped a lacquer bowl and used it to fetch the rice from the right basket to his cloth bag. All of them took the rice back home and ate it. With the passing of time they forgot how much they had stolen. Now they could only hand their carrying tools over. By measurement, a *ma shao* of rice equates 19 *ge*, a wooden shoe 17 *ge* and a lacquer bowl 12 *ge*. Then it is asked that how much rice in total was stolen and how much each of the three thieves got.

To answer the questions is to find the solution to the congruence equations $N \equiv 1(\text{mod}19) \equiv 14(\text{mod}17) \equiv 1(\text{mod}12)$ and Figure out k_1, k_2 and k_3 satisfying the conditions of $k_i \times 14 \equiv 1(\text{mod}19)$, $k_2 \times 7 \equiv 1(\text{mod}17)$ and $k_3 \times 11 \equiv 1(\text{mod}\ 12)$. Based on the *da yan* algorithm, the procedures to find k_1 are illustrated in Fig. 6.13①~⑤:

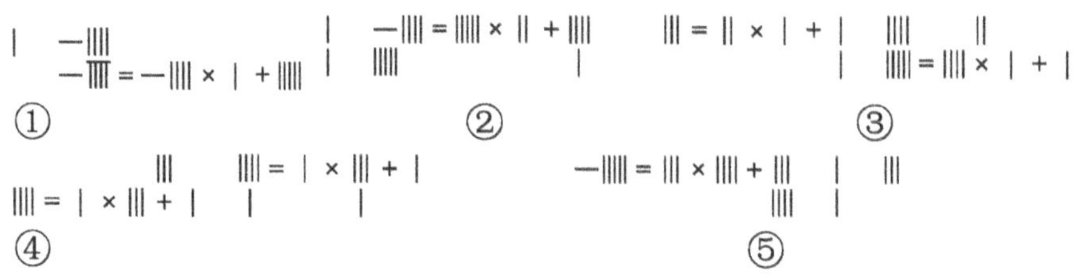

Fig. 6.13 The diagrammatic sketch of procedures to find K_1

Hence $k_1 = 15$. Following the same procedures, we can get $k_2 = 5$ and $k_3 = 11$. Finally, $N \equiv 1 \times 15 \times 204 + 14 \times 5 \times 228 + 1 \times 11 \times 323(\text{mod}3876 \equiv 22573(\text{mod}3876) = 3193$. To sum up, there were 3,193 *ge* of rice in each basket, from which A and C, respectively, stole 3,192 *ge*, and B 3,197 *ge*. Therefore, there were 9,563 *ge* of rice in total.

6.8 The Algorism and Rod Calculus: Characteristics of Traditional Chinese Mathematics

It is fair to say that rod calculus could not help to solve all the ancient Chinese mathematic problems like the methods of limit process and infinitesimal segmentation used in proving the circular area formula and Liu Hui theory.[2] Liu Hui himself also said, "To solve those problems of seeking limit numbers, we cannot use rod calculus but only to infer according to different situations." However, in other fields of ancient Chinese mathematics, the algorism and rod calculus together made it climb to the top of the world's mathematics time and time again, and made it distinguished from other cultural traditions.

First, the algorism and rod calculus have determined that the advantage of traditional Chinese mathematics lies in calculation. The study of quantitative relations particularly attracted scholars' attention and several algorithms were inducted. The achievements mentioned above and those unmentioned were all presented in the forms of formulas, solutions and computational procedures. Even in solving the geometric problems, what was taken into account was the length of a line segment, the area of a plane or the volume of a solid, whereas the nature of the geometrical figures apart from the quantitative relationships was rare in people's consideration.

Secondly, the place-value notation is embodied not only in the numeration but also in mathematical expressions. In the expressions of *tian yuan* (polynomial) and *kai fang* (extraction of square and cube roots), the powers of each term are entirely determined by their relationships with the positions of the Chinese character "*yuan*" (primary term) or "*tai*" (constant term). In *fang cheng* (equation) and *si yuan shi* (multivariate equation of higher order), the unknown number and its power are also determined by its position. It is particularly convenient to use this expression for the two-line operation and the addition, subtraction, multiplication and division of polynomials when eliminating variables in equations. You need to do nothing but to move the Chinese character "*yuan*" up and down when you do the operation of multiplying or divide a polynomial by the power of an unknown number. The place-value notation also runs through arithmetic operation. For example, in division, the dividend is termed "*shi*" (real) and the divisor "*fa*" (method) whose positions are still called "*shi*" and "*fa*" in operation regardless of what values they have changed into. In addition, *kai fang* method is derived from division, so, in *Jiu Zhang Suanshu* (*Nine Chapters on the Mathematical Art*), "*shi*" and "*fa*" in *kai fang* method actually mean the same as those in division. Later, it developed into a method of extracting the cubic root and solving higher order equation, of which the constant term is named "*shi*," the monomial coefficient in *Jiu Zhang Suanshu* is expressed as "*fa*" or "*fang fa*" (method) which is called "*fang*" (square) by Liu Hui, and the highest coefficient of a polynomial in *Jiu Zhang Suanshu* is "*jie suan*" and "yu" (corner) by Liu Hui which was later used mostly. The coefficients of the terms between the higher order one and the primary one are named "*zhong hang*" (middle lines) in *Jiu Zhang Suanshu* and "*lian*" (the square and straight corner of a central room in ancient Chinese architecture) by Liu Hui. Later, names like *shang lian* (the upper corner of a central room), *xia lian* (the lower corner of a central room), and even those such as *yi lian* (the first corner), *er lian* (the second corner) and so on are all employed to distinguish different coefficients in equations of the fourth and/or higher degree. In fact, these names were used not only in the initially established equations but also during the process of solving them. No matter what numbers the coefficients have changed into, their original names always remain. In other words, it is still the position of a number in the arithmetic formula that can determine the power of the unknown number to which it attaches.

Therefore, the algorism has played an extremely important role in traditional Chinese mathematics.

Lastly, the algorism and rod calculus made traditional Chinese mathematics dominant in algorithmic research and characterized by the algebraization of geometry or the combination of geometric problems with both arithmetic and geometry. Liu Hui remarked that "the mathematic methods can be used in solving problems of circles, rectangles, length and capacity", which means the geometric problems can be solved with the arithmetic and algebraic methods. Wu Wenjun attached particular importance to the algebraic characteristics of the geometric problems in traditional Chinese mathematics. He pointed that "The algebraization of geometric problems and the systematic solution with algebraic methods were one of the major achievements of Chinese mathematicians at that time." [27] After the time when *Jiu Zhang Suanshu* came out, all the problems of measuring area, volume and right-angled triangles, which we attribute to geometry today, had to be transformed into arithmetic and algebraic problems.

More importantly, the rod calculus of the traditional Chinese mathematics has the characteristics of constructivity and mechanization. The so-called constructive mathematics refers to the mathematical theory that starts from some initial objects constructively and develops by explicitly regulated operations. All of the following achievements in traditional Chinese mathematics can be categorized into typical constructive methods: four fundamental operations of arithmetic related to fractions, method of *jin you* (now there is), method of *cui fen* (distribution by proportional parts), method of ying bu zu (surplus and deficit), method of *kai fang* (extraction of square and cube roots), method of *zheng fu* (positive and negative), method of *fang cheng* (equations), formulas of area and volume and solutions to problems of right-angled triangles recorded in the bamboo and wooden slips and *Jiu Zhang Suanshu* in Qin and Han Dynasties (221 B.C. to 220); Liu Hui's arithmetic proof to the formula of circular area and to Liu Hui theory and the procedures of seeking *pi*; Jia Xian and Qin Jiushao's *zeng cheng* method for the extraction of roots and *zheng fu* (positive and negative) method for the extraction of roots to seek the positive roots in equations of higher degree; Qin Jiushao's *da yan* method, i.e., a solution of a system of linear congruence equations; *tian yuan* method used by Li Ye and Zhu Shijie as an astronomical technique; *er yuan* method, *san yuan* method and *si yuan* method, i.e., solutions to equations of two unknowns, three unknowns and four unknowns created by Li Dezai, Liu Dajian and Zhu Shijie in Jin (1115–1234) and Yuan (1271–1368) dynasties. "The so-called mechanization is nothing but being stereotyped and standardized." Said Mr. Wu Wenjun, "Mechanization of mathematical problems requires that, in the process of computation or proof, there must be a definite and necessary 'next step' following the former one which has moved forward, so that the conclusion can be reached after these regular and rigid operations." In all the above-mentioned methods, solutions and proofs, there are standardized procedures, which definitely belong to typical mechanized ones. Therefore, Mr. Wu Wenjun concluded, "Ancient mathematics in China, generally speaking, is such a kind of mathematics: constructive and mechanized, which are also its two major characteristics."

It is unfair that some Western and Chinese domestic scholars with Euro-centrism often exclude traditional Chinese mathematics from the mainstream of world mathematics development. Wu Wenjun remarked, "Along the long river of history, the two mathematic systems of mechanized algorithm and axiomatized deduction have always been, alternately and repeatedly, the mainstream of mathematical development." Traditional Chinese mathematics together with later Indian and Arabic mathematics is a typical representative of the mechanized algorithm system, so it is in the mainstream of world mathematical development. It is also the main aspect in the mainstream especially from the time after the ancient Greek mathematics declined in 200–300 B.C. to the fourteenth century.

Translators' Notes

1. A number of scholars believes that Liu Hui determined the ratio of the circumference to the diameter with the formula $100\pi = 314$ after he obtained the approximation 314 *cun* of the circle's area. However, this argument deviates from Liu Hui's commentaries on *Nine Chapters* and places him in the dilemma of circular reasoning.
2. To understand Liu Hui Theory, we should first understand some Chinese terms used in the theory. *qian du* (堑堵) is a kind of triangular cylinder produced by cutting through a cuboid along its two diagonals. Cut *qian du* through along its apex and the opposite ribs, and a quadrangular pyramid called *yang ma* (阳马) and a triangular pyramid called *bie nao* (鳖臑) are obtained. Thus, to be more specific, yang ma is a kind of quadrilateral pyramid with two right-angled triangles vertical to a rectangular base and *bie nao* is a tetrahedron consisting of four right-angled triangles.
3. To prove the volume formula of *yang ma* and *bie nao*, Liu Hui put forward an important theory which says: "Cut through *qian du* along its apex and the opposite ribs, then a *yang ma* and a *bie nao* are obtained. The volume of *yang ma* doubles that of *bie nao*, which never changes." If the volume of *yang ma* is V_{ym} and *Bienao* is V_{bn} then V_{ym}: V_{bn}= 2:1. After his proof of in the theory, Liu Hui added that, "Therefore, without *bie nao*, there is no way to know how big the volume of *yang ma* is; without *yang ma*, there is no way to know things like pyramids. All of this is for practical use." It is consistent with the volume theory of modern mathematics to establish the polyhedron volume theory on the infinitesimal segmentation.

References

1. Zhong, Song. 1957. *Book of Origins*, 36. Beijing: Commercial Press.
2. Book of Changes. 1979. See *Annotated Thirteen Classics*, vol. 87. Beijing: Zhonghua Press.
3. Yirang, Sun. 2001. *Annotated Classics of Mozi*, vol. 326. Beijing: Zhonghua Press.
4. Yirang, Sun. 2001. *Annotated Classics of Mozi*, vol. 383, Beijing: Zhonghua Press.
5. Qianzhi, Zhu. 1984. *Collated Canon of the Dao and Its Virtue*, vol. 108. Beijing: Zhonghua Press.
6. Bojun, Yang. 1990. *Annotated the Commentary of Zuo*, 1171. Beijing: Zhonghua Press.
7. Ban, Gu. 1962. *History of the Han Dynasty*, 956. Beijing: Zhonghua Press.
8. Museum of Baoji City, Cultural Center of Qianyang County, Institute for the History of Natural Sciences of Chinese Academy of Sciences. Counting-rods Unearthed from the Tombs of the Western Han Dynasty in Qianyang County. *Chinese Archaeology*. 1982 (3):85–88, 108.
9. Shengwu, Li, and Guo Shuchun. 1982. Tombs of the Eastern Han Dynasty and Rods Unearthed in Shijiazhuang city of Hebei province. *Chinese Archaeology* 3: 255–256.
10. Shuchun, Guo. 2014. *New Collated Nine Chapters on the Mathematical Art*. Beijing: University of Science and Technology of China Press.
11. Shuchun, Guo. 2001. *The Mathematical Classic of Sun Zi. See Ten Mathematical Classics*. Taipei: Jiuzhang Publishers.
12. Shuchun, Guo. 2001. *Yang's Mathematics Manual. See Ten Mathematical Classics*. Taipei: Jiuzhang Publishers.
13. Ye, Li. 1993. Sea Mirror of Circular Measurement. See *Chinese Classics of Science and Technology Mathematics*, vol. 1, 729–869. Kaifeng: Henan Education Press.
14. Jiushao, Qin. 1993. Mathematical Treatise in Nine Sections. See *Chinese Classics of Science and Technology Mathematics*, vol. 1, 439–648. Kaifeng: Henan Education Press.
15. Dunjie, Yan. 1982. History of Numerals in China. See *Collection of Science & Technology in History Vol. Mathematics*, 32–34. Shanghai: Shanghai Scientific & Technical Publishers.
16. Shuchun, Guo. 2010. *History of Chinese Science and Technology Mathematics*, 34–36. Beijing: Science China Press.
17. Hui, Yang. 1993. A Detailed Analysis of the Methods of Computations in *Nine Chapters*. See *Chinese Classics of Science and Technology Mathematics*, vol. 1, 1414–1427. Kaifeng: Henan Education Press.
18. Shijie, Zhu. 1993. Jade Mirror of Four Unknowns. See *Chinese Classics of Science and Technology Mathematics*, vol. 1, 1205–1280. Kaifeng: Henan Education Press.
19. Hui, Yang. 1993. A Detailed Analysis of the Methods of Computations in *Nine Chapters*. See *Chinese Classics of Science and Technology Mathematics*, vol. 1, 949–1043. Kaifeng: Henan Education Press.
20. Book of Changes. 1979. See *Annotated Thirteen Classics*, vol. 52. Beijing: Zhonghua Press.
21. Qianzhi, Zhu. 1984. *Collated Canon of the Dao and Its Virtue*, vol. 176. Beijing: Zhonghua Press.
22. Al-khwarizmi. 2008. *Algorithm and Algebra*, 1–116. Beijing: Science Press.
23. Smith, D.E. 1925. *History of Mathematics*, vol. II, 382. Dover Publications.
24. Yi, Zu. 1993. The Epilogue of Jade Mirror of Four Unknowns. See *Chinese Classics of Science and Technology- Mathematics*. vol. 1, 1206. Kaifeng: Henan Education Press.
25. Ye, Li. 1993. *Yi Gu Yan Duan*. See *Chinese Classics of Science and Technology-Mathematics*, vol. 1, 873–941. Kaifeng: Henan Education Press.
26. Shijie, Zhu. 1993. Jade Mirror of Four Unknowns. See *Chinese Classics of Science and Technology Mathematics*, vol. 1, 1209. Kaifeng: Henan Education Press.
27. Wenjun, Wu. 1995. *On the Mechanization of Mathematics by Wu Wenjun*. Jinan: Shandong Education Press.

Shuchun Guo Born in Jiaozhou, Shandong in 1941, graduated from School of Mathematics of Shandong University, now serves as the researcher and doctoral supervisor of the Institute of Natural Science History of the Chinese Academy of Sciences. He once served as deputy director of the Academic Committee of the Institute of Natural Science History of the Chinese Academy of Sciences, director of the History of Mathematics Astronomy Research Office, and chairman of the National Mathematical History Society.

Research interests: History of Chinese mathematics.

Major works: Collecting and Proofreading the *Jiu Zang Suan Shu* and its supplement; *The new Proofreading version of the Jiu Zang Suan Shu; LIU Hui: World Mathematics Master in Ancient Times*; Collating *Suan Jing Shi Shu; Translation and Annotation of Jiu Zang Suan Shu; Chinese-French version of Jiu Zang Suan Shu; Chinese-English version of Si Yuan Yu Jian; Chinese-English version of Jiu Zhang Suan Shu; chief editing Completed Works of Li Yan and Qian Baocong on History of Science; History of Chinese Science and Technology: Mathematics; History of Chinese Science and Technology: Dictionaries.*

Awards: *Chinese Science and Technology Classics: Mathematics* won the Nomination of the Third National Book Award in 1997;

Completed Works of Li Yan and Qian Baocong on History of Science won the 4th National Book Award Honor Award in 1999;

Chinese-French version of Jiu Zang Suan Shu won the Bachelor of French Academy Award in 2006;

History of Chinese Science and Technology: Mathematics won the first prize of the Fourth Guo Moruo History Award in 2012.

Chapter 7
Bronze Metallurgy

Benshan Lu and Rongyu Su

Bronze smelting and casting technology is a technological system consisting of copper mining, smelting and casting, which helped the ancient Chinese people to create a brilliant bronze civilization in the Shang and Zhou dynasties that lasted as long as 2000 years.

7.1 Copper Mining and Smelting

Developed from identifying copper ore, copper mining and smelting is a great epoch-making invention in human history. The earliest known copper products were found in Iran and they belong to native copper products made in 7000 B.C. or 8000 B.C., while mined-copper products, dug out in Zagers of the western Iran, can be traced back to 7,000–6,000 years ago, indicating that extracting copper from ore was much later than using natural copper. The earliest copper alloy was smelted from paragenesis, and alloy smelted from domeykite was found in both West Asia and China, while arsenious copper axes made in 3500 B.C. were discovered in ancient Egypt. A good many of copper products containing a little arsenic and nickel were excavated in the painted pottery cultural site (3000–2000 B.C.) in Ganges River of India. In China, people could smelt copper zinc alloy during the Yangshao Culture period (5000–3000 B.C.), and brass sheets and tubes were found in Jiangzhai village site (4020 B.C.) of Shanxi [1].

B. Lu (✉)
Guangdong Beauty and Cosmetic Museum, Guangzhou, China
e-mail: 605647582@qq.com

R. Su
The Institute for History of Natural Sciences, Chinese Academy of Sciences, Beijing, China

J. Hua and L. Feng (eds.), *Thirty Great Inventions of China*,
https://doi.org/10.1007/978-981-15-6525-0_7

Tin bronze was the earliest resultant product from alloying, the production of which first began in West Asian Ur Dynasty (about 2800 B.C.) and Majia Kiln Culture (2780 B.C.) in Gansu, China [2].

Most of the earliest known copper relics in China were in the Yangshao Culture period. It is also almost during the same period that Huang Di (Emperor Yellow, 2717–2599 B.C., one of Chinese ancient ancestors) and Chi You (2730–2689 B.C. one of Chinese ancient ancestors) mined and used copper, as recorded by ancient history of China. Thus based on these records, we can find the historical relics of copper mining and smelting.

7.2 Copper Mining

The locations of copper ore resources in China are relatively concentrated, including the copper ore zone in the middle and lower reaches of the Yangtze River, Chuan-Dian the copper ore zone in Sichuan and Yunnan, Zhongtiaoshan copper ore zone in Shanxi, Baiyun copper zone in Gansu. Consequently, the already found copper mining ruins are identical with the present distribution of copper ore resources. In Hubei, a province along the middle and lower reaches of Yangtze River, there are over 100 ruins of copper mining scattered in Tieshan (meaning iron mountain) District of Huangshi, Daye City and Yangxin County, among which the biggest and richest historical site, the ancient Tonglu Mountain Copper Mining (in Daye) had been exploited from the late half of the Shang Dynasty (1600–1046 B.C.) to the Han Dynasty (202–263 B.C.). The ancient Tongling (meaning copper mountain) Copper Mining in Ruichang County of Jiangxi was exploited from middle Shang Dynasty to Warring States Period (500–221 B.C.). The ancient Wannan (meaning southern Anhui) Copper Mining was exploited during the Western Zhou Dynasty (1046–771 B.C.) and Han Dynasty. Through archeological exploration and investigation, there are some other copper mining ruins such as Niuheliang Copper Mining Ruins (3500–3000 B.C.) in Lingyuan County in Liaoning, Dajing Copper Mining Ruins (about 2800 years ago) in Linxi County of Inner Mongolia Autonomous Region, Nulasai Copper Mining Ruins (about during the Spring and Autumn period, approximately 770–476 B.C., the former half period of Eastern Zhou Dynasty) in Nulasai Copper Mining Site of Xinjiang Uygur Autonomous Region, Zhongtiaoshan Copper Mining Ruins (the Zhou Dynasty, 1046–256 B.C.) in Shanxi, the copper mining ruins, respectively, in Zhongwei in Ningxia Hui Autonomous Region during the Eastern Zhou Dynasty (770–256 B.C.), in Ludian, Yunnan during Eastern Zhou Dynasty and in Mayang County of Hunan during Warring States Period, and so on.

In the twenty-first century B.C., China entered the Bronze Age. Then during the later period of the early Shang Dynasty (1600–1046 B.C.), with the rapid development of bronze civilization and wide use of bronze ritual vessels, the increasing need for copper material catalyzed the first boom period of copper mining and smelting, hence making such large-scale mining producers as Tonglu Mountain Copper Mining and Tongling Copper Mining possible to come into being, of which ruins and relics are the solid evidence of exquisite workmanship with great originality in copper mining and smelting of that time.

7.2.1 Opencast Mining and Underground Mining [3]

In the early stage of mining industry, people did opencast mining first and then underground mining, following the principle of mining from the surface to the depth, which is also a general rule of mining and matched the technological condition of ancient times. In most of the above-mentioned ruins, vestiges of opencast mining were found. Along the middle and lower reaches of the Yangtze River, most of the copper mines are geologically made of soft rocks. Thus two types of opencast mining were operated: hillside mining and pit mining. In pit mining, a ditch or a big pit of a low slope angle was dug (Fig. 7.1). While in Dajing of Linxi County (Inner Mongolia) and Nulasai of Neleike County (Xinjiang), the geological strata are firm rocks. Thus most of the ancient stopes in those places were trenches with high slope angles and drives and chambers on the bottom, using a combination of both opencast mining and underground mining.

The Dajing Copper Mining Ruins was discovered in 1974, which was a fissure filling copper deposit with hundreds of mineral veins and a mining area of about 2.5 km^2. On the surface, there are 47 visible ancient pits, which means open pit development was operated along the strike of mineral veins. The longest contour line is over 500 m long and 25 m wide. The mining depth is 7–8 m and the deepest amounts to 20 m. All these figures indicate the large scale of mining was practiced there at that time. Because of the steeply pitching mineral veins, in order to reduce overburden amount, miners adopted steep slope development approach with the ultimate pit slope angle of 70–90°. They combined trenching and pit mining together, namely digging an adit at the bottom of an open pit stope. As a result, rich ore in the depth was mined, and overburden amount caused by additional pit mining was avoided. In those ancient

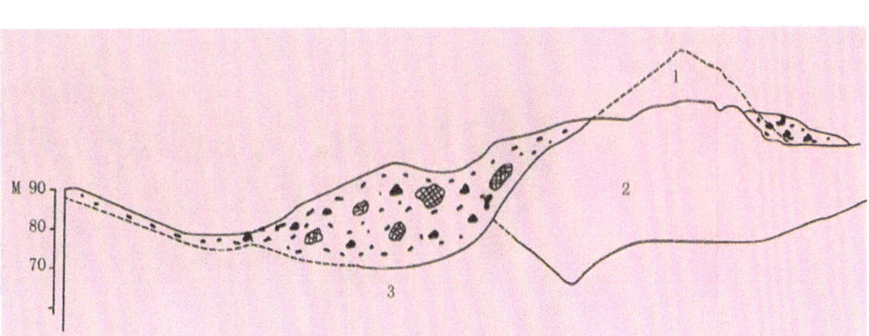

Fig. 7.1 The geological cross-section of Line 0 in Tongling copper mining. 1. The ancient opencast mining site, 2. Malachite, iron clay, 3. Limestone

pits, over 1,060 stone mining tools were excavated, including hammers and borers roughly made from gravels of granite and basalt. Miners were able to trace rich ores, and mine hard ores with "hammer and wedge". Actually, this underground development technique was mastered quite well by ancient craftsmen at least 6,000 years ago. The good example is from the Xiqiaoshan Quarry Ruins in Nanhai District of Guangdong, in which there are seven caverns and the deepest one is of 37 m [4]. After several archeological investigations in Nulasai Copper Mining Ruins, such relics left by pit mining and copper smelting as stent wood, hammer stones, ores and copper ingots were discovered, among which stent wood is about 2650 ± 170 years old by 14C dating. In 1989, I led a research group to make a detailed investigation of the site and check up on the relics. The result proves that there are two ancient opencast mining pits and two underground stopes. On the north slope of the mining area, there are piles of slag by which we detected that copper sulfide ore was mined. The scale of hard rock mining here was so large that we had rarely seen the same kind anywhere else in China before. No. 1 and No. 2 main mineral veins are hidden under the ridge along which the open pit mining was developed, identically with the strike of the veins. No. 1 opencast stope is 90 m long and 50 m deep. An almost 90-degree slope angle was employed so as to reduce overburden amount. No. 2 opencast stope is 123 m long and 4–5 m wide. And, on the southeast side of the trench, adits were developed near the bottom so that several large and high rooms, 47 m away from the surface, came into being. In underground mining, two approaches, room-and-pillar mining and horizontal bracing mining (Fig. 7.2), were adopted. In No. 1 stope, after open-pit mining reached the bottom, a drive was chiseled through glutenite into the rich vein copper deposit, and then the ore body was split along its strive. On the roofs of the large rooms, there were many horizontal bracing wooden platforms, positioned according to mining directions. The ends of the bracing woods were embedded into the supporting holes of the surrounding rocks, which made the platforms quite steady. The function of those platforms was for stoping and artificial ore breaking, an up and down filling mining method. The magnificent open and

Fig. 7.2 The horizontal bracing in Nulasai copper mining site of the spring and autumn period

underground stopes along with the surface waste rock dump obviously reflect the reasonable procedures of excavating, loading, transporting, dumping, and so forth.

Mayang (in Hunan) Copper Mining Ruins, a sandstone-type copper deposit with native copper as its major ore, is located in the hilly terrain of West Hunan. In 1982, 14 surface and underground stopes together with some wooden and iron tools were discovered and dug out in the ruins. By 14C dating, the excavated mallets can be dated back to 2730 $\pm$ 90 years ago. Owing to the pitching ore body, underground development is characterized, to a large extent, by inclined drives while most of goafs are long-bag-shaped. The dip length of No. 2202 drive is 140 m and its dip angle is 36° with the deepest point being over 80 m away from the surface, all of which is sound evidence for the relatively high-level techniques of deep ore mining during the Warring States Period. No. 1203 drive was developed along the mineral vein strike and is about 400 m long, much longer than the earlier drives.

7.2.2 Underground Mine Development [5]

Opencast mining embraces several advantageous features, including fully utilization of the copper mine, high percent recovery and low dilution rate. However, the removal of large amount of the ore waste stands in the way. For a lower overburden rate, underground mining was proposed by ancient miners after studying the features of an ore's topography and mineral occurrence.

Archeological excavations indicate that in middle of the Shang Dynasty two underground mine development methods exist in Tongling, Jiangxi and Tonglu Mountain, Hubei: (1) Single development methods: shaft, inclined shaft and drift; (2) Joint mine development methods: Any combinations of the stated single development method (Fig. 7.3). For example, shaft $\rightarrow$ drift $\rightarrow$ blind shaft; shaft $\rightarrow$ inclined shaft $\rightarrow$ drift (Fig. 7.4). It is found that the depth of mines in the Shang Dynasty reaches over 10 m and the longest drift measures more than 10 m.

The feedback of drilling and radar detection demonstrates that the depth of spot VII3 on the Tonglu Mountains formed in the Western Zhou Dynasty has been recorded as 36 m, and the size of the underground mining area was 85 m by 26 m. The underground mining area of the Tonglu Mountains back in the Western Zhou Dynasty covered around a unit area of 2,000 m^2 and comprised shafts and drifts of larger cross dimensions than that in the Shang Dynasty. Later a significant breakthrough has been achieved in the Spring and Autumn Period, making possible larger depth, width, cross-dimension and a sounder layout of shafts and drifts. Line 24 of the Tonglu Mountains is a rather complete mine of the Warring States Period. Within 100 m^2 lays 5 drifts and 11 (inclined) shafts with the help of joint mining method. An inclined shaft intersects straight from the top to the bottom of the seam (the third layer) and the other 10 shafts are separated into the top, middle and the bottom layer to the south, north and west of the inclined shafts (Fig. 7.5).

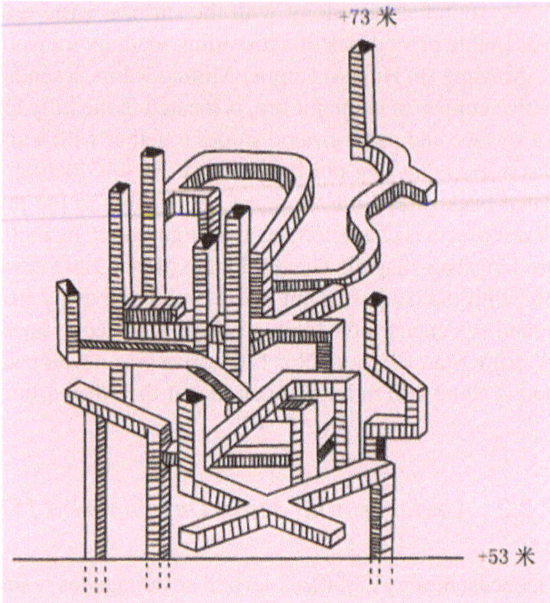

Fig. 7.3 The mine development of the Tonglu Mountains in the Shang Dynasty

Fig. 7.4 The joint mine development of Tongling in the Shang Dynasty. 1. Shaft, 2. Inclined shaft, 3. Drift, 4. Ore body

7.2.3 The Shaft and Drift Supporting [6]

The ore zone along the middle and lower reaches of the Yangtze River is mostly contact-metasomatic deposit of copper and iron, and years of weather eluviation leads to the fact that the secondary-enriched cupric oxide lies mostly in the crushed zone. The soft and broken geological structure of this kind is undoubtedly beneficial to drilling and tunneling, but compounds the difficulties of ground pressure management. Faced with this problem, ancient Chinese brought forward the technology of wooden support and kept innovating other support methods.

Fig. 7.5 The application of shafts and drifts in the mine development of Tongling in warring states period

The wooden support is designed as an underground structure which forms the frame and the backplane along the shaft/drift wall with wood, bamboo, wattle and so forth. Premade on ground, all components would be used to set up support while drilling and tunneling downwards, making easy the construction and saving a considerable amount of working hours. The word "pit" in Chinese was written as 井 and the original form of character井 is ┤ ├. The two horizontal lines in ┤ ├ represent the ground above the pit and the two vertical lines stand for rockshafts and wooden support backplanes, which are higher than the surface to prevent water and the falling rocks.

7.2.3.1 Shaft and Drift Supporting Structure of Tongling in the Shang Dynasty

Shaft: A rectangular pit frame was made of two logs as crossbeam with ends embedded into surrounding rock, and the other two logs' ends were made into a bowl-shaped bracket to support the crossbeam. The pit frame was fashioned to support the rockshafts at intervals, and at each of the four corners erected a column, tied with the frame by rattan to form an integral skeleton. Wooden sticks were inserted between the support and surrounding rock, and a thatch was placed on the outside to form a wall (Figs. 7.6 and 7.7). Moreover, the cross-section of the shaft was 126 cm by 116 cm.

Drift: The trusses consisted of the top beam, column and mantle were arranged at intervals (Fig. 7.8). There were two forms of joint constructions and combinations of the trusses: the bowl-shaped end which is the same as that in the shafts and the open-ended column connecting the upper and lower logs with tenons. The supporting logs were placed between the trusses and the surrounding rock. The discovered drifts

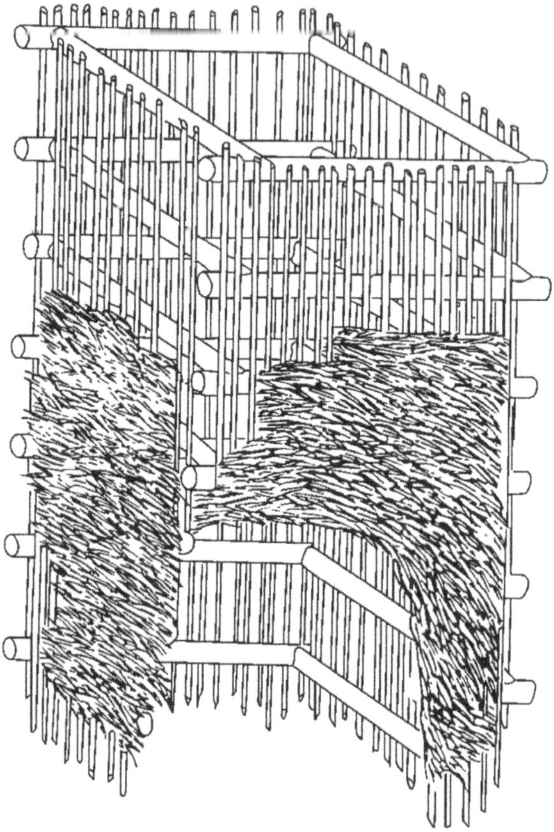

Fig. 7.6 The J72 supporting structure of Tongling in Shang Dynasty (Jiangxi Institute of Archeology, Museum of Ruichang City: *The Discovery and Research of Ancient Copper Ore in Tongling*, Jiangxi Technology and Science Press, 1997:16)

Fig. 7.7 The supporting structure for the shafts in Tongling in the Shang Dynasty

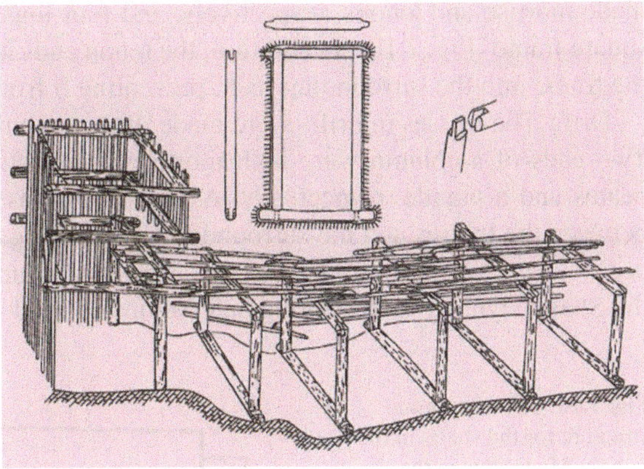

Fig. 7.8 The X1 supporting structure of Tongling in the Shang Dynasty (*The Discovery and Research of Ancient Copper Ore in Tongling*, 21)

were measured as around 1 m by 1–1.3 m. If the bottom of the drift is hard limestone ground, a half-frame support might come into application without any mantle.

7.2.3.2 The Supporting Structure for the Shafts and Drifts in the Tonglu Mountains in the Middle and Late Shang Dynasty

Shaft: Mostly, the pithead was a square. As for the joint structure, the two ends of two logs were used as mortises and the ends of another two logs served as tenons, joining together to form a frame (Fig. 7.9). Between the frame and the surrounding rock wooden supports were set. And the cross-section of the vertical shafts measured 50 cm by 50 cm. Improvements were made later on. The two ends of one log were

Fig. 7.9 Flat-end mortise and single tenon joint structure

made mortise and tenons, respectively, and four logs of this kind joined to form a square frame (Fig. 7.10). In addition, the tenon ends helped to insert four corners of the frame into the surrounding rock, preventing it from falling.

Drift: The trusses in drifts were made of two columns, a mantle and a top beam. Two ends of a column were both mortises, connecting the tenon boards of a top beams and a mantle, respectively. A support was established with placing boards between the trusses and the surrounding ground (Fig. 7.11). The measurement of a drift is recorded as 76 cm by 46 cm. This type of support for drifts was used from the Shang Dynasty to the Spring and Autumn Period.

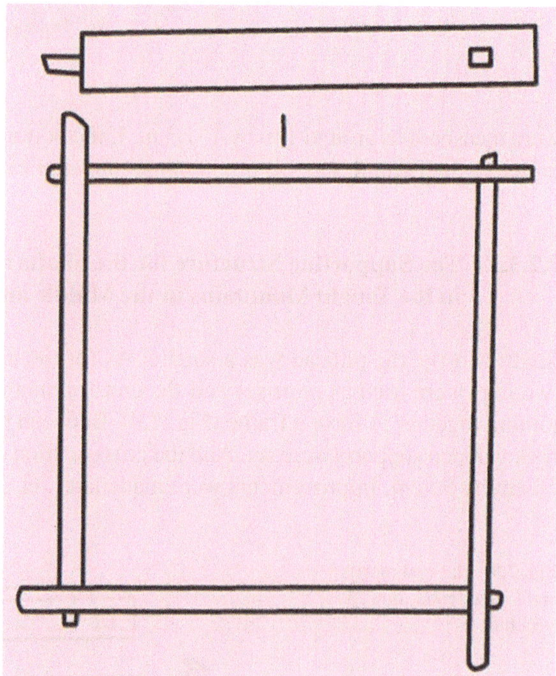

Fig. 7.10 The supporting structure for the shafts in the Tonglu Mountains in the Shang Dynasty

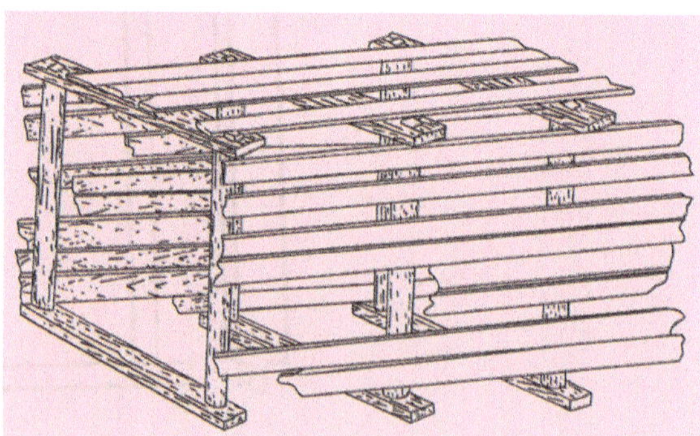

Fig. 7.11 The shafts and drifts featured with mortise and tenon in the Tonglu Mountains in the Shang Dynasty

7.2.3.3 The Supporting Structure for the Shafts and Drifts in Tongling in the Western Zhou Dynasty

Shaft: The structure for the shafts in this period was featured by alternate bowl-shaped joint inner support (Fig. 7.12). Four corners were placed alternatively in a vertical support, which was beneficial to compression resistance in the rockshaft. The square pithead measured 80 cm of its side length. And the support is 20 cm above the surface with green paste mud covering around the pithead to prevent the permeation of the surface water. There was also an enhanced support formed by two overlapping frames, with four corners embedded in the surrounding rock (Fig. 7.13), which endowed the structure with multiple supporting points for a significant improvement in compression resistance. Support of this kind was also found in the mine relic in Gangxia, Yangxin of the Western Zhou Dynasty. It is understood that No. 37 shaft in Tongling (Fig. 7.14) was the main pit of the mine of the Western Zhou Dynasty with a cross-section of 259 cm by 178 cm. At the one-third height of the rockshaft existed several logs, which were used to set up a ladder with two columns.

(Inclined) Drift: Its support structure was about the same as that in Tonglu Mountain of the Shang Dynasty, but was of larger scale: 72–90 cm in height and 65 cm in width (Fig. 7.15).

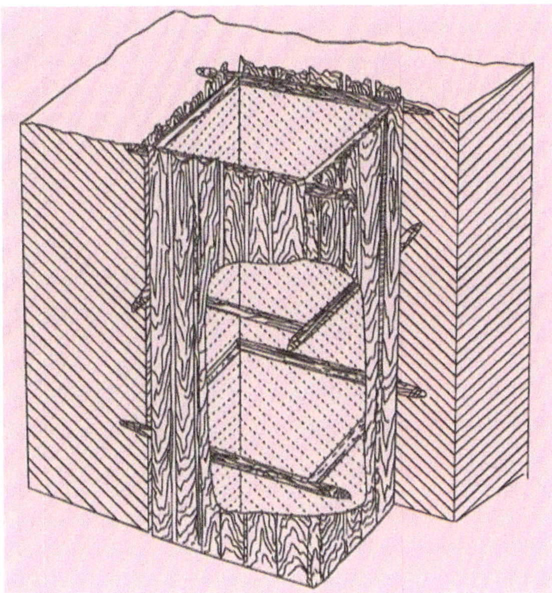

Fig. 7.12 The J23 supporting structure for the shafts in Tongling in the Western Zhou Dynasty (*The Discovery and Research of Ancient Copper Ore in Tongling*, 39)

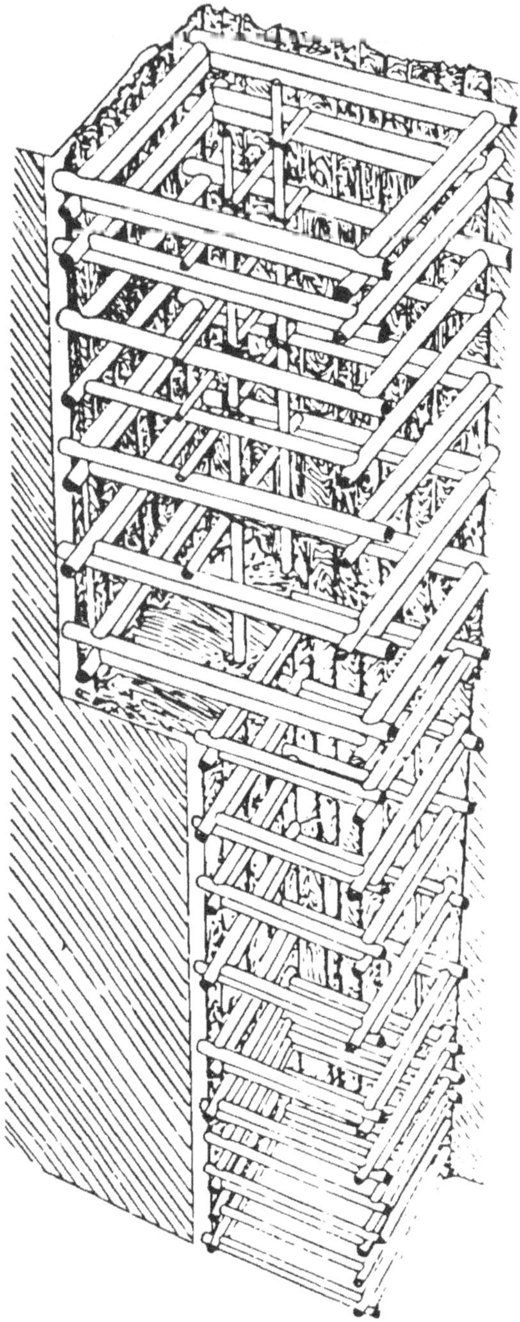

Fig. 7.13 The J37 supporting structure for the shafts in Tongling in the Western Zhou Dynasty (*The Discovery and Research of Ancient Copper Ore in Tongling*, 41)

Fig. 7.14 The vertical shafts and wooden ladder found in Tongling in the Western Zhou Dynasty

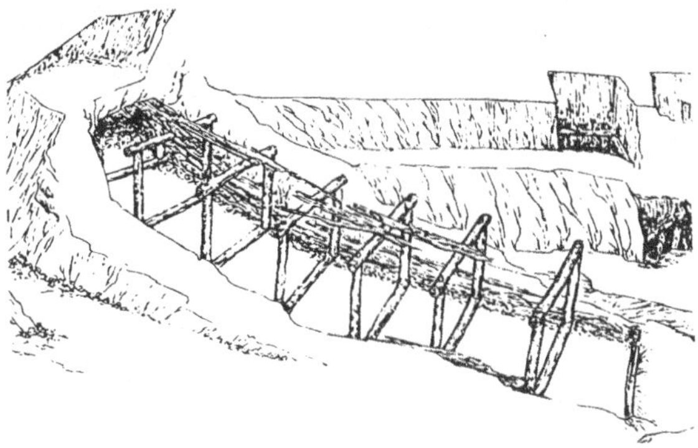

Fig. 7.15 X2 support for shafts and drifts in Tongling of the Western Zhou Dynasty (*The Discovery and Research of Ancient Copper Ore in Tongling*, 42)

7.2.3.4 The Supporting Structure for the Shafts and Drifts in the Tonglu Mountains in Western Zhou Dynasty

Shaft: Sword-shaped mortise–tenon joints were utilized in the supporting structure in Tonglu Mountain and Tongling of the middle Western Zhou Dynasty. The pithead was also a square. And the cross-section of the mine in Tonglu Mountain was recorded 50 cm by 50 cm, while that in Tongling 98 cm by 98 cm. Embedded into the surrounding rock, the four corners of the support were four sword-shaped logs, which could prevent the frame from sliding (Fig. 7.16). Furthermore, in the late Western Zhou Dynasty, the sword-shaped logs of the upper frame and the lower frame were placed alternatively to strengthen the robustness of the frame (Fig. 7.17).

(Inclined) Drift: the supporting structure for drifts of this period was the same as the Shang Dynasty.

In summary, from the middle of the Shang Dynasty to the Western Zhou Dynasty, the form of support for either shafts or drifts had been continuously improved for

Fig. 7.16 The shafts in the Tonglu Mountains in the Western Zhou Dynasty

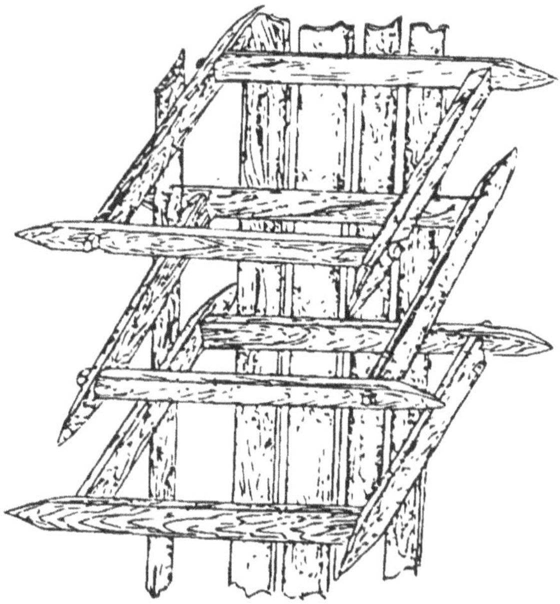

Fig. 7.17 The J10 supporting structure for the shafts in Tongling in the Western Zhou Dynasty (*The Discovery and Research of Ancient Copper Ore in Tongling*, 38)

simpler structure and stronger support. Local broad-leaf woods such as callery pear and European beech were basically the providers of supporting logs, since they had no knots and twisted defects, and had excellent pressure bearing capacity. In addition, the components were of uniform specifications. What's more, the premade components and knockdown support technique helped to get rid of limitations of underground space and greatly improved work efficiency. The supporting logs processing plant of this period was found in both Tongling and Tonglu Mountain, and there were remains of chopped wooden piers, copper axes, sawdust and slag deposits.

Fig. 7.18 The shafts and drifts in the Tonglu Mountains from the Western Zhou Dynasty to the spring and Autumn period

7.2.3.5 The Supporting Structure for the Shafts and Drifts in Eastern Zhou Dynasty

The scale of mining in the Spring and Autumn Period was huge (Fig. 7.18). The mining area in Tonglu Mountain of the Spring and Autumn Period comprised nine ore bodies and was connected to the adjacent Shitouzui Mine and Huangniushan Mine, covering a few square kilometers. In the late Spring and Autumn Period, the support for shafts featured by flat-end and mortise–tenon frame and dense curb was invented and put into effect in Tonglu Mountains. During the Warring States Period, this technique was promoted to Tongling, Jiangxi, and then to areas such as Tongling, Anhui in the Han Dynasty.

The supporting structure for the shafts in the Tonglu Mountains in early Spring and Autumn Period was similar to that of the Western Zhou Dynasty, but there were many improvements with the application of hanging frame structure by bamboo ropes. At this period, the side of the rockshaft was longer than 60 cm (Fig. 7.19). The interior of the frame was usually coated with grass stems and green paste mud, and some were surrounded by bamboo weaving to form a closed space, which contributed to preventing the wall from collapsing and to facilitating air circulation. At the bottom of the shaft, there is often an ingate (Fig. 7.20) connecting the drifts. From the late Spring and Autumn Period to the Warring States Period, the support technology

Fig. 7.19 Shafts in the Tonglu Mountains in the spring and autumn period

Fig. 7.20 Shafts of the spring and autumn period and ingrates

has undergone revolutionary innovation, which gave birth to the drifts or inclined drifts with the characteristics of duck-billed trusses support. The basic features of the support for shafts during this period were that the ends of the four logs were cut into stair-stepping joints, and were joined to form a frame; frames were stacked into dense curb to compose a complete wooden rockshaft.

Within 50 m^2 of the Line 12 mine site in the Tonglu Mountains in late Spring and Autumn Period, 8 shafts and 1 inclined drifts were excavated with a mining depth of at least 50 m. The square pitheads of the mine measured lengths of 90, 105, 120, 210 and 230 cm (Fig. 7.21). And the stacking joints were simple and placed close to each other. Those components were of great compression resistance and were easy to manufacture and erect, which significantly deepened the shaft. The Line 24 mine site in Tonglu Mountain of the Warring States Period was recorded more than 80 m above the surface. More than 70 pieces of iron, wood, rattan, bamboo and ceramic items were unearthed from the mine. The iron tools included axes, drills, hammers, rakes, hexagonal shovel and "凹"-shaped shovel (Fig. 7.22). This period owned a

Fig. 7.21 The support for shafts in Tongling in the warring states period

Fig. 7.22 The mining tools found in the Tonglu Mountains in the warring states period

complete and wide array of mining tools which could not be found in any other previous periods.

In the early and middle stage of Spring and Autumn Period, a section of the (inclined) drifts was enlarged. VII (2) X10 site in the Tonglu Mountains of this period measured 120 cm in height and 80 cm in width. The support for drifts in No. IV ore body is illustrated in Fig. 7.23. The trusses found in the X5 drift in Tongling

Fig. 7.23 The support for shafts and drifts in the Tonglu Mountains in the spring and autumn period

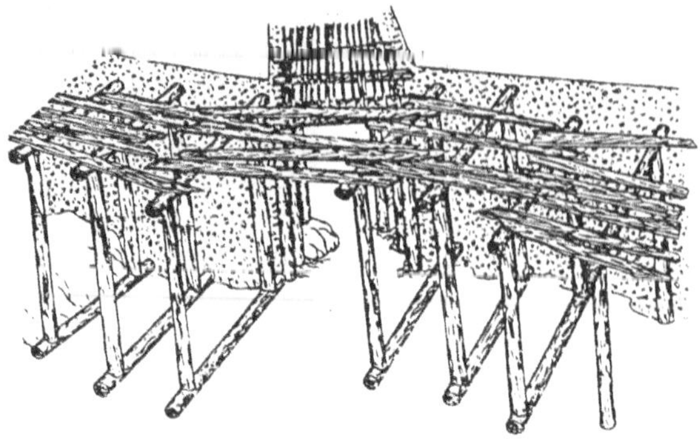

Fig. 7.24 X5 and X6 Tunnels in Tongling in the spring and autumn period (*The Discovery and Research of Ancient Copper Ore in Tongling*, 59)

Fig. 7.25 The support for the vertical drifts in the Tonglu Mountains in the warring states period

were 108 cm by 88 cm (Fig. 7.24), while duck-billed trusses support was found at the mine sites in the Tonglu Mountains in the Warring States Period (Fig. 7.25). Each row of the truss consisted of five components, namely two columns, a top beam, a mantle and an inner strut.

Both ends of the mantle were cut into flat-end tenons, connecting the column. The top beam was supported by a duck-billed structure at the top end of the column and the two pitchforks under the top beam were chiseled into tenons, with inner struts embedded into form a mixed frame. On the beams and off the columns were backplanes ceilings fashioned by wooden sticks or wooden boards. Studies have shown that this kind of support had an advantageous overall compression resistance. The mantle was designed with flat-end tenons, which embraced greater compression resistance than the mortise–tenon joints did. The column served as a strut and was stabilized by a top beam. The inner struts under the beam not only strengthened the bending resistance of the top beam but also supported the column to resist the side pressure. This type of truss combined the superiorities of various joint structures, and was an agglomeration of ancient support techniques. It was simple in process and efficient in both time and labor. Up to now, should there be any shafts or drifts with traditional support, this structure would stand out.

7.2.4 Underground Mining Methods [7]

Modern underground mining method is a systematic engineering which includes preliminary mining, segmentation and stoping before the successful extraction of the ore blocks. There was no such modern tunnel for preliminary mining in the Shang and Zhou dynasties. However, tunnels for pedestrians, ventilation and drainage did exist. Back then, shafts and drifts could be arranged for segmentation in a reasonable way according to various geological conditions, and different approaches had been adopted to control the compression and the ore break. The underground mining methods used were mainly as follows:

(1) Square set method. It was characterized by the cuboid made of wooden frames to fill the space and placed along the stoping. Ancient Chinese would take into account of the strike of rich ore and decided to dig vertically to form a shaft, and then to construct a blind shaft after enlarging, or they went for drifts or inclined drifts at the bottom of pit. Examining the sites in the Tongling and Tonglu Mountains, one may find that this method had been formed in the Western Zhou Dynasty, and was suitable for the mining of granular malachite in earthy surrounding rocks.

(2) Square set and fill stoping method. It is featured with horizontal stratification by two or more drifts, and the mining proceeded from bottom to top, letting waste rock on the top partially or completely fell and filled the lower layer. No. 2 drift and No.3 drift of VII2 site in the Tonglu Mountains in the Shang Dynasty were layered up and down, which was the prototype of this method. During the Western Zhou Dynasty, this method was found in the Tonglu Mountains and in Gangxia, Yangxin, and there were two layers of horizontal stratified sheds in the southeast section of No. XI ore body in the Tonglu Mountains. The upper drifts are connected to the vertical shafts, the lower drifts ran either vertically or horizontally, and the lower layer was partially filled with waste rocks from the upper part. During the Warring States Period, this method was rather well established in the Tonglu Mountains (Fig. 7.26).

(3) The mining method featured by vertical trusses (Fig. 7.27): It was characterized with the upward stoping, and the final stope resembled a cave with relatively hard surrounding rocks. This method is found at the VII6 site in the Tonglu Mountains in the middle and late Spring and Autumn Period and at the Nulasai Site in Xinjiang. The cave of the former was 8.6 m long and 4.5 m wide, forming a three-layer platform for ore breaking. The mining was started with the enlargement of the bottom to all sides, and at a certain height erected a T-shaped transverse bracket, which not only supported the goaf but also served as a platform for drilling, ore breaking and transporting.

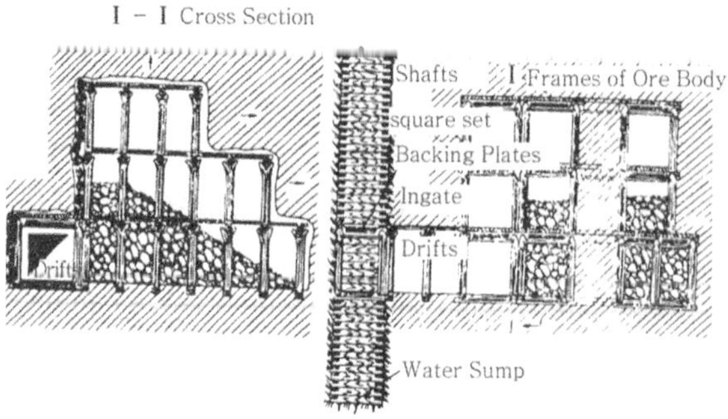

Fig. 7.26 The square set and fill stoping method

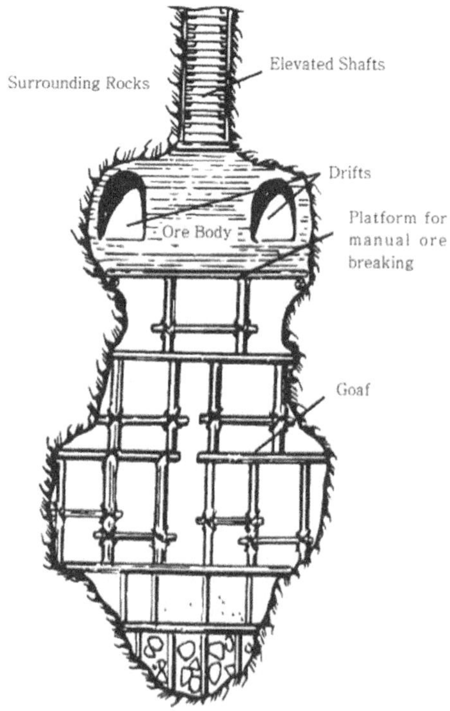

Fig. 7.27 The mining method featured by vertical trusses

(4) Open stoping method with support. This method was used for the mining of horizontal lode. On the side of the tunnel lays, at the same time, the Xianrenzuo Ancient Stope in the Tonglu Mountains and other similar stopes (Fig. 7.28). The depths of the stopes were 1.8–2.4 m, and the open stope was of 20–30 m^2. The goaf was a fractured zone, and the surrounding rock was soft and contained water. Therefore, the support was made of wood and logs.

Fig. 7.28 The open stoping method with support

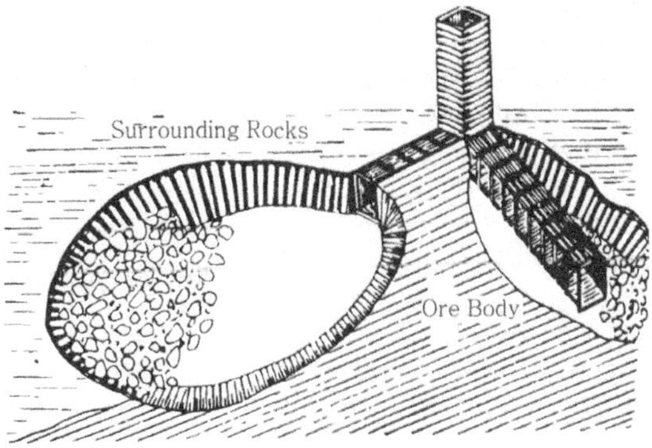

7.2.5 Tools and Appliances for Mining and Loading [8]

Bronze tools for mining have been unearthed at ancient pit sites like the Tonglu Mountains in Daye, Shitouzui, Tongshankou, Guojialong (Yangxin), Fengshan Cave, Tongling in Jiangxi and Tongling in Anhui. There were adzes, chisels and choppers during the Shang Dynasty, while in the Western Zhou Dynasty, hoes, picks and axes were added. 12 pieces of copper axe were excavated from the tunnels in Tonglu Mountain of the Spring and Autumn Period. During this period, types and weight of the tools increased significantly compared with that in the Western Zhou Dynasty (Fig. 7.29). The largest piece of copper axe was 47 cm long with a blade of 41 cm wide and weighed 16.3 kilograms. Judging from the wear condition of the axe blade, the suspension operation is performed, and the holder used the falling inertia of the axe to hit the work plane (Fig. 7.30).

At the early stage, stone tools and wooden tools were not completely abolished. There were still stone hammers, wooden adzes, wooden mallets, wooden pry bars, etc. and wooden spades, wooden scoops, wooden shovels, wooden ladles, etc. for ore shoveling. Wood tools were easy to obtain, simple to make, light in weight and low in price, and suitable for shoveling of mineral particles and waste rocks, which was the reason why they have been used for a long time.

Iron tools were applied in copper mining. The earliest known iron appliance was the iron adze found in the Tonglu Mountains in the Spring and Autumn Period. During the Warring States Period, iron tools, such as axes, hammers, rakes, hoes, hoes, chisels, bits, etc. were put into full use for mining. The tools at the time were made of high strength ductile cast iron with high temperature heat treatment, which were developed from wrought iron and pig iron casting. And the extraordinary performance of the iron made possible the great improvement in mining efficiency. Furthermore, from the Shang Dynasty to the Warring States Period, the containers for ore in the South were mostly bamboo baskets whose ingredients could be obtained locally and which were rather handy.

Fig. 7.29 The mining tools found in the Tonglu Mountains of the spring and autumn period—large copper axe

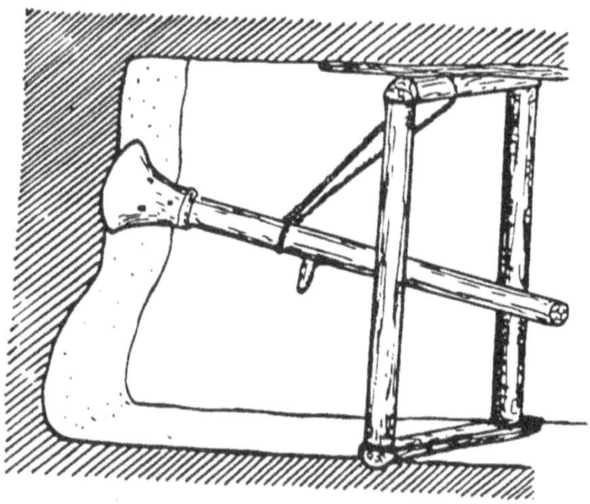

Fig. 7.30 The application of large bronze axe found in the Tonglu Mountains

7.2.6 Drainage, Ventilation and Lighting [9]

Underground waterproofing and water drainage are major issues to ensure mine safety. The drainage tools used in the middle of the Shang Dynasty included wooden troughs, wooden barrels and bamboo tubes. The underground drainage was designed to adapt to the local conditions, and the water should be discharged directly from the drifts to the hillside, or the water at the bottom of the pit (water sump) would be discharged by wooden barrels. Abundant rainfall in the South and a number of lakes brought considerable amount of surface water and ground water which sabotaged the mining safety. As the mining depth increased during the Western Zhou Dynasty, the ground waterproofing and underground drainage systems were quite complete at the time. Ground waterproofing included the following procedure: first, the support is above the surface with green paste mud covering around and a shed set overhead for some pits; the second is to dig a ditch, and to establish wooden troughs for water drainage and diversion. As for underground drainage, system was as follows: first, set up special drainage tunnels and water sum. The drainage tunnels could be constructed into butted board type, wooden partition type or stick partition type. The stick partition type was featured with drains whereas wooden sticks were densely packed with at the bottom and the sides of the tunnel and green plaster were applied on the top. When the drains ran through the stope, board covers would be added to make them covered ones.

The second is to set up the lock wall to cut off water in the abandoned or temporarily unused tunnels. Since the Western Zhou Dynasty, depth of underground mining has been developing. As a result, the water sump should be set in different sections, enabling staged drainage, carrying relay and final discharge. The drainage wooden troughs seen in Daye, Yangxin and Tongling were all carved out of whole wood, and a single trough measured 3 m long, 54 cm wide and 20 cm deep, with a large displacement. These complete drainage systems were the technical prerequisite for a successful underground mining in ancient times.

The shafts and drifts of the Shang Dynasty were not deep, and there was natural ventilation. While in the Western Zhou Dynasty, the mining depth had exceeded 50 m, which means that the ventilation depended on the wind pressure generated by the man-made temperature difference. And there were traces of bamboo burning at the bottom of the tunnels, both in the Tonglu Mountains and Tongling. Some tunnels were also blocked by earthy rock to form windbreak to control the wind direction. When it comes to lighting underground at the time, bamboo torches, bamboo firebrand devices and grease would be mentioned.

7.2.7 Transportation of Ore [10]

The use of simple mechanical devices to lift and transport materials has emerged in the middle of the Shang Dynasty. Wooden blocks (Figs. 7.31 and 7.32) unearthed

Fig. 7.31 Wooden blocks found in Tongling in the Shang Dynasty

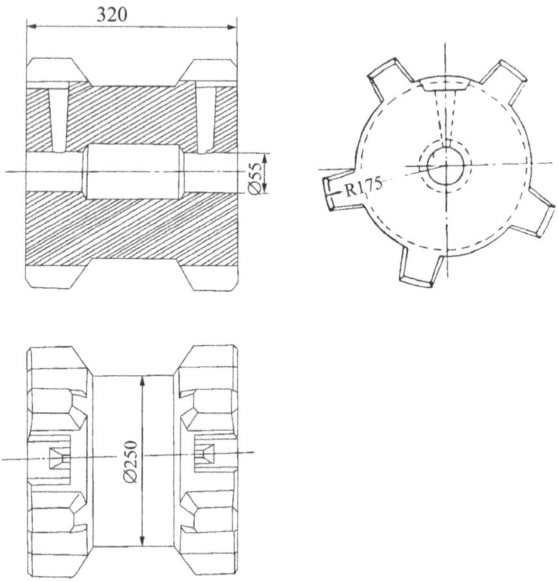

Fig. 7.32 The restoration of the wooden blocks found in Tongling in the Shang Dynasty

from the Tongling site were consistent with the structure of modern sliding bearings. Radial holes connecting with the bores were designed on each sides of the block and grease could be filled into reduce friction. A sliding bearing was formed between the larger diameter end and small diameter end of the bore, which would reduce the size of rubbing surface and friction between the bearing and the shaft. Referring to the analysis of the remains, the blocks were placed above the pithead with a shed overhead. The wooden steering struts were discovered at the junctions between the drifts and the blind shafts in the Tonglu Mountains of the Shang Dynasty. And

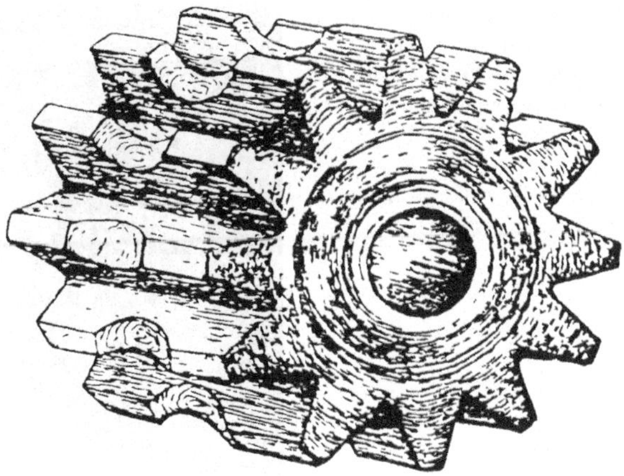

Fig. 7.33 Wooden blocks found in Tongling in the Spring and autumn period (*The Discovery and Research of Ancient Copper Ore in Tongling*, 70)

the struts functioned like fixed pulleys. Moreover, there were wooden ladders with handrails for craftsmen to climb up and down.

In the Spring and Autumn Period and the Warring States Period, the mine transportation technology had made great progress. There were many pieces of wooden blocks unearthed in Tongling (Fig. 7.33). Some of the blocks were found at the turning of the open pits and the drifts. The installation of the blocks at this position could change the direction of the loading baskets (Fig. 7.34) for easier ore extraction.

The large wooden axle shaft of the winch (Fig. 7.35) of the Warring States Period excavated from the Tonglu Mountains, and Hongwei Copper Mine sites was 250 cm long and 26 cm in diameter. The top of the shaft was equipped with wooden bars

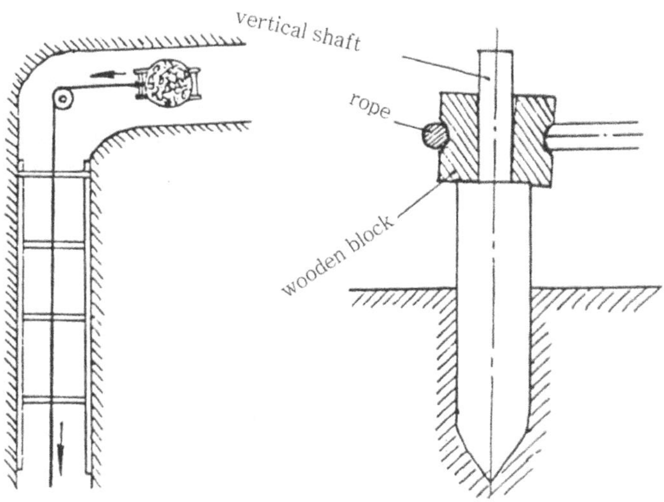

Fig. 7.34 The positions of the wooden blocks in the tunnels in Tongling in the spring and autumn period and the assembly of the blocks

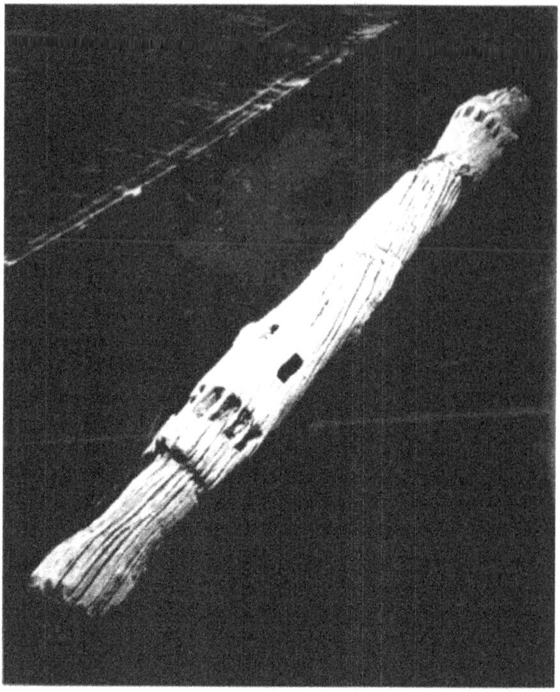

Fig. 7.35 The wooden axle shaft of the winch found in the Tonglu Mountains in the warring states period

and ratchet brakes. As to the winch, it was found in more than 60 m underground, designed for staged lifting.

7.2.8 *Ore Dressing [11]*

The ore dressing tools and equipment found at the sites of the Tonglu Mountains, Hongwei and Tongling were mostly relics of gravity concentration (panning or sluice concentration). The method of panning began at least in the middle of the Shang Dynasty, while the sluice concentration started in the Western Zhou Dynasty. Tools for disintegration consisted of wooden mallets, pestles and mortars; the crushing tools included stone anvils, balls and hammers; the ore dressing tools were composed of panning trays (boat shape, rectangular shape, peach shape, etc.) and panning baskets. In regard to dressing equipment, a wooden sluice was found and there were also relics of the dressing site. The copper ore produced in the southern remains was mostly cemented by viscous gangue or ferrous clay, which required disintegration treatment. For massive symbiotic copper ore, crushing was desired. The gravity concentration was a separation process based on different proportion of various mineral and rock particles in aqueous medium. These two types of gravity concentrations have been tested by the author, and were of decent performance with a high-grade concentrate.

Fig. 7.36 The ore dressing tools in Tonglu Mountains

The crushing tools (Fig. 7.36) were made of wood and had a certain toughness, which could crush the clay without smashing the copper oxide mineral to meet the size requirement for copper ore particle during the gravity concentration.

Dressing plant in Tongling in the Western Zhou Dynasty consisted of a canal, a wooden sluice, a tailings pond, a water filter and a shed. The wooden sluice (Fig. 7.37) was cut out of logs and measured 45 cm in diameter and 343 cm in length. The front and rear rims were 34 cm and 42 cm wide, respectively, and were 20 cm deep. Moreover, there were movable concentrate intercepting plates with a sluice at the tail. The simulation experiment showed that the sluice was advanced in structure, could process large ore dressing and was easy to operate. It is the earliest and most advanced mineral dressing equipment in China.

During the Warring States Period, the sluice concentration was developed toward underground operations, such as that in Tonglu Mountain concentrate separation was completed underground, which greatly reduced the transportation of waste rocks, and also reflected the ancient Chinese superior ability to control underground water.

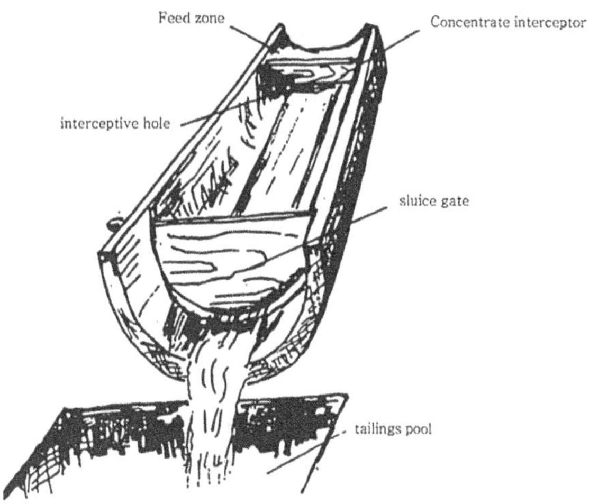

Fig. 7.37 The structure of the wooden sluice in Tongling in the Western Zhou Dynasty

Generally speaking, the mining and dressing technology of the Pre-Qin Period copper mine can be divided into three stages, namely the formation period of the middle Shang Dynasty, the development period of the Western Zhou Dynasty and the innovation period of the Eastern Zhou Dynasty. The joint development methods and support technology for shafts and drifts in the middle of the Shang Dynasty have reached a certain level, which made possible the advancement in the Western Zhou Dynasty, and the innovation in the Eastern Zhou Dynasty. Ancient China has established her own mining system, which is rare in the world mining history.

7.3 Copper Smelting

The highly developed pottery of the Neolithic Age allowed China's copper smelting technology to grow from a high starting point. The birth of copper smelting technology is closely related to the production of pottery. The gray pottery is obtained in the reducing atmosphere, and this method was mastered by ancient Chinese in the early Neolithic Period. More than 6000 years ago, sintering temperature for pottery from Yangshao Culture could reach 950–1050 °C, and the actual kiln temperature was higher. This means that the high temperature for copper smelting and the technical conditions for obtaining a reducing atmosphere have already been obtained during this period.

There are more than 240 kinds of copper minerals in nature, but only pyritum, copper sulfide or copper oxide ore have substantial industrial values.

Sulfide minerals include chalcopyrite, chalcopyrite, bornite, covellite, tetrahedrite and enargite, and oxidized minerals malachite, chrysocolla, hematite, azurite and tenorite. Copper ores are composed of several minerals. Before the Qin Dynasty, the ancients obtained copper ores locally based on their appearance and physical properties and the experience in identifying them. The kind of copper to be smelted is decided by the kind of ores mined, and the kind of copper smelting technology is developed based on the kind of copper ores to be smelted. Four pyrometallurgical copper smelting processes were invented before the Qin Dynasty as follows:

a. Reduction smelting of oxidized ore into copper, referred to as "oxidized ore-copper" process.
b. Direct reduction smelting of copper ore.
c. The sulfide ore is converted into matte, the high-grade matte is roasted by desulfurization for many times and the matte is finally melted into copper by dead roasting, referred to as "sulfide ore-matte-copper" process.
d. Sulfide ore is desulphurized by dead roasting and then reduced to copper, referred to as "sulfide ore-copper" process.

The so-called "direct reduction smelting of copper ore" process, i.e., the copper ore into the furnace contains both copper oxide and copper sulfide. Can copper and sulfur-mixed ores be directly reduced and smelted to extract some pure copper? The answer is yes. From 1958 to 1959, there were many examples of copper smelting by indigenous method in many places in China, including "crude copper is produced by direct reduction smelting of raw ore without roasting", which is under the condition that the copper content of the ore in the furnace is four times that of the sulfur content [12, 13]. The ore of Dongchuan District, Yunnan contains 10.74% copper, 2.78% sulfur and the ratio of copper to sulfur is 4:1. The ore is directly melted without roasting, producing crude copper and matte. The molten slag, matte and crude copper are stratified from top to bottom according to different specific gravity in the hearth. There are also examples of the low copper grade ores. The grade of copper-bearing shale in Mouding County, Yunnan is 1.1% copper, 0.12% sulfur, and the ratio of copper to sulfur is 9:1. Chimashan ore in Daye contains 2.36% copper, 1.28% sulfur, and the copper/sulfur ratio is only 2:1. Because of the low coke rate and high desulphurization rate of the earth furnace, the copper/sulfur ratio changes during smelting, and this destroys the formation conditions of Cu2S and produces crude copper and matte.

Abroad, in the second millennium B.C., Oman mixed a small amount of sulfide ore into the oxidized ore and produced copper and white matte in a reducing atmosphere without roasting [14].

7.3.1 Early Copper Smelting Technology

About 6,000 years ago, there were brass sheets and pipes in the late Yangshao Culture of Jiangzhai Village, Shaanxi and the smelting temperature then was 950–1200 °C. Brass can be obtained by reducing copper-zinc mixed ore or symbiotic ore with carbon. Early brass ware, like early bronze ware, is a product accidentally obtained under original smelting conditions.

Malachite is the earliest mineral used to smelt red copper in China. Copper residue was unearthed in the ash pit of Linjia Site in Dongxiang County, Gansu. Petrographic identification shows that the small pieces consist of malachite, and the large ones 30% malachite, 45% iron ore and 5% copper [15]. Malachite was also found in the three sites of Shijiahe Culture (about 4,100 years ago) in Tianmen, Hubei [16], and smelting slag and malachite were found on the five pieces of bronze sheets in Luojiabailing Site.

The discovery of Niuheliang Site in Lingyuan, Liaoning is of great significance to the exploration of early copper smelting technology in China. Li Yanxiang studied the unearthed debris of furnace wall [17] and the adherent slags in 3,500–3,000 B.C., and found that the slags were copper smelting slag. The inner diameter of the upper part of the furnace is 18–20 cm, the wall thickness is 1.5–3.0 cm and the height is about 35 cm. It is made of straw mixed mud. There are small holes for manpower blast on the furnace wall, with an inner diameter of about 3.7 cm, an inclination of about 35 degrees inward, and 6 holes in each of two rows, a total of 12 holes. When the liquid copper sinks at the bottom of the furnace, the lower part of the furnace must be smashed to extract copper. Residual stoves and copper smelting slag were also found at Longshan cultural sites (2800–2300 years ago) in Huaiyang County and Linru County, Zhengzhou. The copper layer on the remnant of a crucible has been tested to contain about 95% copper, and there is no burn mark on the outer wall of the crucible [18]. It seems that in the initial stage of copper smelting, the internally heated crucible was used for copper smelting or smelting.

About 4,000 years ago, the development of the copper ore belt in the middle and lower reaches of the Yangtze River has begun to take shape. According to the author's survey and statistics, there are nearly 100 copper smelting relics in Huangshi area, Hubei, most of which can be traced to the Neolithic Age. From 1984 to 2003, the Institute of Cultural Relics and Archaeology of Hubei excavated the Dalupu Site in Yangxin County and Xizidi Site in Daye and unearthed a number of copper smelting and casting materials [19, 20].

The Dalupu Site is 15 km northwest of the Tonglu Mountains. In its eighth layer, which belongs to the late Shijiahe Culture, about 4,100 years ago, slag and remnants of furnace wall were unearthed [21], and in the seventh layer, which belongs to the post-Shijiahe culture, about 4,000 years ago, bronze debris, ores, slags and burnt soil deposits were excavated. The ore composition is shown in Table 7.1. It is a typical local copper-iron ore and iron-copper ore. The slag has a good fluidity, and its composition is shown in Table 7.2. In this site, more smelting remains of the Shang and Zhou Dynasties were found, such as stone drills, hammers and burning pits, in addition to a large number of slags, ores and furnace walls.

The Xiezidi Site is 6 km south of the Tonglu Mountains. In its fourth floor are relics of the early Xia Dynasty, including malachite, stone anvil and stone hammer. The smelting remains of the early Western Zhou Dynasty are many, including refining

Table 7.1 Ore composition of layer 7 of the Dalupu site

Sample number	Ore	Petrographic identification	Composition				
			Cu	Fe_2O_3	SiO_2	CaO	MgO
T2307 ⑦: 16	Copper-iron ore	Malachite, quartzite, kaolin	40.12	2.35	9.04		
T2506 ⑦: 15	Iron-copper ore		3.18	38.56	28.46		
T2506 ⑦: 18	Iron-copper ore	Hematite, quartzite	1.53	30.27	39.03		4.86
T2606 ⑦ C: 21	Iron-copper ore		3.18	38.56	28.46	16.03	
T2506 ⑦: 19	Iron ore		Iron content 97.84%				

Table 7.2 Quantitative analysis of slag XRF powder

Sample number	Cu	S	Fe_2O_3	SiO_2	CaO	Al_2O_3	K_2O	P_2O_5	MgO	MnO	TiO	NaO_2
T2507⑦: 24	0.03	0.03	14.74	47.78	5.00	20.07	7.84	0.87	0.44	0.17	0.85	2.61
T2506⑦: 20	1.19		90.27	2.53	4.68	0.25	0.21	0.19		0.15	0.08	

slags and copper granules; the copper content of the latter is 83.68%, 93.47% and 96.23%.

Li Yanxiang analyzed the slag on the seventh layer of Dalupu Site and found that the apparent diameter of 9 matte particles was small, in the range of 3–100 μm, and the irregular copper particles were more than 2000 μm in length. Metal copper particles or metal copper coexisted with matte. The conclusion was that the smelting technology was mainly the "oxidized ore-copper process" or "sulfide ore-copper process" which used oxidized ore or sulfide ore to directly smelt copper [22].

7.3.2 Copper Smelting Technology in the Shang and Zhou Dynasties

The mining remains of Tongling in Jiangxi, which began in the middle of the Shang Dynasty, are distributed within the 70,000 m^2 from the Tie Mountains to the He Lian Mountains. The remains of smelting copper are distributed in the south, west and north, with an area of about 170,000 m^2. The slag accumulation is thicker, mostly in the form of flakes, black in color, smooth in surface and good in fluidity. The surface of some slags is flat and the bottom is curved, which seem to be formed by the solidification of the pot-shaped slag. The slag was tested to show that copper content was only 0.334% (Table 7.3), indicating that copper and slag were separated well.

Tongling was a copper-iron symbiotic deposit. After weathering and leaching, it was downward enrichment of malachite, azurite and limonite, and the average copper content was more than 10%. In 1993, the author conducted an ore dressing simulation experiment at the mine. The selected concentrate's chemical composition is shown in Table 7.4 [23].

Copper smelting remains in Tongling indicated that the copper smelting ore was copper oxide ore, the copper/sulfur ratio of the slag was 16.7, and the smelting technology was the "oxidized ore-copper" process.

Table 7.3 Chemical composition of Tongling Slag

Cu	S	Fe_2O_3	FeO	SiO_2	CaO	MgO	Al_2O_3	Zn	Pb	Sn
0.334	0.02	2.68	40.65	43.50	1.31	0.41	5.57	0.452	0.0065	<0.001

Testing Department: Department of Minerals, China University of Geosciences

Table 7.4 Chemical composition of selected concentrate in Tongling of the the Western Zhou Dynasty

Cu	Zn	Pb	Sn	S	SiO_2	Fe_2O_3	FeO	CaO	MgO	Al_2O_3
20.48	0.212	0.0115	<0.001	0.02	38.71	8.77	0	0.56	0.31	6.91

Chemical Analysis Department: Analysis Center of China University of Geosciences (Wuhan)

By the late Western Zhou Dynasty, China had used the blast shaft furnace to smelt copper. A total of 10 furnaces of the late Western Zhou Dynasty (Table 7.5) were cleaned up in the XI site of the Tonglu Mountains. These 10 furnaces are composed of furnace base, hearth and furnace body (Fig. 7.38) [24]. When unearthed, the furnace body has collapsed, but auxiliary facilities were still there, such as workbench, crushing table (Fig. 7.39), screening field, mud floor and slag pit.

Furnace foundation: The foundation pit was paved with clay and stone. The T-shaped or "+"-shaped air channel was built in the middle, and the channel wall was made of kaolin.

The air channel was about 2 m long, 40–80 cm wide and 40 cm high. The inner stone supported the hearth. The term "air channel" came from the traditional copper industry, or "fire channel". Its function was to ventilate and protect against moisture and heat. Therefore, the surface was covered with baking marks, and charcoal and ash were left at the bottom of the channel. This structure was inspired by the practice to prevent freezing cylinders in long-term copper smelting process. Afterwards, the air channel became a major feature of ancient Chinese furnace. Ke Xi Tai copper smelting furnace in the Tonglu Mountains in the Warring States Period, Lujianao copper smelting furnace in Daye of the Han Dynasty, iron and copper blast furnace in Nanyang City of Henan Province and Acheng Heilongjiang of the Han, Ming and Qing dynasties were equipped with air channel. In the late Ming and early Qing dynasties, Gu Yanwu's *Tian Xia Jun Guo Li Bing Shu* recorded that "the bottom of

Table 7.5 Shaft furnace chronological data

Furnace No.	Thermoluminescence age (year)	14C dating (year)	References
3	2856 ± 295		Archaeology, No. 6, 1981, P. 551
6		3205 ± 400 3110 ± 400	Archaeology, No. 6, 1981, P. 376
10	2895 ± 305 3014 ± 320		Archaeology, No. 6, 1981, P. 551

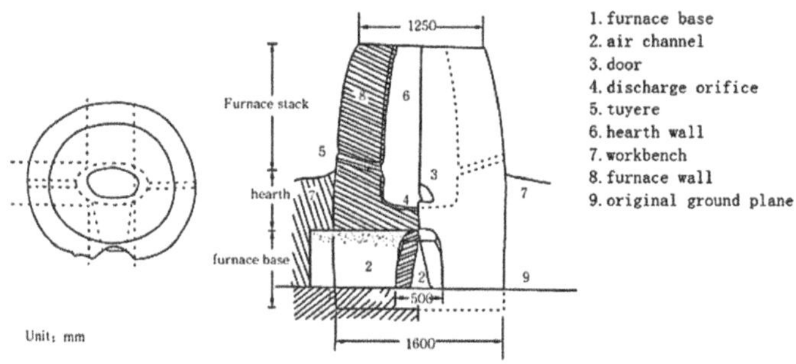

Fig. 7.38 Restoration of copper smelting shaft furnace

Fig. 7.39 The crushing tools of Late Western Zhou Dynasty in Tonglu Mountains: stone anvil and stone ball

the kiln is a channel" in Fujian iron-making furnace. It can be seen that the traditional Chinese furnace structure was derived from the same origin.

Hearth: Take No. 4 furnace as an example, it was oval and had a burning flow. According to petrographic identification, it was a mixed refractory material, in which silicified igneous rock debris (including quartz sand) accounted for 73%, kaolin 10%, and glass phase precipitated after high temperature 15%. The outer wall was composed of 34% stone sand, 30% iron ore debris, kaolin and garnet chips. The bottom of the hearth was divided into four layers from bottom to top: the bottom layer is the same as the outer wall; the second layer is a mixed layer of quartz sand, garnet, iron ore coarse crumb and kaolin; the third layer is mainly kaolin; and the fourth layer is patched with dark brown refractory loach.

The front wall of the hearth had an arched door, which was high and wide in the outside, low and narrow in the inside. The door would be open when charcoal needed to be ignited. After the hearth was filled with charcoal, the door would be closed. In case of failure, the door could be dismantled to be fixed when the furnace was shut down. When the door wall was pressed outward by the side pressure of the charge, the arched door would be tightly blocked. It is commendable that the bottom of the hearth was 6 cm below the outer ridge of the door. It means that the copper liquid with a depth of 6 cm would not be discharged out of the hearth, which was convenient for continuous feeding of ores. The above mentioned is recorded in Wu Qichen's *Dian Nan Kuang Chang Tu Lue* of the Qing Dynasty.

The air outlet of No. 4 furnace was located on the side wall of the long shaft of the hearth, leaving only one air outlet with a diameter of 5–7 cm and a downward inclination of 19°. In 1981, Mr. Hua Jueming and the author made a restoration study [25] and got the following conclusions: First, the shaft furnace was about 1.5 m high in the shape of waist drum. The furnace wall was inclined with the angle of 79 degrees, and the effective volume was 0.3 m^3. Second, the furnace should have two symmetrical blast outlets, the cross-sectional area of the tuyere was 0.23 m^2, and the center line of the tuyere was 30 cm away from the bottom of the hearth, which was beneficial to increase the temperature of the hearth, making full use of the wind pressure to balance the wind. The blower might be a leather bag. Third, the charge distance was about 1.2 m with the large volume and high thermal efficiency. Fourth, the furnace structure was reasonable, easy to operate, sturdy and durable, and more importantly the furnace was old. The Tonglu Mountains blast copper smelting shaft furnace was of significance in the world metallurgy history. Professor Madin from the University of Pennsylvania in the United States said after the on-site inspection, "There are many mine smelting relics in the world, among which the Tonglu Mountains is the first-class. Although the Middle East started copper smelting very early, there is no such large-scale underground excavation sites. Such furnaces with high slag temperature, good fluidity and low copper content are rare" [26].

The characteristics of the Tonglu Mountains copper ore determined the long-term existence of the oxidized ore-copper process. This unique copper mining area was particularly developed due to the secondary oxidation-enriched zone. Natural copper, copper-bearing ore, malachite and azurite were more exposed or close to the surface, which was suitable for development by the ancients in the following three aspects:

(1) The Excavation Report of the Tonglu Mountain Copper Mine Site clarified that the eight excavation areas were all copper oxide mines and many bamboo baskets filled with malachite were unearthed [27]. The author once carried out a panning and dressing simulation experiment in the No. VII ore body, and obtained malachite, chrysocolla and cuprite. The composition analysis is shown in Table 7.6. See Tables 7.7 and 7.8 for the ore composition of the No. XI ore body shaft furnace and adjacent old mines.
(2) Many ancient mine shafts separated from modern mines were found in the copper oxide ore in the marble fracture zone or the igneous mica belt.

Table 7.6 Composition of washed copper ores(%)*

Cu	Fe_3O_4	Fe	SiO_2	CaO	Al_2O_3	MgO	S
40.40	0.60	2.00	20.00	0.64	4.41	0.37	0.53

Chemical Analysis Department: Central Laboratory of Daye Nonferrous Metals Company

Table 7.7 Ore composition (%) of 12-line mines during the spring and autumn period (%)*

Composition	Cu	Fe	SiO_2	CaO
In-Mine sampling	30–40	3–5	25–30	2–3
In-Mine Surface sampling	10–16	42–48	9–18	1–2
Average value	24	24.5	20.5	2

Chemical Analysis Department: Central Laboratory of Daye Nonferrous Metals Company

Table 7.8 Ore composition

Name	Unearthed Place	Composition (%)										Sample No.
		Cu	FeO	Fe_2O_3	Fe_3O_4	Fe	SiO_2	CaO	Al_2O_3	MaO	S	
Malachite	Beside furnace No. 6	53.52	0.51	0.43			3.52	0.29	0.33			Ore No. 23
Iron ore	Beside furnace No. 9	0.98			7.6	62.38	25.53	1.18	2.55	0.61	0.18	Ore No. 18
Copper iron ore	Beside furnace No. 6	3.28	2.71	21.35			9.68	0.47	1.63	0.82		Ore No. 19
		2.059	3.33	18.85			17.60	1.22	1.37	1.51		Ore No. 17
	Tonglu Mountain	1.87				54.2	12.44	1.64	2.1	0.48		
Silica malachite	West of Copper smelting factory	20.34			Slight	6.34	48.83	0.23	10.72	0.12	0.59	Ore No. 20
	West Zhou Dynasty Mine	40.40			0.60	2.99	20.00	0.64	4.41	0.37	0.53	Ore No. 14
Malachite and iron ore inclusions	West of Copper smelting factory	31.88			1.40	17.29	15.72	1.27	2.50	0.22	0.53	Ore No. 19
Malachite	Spring and Autumn- Period Mine	51.82			Slight	1.90	2.88	0.073	0.77	0.14	0.57	Ore No. 23
Melaconitc	Spring and Autumn- Period Mine	22.45				21.29						IV-1

*Chemical analysis department: Except sample IV-1 is inspected by Tonglu Mountain Mine Laboratory, other samples are inspected by Central Laboratory of Daye Nonferrous Metals Company

(3) Modern mine production and geological exploration data reflected that the copper oxide ore of several ore bodies was fully mined by the ancients [28, 29].

The History of Da Ye County notes that malachite would be seen after a heavy rain in the Tonglu Mountains in Qing Dynasty. According to the stone anvils, stone balls and screening fields unearthed in the Tonglu Mountains in late Western Zhou Dynasty, ores were broken into blocks of proper sizes before entering the furnace to reduce the resistance of the furnace gas and increase the contact area between the reactants, which accelerated the smelting process. This granulation technology is a very important initiative. The fuel for copper smelting was charcoal, which had been identified as a hard-burning eucalyptus charcoal with high heat power burned from locally produced eucalyptus. After the shaft furnace is built, it needed to be naturally dried. Add wood for ignition and preheating before smelting, and add charcoal and ore layer by layer from the furnace mouth (sometimes may add flux), and then increase the air volume. When slag and copper liquid in the hearth have a certain amount of accumulation, poke the discharge hole of the arched door, release the slag from the copper and then block the discharge hole with peat to continue the smelting. Then repeat the whole process until the furnace is stopped.

The smelting of copper ore was essentially a process of controlling and operating slags. The texture and characteristics of the slag could show whether the operation of the smelting process was normal or not. The slags in the Tonglu Mountains (Fig. 7.40) were mostly black-gray flakes, with smooth upper surface. Some had corrugated folds, and the fractures were dense and had pores, indicating that the slag had good fluidity and the furnace run normally. The reductive smelting slag of copper oxide ore in modern blast furnace required that the copper content was 0.70–1.0% [30]. The slag sample analysis of the four ancient furnaces, No. 3 to No. 6 (Table 7.9), showed that the copper content was 0.20–1.26%, indicating that the slag and copper were well separated and the copper recovery rate was high (Table 7.10). A piece of crude copper was unearthed on the west side of No. 3 furnace, weighing 94 grams, containing 93.32% copper; the residual copper in the middle of the No. 4 furnace was 2.3 kilograms (Fig. 7.41), containing 93.99% copper, 1.33% sulfur, and it was

Fig. 7.40 Slag accumulation in Tonglu Mountains No. 11 ore body during the spring and autumn period

Table 7.9 Slag +lts

Furnace No.	Sampling point	Description	Composition (%)										Silicon acidity (K)
			Cu	FeO	Fe_2O_3	Fe_3O_4	Fe	SiO_2	CaO	MgO	S	Al_2O_3	
No. 3	First layer	Lumpy	0.58	41.92	7.15			24.11	0.96				
	Second layer		0.67	46.70	13.71			25.08	2.58				
	Third layer		0.20	46.83	14.33			26.90	1.68				
	Average		0.48	45.15	11.73			25.30	1.74				
No. 4	Tailings		1.26			10.00	49.54	6.89	0.46	0.43	0.74	5.70	
No. 5	Tailings		0.625			800	43.88	31.92	1.86	0.68	0.58	6.66	
No. 12	Besid the ditch of furnace No. 6	Platy	0.955	44.73	5.61			36.74	3.17	0.91	0.39	7.807	1.8
No. 13	In the earth of No. 6	Crude	0.798	36.02	1.84			22.22	1.34	1.33	0.84	4.355	1.4
No. 1.4	Besid the ditch of furnace No. 6	Platy aerugo	1.120	51.81	Slight			28.60	3.90	1.62	0.57	6.410	1.2
No. 15	Besid the ditch of furnace No. 6	Schistose	0.858	50.46	0.53			25.96	1.95	1.51	0.93	6.30	1.3

Table 7.10 Major components of raw copper*

Unearthed place	Cu	Sb	Pb	Zn	Sn	Fe	S
Furnace No. 3	93.32	0.0075	0.038	0.014	0.023	3.35	
Furnace No. 4	93.99			0.66		3.99	1.33

*Furnace 3 was tested by Central Laboratory of Daye Nonferrous Metals Company, and Furnace 4 was examined by SEM of General Institute of Iron and Steel Research, the Ministry of Metallurgy

Fig. 7.41 The crude copper unearthed from Tonglu Mountain No. 4 furnace in the late Western Zhou Dynasty

quite close to the copper content required by modern blister copper (92–95%). The smelting temperature of 13 slag samples was measured at 1,200 °C or slightly higher under argon gas protection. This temperature facilitated the complete reduction of the metallic copper and separation from the slag without causing the reduction of the iron into the copper. In smelting copper, the slag silicate degree was required between 1 and 1.5. If the slag silicate degree was too high, the viscosity of the slag would be high, which would affect the separation of slag and copper as well as normal discharge. The slag of No. 6 furnace has a silicate degree of 1.4. The specific gravity of multiple slags was estimated to be 3.6–4, and the high difference in specific gravity between copper and slag was favorable for their separation.

To obtain a suitable slag type, the batching technique was used in late Western Zhou Dynasty. According to the calculation of the composition of the Tonglu Mountains ore and type of slag, iron ore was also used as a flux, which was consistent with the remains of the crushing table and workbench. Analysis showed that the slag was a high iron slag, a typical copper smelting slag.

7.3.3 The Adoption of Copper Sulfide Ore in Copper Smelting

It is an important issue in the study of metallurgical history that when China began to use copper sulfide ore to smelt copper. Due to the complexity of this smelting process, the record of preroasting process was only found in the literatures of the Song Dynasty, and some scholars previously thought that this technology appeared later.

From 1976 to 1995, the sites of the copper mines, such as Dajing Town of Inner Mongolia, Tongling, Anhui, Zhongtiao Mountains in Shanxi and the Tonglu Mountains in Hubei, kept revealing the information on smelting copper with copper sulfide ore, most of which was recorded in early Spring and Autumn Period. It can be inferred that China began to use copper sulfide ore to smelt copper no later than earlier the Spring and Autumn Period.

The Dajing Site belongs to the upper Xiajiadian culture, which appeared in about the early Spring and Autumn Period. It is the only mining and smelting site known in the world to directly smelt bronze with symbiotic ore. In the Tabuaobao smelting copper site covering more than 10,000 m^2, 10 km away from Dajing Site, 4 roasters, 5 furnaces, and 1 blast pipe were unearthed, and remains of furnace walls, slags and charcoals were left. The roasting furnace is a porous furnace with a diameter of 1.5–2 m and above diameter of 8–10 cm. The cross-section of the furnace is elliptical with a long axis of 120 cm and a short axis of 80 cm. The wall of the furnace is made of grass mud, 10–12 cm thick, and the arched door is about 20 cm high and about 10 cm wide. The outer diameter of the blast pipe is 6 cm [31]. The ore at the bottom of the ancient mining pit was identified by the geological department: the R1 pit is mainly chalcopyrite, and some pyrite and copper blue, and the R2 pit is mainly blue copper, cyanite, cassiterite and limonite. Li Yanxiang believed that the ores in oxidized zone, mixed zone and primary zone had been mined by the ancients. After analysis, the slag should have gone through the process of "sulfide ore-copper" and it might be baked with matte [32].

The copper ruins of the Qin Dynasty in southern Anhui are mainly distributed in Nanling County and Tongling County. By 14C dating, the age of the Muyu Mountain Site in Tongling County is 2,882 $\pm$ 55 years old, where quantities of slags, red burned soil, ores and charcoals were found. No.1 furnace of the Jiangmuchong Site in Nanling County is 2725 $\pm$ 115 years ago, and the slag distribution range is about 1.5 square kilometers, where a total of nine copper smelting furnaces were cleaned up. The furnace type and building materials are similar to those in the Tonglu Mountains.

A few copper mines in southern Anhui are large ore bodies, most of which are nest-like ore bodies. These ore bodies, mostly copper and iron type, were well developed and buried deep, so they were easy for the ancients to mine. Sulfide ores were dominated by chalcopyrite. The ancient slags in southern Anhui, characterized by the shape of inverted mushroom, are brownish black and have flow marks on the surface. Mu Rongping made analysis of chemical composition, scanning electron microscopy and X-ray diffraction on the slags of the Jiangmuchong Site, and believed that the "sulfide-matte-copper" smelting process had been adopted. Chen Rong and Zhao

Yuhua analyzed the four copper ingots of the Zhou Dynasty from the Muyu Mountain Site and Fanchangli Mountain Site, which were imaged by electron microscopy. The dark phase was white metal [$(Cu_2S)_2FeS$], which was formed by the insufficient calcination and reductive smelting of copper sulfide ore. Accordingly, it was concluded that chalcopyrite had been used in copper smelting [33].

The diamond-shaped copper ingot was a major feature of the copper smelting in Qin Dynasty. This form of copper ingot was currently only found in this area. In the 1970s, more than 100 kilograms of diamond-shaped copper ingots from the Spring and Autumn Period were unearthed in the Muyu Mountain Site, and from 1984 to 1994, 60 pieces of the Spring and Autumn Period or earlier in the Jiangmuchong Site. Mr. Hua Jueming and others studied the diamond-shaped copper ingots of the Eastern Zhou Dynasty unearthed in Guichi [34] and found that no residues were found in all the ingots. It could be known that copper liquid was released from the furnace and solidified in a prefabricated diamond-shaped sand bed. Tests and analysis showed that the main composition of copper ingot was copper solid solution and iron solid solution, which were the mechanical mixture of copper and iron, containing about 63% copper and about 34% iron. It was neither matte nor pure copper but an intermediate product like sulfonium. If it was crushed and remelted in crucible, pure copper could be obtained by removing the iron and sulfur, therein after repeated segregation and slagging were carried out by the difference in the specific gravity of copper and iron.

Li Yanxiang researched on the smelting technology of the Spring and Autumn Period in the Nulasai Site of Xinjiang Uygur Autonomous Region [35], and believed that during the process of "sulfide ore-matte-copper", a matte smelting and a reductive smelting were engaged with the participation of arsenic. This is the only ancient mining and smelting site known in Eurasian Continent to smelt high-arsenic copper alloy by adding arsenic minerals, which is of great significance in the history of metallurgy.

The development of copper mining and copper smelting technology before the Qin Dynasty shows that China's unique technological culture and copper deposit characteristics have fostered a complete system of copper mining and smelting technology. Since the middle and late Yangshao Culture Period (about 7000 or 5000 years ago), a series of innovative mining techniques and copper smelting techniques had emerged from the middle of the Shang Dynasty to the Warring States Period after more than 3,000 years of accumulation. It is these original major inventions that created China's splendid bronze civilization.

7.4 Bronze Piece-Mold Casting

Process of natural copper, artificial copper smelting and bronze were invented in the Near East, and the traditional process of forging and casting copper was formed in the Chalcolithic Age. Among the cast copper wares, many were formed by the stone mold method and the lost-wax casting method, and also by the mud mold method.

Therefore, the forming process of copper and bronze was invented in the Near East over 2000 years earlier than in China.

When copper smelting first took place in China was still unclear, but it was practiced in the Neolithic Age, gained progress in the Chalcolithic Age, and achieved breakthroughs during the Erlitou Culture period. The production of bronze sacrificial utensils marked the beginning of Bronze Age.

7.5 *The Erlitou Culture: Establishment of the Mold Method

In the second phase of Erlitou Culture, bronze wares increased significantly in number and variety. Most were made of copper-tin bronze or copper-tin-lead bronze by the mud mold method, and no forging ones were found. In other words, in the Chinese Bronze Age when the Erlitou Culture began, the mud mold method was formed as a technological tradition in bronze wares manufacturing, which was different from other civilizations.

Many cast copper relics were unearthed in the copper casting workshop of Erlitou Site, including mud molds, furnace fragments, slags, charcoal and so forth. Among them, many were used and broken mud molds, and only two mud cores could be identified. Most of the mud molds were thick, with brick red, yellowish and light gray colors and loose texture. They were mixed with plants and had a low calcination temperature and a dense and smooth parting face. The mud mold in later period was carved with patterns which were directly portrayed on the surface, and was made of fine mud mainly composed of silica, alumina, ferric oxide and calcium oxide. The main components of the mud are quartz, followed by feldspar and mica, indicating that the material was processed less, and judging from the proportion of phytolith contained in it, reed and grass fiber or ash were mixed. The mud mold can withstand temperatures above 1050 °C [36].

The representative bronze wares of the Erlitou Culture were cast in complex molds by the mold method. The mold for casting bronze cup even needed six or seven mud molds and one mud core. The wall of the Erlitou bronze wares is thin and even in thickness, usually 1–2 mm [37], indicating that the molds and core assembly was very precise and stable, showing superb casting skills.

The mold method was a complex technical system, and the framework of it was shown in Fig. 7.42. The technical elements of the above framework were gradually developed on the basis of the Erlitou Culture, and a series of technological inventions made the mold method complete.

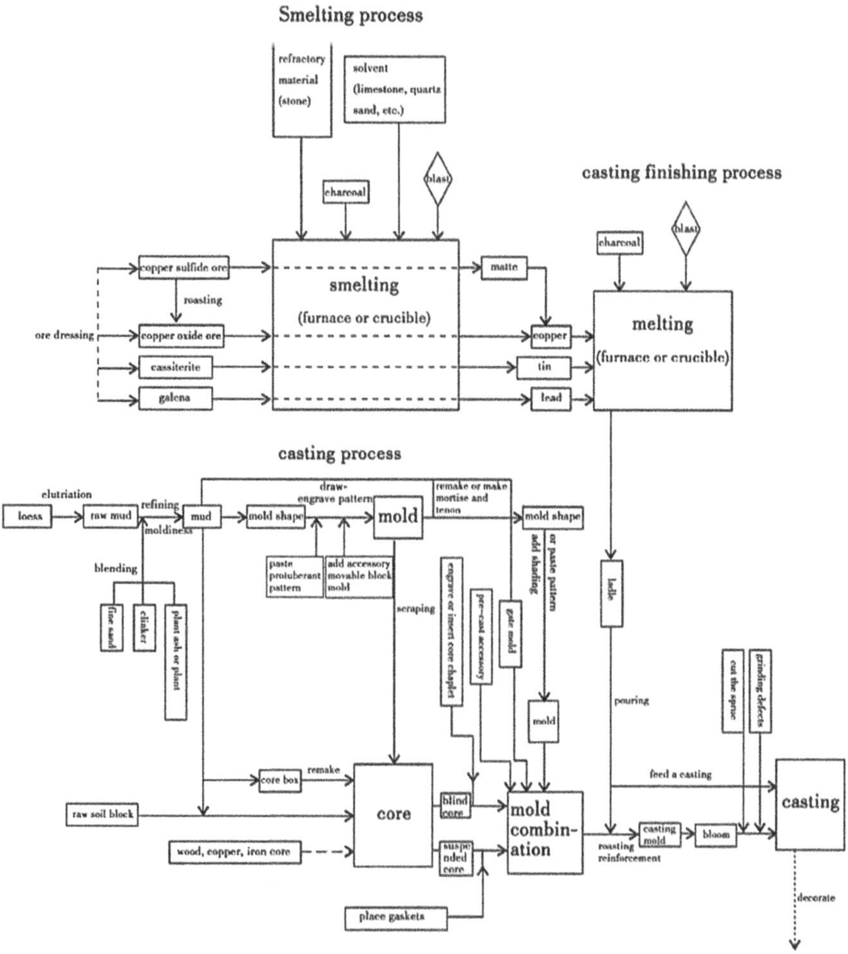

Fig. 7.42 Flowcharts of copper casting by the mold method (dotted lines represent possible correlations)

7.6 *Invention and Use of Core Chaplet and Spacer

Bronze containers were also political and religious symbols. One of the keys to the technology of casting bronze wares is the treatment of clay cores. Most of clay cores are suspended cores, which are difficult to fix in casting and assembly. Due to lack of research on bronze wares in the Erlitou Culture, it is still unknown if there are clay cores in them. However, during the ensuing Erligang Period, chaplet and clay cores were found in Panlong City bronze wares in Huangpi, Hubei.

7.6.1 *Invention and Use of Core Chaplet

There are patches at the bottom of Yangjiawanjue YWM6:1, a Panlong City bronze ware in the early Erligang period; there is a patch at the bottom of Ge YWM6:2 (Fig. 7.43). Four patches and a quadrilateral hole are found at the same height of the feet of the ware. They are all original technological holes. The post-casting parts have fell off, but the other technological holes are still on the walls. The original technological hole is formed by setting core chaplet, which is used to support the ventral core and maintain the size of cavity. There are three patches at the bottom of Lijiazui LZM: 55 tripod in Panlong City: two casting-on patches and one large patch in the middle of the bottom. No copper spacer was found on the X-ray film. The function of the patch is the same. There are three patches in symmetrical distribution and uniform shape under LZM: 20 crotch. They should be made on purpose, and the later recasting shows that it is not an end but a means.

A breakthrough of the core chaplet technology was achieved in the late Shang Dynasty, as blind cores could be set in the ears or feet of the wares, making the casting structure more reasonable and the wall thickness more even. Usually a number of small protrusions are projected from the blind cores, and they are inserted into the mold in casting combination. So these protrusions are self-contained core chaplets. There are also independent mud nails (core chaplets), one end inserted in the blind core with the other end connected in the adjacent mold to fix it. The tripod foot clay core (03AXST2006 H232: 35) unearthed from Xiaomintun Casting Copper Site in Anyang, Henan belongs to the category of self-contained core chaplets, and it is used to cast hollow foot [38]. On the tripod of the early Western Zhou Dynasty unearthed from the Zhuyuan Valley in Baoji, the core chaplets in the blind cores of the feet fell off and showed round holes (BZM13:5); the core chaplets in the ear of the round Ding (BZM13:105) were exposed, indicating that the tripod's ears are hollow [39]. In order to increase the strength of some blind cores, there were wooden core bones in the blind cores at the latest in Western Zhou Dynasty, and iron or copper core bones might be used in the Spring and Autumn Period.

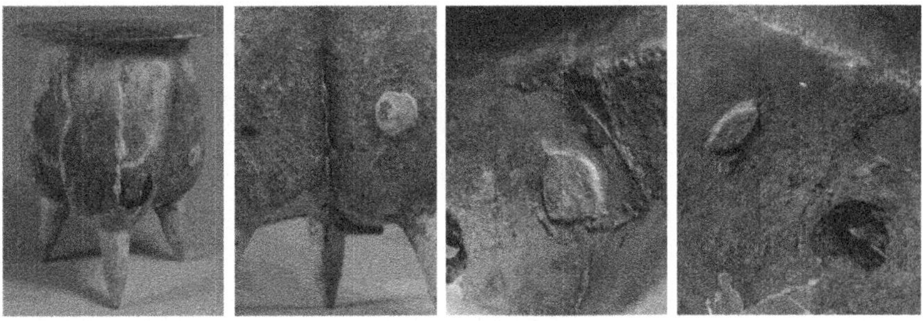

Fig. 7.43 Ge YWM6: 2 and abdominal core hole of Panlong

7.6.2 *Invention of Copper Spacer

Spacer refers to a kind of bronze sheet which is placed between mold and mold, or between mold and core, or between core and core in casting the mold of bronze wares, and a device to ensure the size of mold and support the suspended core, blind core and clay mold. Clearly, spacers are unique to the clay mold method, different from the chaplets used in the lost-wax method. Waste bronze fragments were often used as the spacers, and they were as thick as the wall of bronze ware to be cast.

The invention of spacer can be traced back to the Erligang Period. The spacer on Panlong City bronze wares first appear in the middle of the Erligang Period. Two spacers can be seen in the abdomen of Ge LZM2: 38, and two spacers at the feet. The X-ray film of Jia LZM2: 22 (Fig. 7.44) shows that there is a spacer at the Y-shaped fin intersection point at the bottom, with unclear boundaries and markings of fin on it, which were probably caused by the penetration of copper liquid outside the spacer. There are two spacers at the bottom of Pan LZM2:1. There are three spacers and three patches at the bottom of Lei LZM2:75, each patch covered by a spacer. Two spacers are visible at the bottom of Gu LZM1: 20, and three spacers may be used. There are four spacers at the bottom of Hu LZM1: 9 (Fig. 7.45).

Both spacers and core chaplets are used on the bronze wares in the fourth phase of Panlong. For example, there is a spacer at the center of the bottom of Jia LZM2:22, and rectangular chaplet holes, each of three feet, which reflects the transition from

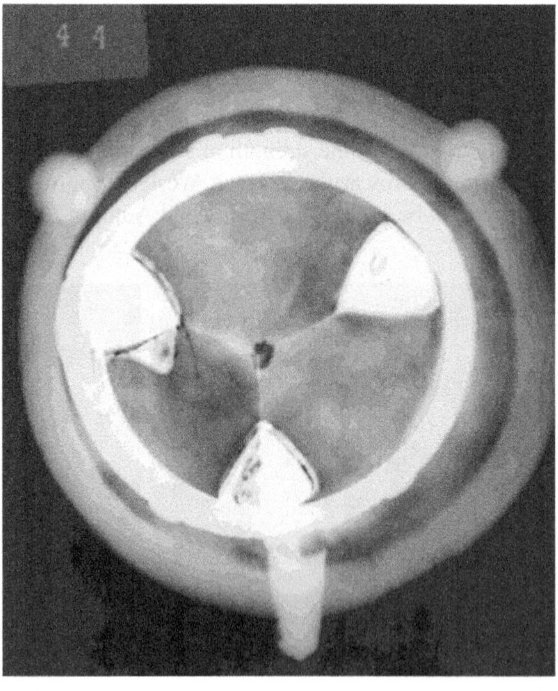

Fig. 7.44 Spacer at the bottom of Panlong City LZM2: 22 (X-ray)

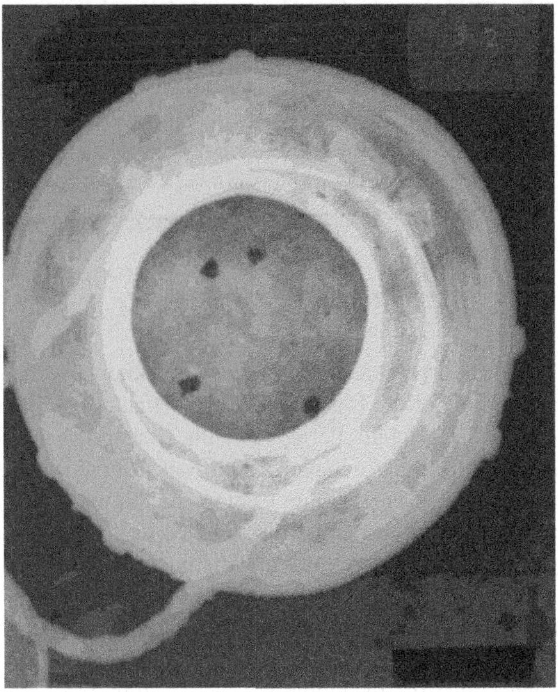

Fig. 7.45 Spacer at the bottom of Panlong City Pot LZM1: 9 (X-ray)

core chaplet to spacer. It can be considered that spacer was invented in this period. Spacers were mostly used in large wares among Panlong City bronze wares, such as Zun and Lei. Recent studies have revealed that copper spacers might originate in the south of China, and then spread to the Central Plains, becoming a key technology in casting bronze wares there [40, 41].

Until the Yinxu period in Anyang, spacers were not used on bronze containers, such as some unearthed from the Heihe Road in Anyang. The bronze wares found in the Baoji Cemetery of the State and Cemetery of State Yan at Liulihe in Fangshan District of Beijing prove that spacers were widely used in early Western Zhou Dynasty [42], tended to be abused in the middle and late Western Zhou Dynasty and got used rationally in late Spring and Autumn Period. In the development of the technology of casting bronze by the mold method in ancient China, the invention of spacer was inevitable. By using spacers, the craftsmen could effectively control the wall thickness of bronze wares and prevented some defects such as mold collapse, movable molds and dislocation of movable mold core, but later they tended to rely on it and even abuse it [43].

7.7 *Separate Casting and Attaching

In the technical system of mold casting, separate casting is the best method for casting large or complex bronze wares. The post-casting method was invented earlier than the pre-casting method; the former appeared in the middle Erligang Period and the latter in the late Erligang Period.

*7.7.1 *Invention of Post-Casting*

The earliest post-casting bronze wares known so far were found in Tomb Lijiazui No.1 Tomb of the Longcheng City—Huangpi. All the four bronze wares belong to the middle Erligang Period. The flat C-shaped handle of Jia LZM1:12 has a ridge-shaped fin on the outside, which looks obviously separate from the abdomen, so it was formed by separate casting. At the joint of the handle and the abdomen, a rivet head similar to patch is attached to the inner surface of the abdomen, with a gap between the rivet and the abdomen. At the joint of the handle and abdominal wall inside the abdomen, there is also a protrusion similar to patch. It is big upper and small lower, irregular in shape and distinctly separated from the abdominal wall (Fig. 7.46), which shows that the handle is cast to the abdomen later, and to make this connection firm, the method of cast-riveting post-cast is adopted, since the abdomen wall is only 1.5 mm thick. The ribbon under the handle is broken and forms the blank, which indicates that the post-cast handle is not the initial design; but due to its being broken or other defects, it was replaced by post-casting.

The flake C-shaped handle of Jia LZM1:13 also has a ridge-shaped fin on the outside. The residue of sprue on it shows that the sprue was not polished after being knocked off. There are obvious casting marks at the bottom of the handle (Fig. 7.47), indicating that the handle is cast separately and post-cast. The complete

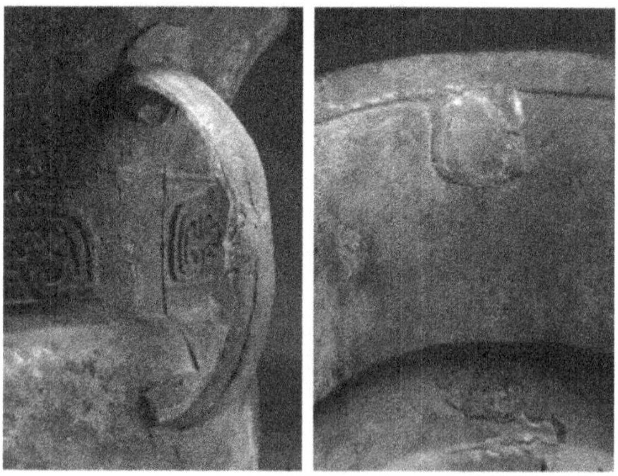

Fig. 7.46 Cast riveting-style post-casting of Panlong City LZM1:12

Fig. 7.47 Separate casting of Panlong City LZM1:13

convex strings at the bottom of the handle indicate that the separate casting of it is deliberately designed.

Separate casting of the Hu LZM1:9 is embodied in the casting seal of the ring gap connecting the cover button and the beam button (Fig. 7.48). The chain section was

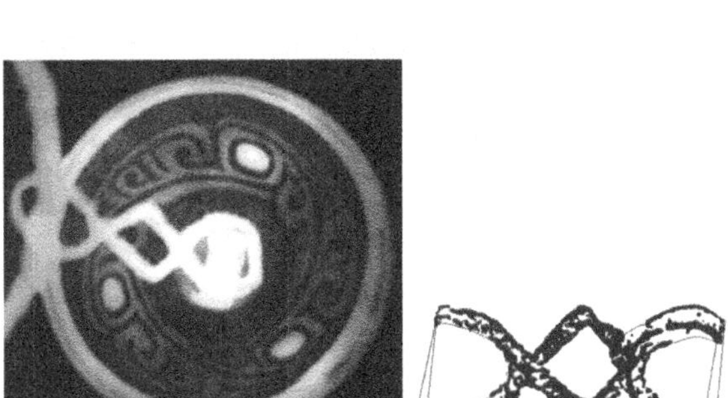

Fig. 7.48 Cover structure and chain section of Panlong City Pot LZM1:9

cast twice. It was first cast into a hollow piece with two forks at both ends. Then it was inserted into the cover button and made into a casting mold to seal the incision. The vertical casting fin on the chain joint is actually the trace of casting, which is closely related to recasting. The two ears of Gui LZM1:5 were separately cast and riveted into the ware [44].

7.7.2 *Invention of Pre-Casting

In the winter of 1973, Phoenix Jia of the late Erligang Period was unearthed from a tomb in Hejiacun Village, Qishan County. Its handle was post-cast in the method of riveting. The inner abdomen of Phoenix Jia is very special, and there are convex ridges with different dimensions. They are larger than the ridges (*fei leng*). There are no convex ridges on the inner abdominal wall corresponding to the ridges (*fei leng*) as the boundary of the animal facial marking unit. The six three-groups of convex ridges on the inner wall of the abdomen are special designs of casting and attaching the three-group of ridges (*fei leng*) as the beast's nose. Because the wall thickness is less than 2 mm, it is difficult to successfully cast the ridge (*fei leng*), hence such a special design (Fig. 7.49) [45], which is the earliest example of pre-casting known so far.

It was the invention and development of separate casting and attaching technology that became an important technical support for the booming of bronze technology in the Erligang Period and cast such a big square tripod that is 1 m high and 86 kilograms heavy earthed in Zhengzhou Shang City Site.

Fig. 7.49 Distribution and convex ridge of Phoenix Jia and its abdomen in Hejiacun Village, Qishan County

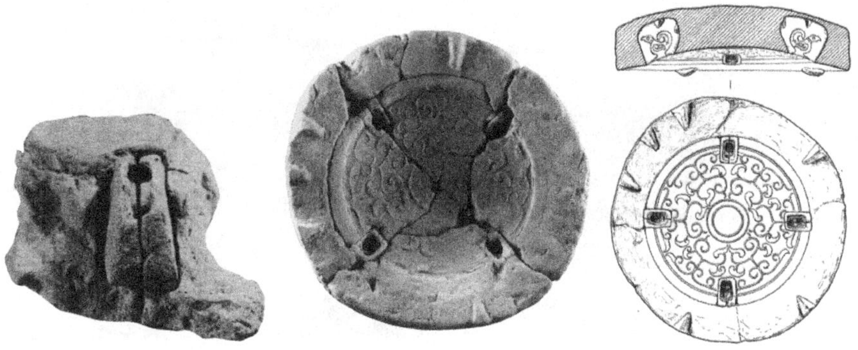

Fig. 7.50 Houma tripod cover casting IIT 24H 24: 7 (left, cited from Shanxi Institute of Archaeology: Houma Copper Casting Site. Cultural Relics Press, 1993) and tripod cover mold IIT 31F 13:11 (right, cited from Shanxi Provincial Institute of Archaeology: Houma Pottery Mold Art. Princeton University Press, 1996: p. 958, 959)

7.8 *Invention and Development of Movable Mold

The movable mold method is a technique of making separate molds and then combining movable molds with other molds and cores to form casting molds. It is also an important invention in the system of the mold method, which embodies the technological thought of "handling complexity with simplicity".

It is still unknown when this technology was invented. The bronze cup of the Erlitou Culture and the round carved animal heads on the shoulder of bronze Zun in the Erligang Period should be made in such a way. The *Fangzuo Gui* earthed from the Baoji National Cemetery at Zhifangtou Site, and *Boge Zun, You,* raised horns and round carving animal heads at the turn of handle excavated in Zhuyuangou Site were all made by movable molds [46].

The molds unearthed from the Houma Copper Casting Site (Fig. 7.50) show that the ring button on the tripod cover was also formed by remaking the movable mold into the movable casting mold and then inserting it into the casting mold of cover for pouring.

7.9 *Welding

Welding is an important metal processing technique, which is usually divided into large welding with copper as flux and small welding with lead-tin alloy as flux.

In the technical system of the mold method, welding was rare in the Shang and Western Zhou dynasties because of the invention of casting. Some scholars believed that welding appeared in the Shang and Zhou dynasties, though it was rare [47].

In the middle and late Spring and Autumn Period, the production of bronze wares underwent a transformation, and most accessories of bronze wares were separately

Fig. 7.51 Wangziwu Tripod SQUARE METERS: 38 of Xichuan Xiasi Temple and structure of its bottom and accessories (cited from Henan Institute of Cultural Relics, Xichuan County Museum, etc.): Chu Tomb of Spring and Autumn in Xichuan Xiasi Temple. Cultural Relics Publishing House, 1991, p. 114, p. 384, Fig. 1

cast, which was most represented by the bronze wares of Chu tombs in Xichuan Xiasi Temple. Among them, the Crawling Beast and accessories of Wangziwu tripod (Fig. 7.51) consisted of 38 castings (including 6 beast horns cast by the lost-wax method), which were cast 20 times and welded 14 times [48]. The accessories of the Sheng tripod and bell frame unearthed from the Tomb of Zeng Houyi of the early Warring States Period were welded to the main body [49, 50].

Because of its special function and special technical background, the casting technology of ancient Chinese bronze wares, with a nearly exclusive clay mold casting system and through a series of inherent inventions like clay chaplet, copper spacer, movable mold, casting and attaching (post-casting, pre-casting), welding and complex ways of combining casting with welding, has created brilliant Chinese bronze culture and became an immortal masterpiece in the history of human technology and art.

References

1. Xi'an Banpo Museum. 1988. *Ginger Village—Report on the Excavation of the Neolithic Age Ruins*, 148. Cultural Relics Press.
2. Rufen, Han, and Ke Jun. 2007. *The History of Chinese Science and Technology, Volume of Mining and Smelting*, 175–176. Science Press.
3. Benshan, Lu. 2007. *The History of Ancient Chinese Metal Mining and Coal Mining Technologies, Volume of Metal Mining*, 56, 57, 114, 125. Shanxi Education Press.
4. Benshan, Lu. 2007. *The History of Ancient Chinese Metal Mining and Coal Mining Technologies, Volume of Metal Mining*, 5. Shanxi Education Press.
5. Benshan, Lu. 2007. *Technological History of Ore and Coal Pit Mining—Mineral Ore*, 26, 58, 124. Shanxi: Shanxi Education Press.
6. Benshan, Lu. 2007. *Technological History of Ore and Coal Pit Mining—Mineral Ore*, 28–39, 58–71, 114 –121. Shanxi: Shanxi Education Press.

7. Benshan, Lu. 2007. *Technological History of Ore and Coal Pit Mining—Mineral Ore*, 74–75, 125–129. Shanxi: Shanxi Education Press.
8. Benshan, Lu. 2007. *Technological History of Ore and Coal Pit Mining—Mineral Ore*, 40, 41, 72, 73, 111, 112. Shanxi: Shanxi Education Press.
9. Benshan, Lu. 2007. *Technological History of Ore and Coal Pit Mining—Mineral Ore*, 43, 44, 76–78, 133, 134. Shanxi: Shanxi Education Press, 2007.
10. Benshan, Lu. 2007. *Technological History of Ore and Coal Pit Mining—Mineral Ore*, 41, 42, 78–80, 130–132. Shanxi: Shanxi Education Press.
11. Benshan, Lu. 2007. *Technological History of Ore and Coal Pit Mining—Mineral Ore*, 82–91, 135, 136. Shanxi: Shanxi Education Press.
12. Dongchuan Mining Bureau. 1959. *Dongchuan Local Copper Smelting Experience*, 5. Metallurgical Industry Press.
13. Zhidao, Zhang. 1959. *Slag and Ingredients for Copper Smelting*, 16–19. Metallurgical Industry Press.
14. Yanxiang, Li. 1995. *Research on Ancient Copper Smelting Slag in Tonglu Mountain and Jiuhua Mountains*, Ph.D. Dissertation, 26.
15. Rufen, Han, and Ke Jun. 2007. *The History of Chinese Science and Technology,•Volume of Mining and Smelting, Mining and Metallurgy Volume*, 191. Science Press.
16. Hubei Institute of Cultural Relics and Archaeology, et al. 1999. *Dengjiawan*. See *Xiaojia Roof*, 23. Cultural Relics Publishing House.
17. Rufen, Han, and Ke Jun. 2007. *The History of Chinese Science and Technology, Volume of Mining and Smelting*, 275–279. Science Press.
18. Henan Second Team of Institute of Archaeology of Chinese Academy of Social Sciences. Report on the Excavation of Linru Coal Mountain Site in Henan Province, *Journal of Archaeology* 1982 (4): 427–475.
19. Hubei Institute of Cultural Relics and Archaeology, etc. 2013. *Dalupu Site of Yangxin County*. Press of Cultural Relics.
20. Hubei Institute of Cultural Relics and Archaeology. 2010. Report on the Excavation of Crab Site in Daye City, Hubei Province in 2009. *Jianghan Archaeology* 4: 18–62.
21. Ying, Qin, et al. 2013. Detection and Analysis of Mining and Metallurgical Relics at Dalupu Site in Yangxin County, *Dalupu Site of Yangxin County*, compiled by Hubei Institute of Cultural Relics Archaeology and Huangshi Museum of Hubei Province, Cultural Relics Publishing House, 861–868.
22. Yanxiang, Li, et al. 2013. Preliminary Study on the Slag of Dalupu Site, Yangxin County. See Hubei Institute of Cultural Relics Archaeology and Huangshi Museum of Hubei Province. *Dalupu Site of Yangxin County*, 786–800. Cultural Relics Publishing House.
23. Benshan, Lu. 1994. Mineral Processing Technology in Shang and Zhou Dynasties and Its Simulation Experiment. *Chinese Science and Technology History* 4: 55–64.
24. Huangshi Museum. 1999. *Tonglu Mountain Ancient Ore—Smelting Site*, 145–154. Cultural Relics Publishing House.
25. Benshan, Lu, and Hua Jueming. 1981. Research on the Restoration of the Bronze Furnace of the Copper-Copper in Tonglu Mountain of Spring and Autumn. *Cultural Relics* 8: 40–45.
26. Protection and Management Committee of Tonglu Mountain Bronze Mine Site in Daye City. 2013. *The Memory of Preface to Tonglu Mountain Bronze Mine Site*, X-VII. Science Press.
27. Huangshi City Museum. 1999. *Tonglu Mountain Ancient Mining and Metallurgy Site*, 54, 103, 106, 124, 136. Cultural Relics Publishing House.
28. Yongguang, Yang, et al. 1980. Study on Mining Methods of Bronze Mine in Tonglu Mountain. *Nonferrous Metals* 4: 87.
29. Yuanchang, Yu, et al. 1985. Copper-iron Deposit in Tonglu Mountain in Daye County, 68–69. Hubei Province. Hubei Provincial E-Southeast Geological Brigade.
30. Авеисян, Х.К. 1957. *Bronze Metallurgy*, 347. Metallurgical Industry Press [Russia].
31. Rufen, Han, and Ke Jun. 2007. *The History of Chinese Science and Technology, Volume of Mining and Smelting*, 282. Science Press.

32. Rufen, Han, and Ke Jun. 2007. *The History of Chinese Science and Technology, Volume of Mining and Smelting*, 285–289. Science Press.
33. Rong, Chen, and Zhao Kuanghua. 1994. Study on Sulphur and Copper Smelting in Tongling Area in the Pre-Qin Period. *Research on the History of Natural Science* 2: 139–144.
34. Jueming, Hua. 1994. *Ancient Metal Technology in China*, 74–79. Elephant Press.
35. Rufen, Han, and Ke Jun. 2007. *The History of Chinese Science and Technology, Volume of Mining and Smelting*, 304–308. Science Press.
36. Haiping, Lian, Tan Derui, and Zheng Guang. 2011. Study on Copper Casting Technology in Erlitou Site. *Journal of Archaeology* (4): 569, 572–574.
37. Haiping, Lian, Tan Derui, and Zheng Guang. 2011. Study on Copper Casting Technology in Erlitou Site. *Journal of Archaeology* (4): 569–570, 567–568.
38. Yu, Liu., et al. 2008. The Manufacturing Technology of Bronze Ritual Wares Casting Molds Unearthed in Yinxu. *Archaeology* 12: 86.
39. Rongyu, Su, et al. 1988. Investigation of Bronze Wares Casting Technology and Detection of Metal Artifacts in Cemeteries of the State. See Lu Liancheng and Hu Zhisheng. *Baoji Cemetery of the State*, 597–605. Cultural Relics Publishing House.
40. Rongyu, Su, et al. 1997. Research on Casting Technology of Bronze Wares of Shang Dynasty Tomb at Dayangzhou Township, Xingan County. See Jiangxi Institute of Cultural Relics and Archaeology, etc. *Shang Dynasty Tomb at Xingan County*, 294–296. Cultural Relics Publishing House.
41. Rongyu, Su. 2012. Research on Casting Technology of Shang Dynasty Bronze Wares at Dayangzhou Township, Xingan County. See *Moji - Selected Collection of Su Rongyu*, 110–114. Shanghai People's Publishing House.
42. Rongyu, Su, et al. 1988. Investigation of Bronze Casting Technology and Detection of Metal Articles in National Cemetery. See Lu Liancheng and Hu Zhisheng. *Baoji National Cemetery*, 564–566. Cultural Relics Publishing House. Zhou Jianxun. Several Discussions on the Casting Technology of Bronzes in Shang and Zhou Dynasties. See *Cemetery of State Yan at Liulihe.* Cultural Relics Publishing House, 1995: 260–264.
43. Rongyu, Su, and Hu Dongbo. 2014. The Use and Abuse of Spacers in Casting Golden Tripod in Shang and Zhou Dynasties. *The Journal of Rao Zongyi College*, 101–134.
44. Rongyu, Su, and Zhang Changping. 2014. Research on Casting and Attaching Technology of Panlong City Bronze Wares. *Paper of International Symposium on Panlong City and Yangtze River Civilization in Shang Dynasty*. Huangpi, Wuhan.
45. Rongyu, Su. 2014. Analysis of Casting Technology of Shangfeng Column Unearthed in Qishan County and Discussion of Related Issues. See Shanxi Institute of Archaeology and Shanghai Museum. *Collection of Papers of International Symposium on Rui State Treasures from Hancheng, Shaanxi Province and Feudal States Archaeology Research in the Zhou Dynasty*, 550–563. Shanghai Ancient Books Publishing House.
46. Rongyu, Su, et al. 1988. Investigation of Bronze Casting Technology and Detection of Metal Articles in National Cemetery. See Lu Liancheng and Hu Zhisheng. *Baoji National Cemetery*, 534–548. Cultural Relics Publishing House.
47. Pope, John A., and Rutherford John Gettens, James Cahill et al. 1967. *The Freer Chinese Bronzes, Volume I, Catalogue*, 284–289. Washington: Smithsonian Publication.
48. Shigang, Zhao. 1991. Bronze Casting Technology of Chu Tomb in Spring and Autumn Period in Xichuan Xiasi Temple. See Henan Institute of Cultural Relics, Xichuan County Museum, etc. *Chu Tomb in the Spring and Autumn Period in Xichuan County*, 385. Cultural Relics Publishing House.
49. Jueming, Hua. 1981. Smelting and Casting Technology of Bells' Components of Zeng Houyi. *Jianghan Archaeology* 1: 11–13.
50. *Tomb of Zeng Houyi (I)*. Cultural Relics Publishing House, 1989: 192–193.

Benshan Lu Born in Wuhan, Hubei in 1949. He has been engaged in archaeological excavations and archaeological research on mining and metallurgy in Tonglushan, Daye, Hubei Province and Tongling, Ruichang, Jiangxi Province for many years. Former curator and deputy researcher of the Tonglvshan Ancient Copper Mine Site Museum, the curator of the Fanyu District Museum in Guangzhou, and the current director of the Guangdong Beauty and Cosmetics Museum.

Research interests: Mining and Metallurgical Archaeology, History of Mining and Metallurgy.

Major works: He has published many papers, and participated in the writing of several monographs, such as *Mining Report of Tonglvshan Ancient Mining and Metallurgical Site, History of Chinese Science and Technology: Mining and Metallurgy, Research on Mining and Metallurgy Technology in Ancient China.*

Rongyu Su Born in Shanyang, Shanxi in 1962, graduated from Xi'an Jiaotong University with a bachelor degree in engineering in 1983. In the same year, he was admitted by Research Center on the History of Natural Science at China University of Science and Technology, and supervised by Professor Huajueming and Professor Lizhichao. In 1986, he got a Master of Science degree and worked at the Institute of Natural Science History of the Chinese Academy of Sciences. He was successively designated as Assistant Researcher, Associate Researcher, and Researcher. He is the former deputy director of the Research Office of Technology History. In 2003, he was appointed as a doctoral supervisor. In 2009, he was appointed as director of the Institute of Cultural Heritage Protection and Management of the Nanjing Academy of Arts. In 2013, he served as vice president of the Chinese Traditional Arts and Crafts Research Association. From 2000 to 2001, he was a Forbes researcher at the Fleur Museum of the Smithsonian Institution. Since then, he has been a visiting professor at the Eberhard Karls Universität Tübingen and the Technische Universität Berlin in Germany. From 2000 to 2004, he served as executive director and secretary general of the Chinese Academy of Science and Technology History. In 2004, he was the director of the Center for Traditional Arts and Cultural Relics Technology of the Chinese Academy of Sciences.

Major works: Investigation of the Bronze Casting Technology and the Detection of Metal Utensils in Chinese Cemetery; Study on Bronze Casting Technology of Zhongshan King Tomb in Pingshan; Chinese Ancient Metal Technology; Study on the Casting Technology of Bronze Wares in the New Shang Dynasty Tomb; Preliminary Study on the Bronze Casting Technology of Sanxingdui in Guanghan; Chinese Early Wax Loss Method and Its Possible Origin; Grinding- A Collection of Su Rongyu's Works; On the Bronze Casting with Revit Structure.

Awards: The Outstanding Science and Technology Book Award in 1996; the second prize for the cultural relics science and technology achievements granted by the State Administration of Cultural Relics.

Chapter 8
Pig Iron-Based Steel Smelting Technology

Jueming Hua and Xing Huang

Pig iron smelting and casting technology was first invented in China, and then a series of relevant major inventions and creations emerged, such as iron mold, softening cast iron technique, co-fusion process of steel-making and giant iron castings. The advanced steel smelting technology of China with its own characteristics has created a splendid history of over 2000 years.

8.1 Iron Smelting as an Epochal Invention

Iron ore deposits are widely distributed on the earth, and they account for about 5% of the total crustal weight, second only to aluminum among various metals. Pure iron is soft, while pig iron is brittle and hard. Through carburization and decarbonization, iron can be transformed into steel or malleable iron, which have much better strength and toughness than bronze and can be adjusted in a wide range through quenching and tempering.

These characteristics and advantages of iron such as high quality, low price and strong adaptability, which were incomparable to other metal and non-metal materials in ancient times, enabled iron to become a good material for making various tools, blades, components and daily appliances. Only iron eventually replaced stone and bronze in ushering in the Iron Age for human beings, a time of greater development of social economy and culture.

Iron was the most important raw material used in early human life. The invention of iron smelting is an epochal event across the world.

J. Hua (✉) · X. Huang
The Institute for History of Natural Sciences, Chinese Academy of Sciences, Taiwan, People's Republic of China
e-mail: huajueming@vip.sina.com

J. Hua and L. Feng (eds.), *Thirty Great Inventions of China*,
https://doi.org/10.1007/978-981-15-6525-0_8

8.2 Chinese Ancient Iron Smelting Techniques Based on Pig Iron

Based on the technical conditions of the ancient times, there were two common ways to reduce iron ore to iron. One is low-temperature solid reduction, which uses wood or charcoal as fuel and reducing agent to reduce iron oxide to loose sponge iron at a lower temperature (about 1,000 degrees Celsius) in a bowl furnace or a lower shaft furnace. Through forging and slag extrusion, sponge iron was transformed into wrought iron, and steel was made from bloomery iron by repeated forging and carburizing.

The other method is high-temperature liquid reduction, which uses charcoal or firewood and coal as fuel and reducing agent to reduce iron oxide to liquid pig iron at a high temperature. The liquid pig iron is then discharged from the furnace to be cast into ingots or wares. Steel can be obtained by decarbonized heat treatment of pig iron, or by stir-frying pig iron into wrought iron and fusing the two kinds of irons together.

The Hittites living in Asia Minor were the first to invent the technique of iron smelting in the early 2,000 B.C., and their method of smelting iron is bloomery iron process. Later, this method spread to Europe, North Africa and other places, and lasted until the thirteenth century.

Bloomery iron was also used in China in the early years. Shortly afterwards, the ancient Chinese invented the technique of pig iron smelting and casting, which then became the mainstream. Pig iron-based smelting is the greatest feature and advantage of ancient Chinese iron and steel smelting technology, and also the source of a series of major inventions and technological achievements which ensued.

8.2.1 Pig Iron Smelting

Jade handle iron sword (Fig. 8.1), copper dice iron leaf spear and copper inner iron dagger unearthed in the Cemetery of the Nation Guo in Sanmenxia are the earliest known artificially smelted iron products in China; they were made of bloomery iron and bloomery carburized steel in the late Western Zhou Dynasty (about the eighth century B.C.). The earliest known pig iron products are two fragments of iron wares

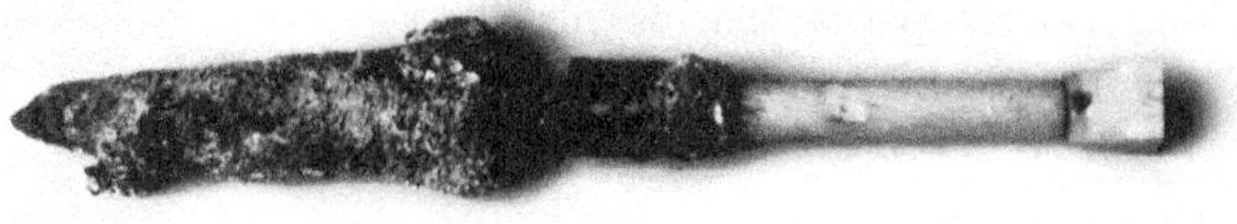

Fig. 8.1 Jade handle iron sword (the Cemetery of the Nation Guo in Sanmenxia, Henan, Late Western Zhou Dynasty)

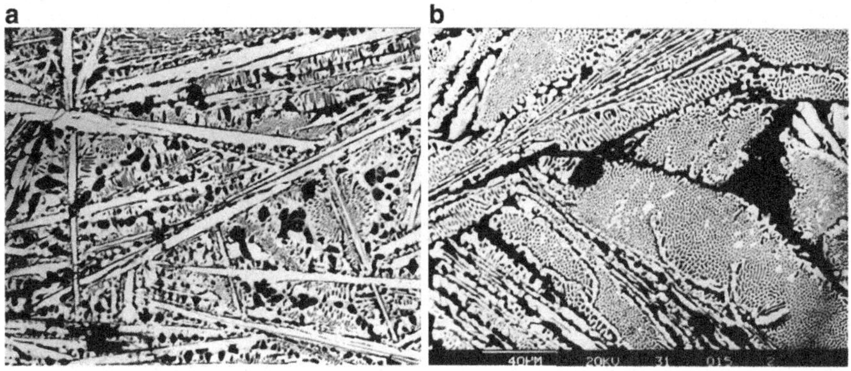

Fig. 8.2 Metallographic structure **a** and hypereutectic white iron **b** (the Cemetery of the Nation Guo in Sanmenxia, Henan, Spring and Autumn Period)

unearthed in the Tianmaqu Village, Shanxi, which were made of hypereutectic and eutectic white iron (Fig. 8.2) in the early and middle Spring and Autumn Period (about the eighth and seventh century B.C.) [1].

At present, the earliest shaft furnace for pig iron smelting has been found in the Jiudian Village of Xiping County, Henan, and it is a relic of the late Warring States Period. After research and recovery, it is known that the furnace body is about 4 m high and its capacity is about 5 cubic meters. Above the hearth, there is a distinct belly angle, and the tuyere is located in the middle of the hearth. The bottom of the furnace is equipped with air channel (Fig. 8.3). Such a furnace design and horizontal blast mode could effectively reduce erosion on the wall, prevent the furnace charge from being too compact, enable air flow to reach the central of the furnace and keep the hearth active. The setting of air channel could help hot metal insulation and prevent hearth accidents. Similar furnace design was also adopted in the western shaft furnaces of the modern times [2].

In order to increase the output of iron, large oval shaft furnace was built in the West Han Dynasty. For example, the hearth of the furnace unearthed in Guying Town has a diameter of 2.3– 3.1 m and a capacity of over 20 cubic meters (Fig. 8.4) [3]. However, due to various constraints, medium and small circular shaft furnaces with a height of 35 m and an inner diameter of 1.5–2 m were still used in Han Dynasty and a short period after it.

The Song and Liao Dynasty was another peak of shaft furnace smelting. The furnace body was built with stone blocks; its shape curve was more distinct, and the furnace capacity increased. The shaft furnace of the Northern Song Dynasty unearthed in Kuangshan Village, Wu'an, Hebei is 6.4 m high, and it is the highest existing ancient shaft furnace in China (Fig. 8.5) [4].

The industry of iron smelting kept increasing in size during the Ming and Qing dynasties, when the large shaft furnace had a height of 6–10 m and a daily of iron output of 20 tons. In the late 1950s, Huang Zhanyue and Wang Daizhi inspected the traditional iron smelting furnaces in Luoci, Yunnan (Fig. 8.6). The traditional

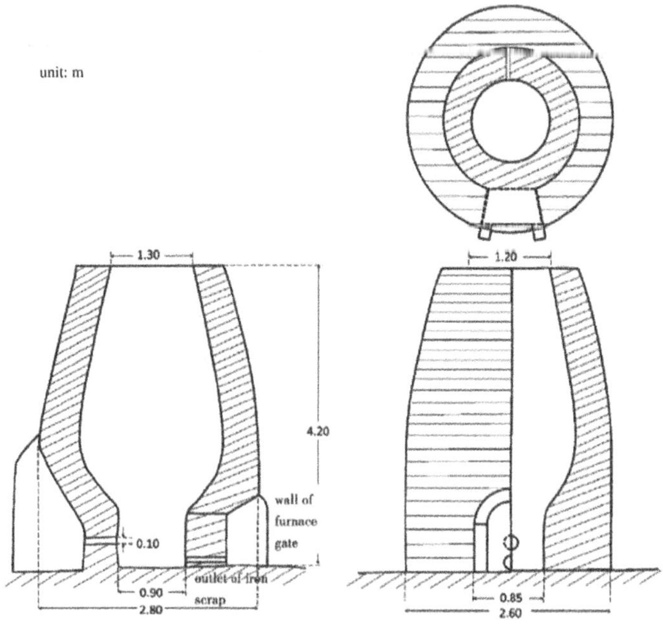

Fig. 8.3 Restoration of shaft furnace of the Late Warring States Period in Jiudian Village, Xiping County

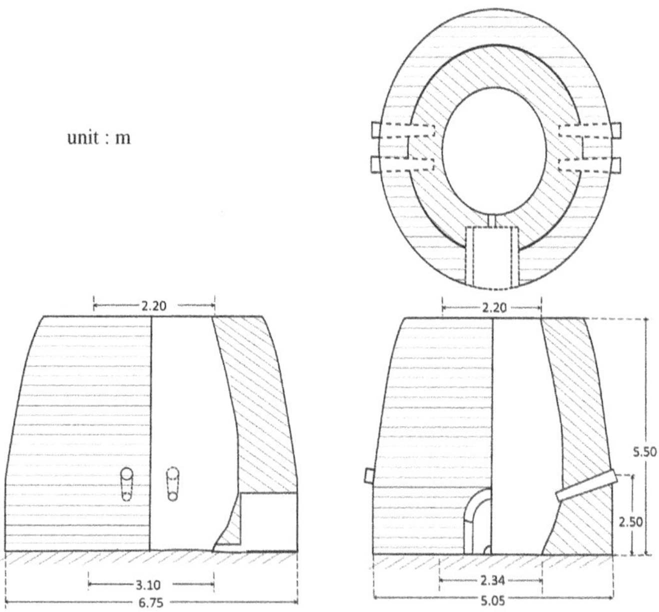

Fig. 8.4 Restoration of No.1 furnace in Guxing Town, Zhengzhou

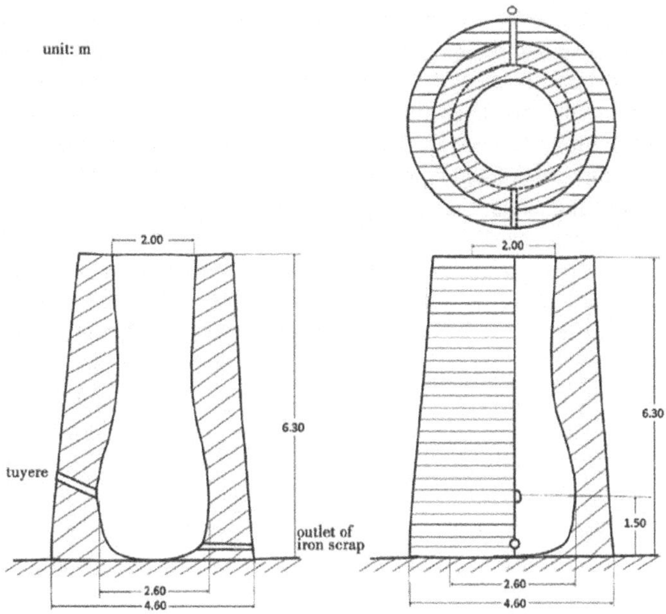

Fig. 8.5 Restoration of shaft furnace of the Song Dynasty in Kuangshan Village of Wu'an

Fig. 8.6 Shaft furnace in Luoci, Yunnan (cited from *Archaeology,* No. 7, 1962)

iron smelting furnace was square cone-shaped, with a total height of 6 m and a diameter of 2 m. The hearth and door were built with refractory stone, and at the bottom of the furnace there was a "十" shaped wind channel. Hydraulic blast was used, and the length of cylindrical bellow was 4 m. The furnace body was made of rammed earth, with salt-added refractory soil inside and large frame outside. Iron was produced every four hours, with an output of 200–300 kg and an annual output of 400–600 tons [5]. There were many small shaft furnaces with the function of dumping, commonly known as "tilting furnace." Since the end of the Ming Dynasty at the latest, the tilting furnace at Yangcheng County, Shanxi had been able to smelt pig iron to directly cast ploughboard in the iron mold. As a masterpiece of pig iron smelting and casting technology, this project was approved by the State Council to be listed into the first group of national intangible cultural heritages in 2006.

Reasonable design and structure of the furnace type were the precondition of shaft furnace running smoothly. In order to liquefy slag and iron and carry out continuous production, strong blast was necessary. The bosh temperature should be kept between 1,200 and 1,400 degrees Celsius. Soft melting zone should be formed in this position, so that the slag and iron droplets fall into hearth and gather. As the folk proverb says, "Where there is wind, there is iron." Blast air bags and multiple blowers might have been used as early as in the sixth century (Figs. 8.7 and 8.8). In the Han Dynasty, driven by water and animal power, the blowers could greatly raise air pressure and volume to increase furnace capacity and the output of iron and reduce the cost.

The piston wooden blower is more efficient, and it can be made bigger and stronger to produce higher air pressure and larger air volume. Wooden fans first used in the Tang Dynasty were the earliest among such blowers. Their shapes were first seen in *Wu Jing Zong Yao* of the Northern Song Dynasty, *Ao Bo Tu* by Chen Chun in the Yuan Dynasty and *Agriculture Book* by Wang Zhen (Fig. 8.9) [6].

The dual-acting piston bellow, which originated in the Song Dynasty, was a major improvement of blast equipment. Characterized by continuous air supply during the

Fig. 8.7 Iron smelting on the stone relief of the Eastern Han Dynasty in Hongdaoyuan of Teng County

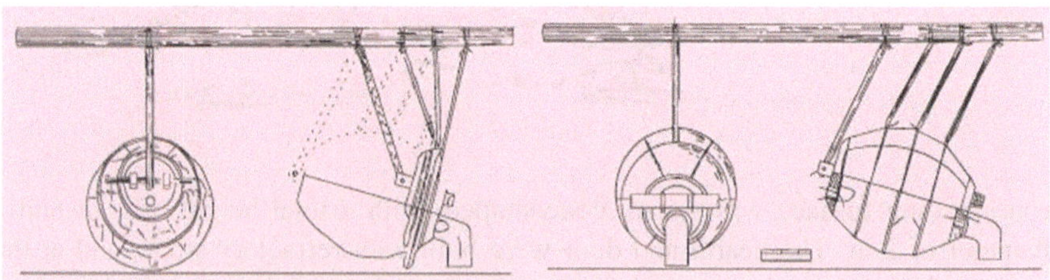

Fig. 8.8 Restoration of blast air bag of the Eastern Han Dynasty

Fig. 8.9 Wooden fans

push–pull process, this kind of bellow greatly improved the working efficiency. It includes square bellow and cylindrical bellow. The cylindrical bellow could withstand greater pressure, and there was no cracks on the whole cylinder body except the interfaces at the two ends. So with guaranteed air tightness, the cylindrical bellow is more used in metallurgical sites (Fig. 8.10) [6].

Charcoal was used as fuel and reducing agent in shaft furnace in ancient times. In his *Shui Jing Zhu,* Li Daoyuan quoted a saying that "The mountain which is 200 *li* to the north of Quci (name of an ancient place) could see fire at night and smoke in the day; the stones and charcoals there were taken by people to suffice the use of thirty-six countries." Many scholars cited this quotation as the basis for use of coal in iron smelting during the Northern Wei Dynasty. Coal mining and utilization were booming in the Song Dynasty. During the reign of Emperor Zhao Xu, coal mine was discovered in Xuzhou, and this event was specially recorded in a poem composed by Su Shi, a renowned poet in Chinese history. The poetic lines in this poem told people that it was vital to protect forest resources and alleviate fuel shortage by using coal in iron smelting. The Song Dynasty iron ingots unearthed in the Tangpo Site, Anyang, Henan contain 1.075% sulfur, which is probably smelted with coal [7]. Coke was used in iron smelting at the latest in the Ming Dynasty. Fang Yizhi's *Wu Li Xiao Shi,* Li Yi's *Jie An Man Bi* and Sun Tingquan's *Yan Shan Za Ji* describe the procedures of

Fig. 8.10 Cylindrical double-acting wooden bellows (the Qing Dynasty)

coke-making and its advantages in use. With high mechanical strength and calorific value, coke is the best fuel for smelting iron by shaft furnace, so coke-making is an invention of great value in the Ming Dynasty.

The shaft furnace smelting technology had matured by the Han Dynasty. The ore grade used was mostly 45–60%, the particle size was 1–5 cm and limestone or dolomite is used as flux. At high temperatures, calcium, magnesium ions and silicon dioxide in ores combined to form silicate with low melting point, forming liquid slag, which was discharged from the slag hole, thus realizing the continuous production of pig iron. The remains of iron slag were mostly flaky, glassy and low alkalinity, proving its good fluidity for the easy separation of slag and iron. Fluorite (CaF_2) was also used as flux in iron smelting in Zunhua, Hebei, which could lower the melting point and help the furnace work smoothly. For details, see Zhu Guozhen's *Yong Zhuang Xiao Pin* and Sun Chengze's *Chun Ming Meng Yu Lu*.

Crucible iron-making is another form of high-temperature liquid reduction method. Originally invented in China, this technique became popular in Shanxi, Henan, Hebei, Shandong, Liaoning and other provinces in Qing Dynasty, especially in Jincheng of the Taihang Mountainous Area. *Xianfeng Qingzhou Prefecture Annals* said that in 1663, the second year during the reign of Emperor Kangxi, Sun Tingquan summoned the ironworkers of Shanxi to practice the method of crucible ironmaking. Field investigation showed that the furnace body was square (Fig. 8.11), each crucible had a height of 30 cm and contained rich iron ore, anthracite and black soil (inferior pulverized coal used as flux), and about 100 crucibles could be placed in each

Fig. 8.11 Square furnace iron-making scene (quoted from Richthofen's *China*, 1882)

furnace, using anthracite as fuel. The furnace temperature was up to 1450 degrees Celsius, and the air blasting time is one day and one night [8]. In 1916, the Institute of Geological Investigation counted a total of 170,000 tons of pig iron smelted by traditional methods in China within that year, of which 70,000 tons (mostly crucible iron) were produced in Shanxi Province, accounting for 41% of the total.

8.2.2 Iron Mold Casting

Clay mold was used mostly in early iron casting. Then during the Warring States Period, iron mold was applied for iron-making in batches, hence a major innovation in the technology of iron smelting. The representative works are "Xinglong Iron Molds" produced in Xinglong County, Hebei (Fig. 8.12). There are 42 pieces in six categories, including molds for axes, adzes, hoes, double chisels, double sickles and vehicles cast by the smelters of the official workshop "Zuolian" in State Yan (1044–222 B.C.) [9]. Similar molds were also found in Cixian County, Shijiazhuang in Hebei and Xinjian District of Nanchang in Jiangxi.

During the Han Dynasty and the Wei, Jin and Southern and Northern Dynasties, the range of both the categories and the application of iron molds had been continuously expanded. They shared the same advantages: the contour of the iron mold was consistent with the shape of the castings; the wall thickness was even, which was conducive to heat dissipation, thus prolonging the service life; there was a handle on the back of the mold, which was easy to hold and increases the

Fig. 8.12 Double sickles mold (unearthed in Xinglong County, Hebei)

rigidity; and the setting of iron cores was of high craftsmanship, which was difficult to achieve even in modern times. A good example can be found in an ironwork cellar of the Northern Wei Dynasty in Mianchi County, Henan, where 152 iron molds were stored for ploughs, double-handled ploughs, shovels, hoes, axes, sickles, arrowheads, hammers and plates [10]. More importantly, during this period, the material of iron mold was greatly improved, ranging from Xinglong's hypereutectic white iron, to Laiwu's tinplate in the early Western Han Dynasty and to Mianchi's gray iron material with better property (Fig. 8.13). As early as in the Western Han Dynasty, gray

Fig. 8.13 Metallographic structure of iron mold, unearthed in Zhenping County, Henan (gray iron, the Eastern Han Dynasty, cited from *Archaeology* No. 3, 1982)

iron, which is superior to white iron in properties, was obtained by people, and later was used for making molds, which is another important achievement of iron casting technology.

8.2.3 Cast Iron Softening

The iron produced in shaft furnace was mostly white iron in the initial stage, and the carbon in it existed in a combined state (iron carbide), making the iron very brittle and hard. If the carbon is oxidized and migrated outward under continuous high temperature and oxidizing atmosphere, or if it is precipitated in the form of graphite under a neutral atmosphere, the white iron can be denatured into white-heart or black-heart malleable cast iron with good toughness and strength.

It is based on the above principle that as early as in the Warring States Period, shortly after the invention of pig iron smelting, Chinese casters obtained white and black core malleable iron through the softening treatment technique (Figs. 8.14 and 8.15). Inspections to the excavated objects show that the iron wares of this period, especially the farm tools, were unearthed in such cities like Luoyang, Nanyang and Huixian of Henan, Changsha of Hunan, Yi County, Wuan, Shijiazhuang of Hebei and Huangshi of Hubei. In the Han and Wei dynasties, thanks to the continuous improvement of softening cast iron technique, white-heart malleable cast iron without being completely decarbonized was seldom seen in ironmaking. What's more, this technique could help to obtain malleable cast iron with spheroidal graphite, which was good in mechanical properties and comparable to the modern nodular cast iron in terms of graphite properties (Fig. 8.16) [11].

The process of softening the cast iron is carried out in batches in an annealing kiln. Carbon content of charcoal pig iron in early stage was mostly near eutectic point (C4.3%), and other elements were low in content, which was helpful to graphite precipitation. The blank of malleable cast iron must be white iron. If graphite is

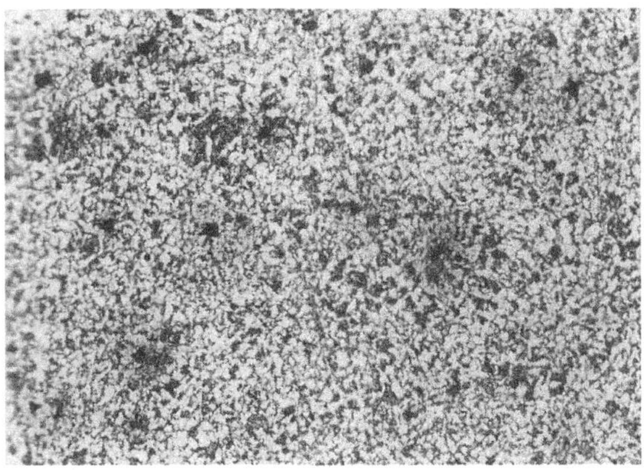

Fig. 8.14 The metallographic structure of ring-handled long sword from the Tunliuhe Tomb, Changzhi, Shanxi (white-core malleable cast iron, mid-late Warring States Period, edited by Han Ruyu and Ke Jun. *Chinese Science and Technology History, Mining and Metallurgy*. Science Publishing House, 2007, p. 433, Fig. 7-3-21)

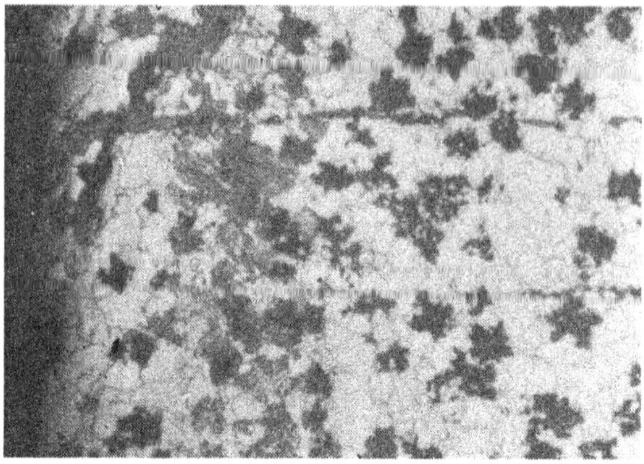

Fig. 8.15 The metallographic structure of iron from the smelting and casting site of Wafangzhuang, Nanyang, Henan (malleable cast iron with spherical graphite, Eastern Han Dynasty, cited from *Natural Science Research*, No. 1, 1982, Fig. 4)

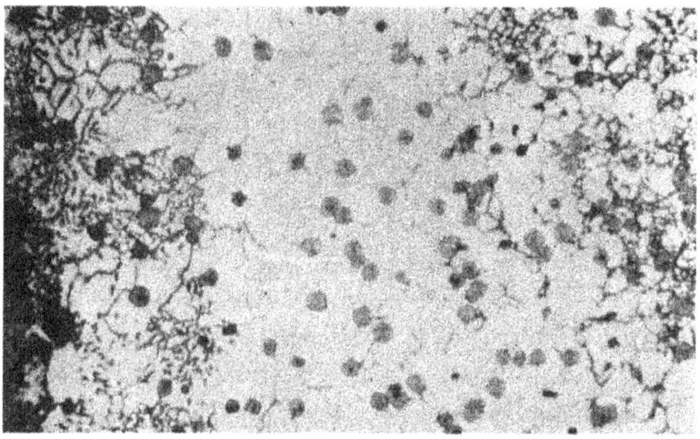

Fig. 8.16 The metallographic structure of technetium from the iron smelting site of Tieshenggou Village, Gongyi, Henan (malleable cast iron with spheroidal graphite, cited from Discussion on Metallographic Inspection and Manufacturing Process of Iron Shovel in Han Dynasty, *Ductile Iron*, No. 1, 1979.)

precipitated from the blank, it will grow up after annealing, deteriorating the performance of the work piece or even scraping it. White iron was guaranteed by iron mold casting, and the process of graphitization could be accelerated and improved because of good heat conduction, which promoted the chilling of castings, grain refinement and strong decomposition tendency. It can be seen that the widespread use of iron mold is closely related to the popularization of cast iron softening. Tests to these castings showed that more than half of the cast iron farm tools in the Warring States Period and the Western Han Dynasty had gone through the softening treatment, proving that this technique played a huge role in the economic development of that period. Some foreign scholars ever questioned how China could enter the Iron Age in the Warring States Period when brittle and easy-to-break pig iron devices were used. This question has been answered by the above discussions on iron mold casting and cast iron softening technique.

8.2.4 Decarburization of Cast Iron into Steel

The formed plate-like and strip-shaped cast iron pieces were subjected to high-temperature decarburization treatment in an oxidizing atmosphere to obtain a metal structure of the steel, without precipitation or little precipitation of graphite. At the very beginning of the Warring States Period, China invented decarburization of cast iron into steel, a novel and unique steel-making technique. Although the heat treatment process was similar to the softening treatment of white-heart malleable cast iron, the processed objects were plates or strips instead of formed castings; the steel must be repeatedly forged to eliminate casting defects such as shrinkage cavity and shrinkage porosity to make metal structure more compact, resulting in high-quality steel castings. These two processes are different in forming concept and technique. The earliest known pottery mold for casting plates and strips was from the Yangcheng Site of the early Warring States Period in Dengfeng, Henan. It has been identified that 8 of the 10 pieces of iron castings produced by the site were decarburized into wrought iron and low-carbon or medium-carbon steel. Batches of plates and strips (Fig. 8.17) unearthed in Nanyang and Guying Town, iron shears found in Dongshima Village of Zhengzhou, and plates, steel farm tools and other tools excavated in Mianchi County all were made of this material, indicating that this invention played a major role from the fifth to sixth century B.C. [12].

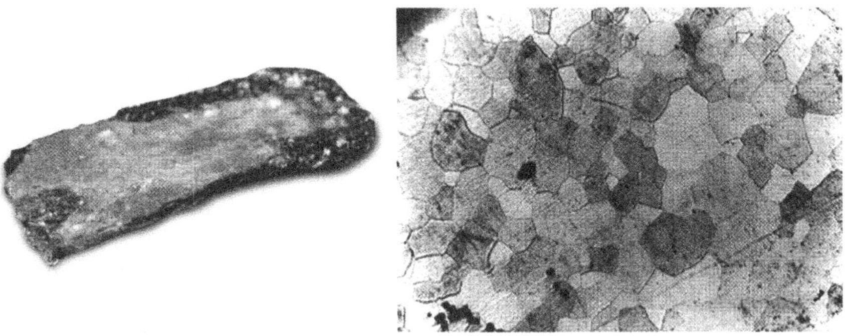

Fig. 8.17 The plate and its metallographic structure from the Guying Site of the Han Dynasty (cast iron decarburized steel, cited from *The History of Chinese Science and Technology, Volume of Mining and Smelting*, p. 611, Fig. 10.2.6)

8.2.5 Iron and Steel Stir-Frying

Using pig iron sheets and blocks as raw material, traditional method of frying iron is to heat them in a furnace until they smelt in a semi-solid state. After stir-frying, carbon in the iron is oxidized and the temperature rises; then silicon, manganese and other components are oxidized to form inclusions. With the decrease of carbon content and the increase of smelting point of iron, pig iron becomes solid wrought iron which can be usable after it being forged and slag being extruded [13]. Some scholars believe that low, medium or high carbon steels can be obtained if the technology of stir-frying is appropriately practiced [14].

As early as in the mid-Western Han Dynasty, China invented the technique of iron stir-frying, which was the result of its unique development of iron and steel technology. The long-term practice of softening and decarbonizing cast iron into steel led to the intended action of stirring pig iron at high temperature, making it oxidized and decarbonized into wrought iron. The earliest iron-frying furnace discovered is located at the Tieshenggou Smelting and Casting Site in Gongyi, Henan.

There are three traditional types of stir-frying furnaces: underground furnace, reverberatory furnace and pig-and-wrought-iron-making furnace, which were still used all across China until the first half of the twentieth century. The last one can be found in Song Yingxing's *Tian Gong Kai Wu·Hardware*: "If we want to make wrought iron, first we need to build a square pond connecting to the furnace, within several feet far and several inches lower, and with short brick walls against the two farthest sides. Getting some mud, we can dry it and pound it into powder so that some workers can hold the willow sticks standing on the wall in a row. When pig iron flows out into the pond, one of the workers is responsible for splashing mud powder rapidly on the flow, while the others keep stirring the stuff quickly. Wrought iron will be obtained immediately (Fig. 8.18)." At the end of the eighteenth century, Europeans made wrought iron out of liquid pig iron in a reverberatory furnace by using the technology of puddling. They puddled pig and wrought iron together in an open-hearth furnace to make steel. As recorded in *Tian Gong Kai Wu*, the obtained wrought iron, together with pig iron, was used to make steel by co-fusion method, which is similar to the European puddling method. Both methods can be categorized into the two-step steel-making process, and the ideas embodied in them were sophisticated in their times.

The method of iron frying is an epoch-making invention. It not only helped people to get a large number of cheap and easily accessible raw materials for making steel, but also directly led to the invention of co-fusion steel-making method, which is of great significance to the transformation of China from the early Iron Age to the complete Iron Age.

In the Jin and Sui dynasties, pig or wrought iron frying was recorded in books like *Xiahou Yang Cheng Jing*, *Ming Hui Dian*, *Wu Bei Zhi*, *Wu Bian*, *Yong Zhuang Xiao Pin*, *Shen Qi Pu* and *Guangdong Xinyu*, which is evident enough to show that it has a far-reaching influence as many forms of iron smelting derived from it through 2000 years of development and variation from the beginning of the first century.

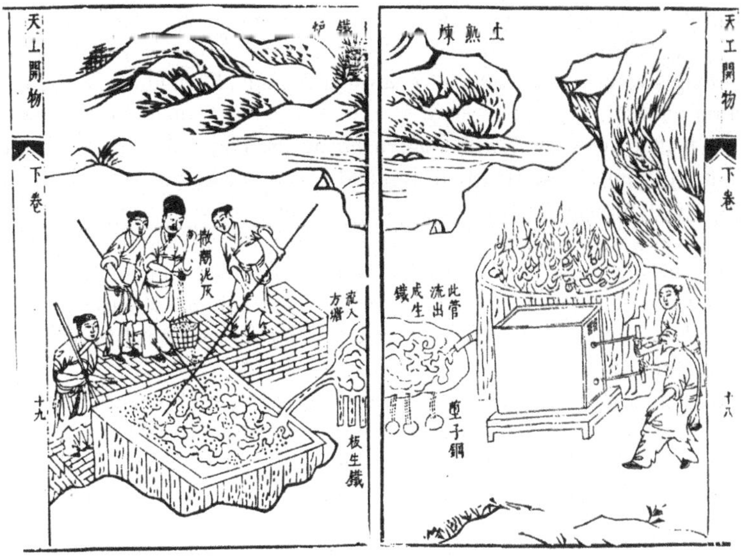

Fig. 8.18 The pig and wrought iron smelting furnace from *Tian Gong Kai Wu Hardware*

8.2.6 Well-Tempered Steel

Among many kinds of steel, the well-tempered steel is the most famous. Liu Kun (271–318) in Western Jin Dynasty wrote a well-known poetic line "Who could imagine that once the well-tempered steel girder would one day turn to be a soft thread twining the finger?" The phrases "steel-making by a thousand hammers" and "steel-making by well-tempering" have become popular idioms. Then "well-tempered steel" was naturally regarded as the best one.

Steel made by carburizing was improved in quality because of the increase of forging times. Thus it can be regarded as the predecessor of the well-tempered steel. After the invention of iron frying method, the well-tempered steel has made great progress by using the fried iron as material, and, accordingly, the forging times and folding levels increased. In his work *Wu Ku Fu*, Chen Lin (?-217) said "Talking about armors, the best ones, which are made of well-tempered steel, must be from the East Xiongnu and Quegong." This was in the late Eastern Han Dynasty. Cao Cao's *Bai Jie Ling* mentions "Weapons that cut everything." Cao Pi's *Dian Lun Jian Ming* also says, "Choose good metal, instruct craftsmen of their country to work, and refine it until it can cut through everything." *Taiping Yu Lan* quotes Tao Hongjing, "The governor of the Shu Area (today's Sichuan), Liu Bei (161–223), demanded Puyuan (a famous sword-maker) to make 5,000 swords (similar to European falchion), with small rings at the end of handles and inscription on the blade." *Jin Shu·He Lian Bo BoZaiji* records that the general ordered craftsmen forge swords out of well-tempered steel. To prove with physical items, there is the Sa Lian sword (thirty-tempered sword, lit. meaning the sword being tempered at least 30 times) forged in officially-built craftsman workshop in Guanghan Prefecture, Sichuan in the year of 104, as recorded in Luo Zhenyu's *Zhensong Tang Ji Jin Tu*. In 1974, the ring-handled long sword was unearthed in Cangshan County, Shangdong, with the gilded inscription "永初六年五月丙午造卅湅大刀吉羊宜子孙" on it. After examination, it is known that the sword, with martensite on the edge, was made by repeated folding and forging of the stir-fried iron and was processed by quenching [15]. In addition, fifty-tempered swords (Fig. 8.19) were unearthed in Tongshan of Xuzhou in Jiangsu. They were made by craftsmen from official workshop in Shu Prefecture in the year of 77, the second year of Jianchu Period. In Kumamoto, Japan, there are eighty-tempered swords unearthed, which were made in the early fifth century. Seven hundred-tempered swords of the late fourth century from Baekje are kept in Isonokami Shine, Japan.

To sum up, from the mid-Eastern Han Dynasty to the Southern and Northern Dynasties, the number of forging times was gradually increasing. The so-called "hundred refinements" is the essence of many refinements and forgings. The technology of making well-tempered steel was quite mature in the Eastern Han Dynasty and was widely used and highly praised in the following dynasties. It is recorded in many classic books like *Meng Xi Bi Tan*, *Ben Cao Gang Mu*, *Tian Gong Kai Wu* and *Hai Guo Tu Zhi*. The "Nine-tempered steel" recorded in the book *Ce Fu Yuan Gui* in the Five Dynasties is also categorized to this technology.

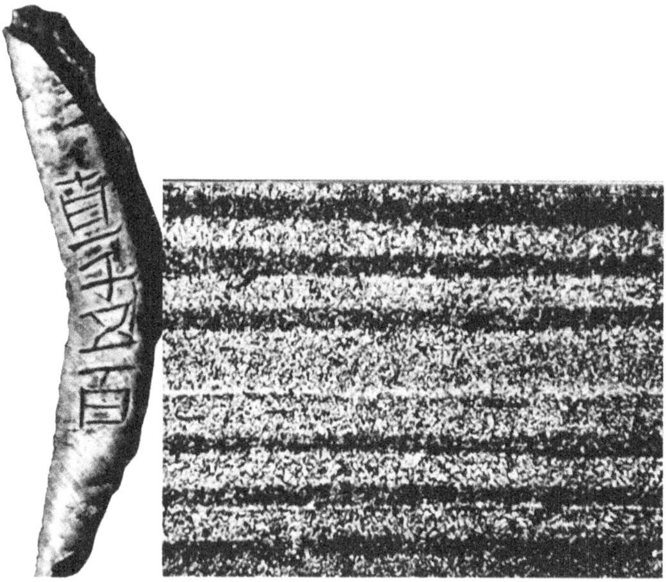

Fig. 8.19 Fifty steel knives and their metallographic structure from Tongshan, Xuzhou (cited from *History of Science and Technology in China*, Mine and Metallurgy Volume)

8.2.7 Co-Fusion Process of Steel-Making

Steel-making was weak in the early Iron Age in China. The invention of frying iron provided an opportunity to break through this bottleneck. According to the records in Wang Can's *Dao Ming,* the iron had been smelted and cast for many times before it was made into knives, which indicated that the technology of steel-making by fusing pig iron and wrought iron together had come into being in late Eastern Han Dynasty. Similar records could also be found in Zhang Xie's *Qi Ming. The Book of Northern Qi Dynasty* recorded how Qimu Huaiwen, an expert in iron and steel smelting in the Eastern Wei Dynasty, made a sword by repeatedly pouring liquid pig iron onto unforged wrought iron. It only took several days and nights for the formation of steel, which showed that this technique had become ever more mature.

The terms like "Tuangang" and "Guangang," referring to the co-fusion process of steel-making, are mentioned in *Meng Xi Bi Tan* written by Shen Kuo in the Song Dynasty. This is the first time that these terms appeared in ancient books. This technique could not only reduce the times of perfusing, but also use mud sealing for slag-making and protection, which was much better than the method adopted by Qimu Huaiwen. *Wu Bian* of Tang Shunzhi in Ming Dynasty has already recorded that pig and wrought iron sheets can be smelted and refined together, or pig and wrought iron are heated together and when the pig iron is melted, it will be rubbed into the wrought iron. *Tian Gong Kai Wu* and *Wu Li Xiao Shi* also have similar records about this technique, which has been prevalent from the Qing Dynasty to modern times in places like Jiangsu, Anhui, Hubei, Hunan, Sichuan and Fujian, where it is known as "Mogang" and "Sugang."

The invention, evolution and promotion of co-fusion process, which is a unique steel-making technology in China, has made up for the deficiency of steel-making technology in the past. It signifies the availability of a whole range of steel products by the handicraft production mode in the natural economic structure and completes the traditional steel-making system based on pig iron.

8.2.8 Inlaid Steel, Added Steel and Pig Iron Drenching

The main body of the Huanshou Sword, which was unearthed from the tomb of Han Dynasty in Laoheshen Village, Yushu, Jilin, was made of low carbon steel while its edge was made of steel with 0.7% carbon. On the sword, there is a clear boundary between the two kinds of steel. It is an early inlaid steel product [16]. *The History of the Northern Dynasties* records that Qimu Huaiwen made the ridge of a sword with pig iron and the blade with steel, which is the earliest known documentary record of inlaid steel. *Meng Xi Bi Tan* says, "The ancients usually used hard iron as the blade while pig iron as the body to avoid its being broken." *Tian Gong Kai Wu* describes the parts of swords, axes, planers and chisels made of steel, and also depicts the methods. Added steel and inlaid steel are both synthetic materials, which are tough

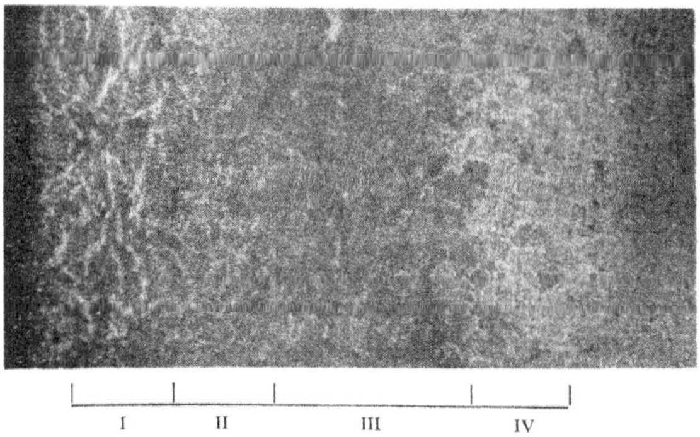

Fig. 8.20 The metallographic structure of transition layer of the hoe drenched with pig iron (from right to left, the carbon content increases gradually. *The Collection of History of Science, 1966(9):75)*

and flexible, so they were widely used to make knives in ancient China due to their adaptability and durability.

Pig iron drenching refers to a kind of heat treatment technique by drenching ironworks with liquid pig iron. *Tian Gong Kai Wu* records that "all the tools used to plough the fields are made of wrought iron melt and drenched with liquid pig iron, which becomes hard once it is quenched. Based on its own weight, each tool should be rubbed with the appropriate amount of solid pig iron in case that it will not be hard enough or too hard to be broken." This is a unique surface carburizing process still widely used in the 1960s, which is also known as "paving raw" and "boiling raw." The farm tools made by this process are labor-saving, self-sharpening and durable, and are very popular with the masses (Fig. 8.20) [17].

8.3 Splendid Iron and Steel Smelting Civilization of China

The first kind of iron used by humans was meteorite iron, and then came the invention of iron smelting technique. It is known that the earliest meteorite tools in China are copper-indigo iron blades unearthed in Gaocheng, Hebei and Pinggu District, Beijing, dating from the middle and late Shang Dynasty [18]. Iron weapons of Nanzhao State recorded in Tang Fanchuo's *Man Shu* and Duan Chengshi's *You Yang Za Zu* were also made of meteorite. It is believed that the production and use of meteorite instruments might have contributed to the invention of iron smelting.

Up to the beginning of this century, about 265 iron wares of the fifth century B.C. were unearthed by archeological excavation, and 42 of them, including 26 pig iron wares, were identified by metallographic examination [19]. This proves that, as mentioned above, the artificial iron wares originally used in China also came from the bloomery iron process, but the pig iron-based smelting technology was invented very early and then became the mainstream.

At the turn of the Spring and Autumn Period and the Warring States Period, China began to enter the Iron Age. Major breakthroughs occurred in the middle Warring States Period. According to the statistics, by the end of the 1970s, nearly 1200 pieces of iron wares in the mid-late Warring States Period were unearthed. Distributed in 41 counties and cities of 12 provinces, these ironworks were of complete variety and could be used in complete sets. Farm tools, other tools and containers were mainly castings while most of the weapons were forgings [20]. This shows that the iron smelting industry with certain scale and production capacity had been established in the seven countries at that time, and the technology level was roughly the same. A unified iron culture was spreading to Baiyue and other regions.

In *Chapter Haiwang of Guan Zi*, it is claimed that, "A woman must have a needle and a knife," "A farmer must have a plough, a rake and a hoe," "A handicraft manufacturer of carriages must have an axe, a saw, a drill and a chisel," and "There is no one in the world that has achieved something without one of them." It is the true portrayal of the wide use of iron wares at that time, especially the high-end production and popularization of farm tools (Fig. 8.21). This is a great event in the history of ancient China, which greatly promoted the development of economy and culture and improved the society as a whole. *Strategies of the Warring States on State Qi* depicts the commercial development of Linzi, the capital of State Qi, and notes that on the roads of Linzi, there were a lot of carriages and numerous people, who were rich and ambitious. During the Warring States Period, the booming economy and the contending among a hundred schools of thoughts were just based on the flourishing iron production.

Fig. 8.21 The iron farm tools of the Warring States and Han Dynasties

① Hex-angular Hoe (The Warring States Period)
② Pickax (The Warring States Period)
③ (The Western Han Dynasty)
④ Spade (The Eastern Han Dynasty)

As early as the Spring and Autumn Period (about seventh century B.C.), Guan Zhong, the prime minister of State Qi, carried out the national monopoly of salt and metal in the country. Soon after, a government sector specialized in iron-working was set up in Nanyang County, State Qin. Those policies were followed in the Han Dynasty and 49 iron administrations were found consecutively all over the country. The rapid development of iron industry made a great contribution to the industrial and agricultural production and the national power of Han Dynasty (Fig. 8.22). The frying process to fine molten pig iron was invented in the middle Western Han Dynasty (about second century B.C.) and soon provided the steel industry with plenty of qualified and affordable raw materials easily and efficiently. Subsequently, well-tempered steel flourished.

The steel-making technology experienced booming development in the Wei, Jin, Southern and Northern Dynasties period. According to the records in Wang Can's *On Blade*, the process of melting wrought iron and pig iron together for steel came into being in the late Eastern Han Dynasty, and then evolved to maturity in the subsequent 400 years. The iron-working techniques in *History of Northern Dynasties* by Li Yanshou in the Tang Dynasty, verified by the texts in *Book of Northern Qi Dynasty*, were claimed to be applied in national defense and daily livelihoods. By that time,

Fig. 8.22 The cattle farming painting of the Eastern Han Dynasty,Mizhi County, Shaanxi

the co-fusion process of steel-making had been standardized and widely used in production.

After a long period of division in Northern and Southern dynasties, the great changes had taken place in iron industry along with the economic and cultural prosperity in Sui and Tang dynasties. The southern China shared more than 50% of the iron output in the late Tang Dynasty and the steelworks in the Song Dynasty. One symbol of progress at the time was the cast iron replacing the wrought iron in farm tools making. The historical improvement in the quality of farm tools played a great role in the leap forward of agricultural production in Tang Dynasty (Fig. 8.23 and 8.24). Featured by those strikingly different techniques like casting with mold in *The Book of Rite*s and smelting in crucible in *Tian Gong Kai Wu,* the iron industry of ancient China came to a summit with the completion of distinguishing technical system of iron-working in Tang and Song dynasties. Based on the existing techniques and the long-term improvement, this system had been adopted for over a thousand years until the modern times, including streaming stone for iron, frying pig iron, co-fusing pig iron and wrought iron, and the processing techniques of carburizing, inlaying and so forth.

From the beginning of Iron Age (about fifth century B.C.) to the twelfth century, no other countries than China could run the iron industry for 1,700 years and process wrought iron from pig iron. The annual iron output of a bloomery furnace in the Roman Empire was 50 lb (22.68 kg), while that of a big shaft furnace in the Han

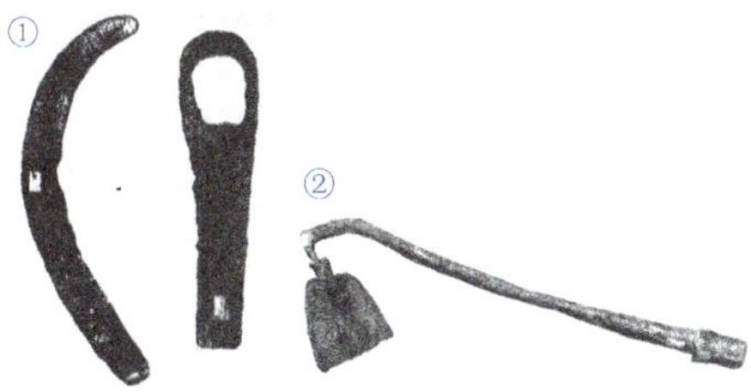

① iron shovel and sickle, unearthed in Sanmenxia (Hua Jueming. *Essays in History of Smelting and Casting in China.* Cultural Relics Press, 1986: 8)
② iron hoe, unearthed in Yi County, Hebei (*Cast Iron in China.* Metallurgical Industry Press, 2011: 34)

Fig. 8.23 The farm tools made of cast iron in Tang Dynasty

Fig. 8.24 Iron hoe: farm tools made of cast iron in the Northern Song Dynasty (Hua Jueming. *Metal-making Technology in Ancient China.* Zhengzhou: Elephant Press, 1999: 417)

Dynasty was 1 ton. It would take 6–8 h to make the bloomery iron carburized only for 2 mm, while 10–15 kg of wrought iron could be made in just an hour by frying pig iron in a pit furnace and then be confused with pig iron. The output of this technology is several times that of carburizing bloomery iron. Nowadays, people produce pig iron by fining ore in shaft furnace, wrought iron by frying pig iron in reverberator and steel by fusing the two kinds of irons together in open-hearth furnace. The traditional Chinese iron smelting technology was based on the production of pig iron and consistent with the modern map. Thanks to the advanced technology, especially pig iron smelting and iron frying, Chinese people created a leading iron civilization in the world. Throughout the ancient times, China had never been in shortage of iron and steel, but enjoyed a good reputation with a series of technical inventions and some giant iron castings like iron ox in Pujin Ferry (Fig. 8.25), iron lion in Cangzhou (Fig. 8.26) and iron tower in D^3angyang County. Joseph Needham pointed out that only Chinese people could get as much iron as they wanted and produce steel in advanced ways from fifth to seventeenth century [21], while H. H. Coghlan regarded that China was the first country in the world to produce giant and heavy iron castings, and they had great power and advanced technology in doing this [22].

China's iron smelting technology spread to the Korean Peninsula as early as in the late Warring States Period, to Central Asia during the reign of Emperor Hanwu of the western Han Dynasty and to Japan in the third century. Blast furnaces were used in pig iron smelting in Sweden and Germany in the thirteenth century [23], and the origin of this technology was believed to be China by Joseph Needham and Tylikot [24]. Whether this statement is true or not remains uncertain, but technological transfer is always two-way. In 1000 B.C., the bloomery iron process appeared in Xinjiang Hui Autonomous Region; scholars thought it might have been introduced from Central Asia. The wrought iron, which originated in Iran and India, was introduced to China in the Wei-Jin, Southern and Northern Dynasties periods, and its manufacturing

Fig. 8.25 The iron ox in Pujin Ferry (cast in the Dynasty)

Fig. 8.26 The iron lion in Cangzhou (cast in the Five-Dynasties Period)

method and pattern characteristics were described in *Yi Qie Jing Yin Yi* by Hui Lin in the Tang Dynasty and *Gegu Yao Lun* by Cao Zhao in Ming Dynasty.

In the development course of the iron and steel smelting technology in ancient China, great inventions and creations were more witnessed before the Sui and Tang dynasties. In the Tang and Song dynasties, this technology underwent a process of being absorbed, selected and fixed. After the Song Dynasty, due to the great pressure of political and economic reasons, the industry of iron smelting lost the driving force to move forward and saw no more achievements, though it kept increasing in Ming and Qing dynasties because of population growth and people's livelihood.

8.4 The Causes and Characteristics of the Iron and Steel Smelting Technology in Ancient China

For the study of history of technology, it is much easier to verify "what it is" than "why it is." It is a puzzle for many people why the pig iron smelting and casting technology was invented by Chinese so early in ancient times. A tentative explanation will be made as follows.

As is known to all, the industry of iron smelting and casting was so developed that its exquisite techniques were rare in the world in Shang and Zhou dynasties. In classical books, iron smelting was listed together with pottery smelting and pottery casting, and bronze smelting was just developed on the basis of the ceramics industry. Pottery clay smelting, pottery making and burning made full preparations for the high temperature technology, molding materials and molding technology needed for bronze casting. From the Xia Dynasty to the Warring States Period, casting had been dominating the metal technology, and most tools and wares were formed by it. Thus a casting-based technological tradition, which exerted a far-reaching influence on the later generations, was gradually developed.

As early as in the period of the Clan Society, Chinese handicraft had a special division of labor among tribes, such as the Kunwu clan, which was known for being good at smelting. In Western Zhou Dynasty, the handicraft industry was controlled by the government. The craftsmen followed their careers and were attached to their personal status for their whole life; their heirs studied their skills in childhood. As is said in *The Book of Rite*s, "The son of an excellent blacksmith should first learn how to make leather coat." Therefore, it is just based on the casting-based technological tradition and the family carry-on system of craftsmen that the ancient Chinese casters possessed superb skills that ordinary people could hardly reach. When iron came into being as a new metal material, the casters fuse it into molten iron in furnace and then cast into ingots or wares by the method passed on from the old generation. Pig iron was also acquired in the West in Roman times, but was abandoned as waste by blacksmiths.

Chinese casters, however, did not throw away the brittle and hard pig iron but reformed the properties of pig iron by the method of iron mold casting and technique of softening cast iron. And then the wide use of pig iron pushed China into the Iron Age. Under the technological conditions in ancient times, white iron became malleable iron only after 8–10 days of high-temperature treatment. Initially, incomplete white-core malleable iron was obtained. Later, the process was improved, and the production cycle was prolonged from 3 to 10 days. Through their persistent efforts, pig iron was softened, and they succeeded in finding a new iron smelting and casting method. Just imagine how much courage, perseverance and wisdom it took them to achieve this!

The advanced and colorful iron and steel smelting technology in ancient China also benefited from the technological thought with "harmony" as the core. In the view of the sages in Chinese history, this thought is ubiquitous. The core concept of all fields, ranging from music, cooking, medicine to artistry, is just the word "harmony," and the terms used are also similar. For example, the terms such as "steamed," "boiled," "fried" and "stir-fried" used in the field of iron smelting are also used in cooking. As is noted in *Yan Tie Lun*, "if several families cooperate to smelt iron, they will be in shortage of time and funds and the irons smelted tend to be too hard or too soft." Being too hard or soft means violation of harmony, so harmonious methods should be adopted to make smelted iron "harmonious," such as carburization, softening and co-fusion of pig iron and wrought iron [25].

From this point of view, the technology of pig iron smelting and casting, which emerged in the early period of China's history and developed to a great height after that, is not related to the characteristics of ores. Rather it is the result of comprehensive interaction of craft inheritance, technological tradition, technical thoughts and many other factors under the special conditions of ancient China.

Technology is something material and spiritual, and also something historical and realistic. As mentioned above, it is the inheritance and renewal of the established pattern under the guidance of technological thought that leads and supports the emergence, formation and development of technology. The metaphysical and physical aspects of technology are mutually causal and complementary to each other, while the role of the former is decisive.

References

1. Henan Institute of Cultural Relics and Archaeology, Sanmenxia Municipal Cultural Relics Workstation. *The Tomb of the Twin Kingdom in Sanmenxia*. Cultural Relics Publishing House, 1999: 539–573.
2. Huang Xing. *Research on Shaft Furnace Type of Ancient Iron Smelting in China*. Doctoral Dissertation of Beijing University of Science and Technology, 2014: 50–53, 89–91, 124–132, 200–203.
3. Yuncai, Liu. 1978. Origin and Evolution of Blast Furnace in Ancient China. *Cultural Relics* 2: 18–27.
4. Huang Xing. *Research on Shaft Furnace Type of Ancient Iron Smelting in China*. Doctoral Dissertation of Beijing University of Science and Technology, 2014: 28, 57–60, 67–69, 97, 141–146.
5. Zhanyue, Huang, and Wang Daizhi. 1962. An Investigation of Local Iron-making in Yunnan. *Archaeology* 7: 368–374.
6. Xing, Huang, and Qian Wei. 2013. A Comparative Study of Ancient Blowers in the World. *Research on the History of Natural Science* 1: 84–111.
7. Han Rufen, Ke Jun. *The History of Chinese Science and Technology, Volume of Mining and Smelting*. Science Press, 2007: 589.
8. Baisheng, Fan. 1983. *Investigation on Crucible Iron-making in Jincheng*. Collection of History of Metals: Shanghai Science and Technology Press.
9. Shaozong, Zheng. 1956. The Warring States Tool Casting Model Discovered in Xinglong. *Archaeological Newsletter* 1: 29–35.
10. Museum, Mianchi County Cultural., and Henan Museum. 1976. Ancient Iron Wares Discovered in Mianchi County. *Cultural Relics* 8: 45–51.
11. Jueming, Hua. 1982. Discussion on High-strength Cast Iron in Han and Wei Dynasties. *Research on the History of Natural Science* 1: 1–12.
12. Han Rufen, Ke Jun. *The History of Chinese Science and Technology, Volume of Mining and Smelting*. Science Press, 2007: 609–613.
13. Hua Jueming. *Ancient Metal Technology in China, the civilization created by copper and iron*. Elephant Press, 1999: 383–385.
14. Han Rufen, Ke Jun. *The History of Chinese Science and Technology, Volume of Mining and Smelting*. Science Press, 2007: 612–615.
15. Zhong, Li. 1975. Discussion on the Development of Iron and Steel Smelting Technology in the Early Period of Chinese Feudal Society. *Journal of Archaeology* 2: 1–22.
16. Han Rufen, Ke Jun. *The History of Chinese Science and Technology, Volume of Mining and Smelting*. Science Press, 2007: 809.
17. Xianbao, Zhang, and Zhang Xianlu. 1990. Analysis and Identification of Copper and Iron Blades of the Shang Dynasty Unearthed in Liujiahe, Pinggu. *Beijing. Cultural Relics* 7: 66–71.
18. Cultural Relics Administration Office of Hebei Provincial Museum. 1973. The Relics of the Shang Dynasty in Taixi Village, Gaocheng. *Hebei. Archaeology* 5: 266. Yeqin, Ling. 1966. Origin, Spread and Function of the Technology of Pig Iron Drenching. *The Collection of History of Science* 9: 71–77.

19. Rufen, Han. 1998. Metallographic Studies of Early Iron Wares in China. *Cultural Relics* 2: 87–96.
20. Wenbin, Zhang, and Meng Fanren. 1980. The Development of Metallurgical Industry in Warring States and Western Han Dynasty from Archaeological Perspective. *Journal of Zhengzhou University (Social Science Edition)* 1: 33–41.
21. Needham, Joseph. 1958. *The Development of Iron and Steel Technology in China*. London: Newcomen Society.
22. Coghlan. H. H. *Notes on Prehistoric Copper and Early Iron in the Old World*. Oxford, 1956: 85.
23. Jockenhovel, A. 1997. The Beginning of Blast Furnace Technology in Central Europe. *Historical Metallurgy Society News* 37: 4–5. Magnusson. G. Iron Production, Smelting and Iron Trade in the Baltic during the Late Iron Age and Early Middle Ages (5th-13th Centuries). *Archeology East and West of the Baltic*. Stockholm: Stockholm University, 1995: 61–70.
24. Joseph Needham. Chinese Priority in Cast Iron Metallurgy. *Technology and Culture*, 1964, (5).
25. Hua Jueming. *Ancient Metal Technology in China, the Civilization Created by Copper and Iron*. Elephant Press, 1999: 555–559.

Jueming Hua Born in Wuxi, Jiangsu, in 1933, graduated from the Department of Mechanical Engineering of Tsinghua University in 1958. In 1967, he got the master degree in the Institute of Natural Science History of the Chinese Academy of Sciences. In 1986, he served as a researcher at the Institute of Natural Science History of the Chinese Academy of Sciences, served as deputy director from 1988 to 1993, and retired in the same year. From 1993 to 2003, he served as the director of the Institute of Science History and Ancient Literature, Tsinghua University.

Research interests: History of technology; the philosophy of technology; traditional crafts Major works: A Collection of Papers on History of Chinese Metallurgy, Ancient Chinese Metal Technology: Civilization Made by Copper and Iron, Complete Works of Chinese Traditional Crafts, Chinese Handcrafts and Collections of Hua Jueming's Works Awards: The first prize of scientific and technological achievements granted by the Ministry of Culture;the first prize of the National excellent publications;the Lifetime Achievement Award for the Foundry industry.

Major works: A Collection of Papers on History of Chinese Metallurgy, Ancient Chinese Metal Technology: Civilization Made by Copper and Iron, Complete Works of Chinese Traditional Crafts, Chinese Handcrafts and Collections of Hua Jueming's Works.

Xing Huang Born in Xuanhua, Hebei in 1981 Ph.D. assistant researcher at the Institute of Natural Science History, Chinese Academy of Sciences.

Research interests: History of metallurgy; history of Compass; Mechanical history

Major works: Monographs: An Empirical Study of Ancient Chinese Compass

Papers: An Empirical Study on Sinan of Natural Magnet Spoon; Literature Review on the History of Chinese Compass; Study on Numerical Simulation of Air Flow Field in a Single Air Outlet Iron Smelting Shaft Furnace—Taking Shuiquanggou Iron Smelting Site in China in the 9th and 13th centuries as an example; A Comparative Study of Ancient Positive Blower in the World; Research on Type and Performance of Ancient Chinese Fan Car.

Awards: The 9th Special Grant from China Postdoctoral Science Foundation;The Second Prize of the Second Beijing Science Innovation Competition granted by the Chinese Academy of Sciences; Excellent Doctorate Thesis of Beijing University of Science and Technology.

Chapter 9
Canal and Navigation Lock

Weibing Zhang and Kuiyi Zhou

The construction of the great canals of China enjoys a long history. It started from the excavation of canals between adjacent watersheds, and finally formed nowadays national water transport network with the Beijing-Hangzhou Great Canal as the backbone to connect all the major water systems in China. The emergence and development of the great canals of China not only manifests the world-leading achievements in hydraulic science and technology made by ancient China, but also relates to its national historical development, always producing a significant impact on the development of the country.

China ranks as the first country in the world to build navigation locks. Its navigation locks of ancient canals successively experienced three developments, namely, that of single-gate navigation locks, multi-gate navigation locks, and navigation locks with pond. The Song Dynasty saw the peak of ancient China's navigation lock technology, and the new kind of double lock that was produced in 984 was the prototype of modern navigation locks, which was about 400 years earlier than the first navigation lock that appeared in Europe.

9.1 The Great Canal

The term "canal" is commonly defined in international dictionaries as a man-made navigable waterway, such as the Suez Canal, the Panama Canal, etc. In ancient China, a canal could be defined in narrow and broad senses, respectively. In the narrow sense,

W. Zhang (✉)
Institute of Water Resources and Hydropower Research, Beijing, China
e-mail: zwb9707@126.com

K. Zhou
The Society for Research on Hydraulic Engineering History, Chinese Hydraulic Engineering Society, Beijing, China

J. Hua and L. Feng (eds.), *Thirty Great Inventions of China*,
https://doi.org/10.1007/978-981-15-6525-0_9

it was in line with the international definition, while in the broad sense it also included the waterways [1] excavated by use of natural rivers and lakes. In China, most of the main river systems flow from west to east, between which run in the same direction the dividing crest(s), and there lacks connection between the north and the south. This geography made it tough for ancient traffic which took waterways as its backbone. In the long historical development of China, in order to expand the space of activities and meet the needs of political, economic, military, and socio-cultural exchanges, a great number of man-made canals or waterways had been excavated either by use of convenient conditions of natural rivers and lakes or by overcoming the difficulties in topographical drops and insufficient water resources, finally forming a national water transport network connecting the major water systems, which is called the great canals of China. The formation of the great canals of China was surely not simple and easy at all, and had undergone many expansions and reconstructions. Among them, two large-scale reconstructions and expansions happened in the Sui Dynasty and the Yuan Dynasty were the most important, which finally led to the North–South great canals with the Beijing-Hangzhou Great Canal as the backbone. In the history of water transportation in the world, the great canals of China have been well-known for their earliest excavation, longest running distance, greatest shipping capacity, most exquisite engineering design, and longest-term efficiency. The great canals of China together with the Great Wall, which look like a herringbone set on the land of China, have become a symbol of the spirit of Chinese nation.

9.2 Excavation of Canals Connecting Adjacent Watersheds

According to *Yugong* in *The Book of Documents*, there had existed water transportation on the natural rivers in the present Reaches of Yellow River when King Da Yu controlled the flood. In the Spring and Autumn Period (770 B.C.–476 B.C.), there had been canals recorded, and Sima Qian made a systematic description of the grand occasions of building canals during the Spring and Autumn Period and the Warring States Period (475 B.C.–221 B.C.) in *Rivers and Canals* in his famous book *Historical Records*. In ancient China as an agricultural society, taxes in kind were the main financial source of the government then, and mostly grain was paid. And the central government, aristocrats, and troops were mainly on the capital area of each dynasty, and the materials collected by local governments were required to transport to the capital. Of course, water transportation is most suitable for the bulk of materials. In topography, the main river systems in China generally flow from west to east. They, together with the artificially dug north–south canals, constituted a huge water transport network. Consequently, the construction of canals has been necessary for social development, and also a key task for flood control and farmland irrigation to achieve national and social security. It's the case in times of peace, and the canals even played an irreplaceable role during wars by securing military supplies. At the earlier time of canals, their important value had already been demonstrated. Now we choose the Hangou Canal, the Honggou Canal, the Xuxi Canal, the Pogangdu Canal, and the Lingqu Canal as the representatives of ancient canals to discuss.

9.2.1 The Hangou Canal and the Honggou Canal

The Hangou Canal connecting Yangzhou and Huai'an in Jiangsu province is the first canal that was exactly recorded in Chinese history. In 486 B.C., the then State of Wu, in order to proceed northward and fight for hegemony, built a water transport connecting the Yangtze River and the Huaihe River to transport "grain" (see *The Ninth Year of King Ai of the State of Lu* in *Zuo Zhuan*). This waterway was just the Hangou Canal, which could support the military practice. According to *The Huaihe River* in *Shui Jing Zhu* (*Commentary on the Waterways Classic*), the Hangou Canal ran northward from today's Yangzhou city, and went through many lakes along the way into the Huaihe River in today's Huai'an city. As its construction naturally used the local dense waterway network of rivers and lakes, it inevitably came across many twists and turns as well as the danger of wind and waves. It met its biggest twist in the Sheyang Lake, where it first turned to the northeast after running north over Gaoyou County, and then went out of the northwest of the Sheyang Lake into the Huaihe River. It was not until the years of 345–356 that the Hangou Canal abandoned the route of passing the Sheyang Lake, and directly fed into the Huaihe River by way of a northward canal that was newly dug on the west side of the lake. It was a common practice to use natural rivers and lakes to reduce canal construction in order to adapt to the productivity and military operations at that times. The projects of curve cut-off showed the historical progress in canal construction (Fig. 9.1).

More than a hundred years after the completion of the Hangou Canal, the Honggou Canal (also known as the Langtang Canal in the Western Han Dynasty, and the Bianqu Canal in the Eastern Han Dynasty when it was diverged to the east) linking the Yellow River and the Huaihe River also emerged as the times required. In 361 B.C., King Hui of the State of Wei moved the capital to Daliang (now Kaifeng, Henan Province), and began to build the Honggou Canal in sections in the following year, which was eventually accomplished in 339 B.C. The canal drew the water of Yellow River in Daliang, then turned to the south from the east, and connected the north tributaries of the Huaihe River, namely, the Danshui River, the Suishui River, the Woshui River, and the Yingshui River, therefore forming a water transport network between the Yellow River and the Huaihe River. The network provided great support to the social and economic development of the region, and promoted the prosperity of the cities along the rivers involved (Fig. 9.2). Tao (now Dingtao, Shandong province) at the junction of the Jishui River and the Heshui River, Chen (now Huaiyang, Henan Province) at the intersection of the Shashui River and the Honggou Canal, Shouchun (now Shouxian County, Anhui Province) at the entrance of the Xifei River and the Yingshui River into the Huaihe River, Suiyang (now Shangqiu, Henan Province) on the bank of the Suishui River, and Pengcheng at the intersection of the Danshui River and the Sishui River, all became the metropolises then. After the unification

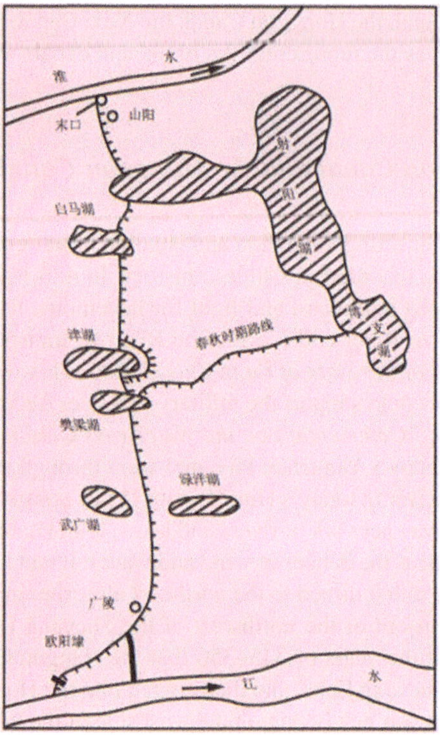

Fig. 9.1 A diagram of the waterways of the Hangou Canal in the Eastern Jin Dynasty (Selected from Yao Hanyuan. *Outline of the History of Water Conservancy in China*. Beijing: Hydraulic and Hydroelectricity Press, 1987: p. 127, Fig. 4-7)

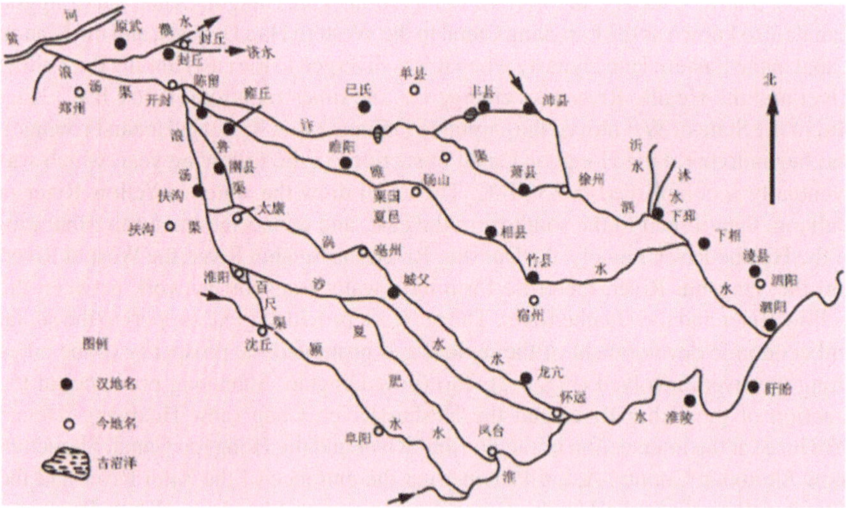

Fig. 9.2 A diagram of the Langtang Canal and the Bianqu Canal in the Han Dynasty (repainted according to Yao Hanyuan. *Outline of the History of Water Conservancy in China*. Beijing: Hydraulic and Hydroelectricity Press, 1987: p. 82, Fig. 3-7)

of the Qin and Han dynasties, the canals of China met greater development, and the Hangou Canal and the Honggou Canal became the backbone canals linking the capital Chang'an (now Xi'an, Shaanxi Province) and the south of the lower reaches of the Yangtze River.

9.2.2 The Xuxi Canal and the Pogangdu Canal

Both the Taihu Lake and the Shuiyang River areas in the south of the Yangtze River had long been economically developed, but they were blocked by Mount Maoshan which extends from southwest to northeast. When King Helv (?–496 B.C.) of the State of Wu launched an attack on the State of Chu, one of his generals Wu Zixu suggested to dig the Xuxi Canal to connect the two water systems. The key project of building the Xuxi Canal was to cut through the water-deficient Mount Maoshan, which is 20 m above the sea level on the highest point, reduced to 8 m in the west and 6 m in the east. The canal newly dug out through it took advantage of five weir dams to block waterways, and between any two dams natural rainwater or the water by waterwheel were poured in and stored. The Xuxi Canal is thus known as the Yandu Canal, also called the Five-weir Canal. In this way, the canal connects Jingxi in the upper reaches of the Taihu Lake and pours into the Yangtze River through the Shuiyang River, becoming the de facto canal which cuts through mountains.

To the north of Mount Maoshan, there also runs a second canal which cuts through Mount Maoshan and links the Taihu Lake basin to the Yangtze River. It is named as the Pogangdu Canal, built in 245 during the three Kingdoms period, it runs from Jurong in the west to Danyang in the east, with a total length of 40–50 *li*. The canal is also a waterway dug out on hills, and its navigable water source is mainly rainfall and the water pouring down the slopes which are blocked and stored by weir dams, so it is necessary to build multi-level weirs to store water step by step in order to maintain a navigation depth. According to *Jiankang Records*, there were 14 dams in total built along the Pogangdu Canal, with an average of one dam per 3-*li* waterway, which indicates that the canal has a very steep longitudinal slope. The practice of the canal's building of weir dams as well as step-type waterways is first recorded in historical documents, and has pioneering significance in the history of canals.

9.2.3 The Lingqu Canal

The Central Plains region had been blocked from the areas of Guangdong and Guangxi by the Nanling Mountains which extend from the east to the west, thus making the traffic between them rather tough. The Hunan-Guangxi corridor, a south–north geographical corridor seated between the Yuecheng Ridge in the northwest and the Dupang Ridge in the southwest of Xing'an County, Guangxi province, had been the only road access from the south of the Five Ridges to the Central Plains. Immediately after unifying China, First Emperor of Qin (259 B.C.–210 B.C.) launched a military attack on the south of the Five Ridges, but only found that his large number of military supplies was restricted by the mountains, which were truly the obstacle to the military action. In 219 B.C. [2], First Emperor of Qin sent Shi Lu to preside over "the project of digging a canal for transporting food" (*Huainan Zi: Ren Jian Xun*). Both the sources of the Xiangjiang River as a tributary of the Yangtze River and the Shi'anshui River as the origin of the Lijiang River, which is the tributary of the Zhujiang River, originated from Mount Haiyang in Xing'an County. The Xiangjiang River flows from south to north, while the Shi'anshui River from north to south, and the two are only 1.6 km far away at the nearest point when they both run in the saddle of the mountain. But as the water level here of the Xiangjiang River is 6 m lower than that of the Shi'anshui River, even if the canal is built, the possible rapid current would make it impossible for any ship to sail. To solve this problem, the craftsmen of the State of Qin decided to divide the Xiangjiang River into two branches at the distance of 2.3 km from the nearest point of the two rivers along its upstream side. The southern branch goes toward the Shi'anshui River and finally down into the Lijiang River, known as the South Canal. The northern branch was connected by a newly built canal, known as the North Canal, and its lower reach still feeds to the Xiangjiang River. Therefore, the Lingqu Canal connecting the Xiangjiang River and the Lijiang River realized the water transportation between the Yangtze River and the Zhujiang River, becoming the only waterway connecting the south of the Five Ridges at that time. The Lingqu Canal's cross over the Nanling Mountains made it the first de facto canal which cuts through mountains in China.

The scientific and technological advantages that the Lingqu Canal obtained cover: firstly, its best water distribution location chosen by meticulous measurement not only reduced its workload, but also made the artificial canal smoothly connect with natural waterways, which thus guaranteed suitable velocity of flow and depth of navigation for both the South Canal and the North Canal to sail; secondly, the construction of a fish mouth-type dividing dike (similar to that of the Dujiang Weir) divided the Xiangjiang River into two, thus ensuring sufficient water supply to the two canals; thirdly, one large and one small sluices were constructed, with the large linking to the North Canal and the small to the South Canal, both of which not only ensure a suitable navigation depth for the two canals, but also directly discharge excess water of the Xiangjiang River to the Xiangjiang channels in order to guarantee the projects of the two canals; fourthly, due to the possible hindrance of many large rocks in the Xiangjiang River, the North Canal abandoned the course of the Xiangjiang River and was specially built into a tortuous river way, which reduced the slope drop of the Xiangjiang River from 3.75‰ to 1.7‰, and meanwhile the slope drop of the South Canal to 0.9‰, thus making both sail by manpower; fifthly, a number of sluice gates (temporary navigation locks) were built to ensure sufficient depth of navigation for the South Canal. The materials used for building sluice gates had been improved between the Tang and Song dynasties (Fig. 9.3).

As a magnificent construction in the history of ancient canals in China, the Lingqu Canal connects the two major water systems of the Yangtze River and the Zhujiang

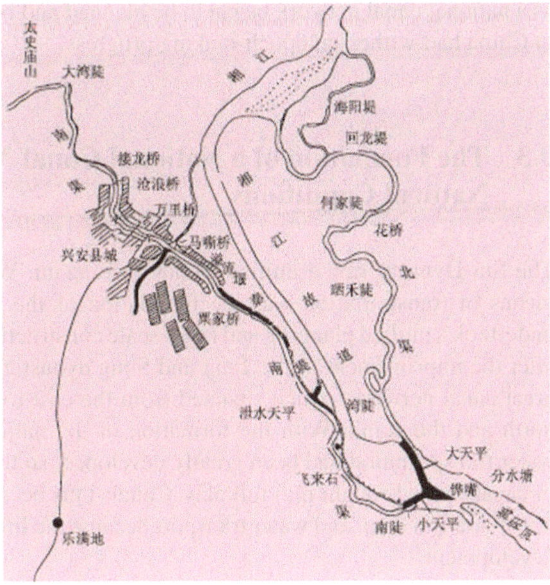

Fig. 9.3 A diagram of the Lingqu Canal's water diversion project (Selected from Zhou Kuiyi. *History of the Science and Technology in China: Volume of Water Conservancy*. Beijing: Science Press, 2002: p. 238, Fig. 3-36)

River, so that the Central Plains and the south of the Five Ridges could be directly communicated by water. Since the completion of the Lingqu Canal, large numbers of reinforcements and grain of the State of Qin had been transported to the south, so the south of the Five Ridges was eventually unified. Afterwards the three counties of Guilin, Xiangjun, and Nanhai were set up, and the migration of migrants from the Central Plains to the south of the Five Ridges was carried out many times, which made the south of the Five Ridges integrated into the territory of China. This indicated that the unification of China was finally completed. After the Han Dynasty and before the construction of the Hunan-Guangxi railway, the Lingqu Canal had long been the main artery of transportation between the North and the South of China. Chen Yuanlong, the governor of Guangxi in the Qing Dynasty, once remarked that small as it was, the Lingqu Canal was the strategic passage among the areas around the State of Chu, Guangdong and Guangxi, and the only way that could be relied on for the transport of troops as well as the circulation of grain and general merchandise.

In today's Lingui County, southwest of Guilin, there is another canal similar to the Lingqu Canal in terms of role and constructional detail, named the Xiangsidai Canal, also known as the Guiliu Canal. It together with the Lingqu Canal is called the East–West Dou Rivers. It was built in the Tang Dynasty, and shared similar layout with the Lingqu Canal.

After the Han Dynasty, during about 400 years in Wei, Jin, Southern and Northern dynasties, the political division, the lack of long-term planning of canal construction, the hasty completion of most of the canals, and the inadequate maintenance, all made the role of canals as a means of water transport rather limited. Since the Sui Dynasty, as a national canal network began to be planned and built, the construction of canals in China had witnessed epoch-making progress.

9.3 The Formation of a National Canal Network Fitting Natural Conditions

The Sui Dynasty saw a unified China once again. Water transportation as a main means of transportation was the foundation of the unified empire. This dynasty undertook a unified planning and large-scale construction of national canals, and later after the improvement in the Tang and Song dynasties, there had formed a national great canal network, which spanned from the east to the west and ran through the north and the south. With the formation of the national water transport network, water transportation had been greatly developed, so the economic and social effects of canals were brought into full play. Canals thus became the lifeline of the national transportation then, and water transport became the lifeblood of social and economic development.

9.3.1 The Planning and Layout of National Canal Network

With the capitals settled down in Chang'an and Luoyang, the Sui Dynasty carried out a comprehensive planning and large-scale construction of national canals in order to improve the link of the capitals to the rich areas in the south, to strengthen the central control over local authorities, and consolidate the stability and unity of the regime. In particular, during the years of 605–610 under the regime of Emperor Yang of the Sui Dynasty, by taking advantage of the favorable natural conditions in topography and the water source of rivers and lakes, a planned construction of four conjoint canals, that is, the Tongji Canal, the Shanyangdu Canal, the Yongji Canal, and the Jiangnan Canal, had been carried out along the 2,700-kilometer-long line from Zhuojun to Yuhang on the basis of the scattered and intermittent canals dug between adjacent watersheds by previous generations. A national network of canals was accordingly formed. Since then, the canal construction of China had entered a new era.

The planning and construction of canals in the Sui Dynasty began with Emperor Wen. The dynasty established its capital in Chang'an, but the Guanzhong Plains were long and narrow with limited area, where the grain and materials produced could never meet the needs of the capital at all. To this end, Emperor Wen began the construction of water transport in Guanzhong after ascending to the throne. In 584, Emperor Wen "ordered Yuwen Kai to dig a canal to draw the water of the Wei River, which, named Guangtong Canal, extends from the east of Chang'an city to Tongguan for more than 300 *li*" [3].

This is the first canal project built after the establishment of the Sui Dynasty, and also the westernmost section of the national canal network planned by the Sui Dynasty. In 587, for the reunification of the regions south of the Yangtze River, Emperor Wen began to build the Shanyangdu Canal on the basis of the ancient Hangou Canal. The Shanyangdu Canal runs from Jiangdu (now Yangzhou) in the south to Shanyang (now Huai'an) in the north, about 300 *li* long, connecting the Yangtze River and the Huaihe River.

After ascending the throne, Emperor Yang of the Sui Dynasty carried out the construction of the eastern capital Luoyang, and began to plan building a canal network with Luoyang as the center. The water route between Luoyang and the rich areas of the south of the Yangtze River began to be dug at first. In 605, Emperor Yang ordered to build the Tongji Canal on the basis of the Bianqu Canal dug in the Han Dynasty. The Tongji Canal was divided into two sections. Starting from Xiyuan of Luoyang and roughly by use of the Yangqu Canal built by Zhang Chun in the Eastern Han Dynasty, the east section drew water of the Gushui River and the Luoshui River into the Yellow River, while the west section drew water of the Yellow River at Banzhu (now north of Yingyang, Henan Province) to flow eastward through Kaifeng, then turn to the southeast and eventually into the Huaihe River. The Tongji Canal was more than 1,300 *li* in length and 40 steps in width, with royal roads built and willow trees planted on both sides. In order to stay away from the dangerous beaches of Lvlianghong and Xuzhouhong in the south of Xuzhou, the canal did not follow the flow direction of the Bianqu Canal in the Eastern Han Dynasty, but was rechanneled to the west and directly into the Huaihe River on the basis of the ancient Qishui River. Running across the Central Plains, the Tongji Canal made full use of the favorable conditions of water source and topography in the process of excavation. The tributary in the north side of the Huaihe River flowed from northwest to southeast, which catered to the specific drop and flow direction suitable to excavate the artificial channel. At the same time, this region was formed by siltation of the Yellow River and Huaihe River, so easy to excavate, which ensured the project smooth to carry out. "Should it be the call of Heaven to excavate the Bianshui River, thus there are no mountains within more than a thousand *li*" [4]. This poem written by poet Pi Rixiu in the Tang Dynasty vividly reflects the topographic condition along the Tongji Canal. Along with the construction of the Tongji Canal, Emperor Yang also carried out large-scale dredging of the Shanyangdu Canal, which could thus sail huge dragon boats and grain ships. Since then, people at that time could travel from Luoyang to Sizhou through the Tongji Canal, go along the Huaihe River down to Shanyang, then cross over the Hangou Canal to Yangzhou, and finally arrive at the regions south of the Yangtze River after passing through the Yangtze River, which meant that an arterial water carriage connecting the Central Plains and the rich areas of the south had been formed.

In the first month of the year of 608, Emperor Yang "sent out imperial decree to call more than a million of men and women in various counties of Hebei to open the Yongji Canal, aiming to draw water of the Qinshui River southward to the Yellow River, and northward to Zhuojun County" [5]. The Yongji Canal exploited many other natural rivers such as the Weihe River, the Qingshui River, the Qishui River, and the Baigou Canal along the way, linking the Yellow River to the Haihe River for the first time in history. More than 2,000 *li* in length, it lies completely to the north of the Yellow River, ranking among the backbone canals in the northern canal network of ancient China. Along its line, people could travel from Luoyang to reach Jicheng (now Beijing), a major military city in the north, which made it easier to deploy military forces to northeast China and control the situation in the north.

In 610, Emperor Yang rebuilt the Jiangnan Canal, "which extends from Jingkou (now Zhenjiang) to Yuhang (now Hangzhou), more than 800 *li* in length and more than 10 *Zhang* (=3 13 m) in width, suitable for dragon boats to sail."[6] The Jiangnan Canal flowed through flat terrains with more lakes, which guaranteed a steady water source of the canal. After the Sui Dynasty, except for partial renovation, the river way of the canal had almost experienced no big changes. The Jiangnan Canal links the Yangtze River to the Qiantang River.

Thus, the beginning of the seventh century had seen the formation of a vast canal network (Fig. 9.4) in ancient China. More than 2,700 km long, this network took Luoyang as the center, and went as far as the Guanzhong Basin in the west, the Hebei Plains in the north, and the south of the lower reaches of the Yangtze River in the south, connecting the five major water systems of China, namely, the Haihe River, the Yellow River, the Huaihe River, the Yangtze River, and the Qiantang River. This canal network running through the major rivers was characterized by reasonable layout and long route, and covered the main developed areas, greatly contributing to the economic and cultural prosperity in the east of China by convenient water transport. In his *Science and Civilization in China*, Joseph Needham made an evaluation of the canals built in the Sui Dynasty: "Among the constructions by the Sui Dynasty, the

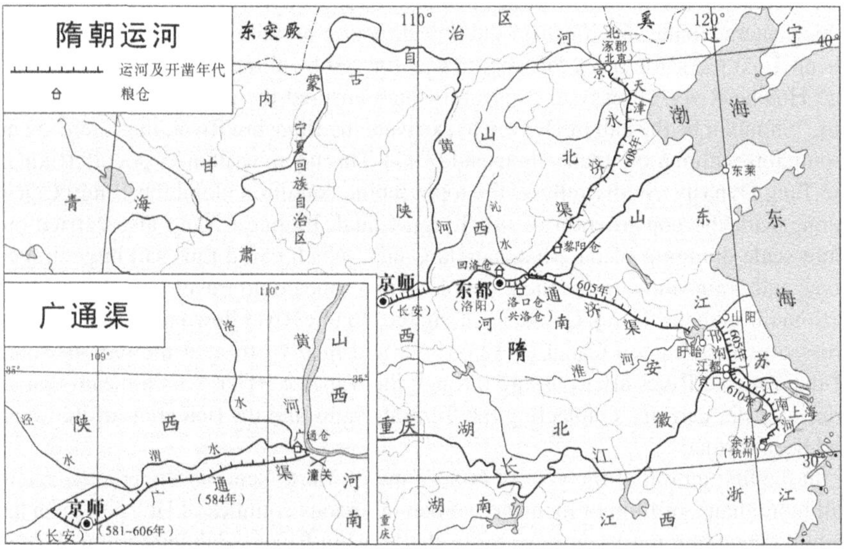

Fig. 9.4 A map of canals of the Sui Dynasty

Great canal, which is about 1,800 km long and links the north to the south of China, is absolutely the largest and most influential one. Between the end of the Sui Dynasty and the rise of modern highway and railway, this canal had become the traffic artery connecting the north and the south of mainland China, greatly contributing to the political, economic and cultural development of the country" [7].

9.3.2 The Replenishment and Improvement of the National Canal Network

The formation of the national canal network of the Sui Dynasty brought great convenience to the later generations and resulted in the social economic development and cultural prosperity of the Tang Dynasty. In *An Epigraph in Praise of the Bianhe River*, Pi Rixiu, the poet of the Tang Dynasty, said: "Without canals, people of the Sui Dynasty suffered a lot; with canals, people of the Tang Dynasty benefited a lot." In the Tang Dynasty, the central government had attached great importance to the maintenance of canals. In the Northern Song Dynasty, the convenience of water transport even made the imperial court establish the capital in Bianjing (now Kaifeng). In order to closely connect the north as the military and political center with the south as the economic center, the Northern Song Dynasty mainly dredged and managed the Bianqu Canal and the Huaiyang Canal, especially the former. Since the Southern Song Dynasty established the capital in Lin'an (now Hangzhou), the Zhedong (east of Zhejiang) Canal came to be the backbone transportation at that time. It was directly managed by the central government, and gradually integrated itself into the national canal network.

9.3.2.1 The Bianqu Canal

In the Northern Song Dynasty, the then canal network centered on the capital Bianjing covered four main man-made waterways, namely, the Bianhe River (or Bianqu Canal), the Huimin River, the Guangji River, and the Jinshui River, collectively known as the "Four Canals of Water Transport (of grain to the capital)." Among them, the Bianhe River was the most important, and had been highly appreciated at that time [8]. Its average transportation capacity could reach 5 or 6 million *dan* (a unit of dry measure for grain, = 100 *sheng*) annually, even 8 million *dan* at most, which was the highest record of annual water transport of grain in the Northern Song Dynasty as well as the highest in this regard in history.

The Bianhe River was rebuilt almost on the basis of the Tongji Canal of the Sui Dynasty. With the route roughly starting from the south of Heyin County of Mengzhou (now the west of Yingze County, Henan Province), the Bianhe River drew water from the Yellow River and flowed eastward to the gate of Kaifeng City. Then it flowed to the east and was divided into two branches: one flowed eastward and

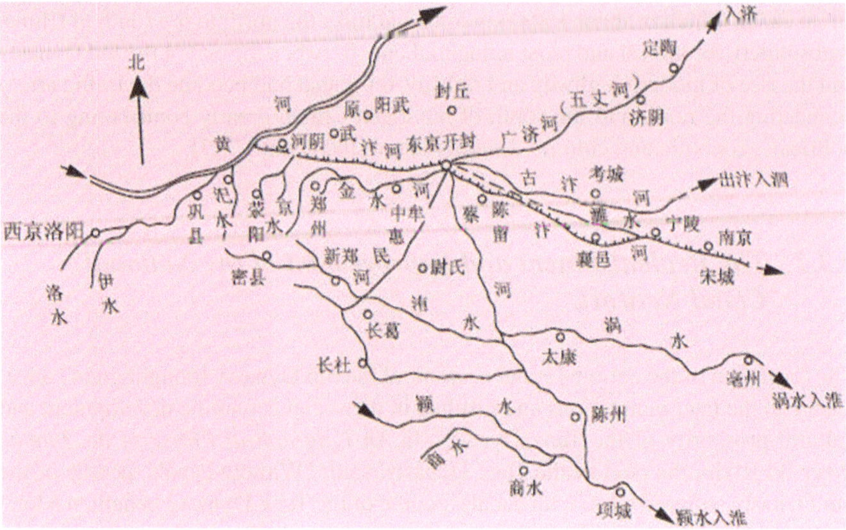

Fig. 9.5 A diagram of waterways neighboring Kaifeng in the Song Dynasty (Selected from Yao Hanyuan. *Outline of the History of Water Conservancy in China*. Beijing: Hydraulic and Hydroelectricity Press, 1987: p. 261, Fig. 5-5)

passed through Caozhou (now Cao County, Shandong Province), Jizhou (now Jining, Shandong Province) to Liangshan (now Dongping, Shandong Province), joining the water transport of grain in Shandong; the other flowed to the southeast, passed through Songzhou (now Shangqiu, Henan Province) and Suzhou, and finally fed the Huaihe River at the Sishui River, joining the water transport of grain in the areas of the Yangtze-Huaihe Rivers (Fig. 9.5). The Bianhe River in the Northern Song Dynasty reached all directions, and created the prosperity of Kaifeng, the capital. The famous ancient painting *Qingming Shang He Tu (Riverside Scene at Qingming Festival)* (Fig. 9.6) just depicts the prosperous look along the Bianhe River in Kaifeng at that time.

Being lifeline of the Northern Song Dynasty, the Bianhe River was connected with the Yellow River. The Yellow River had been well-known for its sudden rise and fall of water as well as large sediment concentration, which always brought about a series of problems. In order to solve the problems concerning water source and sediment, people of the Northern Song Dynasty creatively carried out the project of river narrowing and the project of clearing the Bianhe River, creating a glorious chapter in the history of the canals of China.

The project of river narrowing means to narrow the body of a river by binding up its banks with wooden piles and plates, in order to increase the flow speed of the river, raise its water level, and reduce its siltation. As the main engineering measure to control the Bianhe River in the Northern Song Dynasty, this project was first put forward by Emperor Zhenzong of the Song Dynasty in 1015. *A Sequel to History as a Mirror* makes a record of it: In the broad but shallow positions of the Bianhe River,

Fig. 9.6 *Riverside Scene at Qingming Festival* (partial): Bustling Scene on the Wharf of the Bianhe River (Selected from Yu Hui. *A Guide to Riverside Scene at Qingming Festival by Zhang Zeduan*. Beijing: Peking University Press, 2015: p. 59, Fig. 3-1)

saw tooth-shaped banks were built to narrow the width of the river and raise the water level to speed up its current speed [9]. This measure of river regulation by changing the width of a river was put forth for the first time in history, being an innovation made by the people of the Northern Song Dynasty after a long-term practice. In 1056, the project was further defined as "narrowed river with wooden banks" [10]. According to *The History of the Song Dynasty: Biography of Fu Weizhong*, Fu Weizhong who presided over the Bianhe River at that time pointed out once: "As for a river, it is sometimes wide, sometimes narrow. When it is wide and slow in current speed, the siltation occurs, which makes it difficult to sail boats. At this point, the river needs to be narrowed by binding up its banks with wooden piles and plates."[11] This was actually the earliest proposal and preliminary explanation of the theory of "clearing sands with converging flow" in later generations.

Another great achievement in controlling the Bianhe River made in the Northern Song Dynasty was the implementation of clearing the Bianhe River. In the late Northern Song Dynasty, the Bianhe River became a river on the ground. In 1079, in order to ensure that the transportation dry line was clear of traffic, the government appointed Song Yongcheng to carry out the project of clearing the Bianhe River (Fig. 9.7). The main steps taken in the project included: (1) Building a canal, namely, blocking Luokou and Biankou, and creating a new 51-*li* diversion canal to connect the Bianhe River; (2) Retaining water, namely, by constructing three small reservoirs of Fangjia, Huangjia, and Mengwang in the higher upper reaches of the Suoshui River, and drawing water of the Suoshui River into them, respectively, as regulating

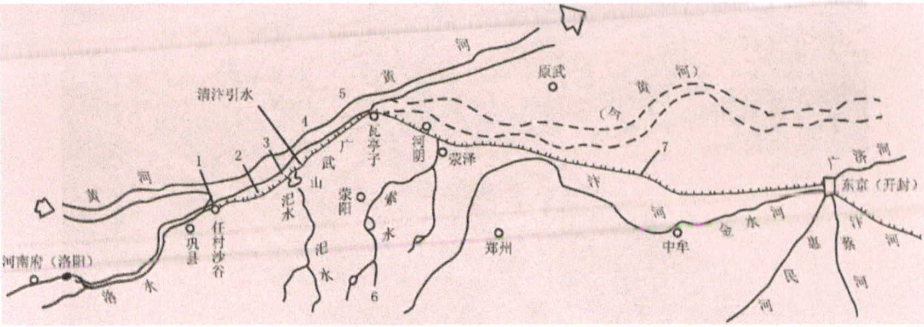

Fig. 9.7 A diagram of the project of clearing the Bianhe River (Selected from Guo Tao. *The History of Hydro Science and Technology in Ancient China*. Beijing: China Construction Industry Publishing House, 2013: p. 150, Fig. 7-3)

reservoirs of the new canal; (3) Building embankment, namely, building a great embankment which ran from Mount Shenwei in the west to Dike Tujia in the east, with a total length of 47 *li*; (4) Renovating the riverbed of the Bianhe River, namely, driving a bundle of piles every 20 *li* and a water lock every 100 *li* to control water flow and increase water depth; (5) Renovating the former mouth of the Sishui River into the Yellow River, and constructing a water gate in the upper and lower reaches of the Sishui River, respectively, to create a new passage between the Yellow River and the Bianhe River [12]. After the completion of the project, the Bianhe River became a river with clear water as its source, "with the flow gentle, the banks straight and the navigational speed doubled. Since then, all kinds of boats, canoes or ships have traveled on it day and night, with least risk but most convenience," [13] which just demonstrates the remarkable benefits of the project.

The project of clearing the Bianhe River had not only changed the source of the river itself, but also was a comprehensive application of various techniques in canals, such as surveying, building channels, installing water gates, flood controlling, water tank, and so on. In the project, the then advanced navigation lock technology was adopted, the scientific concept "flow" of modern hydrology was preliminarily understood, and the natural characteristics of the river were taken into account in the planning and design. All these show that the project was just representative of the highest-level ancient canal technology in the eleventh century, and in a sense a comprehensive reflection of the achievements of science and technology made in the Northern Song Dynasty, which could be safely said to be a heyday of the development of science and technology in ancient China. In *History of Scientific Thought* in *Science and Civilization in China* (Vol. 2), Joseph Needham remarks on the development of science and technology in the Northern Song Dynasty, "The Song Dynasty indeed met the most prosperous period of science in the mainland of China," "the golden age of natural sciences" of China [14]. "The development of science and technology in China has reached its peak in the Song Dynasty, actually surpassing in many respects the level of Britain or Europe before the industrial revolution in the middle of the 18th century" [15].

9.3.2.2 The Huaiyang Canal

The Huaiyang Canal is just the ancient Hangou Canal, also known as Chuyang Canal or Chuzhou Canal by people of the Song Dynasty. It was mainly open to shipping with the help of the tide of the Yangtze River. Since the Six Dynasties, the canal had been controlled by water gates and weirs to accept water of the Yangtze River and prevent its own water from releasing. *Lai Nan Lu* written by Li Ao of the Tang Dynasty records: "(The Huaiyang Canal) there are 350 *li* from Huaiyin to Shaobo, against the current, while 90 *li* from Shao Bo to the Yangtze River, 800 *li* from Runzhou to Hangzhou. Though the canal rises and falls, its water does not flow away." The so-called "not flow away" was just due to the control by a series of dams and water gates. The famous cases like this include the Yilou Dam, the Shaobo Dam, and so on.

In the Song Dynasty, the port of the Huaiyang Canal was built at Mokou of Chuzhou, and the entrance of the Bianhe River into the Huaihe River was within Xuyi County. The two stood almost 200 *li* apart. If a boat traveled between them, it had to go through a section of the Huaihe River, where the wind was strong and the waves high, taking a risk of shipwreck. In order to reduce the travel within the Huaihe River, new canals had been built for three times successively to improve the navigation condition in almost one hundred years since the beginning of the Song Dynasty.

The first time happened during 984–987, when Qiao Weiyue, the official in charge of transportation of Huainan, built the 40-*li* Shahe Canal (Fig. 9.8) which ran from Mokou to Mopankou of Huaiyin, and deliberately kept away from the dangerous Shanyangwan of the Huaihe River. The second happened during 1041–1048, when Xu Yuan, the official in charge of shipping and dispatching of Jianghuai continued to build a new river westward from Huaiyin, called the Hongze Canal, which was 49 *li* long and kept away from the dangerous Mopankou in the northeast of Huaiyin. The third time happened in 1083, when the Guishan Canal was built in further west, which ran from Hongze Town to Guishan Town of Xuyi, and was about 60 *li* long. In this way, the estuary of the Huaiyang Canal was basically opposite to that of the Bianqu Canal, so ships could pass through the Huaihe River and enter the Huaiyang Canal after traveling out of the Bianqu Canal.

The Song Dynasty had made two important achievements in its renovation of the Huaiyang Canal. First, during the construction of the Shahe Canal, a navigation lock was built, which was the earliest in both China and the world, about 400 years before the first appearance of western navigation locks. Second, in the building of the Guishan Canal, a scientific method similar to modern geological drilling of soil samples began to be adopted. *Bei Xing Ri Lu* written by Lou Yue of the Southern Song Dynasty has the following records: in 1170, "from Hongze to Guishan, a well was dug every one to two *li* to measure the earth and stone underground. After achieving the active data, a canal was requested to build". This was a new technology used to know the soft and hard nature of the water and soil underground in order to ensure the quality of a project, which reflects the technical level in canal construction in the Northern Song Dynasty.

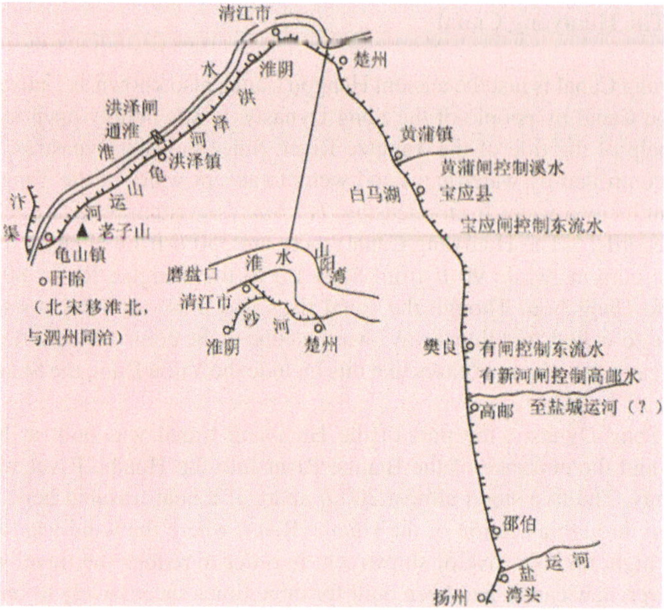

Fig. 9.8 A diagram of the Huaiyang Canal of the Song Dynasty (Selected from Yao Hanyuan. *History of the Beijing-Hangzhou Great Canal*. Beijing: Hydraulic and Hydroelectricity Press, 1998: p. 54, Fig. 2-6-2)

In the Northern Song Dynasty, the Huaiyang Canal used the original pools to enrich its water tank, and such facilities as overflow weirs, water locks, and others were built along it to control its water. By 1119, there had been 79 sluice gates or navigation locks built on the banks of the canal of Zhen, Yang, Chu, Si, Gaoyou, and so on [16].

9.3.2.3 The Zhedong Canal

The Zhedong Canal, or the East of Zhejiang Canal, was initially built in the State of Wuyue (907–978) in the Spring and Autumn period. During the Western Jin Dynasty, He Xun who was in charge of Kuaiji presided over the dredging and opening of the Xiling Canal, which thus became a canal of more than 200 *li* long extending from Xiling (now Xixing Town of Xiaoshan) in the west to Cao'ejiang in the east. It was renamed Xixing Canal in the State of Wuyue in the Five Dynasties (907–960). After the Southern Song Dynasty established the capital in Lin'an, the canal became lifeline of the dynasty. In addition, the dynasty attached importance to foreign trade, and Mingzhou (now Ningbo) was then an important foreign trade port. In order to strengthen the ties between the capital and the rich areas of eastern Zhejiang, the dynasty carried out quasi-militarization management of the canal, each section of which was specially maintained and dredged by the forces. According to *Jiatai Annals of Kuaiji* (Vol. 12), at that time, the Zhedong Canal could be passable for vessels carrying 200 *dan* within Xiaoshan County and Shangyu County, and for vessels carrying 500 *dan* within Shanyin County and Yaojiang. The shipping conditions and prosperity of the canal had reached the peak of its history.

Eastern Zhejiang is high in the south and low in the north in topography, so most of its rivers run north–south. The east–west Zhedong Canal, therefore, had to pass through many natural rivers (Fig. 9.9). In order to maintain the water level in different river sections and make the vessels pass through the river sections with different water level, the Zhedong Canal had been built with many navigation locks as well as dams and weirs, the famous of which include the Qianqing Weir, the Cao'e Weir, and so on. In 1072, Cheng Xun, a Japanese monk, came to China to visit the Buddhist resorts of Mountain Tiantai and Mountain Wutai. He described the situation in his *Records of Visiting Mountain Tiantai and Mountain Wutai* when he passed through the Qianqing Weir by boat: "(May 6) At 1 p.m. to 3 p.m., we passed through the Qianqing Weir. It's rare to find that our boat was pulled over the land by two oxen, while most of us passengers had to walk through from a floating bridge made of ten small boats…" [17]

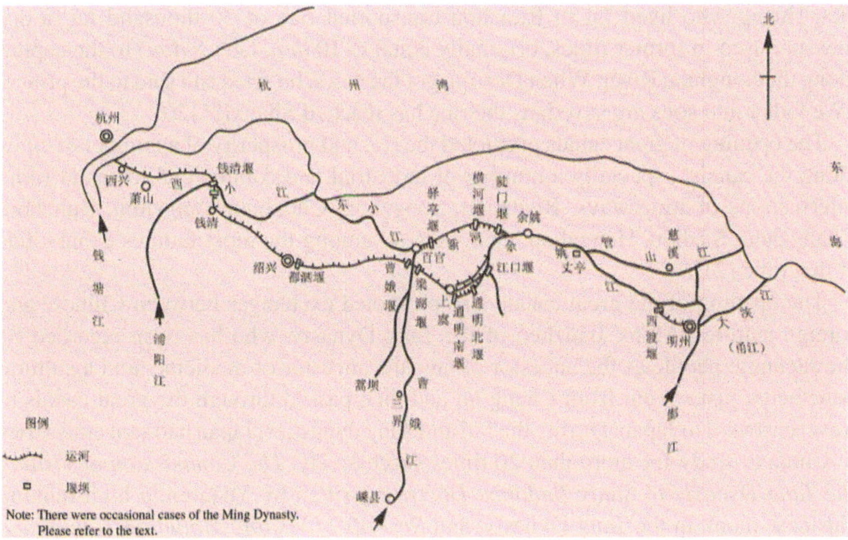

Fig. 9.9 A diagram of the Zhedong Canal of the Song Dynasty (Selected from Yao Hanyuan. *History of the Beijing-Hangzhou Great Canal*. Beijing: Hydraulic and Hydroelectricity Press, 1998: p. 739, Fig. 1-1)

9.3.3 Great Social Value of the Formation of the National Canal Network

During the Sui, Tang, and Song dynasties, the formation and continuous improvement of the national canal network had produced a far-reaching impact on the development of history, and played a more and more important role in both the transfer of properties and the economic and cultural links between the North and the South. In the early years of the Tang Dynasty, the grain transported to the capital Chang'an along the Tongji Canal was "no more than two hundred thousand *dan* every year" [17] at the start, but increased to 4 million *dan* [18] in 743, which was the highest record of water transport of grain in the Tang Dynasty. Pi Rixiu, a poet of the late Tang Dynasty, wrote it in his *Redolent of the Bianhe River* that though this river was said to be cause of the subjugation of the Sui Dynasty, it still stretched green and vast. He even compared the building of canals in the Sui Dynasty to the controlling of flood by King Dayu in terms of contribution [19]. The great canals were also of great significance for the Tang Dynasty to maintain its regime. In the early reign of Emperor Dezong, the water transport of grain was interrupted due to the segmentation by military governors, so the imperial center was in jeopardy. Because of the lack of food, the forbidden forces guarding Chang'an city mutinied and surrounded the palace. In this critical moment, Han Huang who lived far in Runzhou transported rice of 30 thousand *hu* (a dry measure used in former times, originally equal to 10 *dou*, later 5 *dou*) to the capital along the Jianghuai River. When Dezong got the news, he cried and said to the prince, "We father and sons are saved, as the rice has reached Shanxi!" [20].

The opening of great canals promoted the rise and prosperity of a number of cities along the canals, especially a number of industrial and commercial towns at some intersections of waterways. Bianzhou, Songzhou, Chuzhou, Yangzhou, Runzhou, Changzhou, Suzhou, Hangzhou, etc., were all among the most famous canal cities at that time [21].

The opening of the great canals also promoted exchanges between Chinese and foreign cultures. Monk Jianzhen of the Tang Dynasty, who has been regarded by the Japanese people as the ancestor of law, the ancestor of medicine, and a cultural benefactor, just set out from Chang'an and then passed through the great canals to travel eastward to Japan. During the Sui and Tang dynasties, Japan had sent emissaries to China to study for more than 20 times successively. *The Commission of Visiting the Tang Dynasty to Study Buddhist Doctrine* written by Yuanren, a high-ranking Japanese monk in the Tang Dynasty, and *Records of Visiting Mountain Tiantai and Mountain Wutai* written by Cheng Xun of the Northern Song Dynasty both gave a full account of their journeys along the canals, in which they saw and heard on the spot the peaceful and polite society of the Tang and Song dynasties, respectively. In 1265–1274 of the Southern Song Dynasty, Puhaddin, an Arab missionary, had been in Yangzhou and preached. According to legend, he was the XVI-born grandson of Mohamed, the founder of Islam. He finally died of illness in a ship traveling in the Beijing-Hangzhou Great Canal. According to his will, Yangzhou government then buried him on a higher land on the east bank of the Great canal in Yangzhou.

9.4 The Completion of the North–South Great Canal that Could Cross the Mountains and Rivers

As the Yuan, Ming, and Qing dynasties all established the capital in Beijing, the northward transfer of political center greatly changed the layout of the main canals. In the Yuan Dynasty, as the section within Shandong was carried out the projects of curve cut-off, and the section within Beijing was redesigned, the Beijing-Hangzhou Great Canal finally runs straight instead of arcuate, which made the whole north–south line joined up. Setting out from Hangzhou, the grain ships of the south of the lower reaches of the Yangtze River could travel northward across the Yangtze River, the Huaihe River, and the Yellow River, all the way to Beijing. In the Ming and Qing dynasties, in order to overcome the difficulties in topographical drops along the Beijing-Hangzhou Great Canal, especially in the insufficient water sources when the canal crosses the horsts in Shandong, the ancient Chinese made constant innovations in engineering technology and basically ensured the safety of navigation and the smooth flow of the canal. The canal thus became the backbone running through the North–South economic artery of the unified empire.

9.4.1 The Feasibility Planning of the Beijing-Hangzhou Great Canal Made by Guo Shoujing

As for the whole north–south line joined up of the Beijing-Hangzhou Great Canal in the Yuan Dynasty, Guo Shoujing is obviously an important figure who has to be mentioned. Modern people have highly regarded Guo Shoujing as the founder, general planner, and designer of the canal.

Guo Shoujing (1231–1316), with the courtesy name Ruosi, came from Xingtai, Hebei Province. As an outstanding scientist of the Yuan Dynasty, he specialized in water conservancy and astronomical calendars. Among his achievements, the one of the most far-reaching influence on the future is that he was fully responsible for the planning and design of the Beijing-Hangzhou Great Canal. In 1275, Prime Minister Bo Yan felt the importance of a North–South waterway and thus suggested building the later Beijing-Hangzhou Great Canal, which was accepted by Kublai, the fifth emperor of the Yuan Dynasty, so sent Guo Shoujing, who was then in charge of water administration, to carry out an investigation. Guo made a survey of the area of 800 km surrounding the capital, "acquiring that Jizhou, Daming, Dongpingsi, Wen and the Yuhe River are actually connected, so he accordingly made a map and presented it to the emperor" [22]. The brief biographical sketch of Guo Shoujing records:

(In 1271) Guo began to be in charge of water administration. In 1275, Prime Minister Bo Yan fought southward, and conferred to set up water courier stations in order to make him easier to make inspection tours. Bo Yan ordered Guo to conduct a corresponding survey of it. Guo successively surveyed the areas from Lingzhou to Daming, from Jizhou to Pei County and further to Lvliang in the south, from Dongping to Gangcheng, across the old way of the Yellow River from Dongping and Qinghe to the Yuhe River, from the section of the Yuhe River in Weizhou to Dongping, and from Shuibo (now Dongping Lake) southwest of Dongping to the Yuhe River, and acquired that Jizhou, Daming, Dongpingsi, Wen and the Yuhe River were actually connected. He accordingly made a map and presented it to the court. The next year, the office in charge of water administration was incorporated into the Ministry of Works in feudal China... Then Guo and King Zanshan led the offices of astronomical phenomena in the north and south, respectively, to survey and draw independently [22].

Among them, Jizhou (now Jining, Shandong Province) was located at the southern end of the horsts of Shandong, connecting with the Sishui River drawn from the northeast, Daming (now Daming, Hebei Province) on the northern bank of the Yuhe River (now the Weihe River), and Dongping (now Dongping, Shandong Province) close to the northern shore of the Wenshui River. It can be seen that in the feasibility study of the Beijing-Hangzhou Great Canal, Guo Shoujing specially circled the key places that interconnected with the Sishui River, the Wenshui River, and the Yuhe River. On topography, the Yuhe River and the Sishui River at that time were on the one side of the horsts and ran northward and southward, respectively, so the watershed between them could only draw water from the Wenshui River which is high and situated in the middle. That is, the Wenshui River should be drawn to an appropriate location into the canal, then diverged to the north and south and connected with the Sishui River and the Yuhe River. According to the six measured lines (Fig. 9.10), Guo Shoujing intended to take Dongping as the center and then to diverge the Wenshui River into the canal. The surveying and planning carried out by Guo Shoujing provided a scientific and technological guarantee for the canal's successful crossing over the horsts of Shandong. Unfortunately, as Guo was moved in 1276 to host astronomical observations and the draft of official calendar, his survey results of the Beijing-Hangzhou Great Canal were buried, which caused the dilemma of water source of the canal in the Yuan Dynasty. It was not until 1411, more than 130 years later, that it was solved by the old man Bai Ying. Besides, Guo Shoujing presided over the survey of other places. At that time, his survey was more advanced, over 600 years earlier than the European benchmark built by sea-level elevation.

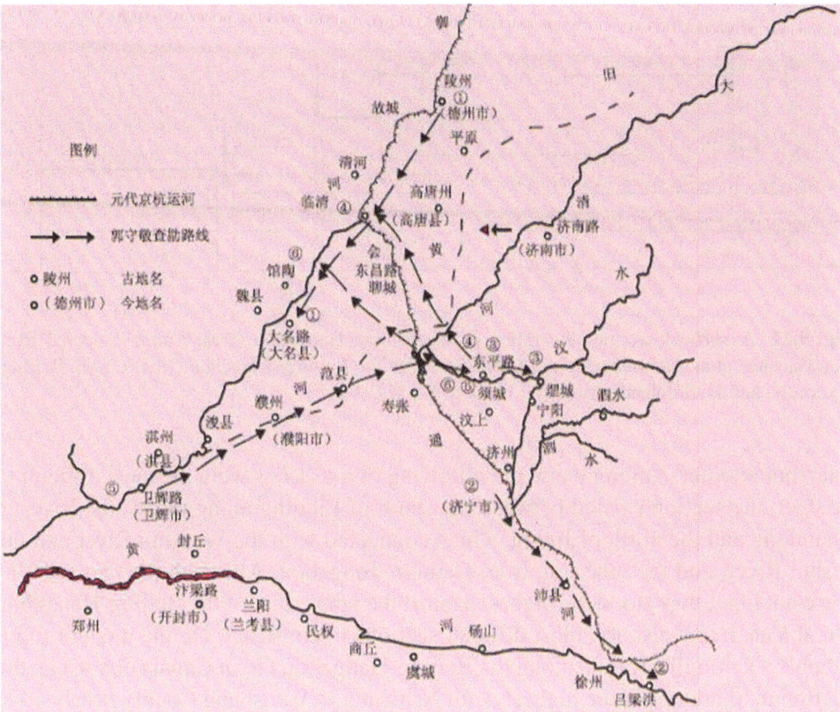

Fig. 9.10 A diagram of the routes Guo Shoujing surveyed along in the feasibility study of the Beijing-Hangzhou Great Canal (redrawn according to Fig. 2-37 in *History of the Science and Technology in China: Volume of Water Conservancy*)

Notes: The routes Guo Shoujing surveyed along

1. From Lingzhou to Daming 2. From Jizhou to Lvliang 3. From Dongping to Gangcheng 4. From Dongping and Qinghe to the Yuhe River (now the Weihe River) 5. From the section of the Yuhe River in Weizhou to Dongping 6. From Shuibo southwest of Dongping to the Yuhe River.

9.4.2 *The Whole North–South Line Joined up of the Beijing-Hangzhou Great Canal*

At the beginning of the Yuan Dynasty, the Beijing-Hangzhou Great Canal was intended to be replanned and redesigned with the aim of fundamentally changing its arcuate way developed in the Sui, Tang, and Song dynasties into a north–south straight line. It would no longer bypass the Central Plains, but directly pass from Huaibei (north Anhui Province) through Shandong into the North China Plain, finally reaching Dadu. As a result, during 1260–1264, each section of the canal had been dredged

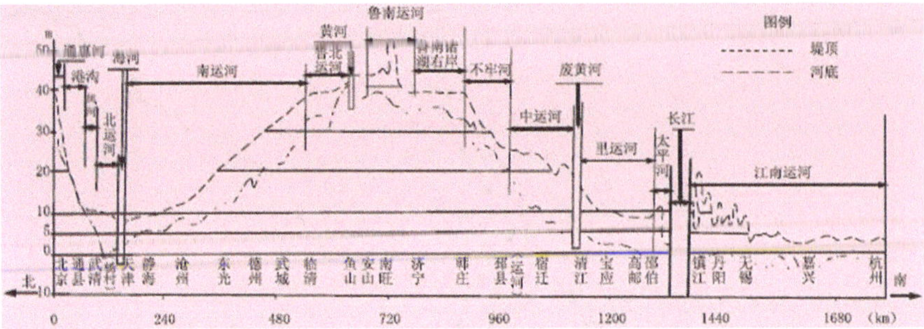

Fig. 9.11 A vertical section illustration of the Beijing-Hangzhou Great Canal (Selected from Zou Baoshan et al. *Management and Development of the Beijing-Hangzhou Great Canal*. Beijing: Hydraulic and Hydroelectricity Press, 1990: p. 52, Fig. 3-1)

one after another. Among them, the following two sections were the most difficult to dredge: one section existed between the south of Linqing along the Weihe River in Shandong and the north of Jining, which connected with the Wenshui River and the Sishui River, and the other ran from Dadu to Tongzhou. Although the two sections were not long, they stood the highest against the whole line of the Beijing-Hangzhou Great Canal, and also the most difficult sections to construct. On the modern topographic section illustration along the Beijing-Hangzhou Great Canal (Fig. 9.11), the section in Shandong is the highest, with Nanwang in Wenshang County nearly 40 m high, the highest point along the whole canal. It's followed by the Tonghui River, the bottom of which is about 30 m above sea level. The difficulty of dredging the two sections mainly rested on the difficulty of water supply.

9.4.2.1 The Jizhou River and the Huitong River

The section of the Beijing-Hangzhou Great Canal in Shandong passes through the horsts of Shandong, which has the largest topographical drop along the whole canal. The construction of the long mountain-crossing canal with larger topographical drops needed to overcome the difficulties in both topographical uplift and water shortage. The project was completed after two constructions.

The excavation of the Jizhou River was firstly carried out. The project of the over 130-*li* section extending from Jining to Anshan of Dongping in Shandong started in 1276, and was completed in 1283. In order to ensure that the passage was clear and the crossing over ridges were smooth during the project, water gates were also set up along the Jizhou River to save water. The opening up of the Jizhou River confirmed that the diversion and distribution of water across river basins was rational, which was a key step for the canal to finally join up the Yuhe River, the Wenshui River, and the Sishui River, and smoothly cross the areas with poor water resources [23].

After the opening up of the Jizhou River, there still remained a section not connected between the Sishui River and the Yuhe River. Yang Wenyu from the

Yuan Dynasty once made a record of the section through which land transportation could only be relied on to pass at that time [24]. The continuous construction of the canal to the north of the Jizhou River thus became an urgent matter. In the first month of 1289, the construction of the Huitong River began and was completed in June. As a result, the Huitong River could connect with the Jizhou River in the south and Linqing as well as the Yuhe River in the north, running more than 250 *li* in length (Figs. 9.12 and 9.13).

The connection of the Huitong River with the Jizhou River could solve the problem of ships to cross hillsides and ridges when traveling along the Beijing-Hangzhou Great Canal. Later, the Huitong River and the Jizhou River were joined up together, commonly known as Huitong River, which was the most crucial project of the Beijing-Hangzhou Great Canal. In *A History of Technology*, Charles Singer made a remark to the history: "The section of the canal completed in 1283 crosses the mountains of Shandong, and is the earliest 'mountain-crossing' canal... Building a canal at the top of the watershed that separates two rivers requires both bold imagination and proper construction skills to provide sufficient water at the top of a mountain" [25].

The Huitong River, however, also had a deficiency, namely, it had been troubled by the two major problems of water source and the erosion and siltation of the Yellow River since its opening up. To this end, the Yuan Dynasty took a series of measures. The first was to draw water into the river. After the opening up of the Huitong River, water of the Wenshui River and the Sishui River was drawn to Jizhou as a water source; at the same time, "water locks and weirs were set up in Yanzhou to lead the Sishui River to flow westward, and water locks and weirs were set up in Gangcheng to diverge the Wenshui River into the Huitong River, and finally flow southward to Jizhou" [26]. The second was to adjust by means of a chain of navigation locks. In the Yuan Dynasty, 31 navigation locks were built over the Huitong River, and 13 navigation locks over the natural rivers such as the Sishui River, the Wenshui River, the Guanghe River, the Fuhe River, the Yanhe River, etc., respectively, finally forming a chain of navigation locks (Fig. 9.12). These navigation locks, the earliest of which was built in 1289 and the latest in 1341, were the first chain of navigation locks used in the world, about 350 years earlier than similar locks used in the west.

9.4.2.2 The Tonghui River

With the opening up of the Jizhou River and the Huitong River, ships of grain in the south could reach Tongzhou. There were only 50 *li* from Tongzhou to Dadu, but as there was no water transportation, it was very difficult to go between them by land transport. In 1292, the court appointed Guo Shoujing as the official in charge of water administration, responsible for the design and construction of the Tonghui River.

The construction of the Tonghui River began in 1292 and was completed in 1293 (Fig. 9.14). Kublai, the fifth emperor of the Yuan Dynasty, attached great importance to the project and personally presided over the opening ceremony. He followed the ceremony Emperor Wu of the Han Dynasty once performed and ordered "all officials below the Prime Minister to carry dustpans and spades" to open up the river on the

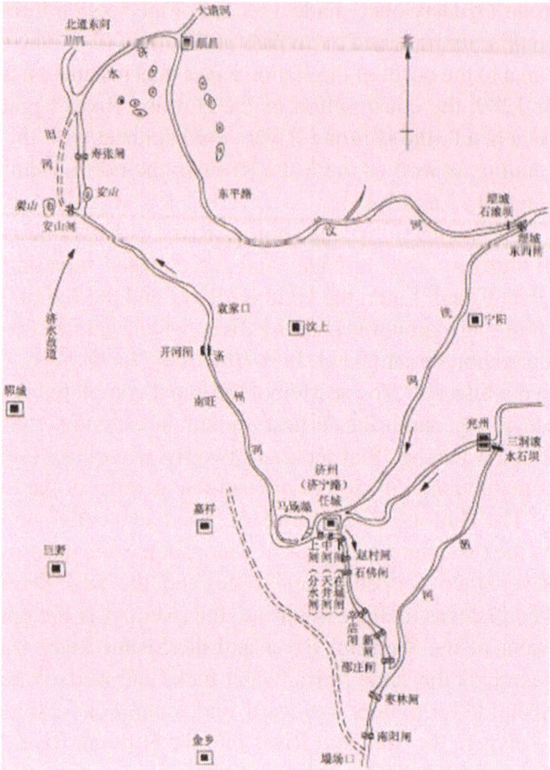

Fig. 9.12 The Huitong River (the section between Dongping and Jining) during the Yuan, Ming and Qing dynasties. (Selected from Yao Hanyuan. *History of the Beijing-Hangzhou Great Canal*. Beijing: Hydraulic and Hydroelectricity Press, 1998: p. 114, Fig. 3-12-2)

construction site. After the opening up of the Tonghui River, ships of grain could sail directly into the Jishuitan Pool in the city of Dadu, finally making the whole north–south line joined up of the Beijing-Hangzhou Great Canal. At that time, there appeared an unprecedented gathering of ships in the Jishuitan Pool. Kublai got there in person and experienced the scene of a convoy of ships "covering the surface of the river." He felt a tremendous sense of exultation, and named it Tonghui River.

During the construction of the Tonghui River, it was also necessary to solve the problems in both topographical uplift and water shortage. On the basis of his on-the-spot investigation and careful survey, Guo Shoujing selected an ideal diversion route: to open up a channel starting from Baifu Village to lead the Baifuquan River to go westward at first, and then head towards the south roughly along the 50-m contour line. This route could avoid those valleys and lowlands, and intercept along the way the water from the Shahe River, the upper reaches of the Qinghe River, and the rivers at the foothills of the Western Hills into the Kunming Lake. After the water flew into Dadu City, "it enters the city from the West [Water] Gate, joins up the Jishuitan Pool in the south, goes out of the Wenming Gate in the southeast, and reaches Gaolizhuang Village of Tongzhou into the Baihe River in the east. It is 164 *li* and 104 *bu* (an old unit for measurement of length, equivalent to five *chi*) in length"

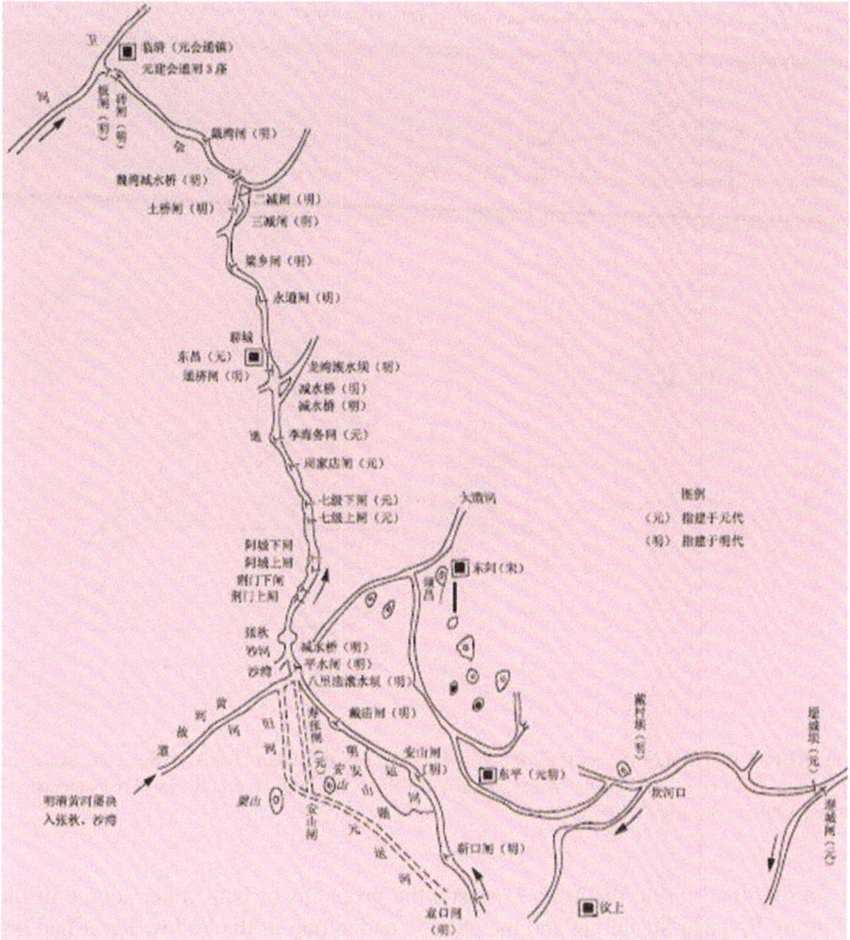

Fig. 9.13 The Huitong River (the section between Linqing and Dongping) during the Yuan, Ming and Qing dynasties. (Selected from Yao Hanyuan. *History of the Beijing-Hangzhou Great Canal.* Beijing: Hydraulic and Hydroelectricity Press, 1998: p. 114, Fig. 3-12-2)

[27]. After the opening up of the Tonghui River, 24 sluices were built to save water for boat-going in order to ensure that the passage of water transport of grain was smooth.

The Great canal of the Yuan Dynasty (Fig. 9.15) extended from Dadu in the north to Hangzhou in the south, passing through such sections as the Tonghui River, the Yuhe River, the Huitong River, the Jizhou River, the Huaiyang Canal, the Jiangnan Canal, etc., and made use of the natural rivers such as the Luhe River, the Guangshui River, the Sishui River, the Yellow River, and the like. It was about 2,000 km in total length, [28] and as a great artery through the north and south, it established the foundation of the Beijing-Hangzhou Great Canal in the Ming and Qing Dynasties.

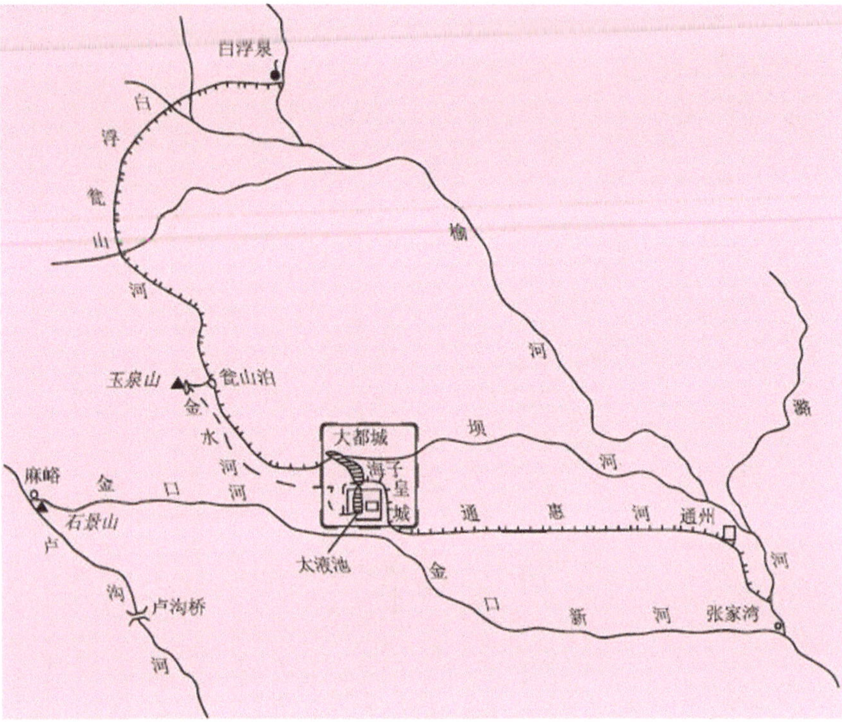

Fig. 9.14 A diagram of the route of the Tonghui River in the Yuan Dynasty. (Selected from Yao Hanyuan. *History of the Beijing-Hangzhou Great Canal*. Beijing: Hydraulic and Hydroelectricity Press, 1998: p. 88, Fig. 3-10-2)

It's a pity that during the Yuan Dynasty, the problems of both water source of the Huitong River in Shandong and the erosion and silting of the Yellow River had not been completely solved. This made the Beijing-Hangzhou Great Canal difficult to fully produce beneficial result, and the sea transportation remained to be the main way of north–south transport. At that time, the annual amount of transport of grain was 3 million *dan*, among which an amount of about 300,000 *dan* was transported by the Beijing-Hangzhou Great Canal.

9.4.3 Treatment of the Beijing-Hangzhou Great Canal

The Ming and Qing dynasties established the capital in Beijing, so the Beijing-Hangzhou Great Canal became the lifeline connecting the north and south. During 1403–1424 of the Ming Dynasty, the problem of water source of its section in Shandong was initially solved, which made the water transport of grain gradually develop steadily. This period had witnessed the repeated breaking down of the Yellow River,

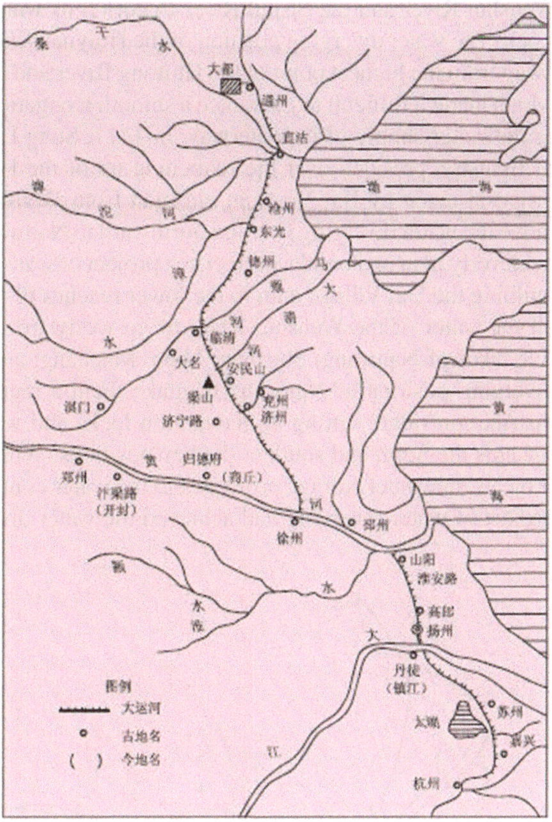

Fig. 9.15 A diagram of the course of the Beijing-Hangzhou Great Canal in the Yuan Dynasty (Selected from Zou Baoshan et al. *Management and Development of the Beijing-Hangzhou Great Canal*. Beijing: Hydraulic and Hydroelectricity Press, 1990: p. 11, Fig. 1-3)

which posed a serious threat to the canal. In order to solve it, the New Nanyang River, the Jiahe River, and the Central Canal had been dug successively since the reign of Emperor Jiajing (1522–1566) of the Ming Dynasty, finally achieving the separation of the canal from the Yellow River. In addition, since the canal parallel intersected the Yellow River, the Huaihe River, etc., during the Ming and Qing dynasties, especially in Qingkou where the Yellow River, the Huaihe River and the canal intersected, it was a world-class difficult problem to keep the passage clear, which involved extremely complicated engineering technology. Therefore, the construction of the Beijing-Hangzhou Great Canal during the Ming and Qing dynasties focused on the solution of the water source of the Huitong River, the separation of the canal from the Yellow River, and the treatment of the key project of Qingkou.

9.4.3.1 Nanwang Key Project of Water Diverging

In the Yuan Dynasty, the north–south water diverging of the Huitong River crossing the horsts in Shandong was mainly achieved by means of drawing water from the

Wenshui River and the Sishui River to enrich its water through the Huiyuan Water Lock. The water diverging position of the Huiyuan Water Lock, however, was 40 km away from the highest point of the Huitong River and about 8 m in topographic drop, which made it difficult to guarantee a smooth navigation of the Huitong River as well as the water supply of its waterway. In 1411, Song Li, the minister of the Ministry of Industry, presided over the reopening up of the Huitong River. He adopted the proposal made by Bai Ying, an old man from Wenshang County of Shandong, to move the water diverging position northward to Nanwang, and centered on Nanwang to cleverly plan and build a perfect key project of water diverging (Fig. 9.16). That is, building the Dai Village dam in the lower reaches of the Wenshui River to intercept all the water of the Wenshui River to go westward into Nanwang; renovating the five lakes of Nanwang, Shushan, Mata, Machang, and Anshan situated around the diverging position of Nanwang to make them a regulation reservoir on the water distribution ridge; setting up navigation locks and water gates for diverging water towards the north and south at the key positions. With the roles of the water source projects, the water storage projects and the water control projects, the Nanwang key project of water diverging had achieved the water diversion control target of "70%

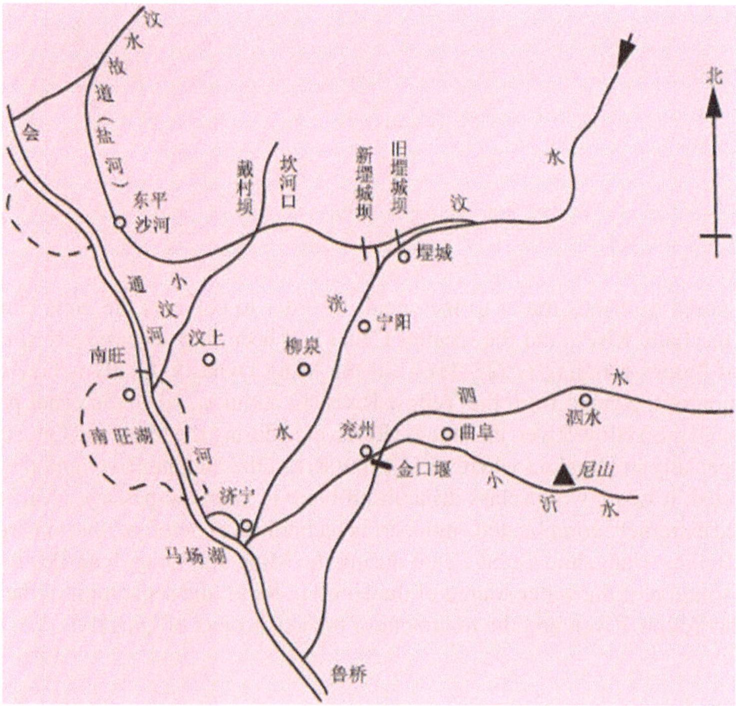

Fig. 9.16 A diagram of the Nanwang key project of water diverging (Selected from Yao Hanyuan. *History of the Beijing-Hangzhou Great Canal*. Beijing: Hydraulic and Hydroelectricity Press, 1998: p. 198, Fig. 4-18-2)

of the whole amount of water flow to the north and the rest to the south" in order to cater to the characteristics of less water in the north and more water in the south, which accordingly solved the problem of water shortage of the mountain-crossing canal to the north of Jining and ensured the successful crossing (over) mountain after mountain of the fleet of grain. The Nanwang key project of water diverging reflected that China at that time had taken the lead in the planning of key projects of water conservancy and in the engineering technology of canals, as no other countries than in the world had built artificial canals that could climb and cross mountains. [29] Later, the number of navigation locks that were built over the Huitong River increased to 38, the use of which solved the problems of both topographic drops and water control of the passage of canals, and improved the navigation conditions of the Huitong River. This river ran about 200 km in length, and had an average topographic drop of about 2‰. In addition to navigation locks, water impounding gates, water intake sluices, water release sluices, water leveling sluices as well as weirs and dams had been built over the Huitong River. There were more than 100 sluices and dams altogether that had been built along, so the Huitong River has also been called "Sluice Channel."

After the completion of the Nanwang key project of water diverging, the Huitong River began to enjoy a reliable water source. After 1412, the water transport of grain northward along the canal soon exceeded 4 million *dan*.

9.4.3.2 The Separation Project of the Yellow River and the Beijing-Hangzhou Great Canal

From the Yuan Dynasty to the reign of Emperor Jiajing of the Ming Dynasty, the Beijing-Hangzhou Great Canal had intersected the Yellow River in Xuzhou, from which to Huaiyin the Yellow River had been used to travel along. This route would pass through the two dangerous beaches in Xuzhou Hong and Lvliang Hong, respectively. The two positions were often silted or interrupted due to the burst of the Yellow River. In addition, the Yellow River often flooded from Henan to the north and broke through the Huitong River. Since the early reign of Emperor Jiajing, the route of the Beijing-Hangzhou Great Canal had been proposed to change to realize the separation of the Yellow River and the canal. Until the middle reign of Emperor Kangxi of the Qing Dynasty when Jin Fu opened up the Central Canal did the canal separated from the Yellow River. Three important projects were then carried out, namely, the New Nanyang River and the Jiahe River of the Ming Dynasty, and the Central Canal of the Qing Dynasty.

The construction of the New Nanyang River was first proposed in 1527, whose plan was to change the canal between the south of Nanyang and Liucheng from the west of the Zhaoyang Lake to its east in order to avoid the erosion and deposition of the Yellow River and made the Zhaoyang Lake storage of flood. The following year, the construction began, but was forced to shut down halfway as being attacked by people in drought. Zhu Heng presided over the project in 1566, and the next year a new river was completed with a total length of more than 140 *li*, called New

Nanyang River (Fig. 9.17). The new river completely separated the section of the Beijing-Hangzhou Great Canal in Shandong from the Yellow River, eliminating the threat of siltation of the Yellow River. Weng Dali, the general official in charge of rivers in the early reign of Emperor Longqing in the Ming Dynasty, once said: "Compared with the old river, the new river has five benefits: first, as the new river slightly faces upward in topography, it is difficult to be burst by the Yellow River; second, it flows calmly, so not necessary to be on guard against it; third, it is not steep as the old river; fourth, there is always reserved space along the river, which could be used to plant crops; fifth, it is quite open to all vessels" [30].

In 1569, in view of both the superiority of the New Nanyang River and the harm of the silting of the canal in Chacheng near Xuzhou, Weng Dali, the imperial official of the capital, put forward to build again the Jiahe River so that the canal could avoid the section of the Yellow River within Xuzhou from the south of Xiazhen and

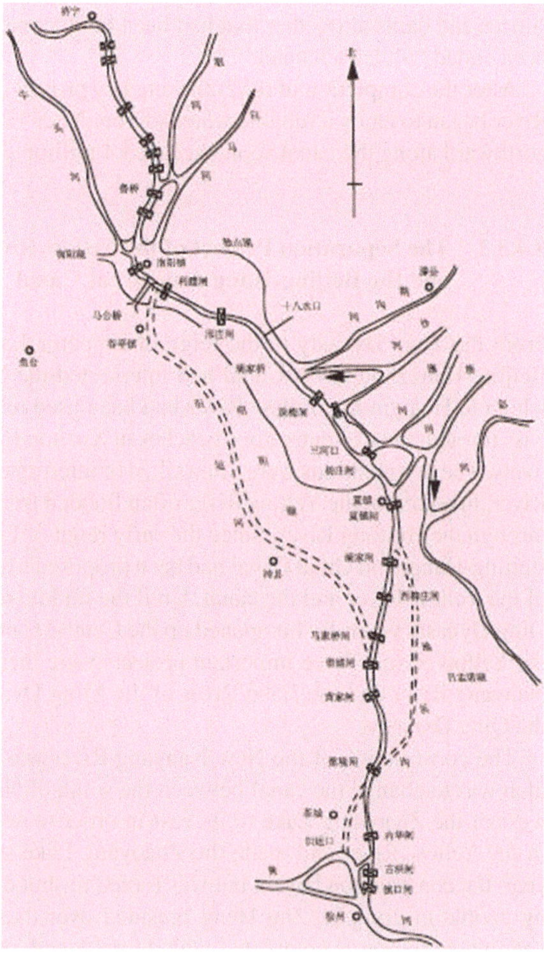

Fig. 9.17 A diagram of the New Nanyang River (Selected from Yao Hanyuan. *History of the Beijing-Hangzhou Great Canal*. Beijing: Hydraulic and Hydroelectricity Press, 1998: p. 245, Fig. 4-20-2)

flowed directly into Pizhou. The plan was not completed until the year of 1605 due to the opposition of Pan Jixun and others. This new river, which joined the New Nanyang River in the upper reaches and ran directly into the Yellow River beside the Luoma Lake in the lower reaches, made the canal separated from the Yellow River between Xuzhou and Pizhou (Fig. 9.18). The opening up of the Jiahe River could avoid the hidden danger of the burst of the Yellow River and the dangerous sections of the Yellow River in Xuzhou and Lvliang, making it the main channel within the middle section (the section between South Shandong and North Jiangsu) of the Beijing-Hangzhou Great Canal. Jin Fu, governor in charge of rivers during the reign of Emperor Kangxi of the Qing Dynasty, once said: "As for the management of rivers during the Ming Dynasty, the construction of the Jiahe River enjoyed the greatest achievement" [31].

After the completion of the Jiahe River, the passage extending over 200 *li* between the south of the mouth of the Zhihe River in Pixian County and Qingkou was still needed to travel by the Yellow River. On the basis of the Jiahe River of the Ming Dynasty, the project of a new canal extending from Zhangzhuang Port and through the mouth of the Luoma Lake started from 1686 to be carried out under the management of Jin Fu, which would pass through Suqian, Taoyuan, and flowed into the Yellow River at Zhongjiazhuang of Qinghe. The project was completed in the first month of 1688, known as the Central River (Fig. 9.19). It's just here that the Beijing-Hangzhou

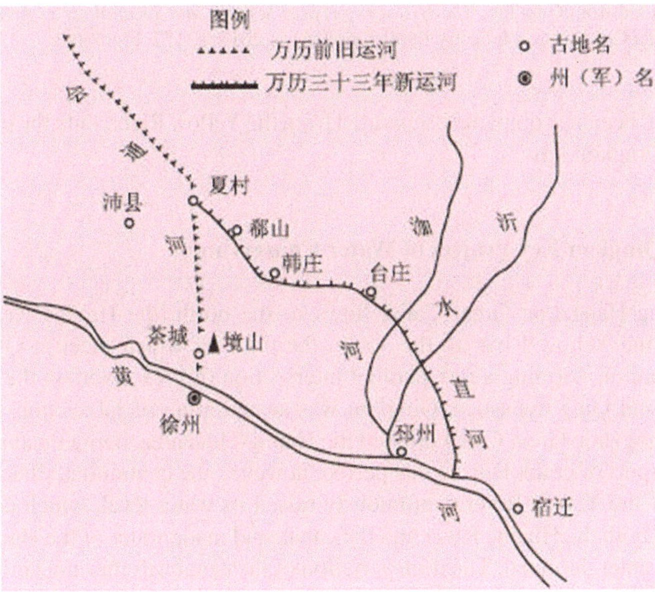

Fig. 9.18 A diagram of the Jiahe River (Selected from Guo Tao. *The History of Hydro Science and Technology in Ancient China*. Beijing: China Construction Industry Publishing House, 2013: p. 156, Fig. 7-6b)

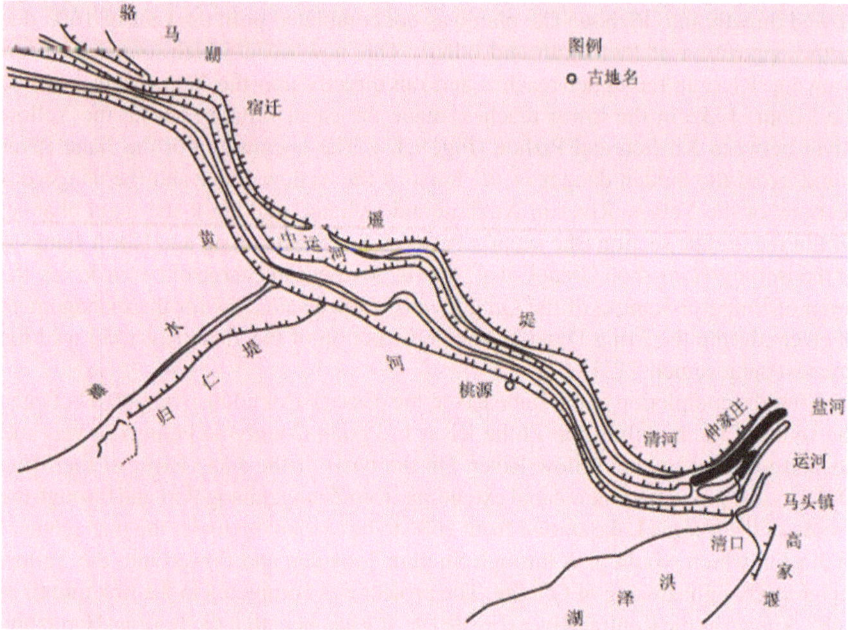

Fig. 9.19 A diagram of the Beijing-Hangzhou Great Canal after the Central Canal was built by Jin Fu (Selected from Guo Tao. *The History of Hydro Science and Technology in Ancient China*. Beijing: China Construction Industry Publishing House, 2013: p. 157, Fig. 7-6c)

Great Canal became completely separated from the Yellow River, only the intersected region in Qingkou left.

9.4.3.3 Qingkou Key Project of Water Conservancy

The Beijing-Hangzhou Great Canal flows to the north, the Huaihe River to the west, and the Yellow River to the south, the three of which meet at Qingkou of today's Huai'an, forming a rare parallel intersection of great rivers in the world. In the Ming and Qing dynasties, Qingkou was seated at a crucial section where the Beijing-Hangzhou Great Canal crossed the Huang-Huai area, being the throat of the water transport of grain. During this period, however, the continuous siltation on the riverbed of the Yellow River continuously raised its water level, which constricted the flow of both the Huaihe River and the canal, and made water of the Huaihe River difficult to enter the canal. The transportation of the canal was thus not smooth at all. Especially in flood season, the flood and silts of the Yellow River made it rather difficult and risky for any water transport fleet of grain to pass through Qingkou smoothly. Therefore, during the Ming and Qing dynasties, a large number of projects had been built in Qingkou, which basically maintained a smooth transportation of the canal (Fig. 9.20). There were three major engineering projects involved: the first was to

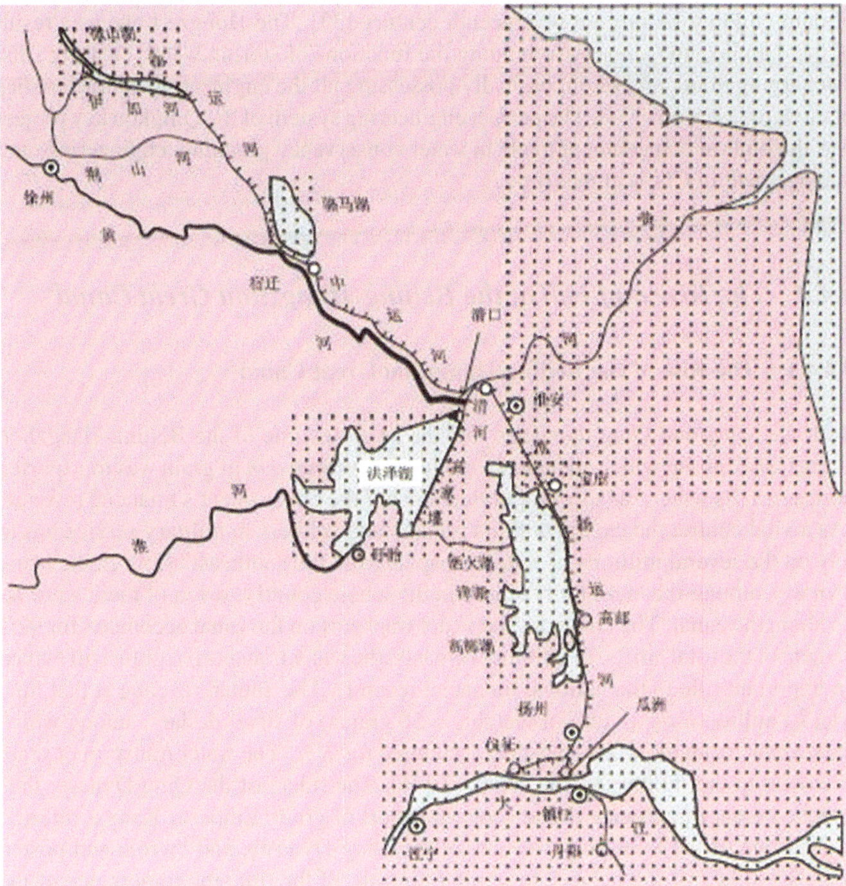

Fig. 9.20 Qingkou Key Project of Water Conservancy: an intersected project among the Beijing-Hangzhou Great Canal, the Yellow River, and the Huaihe River (Selected from Zhou Kuiyi, etc. *General Annals of Chinese Culture: Water Conservancy and Transport*. Shanghai: Shanghai People's Publishing House, 1998: p. 222, Fig. 7-10

move away and rebuild all river ports along to avoid possible burst, surface flooding, or silting of the Yellow River; the second was to construct water-controlling projects and spur dikes, and draw clear water to clear away the silts at Qingkou and protect the river channel; the third was to clear the intersected channel at Qingkou in order to make the sluice smooth [32].

With the continuous construction in the Ming and Qing dynasties, a huge hydro-engineering complex had been formed at the area of Qingkou, which has been the most complex engineering system in the world so far. The Gaojiayan Dam, the main body of the key project, was about 60 km long, with some dyke-dams 15–20 m high. Before steel reinforced concrete appeared, the dam had been one of the tallest masonry dams in the world, and also a worldwide landmark in the history of dam building before the seventeenth century [33]. The Hongze Lake as a result of the dam became a reservoir holding the functions of storing water, clearing silts, discharging flood, and so on, basically possessing all the engineering characteristics of modern reservoirs [34]. The perfect engineering system of the Qingkou key project was the highest achievement made in water conservancy planning, engineering, and management in China at that time.

9.4.4 The Rise and Fall of the Beijing-Hangzhou Great Canal

9.4.4.1 The Rise of the Beijing-Hangzhou Great Canal

Both the Ming and Qing dynasties saw the glorious time of the Beijing-Hangzhou Great Canal, along which an annual amount of 4 million *dan* of grain was transported northward. Its water transport supported most of the government's financial revenue. "As a vital centre, the capital with its officers and soldiers in military service has to rely on the several millions of water transport from the southeast" [35]. In the Ming Dynasty, among the eight tariff offices in the whole country, seven of them were set up along the canal. The tariffs for the water transport on the canal accounted for 92.7 percent of the total tariffs. In the Qing Dynasty, the water transport of grain still played an important role in the national fin ancial revenue. The annual revenue at that time was 70 million *liang* (a unit of weight, = 50 grams) of silver dollars, among which the revenue from the water transport accounted for 2/3. "The water transport of grain is crucial to the military of the country" [36]. The rulers of the Qing Dynasty thus attached great importance to the water transport of grain. Emperor Kangxi referred to it as one of the three major events of the Qing Dynasty, and "wrote and posted it on the pillars of the palace" to remind himself all the time the importance of the canal to maintain the stability of the Qing regime.

Besides, the Beijing-Hangzhou Great Canal promoted the rise and prosperity of towns along its route. During the Ming and Qing dynasties, there were more than 30 large and medium-sized cities with developed industry and commerce, almost half of which were situated along the canal. Among them, the famous towns included Linqing, Jining, Xuzhou, Huai'an, Gaoyou, Yangzhou, Hangzhou, and so on.

These developed commercial and industrial towns along the canal attracted the attention of foreigners from far away, and promoted the cultural exchange and integration between China and foreign countries. Marco Polo, a European traveler of the Yuan Dynasty, came to China with his father. He had been in charge of Yangzhou for three years and visited Zhenzhou, Guazhou, Yangzhou, Gaoyou, Baoying, and Huai'an along the canal. After returning to his own country, he wrote the popular *The Travels of Marco Polo*, which vividly introduces the local conditions and customs along the Beijing-Hangzhou Great Canal.

During the reign of Emperor Yongle of the Ming Dynasty, the three kings on the Sulu Islands (now the Philippines) led their families of 340 members to China for a friendly visit, and received grand reception of Emperor Yongle. When they went southward to return along the Beijing-Hangzhou Great Canal by boat, the East King died of illness. He was finally buried in Dezhou, and his family members were allowed to come to Dezhou to sacrifice him. Besides, certain persons were specially assigned to guard his tomb. All these expenses were paid by the court. In the early reign of Emperor Hongzhi, Cui Pu from North Korea wrote in his *Piao Hai Lu* and praised the benefits of the canal. During the reign of Emperor Wanli, Italian Jesuit Matteo Ricci came to China and traveled from Nanjing to Beijing by boat. Later, he wrote *Regni Chinensis Descriptio* according to what he had seen along the canal.

In the Qing Dynasty, many foreign envoys, missionaries, and tourists also traveled by way of the Great canal. The water conservancy projects, towns and villages, and local conditions and customs along the canal impressed them deeply. *Record of the British Envoys' Visit to Emperor Qianlong* describes what the British mission saw of the Great canal from Tongzhou to Hangzhou. *The First Visit to China of the Envoys from the Netherlands* records what the Dutch mission to China saw of the flourishing canal within Yangzhou during the years of Emperor Shunzhi.

During the Yuan and Ming dynasties, the Beijing-Hangzhou Great Canal linked China to the world, becoming a window for foreigners to observe the material civilization and regional culture of China. They took a novel look at the Beijing-Hangzhou Great Canal, and their vivid description of it reflected the exchange and collision between Chinese and foreign cultures, becoming an important supporter to spread Chinese culture to the outside world [37].

9.4.4.2 The Fall of the Beijing-Hangzhou Great Canal

In 1855, the Yellow River burst in Tongwaxiang of Lankao, Henan, and changed its course into the sea in Lijin, Shandong. This flooding dashed against the banks of the Great canal in Shandong, and caused the canal to completely suspend shipping service for more than 10 years. It was a turning point to fall in the history of the canal. Since then, although there had been plans to resume navigation, the continuous flood of the Yellow River and the decline of the country made the canal gradually silt up. During 1875–1908, the Qing Dynasty finally abandoned the construction of the Great canal, which finally completed its 1,000-year journey, reaching the lowest point of its destiny.

With the silting of the canal, the mode of transportation had changed throughout the country. In 1826, trial sea transportation was successfully carried out. In 1852, the water transport of grain in Jiangsu and Zhejiang was changed to sea transport. The following year, the water transport of grain in the four provinces of Hubei, Hunan, Jiangxi, and Anhui changed their courses, which meant that the water transport of grain by way of the canal had ceased to exist except in name. In 1872, China Merchants Bureau was established in Shanghai, sea transport thus becoming smoother. In 1900, the Qing Dynasty collected tax by ready silver dollar instead of material object, and the water transport of grain was abolished. In 1911, the Jingpu Railway opened to traffic, which replaced the shipping service. As only the section between Jining of Shandong and Hangzhou as well as the Xixing Canal south of the

Qiantang River was preserved, the Beijing-Hangzhou Great Canal had evolved into an interval canal. Its glorious history and cultural values certainly remain forever.

In its development of more than 2,000 years, in order to overcome the problems of topographic drops and water shortage, the Great canal of China had constantly adjusted its routes and perfected the engineering measures, so as to achieve smooth water transport of grain. In the face of extremely complex natural conditions, Chinese people of successive dynasties carefully designed and skillfully arranged the transporting projects along the canal. They unceasingly innovated and solved many worldwide engineering and technical problems at that time, demonstrating the world's leading achievements made by ancient China in water conservancy technology. No other famous canals in the world could match it then. In addition, the emergence and development of the Great canal of China closely related to the historical development of the country, greatly influencing the country's politics, economy, military, society, culture, ecology, and others.

The history of the great canals of China for more than 2,000 years has been accompanied by the growth of Chinese nation, showing the spirit of hardship and excellence in which Chinese nation grows. The people of all ages have achieved remarkable achievements through their hard work, which embody the dignity of our nation and greatly contributed to both the history of canals and the history of cultures in the world. In the last 150 years, the great canals had evolved into interval canals, but they are still playing a continuous role today. The former channel of the Beijing-Hangzhou Great Canal is used by 90% of the eastern route of the South-to-North Water Transfer Project. From this point of view, the great canals of China are in nature living cultural heritage. As time progresses, both the historical and cultural value and the socio-economic value of the great canals of China will be more and more highlighted. *The Book of Songs: Greater Odes* says: "Although the Zhou Dynasty was an old state, its fate relied on its reform." The history of the great canals of China tells us that nowadays, if we intend to be a great country that other countries in the world feel deeply convinced of and whose people could live with dignity, we have to foster our cultures and reform the systems in addition to developing science and technology.

9.5 The Development and Innovation of Canal Navigation Lock

Since ancient times, human beings had made boats as a means of vehicle and started original water transportation. Chinese classic *Huainan Zi* written by Liu An has a record of it. In the Han Dynasty when Liu An lived, water transportation not only referred to that on natural rivers and lakes, but also on artificial canals that were over 100 km long and built to connect with natural rivers and lakes. At the same time, shipping science and technology were also improving, so Liu An pointed out that "appliances will turn to be suitable as time varies." In other words, along with the progress of the times, waterway engineering facilities should not follow the old routine, but conform to the times and make constant innovations. It is with the development of productivity that the opening up of the canals and the navigation locks on them had been constantly innovated. Generally, a navigation lock is built on the section of a canal with steeper slope, especially with intensive topographic drops. It is used to adjust the depth of a canal, the gradient of water surface, and the velocity of flow, in order to facilitate the navigation of boats.

The navigation lock, formerly known as weir dam, is a dam that holds back water of a canal. When a ship passes through a mountain, as there is a larger drop between the mountain and flat ground, it needs one or multi-level weirs as a transition to go smoothly. The slopes in the upper and lower reaches of each weir dam are rather gentle, which is easy for tugboats to go up and down. Each weir dam is gradually reduced in elevation, which thus divides the whole drop into several levels. The water between weir dams is stored to ensure the depth of navigation. When a boat passes the weir dam, the heavy-duty one is usually unloaded, while the empty one is dragged over the dam by manpower or buffalo. After crossing the dam, boats will be then reloaded and continue their voyage. For larger boats, simple machineries such as winches, which are similar to beveled ship lifters still visible in some areas nowadays, will be used. In order to reduce the friction between ship bottom and the surface of weir dam, some lake mud and water weeds to be got nearby are generally applied on the surface of weir dam. This method is simple but effective. This way of crossing dams is generally referred to as "Pan Ba (coil the dam)" or "Zhuan Ban (transfer)." As a kind of inoperable physical building, a weir dam can not be called navigation lock, but has its main functions, hence the predecessor of navigation lock. Canals with weir dams had been recorded in the late Spring and Autumn Period, among which the Xuxi Canal was the most famous.

The Taihu Lake area and the adjacent Shuiyang River basin are distributed with a wide lake network, which had been economically developed in the Spring and Autumn Period, but blocked by the Maoshan Hill extending from southwest to north-east. When King Helv of the Wu State sent armed forces to suppress the Chu State, one of his generals Wu Zixu suggested opening up the Xuxi Canal to connect the two waters. The Maoshan Hill is 20 m at the highest elevation, 8 m in the west, 6 m in the east. The section of the Xuxi Canal flowing within the hill was helped to block the waterway by five weir dams, so the canal got the name of weir ditch, also known as Wu Yan (Five Weirs). Between the weirs, natural precipitation or slope water was stored, and the water surfaces decreased level by level. Any vessel had to cross the hill from here, which made it possible to travel from Jingxi in the upper reaches of the Taihu Lake to the channel pouring the Yangtze River through the Shuiyang River. The Xuxi Canal still undertook a large number of transport tasks in the Song Dynasty, and it still exists nowadays.

The development of navigation locks of canals in China can be divided into three stages: the stage of single-gate navigation locks, multi-gate navigation locks, and navigation locks with pond, respectively.

9.6 *Single-Gate Navigation Lock: A Facility for Adjusting the Gradient of Water Surface of Canals

*9.6.1 *The Navigation Locks Technique of the Lingqu Canal*

The earliest single-gate navigation lock could date back to the Lingqu Canal, which is located in now Xing'an County, Guilin, Guangxi, and was built at the beginning of the third century B.C. The Central Plains and the areas of Guangdong and Guangxi are blocked by the east–west Nanling Mountains, so it was rather difficult to travel between them. Between the Yuecheng Ridge in the northwest and the Dupang Ridge in the southwest of Xing'an County, Guangxi, there exists a south–north Hunan-Guangxi Corridor, which was the land route from Lingnan (south of the Five Ridges) to the Central Plains. After the reunification of the six states, First Emperor of Qin (259–210 B.C.) commanded his forces to Lingnan, but his large supply of military supplies was restricted by the barrier of mountains. To overcome the obstacle of military actions, the emperor ordered Shi Lu to preside over "the construction of a canal to open up a passage to transport grain" (*Huainan Zi*) in 219 B.C. The Xiangjiang River and the Lijiang River happen to originate from the Haiyang Mountain in Xing'an County. The Xiangjiang River flows from south to north, while the Shi'anshui River, a branch of the Lijiang runs from north to south. The craftsmen of the Qin State decided to search a proper diversion point in the upper reaches of the Xiangjiang River. They finally found the position of Meitan, 2.3 km away from the nearest point in the upper reaches of the above two rivers, where the Xiangjiang River was 1.1 m higher than the Shi'anshui River, and accordingly built a key project of water diverging here, which divided the Xiangjiang River into two branches. The southward branch flows through the Shi'anshui River and finally into the Lijiang River, called the South Canal. Next to the Xiangjiang River, the northern branch was built a new North Canal nearby, and its lower reaches still flows back to the Xiangjiang River. Therefore, the Lingqu Canal which connects the Xiangjiang River with the Lijiang River helps to link the Yangtze River to the Pearl River, becoming the only water transportation channel connecting Lingnan at that time. As crossing the Nanling Mountains, the Lingqu Canal was the first truly mountain-crossing canal in China.

Precisely because the Lingqu Canal crossed the mountains here, although the South Canal and the North Canal had chosen an appropriate water diversion position, the two canals still had much larger gradient for navigation. The average gradient of the South Canal reached nearly 1‰, and its water flow speed was faster, with its navigation depth also reduced, which brought difficulties to its navigation. Navigation Locks should thus be built to adjust the state. The navigation locks built on the Lingqu Canal were locally called *dou* or *doumen* (sluice gate), which were used to save the canal flow. Building a sluice gate in the desired section of a canal can reduce the gradient and velocity of the canal, and increase the navigation depth, producing a spectacle of boats climbing over mountains. Due to the critical role of sluice gates in the transport of the Lingqu Canal, the Lingqu Canal was also called the Dou Canal (Sluice Gate Canal).

The origin of the construction of sluice gates is lost, but the first document of them could be found in *A Record of the Lingqu Canal* written by Yu Mengwei in 868 [38]. Yu wrote down afterwards what Li Bo saw when he first visited the Lingqu Canal in 825, whose "sluice gates had been damaged with age, whose water was too shallow to ship." Its reconstruction was thus started, the main projects of which included "ploughing its banks to prevent its water from flowing away, and improving its sluice gates to empty directly." It can be seen here that the setting up of sluice gates existed long since the Tang Dynasty. The Song Dynasty saw more records of them. Fan Chengda and Zhou Qufei from the Song Dynasty had served as an official in Jingjiang Fu (now Guilin, Guangxi) at the same time during the reign of Emperor Chunxi, with Fan magistrate of a prefecture, Zhou an official under county magistrate in charge of lawsuit. They had opportunities to visit the Lingqu Canal together. In their writings, they made similar records of the canal and the sluice gates on it. In 1178, Zhou Qufei made such a description of the canal: "There are 36 sluice gates built inside the canal. When a boat passes through a sluice gate, the gate is then locked. After the water within the lock gathers enough to sail the boat, it means that the boat can then sail up and down the mountain. In this way, all boats can go through it" [39]. In the description, the word "lock" (Fan Chengda described it as "sluice gate") shows that the sluice gate of the Lingqu Canal in the Song Dynasty must be operated manually. But then, it is unclear whether the gate to lock the water was surface plate-type, laminated beam-type, or that similar to the steep lever made in the Qing Dynasty.

There had been 36 navigation locks in the Song Dynasty, 14 of which still remain, all reconstructed in the Qing Dynasty. The walls on both sides of a navigation lock are a masonry structure with a general height of 1.5–2.0 m and a width of 5.5–6.0 m. Most of the walls are semi-circular or arc-shaped. The distance between two navigation locks corresponds to the topographic elevation, with the nearest about 60 m, the farthest 2,000 m. The north–south width of the Lingqu Canal is mostly 10–15 m, which is one to two times the width of its navigation locks (Fig. 9.21).

The structure of navigation locks made in the Qing Dynasty is shown in Fig. 9.22. The support components of a sluice plate comprise a small sluice bar, an outside bar, and a bottom bar. When a boat is passing over, the small sluice bar is obliquely inserted into the stone hole of the bottom plate of the navigation lock, with its upper end obliquely embedded into the fillister of the stone sluice base on the other side; one end of the bottom bar is inserted into the hole fillister of the opposite sluice base, with the other end fended off on the lower end of the small sluice bar. The outside bar is also the case. In front of the above support sluice bars there exists a wood tripod dam which is made of three wooden poles tied. The wood tripod dam leans against the outside bar and the bottom bar, and on it the bamboo-made mats are laid. When the sluice gate is opened, the small sluice bar is unclenched, and the vessels move downstream with the collapse of the sluice plate. The sluice plate is obviously not densely made enough to hold all water, but the loss of a small amount of water in a water-rich canal like the Lingqu Canal does not make it difficult for vessels to

Fig. 9.21 Bull-like navigation lock (Selected from Liu Zhonggui. *The Lingqu Canal*. Nanning: Guangxi Science and Technology Press, 2014: p. 53)

travel. Although sluice gates were not made fine, they fully have the function of a navigation lock. They are in nature simple single-gate navigation lock.

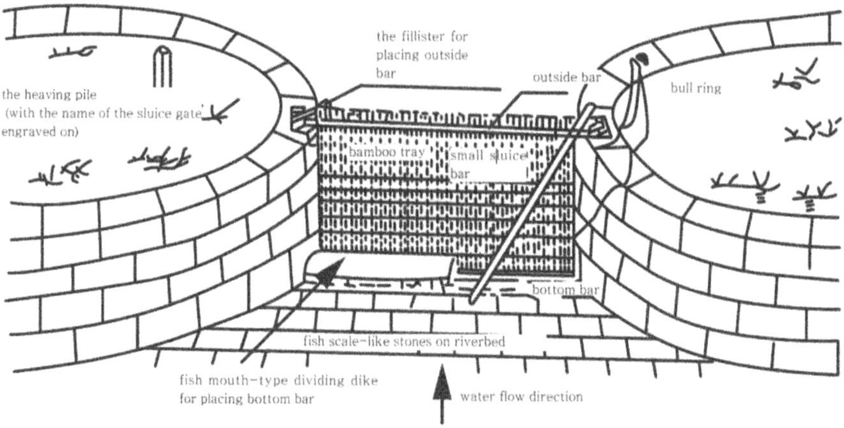

Fig. 9.22 A structural diagram of the sluice gate of the Lingqu Canal (Selected from Liu Zhonggui. *The Lingqu Canal*. Nanning: Guangxi Science and Technology Press, 2014: p. 46)

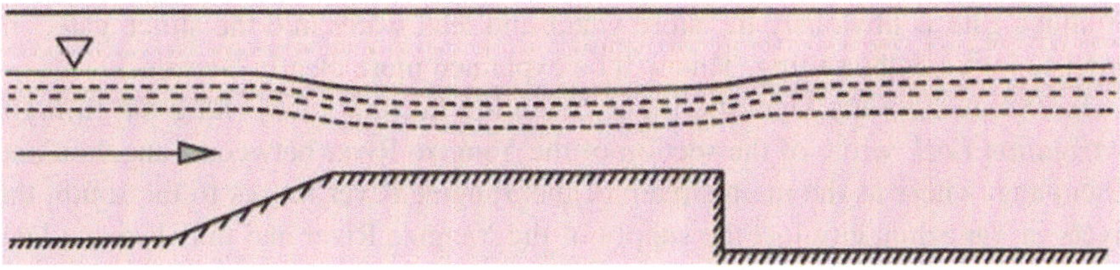

Fig. 9.23 A longitudinal profile of the wide-topped weir flow-like water surface line

As for the Lingqu Canal in the Qin Dynasty, it was also necessary to build facilities with the same function for ships to cross the mountains, although its structure might be relatively simple. As a result of the need to serve the war, perhaps initially there was only the sluice base, but no sluice plate. Even the sluice base might be temporarily constructed, not as regular as those of later periods, but it had to be made in order to sail ships. Even if there was no sluice plate, as the sluice base could narrow the water surface, there would also form a wide-topped weir flow-like water surface line (Fig. 9.23) in the upper reaches of the sluice gate, thus adjusting the water drop in the canal, increasing its upstream navigation depth, and slowing down its velocity of flow. This improved the conditions of sailing in a certain degree, and any boat could sail across with the aid of the dragging by manpower.

*9.6.2 *The Structure and Development of the Conventional Single-Gate Navigation Locks in Ancient Times*

The sluice gates used in agricultural irrigation system to save and distribute water originated much earlier, referred to at least in the irrigation system of the Han Dynasty. The canal navigation lock which was well documented first appeared in Yangzhou, where the Huaiyang Canal connects with the Yangtze River. According to *History of the Song Dynasty: The Biography of Xie Lingyun*, when Xie passed Yangzhou on the way in November of 416, he once saw "the boats crossing over two sluice gates and rising." In 423–424, someone sailed through the navigation lock in Yangzhou, but "fell into the river and died." In the legal document *Shui Bu Shi* revised in 737 of the Tang Dynasty, there also records that "there are two navigation locks in Yangzijin of Yangzhou," as well as the detailed information about the management personnel, material supply, etc., of them [40]. Soon after, Li Bai (701–762), a great poet of the Tang Dynasty, passed through Yangzhou and wrote a poem, in which he made a vivid description of the sluice gate. In the poem, Li praised Qi Huan of the Tang Dynasty for opening up the Yilou River in 738. Along the river, two sluice gates were seemingly built below two bridges, which possibly held the function of opening or closing the sluice plate, with attics built at the head of each one. It can be seen that the structure of a sluice gate had developed fine, absolutely not simple. The function of sluice gate is obviously to "store water and lead water into the sluice gate," in order to improve the sailing. This will be explained more clearly later on.

In 788 of the Tang Dynasty, Liang Su in his *Records of Opening the Aijingpi Navigation Lock* wrote of the section of the Yangtze River between Yangzhou and Zhenjiang, where as the main stream of the Yangtze River moves to the south, the rivers in Yangzhou city lost the supply of the Yangtze River and thus became lack of water. This made the shipping blocked and the urban environment deteriorated. Du Ya, the official in charge of Huainan (the middle part of Anhui Province) at that time, managed to build the Aijingpi (also known as Chengongtang) Navigation Lock which covered an area of a hundred *li* round to supply water to the Great canal, which greatly improved the situation in Yangzhou. It can be seen that the two sluice gates in Yangzhou were built to store tidal water directly for the canal.

Single-gate navigation locks were widely used in ancient times, and had been used for the longest time. Take the Qingfeng Navigation Lock on the Tonghui River in Beijing as an example. The sluice gate had been made of block stones with slurry from the Yuan Dynasty to the Qing Dynasty. The whole base of the sluice gate could be divided into the sluice wall, the foundation, the sluice plate, the device for opening and closing the sluice plate, and so on. The water gate of the base was 6.4 m wide and 6.35 m high. The sluice wall (also known as diamond wall) on each bank was built of block stones with slurry, 13.5 m long. In the middle of a sluice wall, fillisters (also known as sluice slot) were set up from top to bottom (*see* the left side of Fig. 9.24). In the upper side of the sluice gate, the eight-character wing walls, which were 12.5 m long and called water-facing geese wings, were built with slurry along the sluice walls, while the wing walls in the lower side called water diversion dovetails. In the part of the foundation, many wooden piles (also known as ground piles) were set up from bottom to top; on the foundation, there was a layer made of triad soil, about 15 cm thick, on the top of which was covered with stone slabs. Each sluice plate is 6.5 m × 0.3 m × 0.3 m in size, and consists of 13 stoplogs. At each end of each plate there is an iron plate ring for lifting the sluice plate. The device for opening and closing the sluice gate is called gate-entwining stone. There is one such stone on the left and right sides on each bank, thus a total of 4. See the right side of Fig. 9.24. In the center of a gate-entwining stone, there is an axle made of wood or iron, which belongs to fixed pulley kind. When the rope of opening–closing the gate passes through the pulley axle, the sluice plate could open or close under the pull by man or animal power. Ancient ship sluice plates were generally stoplog-kind, rarely whole sluice plate-kind, which might be related to the capacity of the opening–closing device.

Since the Yuan Dynasty, series of single-gate navigation locks had appeared along the Beijing-Hangzhou Great Canal to overcome the shipping difficulties caused by larger topographic drops. Among them, a typical example is the Tonghui River (Fig. 9.25), which extends from Beijing to Tongzhou and was presided over by Guo Shoujing in 1293. It is the northernmost section of the Beijing-Hangzhou Great Canal, about 30 km long between Jishuitan and Tongzhou of Beijing, with the elevation reduced from 30 m to about 10 m. At that year, there were a total of 24 single-gate navigation locks along the Tonghui River, with the separation distance between gates ranging from 500 m to 2,500 m. When adjacent gates open or close one after another,

Fig. 9.24 The gate fillister and the opening-closing device set up on the Qingfeng Navigation Lock in Beijing. The left figure shows the fillister on which the sluice plate is set up, and the right figure shows the stone lifting support for lifting the sluice plate (Selected from [British] Joseph Needham. Colin Ronan (Adapted). *The Shorter Science and Civilization in China (Vol. V)*. Shanghai: Shanghai People's Publishing House, 2003: pp. 240–241. Photo 492 and 493 were taken by Wang Wenbi in 1935)

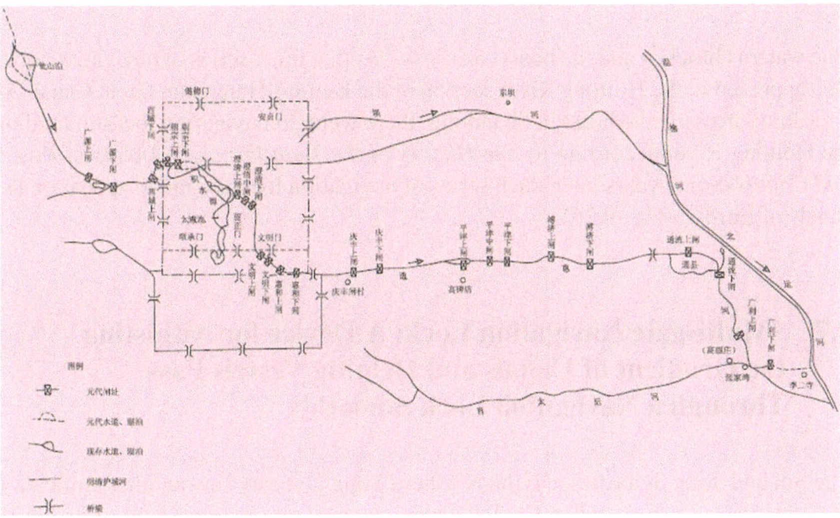

Fig. 9.25 The distribution of the 24 single-gate navigation locks along the Tonghui River in the Yuan Dynasty (Selected from Yao Hanyuan. *History of the Beijing-Hangzhou Great Canal*. Beijing: Hydraulic and Hydroelectricity Press, 1998: p. 91, Fig. 3-10-3)

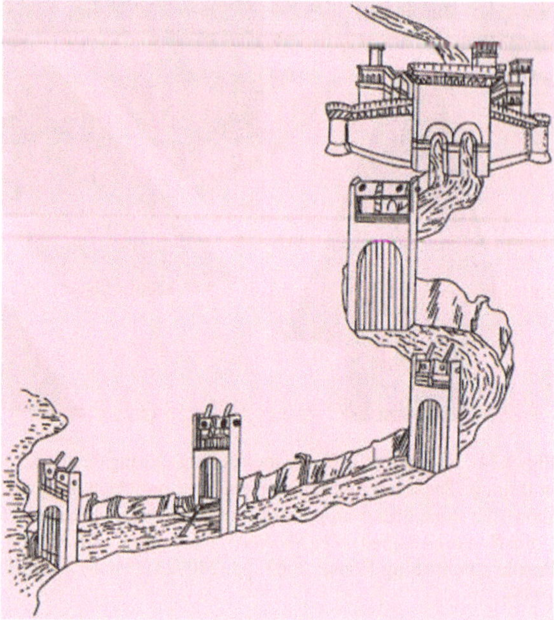

Fig. 9.26 As for series of single-gate navigation locks, when a gate is open, ships must pass through the torrents. This picture is selected from the manuscript by an Italian hydraulic engineer in the fifteenth century (Selected from [India] Charles Singer, E. J. Holmyard, A. R. Hall, et al. (eds.). *A History of Technology (Vol. II)*. Wang Qian, Sun Xizhong (Trans.). Shanghai: Shanghai Scientific & Technological Education Publishing House, 2004: p. 490, Fig. 626)

"the water is blocked and the boats can cross." At that time, series of navigation locks also appeared in the Huitong River section of the Beijing-Hangzhou Great Canal. As a result of crossing the horst in Shandong, there were 26 navigation locks in total on the Huitong River according to *The History of the Yuan Dynasty: Annals of Rivers and Canals*. Some years later, such series of navigation locks began to appear on the canals of Europe (Fig. 9.26).

9.7 *Multi-gate Navigation Lock: A Device for Adjusting the Gradient of Canals and Helping Vessels Pass Through a Navigation Lock Smoothly

The Sui and Tang dynasties and the Northern Song Dynasty saw another unification of China in the history, when China became a world power as a result of its developed economy and prosperous culture as well as its great progress made in internal and external traffic. As for the inland water transportation, a national water transport network centering on the capitals (located in Chang'an, Luoyang and Kaifeng respectively) at that time had formed, connecting the Haihe River, the Yellow River, the Huaihe River, the Yangtze River, and the Qiantang River, with a total length of 2,400 km. In this period, the Huaiyang Canal and the Jiangnan Canal represented the highest achievements made in the science and technology of canal navigation locks, which made great contributions beyond previous generations or even worldwide.

*9.7.1 *Achievements of the Multi-gate Navigation Locks Made in the Northern Song Dynasty*

9.7.1.1 *The Creation of the Xihe River Navigation Lock

A multi-gate navigation lock often has two or more gates. There is a lock chamber between the two gates of a double-gate navigation lock, and two lock chambers of a three-gate navigation lock. It is the embryonic form of modern navigation locks. The multi-gate navigation lock that was clearly documented in China appeared in the first year of Emperor Yongxi (984) in the Northern Song Dynasty. According to *The History of the Song Dynasty: Biography of Qiao Weiyue*:

> …Qiao Weiyue ordered to build two sluice gates on the third weir of the Xihe River, more than 50 *bu* (an old unit for measurement of length, equivalent to five *chi*) apart between the two gates. On the gates houses were built. And some County gates were also set up to store water; as soon as the water was well up, sluice it. Cross-bridges were built ashore, and reinforced its site by stamping earth and piling up stones.

All the disadvantages were thus got rid of, and boats could come and go smoothly.

This document records the reasons why the weir that regulated the flow of water was constructed and the multi-gate navigation locks were built in order to ensure the depth of navigation of the canal, as well as the main structure of original multi-gate navigation locks. The historical books such as *Records of the Song Dynasty* made similar records of this navigation lock [41]. At that time, the Huaiyang Canal was higher than the Huaihe River and the Yangtze River at its both ends. The lock was located at the north entrance of the Huaiyang Canal to the Huaihe River, and the chamber between its two gates is 50 *bu* long, about 80 m. The sluice gate would open when the water level of the canal was close to the tide level of the Huaihe River, so that vessels could pass smoothly. As a result, it's not necessary for ships to wind up the dam and the loss of materials was accordingly saved. Its basic working principle is: when a vessel enters the canal from the Huaihe River, the sluice gate closer to the Huaihe River opens first, and the vessel enters the lock chamber; then the sluice gate closer to the Huaihe River is closed, and water begins to be injected into the lock chamber by water storage facilities to raise the water level of the lock chamber; when the water level of the lock chamber is similar to that of the canal, the second sluice gate opens, and the vessel can sail into the canal, which ends the crossing of the navigation lock. This principle is completely the same as that of modern navigation locks. When a ship passes through a single-gate navigation lock, as any two single-gate navigation locks in a river are generally far apart from each other, the more water stored between the two navigation locks will lose a great deal when the sluice gates open. Compared with it, multi-gate navigation locks with chambers will mainly lose the water stored in the chamber between two sluice gates which are about 100 m far away. For the river section which lacks water resource and has intensive topographic drops, multi-gate navigation locks have obvious superiority over single-gate ones. In addition, if there is only one single-gate navigation lock in a river, it means that the water-head between the upper and lower reaches of the river only rests on the position of the navigation lock. When the sluice gate is opened, the turbulent river is much dangerous for the crossing vessels. A double-gate navigation lock, however, can achieve a smooth passage for vessels to pass through by means of its two sluice gates (one lock chamber) to adjust the water level within. A three-gate navigation lock with two lock chambers can decompose the larger one-stage water-head between the upper and lower reaches of a river into two-stage water-head, so vessels can pass through more smoothly (Fig. 9.27).

In 1072, when Japanese monk Cheng Xun was marshaled into the presence of the Buddhist holy sites of Mountain Tiantai and Mountain Wutai, he traveled northward by boat along the Jiangnan Canal and the Huainan Canal. When he entered the Huaihe River from the canal, he had an experience of crossing navigation locks. He even passed through two multi-gate navigation locks in Chuzhou, where there was only one a hundred years ago. The first one was in Chuzhou (now Huai'an city), where Cheng Xun as well as his boat waited for the rising of the Huaihe River at the first sluice gate on September 17, "and at 7 p.m. to 9 p.m., the first gate was opened in accordance with the tide. The first more than a hundred of ships entered the lock chamber… and stayed there at night." By the next day (18th), "the second gate opened at 7 p.m. to 9 p.m. and the ships sailed out of the chamber." It meant that they passed through the first multi-gate navigation lock smoothly. Then they went on sailing northwestward for "60 *li*, and stopped at Xinkai, Huaiyin County of Chuzhou." On the third day (19th), they continued to sail "for another 60 *li*, and arrived at the gate of the other navigation lock in Shiliang Town at 3 p.m. At 7 p.m.

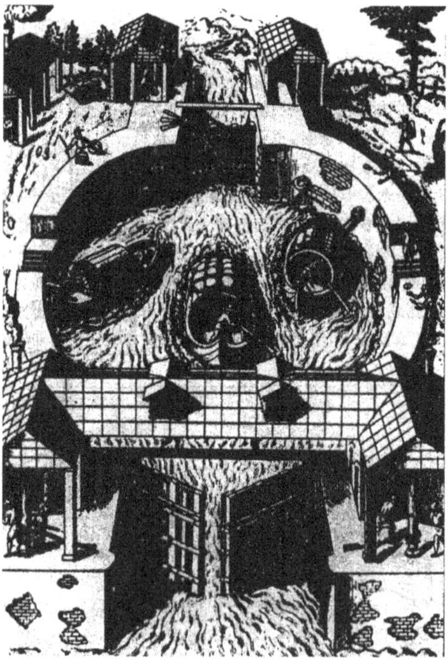

Fig. 9.27 A lock chamber that can move barges from one height to another in a river (Selected from [India] Charles Singer, E. J. Holmyard, A. R. Hall, et al. (eds.). *A History of Technology (Vol. II)*. Wang Qian, Sun Xizhong (Trans.). Shanghai: Shanghai Scientific & Technological Education Publishing House, 2004: p. 489, Fig. 625)

to 9 p.m., the first sluice gate of the navigation lock was opened and the ship entered the chamber. They stopped and rested at the entrance of the Huaihe River." At that time, they were actually in the chamber of the second multi-gate navigation lock. On the fourth day (20th), "they sailed out of the chamber at 3 a.m. to 5 a.m. and into the Huaihe River." The second multi-gate navigation lock they passed through on 19th and 20th was just the Xihe Navigation Lock which was originally built by Qiao Weiyue. The first multi-gate navigation lock and the second one were 120 *li* apart, which conforms to the record of the 120-*li* water course which Lou Yue had traveled from Chuzhou to Huaiyin (60 *li*) and from Huaiyin to Hongze (60 *li*) when he served as an envoy to visit the State of Jin in 1169 [42]. The navigation lock on the Huainan Canal, however, had been damaged as a result of preventing the Jin forces from heading south, so Lou Yue had never seen it in reality.

A History of Technology edited by Charles Singer et al. evaluates the multi-gate navigation lock created by Qiao Weiyue in 984 as "a facility of vital importance in technology." [43] *The Shorter Science and Civilization in China* also assesses the contribution made by Qiao Weiyue as "the first multi-gate navigation lock in the world history… It is clear that the Europeans cannot replace the status of Chinese as forerunner in inventing multi-gate navigation locks." The earliest appearance of multi-gate navigation locks in Europe "was exactly in 1373, in the Netherlands…"

Of course, the European navigation lock had its own characteristics and innovations. A century after the use of multi-gate navigation locks, the famous Italian painter Leonardo da Vinci (1452–1519) designed a rotatable "V-shaped" sluice gate [44] to replace the vertical opening–closing gate. This new gate decreased the power to open and close, and ships could pass through more smoothly. Leonardo da Vinci was said to design a mountain-crossing canal for Florence in 1503, which considered the issue of water supply on top of mountains. More than 100 years later, his advanced design was finally put into practice on the Milan Canal.

9.7.1.2 *The Guazhou Navigation Lock and the Zhenzhou Navigation Lock

Close to the Huaihe River, the Xihe River Navigation Lock was situated at the north end of the Huaiyang Canal. Near the estuary of the Yangtze River, there was also a multi-gate navigation lock at the south end of the Huaiyang Canal, called Guazhou Navigation Lock. In Yangzhou (now Yangzhou city), many weirs or navigation locks were built in the Tang and Song dynasties for navigation. In 1072, Japanese monk Cheng Xun sailed from the Yangtze River to the Great canal and then entered Yangzhou. When passing by Liuzhou Town, he as well as his boat met a navigation lock. The boat entered the lock chamber with the help of the tide of the Yangtze River at 3 p.m. to 5 p.m., and then sailed further for two *li* and stopped at the Guazhou Weir, where their boat was pulled over the weir by 11 oxen on the left and right side, respectively, and went into Yangzi Town.

In this section of the Yangtze River, there was also Zhenzhou Navigation Lock built in the upper reaches of Guazhou connecting the Great canal with the Yangtze

River. The navigation lock was built in 1026 and presided over by the imperial bodyguard Tao Jian. Hu Su (996–1067) once served as supreme government official in charge of the military affairs of Yangzi. He wrote *The Chronicle of the Zhenzhou Navigation Lock* to make a detailed account of it: "In the past, it's rather inconvenient for vessels to be pulled over the weir by oxen". Especially in autumn and winter when the water of the Yangtze River was lowered, it was even more difficult for ships to enter the Huaiyang Canal. After taking office, Tao Jian ordered to build a double-gate navigation lock, which greatly benefitted the waterway transportation. According to *Meng Xi Bi Tan*, the encyclopedia written by Shen Kuo in the Song Dynasty in China, the completion of the double-gate navigation lock not only freed many vessel-pulling workers, but also could sail larger boats: "In the past, a boat could carry rice of no more than 300 *dan*; after the completion of the lock, a boat of the same size could carry 400 *dan*. Later, the carrying capacity of official ships was raised to 700 *dan*, while a private ship could carry more than 800 sacks of rice (2 *dan* per sack)". According to *The Chronicle of the Zhenzhou Navigation Lock*, a new pond was excavated to supply water to the chamber of the Zhenzhou Navigation Lock, which meant that the water level of the chamber could be adjusted within so as to conform to the upstream and downstream water levels. The Zhenzhou Navigation Lock was similar to the Xihe River Navigation Lock in structure, but had its technological superiority in details. Its facilities to supply water to and drain away the water of the lock chamber were crucial to the normal operation of multi-gate navigation locks. In particular, the scene of "circling whirlpools overflowing inside the chamber" more vividly describes the water flow produced when the sluice gates were closed and water was injected from the new pond to the lock chamber. The pond of the Zhenzhou Navigation Lock appeared to have only the function of supplying water to the chamber, slightly different from navigation locks with pond (as mentioned below) which were built 70 years later and could save water.

9.7.2 *Reasons for the Decline of Multi-gate Navigation Locks

Later, the Zhenzhou Navigation Lock and the key positions built along the Huaiyang Canal had been maintained and abandoned many times. Joseph Needham said: "The fact that the foreign visitors of China who had sailed on the Great canal before the 17th century talked about only water locks (i.e. single-gate navigation locks) and sluicing weirs seemed to be sufficient evidence that multi-gate navigation locks had not developed in China... Dr. Joseph Needham believes that he can prove this." This statement made by Joseph Needham is well founded, as we can achieve the confirmation from *Annals of Rivers and Canals (Part Six)* in *The History of the Song Dynasty* (Vol. 96). In 1121 of the Northern Song Dynasty, Xiang Ziyin wrote:

> The Great canal is several *zhang* above the Yangtze River and the Huaihe River. It's hundreds of *li* apart from the Yangtze River to the Huaihe River, which makes it hard to dredge the

passage by manpower. In the Tang Dynasty, Li Jifu abandoned navigation locks, built weirs, harnessed ponds, sluiced and stored redundant water, which kept the water transport of rice open. Zeng Xiaoyun, the then official in charge of shipping and dispatching, strictly performed the system of one available use of the canal per three days, and built a new pond to save water, treasuring water like gold. But in recent years, nonstop travel has been popular, and the transport of salt and tea can earn much profit. Especially, the canal can be open or closed freely by use of undue authority. There is no time for it to store water at all. In addition, certain navigation locks were destroyed, so the water transport in the area of hundreds of *li* from Hongze to Zhaobo is obstructed. To get rid of the disadvantages, it is advisable to build a dam in Port Taizi of Zhenzhou to regain the old passage of the Huaizi River, and to build a dam at the mouth of Guazhou to regain the Longzhou Weir…

He suggested that dams should be built in the north and south end of the Huaiyang Canal in order to solve the problem of water shortage that was caused by privileged interference in both the transportation system reform and the operation of multi-gate navigation locks. The water-saving technology of navigation locks, which had been good and successfully applied, was destroyed under the interference of authoritarian privileges, and had to end up in retrogression. In 1187, the Guazhou Navigation Lock, the Zhenzhou Navigation Lock, and others were repaired. In 1123, it's officially reiterated that "the six navigation locks in the southeast should open and close on schedule. It's prohibited that anyone make improper claim in this regard, or unlock any navigation lock unpunctually, so that the water is sluiced and the key traffic thus obstructed." The effort to restore the original system of the canal, however, turned out to be a useless scrap of paper.

In the early years of the Southern Song Dynasty, the Song and Jin states fought in the Huaihe River basin. "At the beginning, the Jin forces invaded and ravaged Huainan, and did not withdraw. In the fourth year, the Song forces were instructed by imperial decree to burn down the port navigation lock in Wantou of Yangzhou… then ordered to destroy the Zhenzhou Navigation Lock, the key positions built along the Huaiyang Canal and the Chengongtang (i.e. Aijingpi) Navigation Lock in Zhenzhou, in case the enemies use them." This was the destruction caused by war.

9.8 *Navigation Lock with Pond: Multi-gate Navigation Lock with Water-Saving Facilities

A navigation lock with pond was actually a further development on the basis of multi-gate navigation locks. Its additional pond(s) has/have the dual functions of storing water to replenish canals and reuse the water which helps vessels to cross navigation locks. How can the functions be achieved? If a new pond is dug on the low-lying land next to the navigation lock to store water when vessels are heading upstream, the downstream sluice gate is opened; after the vessels enter the lock chamber, the pond with higher water level can supply water to the lock chamber to increase its water level so that the downstream vessels can enter the upstream lock chamber. This operation is the same as that of a multi-gate navigation lock. When vessels sail from the upstream lock chamber to the downstream one, it is bound to let out part of the water volume in the upstream lock chamber in order to line up with the water level of the downstream lock chamber, so that the vessels can smoothly enter the lower lock chamber. Then, with the intention of temporarily storing the abandoned water of the upstream lock chamber in the pond for reuse, the ancients invented the navigation lock with pond.

9.8.1 *Technical Achievements Made of the Jingkou Navigation Lock

Zeng Xiaoyun, nephew of Zeng Gongliang who was the author of the famous military book *The Pandect of Military Classics*, was the first to propose the construction of navigation locks with pond. In 1094–1097, when in charge of the Department of Shipping and Dispatching, Zeng Xiaoyun submitted a written statement *The Advantages and Disadvantages of Navigation Locks with Pond* and suggested that water-saving navigation locks with pond be built in "Guazhou of Yangzhou, Jingkou of Runzhou and Benniu Changzhou" which were extremely short of water to replace the original weirs. He was ordered to preside over the construction, and completed it in 1099. Moreover, he established the rules and regulations for the opening and closing of navigation locks with pond. Among these navigation locks, the records of the Jingkou Navigation Lock located in now Zhenjiang city are relatively complete.

The Jingkou Navigation Lock is located in Dantu County, Runzhou (now Zhenjiang in Jiangsu), facing the Yangtze River in the north. Topographically low in the north and south but high in the middle, this region is the watershed between the Yangtze River and the Taihu Lake basin. In order to keep enough water for ships to pass through Runzhou, weirs were built here in the Tang Dynasty and the Northern Song Dynasty. In 1089 when it met a great drought, Lin Xi, the magistrate of Runzhou, presided over the construction of the upper, middle, and lower navigation locks, which formed two lock chambers. Later in 1099, two ponds were built near the two chambers to store water. The overall plan of the Jingkou Navigation Lock built during the Northern Song Dynasty was repeated in the reign of Emperor Ningzong of the Southern Song Dynasty as follows:

> Nanxu is high in topography, and the Great canal for transport of rice runs through the middle of it. The Xijin Navigation Lock was built here to link to the Yangtze River in order to strengthen the transport of rice. Those who planned to build the canal in the past feared that when the navigation lock opened, the water would run off. They thus built two ponds in Jishui and Guishui respectively to supply water to the canal. Jishui is situated in the east, while Guishui in the north. Both regions were built with navigation locks. When the canal is full of water, the navigation locks will be closed. Otherwise they will open. To supply with saved water when in deficiencies, the canal is thus always open to shipping without the danger of silting-up.

The above quotations, in particular this part "feared that when the navigation lock opened, the water would run off. They thus built two ponds in Jishui and Guishui respectively to supply water to the canal", succinctly pointed out the essence of the innovation made of navigation locks with pond. At the same time, in order to facilitate the connection between a navigation lock with pond and its lock chamber, channels and regulating gates on them were built to “store water for navigation locks with pond, supply water for ditches, and return water to navigation locks”. In this way, the joint operation of a canal and the pond(s) could store some of the abandoned water in the upper lock chamber into the lower one, which would be returned to the navigation lock from the pond(s) if needed. In addition, the water in a low-water pond may be carried into a high-water pond by means of a waterwheel. The supply and storage of water through ponds could accordingly realize the reuse of part of the water, thus achieving the purpose of saving water. This is particularly important for a canal built over ridges where a desperate shortage of water is suffering (Fig. 9.28). Ten years later, “in September of 1099, the Jingkou Navigation Lock with Pond in Runzhou and the Benniu Navigation Lock with Pond in Changzhou were completed”, and special officials were ordered to “go to the Guazhou Navigation Lock with Pond from Hangzhou and manage the new and old navigation locks built in Changzhou, Runzhou, Yangzhou, Xiu and Hangzhou, and have special responsibility for navigation locks with pond according to the issued system”.

After the Southern Song Dynasty moved its capital to Lin’an (now Hangzhou, Zhejiang), the transportation from Anhui, Jiangsu, Hubei, Hunan, Sichuan, and other places to Hangzhou through Zhenjiang was going up every day as a result of the rich supply to the capital. Prefecture Chief Shi Mijian knew the profound mystery of the history that “the ancient people built ponds to store water, which would be used to replenish the canals. That’s the reason why a navigation lock is opened every day,

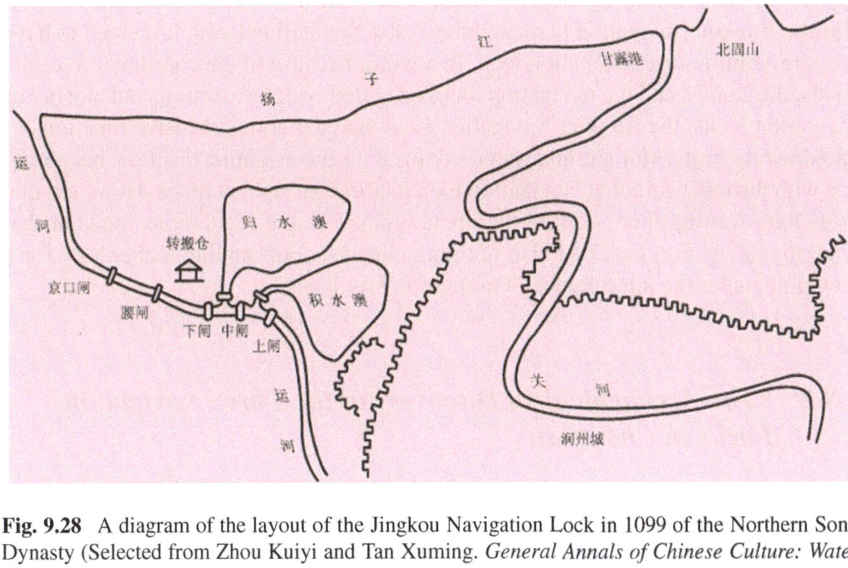

Fig. 9.28 A diagram of the layout of the Jingkou Navigation Lock in 1099 of the Northern Song Dynasty (Selected from Zhou Kuiyi and Tan Xuming. *General Annals of Chinese Culture: Water Conservancy and Transport*. Shanghai: Shanghai People’s Publishing House, 1998: p. 213, Figure above in Fig. 7-7)

but the canal is never emptied". He, therefore, presided over the renovation of the Jingkou Navigation Lock with Pond in 1213. The navigation lock had five sluice gates altogether. "The Jingkou Navigation Lock is several *li* away from the Yangtze River, with the Waist Navigation Lock in the south, a lower, middle and upper sluice gate in the east in succession… connecting with the spring tides, and being opened and closed on schedule to retrench water". The distance between the Yangtze River and the south navigation lock is 1,869 *zhang*, and it was completed in June the next year. At that time, the pond in Jishui had been destroyed. The pond in Guishui was recovered, and connected with the trench in the southwest built to protect the silos of transferred grain, which not only enhanced the safety of the grain silos, but also expanded the capacity of the pond in Guishui. Besides, the canal was drawn to the northeast and connected with the Yangtze River at Ganlu Port, which would be easy to draw the tide of the Yangtze River.

The pond in Guishui had storage of 200 cubic *zhang*, the trench to protect grain silos 200 cubic *zhang*, and Ganlu Port connecting with the Yangtze River in the east 120 cubic *zhang*. The total storage together with that of the newly built lakes and pools was three times the volume of the previous pond in Guishui. Among these rivers, navigation locks were built to control the water. "As vessels pass through the channels, enough water will be needed to support, to which the lower ponds always contribute. If the vessels are too many for a navigation lock to accommodate them, our Ganlu Port will open and close to help regulate it". According to *Annals of Zhenjiang*, the lock chamber between the middle sluice gate and the upper one was 39 *zhang* long and *27 zhang* wide in 1214. The size and capacity could be seen clearly. There was the other lock chamber between the middle sluice gate and the upper one. The three navigation locks cooperated into a two-stage navigation lock; if plus the Jingkou Navigation Lock and the Waist Navigation Lock, it seemed to have become a multi-stage navigation lock. In a word, because of the coordination of the ponds, the canals, and the navigation locks, assorted with the opening and closing of navigation locks, the Jingkou Navigation Lock had the comprehensive functions of tide-drawing, water storage, and water-saving as a key navigation position, becoming one with the most complete navigation lock facilities in ancient times. However, due to its long waiting time, despite the opening and closing regulations made at that time, the navigation lock had also not maintained its outstanding technology for a long time under the interference of autocratic privileges.

9.8.2 *The Archaeological Discovery of the Three Navigation Locks in Chang'an

The Chang'an navigation locks, located in the northeast of now Hangzhou city, Zhejiang Province, are another typical kind of navigation locks with pond. It is recorded in literature that "the three Chang'an navigation locks are situated at 25 *li* northwest of Yanguan County, said to be handed down from the Tang Dynasty".

They were renovated in 1191. The upper source of the canal section is in the West Lake of Hangzhou, and it flows down to Chongchang Port in Dongtiaoxi. At present, the measured water-head here is 1.5–2.0 m, which means navigation locks need to be built to adjust water for the sake of sailing.

The three Chang'an navigation locks (Fig. 9.29) are composed of the upper, middle, and lower navigation lock. The distance between the lower navigation lock and the middle navigation lock is 90 *bu*, the middle one and the upper one more than 80 *bu*. Besides the three navigation locks, two ponds were built: the upper pond covers 98 *mu*, the lower one 132 *mu*. "The redundant water will be stored in the two ponds, and be injected into the navigation locks in the drought". The locations of the three navigation locks and the distance among them recorded were confirmed by the archaeological discovery in 2012. According to the measurement, "the lock chamber between the lower navigation lock and the middle one is about 140 m long, the lock chamber between the middle one and the upper one 125 m long. The base of each navigation lock is about 6.9 m wide". The walls on both sides of the base of each navigation lock are usually V-shaped. The width of the lock chambers should be several times the width of the base of the navigation locks. It can be seen that each lock chamber could accommodate dozens of rice-carrying vessels.

In 1072, Japanese monk Cheng Xun passed by this region by boat, and wrote: "At 11 a.m. to 1 p.m., we arrived at the Chang'an Weir in Yanguan County. At 1 p.m. to 3 p.m., the magistrate came and served us tea at the Chang'an Pavilion. At 3 p.m. to 5 p.m., two navigation locks were opened and all boats sailed out. After the boats were out, the first wood sluice gate was closed, and at the same time the third navigation lock opened. Then the second wood sluice gate was closed, and all boats sailed out. The next day, the river was formerly five *chi* below, but after the crossing of the sluice gates, the upper river fell and was at the same level. Then the vessels sailed out". It's smooth for vessels to pass through the navigation locks, and especially didn't take much time. Compared with the Zhenzhou Navigation Lock in

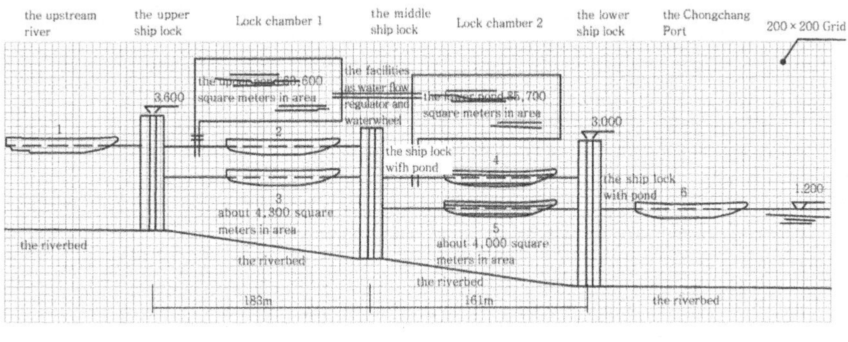

Fig. 9.29 A presumed diagram of the operational principles of the three Chang'an Navigation Locks (Selected from Fu Zhengrong and Zheng Jiali. A preliminary study on the evolution of the Chang'an Navigation Lock on the Jiangnan Canal. Qiu Zhirong, Li Yunpeng (eds.). *The Symposium of Canals*. Beijing: Chinese Literature and History Press, 2014: p. 120, Fig. 12)

Yangzhou, the Jingkou Navigation Lock (which connects with the Yangtze River) in Zhenjiang or the Xihe River Navigation Lock (which connects with the Huaihe River) in Huai'an, the time used was greatly shortened.

The Yuan, Ming, and Qing dynasties all set its own capital in Beijing. As the political center moved northward, the layout of the key canals changed accordingly. Extending 1,700 km and linking Beijing to Hangzhou and further to Ningbo, the Beijing-Hangzhou Great Canal has been open to shipping for 600 years and as the artery of water transportation in China, stretches across the Haihe River, the Yellow River, the Huaihe River, the Yangtze River, and the Qiantang River basins. The unbroken canal crosses complex landforms, especially the horst in Shandong, so its construction took rather long time. As it passes over undulating mountains, the difficulties in both terrain uplifting and the shortage of water resources had to be overcome. Moreover, as the canal parallel intersects with the Yellow River and the Huaihe River in present-day Huai'an city, navigation locks were also required to control water. Remarkable achievements had been made in all these aspects. However, as far as the science and technology of navigation locks is concerned, the Song Dynasty had reached the highest level in ancient times, while the Yuan, Ming, and Qing dynasties had never achieved significant and substantive breakthroughs at all.

References

1. Wang Jian. 2012. Some noteworthy "common senses" about the great canals of China. *China Culture Daily*, Sept. 20, p.14.
2. *See* Zheng Liandi. 1986. *A Brief History of the Lingqu Canal Project*. Beijing: Hydraulic and Hydroelectricity Press, p. 64.
3. Shi Huo (the 19th). 1973. *Records of the Sui Dynasty* (Vol. 24). Beijing: Zhonghua Book Company, p. 684.
4. (The Tang Dynasty) Pi Rixiu. 1981. *A Mélange of Essays*. Xiao Difei, Zheng Qingdu (sorting). Shanghai: Shanghai Classics Publishing House, p. 219.
5. A Chronicle of Emperor Yang of the Sui Dynasty. 1973. *Records of the Sui Dynasty* (Vol. 3). Beijing: Zhonghua Book Company, p. 70.
6. The Sixth Year of the Reign of the Sui Dynasty. 1956. *History as a Mirror* (Vol. 181). Beijing: Zhonghua Book Company, p. 5652.
7. British] Joseph Needham. 1977. *Science and Civilization in China: Civil and Water Conservancy Engineering (Book 10)*[M]. Zhang Yilin, Shen Baixian (Trans.). Taibei: Commercial Press, p. 492.
8. *Annals of Rivers and Canals: The Bianhe River* (Part One). 1977. *The History of the Song Dynasty* (Vol. 93). Beijing: Zhonghua Book Company, p. 2321.
9. *A Sequel to History as a Mirror*. 1995. (Vol. 85). Beijing: Zhonghua Book Company, p. 1959.
10. *A Sequel to History as a Mirror* (Vol. 184). Beijing: Zhonghua Book Company, 1995: p. 4448.
11. *The History of the Song Dynasty*. 1977. (Vol. 463)*: Biography of Fu Weizhong*. Beijing: Zhonghua Book Company, p. 13555.
12. Guo Tao. *The History of Hydro Science and Technology in Ancient China*. Beijing: China Construction Industry Publishing House, 2013: pp. 148–152.
13. *A Sequel to History as a Mirror* (Vol. 297). Beijing: Zhonghua Book Company, 1995: p. 7226.
14. [British] Joseph Needham. *Science and Civilization in China (Vol.2): History of Scientific Thought*. Beijing: Science Press, 1990: pp. 525–527.

15. Cited from Li Rongrong. The great Song Dynasty in the eyes of foreign sinologists: the golden age of prosperity and innovation. *Ke Xue Da Guan Yuan*, 2013, (9): pp. 68–69.
16. *Annals of Rivers and Canals: Rivers in the Southeast* (Part One). *The History of the Song Dynasty* (Vol. 96).
17. [Japan] Cheng Xun. *Records of Visiting Mountain Tiantai and Mountain Wutai*. Bai Huawen, Li Dingxia (checked and annotated). Ma'anshan: Huashan Wenyi Press, 2008: p. 20.
18. Shi Huo Zhi. *New History of the Tang Dynasty* (Vol. 53).
19. [The Tang Dynasty] Pi Rixiu. *A Mélange of Essays*. Xiao Difei, Zheng Qingdu (sorting). Shanghai: Shanghai Classics Publishing House, 1981: p. 220.
20. [The Song Dynasty] Sima Guang. *Annals of the Tang Dynasty. History as a Mirror* (Vol. 232).
21. Dong Wenhu et al. *The History and Future of the Beijing-Hangzhou Great Canal*. Social Sciences Academic Press, 2008: pp. 156–157.
22. [The Yuan Dynasty] Qi Lvqian. Brief biographical sketch of Guo Shoujing in the office of court historian. *Guo Chao Wen Lei* (Vol.50). *Si Bu Cong Kan* (*The Imperial Collection of Four*) (Preliminary edition), p. 544.
23. Zhou Kuiyi. *History of the Science and Technology in China: Volume of Water Conservancy*. Beijing: Science Press, 2002: p. 240.
24. [The Yuan Dynasty] Yang Wenyu. Tablet in memory of the successful opening up of the Huitong River. Cited from (The Ming Dynasty) Wang Qiong. *A Map of Canals* (Vol. V). Yao Hanyuan, Tan Xuming (Arranged and Edited). Beijing: Hydraulic and Hydroelectricity Press, 1990: p.220.
25. [British] Charles Singer, E. J. Holmyard, A. R. Hall, et al. (eds.). *A History of Technology* (Vol. III). Wang Qian, Sun Xizhong (trans.). Shanghai: Shanghai Scientific & Technological Education Publishing House, 2004: p.300.
26. A record of rebuilding the stone dams in Gangcheng. *An Overview of the Canals in Shandong* (Vol. IV).
27. The Tonghui River. *The History of the Yuan Dynasty: Annals of Rivers and Canals* (Part One). Beijing: Zhonghua Book Company, 1976: p. 1588.
28. Zou Baoshan et al. *Management and Development of the Beijing-Hangzhou Great Canal*. Beijing: Hydraulic and Hydroelectricity Press, 1990: p. 43.
29. Guo Tao. The Great canal: The Live Cultural Heritage of the Civilization of Chinese Water Conservancy. *The Three Gorges of China*, 2012, (8): pp.5-13.
30. *Memoir of Emperor Muzong of the Ming Dynasty* (Vol.31). April of the second year in the reign of Emperor Longqing.
31. [The Qing Dynasty] Lu Yao. *An Overview of the Canals in Shandong* (Vol. III).
32. Guo Tao. *The History of Hydro Science and Technology in Ancient China*. Beijing: China Construction Industry Publishing House, 2013: p.161.
33. Tan Xuming, Yu Bing, Wang Yinghua & Zhang Nianqiang. Characteristics and principal components of the heritage of the Beijing-Hangzhou Great Canal. *Journal of Water Conservancy*, 2009, (8): pp.1219–1226.
34. Zhang Weidong. *The Construction of the Hongze Lake Reservoir: Water Conservancy of the Hongze Lake in the 17th Century and Before*. Nanjing: Nanjing University Press, 2009: pp.32–33.
35. Xu Xing. Qing Gai Yun Fa Shu. *Complete Works of Wei Yuan (Vol.15)*. Changsha: Yuelu Press, 2011: p.549.
36. Fan Chengmo. Qing Gai Zhe Cao Liang Shu. *Complete Works of Wei Yuan (Vol.15)*. Changsha: Yuelu Press, 2011: p.509.
37. Hu Mengfei. The Beijing-Hangzhou Great Canal in the eyes of foreigners in the Ming and Qing Dynasties. *Journal of Huaiyin Institute of Technology*, 2014, (2): pp.1-5.
38. Tang Zhaomin. *Selected Collection of Documents of the Lingqu Canal*. Beijing: Zhonghua Book Company, 1982: p.148.
39. Zhou Qufei. *Ling Wai Dai Da (Vol.1): Entry of the Lingqu Canal*. Series Edition of Collections (Vol. 3118): p.7.

40. Yao Hanyuan. *History of the Beijing-Hangzhou Great Canal*. Beijing: Hydraulic and Hydro-electricity Press, 1998: pp.12–15.
41. *Compiled Records of the Song Dynasty: Shi Huo 8: Chapter 31* written by Xu Song records that "in October, 1026, a navigation lock was built on the Beishen Weir of Chuzhou and the Jiangkounan Weir of Zhenzhou respectively" and that "in 1165... the two navigation locks in Hongze were fixed". As for the names of the navigation locks, Yao Hanyuan made a clear discrimination in the notes on page 56 and page 57 in his *History of the Beijing-Hangzhou Great Canal*.
42. [The Southern Song Dynasty] Lou Yue. *Bei Xing Ri Lu. Gong Kui Ji* (Vol.111). The edition of *The Imperial Collection of Four*.
43. [India] Charles Singer, E. J. Holmyard, A. R. Hall, et al. (eds.). *A History of Technology (Vol. III)*. Wang Qian, Sun Xizhong (Trans.). Shanghai: Shanghai Scientific & Technological Education Publishing House, 2004: p.300.
44. See *Encyclopedia Britannica* (International Chinese Edition) (Vol.10). China Encyclopedia Press, 2002: p.160.

Weibing Zhang Born in Hongdong, Shanxi in 1974, PhD, senior engineer of the China Institute of Water Resources and Hydropower Science, standing director and deputy secretary-general of the Disaster History Committee of the China Disaster Defense Association.

Research interests: History of water conservancy and history of disasters.

Major works: Editor-in-chief of Chinese Water Conservancy - Today in History (1949–2012);

Papers: Discussion on the Evaluation Indicator System and Method of Regional Extreme Drought; Cultural Background of the Distribution of River Circulation Names in China; Study on Water Safety in the War of Ming Dynasty

Awards: The first prize of Outstanding Paper in the first National Water Culture Forum of the Ministry of Water Resources; the Outstanding Paper Award of the China Water Conservancy Society.

Kuiyi Zhou Born in Liaoyang, Liaoning Province in 1938, Professorate senior engineer in China Institute of Water Resources and Hydropower Research. He once served as the president of the Hydraulic Engineering Society of China Water Conservancy History Research Association, and a member of the International Committee on Irrigation and Drainage (ICID). He is currently the honorary president of the Hydraulic Engineering Society of China Water Conservancy History Research Association.

Research interests: Water Conservancy History and Flood Prevention and Mitigation.

Major works: *Monographs: the History of Chinese Science and Technology: Water Conservancy; Review and Prospect of Chinese Water Management Strategy*.

Papers: "Study on the gains and losses of Jianhu Lake reclamation"; "Historical Research on the Evolution and Flood Control Planning of Jingjiang River and Dongtinghu Lakes"; "'Historical Model' and Disaster Research".

Awards: The History of Chinese Science and Technology: Water Conservancywon the 3rd Prize of the 3rd Guo Moruo Chinese History Award;Research on China's Major Natural Disasters and Disaster Mitigation Strategies won the 2nd Prize of the 1999 National Science and Technology Progress Award;Study on Drought and Floods in China won the 2nd Prize of the 2001 National Science and Technology Progress Award.

Chapter 10
Plows and Seed Ploughs

Lisheng Feng

Plows and seed ploughs are related agricultural machinery, which as important inventions in the history of human civilization have played an important role in the history of agriculture and machinery.

Plows are a kind of farm tool used to plough or deeply loosen the soil, with mouldboard-plow the most commonly used. A plow is mainly composed of a plough share, a mould-board, a plough beam, and the like. The working surface of a plough share and a mould-board are both a continuous and smooth curved one, whose shape and size are selected according to different soil and cultivation requirements, and related to the traveling speed of the whole unit. Different curved surface of a plow body has different functions such as soil-turning, soil-loosening, soil-breaking, weeds or stubble-covering, etc. Chinese traditional plow is the mouldboard plow, and the mould-board had been used in the Han Dynasty, becoming the most advanced farming tool in the world at that time. The traditional plow was improved by Lu Guimeng in the Tang Dynasty and turned to be more perfect in structure, thus finalizing the basic form of curved-beam plows in ancient China.

As a sowing kind of farm machinery, a seed plough can sow crop seeds in the soil according to certain agronomic requirements. In ancient times, the seed plough was also known as seed plough cart. It can be divided into two kinds: one with one leg and one with four legs, widely used in the northern region of China. The seed plough can accurately control the sowing amount of seeds, row spacing, and seed level. As an important invention in ancient China, seed ploughs had been used to sow in the drill as early as the Han Dynasty. Thereafter seed ploughs had been used for a long time, and got improved and perfected. In the West, drill seeders were not used to sow till the eighteenth century.

L. Feng (✉)
Institute for History of Science and Technology & Ancient Texts, Tsinghua University, Beijing, China
e-mail: fls@tsinghua.edu.cn

J. Hua and L. Feng (eds.), *Thirty Great Inventions of China*,
https://doi.org/10.1007/978-981-15-6525-0_10

10.1 The Emergence and Early Development of Plows

China had invented the ploughs and plowshares for a long time to turn and dig the soil, which were then developed into plows. The appearance of plows can be traced back to the Neolithic Age. Stone plow (Fig. 10.1), soil-breaker (Fig. 10.2), and soil-weeding utensil commonly found in the Liangzhu Culture were all developed to satisfy the needs of paddy field farming. The stone plow was first discovered in China. It is shaped like a flat and thin isosceles triangle, with the closed angle 40–50°, the two waists edged, and 1–3 holes in the middle. The small one is only 15 cm long, while the older one nearly 50 cm long, the rear end slightly flat or concaved. Strictly speaking, the stone plow is only plough share, a part of a whole plow, which needs to be fixed on the plow bed to work. The emergence of plows not only improved the efficiency and quality of farming on soil, but also provided the possibility of utilizing animal power.

During the Shang and Zhou Dynasties, the production of stone plows was obviously improved. The stone plows made in this period unearthed in Changxing County, Zhejiang Province (Fig. 10.3) are different in shape from those made in the Neolithic Age. According to the analysis, these stone plows are in nature soil-loosening tools which can continuously move along the horizontal direction by manpower. Between their upper and lower parts, a plank is sandwiched, only with the edge shown, which aims to avoid being broken off because of its fragile texture. The later copper and iron plough shares shaped like isosceles and equilateral triangles all evolved from

Fig. 10.1 The stone plow made in the age of Liangzhu Culture unearthed in Yuhang, Zhejiang Province (Photo by Feng Lisheng)

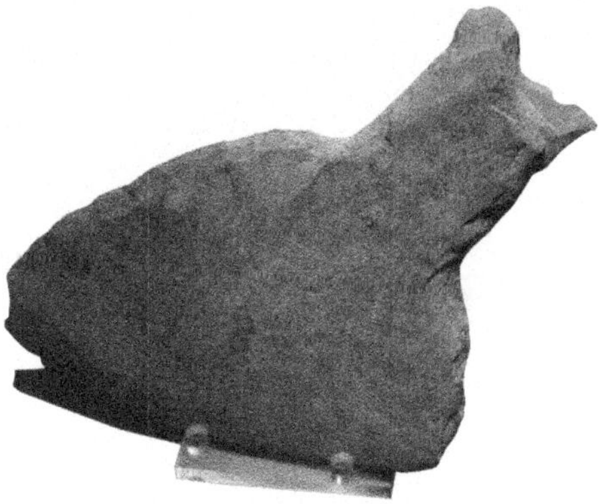

Fig. 10.2 The soil-breaker made in the age of Liangzhu Culture unearthed in Yuhang, Zhejiang Province (Photo by Feng Lisheng)

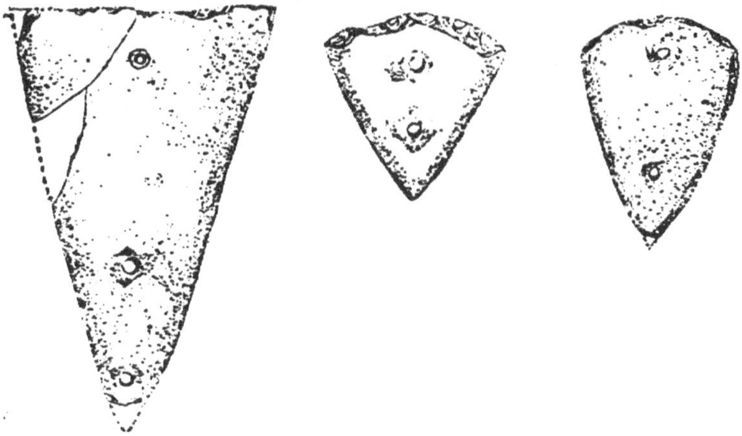

Fig. 10.3 The stone plows unearthed in Changxing County, Zhejiang Province

the narrow, wide stone plows made in this period. Copper plows had appeared in the Shang and Western Zhou Dynasty (Fig. 10.4) [1]. However, the number of them unearthed is very small, which possibly means that they had not been widely used. The improvement of stone plows and the appearance of bronze plows in the Shang and Zhou dynasties were a great development of agricultural machinery in ancient China. Due to the need for agricultural farming and the significant changes in materials, the types and forms of farming tools in the Spring and Autumn and Warring States Period had also been greatly developed. At present, the iron farming tools unearthed in this period cover jue (a pickaxe-like earth-digging tool), sickle, spade, hoe, shovel, rake, plough share, and so on, which could basically cater to the requirements of different sessions in agricultural production such as reclaiming, ploughing, leveling, weeding, soil loosing, harvesting, etc. By the time of the Warring States Period, complete sets of small and efficient farm tools and agricultural machinery

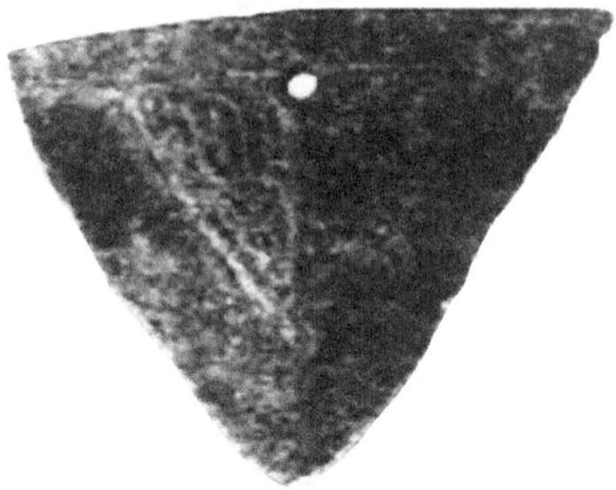

Fig. 10.4 The copper plow made in the Shang Dynasty

had been produced. The production materials of farm tools had also changed, especially iron wares widely used. The formation of ploughing was the most influential. In the unearthed farm tools made in the Warring States Period, iron plough shares are included. In the former Wei, Yan, Zhao, Qin States, and other areas of the Warring States Period, iron plough shares were unearthed, and cattle's farming was proved to be gradually adopted.

Iron plough shares mostly have a V-shaped cap (Fig. 10.5) [2]. A wooden plow embedded with this kind of plough share cap could be used to loosen soil and dig trenches. However, it has limited effect in ploughing and making ridges, though much higher than the "farming by ploughs and plowshares."

Figure 10.6 shows the restored wooden plow with V-shaped iron plough share cap [3]. Since the iron plough share cap can be replaced in order to keep the plough share sharp, it is of great significance to improve the efficiency of farming.

The widespread use of ironware, the finalizing of farm plows, and the popularizing of cattle's farming were three important signs of the development of social productivity since the Qin and Han dynasties. During the Qin and Han dynasties, iron plough shares were promoted throughout the country. The iron plough shares made in the Han Dynasty have been unearthed in Shaanxi, Henan, Shandong, Hebei,

Fig. 10.5 The iron plough share cap made in the Warring States Period (unearthed in Guwei Village, Huixian County, Henan Province)

Fig. 10.6 The restored picture of the wooden plow with V-shaped iron plough share cap made in the Warring States Period

Gansu, Fujian, Liaoning, Sichuan, Inner Mongolia, and other places. In 1970, an all-iron plough share was unearthed outside the north gate of the First Emperor of Qin (259–210 B.C.) Cemetery. It is 25 cm long, with the two wings 25 cm apart; at the intersection of the two wings there is a ridge, which is 5 cm long and about 1 cm wide [4]. Later in 1980, another all-iron plough share made in the Qin Dynasty was discovered at the site of Chenjiagou, Lintong County. The two plough shares made in the Qin Dynasty were larger in size and deeper in ploughing than the V-shaped plough share caps widely used in the Warring States Period. During the reign of Emperor Wudi of the Han Dynasty, iron farming tools were greatly promoted, and cattle's farming got further popularized. According to *On Salt and Iron*, "Ironware has great use by the people," and "whether the tools are sharpened or not, their effects could be a far cry from each other." This reflects the understanding of the importance of the use of ironware and advanced agricultural machinery in the Han Dynasty. The plough shares made in the Han Dynasty were unearthed in as many as 40 or 50 places, mostly all-iron plows including tongue-shaped big plough shares, small plough shares, and giant plough shares. Tongue-shaped big plough shares are more than 30 cm in both length and width, weighing 7.5 kg or so. Small plough shares have similar length and width, 10.8–17.5 cm in length, 7–11.8 cm in front width, and 9–14 cm in rear width. Giant plough shares are generally about 40 cm long and weigh 9–15 kg. For example, the triangular plough share unearthed in Tengxian County, Shandong Province, is 48 cm in length and 45 cm in top width.

The adoption of the device of mould-board in farm plows in the Han Dynasty was a more important innovation in the history of agricultural machinery. Mould-board, also known as plough ear or plough mirror, is set above the plough share, so that plows can not only be used for loosening and breaking soil, but also for turning over the soil to one side, burying weeds under the soil as fertilizer, and killing insects as well. Therefore, the plows with the device of mould-board are said to be very efficient agricultural machinery. The iron mould-boards made in the Han Dynasty have been unearthed in many places such as Anqiu in Shandong, Zhongmou in Henan, and Xi'an, Xianyang and Liquan in Shaanxi.

In Shaanxi, tongue-shaped big plough shares made in the Han Dynasty were unearthed, as well as V-shaped ploughshare caps and mould-boards (Fig. 10.7). When

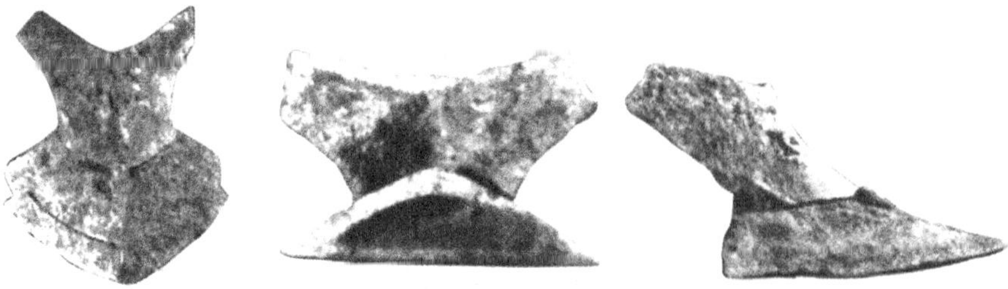

Fig. 10.7 Embedded iron ploughshare in mould-board made in the Western Han Dynasty

unearthed, the V-shaped plough share caps were embedded in the tip of plough share, or stored separately. A mould-board is set above the plough share, and the curved surface of the mould-board and the back curved surface of the plough share together make up a nearly unbroken curved surface. As the bending and torsion aside of the mould-board is not big at all, the soil could be pushed forward and upward to a certain extent to be broken, and then overturns to the right front and falls down, forming discontinuous soil strips. Its ploughing depth is seldom restricted by the ploughing width, and the former is usually larger than the later. At low speeds, the soil can also be broken into ridges.

The mould-boards of the Han Dynasty unearthed can be divided into two categories: the one which is rhombus-shaped, tile-shaped, or square-shaped but with an unfilled corner, with an average length of 45.8 cm and a width of 23.1 cm, and only turns the soil over to one side, and the one which is saddle-shaped, with an average length of 18.3 cm and a width of 20.8 cm, and whose two sides can turn over soil. Figure 10.7①, ② are the front view and rear view of the saddle-shaped embedded iron plough share in mould-board made in the Western Han Dynasty respectively, and Fig. 10.7③ shows its side view. According to the pictures of plow farming unearthed from the Han Dynasty tombs, the plows then had been basically shaped, generally composed of plough beam, plough tip, plough base, plough share, and mould-board, with plough arrows used to adjust the depth of ploughing. After the Eastern Han Dynasty, cattle's farming was also gradually promoted in the south, spread into some of the northeast, northwest, and southwest regions.

At that time, the vertical plough beams were divided into two kinds: single vertical beam and double vertical beam. The former adopted the farming method of two cattle carrying the bar together to pull the plow, that is, "two cattle carrying the bar," while the latter of one cattle to pull the plow, whose plough arrow could adjust the depth of ploughing. The use of the former began after Zhao Guo, an agriculturalist in the middle of the Western Han Dynasty, carried out the "method of rotating field," and was related to the use of both animal power and large-scale plough shares, becoming an important symbol of the development of productivity. The stone relief of the Han Dynasty unearthed in Shuanggou, Suining County, Jiangsu Province (Fig. 10.8) vividly depicts the farming scene of peasants adopting the method of two cattle carrying the bar. The method of one cattle pulling the plow was born later than that

Fig. 10.8 The stone relief of cattle's ploughing in the Han Dynasty (Collected by the National Museum of China)

of two cattle carrying the bar, and since the Han Dynasty, the two had coexisted for a long time.

On the stone relief about cattle's farming of the East Han Dynasty unearthed in the tomb of Wang Deyuan in Suide, Shaanxi Province, there was carved a vivid document of the farming method of one man driving one cattle (Fig. 10.9) [5]. The rubbing of the stone relief is 135 cm high and 29 cm wide. Its general appearance is divided into four grids, the third of which reflects the farming of one man driving one cattle: under a lush mulberry tree, a farmer is holding the plow with one hand, with a whip lifted to drive the cattle to walk forward and plough. The plow in the farming was actually set with a plough arrow, which could adjust the depth of ploughing. This was an advanced ploughing method and had important technical significance. After the Han Dynasty, the pulling by two cattle tended to be gradually decreased, while that by one cattle gradually increased, but it was not until the Wei, Jin, Southern, and Northern dynasties that the pulling by one cattle was popularized. In the Sui and Tang dynasties, it was still in use. According to *Jin Shu* (*History of the Jin Dynasty*) (Vol. 109), Huang Zeng once ordered that "all enclosed ground for raising animals should be given up to the peasants without land. The poor have no assets and cannot live

Fig. 10.9 The stone relief of cattle's ploughing in the Eastern Han Dynasty

their lives, so each of them should be given one cattle. If any one has strength to spare and is willing to reclaim the official land with official cattle, the old provisions made in the Wei and Jin dynasties should apply." If the people without assets were given one cattle to farm, the farming of one plow used with one cow should be adopted.

By the reign of Emperor Wudi of the Han Dynasty, the three important components of a farming plow, that is plough frame, plough share, and plough pattern, had been preliminarily shaped, realizing the fundamental transformation from ploughs and plowshares to plows. And this transformation started with the promotion of the ploughing side by side when Zhao Guo carried out the "method of rotating field."

By the Wei, Jin, Southern, and Northern dynasties, the plough share and the beam of a plow had witnessed some changes. The shape of plough share was changed from the isosceles triangle of the Han Dynasty to the long and narrow tongue shape of cattle, which could meet the needs of both one cattle pulling the plow and ploughing muddy fields. As the plow with long vertical beam was inconvenient to operate, especially in the mountainous or hilly areas, "it is difficult for it to turn around." In the Northern Wei Dynasty, wei plow was invented, which was more convenient to operate than the plow with a long vertical beam [6]. According to *Qi Min Yao Shu* (*Important Arts for the People's Welfare*), "Now there is a kind of farming plow in the eastern Liaoning Province, whose beam is four *chi* long, easy to turn around."

10.2 The Improvement of Plows

During the Sui and Tang Dynasties, the emergence of curved-beam plow was one major technological innovation of Chinese plows. On the basis of summing up the folk technologies of plow making in the area of Jiangdong (the lower reaches of the Yangtze River), Lu Guimeng of the Tang Dynasty completed the *Book of Ploughs and Plowshares*, describing in detail the parts, dimensions, and functions of the plows commonly used in the area of Jiangdong. The construction of Jiangdong plow covers a plough trowel and a mould-board made of metal, as well as plough base, controlling trowel, bamboo tablet, plough arrow, plough beam, plough tip, oblong wooden-frame, curved chock, and wooden tub made of wood. The improved curved-beam plow (i.e., Jiangdong plow, Fig. 10.10) by Lu Guimeng had the advantages of

Fig. 10.10 The restored picture of Jiangdong plow (Copied by the Institute of Scientific and Technological History and Ancient Literature, Tsinghua University)

being simple in structure, sturdy, and durable, easy to operate, convenient to adjust, smooth to farm, and the like. Each of its components had special functions and a reasonable form.

The mould-board of a Jiangdong plow is set on top of the plough trowel, forming a composite device with a curved surface to dig and turn the soil. The plough base and the controlling trowel fix the plough share tightly to enhance the stability of the plow. The bamboo tablet is used to defend the mould-board. The plough arrow and the oblong wooden-frame are devices for adjusting the depth of ploughing, by means of enlarging or narrowing the angle between the plough beam and the plough bed to make the plough share dig deeply or shallowly by adjusting the oblong wooden-frame and the plough arrow. The width of cultivated land by ploughing can be controlled by the plough tip. The most important thing was the adoption of short curved beams. In the past, the plows adopted vertical and long beams, which were not flexible enough to turn back and swerve when in ploughing, not easy to turn the soil, and not high in efficiency. Jiangdong plows, on the other hand, greatly shortened the length of the beam, removed the plough arm, and diminished the plough frame, reducing the weight of a plow. At the same time, a plough tray, ploughing ropes and a curved yoke were added, so animal power was saved, only one cattle required to pull, thus changing the arduous method of two cattle carrying the bar since the Han and Wei dynasties. It was surely one of the major reforms of plows in China [7].

The structure of the Jiangdong plow had been quite perfect, though its development also experienced a long process. According to the cattle's ploughing on the mural in Li Shou's tomb unearthed in Shaanxi, long curved-beam plows had appeared in the early Tang Dynasty. But this kind of plow, like the plow with two cattle carrying the bar in the Han Dynasty, was also pulled by two cattle, and did not change much in structure except that its long vertical beam was changed into a long curved beam. This change, however, narrowed the angle between the end of plough beam as well as the plough arm and the shoulders of cattle, which could reduce the labor intensity of cattle and was thus a progress. On the other hand, as the beam of Jiangdong plows was changed into a short curved one, in the front end of which there was set a rotatable tray, and whose ropes tied to the curved yoke were harnessed to cattle, not only animal power was saved, but also it's flexible to turn back. The kind of cattle's ploughing covered a small space, so was particularly suitable for the farming of paddy fields in the south. In fact, it first appeared and got promotion in the area of Jiangdong. It is generally believed that Jiangdong plows were the first to be created by the working people of the regions south of the Yangtze River. As a technical innovation from the folk, it was made about in the middle and late Tang Dynasty.

The emergence of the curved beam and plough tray of Jiangdong plows was a great progress for traditional plows. But as the parts transmitting power and the working machinery were still integrated, the pulling force was still difficult to make full use of, and the operation flexibility needed to be further improved. Since the Song Dynasty, folk plows had used hooks and soft set to separate the plow from the tools controlling cattle, thus making the plow relatively independent. With the separation of cattle and plow, the structure of the plow had also changed, the beam greatly shortened, the body obviously decreased, the number of parts cut down, and the weight greatly reduced. In this way, the structure of plow turned to be more reasonable, more flexible, and convenient to operate. Cattle's farming thus became not only suitable for paddy fields and flat fields, but also extended to mountainous areas, even for terraced fields. The traditional plows had developed perfectly by this time. Figure 10.11①, ② show the cattle's yoke and the wooden tub, respectively, covered in the section of "Atlas of agricultural instruments" in *Nong Shu* (*History of Agriculture*) written by Wang Zhen of the Yuan Dynasty. The yoke, also known as shoulder yoke, was cut two holes at both of its ends in order to "run through the ploughing ropes." At the same time, a short rope was tied under it "to control the nape of cattle." At this point, the wooden tub had been separated from the beam, with holes dug out at both ends, a ploughing rope attached on the top, and a ring set in the middle and connected with the beam through the hooks. The yoke, the ploughing ropes and the ploughing frame were accordingly joined together into a soft set, that

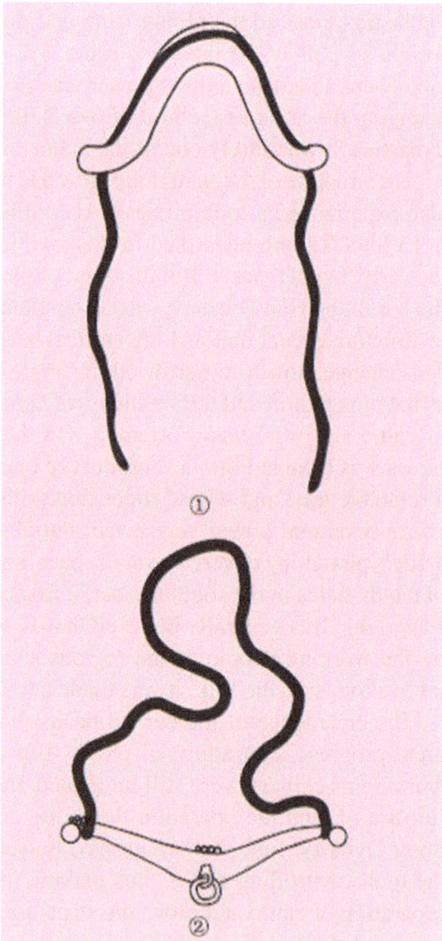

Fig. 10.11 Diagram ①: Cattle's yoke, Diagram ②: The wooden tub

Fig. 10.12 The hooks unearthed in the sites of the Liao and Jin dynasties

is, the driving part of a plow. Figure 10.12 shows the hooks unearthed in the sites of the Liao and Jin dynasties [8].

During the Ming and Qing dynasties, most of the ploughing techniques and methods achieved in the Song and Yuan dynasties had come down, without too many changes. Since the Ming and Qing dynasties, due to the increasing population, the cultivated land per capita was increasingly decreased. The traditional farming tools could fully meet the needs of agricultural production, so had stronger vitality. The increase of agricultural productivity was not mainly manifested in farming tools, but in the folk, there had been people who were committed to the improvement or invention of plows, and appeared some improved agricultural tools. The depth of ploughing by early plows was only a few *cun*, "always three *cun*" as the popular saying in the folk. In the Ming and Qing dynasties, there appeared big plows that could plough deeper, which further improved the technology of farming. According to *Ma Shou Nong Yan*, "The plow chosen to use should be temperate in depth... Those plows which could produce extraordinary depth will make the soil fertility uneven." *Zhi Ben Ti Gang* records: "To plough the soil, the farming of one plow with one cattle can be adopted, or of one plow with two cattle, or of one plow with three or four cattle, or of two plows with one cattle. These plows could produce a shallow ploughing of a few *cun*, or a deep ploughing of more than one *chi*, even of two *chi*." It also points out: "The size of a plow used is dependent on the rigidity of soil. If the soil is rigid, a big plow is used. Otherwise, a small one is used." There were also "firm and heavy plows with two or three cattle" [9]. specially for reclaiming wasteland. In the late Qing Dynasty, the structure of plows had changed due to the use of iron plough beams. For example, the plows used in Jiangsu all adopted iron beams and removed the plough arrow that supported the plough beam. A long hole-slot was dug in the middle of the plough tip, and chocks were used to fix the plough beam and adjust the depth of ploughing. In this way, although the structure of plows was simplified, the efficiency of ploughing was not affected at all. This was actually an improvement, so the plows are still used by peasants. During the Ming and Qing dynasties, there was also a great improvement in the method of connecting the body of cattle with the body of plows. Originally, a circular ring was horizontally set on the end of the beam, and the set hooks bent upward were vertically hung on the ploughing plate. The circular ring was changed into one vertically set, and the set hooks were changed into S-shaped set hooks vertically hung. The soft set of cattle thus could have a greater swing around it, which made it very convenient and flexible to operate.

Short curved beams are still used in the rural areas of China. Compared with vertical beams, they are shorter and greatly lightened in body, thus very light and dexterous. Figure 10.13 shows the structural diagram of the plows widely used in the areas south of the Yangtze River (Shanghai, Jiangsu, etc.) more than 30 years ago [10]. In the diagram, the jack board is actually the plough arrow, one of the main components for pulling, and the joint of the upper part of the jack board and the plough beam is clearance fit; trapezoidal mortises were dug out on the plough beam, through which the upper end of the jack board ran; one high hole and one low hole were drilled at the fixed position of the plough beam, and pinned by wood pins to fix the adjusted position of the plough beam.

As for the production of wooden plough beam, it is very important to select materials as the curve can not be machined freely. The old carpenters who make the

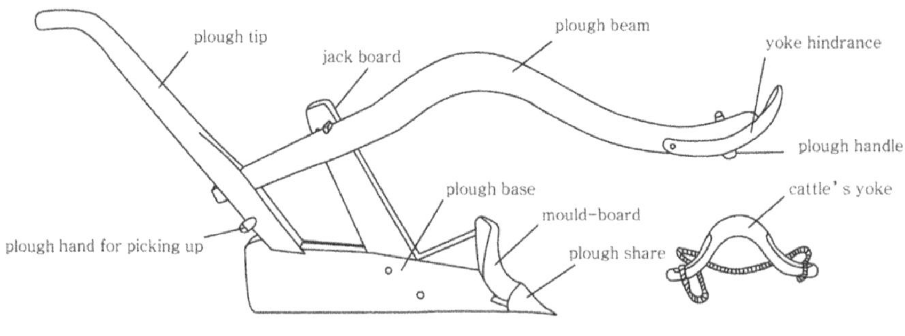

Fig. 10.13 The main structure of a plow

plows often say: "Half of the material, half of the production." Wooden plough beams are usually made of Melia azedarach Linn, wild elm, camphor tree, Chinese scholar tree, etc. If these are not available, black-velvet tree and Japanese ash can also be used. Generally, the bending between the trunk of a tree and one of its branches is mostly made full use of, which is called "making use of the branch, seizing superiority of the trunk." When necessary, artificial pressure should be placed on those suitable trees to make them curve and grow in a fixed shape for a year or two. Therefore, it is no good using ordinary wood to make the plough beam. It's also no good producing the required curve by sawing a large wood, as this kind of plough beam with artificially sawn curve cannot stand the pulling of cattle and will break in use. Accordingly, the trees that grow naturally must be searched. The materials used to make plough beams should not only tend to curve from top to bottom, but also to curve from left to right. When ploughing, the plow made of this kind of materials can turn up the soil blocks, which will not be blocked between the plough beam and the mould-board, thus easier to turn over the soil.

In 2006, "The Achievements Exhibition of China Intangible Cultural Heritage Conservation" was held at the China National Museum, in which the wooden curved-beam plow originally used in Xiaohuang Village, Congjiang County, Guizhou Province as a traditional curved-beam plow was displayed. Figure 10.14①, ② are the pictures of it taken by the author in two different directions. It can be seen that the plough beam not only tends to curve from top to bottom, but also to curve from left to right.

In modern times, one of the important improvements of traditional curved-beam plows was the use of iron-made curved beams. The iron-made curved beams appeared in the late Qing Dynasty, and got promoted in the period of the Republic of China. The curved-beam plows widely used in the rural areas of China in the thirties and forties

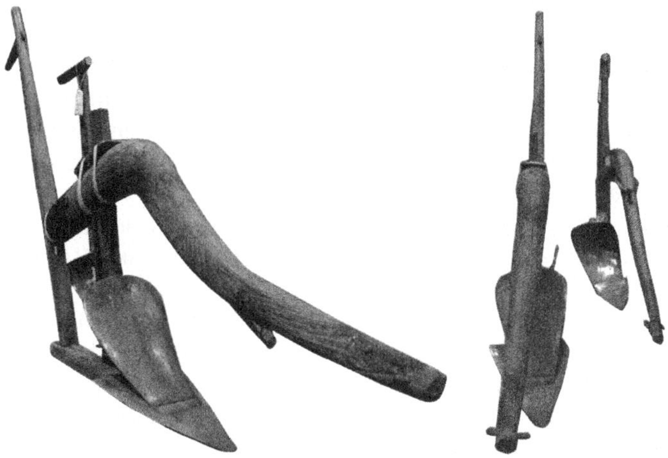

① ②

Fig. 10.14 The wooden curved-beam plow used in Congjiang County (Photo by Feng Lisheng)

of the twentieth century could be divided into iron-beam plows and wood-beam plows according to the material made of plough beams. The former was adopted in the areas where the iron-making industry was developed (such as Wuxi, Shanghai, etc.), and the latter was also rather popular. There were certainly many areas where both the two were used simultaneously.

The use of iron beams not only increased the service life of plows, but also could machine the curve freely, as well as improving the performance of plows. It removed the part of the plough arrow that supports the plough beam and thus simplified the structure of plows. A long hole-slot was dug in the middle of the plough tip, and the inosculated positions between the plough tip and the plough beam were fixed and adjusted by chocks. The chocks could be movable in a certain range, and could lower or raise the plough beam so as to adjust the digging of the plough trowel into the soil and thus control the ploughing depth.

Nowadays, the traditional Chinese plows have a life of their own as usual, as cattle's farming still plays an important role in many areas such as Yunnan, Guizhou, Sichuan, Tibet Autonomous Region, and so on. In particular, on the cold-water terraces of the southern mountainous areas and the dry-land terraces of the northern mountainous areas, where each land parcel is not big at all, the slopes were comparatively large, and the tractors are difficult to arrive, the farming by means of traditional plows is still more common.

10.3 The Invention and Improvement of Seed Ploughs

The (animal-drawn) seed ploughs were invented in the Western Han Dynasty, and thereafter had been further promoted. In the last years of Emperor Wudi of the Han Dynasty, in order to develop agriculture, he ordered to carry out the "method of rotating field," that is each *mu* of field is separated by three ditches and three ridges, with each ditch one *chi* wide and deep; crops are planted in the (low) ditches, and the positions of the ditches and ridges can rotate every year. The method requires that the positions of crops to be planted in should be well-distributed in spacing and every direction. It means that the broadcast sowing and the spaced sowing are impossible to meet the intended requirements, and it is inevitable to use seed ploughs to sow in a drill. Cui Shi of the Eastern Han Dynasty once made a brief introduction to it in his *Zheng Lun* (*On Politics*): "Emperor Wudi appointed Zhao Guo as the official in charge of gathering grain. Zhao taught the people how to cultivate and grow. His method was to use the three-plough with one cattle. One person could operate according to it to sow seeds and pull the seed plough at the same time, and plant one *qing* (a unit of area, = 6.6667 hectares) of field each day. To this day, the three areas of Jingzhao, Zuofengyi and Youfufeng in Yongzhou still benefit from it." "Three-plough" here actually refers to the three-legged seed plough.

In archaeological data, there are also murals of seed ploughs unearthed. For example, in the murals in the tombs of the late Western Han Dynasty unearthed in Zaoyuan, Pinglu, Shanxi Province, there appears a scene in which a person is

Fig. 10.15 The three-legged seed plough in the murals painted in the burial rooms of the late Western Han Dynasty unearthed in Pinglu, Shanxi Province

pulling a seed plough to sow seeds. The seed plough he is using is just a three-legged seed plough (Fig. 10.15) [11]. On the portrait bricks of the Wei and Jin dynasties in Jiayuguan of Gansu, the image of pulling a seed plough with cattle could also be seen. These findings confirm that Cui Shi's records are credible. According to the "Part of Ploughing" in *Qi Min Yao Shu*, "As for sowing on the ridges, two-legged seed plough is not as good as one-legged seed plough." The "Atlas of Agricultural Tools (II)" in the book *Nong Shu* (Vol. 12) written by Wang Zhen of the Yuan Dynasty made a detailed record of seed ploughs (Fig. 10.16): "A seed plough has different kinds, one-legged, two-legged, three-legged…with two handles bent up, three *chi* high, two legs hollow within, as wide as one ridge, four crossbeams. In the middle of it, a dou is set, which loads the seeds. The dou is linked to the hollow legs of the seed plough. Beside the dou, two beams are installed, between which a cattle can accommodate.

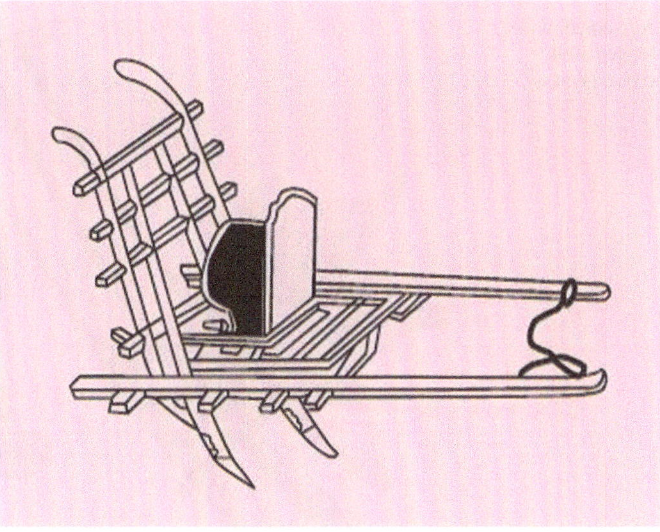

Fig. 10.16 The picture of seed plough in *Nong Shu* written by Wang Zhen

One person pulls the cattle and at the same time holds the seed plough. He walks and shakes the seed plough so that the seeds are dripped out." That is, when sowing, one cattle pulls the seed plough and one person supports it; the seeds are loaded in the dou, which is linked to the hollow legs of the seed plough. The person walks and shakes the seed plough so that the seeds are dripped out. By means of the device, one person can complete the three processes of digging ditches, sowing seeds, covering soil at the same time. It can evenly sow three rows at a time, greatly improving the sowing efficiency. The "Atlas of Agricultural Tools" of Wang Zhen's *Nong Shu* has a picture of the seed plough. The book also explains the distribution of the use of seed ploughs at that time: "In the areas nowadays of the Yan, Zhao, Qi and Lu states, two-legged seed ploughs are mostly used, while in the west of the Shanhai Pass, four-legged seed ploughs are adopted, and one more cattle is used, so it is faster to sow seeds."

As an important link in agricultural production, the seed sowing always has fine timeliness and high-quality requirement. The three-legged seed plough can complete the series work of seed sowing, covering, compacting at the same time, making the seed sowing method greatly improved. The invention was made on the basis of summarizing the experience of folk technology, and played an important role in promoting agricultural production. Liu Xianzhou studied and restored the structure and function of the seed plough [12], in view of which the Chinese Museum of History (now the National Museum) restored the three-legged seed plough made in the Han Dynasty (Fig. 10.17). At the time of the Three Kingdoms, Huangfu Long of the State of Wei acted as prefecture chief of Dunhuang, and had taught the people to make seed plough. Some scholars analyzed that all the seed ploughs used at that time were just three-legged seed ploughs [13]. In the post-Wei period of the Northern

Fig. 10.17 A restored model of the three-legged seed plough in the Han Dynasty

Dynasties (386–581), one-legged and two-legged seed ploughs were invented, which somewhat increased the kinds of seed ploughs [14]. To this day, the three-legged seed ploughs are still used in some rural areas of China.

A seed plough generally consists of three parts: the frame (the supporting part), the dou (an object shaped like a cup or dipper) and the seeding regulator (the part of loading seeds), and the seed plough share (the part of cutting soil and digging ditches). The part of frame is made up of the seed plough tip and its handle, the seed plough plate, the seed plough beams, etc., the part of loading seeds made up of a chamber of dou for loading seeds, a seeding regulator and tubes for dripping seeds, and the part of cutting soil and digging ditches made up of a seed plough share, also known as the trowel for digging ditches. The tip and the legs are integrated as one of the main parts of the frame. The upper part roughly connected with the seed plough beams is the tip and the handle, while the lower part is the seed plough legs so that the seed plough share can be installed. A two-legged seed plough has two tubes for dripping seeds, a three-legged one three. There are two seed plough beams, which are relatively long and can be used directly to harness livestock.

The most important part of a seed plough is its seeding regulator, whose structure and working principle will be further explained below. The structure of an early seeding regulator was simply made. As shown in Fig. 10.18, the regulating and clearing of it were achieved by means of a thin wooden rod or iron bar, one end of which was thinned, sloped into the chamber of dou from the discharge hole, and set against the baseboard of the chamber of dou, while the thick end tied to the handle of the seed plough with a rope. When sowing, the seed plough began to wobble, meanwhile slightly pressing the rope, so that the seeds could pass through the discharge hole smoothly under the swinging of the wooden rod or iron bar. This is an early device of seeding regulator, whose structure was adopted by the restored model of the three-legged seed plough in the Han Dynasty introduced above.

Figure 10.19 shows an improved seeding regulator of seed plough, one of the most common forms at present [15]. A seed discharge hole is opened in the lower part of

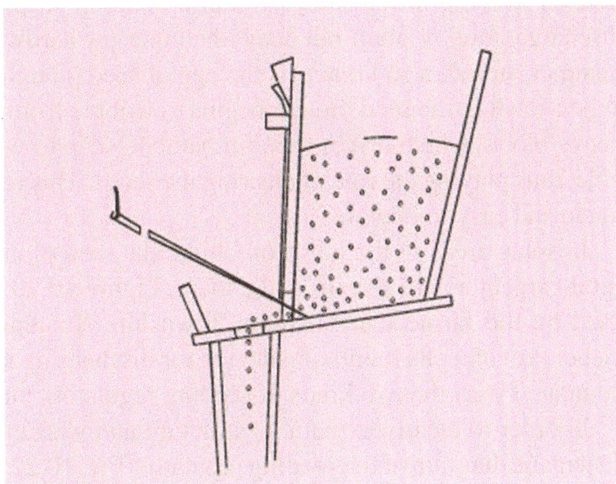

Fig. 10.18 The schematic diagram of the construction of the early seeding regulator

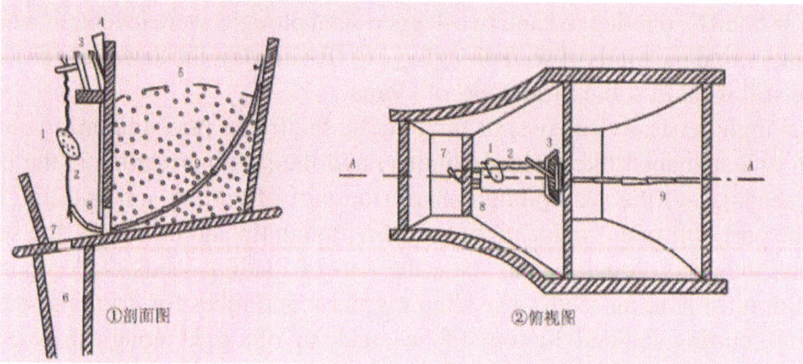

① Sectional view ② Vertical view

1. small heavy hammer 2. hung rope 3. flashboard 4. wedge 5. seeds 6. seed-dripping tube 7. seed-diverting chamber 8. discharge hole 9. thin bamboo strip

Fig. 10.19 The schematic diagram of a modified seeding regulator

the rear wall of its chamber of dou, and a flashboard is installed on the outer side of the rear wall to adjust and open-close the seed discharge hole, with the flashboard fastened with a wedge. When the wedge is pulled out, the flashboard can move up and down as needed to change the effective size of the discharge hole to control the amount of seeds discharged. After the seeds are discharged out of the hole, they enter the tubes for dripping seeds in the seed-diverting chamber. One end of the tubes for dripping seeds is connected to the seed-diverting chamber, and the other end connected to the legs of the seed plough. The seeds are sown into the soil through the tubes for dripping seeds and the hollow legs of seed plough. In case the seeds are blocked up at the discharge hole, a thin bamboo strip made into a movable chain is placed through the hole, one end of which is fixed above the inner wall of the chamber of dou, and the other end connected to a hung rope outside the discharge hole so that the lower end of the thin bamboo strip can wobble freely within the discharge hole. A small but heavy hammer (or hardwood block, iron block, copper hammer, stone), also known as the egg of seed plough, is attached to the hung rope. When sowing, the seed plough begins to wobble from left to right. So does the hung heavy block, which drives the thin bamboo strip to wobble around at the discharge hole, thus playing the role of clearing the seeds. This regulating device is very simply made and easy to operate.

In some areas in the north of China, the seed ploughs are still in use, and some rural carpenters are able to make them. Figure 10.20 is a three-legged seed plough used by the farmers in Shenjing Township, Xuanhua County, Zhangjiakou City, Hebei Province. Its regulating device for discharging seeds was somewhat improved compared with the two kinds of seeding regulators introduced above (Fig. 10.21).

In order to clearly explain its structure and working principle, the author drew a schematic diagram of the seeding regulator (Fig. 10.22). There are two improvements

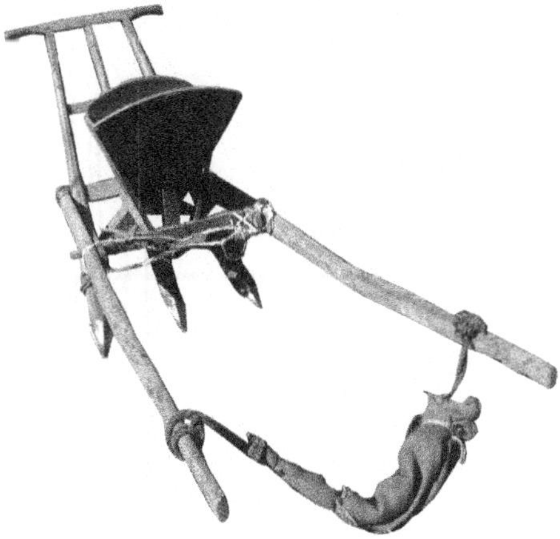

Fig. 10.20 The seed plough used by the farmers of Xuanhua County, Zhangjiakou City, Hebei Province (Photo by Feng Lisheng)

Fig. 10.21 The seeding regulator of seed plough in Xuanhua County (Photo by Feng Lisheng)

in its structure. First, the bottom of the seed-loading chamber and that of the seed-diverting chamber are not in a horizontal line, the former higher than the latter, with the dispersion between them more than 10 cm. In this way, the seeds which come out of the discharge hole will meet a large drop, so it is not easy for them to be blocked up at the outlet, thus improving the working efficiency. As for the first kind of seeding regulator, its seed-diverting chamber and chamber for loading seeds share one baseboard, so there is no dispersion at all, but as the baseboard is a little gradient, the seeds are still easy to flow. Second, in order to facilitate the smooth outflow of seeds from the seed-loading chamber, a thin iron rod is also set up, one end of which is inserted through the discharge hole and fixed in the U-shaped ring on the baseboard of the seed-loading chamber. And in order that the seeds can be evenly distributed among the three dripping tubes in the seed-diverting chamber, a hung rope is fixed on the outer side of the rear wall of the seed-loading chamber. The rope is connected to the other end of the thin iron rod, then attached by a hung object (iron hammer, copper hammer, or stone hammer) to its bottom end, and finally placed under the seed

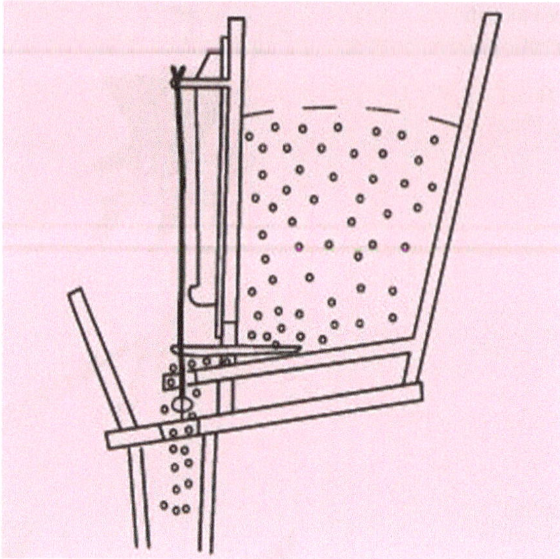

Fig. 10.22 The schematic diagram of the seeding regulator of seed plough in Xuanhua County

discharge hole. When sowing, the seed plough wobbles from left to right. So does the hung object, which drives the thin iron rod to wobble around, thus playing the role of clearing the seeds. And at the same time, when the hung object wobbles, it will strike the seeds flowing out of the discharge hole, and make them fall to the two sides of the seed-diverting chamber to ensure that the seeds can be evenly delivered to the three dripping tubes, thus achieving the effect of sowing evenly. A flashboard and a wedge are installed on the outer side of the rear wall of the seed-loading chamber. When the wedge is pulled out, the flashboard can move up and down as needed to change the effective size of the discharge hole to control the amount of seeds discharged. This is the same as that of the previous two kinds of seeding regulators, whose advantages this kind of seeding regulator combines, and is therefore easier to operate.

10.4 The Western Spread of Chinese Plows and Seed Ploughs and Their Influence

The plows made in China had led the world for a long time in ancient times. The highly efficient plows with iron curved mould-board had been used by the late Western Han Dynasty, and were basically finalized in design. In addition to the curved mould-board, the excellent structure was also a factor for Chinese plows to be superior to those made in other countries in the world. In particular, Chinese plows adopted an adjustable rod to change the distance between the plough share and the plough beam so that the depth of ploughing could be adjusted accurately. The plows with mould-board were not used in ancient India, Persia, and Arab. Chinese plows with iron curved mould-board were spread into Japan in the Tang Dynasty and had had a long-term impact on Japan's agricultural production. In Europe, the plows with wooden vertical mould-board had been used simply for a long time. The kind of plows whose mould-board and share could not be closely combined often carried weeds and soil with them when ploughing, thus hard to be pulled. The plows with iron curved mould-board were not used there until the early eighteenth century, due to the introduction of Chinese plows [16].

The poor efficiency of a traditional European plow was attributed to its large base, heavy wooden wheels, and large wooden mould-board, which could result in great resistance. In addition, as the plow's mould-board and plough share could not be closely combined, weeds and soil were often carried with and stuck in its crevice, so the farmer had to stop every few minutes and use a stick to remove the soil and weeds from it. The plow was far less efficient than traditional plows made in China.

The traditional seed ploughs made in China were the most advanced drill seeder in the world in ancient times. A modern sowing machine can merely perform in succession its whole functions of ditching, seeding, covering, and compacting, but the three-legged seed plough made in the Han Dynasty of China had already been able to finish the working procedures together. In the sixteenth century, the method used in Europe to sow seeds was still broadcast sowing, which had been the only method to sow seeds by the European farmers for a long time. This method was not only inefficient, but also wasting more seeds. In the eighteenth century, Jethero Tull was the first person in Europe who set forth the theory of dry-land grain sowing in the drill in a systematic and complete way. When it came to the shortcomings of broadcast sowing, Tull said: "The seeds cast on the depressed places may be over ten times more than elsewhere, and there may be few or no seed on the projected places. This uneven sowing reduces the number of effective seeds, as the space with 50 seeds can produce not as much as that with only one seed. In a place densely sown, the plants will not be able to get enough nutrients."

Francesca Bray, a famous British expert in technical history, argued that "the key factor in the reform of agricultural technology in Europe in the 18th century was the introduction or development of the plows with iron curved mould-board, drill seed planters and cultivators. Before that, northern China was the only place to use the three together for farming." She further pointed out: "There is strong evidence that modern European plows as well as their lightweight frames and iron curved mould-boards were directly affected by Chinese plows [17]." In the seventeenth and eighteenth century, Christian missionaries were able to travel more widely in mainland China. Their large number of correspondences and publications greatly improved the understanding of the Europeans of China. The agricultural machinery made in China was also introduced to Europe by the missionaries. Most of the European encyclopedias that were introduced into China cover the information about agriculture and farm tools. Xu Guangqi, together with the western missionaries such as Matteo Ricc and Johann Schreck, co-translated a variety of western scientific and technical works, including the ones on irrigation, water conservancy, and machinery. *Nong Zheng Quan Shu* (*An Encyclopedia of Agriculture*) that Xu Guangqi wrote includes chapters on agriculture, irrigation, water conservancy, and agricultural machinery. Its sections of farm tools and machinery have paraphrased and included many of the contents and illustrations covered in *Nong Shu* (*History of Agriculture*) written by Wang Zhen, including the graphics and instructions of plows, seed ploughs and cultivators. The missionaries who had come to China should know much about *Nong Zheng Quan Shu* written by Xu Guangqi.

According to the studies by western scholars, the plows with mould-boards made in China were brought back to the Netherlands by Dutch sailors in the seventeenth century. Some Dutchmen were employed by the British at the time and were responsible for draining off water of the marshes in then East Anglia and of the upland moors in Somerset. The Dutchmen brought with them Chinese plows, later called Rotherham ploughs, alias "own-brand Dutch ploughs." The Dutch and the British were thus the first in Europe to benefit from the highly efficient plows made in China. Chinese plows could work well in paddy fields, and the Europeans soon realized that the farm tool was also very effective in general fields, and soon began to use it in relatively arid fields. The kind of Chinese plows was spread from England to Scotland, and later from the Netherlands to the United States and France. By the 1870s, they were the best plows and sold the cheapest. Over the next few decades, western mechanical designers had renovated the plows, of which the plow that James Small developed in 1784 was a great improvement. And the various plows that J. Allen Ransome produced in the nineteenth century had made new improvements. Modern plows with steel frames can be said to be the result of many improvements to Chinese plows, and also the most important factor that led to the agricultural revolution in Europe [18].

The earliest drill seed planter made in Europe was inspired by the seed ploughs made in China. But as the seed ploughs were mostly used in the north of China, which is far from the ports in the south that the Europeans often visited, the actual samples of the seed ploughs might not have been imported into Europe. The Europeans had read about the planting methods adopted in the north of China, and been impressed by their high efficiency, economy, and appropriateness. In spite of this, as the sources of information available to the Europeans were not sufficiently detailed, they could not directly copy those Chinese farm tools involved. Besides, the accounts and illustrations covered in China's agricultural encyclopedias were not accurate enough to be used as a blueprint. In this way, European machinists had to work on their own, and invented a variant of Chinese drill seed planter. That's why western drill seed planters had little in common with Chinese seed ploughs, but technical historians always regard it as "an example of the spread of excitation [19]."

However, there were obvious deficiencies in the early western drill seed planters. The earliest drill seed planter made in Europe was invented by Camillo Torello, and the Venetian Senate granted him a patent in 1566. But the first drill seed planter that was described in detail was developed by Tadeo Cavalina from the city of Pologna in 1602, though it was rather primitive. The first real drill seed planter made in Europe was developed by Jethero Tull. Soon after 1700, the planter had come into being, and the detailed description of it was published in 1731. In Europe of the eighteenth century, James Sharp also invented a better drill seed planter. The planter could only sow in a single row and was too small in size, so it had not been paid enough attention to. The different drill seed planters made in Europe were not only more expensive and unreliable, but also fragile and uneconomical. It was not until the mid-nineteenth century that Europe had a sufficient number of solid drill seed planters of good quality. That means, Europe had failed to make full use of the inherent principle of Chinese seed ploughs for two centuries [20].

In the late nineteenth century, many officials of the Qing Dynasty visited Europe and the United States, where they came in close contact with and felt the modern science and technology of the West. After seeing the advanced machinery of western agriculture, they were quite surprised to record and comment on them. But the envoys made little account of the plows produced in the West. In 1868, Emperor Tongzhi appointed Anson Burlingame, former U.S. ambassador to China, to lead the Chinese mission to Europe and the United States. Zhi Gang, an official of the Qing Dynasty just in the mission, described in his *Records of the First Diplomatic Corps to Europe and the United States* the farm tools by animal power that they had seen in New York, such as plows, reaping machine, threshing machine, rice milling machine, etc., but made a simple description of plows as "half of the plows are acute angle-shaped [21]." The plows of the West did not seem to have attracted his attention. Guo Songtao, the first envoy of the Qing Dynasty to Britain and France, paid close attention to agriculture there during 1876–1879. In May of 1877, Guo was invited to visit one machine plant situated at one river mouth about 200 *li* northeast of London. He made such a record in his journal: "There are three kinds of ploughing machines: two by wheel and one by horsepower." He made a more detailed description of the steam plows, as they were rather novel and unique to Chinese people [22]. The reason why the envoys made little account of the plows produced in the West was probably that Europe and America were not much more advanced in farming tools than China at that time, so they were not attractive enough [23].

The plows and seed ploughs technology of China had a crucial influence over the agricultural revolution in Europe. Western scholars have had high assessment of this: "When the plows made in China finally spread to Europe, they were copied. Together with adoption of the Chinese cultivation in row and the seed ploughs, they had a direct influence on the European agricultural revolution. It is generally believed that the agricultural revolution in Europe led to the industrial revolution there, and the West thus developed as a world power. It is an irony, however, that all the results depended on the influence of China, not the intrinsic strength of Europe at all [24]."

References

1. Zhang Chunhui. *History of the Inventions of Agricultural Machinery in Ancient China (Suppl.)*. Beijing: Tsinghua University Press, 1998: pp. 16–17.
2. Wenhua, Chen. 1991. *Atlas of the History of Agricultural Science and Technology in Ancient China*, 132. Beijing: China Agriculture Press.
3. Jinghua, Li. 1985. Ancient iron farming tools in Henan Province (Continued). *Agricultural Archaeology* 1: 64.

4. Wenhua, Chen. 1991. *Atlas of the History of Agricultural Science and Technology in Ancient China*, 190. Beijing: China Agriculture Press.
5. Wenhua, Chen. 1991. *Atlas of the History of Agricultural Science and Technology in Ancient China*, 191. Beijing: China Agriculture Press.
6. Zhang Chunhui, You Zhanhong, et al. (eds.) *History of the Inventions of Mechanical Engineering in China (Part II)*. Beijing: Tsinghua University Press, 2004: p. 70.
7. Zhaolin, Song. 1979. A study in the curved-beam plows of the Tang Dynasty. *Journal of Chinese Museum of History* 1: 62.
8. Beijing Brigade of Cultural Relics. The ironware made in the Liao and Jin dynasties unearthed in Beijing. *Journal of Archaeology*, 1963, (3): Plate 10.6-5, 7, 9.
9. [The Qing Dynasty] Yang Shen. "Chapter of knowledge seeking" in *Zhi Ben Ti Gang*. The Chongbenzhai Version in the twelfth year of Emperor Qianlong.
10. The Compiling Group of *Rural Carpentry* in Shanghai Jiading Furniture Factory. *Rural Carpentry*. Shanghai: Shanghai Science and Technology Press, 1979: p. 86.
11. Xianzhou, Liu. 1963. *History of the Inventions of Agricultural Machinery in Ancient China*, 31. Beijing: Science Press.
12. Xianzhou, Liu. 1963. *History of the Inventions of Agricultural Machinery in Ancient China*, 35. Beijing: Science Press.
13. Jinyu, Wang. 1986. The sowing diagram of the three-legged seed plough found in the murals of Grotto 454 in the Mogao Grottoes at Dunhuang. *Agricultural Archaeology* 4: 118.
14. [The Northern Wei Dynasty] Jia Sixie. "Farming first" in *Qi Min Yao Shu* (*Important Arts for the People's Welfare*).
15. Hanzhang, Qin. 1951. *Agricultural Tools*, 108–110. Beijing: Commercial Press.
16. [UK] Francesca Bray. China's contribution to the European agricultural revolution: technological reforms. *See* Li Guohao, Zhang Mengwen, Cao Tianqin (eds.). *Exploration of the History of Science and Technology in China*. Shanghai: Shanghai Classics Publishing House, 1986: pp. 573–581.
17. [UK] Francesca Bray. China's contribution to the European agricultural revolution: technological reforms. *See* Li Guohao, Zhang Mengwen, Cao Tianqin (eds.). *Exploration of the History of Science and Technology in China*. Shanghai: Shanghai Classics Publishing House, 1986: p. 588.
18. [The United States] Robert K. G. Temple. *The Genius of China: 3,000 Years of Science, Discovery, and Invention*. Chen Yangzheng et al. (Trans.). Nanchang: 21st Publish House, 1995: pp. 33–34.
19. [UK] Francesca Bray. China's contribution to the European agricultural revolution: technological reforms. *See* Li Guohao, Zhang Mengwen, Cao Tianqin (eds.). *Exploration of the History of Science and Technology in China*. Shanghai: Shanghai Classics Publishing House, 1986: p. 602.
20. [The United States] Robert K. G. Temple. *The Genius of China: 3,000 Years of Science, Discovery, and Invention*. Chen Yangzheng et. al (trans.). Nanchang: 21st Publish House, 1995: pp. 47–48.
21. Zhi Gang. *Records of the First Diplomatic Corps to Europe and the United States (Vol. 1)*. Changsha: Yuelu Press, 1985: p. 227.
22. Guo Songtao. *The Journal of London and Paris (Vol. 1)*. Changsha: Yuelu Press, 1984: pp. 239–240.
23. Genjin, Ni, and Wei Luling. 2008. The Chinese understanding and thinking of western agricultural machinery before the Sino-Japanese War of 1894-1895. *History of Chinese Agriculture* 4: 18–27.
24. [The United States] Robert K. G. Temple. *The Genius of China: 3,000 Years of Science, Discovery, and Invention*. Chen Yangzheng et. al (trans.). Nanchang: 21st Publish House, 1995: p. 29.

Lisheng Feng Born in Wuhai, Inner Mongolia in 1962 graduated from Inner Mongolia Institute of Technology with a bachelors degree in engineering in 1983 then he got a Master of Science degree in 1987 from the Institute of History of Science. Inner Mongolia Normal University. In 1999 he got his doctor degree of science from the Department of Mathematics Northwestern University.

He was a professor at Inner Mongolia Normal University and currently the director of the Institute of Science and Technology History and Ancient Literature of Tsinghua University, and professor and doctoral supervisor of the Institute of Science and Technology History of Inner Mongolia Normal University. He has been a visiting scholar and visiting professor at Japans Tohoku University, Tokyo University, and Central University. In 2004, he was selected into the "New Century Outstanding Talent Support Program" by the Ministry of Education. He also serves as a member of the National Intangible Cultural Heritage Protection Committee, executive director of the Chinese Academy of Science and Technology History, president of the Chinese Traditional Arts and Crafts Research Association, and vice chairman of the National Mathematical History Society.

Research interests: History of mathematics; History of Technology; Scientific and technological literature; Protection of Science and Technology Cultural Heritage.

Major works: History of Ancient Chinese metrology, History of China-Japan Mathematical Relations.

Awards: The First Prize of Scientific Research Achievements of the Inner Mongolia Normal University; Co-editing Notes of Chou Ren Records won the first prize of the National Outstanding Ancient Books Award;the Encyclopedia of Color Map Science and Technology that he participated in the compilation won the second prize of the 2009 National Science and Technology Progress Award; History of Chinese Science and Technology: Mathematics won the first prize of the Fourth Guo Moruo History Award.

Chapter 11
Water Wheel

Xing Huang and Baichun Zhang

A water wheel is a device that converts the kinetic energy or potential energy of water flow into rotating mechanical energy. In ancient China, water wheels were used as the prime motor of shui dui (water power trip-hammer for husking rice), scoop waterwheel, waterpower roller (for grinding grain), waterpower grinder, boat mill, drainage machinery, water transport astronomical instrument, and so on.

The term "water wheel" has different meanings in ancient Chinese literature. In Fu (descriptive prose interspersed with verse) of *Water Wheels of Quan Tang Wen* (*Collected Literatures of the Tang Dynasty*), a "water wheel" refers to the wheel-like water-lifting machinery. In the Song Dynasty, a scoop water wheel was called "water wheel," and sometimes "water cart." In *Nong Shu* (*History of Agriculture*) written by Wang Zhen of the Yuan Dynasty, a water wheel is explicitly defined as the wheel as a kind of power machine driven by the impact force of water. After that, the term "water wheel" has been used to refer to both the wheel-like machinery as power machine, and the scoop waterwheel. This paper discusses the former, namely, the wheel-like machinery which realizes the conversion of driving forces. This kind of machinery is mainly composed of a shaft, the hub, and blades. The traditional water wheels used in China can be divided into vertical type and horizontal type according to installation mode, up-shot type, down-shot type, and oblique-impact type according to the impact position of water flow, and flat-plate type, inclined plate type, and bucket type according to the structure of blade.

X. Huang (✉) · B. Zhang
The Institute for History of Natural Sciences, Chinese Academy of Sciences, Beijing, China
e-mail: huangxing@ihns.ac.cn

J. Hua and L. Feng (eds.), *Thirty Great Inventions of China*,
https://doi.org/10.1007/978-981-15-6525-0_11

11.1 Up-Shot Vertical Water Wheel

According to Chinese literature, the vertical water wheel was originally used as the prime motor of shui dui. Huan Tan made such a record in his *Huan Zi Xin Lun* in the end of the Western Han Dynasty and the beginning of the Eastern Han Dynasty: "Fu Xi, one of the earliest legendary rulers of China, invented the pestle and benefited the people. Later generations lengthened the arm of the pestle and invented a treadle-operated tilt hammer for hulling rice, which is ten times efficient than the former pestle. Later, new mechanisms were added to the tilt hammer, and donkeys, mules, cows and horses were used to help produce water power to hull rice, which is 100 times efficient than the former pestle [1]." It was inferred that a water wheel had appeared by the end of the Western Han Dynasty [2]. Vertical water wheels were widely used in grain processing machinery, water-lifting machinery, and astronomical instruments.

The term "shui dui" was first found in *Tong Su Wen* (*Popular Essays*) written by Fu Qian and *Rou Xing Lun* (*On Corporal Punishment*) written by Kong Rong in the Eastern Han Dynasty. Later on, the term had been written into literatures of all dynasties. For example, Fu Chang made such a record in his *Jin Zhu Gong Zan* (*Praises of the Princes and Ministers of the Jin Dynasty*): "Du Yu and Yuan Kai made a conjoint shui dui." It can be seen that in the Eastern Jin Dynasty, shui dui had had a more complex structure, and the structure of the water wheel it adopted was likely to be more complex. According to the continuous records of it in literature and the working principle of shui dui, the vertical water wheels should be adopted. In 263 or later, there was a more detailed record of shui dui: "Beside the river, a shui dui is made, and water wheels are set behind it. A crosspiece is installed to run through the wheels. The two ends of the crosspiece are linked alternately to a long wood of about two chi and directly impact the rear wood of the shui dui. When the water is lifted to impact the wheel, the wheel will rotate, which drives the alternate wood to impact the rear wood of the shui dui to hull grain, using no manpower. The device is thus called shui dui [3]." The "crosspiece" here is the horizontal shaft of a water wheel which is certainly vertical, while the "long wood" refers to the convex rod, which means that this shui dui must be a conjoint one.

The up-shot water wheels take advantage of both the impact and the gravity of water flow from above to drive to rotate. *Nong Zheng Quan Shu* (*An Encyclopedia of Agriculture*) (Vol.18) makes a clear description of a vertical water wheel:

It (a vertical water wheel) can be set up beside a river, which is finally chosen according to the rise and fall of its water. If the water is shallow, palisades will be installed. If the water flows gently, wood blocks should be used to block water in order to raise the flow. The wheel, more than one *zhang* (a unit of length, = 3.13 m), should be installed against the shore side and carry water from below to shoot down to drive. This water wheel is called "water-lifting trip-hammer for husking rice." If the water is high and the shore deep, the wheel will be decreased in size but widened, and adopts the boards as steps, with a wooden water channel installed on the top. The water will shoot up to rotate the boards and so drive the wheel. This water wheel is called "dou dui (bucket trip-hammer for husking rice)," or "gu dui (drum trip-hammer for husking rice)." So, the water wheel is set up to suit the measures to local conditions, making full use of the advantages of each place [4].

Obviously, the "dou dui" in the above description is up-shot vertical water wheel, while the "water-lifting trip-hammer for husking rice" down-shot vertical one. According to the description made by Wang Zhen, the blade of dou dui should be bucket type. The drum-shaped water wheel with smaller diameter was widened in order to enlarge the volume of the blade bucket, thus making full use of the momentum and weight of water. Wang Zhen's account of it was included both in *Nong Zheng Quan Shu* in the Ming Dynasty and *Nong Ya* in the Qing Dynasty. The carving copy *Shaoxing Fu Zhi: Water Conservancy* written in the reign of Emperor Kangxi of the Qing Dynasty also mentioned the down-shot vertical water wheel and the up-shot vertical water wheel. Besides, Hommel described the two water wheels made in China in his book written in 1937 [5]. By the 1990s, the up-shot vertical water wheels had been still used in Zhejiang, Yunnan, and Guangxi, whose basic structure and installation were no different from those described by Wang Zhen.

The shui dui driven by an up-shot water wheel is being used in Tongcun Town, Kaihua County, Zhejiang Province to smash small pieces of wood to make wood powder (Figs. 11.1 and 11.2). The water is drawn from mountain streams, and impacts the bucket-shaped blades above the water wheel through the flashboard and the water channel. When the blades turn just below the shaft, the water flows out of the blade bucket. Under the impact of water flow and the gravity of the water in the blade bucket, the water wheel, and the shaft begin to rotate, and the two pairs of convex plates on the shaft press down the two rods of shui dui one after the other, which makes the hammer of shui dui fall and rise. By raising and lowering the flashboard,

Fig. 11.1 Outside view of the hydraulic power workshop in Tongcun Town, Kaihua County, Zhejiang Province (Photo by Zhang Baichun)

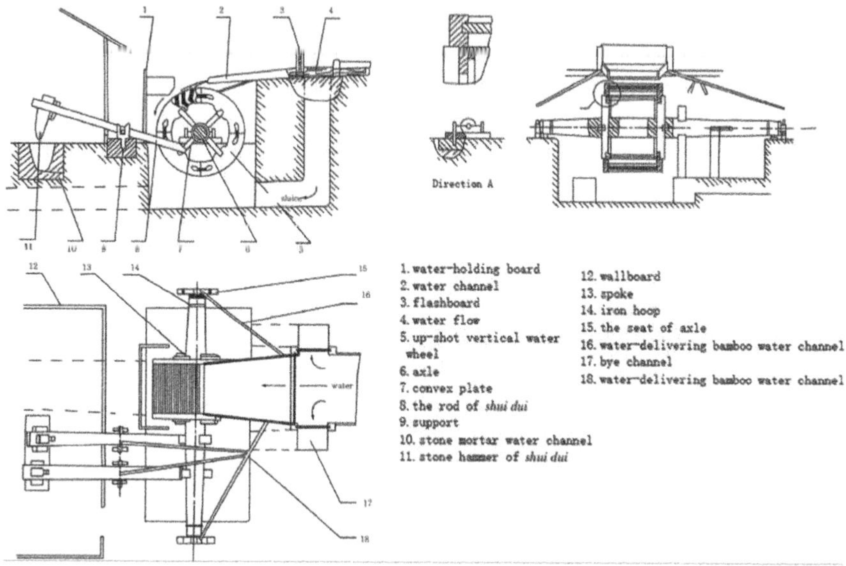

Fig. 11.2 The structure chart of the shui dui in Tongcun Town, Kaihua County, Zhejiang Province (Redrawn according to the article "Chinese traditional water wheels and their driving machinery" by Zhang Baichun)

the flow rate of water can be adjusted so that the speed and torque of the water wheel are changed. When the flashboard cuts off the water flow, the wheel will stop, and the water overflows into the bye channels on both sides of the water channel. Such a shui dui can grind 70 kg of finger-size broken wood into wood flour every day and night [2].

As for the system of shui dui, all parts are made by wood except for the stone hammer, stone water channel, stone seat, iron hoop, iron plate cover, etc. Even the wooden parts are required not to become deformed and rotten easily when immersed in water, so their vertical and horizontal strength should be higher. The important parts as the axle, the rod of shui dui, the spoke, and the rim plate are generally selected with camphor wood. All the parts are connected mainly by hua, latch, wedge, and hoop, and almost no iron nails are used except for the reinforcement bar on water wheel. When the wooden parts become wet, they will expand and the joints will thus be tightened, which is one of the advantages of wooden hydraulic machinery.

11.2 Down-Shot Vertical Water Wheel

There was little description of the structure of water wheels before the Yuan Dynasty. The long scroll of *Picture of the Outspread Rivers and Mountains* drawn by Wang Ximeng (1096–?) in the Northern Song Dynasty depicts the waterpower grinder

Fig. 11.3 The waterpower grinder driven by up-shot vertical water wheel in the *Picture of the Outspread Rivers and Mountains* [6]

driven by up-shot vertical water wheel (Fig. 11.3). *Nong Shu* written by Wang Zhen makes a sketch of the down-shot vertical water wheel, but whose structure of the blade is described very simply. After the appearance of conjoint shui dui and conjoint grinder, water wheels were required to output more power. The basic ways to increase the power include increasing the diameter of a water wheel, the width and the number of its blades, and improving the shape of its blades. When these data reach a certain extent, the connecting difficulty between the axle and the blades will increase, and the space between the blade tips will also be oversize. When the diameter is very large, it is almost difficult to make a practical water wheel without the use of a heavy boss of wheel. To overcome this obstacle, the ancient Chinese installed the rims for water wheels until the spokes separated from the blades. And the spokes turned to be rod-shaped and the blades plate-shaped. Then the number of blades was increased and their shape was improved (Fig. 11.4).

As the early literature is too brief, it is now difficult to determine the initial date of the rims used in Chinese water wheels. However, the wheels that were made up of hub, spoke, and rim had already been produced in China late in the Western Zhou Dynasty. As the wheels were easily spread to use, the makers of water wheels could easily refer to the construction of wheels. A conjoint shui dui required a larger torque for a water wheel to output, so the water wheels with rims were required. According to *Jin Zhu Gong Zan* by Fu Chang, it's inferred that the vertical water wheels with rims had appeared in the Jin Dynasty (317–420). The water wheels included in *Nong Shu* by Wang Zhen, *Tian Gong Kai Wu* by Song Yingxing and *Nong Zheng Quan Shu* by Xu Guangqi were all set up with rims. In the Ming Dynasty, the water wheels that were used to drive conjoint shui dui and conjoint grinder were just made up of more complex spoke, rim, blade, etc.

In Huamin Village, Huabu Town, Kaihua County, Zhejiang Province, there was once a shui dui that could hull, grind, and husk rice driven by a down-shot vertical water wheel (Fig. 11.5). It exploited the water of the Grand Canal to operate, and

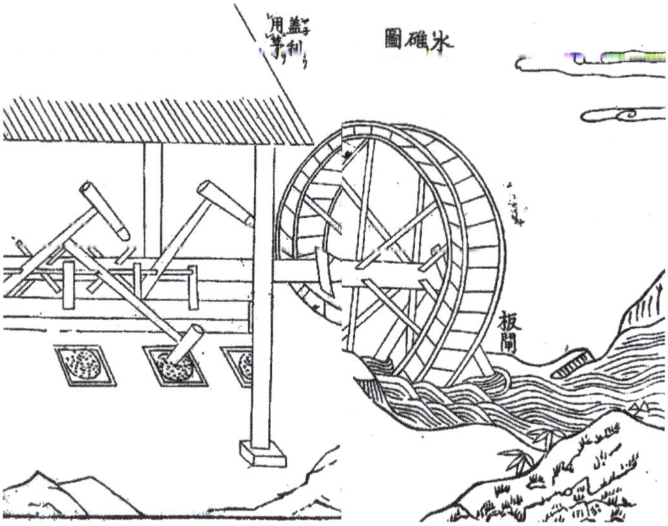

Fig. 11.4 The vertical water wheel with rim (Taken from *Tian Gong Kai Wu*)

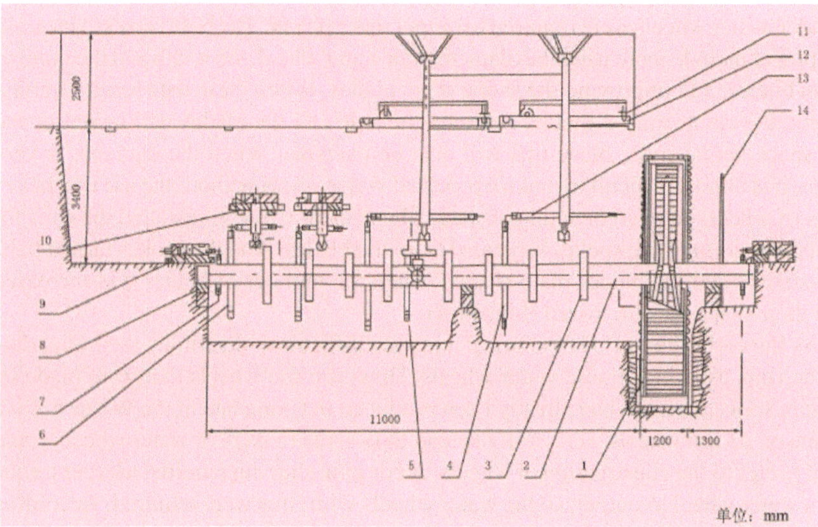

1. Down-shot vertical water wheel 2. main axle 3. convex plate 4. gear 5. gear 6. gear 7. gear 8. the seat of main axle 9. lotus-shaped mill 10. high-table mill 11. runner wheel 12. mill water channel 13. gear 14. water-holding board

Fig. 11.5 Front view of the shui dui in Huabu Town (Redrawn according to *The Research of Chinese Traditional Machinery* by Zhang Baichun) Unit: mm

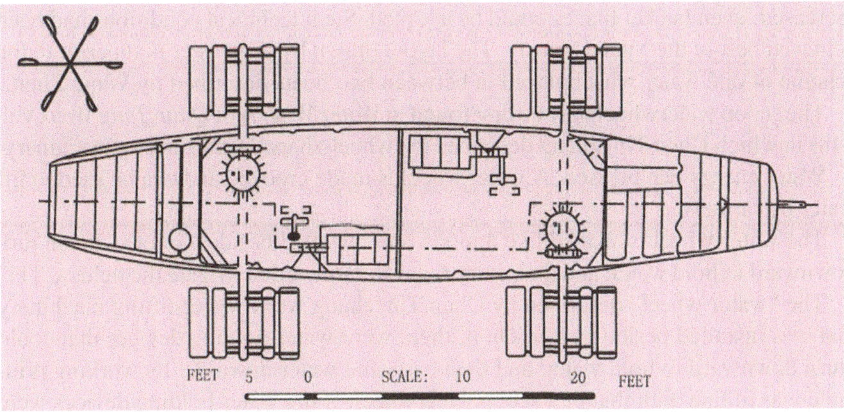

Fig. 11.6 The boat mill and its water wheel in Fuling, Sichuan (Worcester 1940) [7]

had a history of more than 100 years by the 1980s. It had been repaired many times since the late Qing Dynasty. By the 1990s, on its original site, there was a diversion canal, the parts and components such as the seat of stone axle, stone millstone, water channel mill, rolling grinder, and so on, and the remaining buildings. The shui dui had processed grain for more than 2,000 people living in Huabu Town, which could process rice of 1,000 kg, mill flour of 500 kg, or grind rapeseed of about 500 kg per day and night.

In the Tang Dynasty, the water wheel and the rotary grinder had been installed on a boat to make a boat mill. According to *Shui Bu Shi* in the Tang Dynasty, it was forbidden to build any "floating stone mill" on the rivers near Luoyang [8]. Lu You of the Southern Song Dynasty mentioned "the floating stone mill amidst turbulence" [9] in his *Jian Nan Shi Gao* (*Complete Poetry of Lu You*). The "floating stone boat" here should refer to the boat mill. The boat mill used a down-shot vertical water wheel. Wang Zhen made an exact explanation of it in his *Nong Shu* (Vol. 19): "Two boats are tied together, on which four principle columns are established, thus a house built with twitch-grass and bamboo in between the two boats. One grinder is placed on each side to the rapid water by cable. The bow of each boat is aslant inserted with a wood plate to gather water, with an iron claw thrown into the water so that each boat will not lie across. When the water flows across and impacts the vertical wheel, its axle will drive the two grinders beside" [10].

Both *Tian Gong Kai Wu* in the Ming Dynasty and the documents of the Qing Dynasty show that the boat mills had been widely used in the south. However, there is no detailed description or drawing of the structure of boat wheels in ancient literature. Worcester gave a picture of the boat mill mapped in Fuling, Sichuan Province (Fig. 11.6) in a paper published in 1940. In the picture, the vertical water wheel is nearly one meter in diameter, and uses flat-plate blades. In addition, due to the limitation of the boat height, it was not suitable for a boat mill to install a water wheel with larger diameter, but the number and width of the water wheel could be increased; even bucket blades could be adopted. Such technical conditions had been achieved before the Yuan Dynasty. The layout drawn by Worcester is superior to the scheme of one water wheel placed in between two boats described by Wang Zhen.

The scoop waterwheel may be first found in *Water Wheels* of *Quan Tang Wen* (Vol. 948) in which Chen Tingzhang describes the wheel-shaped water-lifting machinery:

Water energy can be used. A water wheel is made crooked and can be used to lift water by farmers…

The water wheel is shaped like a moon and installed beside a river… It can turn downward to hold water, and then carry the water upward to irrigate the fields… [11]

The "water wheel" described by Chen Tingzhang was a water-lifting machinery that was installed beside a river. On it, there were water-holding devices that could "turn downward to hold water, and then carry the water upward." Its working principle was in line with that of a scoop waterwheel. If the water-holding devices were bamboo tubes or wooden tubes, it should be the scoop waterwheel mentioned by later generations.

There were many records about "water wheels" in the Song Dynasty. Li Chuquan of the Northern Song Dynasty wrote in his own collection of works *Yu An Ji*: "In the regions south of the Yangtze River, there is a kind of water wheel that is not driven by man power, really a very smart creation. One water wheel is installed with ten tubes which can decant and inject water, and turns up and down all the time to irritate the fields…" [12] The "water wheel" here should be the scoop waterwheel that was driven by water flows. Su Shunqin in the same dynasty talked about in his works both the spoke and the cup-like bucket of a scoop waterwheel: "The spokes (of a scoop waterwheel) are intersected up and down, and the cup-like buckets circle round… The device that can rotate is not an elegant tool at all, but can help to remove disasters. When it is loaded with water, it will turn upward; when it pours away the water, it will turn downward" [13].

The buckets of a scoop waterwheel are generally made with bamboo as well as wood. The scoop waterwheel is generally produced with lighter materials such as bamboo and wood, and locally made on the wheel rack according to the height of water lifting (Fig. 11.7). Some of the scoop waterwheels installed along the Yellow River in Lanzhou, Gansu Province, are more than 15 m in diameter. Among them, the water wheel located in Xiachuan Village, Xigu District built in the Qing Dynasty (Fig. 11.8) is still in use. The kind of water wheels with braided blades is also used on the sugar cane mills in Manhai Village, Menghan Town, Jinghong City, Yunnan Province [14].

Water power had been used in the astronomical instruments of ancient China for a long time. According to *History of the Jin Dynasty: Annals of Astronomy*, Zhang Heng invented a celestial globe in 130 or so, which could "turn around by water leakage." It's guessed by later generations that Zhang Heng probably used vertical water wheel and the gear train, whose structure could have referred to shui dui [15].

In the Tang Dynasty, astronomical instruments that were driven by water wheels had appeared. According to *The Old-version History of the Tang Dynasty: Annals of Astronomy*, Seng Yixing and Liang Lingzan invented a celestial globe, "that can move automatically if injected by water to impact the wheel." The water wheel

Fig. 11.7 The scoop waterwheel installed in Linna Quarter, Poxin Village, Paoli Town, Fengshan County, Guangxi (Photo by Zhang Baichun)

should be vertical, with the blades bucket type, which could help to improve its power and moving stability. After the Tang Dynasty, the instruments powered by water wheels gradually increased. In the Song Dynasty, the "pivot wheel" of the water transport observatory created by Su Song and others adopted the side-shot vertical water wheel, which rotates by the weight of water rather than kinetic energy. The combination of pivot wheel and the heavenly balance made it come true to use the escapement mechanism to control the rotation speed of pivot wheel.

11.3 Oblique-Impact Horizontal Water Wheel

In China, a horizontal water wheel is often referred to as a horizontal wheel, which is actually an oblique-impact horizontal water wheel. The kind of water wheel possibly appeared in the Eastern Han Dynasty. According to the article written by Fu Xuan that was quoted in *The Records of Three Kingdoms: Biography of Zheng Xie* noted by Pei Songzhi, Ma Jun invented "a kind of acrobatics exploiting turning water" in 227–239, "which was carved with a large wood, wheel shaped, and could be performed parallel to the ground to lift water" [16] to drive a variety of puppets and "mills with pestle." Apparently, Ma Jun made a water wheel. When the ancients classified and

Fig. 11.8 The great water wheel installed beside the Yellow River in Xiachuan Village, Xigu District, Lanzhou, originally built in the Qing Dynasty (Photo by Huang Xing)

called water wheels, the position of the wheel body was adopted as the standard, not the wheel axle. According to this fact, "be performed parallel to the ground" should mean that it was a lying-down water wheel, that is, its water wheel was horizontal type.

According to *History of the Later Han Dynasty: Biography of Du Shi*, when Du Shi was in charge of Nanyang in 31 A.D., he "invented a waterpower blowing machine, used to smelt iron to make agricultural tools" [17]. The waterpower blowing machine Du Shi invented might be horizontal or vertical type. According to *The Records of Three Kingdoms: History of the Northern Wei Dynasty*, Han Ji, prefecture chief of Laoling, improved airing machinery: "When smelting in old times, horsepower blowing machine was used. A large number of horses were thus exploited. Later manpower blowing machine was invented, which also consumed much manpower. Han Ji invented waterpower blowing machine by use of flowing water, whose benefit is three times that of the past methods" [18]. As for the horsepower blowing machine, horses walked around a vertical axle and drove it to rotate, which drove the leather bag to move to and fro through the mechanism composed of a crank, a connecting rod, and so on. Han Ji was from Nanyang, so he might have been influenced by the waterpower blowing machine invented by Du Shi. The waterpower blowing machine invented by Han Ji should accordingly follow the transmission mechanism of horsepower blowing machine, and replace horses with oblique-impact horizontal water wheel.

The existing vertical water wheel that could be used for reference helped horizontal water wheel to develop and turn mature in technology as soon as possible. Its maturity should be equal to that of vertical water wheel, and was reflected in the structure of rims and blades. *Picture of the Rolling Gear at Water-gate* (Painter: Yi Ming) made of the history from the Five Dynasties to the Northern Song Dynasty is the earliest drawing of the horizontal water wheel made in China (Fig. 11.9). The painter described the scene of grain processing by waterpower grinder, that is, the water flow from the oblique water channel impacts the horizontal water wheel with rims, whose vertical axle directly drives the horizontal millstone above the floor slab. The whole device was installed in a building mainly of wooden structure. The painter accurately painted the structure of a waterpower grinder in a reasonable proportion, including the rims, spokes, and blades of its horizontal water wheel, which in the picture appears to adopt the blades with a sharp angle to the horizontal plane. Wang Zhen in his *Nong Shu* (Vol. 19) gave a brief description of waterpower blowing machine and its horizontal water wheel [19].

The earliest horizontal water wheel should have no rims, with the blades just installed on the vertical axle, the flat-type blades perpendicular to the rotating surfaces of wheels. In order to output greater torques, the wheel diameter was gradually increased. When the wheel diameter reached a certain extent, the rims and spokes had to be specially set up, the number of blades also increased. That is just the structure shown in the diagram. After the Jin Dynasty, there came to be more and

Fig. 11.9 The horizontal water wheel in *Picture of the Rolling Gear at Water-gate* [20]

more records of waterpower grinders and waterpower rollers. According to *History of the Southern Dynasties (420–589). Biography of Zu Chongzhi*, Zu Chongzhi "made a shui dui grinder in Leyou Garden and Emperor Xiaowu (of the Song Dynasty in the Southern Dynasties) came to see it in person." [21] For a grinder, the mechanism was made the simplest when horizontal water wheel adopted, and gears were needed when vertical water wheel adopted. If the "shui dui grinder" Zu Chongzhi had made included two independent machineries, then it was likely to use horizontal water wheel. On the contrary, if it was machinery combined with the functions of both hulling and grinding, it should use vertical water wheel, the mechanism of a cam, and gears.

According to *History of the Northern Wei Dynasty: Biography of Cui Liang*, "When Cui Liang worked in Yongzhou, he read the biography of Du Yu who made eight waterpower grinders that could benefit people. He thus taught his people how to make waterpower rollers… Later, he made waterpower rollers in dozens of areas by use of the water of weirs and valleys, that brought ten times of benefit to local people…" [22] It's probably that Cui Liang had made mainly the water grinders and water rollers driven by horizontal water wheels, possibly as well as more complex grinding and rolling mechanical system driven by vertical or horizontal water wheels.

Nong Shu (Vol. 19) written by Wang Zhen illustrates water grinder, water roller, and the "three articles of water wheel" respectively driven by horizontal water wheel. The last one used one water wheel to drive the grinder, the rice huller, and the roller simultaneously [10]. Unfortunately, as for the structure of blades, *Tian Gong Kai Wu*, *Nong Zheng Quan Shu* and other books did not give a more detailed description than *Nong Shu* by Wang Zhen.

The traditional machinery used in today's Yunnan, Guangxi, Sichuan, Tibet, and other places shows the structural details of water wheels. At least in the Han Dynasty, Yunnan had maintained contacts with Sichuan and other places, and become one of the channels for China to communicate with foreign countries. During the Three Kingdoms (220–265), Yunnan was controlled directly by the kingdom of Shu Han (221–263) and got unified, meanwhile dominated by the central government. As a result, the technologies in the southwest of China around Yunnan had something in common with those in the mainland. The treadle-operated tilt hammer for hulling rice and the water wheel is just the case. The former is typical Chinese machinery, and was made a Dongba pictograph character of the Naxi nationality in Yunnan Province. Both in Xizhou, Dali Prefecture, Yunnan Province, and in Tongcun Town, Kaihua County, Zhejiang Province, there has existed a kind of "water-dripping treadle-operated tilt hammer for hulling rice" with the same structure, which indicates that the two have the same origin of technology.

In Songpingzi Village, Shigu Town, Lijiang County, Yunnan Province, a waterpower grinder without rim had been still used in the early 1990s. The current impacted the horizontal water wheel along the water channel to drive the wheel and axle to rotate. The iron key on the upper end of the axle drove the upper millstone to rotate on the lower millstone [23]. There was an angle of less than 90° between the upper edge of the blade of the horizontal water wheel and the axis line. This structure required the water wheel with smaller diameter to have longer blades. The current

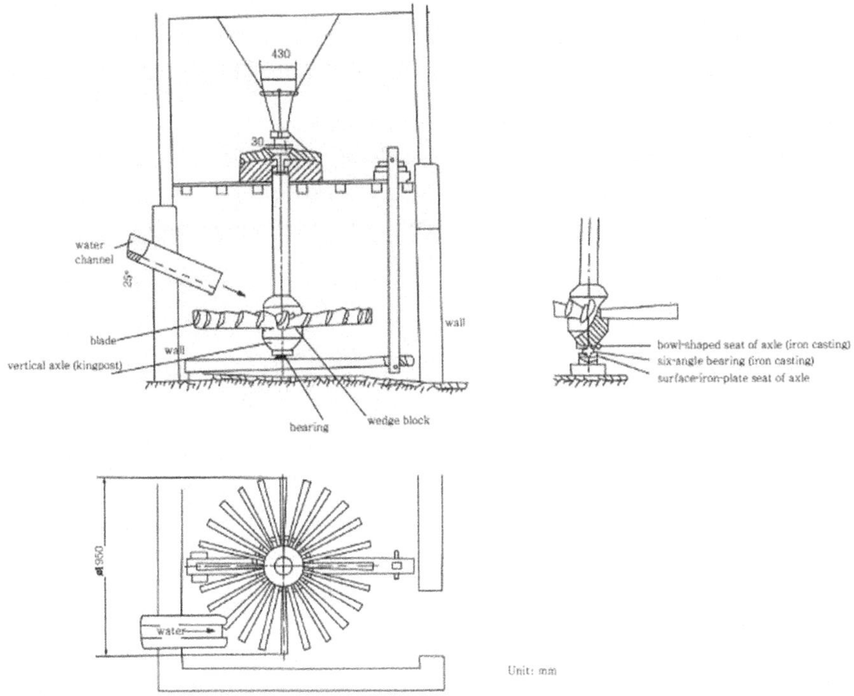

Fig. 11.10 The oblique-impact horizontal water wheel used by the waterpower grinder in Shigu Town (Redrawn according to the article "The research of several traditional hydraulic machineries in Yunnan" by Zhang Baichun)

first impacted the outer end of the blades and then flowed downward along the blades, thus increasing the action time of water flow on the blades (Fig. 11.10). In addition, as there was an intersection angle between the flat surface of a blade and the axis line, the current could impact on the flat surface of blades more effectively. Sometimes, the surface of blades was made curved, that is, slightly bucket-shaped, so that the dynamic performance of water wheels was further improved.

The waterpower grinders without rim that are still in use in Lhasa, Tibet, are similar in structure to that in Lijiang, Yunnan Province. The surface of its blades is flat, and vertically is in the same direction as the axis, thus preserving the most simplified structure (Fig. 11.11). Scholars believe that the technology of waterpower grinder in Tibet was introduced from the mainland when Princess Wencheng of the Tang Dynasty entered Tibet. According to *The Old-version History of the Tang Dynasty: Biography of Tubo* (Tibetan regime in ancient China) (Vol. 207), Srongtsen Gampo, husband of Princess Wencheng, petitioned the then imperial government in 650 for "silkworm seed and craftsmen of making wine, grinder, stone mill, paper and ink, and was approved." There is a clear record of the event, which was also written in Tibetan literature, such as *Xi Zang De Guan Shi Yin* (*Tibetan Guanyin*) (a

Fig. 11.11 The water wheel without rim of the waterpower grinder and its regulating structure in Jiami, Niangre Township, Lhasa, Tibet (Photo by Huang Xing)

Bodhisattva), *Xi Zang Wang Tong Ji* (*A Complete History of the Kings in Tibet*), and *Xian Zhe Xi Yan* [24].

The appearance of water wheels marks a great leap forward in the understanding and application of natural energy. The use of water wheels abroad was obviously earlier than that in China, but different civilized regions might independently invent the same or similar technologies that embodied their own features. Chinese water wheel technology may be invented independently or influenced by other civilizations. From the periods and locations when and where Du Shi, Han Ji, Ma Jun, Cui Liang, and others lived and worked, Chinese horizontal water wheels first appeared in some places in Henan, Shaanxi, and Shandong, then spread to other areas, and took root in the south where there were more rivers and streams.

References

1. [The Eastern Han Dynasty] Huan Tan. *Huan Zi Xin Lun. Si Bu Bei Yao* (*The Remarks of Four*) (Photocopy). Beijing: Zhonghua Book Company, 1920: p. 17.
2. Zhang Baichun. Chinese traditional water wheels and their driving machinery. *History of Natural Science*, 1994, (2): p. 155–163; 1994, (3): pp. 254–263.
3. [The Northern Song Dynasty] Sima Guang. *History as a Mirror*. Beijing: Zhonghua Book Company, 1956: p. 2469.
4. [The Ming Dynasty] Xu Guangqi. *Nong Zheng Quan Shu* (*An Encyclopedia of Agriculture*). Beijing: Zhonghua Book Company, 1956: p. 367.
5. [US] Rudolf P. Hommel. *China at Work: an Illustrated Record of the Primitive Industries of China's Masses, Whose Life is Toil, and thus an Account of Chinese Civilization*. Dai Wusan et al. (trans.) Beijing: Beijing Institute of Technology Press, 2012: pp. 86–89.
6. Zhao Nong. *History of Chinese Art Design*. Xi'an: Shaanxi People's Fine Arts Publishing House, 2004: p. 181.

7. Joseph Needham. *Science and Civilization in China (Vol.4, Part II)*. Cambridge University Press, 1965: p. 412.
8. Research Group of the Scientific and Technological History, the Library of Tsinghua University. *Data Collection of the History of Chinese Science and Technology: Agricultural Machinery*. Beijing: Tsinghua University Press, 1985: p. 307.
9. Research Group of the Scientific and Technological History, the Library of Tsinghua University. *Data Collection of the History of Chinese Science and Technology: Agricultural Machinery*. Beijing: Tsinghua University Press, 1985: p. 306.
10. [The Yuan Dynasty] Wang Zhen. *Nong Shu* (Vol.19, pp. 5B-6A). *Si Ku Quan Shu* (*The Imperial Collection of Four*) by Pavilion of Literary Profundity. Shanghai: Shanghai Classics Publishing House, 2003: pp.543–544.
11. Research Group of the Scientific and Technological History, the Library of Tsinghua University. *Data Collection of the History of Chinese Science and Technology: Agricultural Machinery*. Beijing: Tsinghua University Press, 1985: pp. 149–150.
12. Research Group of the Scientific and Technological History, the Library of Tsinghua University. *Data Collection of the History of Chinese Science and Technology: Agricultural Machinery*. Beijing: Tsinghua University Press, 1985: p. 160.
13. Research Group of the Scientific and Technological History, the Library of Tsinghua University. *Data Collection of the History of Chinese Science and Technology: Agricultural Machinery*. Beijing: Tsinghua University Press, 1985: p. 155.
14. Zhang Baichun. Chinese traditional water wheels and their driving machinery. *History of Natural Science*, 1994, (2): p. 155–163; 1994, (3): pp. 254-263.
15. Xianzhou, Liu. 1959. The full application of the gear train in ancient China. *Journal of Tsinghua University* 4: 1–11.
16. [The Western Jin Dynasty] Chen Shou. *The Records of Three Kingdoms*. [The Southern Song Dynasty] Pei Songzhi (Annotated). Beijing: Zhonghua Book Company, 1959: p. 807.
17. [The Southern Song Dynasty] Fan Ye. *History of the Later Han Dynasty*. [The Tang Dynasty] Li Xian, et al. (Annotated). Beijing: Zhonghua Book Company, 1965: p. 1094.
18. [The Western Jin Dynasty] Chen Shou. *The Records of Three Kingdoms*. [The Southern Song Dynasty] Pei Songzhi (Annotated). Beijing: Zhonghua Book Company, 1959: p. 677.
19. [The Yuan Dynasty] Wang Zhen. *Nong Shu* (Vol.19, pp. 5B-6A). *Si Ku Quan Shu* (*The Imperial Collection of Four*) by Pavilion of Literary Profundity. Shanghai: Shanghai Classics Publishing House, 2003: pp.544–545.
20. The identification group of ancient Chinese calligraphy and painting. (Ed.) *Complete Works of Chinese Painting (3): The Song, Liao and Jin Dynasties (2)*. Hangzhou: Zhejiang People's Fine Arts Publishing House, 1999: p. 16.
21. [The Tang Dynasty] Li Yanshou. *History of the Southern Dynasties (420–589)*. Beijing: Zhonghua Book Company, 1975: p. 1774.
22. [The Northern Qi Dynasty] Wei Shou. *History of the Northern Wei Dynasty (368–534)*. Beijing: Zhonghua Book Company, 1974: pp. 1476-1484.
23. Baichun, Zhang. 1994. The research of several traditional hydraulic machineries in Yunnan. *Ancient and Modern Agriculture* 1: 41–49.
24. Guan Xiaowu & Huang Xing. The waterpower grinder in Jiami of Tibet and the etiquette and custom of eating zanba (roasted qingke barley flour). In *The Technology in Anthropology, Folklore and Industrial Archaeology*. Beijing: Beijing Institute of Technology Press, 2009: pp. 119–137.

Xing Huang Born in Xuanhua, Hebei in 1981, PhD., assistant researcher at the Institute of Natural Science History, Chinese Academy of Sciences.

Research interests: History of metallurgy; history of Compass; Mechanical history;

Major works: Monographs: An Empirical Study of Ancient Chinese Compass;

Papers: An Empirical Study on "Sinan" of Natural Magnet Spoon; Literature Review on the History of Chinese Compass; Study on Numerical Simulation of Air Flow Field in a Single Air Outlet Iron Smelting Shaft Furnace - Taking Shuiquanggou Iron Smelting Site in China in the 9th and 13th centuries as an example; A Comparative Study of Ancient Positive Blower in the World; Research on Type and Performance of Ancient Chinese Fan Car.

Awards:The 9th Special Grant from China Postdoctoral Science Foundation;The Second Prize of the Second Beijing Science Innovation Competition granted by the Chinese Academy of Sciences;Excellent Doctorate Thesis of Beijing University of Science and Technology.

Baichun Zhang Born in Baicheng, Jilin province in 1960, PhD, Research Fellow and Director of Institute of Natural Science of the Chinese Academy of Sciences. He is currently a correspondent member of the International Academy of Science History, an executive member of the International Institute of Institutions and Machines History Committee, and an executive director of the China Institute of Mechanical Engineering.

Research interests: History of Technology and Mechanics; History of knowledge dissemination and comparative history; Strategy for the Development of Science and Technology

Major works: *The Transfer of Soviet Technology to China (1949-1966) (co-author); Communication and Association-Research and Commentary on the" Qi Qi Tu Shuo; on Complete Works of Traditional Chinese Crafts: Investigation and Research on Traditional Machinery; Transformation and Transmission: Chinese Mechanical Knowledge and the Jesuit Intervention (Co-author); Europeanization of sky measuring instruments in the Ming and Qing Dynasties.*

Chapter 12
Traditional Technique of Lacquer

Bei Chang

The term "decoration with lacquer" first appeared in *Zhou Li* [1], a classical book in ancient China, referring to "to decorate or paint with lacquer [2]." That is to say, the term only referred to the use of natural lacquer to paint utensils in ancient times, but now known as "lacquer art."

The ancient people of China were the first to use natural lacquer to paint utensils. Moreover, they had made constant inventions and discoveries in this field, finally developed this technology into a broad and profound handicraft art. Let us start to discuss it with the function of natural lacquer, the main raw material of lacquer art.

12.1 The Discovery and Utilization of Lacquer

As early as tens of thousands of years ago, Chinese ancestors had discovered that the sap of lacquer tree had high viscidity, which could be used to strengthen the adhesion of wooden appliances, production tools, etc. If painted on the woodwork or potteries, it would produce a dense protective coating which was waterproof, solid and wear-resisting, and beautiful and bright enough to prevent the woodwork or potteries from leaking, thus prolonging their service life. Once discovered of the viscidity, protection, and beautification, "natural lacquer" came to be one of the important living materials of our primitive ancestors.

The use of natural lacquer in the preliminary stage was to apply the sap of lacquer tree—natural lacquer—directly on utensils. Natural raw lacquer looks reddish brown, and will turn black after being hardened. Zheng Xuan thus wrote such an annotation in *Zhou Li*: "Lacquer, more red but less black [1]. The lacquered mulberry bow made in about 8,000 years ago (Fig. 12.1), unearthed in the Neolithic site in Kuahuqiao

B. Chang (✉)
Jiangsu Research Institute of Culture and History, Beijing, China
e-mail: zhang_yan44@163.com

J. Hua and L. Feng (eds.), *Thirty Great Inventions of China*,
https://doi.org/10.1007/978-981-15-6525-0_12

Fig. 12.1 The early lacquered wood bow made in Neolithic times (Photo by the Kuahuqiao Site Museum of Xiaoshan)

of Xiaoshan, Zhejiang Province, is 121 cm in residual length. On it, the painted raw lacquer can be clearly seen after being scratched of the raw lacquer ash. The lacquered surface is glossy and has leveling, which is probably the result of being painted with dilute raw lacquer with the tool made of bone or horn.

With the progress of civilization, Chinese ancestors began to plant lacquer trees artificially and collect their sap. The terms of lacquer trees are mentioned many times in *Shan Hai Jing* (*The Classic of Mountains and Rivers*). *Yu Gong* in the historic literature *Shang Shu* includes such a record of "Yanzhou paid tribute to the royal government with lacquer, silk…Yuzhou paid tribute to the royal government with lacquer, hemp, fine ko-hemp cloth and ramie [3]." In the *Book of Songs*, there are such poems: "In the mountains there are planted lacquer trees… [4]" In the Warring States period, Zhuang Zhou "once acted as one of the officials of Lacquer Garden in Meng [5]." In the Western Han Dynasty, "Shandong is rich in fish, salt, lacquer, silk… One thousand *mu* of lacquer trees are planted…A rich and noble family owns one thousand pieces of woodwork painted with lacquer, and one thousand *dou* of lacquer, as rich as a family with a thousand carriages [6]." In other countries of Northeast Asia and the countries of Southeast Asia and South Asia, lacquer trees are also planted, while China enjoys the longest history of planting lacquer trees and is still a major lacquer-producing country.

The earliest record of lacquer cutting can be found in *Gu Jin Zhu* (*Notes to the Ancient and Modern Times*) by Cui Bao of the Eastern Jin Dynasty: "*Nan Yue Zhi* records: lacquer tree, cut it open on the bark with a sharp axe, and catch the drops of sap with a bamboo tube, in which the sap turns to laquer [7]." Choose the lower plano-concave part of trunks where the sap of lacquer is concentrated. Cut a shallow opening to release water, and then cut deeper and meanwhile insert a shell or bamboo tube into the opening; the sap of lacquer will drip into the shell or bamboo tube, finally into the lacquer bucket. The cutting period of lacquer is not more than 140 days from June to September. The "first lacquer" contains more water, but its drying performance is better. The "mid-term lacquer" which is cut out on hot summer days is the best in quality, while the "final-phase lacquer" comes second. The lacquer coming out of the end or branches of lacquer tree is the worst. As a lacquer tree has both the largest flux of sap at dawn and the most abundant sap in hot and foggy weather, lacquer farmers often enter the mountains in the middle of the night and start cutting lacquer at dawn. *Nan Yue Zhi* records: "If starting to cut lacquer trees when cocks crow at dawn, you will certainly harvest. After this time, *yinqi* will go down and *yangqi* rise, so you will harvest nothing [8]." The saying that "you have to walk one hundred *li* and cut one thousand time, you can harvest one *jin* of lacquer" by lacquer farmers reflects the hardships they undergo of harvesting lacquer in deep mountains and forests. Each lacquer tree can be cut 60 openings a year, producing about 250 grams a year, so the sap of lacquer tree is called "liquid gold."

Natural lacquer consists of 40–70% laccase, less than 1% laccol, 5–7% vegetable glue, 2–5% glycoprotein, and 15–40% water. Among them, laccase is the most important and active component of natural lacquer, playing the role of drying and film-forming natural lacquer. The higher the content of laccase is, the better the quality of lacquer is. The one who has the first contact with lacquer often suffers from redness and swollen of the skin, rather itchy, which is commonly known as "ill with lacquer sore." During the Warring States period, Zhao Xiangzi painted the head of Zhi Bo with lacquer and used it as a drinking ware. Yu had his body painted with lacquer as if he was suffering from leprosy. He secretly hid under a bridge and premeditated to assassinate, but was arrested by Zhao Xiangzi. He drew a sword to cut the clothes of Zhao Xiangzi and then killed himself. Both events can be found in *Historical Records: Collected Biographies of Assassinators*. People who have suffered from lacquer sore will be often immune to it.

12.2 The Invention of Preformed Element Forming and the Use of Varnish

Chinese lacquer ware is made from original wooden preformed element or pottery preformed element. The wooden preformed element has been the main one of Chinese lacquer ware for thousands of years. At the ancient Hemudu Cultural Relic of Yuyao, Zhejiang Province of about 7,000 years ago, there were unearthed 10 pieces of wooden lacquer ware such as lacquered wooden bowl, red lacquered wooden barrel bound with bamboo strips, lacquered butterfly-shaped wooden ware, and black lacquered tube. The lacquer ware unearthed at the Liangzhu Culture site in Jiangsu Province was decorated by means of colored drawing and inlay. The black lacquered pottery cup unearthed there proves that as early as the Neolithic times, Chinese ancestors had begun to use raw lacquer to paint wooden work and pottery ware in order to prevent them from leaking, make them easy to clean, and prolong their service life.

Three generations later, the decoration technology with lacquer began to be used in making sacrificial ware [9], and life utensils came to be increased with decorative feature. For example, the pieces of black-red lacquered ware inlaid with turquoise were unearthed at the site of the early Shang Dynasty in Gaocheng, Hebei, and the lacquered wooden work inlaid with teeth and turquoise was found at the ruins of the Yin Dynasty in Anyang. *Zhou Li* recorded a variety of vehicles decorated with lacquer [10, 11]. Many pieces of lacquer ware copied of the sacrificial vessels of the Zhou Dynasty were unearthed in the tomb of the Spring and Autumn period in Dangyang, Hubei. In the Warring States and the Qin and Han Dynasties, the focus of social fashion on "etiquette" in making utensils of the Western Zhou Dynasty was changed to that on "utility," and the lacquer ware whose preformed frames were made of wood, bark, bamboo, rattan, linen and other materials became the main life utensils used in the families of aristocrats and landlords, due to its portability, attraction, durability, resistance to corrosion, and so on. The wooden lacquer ware unearthed so far covers the life utensils such as furniture, cooking utensils, and food utensils, weapons, musical instruments, stationery and toys, funeral utensils, transportation appliance, etc. As for the life utensils, there are table, case, box, bed, tripod, pot, kettle, clock, wine vessel, barrel, tub, stemmed cup or bowl, mirror case, casket, bamboo-plaited basket or suitcase, bowl, spoon, soup ladle, wine goblet, pillow, comb, double-edged fine-toothed comb, eared cup, chamber pot, etc. The weapons include quiver, arrow, arrow shaft, knife scabbard, sword scabbard, dagger scabbard, bow, crossbow, spear, shield, bow and arrow frame, etc. The musical instruments include *qin* (a seven-stringed plucked instrument in some ways similar to the zither), *se* (a twenty-five-stringed plucked instrument), *sheng* (a reed pipe wind instrument), *chi* (an ancient musical pipe), bell frame, inverted bell frame, flying bird with tiger-shaped base, tiger bird rack, bird-like drum with tiger-shaped base, etc. The stationery and toys include lacquer ink stone, Liu Bo chessboard, etc. And the funeral utensils include coffin, bamboo curtain bed, face mask, pillow, tomb beasts, etc. In the tomb of Marquis Yi of the State of Zeng in the Warring States period situated in Suixian County, there were unearthed eight kinds of musical instruments with 125 articles, most of which belong to wooden lacquer ware. According to *Han Shu* (*The History of the Former Han Dynasty*), in the reign of Emperor Chengdi, the Zhaoyang Palace "looked red in the atrium, and the halls were decorated with lacquer [12]," which approves of the fact that the decoration technology with lacquer had been used in wooden architecture in the Han Dynasty.

The forming of wooden preformed lacquer ware includes cutting and digging, carving and sawing, engraving, plate combining, mechanical spinning, wood bending, and so on. The lacquer ware that was made by cutting and digging as well as engraving includes the wood-carved sitting beasts screen decorated with colored lacquer, phoenix, and bird-shaped couple lacquer cups, colored mandarin-duck lacquer beans, etc. unearthed in the Chu tomb of Jiangling, Hubei. The "wood bending," also known as "rolling wooden preformed element," or "rolling semi-finished product" by lacquer workers, refers to the forming technique: first make a piece of pine or China fir wood into flexible thin pieces by splitting and planning, and then make these thin pieces into wooden preformed lacquer ware by bending, slotting, and jointing with glue. In the Han tombs of Yangzhou, a large number of pieces of lacquer ware made of rolling wooden preformed element were unearthed. In the inscriptions on lacquer ware of the Han Dynasty, the original meaning of "   " is to wrap a thin wooden preformed element with a linen cloth with lacquer, with the linen cloth and the lacquer sandwiched layer by layer, to which the words like "Qi Bu Xiao Zhi (small-size lacquered cloth)," "Jiu Zi Zeng Lian (nine-piece toilet case)," and "Bu Zeng Lian (cloth toilet case)" written on the bamboo or wooden slips unearthed in No. 1 Han Dynasty Tomb of Mawangdui could testify. In the last three generations of the Qin and Han Dynasties before the refining technology of raw lacquer was invented, lacquer ware had been mostly produced with raw lacquer, later decorated with the lacquer mixed with oil (Fig. 12.2).

During the Han Dynasty, high-grade lacquer ware was manufactured by state-owned workshops. The inscription carved with needle on the lacquer eared cup of the Western Han Dynasty unearthed in Pingba, Qingzhen, Guizhou Province, includes 70 Chinese characters, recording in detail the strict division of labors in making lacquer ware of the state-owned workshops. The basic-element workers were responsible for manufacturing wooden preformed lacquer ware, the decoration workers for painting the preformed lacquer ware, the copper ear-coating workers for gold-plating the copper ears, the upper workers for decorating with lacquer for the

Fig. 12.2 A wooden lacquer gui (vessel for holding grains in ancient offerings or feast) used in the Warring States period (Photographed by the author in 2011 "Special Exhibition of the Lacquer Ware Unearthed in the Chu Tombs in Jiuliandun, Hubei Province")

finish, the commercial painters for depicting the ornaments, the cleaning workers for the final cleaning, and the cropping workers for the inspection [13]. And there were workers who managed the shaded room. The whole process was supervised by officials at all levels. During the reign of Emperor Zhaodi of the Han Dynasty, the lacquer decorations had been "used by the riches…as well as the middle riches…"; "a drinking vessel made of crooked wood would be made by one hundred workers, while a screen by ten thousand workers," [14] "one minister named Yu once worked in the Eastern Palace, and was granted cups and tables, which were decorated with gold and silver as well paintings…"[15] As the refining technology of raw lacquer had not been invented yet, the lacquer ware of the Han Dynasty was mostly decorated with mixed natural lacquer with dry vegetable oil. The practice of mixing with oil could make raw lacquer diluted, with which the decorated lacquer ware would tend to be leveled and full of luster. The author rubbed the decorated lacquer ware with gold and silver of the Han Dynasty in the warehouse of Yangzhou Museum, whose surface has floating light, without any ground trace. Its gold and silver pattern is level with the surface of lacquer, from which we can know that it was produced by coating the surface with gold and silver pieces when the lacquer was wet after the surface had been decorated with lacquer and leveled. Compared with it, the technology used by the leveled decorated lacquer ware with gold and silver coated with gold and silver pieces made in the Tang Dynasty is different merely in complexity (Fig. 12.3).

Fig. 12.3 The colored set of lacquer toilet case with silver buckles decorated with gold and silver pieces of the Han Dynasty (Unearthed in the Han Dynasty Tombs of Huchang, Yangzhou; selected from *Chinese Culture Pictorial*)

12.3 Maturity of the Refining Technology of Lacquer and Initial Development of the Technology of Polishing

The moisture content of raw lacquer cut from lacquer trees is 20% and 30%, containing impurity of dust. Its molecular structure is loose and rough, and the viscosity is bigger. It can dry in short time, so can't be used in thick coating. Both of its leveling and glossiness are not good at all, and its formed film is thick and hard with strong adhesion. Lacquer workers always describe raw lacquer as "only being scrapped, not being brushed." With the development of civilization, people expect a bright and smooth coating of the utensils they use, so we have invented the technology of adding refined dry vegetable oil to lacquer to improve the brightness of formed films. The technology of refining dry vegetable oil was first seen in the book *Yan Shi Jia Xun* (*The Yan's Traditional Injunctions*) of the Northern Qi Dynasty (550–577), in which there records such a line of "boiling walnut oil to refine tin into silver" [16]. In ancient times, the dry vegetable oil added with the drier lithargite [17] was called "lithargite oil," while the patterns depicted with "lithargite oil" were called "lithargite painting." "In the State of Wei of the Warring States period, someone had talked about the lacquer paintings made with lithargite [18]."

As the further development of civilization, people hoped that the utensils they used should have a smooth and dense surface of lacquer, so the technique of refining and processing raw lacquer was invented. Pour raw lacquer into the lacquer-exposing tray or lacquer-frying pan, and stir it with 15–20% water until the lacquer changes color; when the water content reaches 10–15%, put it in the sun (nowadays in light at about 45 °C) and turn it over with a lacquer-exposing picker at a constant speed. When the original water content reaches 3–5%, and the lacquer looks like translucent red-brown paste, raw refined Chinese lacquer is produced, which was named in ancient times as "bright ointment," "paste lacquer," "heguang (combined light)," "glossy paint," or "bright-sun burnt paint," known as "hongjin lacquer," "translucent paint," "red-bone polishing paint," or "white-base polishing paint" in the folk. When the raw refined Chinese lacquer is mixed with pigments, refined Chinese lacquer or polishing paint is produced. The refining process of the lacquer is the oxidation polymerization of lacquer molecules and the removal of part of water. The refined polishing lacquer becomes low in viscosity, better in leveling, dense in coating, and much brighter in appearance. According to *Bi Jing* by Wang Xizhi in the Eastern Jin Dynasty, "Some people used lvchen to paint a bamboo tube and hollowed it. This has been employed for many years..." [19] Only when the brush penholder is decorated with polishing lacquer and then polished after the lacquer is dry, the surface of lacquer can be as clear as when sunk in water. It can be seen that the Eastern Jin Dynasty was the age when the refining technology of raw lacquer had been mature, and the technology of polishing achieved initial development and became popular. Since then, the decoration technology with polishing lacquer had become the most widely used decoration technology with lacquer. It was introduced to Japan in the late sixteenth century, and was called "colored coating with wax" by Japanese lacquer workers.

12.4 Invention of the Technology of Filling and Inlaying and Excavation of the Potential Performance of Lacquer

The Tang Dynasty gave birth to a brand-new decoration technology with lacquer—the technology of "filling and inlaying," which included such techniques as leveled decorated lacquer art with mother-of-pearl inlay, leveled decorated thick-lacquer art with gold and silver, nonessential gold engraving, rhinoceros-print technique, etc. *A Record of Decoration with Lacquer* by Huang Cheng records it in *The Seventh Chapter of Filling and Inlaying*. The common essentials of filling and inlaying are: Grain, or paste or sprinkle decorative materials on the preformed lacquer ware to make patterns, completely decorate it with lacquer, and then grind it after it turns dry until the patterns come out, and finally polish it. The mature grinding and polishing technologies were necessary for the birth of the decoration technology with lacquer by filling and inlaying. The grinding technology came to be mature from a lower stage to an advanced one, from the grinding with the decorative materials mixed with ash lacquer to the one after pasting or sprinkling decorative materials on the preformed lacquer ware and then decorating it with lacquer. The decorated *qin* (a seven-stringed plucked instrument in some ways similar to the zither) with lacquer in the Tang Dynasty has been regarded highly. The reason involved rests on its production procedure: people of the Tang Dynasty broke antlers into pieces, then burnt them into ash, and mixed it to the lacquer; after it turned dry solid, grind and polish it. As there are a large number of micro-gaps in the molecular structure of antler for raw lacquer to enter, the lacquer and the ash will be highly stuck together, and meanwhile the *qin* can produce more force of penetration of sound. Moreover, after being ground and polished, the finish of the lacquer ware made of the ash of antlers will show tawny dizzy speckles or twinkling color spots, looking very wonderful [20]. In addition to the ash of antlers, the crumbs of gold, copper, or porcelain were also added to the lacquer to make *qin* of thc Tang Dynasty and then to grind and polish. The achievements that the decorated *qin* with lacquer in the Tang Dynasty had made completely depended upon the mastery of painters of the new technologies of grinding and polishing. There are 467 ancient pieces of *qin* of past dynasties collected in Beijing Palace Museum, of which 4 pieces are the famous *qin* produced in the Tang Dynasty. There are more than 30 ancient pieces of *qin* collected in Zhejiang Provincial Museum, including 5 famous pieccs of *qin* of the Tang Dynasty. On the inside of the "Cai Feng Ming Qi" *qin* with the style of Luoxia (evening clouds) made in the Tang Dynasty, there is an inscription of "Da Tang Kai Yuan Er Nian Lei Wei Zhi" (Lei Wei made it in the second year of the Kaiyuan reign of the Tang Dynasty) (Fig. 12.4).

In the Tang Dynasty, the leveled decorated lacquer art with gold and silver developed into its extreme state. The Dongda Temple in Nara, Japan, collects leveled decorated lacquer *qin* with gold and silver, leveled decorated lacquer Hu (the Northern barbarian tribes in ancient China) vase with silver, leveled decorated octagonal diamond-shaped lacquer mirror box with silver, etc. produced in the Tang Dynasty. In the basement of the Famen Temple of Xi'an, there were unearthed leveled decorated lacquer bowls with silver produced in the late Tang Dynasty, which hold preformed lacquer ware of porcelain and were gold-plated with collective flowers (Fig. 12.5). In the Tang Dynasty, there also appeared lacquer ware coated with gold crumbs that was produced mainly by filling and inlaying. The gold and silver lacquer knife with mother-of-pearl inlay of the Tang Dynasty collected in the Dongda Temple in Nara (Fig. 12.6) is just one lacquer ware coated with gold crumbs that was made by sprinkling gold powder, covering bright, grinding and polishing. It is generally accepted by scholars as the forerunner of Japanese Makie.

The leveled decorated lacquer art with mother-of-pearl inlay which is processed mainly by filling and inlaying also reached its maturity in the Tang Dynasty. In the Feiying Pagoda of Huzhou, Zhejiang Province, there was found a black lacquer scripture box with mother-of-pearl inlay made in the Five Dynasties (907–960). On the outside of its bottom plate, such 47 Chinese characters as "Wu Yue Guo Shun De Wang Tai Hou Jin Shi Bao Zhuang Jing Han Si Zhi..." were written in red lacquer. Although its lacquer wooden frame has fallen apart, it is the earliest existing leveled decorated lacquer ware with mother-of-pearl inlay with mature technology. Suzhou Museum also collects a lacquer scripture box with mother-of-pearl inlay made in the Northern Song Dynasty (Fig. 12.7). It has a wooden preformed lacquer ware, with the black-lacquered outer walls adorned with luminous mother-of-pearl patterns of leaves which are completely even with the black lacquer surface. The edge of a partial

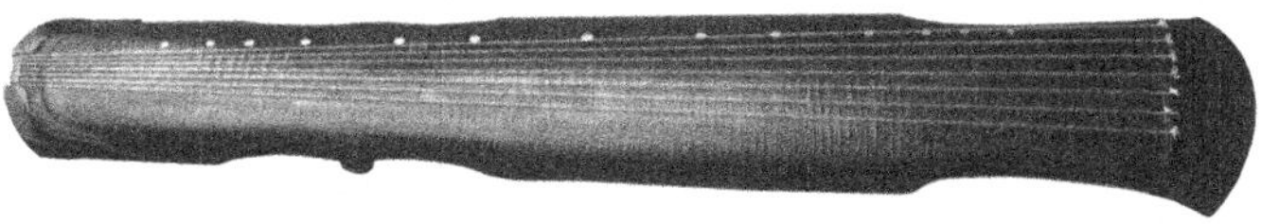

Fig. 12.4 The "Cai Feng Ming Qi" *qin* with the style of Luoxia (evening clouds) made in the Tang Dynasty (Photographed by the author in Zhejiang Provincial Museum)

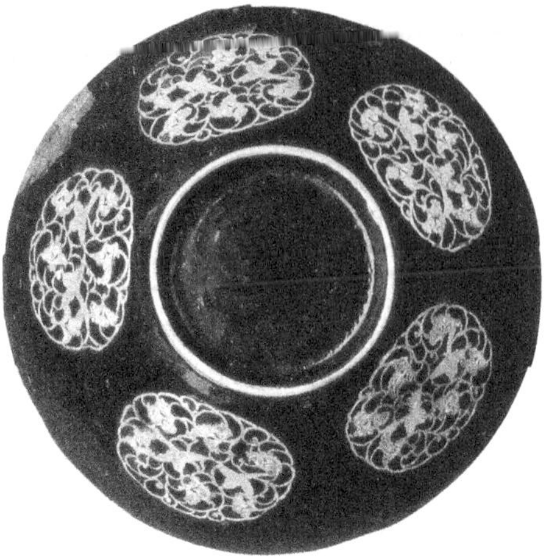

Fig. 12.5 The leveled decorated lacquer bowl with silver produced in the late Tang Dynasty, which hold preformed lacquer ware of porcelain and were gold-plated with collective flowers (Collected by Shaanxi Provincial Museum, and selected from *Complete Collection of Chinese Lacquer Ware: from the Three Kingdoms to the Yuan Dynasty* edited by Chen Jing)

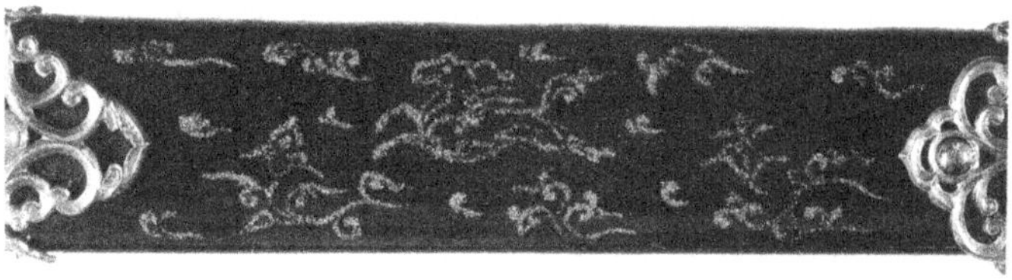

Fig. 12.6 The pattern of the lacquer ware coated with gold crumbs on the gold and silver lacquer knife with mother-of-pearl inlay of the Tang Dynasty (Selected from *Japanese Art 11: Decoration Techniques* by Yoshihiko NiShikawa)

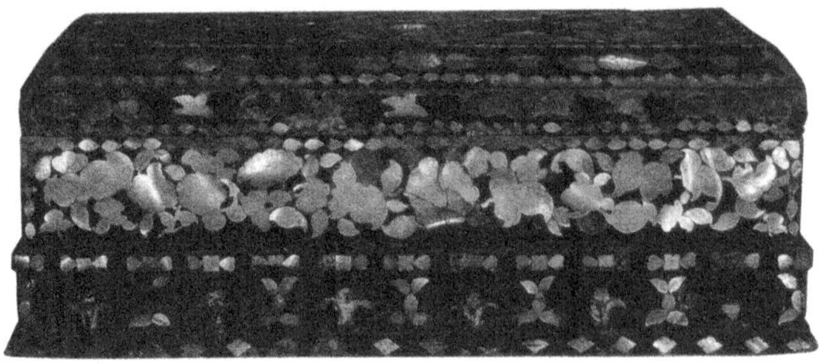

Fig. 12.7 A black-lacquered scripture box with mother-of-pearl inlay produced in the Northern Song Dynasty (Collected by Suzhou Museum, and selected from *Complete Collection of Chinese Lacquer Ware: from the Three Kingdoms to the Yuan Dynasty* edited by Chen Jing)

mother-of-pearl inlay is serrated, and a dropped mother-of-pearl inlay is 1.08 mm thick, whose lacquer surface can be seen as a clear trace of grinding. The leveled decorated lacquer technology with gold and silver and the leveled decorated lacquer technology with mother-of-pearl inlay had both spread to the East Asian countries in the Tang Dynasty, whose colorful decorative style directly promoted the formation of the leveled decorated lacquer art with mother-of-pearl inlay in Korea.

The technical essentials of rhinoceros-print lacquer art also cover graining, color changing, lacquer filling, and then grinding flat into patterns. As a kind of filling and inlaying technology, the rhinoceros-print lacquer art should be born in the Tang Dynasty when the grinding and polishing technologies became mature and the potteries with twistable preformed body were widely used. *Tang Liu Dian* records a "black-lacquered lacquer ware with the pattern of detritus," which just refers to the rhinoceros-print lacquer ware. *Tai Ping Guang Ji* in the Tang Dynasty also writes about a rhinoceros-print lacquer pillow. The filling and inlaying technology was also widely used in the Tang Dynasty, to which a "rhinoceros-print lacquer box inlaid with mother-of-pearl flowers" written in the famous legend of the Tang Dynasty *Huo Xiao Yu Zhuan* (*Biography of Huo Xiaoyu*) can be a side witness. The filling and inlaying technology which covers graining, lacquer filling and then grinding flat into natural patterns, such as "Ruoxia coating," "Jicao coating," "Duizhu coating," "Lingwen coating," and so on, is collectively called by Japanese painters as "Tang coating (coating of the Tang Dynasty)". It can be seen from the name itself that the rhinoceros-print lacquer art had been passed to Japan in the Tang Dynasty.

The thin coating of refined Chinese lacquer is brittle and hard; after accumulated layer by layer and dried, it even turns as hard as iron. With the maturity of the refining technology of dry vegetable oil, Chinese painters found that the addition of refined Chinese wood oil into lacquer could greatly reduce the brittleness of the thin coating of refined Chinese lacquer, and after a preformed lacquer ware was coated layer by layer to a certain thickness, it could then be carved. The potentiality excavation of lacquer and Chinese wood oil led to the creation of the technique of Ti-rhinoceros (scraping-rhinoceros), a carved lacquer ware art to deeply carve clouding in the Southern Six Dynasties. The Ti-rhinoceros lacquer toilet cases made in China in the fifth century are respectively recorded in *Chinese Lacquer Art* edited by the Songtao Art Museum of Tokyo and *The 2,000 Years of Chinese Lacquer Art* edited by the Museum of Cultural Relics of the Chinese University of Hong Kong. The invention of the carved lacquer ware art such as the technique of Ti-red (scraping-red) appeared in the Tang Dynasty when Chinese painters further explored and grasped the potentials of lacquer. According to *A Record of Decoration with Lacquer* by Huang Cheng, "As for the technique of Ti-red…the method adopted in the Tang Dynasty was mainly to use printing board to engrave flat brocades, coloring it red, with the engraving unsophisticated and worth seeing; concave yellow brocades were also engraved." Yang Ming once made an annotation about it: "The method adopted in the Tang Dynasty was written as above, and its engraving is rather fast, beyond the ability of future generations. As for concave yellow brocades, they mostly looked like fine hooked cloud, quite different from the scraping method adopted since the Song and Yuan Dynasties [21]." Nowadays, there has been no carved lacquer ware existed

Fig. 12.8 The Ti-red lacquer saucer with the patterns of peony and grass made in the Northern Song Dynasty (Selected from *Chinese Lacquer Art* edited by the Songtao Art Museum of Tokyo)

which was made in the method adopted in the Tang Dynasty. In the Songtao Art Museum in Tokyo, Japan, there is exhibited a Ti-red lacquer saucer with the patterns of peony and grass made in the Northern Song Dynasty (Fig. 12.8). The patterns look well-developed and graceful, enjoying the pattern style of the Tang Dynasty. Its scraping is very shallow, with six-petal brocades engraved on the grounding of yellowish brown lacquer among the pattern gaps, looking mild and mellow with mentally refreshing beauty. It is the existing Ti-red lacquer ware made in the Song Dynasty most close to the style of the Tang Dynasty.

12.5 Development of Civil Lacquer Art and Prevalence of the Foiled Decoration Technology with Lacquer

In the Song Dynasty, the rise of city construction made the decoration technology with lacquer walk out of the court and become a folk and marketplace craftwork. Civil lacquer art got further development when compared with that of previous generations. A large number of civil plainly decorated lacquer wares were unearthed from the tomb of the Northern Song Dynasty in Huai'an, and the tomb of the Southern Song Dynasty in Yixing, Jiangsu Province (Fig. 12.9), such as petal-shaped bowls, plates, mirror cases, saucers and so on. With light and thin bodies of preformed lacquer ware, these lacquer wares were painted with one color and look clean and elegant in lines, enjoying very high aesthetic value. It can be seen here that the decoration with polishing lacquer had been the most fundamental technology adopted to make civil lacquer ware in the Song Dynasty, and the polishing technology had also been used quite expertly.

Fig. 12.9 Plainly decorated lacquer ware of the Song Dynasty (Unearthed at the tomb of the Southern Song Dynasty in Yixing, Jiangsu Province; photographed in Nanjing Museum by the author)

Archaeologists used to name the gold pieces on lacquer ware produced from the Shang Dynasty to the Han Dynasty "gold foil", *History of the Later Han Dynasty* had recorded something about "gold foil", [22] and the "gold foil" of twisted gold threads unearthed at the tomb of Li Chui (one of the princesses of the Tang Dynasty) is only 8.956 microns thick. But strictly speaking, "flying gold" is the gold foils in the true sense, its thickness only 0.12 microns. It is said that the technology of gold foil forging was initiated by Ge Hong in the Eastern Jin Dynasty. Buddha statues were coated with gold foils in the Tang Dynasty, and the splendid landscapes in green and gold were initiated in the Tang Dynasty and prevailed in the Song Dynasty. Both the covering, plating, and coating with gold foils on lacquer ware and the foiled decoration technologies with lacquer such as tracing a design in gold, heading-on a design in gold and concave tracing a design in gold were popular in the Song Dynasty and should be supported by a mature technology of gold foil forging. As for the written records of the technology of gold foil forging, *Chuo Geng Lu* (*Stopping Farming Record*) by Tao Zongyi at the end of the Yuan Dynasty should be the earliest, and *Tian Gong Kai Wu: The Fourteenth Chapter of Hardware* by Song Yingxing in the Ming Dynasty and *Hui Shi Suo Yan* (*Detailed Talk about Drawings*) by Ze Lang in the Qing Dynasty made much more detailed description.

As for the lacquer ware produced in the Warring States period, the paint mixed with gold powder was often used to flat coat and outline, but it was still difficult to say that there had been a strict design tracing technology in gold at that time. The true maturity of the design tracing technology in foil was in the Tang and Song Dynasties. In the Huiguang Tower in Rui'an County, Zhejiang Province, there were unearthed a lacquer scripture case and a lacquer Buddhist relics case of the Northern Song Dynasty (Fig. 12.10). The walls of the lacquer Buddhist relics case were traced designs in gold of immortals, whose lines are slender and natural like floating clouds and flowing water, absolutely comparable to the contemporary famous painting *The Picture Scroll of 87 Immortals*. Inside the lacquer scripture case, there was set one

Fig. 12.10 The immortals traced in gold in the diamond-shaped light-opened on the lacquer Buddhist relics case of the Northern Song Dynasty, which was produced in both the concave and the convex design tracing technology in gold (Selected from *Complete Collection of Chinese Lacquer Ware: from the Three Kingdoms to the Yuan Dynasty* edited by Chen Jing)

smaller sandalwood lacquer scripture case, which was traced a design in gold on the grounding of red lacquer, simply outlined of the patterns of flowers and birds in the light-opened, and coated with powdered gold out of the light-opened, with twisted branches and flowers painted on the open grounding of red lacquer. The gold images and the red images enhance each other's beauty, exquisite beyond compare. It cost innumerable workers to finish, which no lacquer ware with gold design in later generations could match. It can be considered that the technology of giddying a design in gold in China began in the Song Dynasty. The earliest known object made by the concave design tracing technology in gold is the Zhenzhu Stupa of the Northern Song Dynasty, which was unearthed in the Ruiguang Tower in Suzhou in 1978 (Fig. 12.11). The stupa is 122.6 cm high; its octagonal Sumeru throne was opened the light, and on each corner of the feet and the waist, plastic-like broken branches and flowers, flying Apsaras, and providers were piled with ice-like lacquer and then traced in gold. The Sumeru throne was carved with wood and then traced in gold, supremely elegant. On the wooden case of the splendid stupa, there were the written characters of "Da Zhong Xiang Fu Liu Nian..." The whole stupa is not only a rare masterpiece produced of the concave design tracing in gold and decorating technologies with red lacquer, but also an extremely rare comprehensive art work. On the above-mentioned outside lacquer scripture case in the Huiguang Tower, Five Dhyani Buddhas, strange animals, and flying birds were produced in the concave

Fig. 12.11 The pedestal of the Zhenzhu Stupa of the Northern Song Dynasty which was produced of the concave design tracing in gold and decorating technologies with red lacquer (Photographed in the warehouse of Suzhou Museum by the author)

design tracing technology in gold, while curly grass produced in the convex design tracing technology in gold, looking elegant and serene, absolutely a rare article among ancient lacquer wares. On the bottom of the case, the written characters of "Da Song Qing Li Er Nian (1042)" can still be recognized.

Mr. Wang Shixiang treated the lacquer toilet case of the Han Dynasty unearthed in Guanghua of Hubei Province as the earliest lacquer ware produced in the technology of qiang-jin (heading-on a design in gold). According to *Ci Hai* (a large-scale Chinese dictionary), the Chinese character "qiang" means "heading-on" or "catching up forthwith." Qiang-jin refers to the art that when the lacquer is dry but adhesive, head on to coat it with gold or silver foil powder. The meaning of the phrase "heading-on" thus shows that the technology could only appear in the Tang and Song Dynasties when the flying gold was widely used, as before this time it was the filed gold powder that was used to decorate a lacquer ware. Filed gold powder can only be mixed or buried in lacquer and then ground out, as it is extremely difficult to attach it to anything. In the tomb of the Southern Song Dynasty in Wujin of Jiangsu Province, several pieces of lacquer ware produced in the technology of heading-on a design in gold were unearthed. Among them, the twelve-corrugation lotus petal-shaped red lacquer toilet case with silver buckles, with beautiful women in courtyard headed on in gold, is certainly a national cultural relic. The case is only 21.3 cm high and 19.2 cm in diameter, with the wooden preformed lacquer ware shaped three bumped lotus petals. On the red lacquer surface of its outer walls, some cracks could be seen. On its cover, a "Picture of Beautiful Women in a Garden" was made in the technology of heading-on a design in gold, in which the figures' garments were outlined and

Fig. 12.12 A lacquer scripture case produced in the technology of design heading-on in gold on the grounding of black lacquer of the Yuan Dynasty (Collected in Kyushu National Museum; selected from *Lacquer Tea Sets: Articles of the Tang Dynasty*)

headed on into fine autumn hair-like patterns, the facial features were outlined and headed on slender and vague, the willow trees and the mountain stones thick and strong in lines, the walls into broken branches and flowers, and the petals can be seen fine autumn hair-like lines. Japan has stored many pieces of Chinese lacquer ware of the Yuan Dynasty produced in the technology of heading-on a design in gold (Fig. 12.12). The technology was originally recorded in *Chuo Geng Lu* written by Tao Zongyi during the end of the Yuan Dynasty and the early Ming Dynasty.

1. Ice-like lacquer: In *A Record of Decoration with Lacquer*, the phrase is used to refer to the mixture of oil and lacquer, which contains large amount of glue, delicate and soft, with good plasticity. It is ice-like and used to pile and mold patterns.

12.6 Summary of Chinese Decoration with Lacquer and the Introduction of Japanese Decoration with Lacquer

It was in 1421 of the reign of Yongle that Emperor Chengzu of the Ming Dynasty moved the capital from today's Nanjing to Beijing. It can thus be seen that the official lacquer ware in the reign of Yongle was basically made in the south of the Yangtze River. In the reign of Xuande, the official lacquer workshops were moved northward to the capital, hence the Guoyuan Factory. In the Yuan Dynasty, Zhang Degang and Bao Liang from Jiaxing who were both the inheritors of the Ti-red technology were successively called to Beijing. Accordingly, the Ti-red technology adopted in the Guoyuan Factory inherited the humble but sly style that had developed in Jiaxing of the Yuan Dynasty: concise in carving, fine in grinding, looking smooth and embellished; use multicolor thick lacquer to pile design and color in filling lacquer, and grind the lacquer ware as smooth as a mirror, with the craftsmanship complicated and difficult. In the middle and late Ming Dynasty, the handicraft industry and business in the south of the Yangtze River got prosperous, and the treasure evaluation and storage turned prevailing. The lacquer art thus developed from the pathway of converging to that of diverging, becoming varied in modeling and ingenious in adornment. In the late Ming Dynasty, western missionaries came to China and brought with them western practical instruments, and Chinese scholar-bureaucrats began to reflect on the inveterate idea of denying the practicality of utensils, so they turned their attention from "Dao (logos)" to "qi (utensils)." As a result, Huang Cheng wrote *A Record of Decoration with Lacquer*, Wang Zheng wrote *Illustrated Marvelous Utensils of Europe* according to the oral instruction by missionary Jean Terrenz, Ji Cheng wrote *Yuan Ye*, Song Yingxing wrote *Tian Gong Kai Wu*, and so on, which were all the results of the practical trend and the prevailing treasure evaluation and storage in the south of the Yangtze River in the middle and late Ming Dynasty. Yang Ming, a famous lacquer man from Jiaxing of Zhejiang, made annotations for *A Record of Decoration with Lacquer*, which greatly enriched the content of the book.

A Record of Decoration with Lacquer includes 18 chapters of 220 articles. *Qian Ji* (*Collection of Heaven*) covers 2 chapters, of which *The First Chapter of Utilization* records the materials, tools, and equipments used to make lacquer, and *The Second Chapter of Modeling* emphasizes the importance of decorators with lacquer and enumerates the possible faults involved in all kinds of technologies. *Kun Ji* (*Collection of Earth*) covers 16 chapters: the contents covering from *The Third Chapter of Texture and Color* to *The Sixteenth Chapter of Single Element* record 14 kinds of decorations with lacquer, *The Seventeenth Chapter of Quality* records the technologies of making preformed lacquer ware, and *The Eighteenth Chapter of Valuing the Ancients* records the modeling after an antique, the modeling of contemporary products and the repairing of lacquer ware. *A Record of Decoration with Lacquer* completely records the huge system of the decoration technology with lacquer developed since the middle ancient times, so that later painters can not only trace back to the origin of the ancient art, but also innovate accordingly. The masterpiece has become the only classical work about the decoration art with lacquer in ancient times approved by East Asia and even the whole world (Fig. 12.13).

People of the Ming Dynasty once recorded: "The decoration art with lacquer coated with gold originated from Japan. In the reign of Xuande of the Ming Dynasty, a painter named Yang was sent to Japan to study it. After returning, Yang began to teach it to his son. The son learned it and made some innovations. He employed five-color gold mother-of-pearl, not following its old law. The materials and colors of his art works perfectly matched, looking artless. Japanese thus came to China to see his works. They felt astonished and also sighed, thinking that although their own nation invented the art, they could not achieve such excellence." [23] Japanese decoration art with lacquer was handed back to China, and the exchange of the two countries in

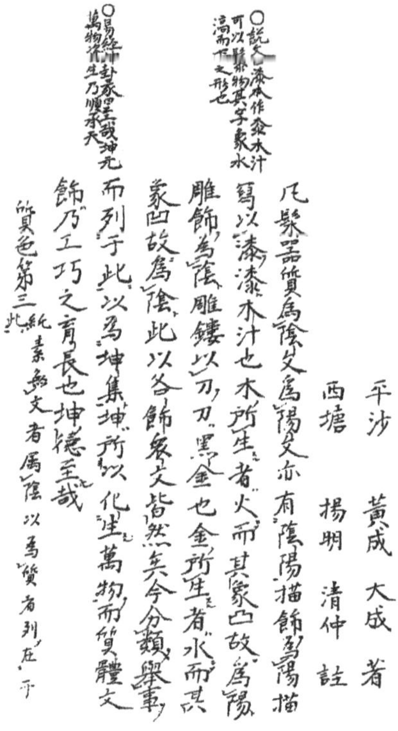
髹飾錄坤集
平沙 黃成大成 著
西塘 揚明清仲 註
凡髹器質為陰文為陽文亦有陰陽描飾為陽描
寫以漆漆木汁也木所生者火而其象凸故為陽
雕飾為陰雕鏤以刀刀黑金也金所生者水而其
象凹故為陰此以各飾衆文皆然矣今分類舉事
而列于此以為坤集坤所以化生萬物而質體文
飾乃工巧之育長也坤德至哉
質色第三 紙素無文者屬陰以為質者列于此
○說文曰漆本作桼木汁可以髹物其字象水滴而下之形也
○易經坤卦彖曰至哉坤元萬物資生乃順承天

Fig. 12.13 A copy of *A Record of Decoration with Lacquer*, a transcript by the Jianjia Hall (Selected from Chang Bei. *A Record of Decoration with Lacquer and East Asian Lacquer Art*)

this field gave birth to a lot of new decorative techniques, such as tracing a design in gold by adding mother-of-pearl, tracing a design in gold by adding mother-of-pearl and grinding with colored lacquer, tracing a design in gold by mixing with placer gold, tracing a design in gold by adding mother-of-pearl and grinding with placer gold, and tracing a design in gold with colored lacquer by adding mother-of-pearl and gold and silver pieces. The last one was introduced into China in the reign of Xuande, and recorded by *A Record of Decoration with Lacquer*. In the late Ming Dynasty, "Recently, those who studied Japanese decoration art with lacquer and returned such as Wu Zhongjiang concluded the technologies of the art, very good at simulating, and tried to affix a seal to the art works with lead. Gold and silver foils, trees and stones inlaid with mother-of-pearl, golden paint, drawing color, all are praised" [24] "As for the techniques such as decolorizing, grinding gold, inlaying with mother-of-pearl, outlining and piling a design with lacquer, and so on, Fang Xinchuan from Xin'an takes the lead." It can be seen that in the late Ming Dynasty, the decoration technology with lacquer coated with gold and silver foil powder mastered by Chinese painters had experienced a qualitative leap. The birth of new technologies in the mid and late Ming Dynasty could not do without the stimulation of the craft of Japanese "Original Makie" [25].

12.7 Engraving and Inlaying on the Decoration of Furniture Screens

During the reign of Emperor Qianlong of the Qing Dynasty, a great deal of lacquer furniture of the palace was produced by the salt administration of the northern and southern areas of the Huaihe River. Lacquer screen is the main variety of Yangzhou lacquer furniture, covering the large-scale kind as hanging screen, the middle-scale kind as stove screen, and the small-scale kind as table screen (ink-stone screen). Ground screen can be divided into the one which is shaped like Chinese character "山," etc. Surrounding screen (a free-standing and mostly folding screen) can have 2 fold, 4 fold, 12 fold, and even 24 fold of its kind, on which literati calligraphy and painting are reproduced. Lacquer screen either stands in a hall to curtain the space off, or hangs on the four walls to decorate, or is placed on a table for the owner to appreciate. [26] Yangzhou painters had inlaid jade, porcelain, bamboo, enamel, marble, etc. on furniture screens and then paid tribute to the court (Fig. 12.14). Emperor Qianlong also gave the tasks of decorating the Cuishang Tower, the Fuwang Pavilion, the Juanqin Study, etc. in the Ningshou Palace within the Forbidden City to the salt administration of the northern and southern areas of the Huaihe River for their workers to engrave and inlay in Yangzhou. The rich and powerful families of Yangzhou also adopted a variety of carved and inlaid screens to produce isolated

Fig. 12.14 Seven pieces of the large-scale engraved lacquer ground screen inlaid with jade shaped like Chinese character "山" made in the middle Qing Dynasty (Selected from Hu Desheng (Ed.). *Masterpieces in the Imperial Palace: Furniture in the Imperial Court of the Ming and Qing Dynasties*)

Fig. 12.15 The engraved gray lacquer screen inlaid with bits of mother-of-pearl set in the outer room of the Shi Kefa Memorial Hall in the late Qing Dynasty (Selected from Shen Huilan. *Yangzhou Eight Carvings*)

indoor space in their big houses and gardens. In today's Yangzhou Shi Gong Temple (Shi Kefa Memorial Hall), a large-scale screen made in the Qing Dynasty, more than one *zhang* high and wide, was set in each of three rooms respectively (Fig. 12.15). As for the screen set in the outer room, on the grounding of black lacquer on the front preformed lacquer ware, there were caved twisted plum branches with blossoms; within the lines, bits of mother-of-pearl were inlaid, and the inscription of "Ji You Liu Nian Li Shi Hu" was engraved. On the opposite side, there were thinly carved the patterns of mountain stones and clouds and rivers on the grounding of eight-treasure gray lacquer, inlaid with colorful enamel flowers. This screen was designed by Li Shihu, a painter of the late Qing Dynasty following the plum painting of Wang Chaolin, and then meticulously made by Yangzhou lacquerers. It can be seen that in the Qing Dynasty, the decoration technology with lacquer by engraving and inlaying was adopted not only by royal buildings, but also by folk halls.

12.8 Patting and Applying with Thin Lacquer and Thickly Decoration by Filling

While court lacquer ware deliberately strived for exquisiteness and over-elaboration, the lacquer workshops in the Qing Dynasty produced a large number of simple but healthy utensils mainly for practical use, forming own distinctive decoration technologies with lacquer from place to place. Among them, the techniques of thin application by patting and thickly decoration by filling could especially demonstrate the creative ability of lacquer artists, contrasted with the extravagant style of both court lacquer ware and lacquer screen couplet which misused the techniques of engraving and inlaying.

Before the invention of the techniques of covering with, applying and coating powdered gold, the method used by lacquer workers to make a piece of lacquer ware contain gold or silver was only to mix filed metal powder into the lacquer and then apply the mixed lacquer on the preformed lacquer ware. In the late Qing Dynasty, Shen Zhenghao, the eldest grandson of Shen Shao'an of a lacquer ware family in Fuzhou, invented the technology of patting and applying with thin lacquer. According to this technology, half of clear oil and half of quick-drying lacquer are mixed, and then the pigment is added; the mixture is then finely mixed on a fine slate, into which lacquer and pigment are added at any time until the pigment is absolutely fine, without particles, enough to be fine dyestuff. Mix with clear oil with quick-drying lacquer in proportion; in 3 *qian* (a unit of weight, = 5 grams) lacquer, the powder of about 1,000 sheets of silver foil is added. The mixture will be slowly rubbed in batches with a stone pestle on the fine slate, turned from time to time with a paint scraper, until it is absolutely fine, without particles, enough to be silver foil powder stuff. Mix with clear oil with quick-drying lacquer and ancient copper foil powder in proportion, and slowly rub the mixture to be fine ancient copper foil powder stuff. And then mix the silver foil powder stuff, the ancient copper foil powder stuff, and the dyestuff evenly enough to be colored thin lacquer with gold and silver, which is finely filtered repeatedly to contain no particles. In a dust-free environment, the lacquer made of silver foil powder stuff, ancient copper foil powder stuff or colored dyestuff is dipped with a very clean hand, and then thinly patted on a very smooth, absolutely dirt-free preformed lacquer ware, from left to right, then from top to bottom, evenly patted for three times or more. During the process, any dust is prevented from falling on the paint surface. Try to use human hair to make a paint brush. First smooth the hair, then put it in a shaded room and dry it for a few days. Put it in a cool and ventilated place, and dry it until the oil in it flies off slowly and thoroughly. The technology of patting and applying with thin lacquer can greatly save the use of lacquer, and make it a reality to produce lacquer ware of high brightness color with gold and silver, not just of the traditional warm or undertone colors like red and black any longer. The coating of golden and silver lacquer makes lacquer ware implicit, no longer dazzling. This is an epoch-making reform in the history of Chinese decoration technology with lacquer, and the most creative contribution to Chinese decoration technology with lacquer made by the lacquer workers of Fuzhou in the Qing Dynasty. Since then, this technology has been widely used in the production of lacquer ware in Fuzhou (Fig. 12.16).

Corresponding to the above technology, the lacquer workers of Fuzhou also invented the technology of "thickly decoration by filling," which is a collective name of the set of techniques of graining (making patterns) on a piece of preformed lacquer ware with lacquer, filling it with lacquer, then decorating it completely, and finally polishing it after drying until the patterns appear. This technology was recorded as "the decoration with lacquer by filling and polishing." The masters in this field include Mr. Li Zhiqing, Mr. Shen Fuwen and so on. Li Zhiqing (1894–1976) began to study at the Fuzhou Arts and Crafts Institute at the age of 19, and then followed Chinese Lin

Fig. 12.16 Modern oblong box decorated with thin lacquer with the pattern of ink landscape (Produced by Shen Shao'an Lacquer Ware Family, selected from the *He Guang Ti Cai* by the "Palace Museum" of Taibei)

Hongzeng and Japanese Harada to study lacquer art. In 1924, he went to Nagasaki Fine Arts School in Japan to continue to study lacquer art, and returned to China in 1926. He began to work in Fuzhou Institute of Industrial Arts in 1956, and then worked in the Second Bodiless Lacquer Ware Making Factory engaged in lacquer painting and design. Combining with traditional Chinese displaying decoration with lacquer, rhinoceros-print technique, and Japanese variable coating, he invented a unique way to produce 100 pieces of decorated sample plates with lacquer by filling and polishing, and produced a number of pieces of thickly decorated lacquer ware by filling (Fig. 12.17). Shen Fuwen (1909–2000) was admitted to the National Academy of Arts of Hangzhou in 1926, and went to Japan in 1935 to study lacquer art in the Matsushita Institute of Lacquer Art. After returning to China, he tried to display the patterns of Dunhuang with the thick decoration technology by filling, and made more than 100 lacquer plates and lacquer bottles (Fig. 12.18). In the 1940 s, his products

Fig. 12.17 Sample lacquer plates produced by polishing and filling (Picture provider: Fuzhou Institute of Industrial Arts)

Fig. 12.18 A modern lacquer disk thickly decorated with the pattern of Dunhuang by filling and inlaying (Author and picture provider: Shen Fuwen of Chongqing)

were given touring exhibitions at home and abroad, which became a heroic undertaking in the art industry. The patting and applying technology with thin lacquer by Shen Zhenghao and the thick decoration technology by filling by Li Zhiqing and Shen Fuwen can be called two flags flying high in the history of the decoration technology with lacquer from the late Qing Dynasty to nowadays. The invention and refinement of the two technologies lay on both the master of the techniques of filling, inlaying, and polishing of lacquer artists, and the further development of the potential performance of lacquer. The latter has become the most common technology used in modern lacquer painting.

12.9 New Creation of Contemporary Lacquer Painting: Covering Bright and Polished Painting with Foil Powder

In ancient China, lacquer painting was dependent on lacquer ware. Since the twentieth century, the stimulation of Vietnam's polished lacquer paintings and the impact of modern art thoughts have pushed Chinese painters to pursue tirelessly. As a result, modern lacquer painting was separated from industrial arts and became a pure art, finally recognized as a new genre of painting in 1984. Modern lacquer painting takes thick decoration with lacquer by filling and inlaying as its main technology, whose most innovative technique is covering bright and polished painting with foil powder.

The ancient painters were powerless to apply colors to a drawing on lacquer board. Lacquer painters thus changed the grounding from polished lacquer into aluminum powder, so changed the lacquer board like glazed printing paper into that like Xuan paper. The lacquer board would look like raw Xuan paper when coated with foil powder in wet in decorating with lacquer and leveling, while like processed Xuan paper when coated with foil powder after the painted surface becomes dry, on which you can freely apply colors to a drawing or draw, and there can even produce the dizzy effect of Chinese ink. Only in the twentieth century when aluminum and aluminum foil emerged in large quantities and became cheaper, rather than the nineteenth century when aluminum was still rare, it was possible in material to make such an invention of covering bright and polished painting with aluminum foil powder (Fig. 12.19). Mr. Zheng Yikun, a lacquer artist from Fuzhou, is absolutely the most excellent grasper of the invention. His "excellence" lies in: by exploiting the translucency of the transparent lacquer, he repeatedly draws a picture with foil powder, repeatedly covers bright and repeatedly polishes painting, some foil powder polished

Fig. 12.19 *A Girl of the Dai Ethnic Group Dressing and Making Up*, a modern lacquer painting made by covering bright and polished painting with dore silver (Author and picture provider: Qiao Shiguang of Beijing)

Fig. 12.20 A modern bodiless lacquer fish tank by polished painting with foil powder (Author: Zheng Yikun of Fuzhou, picture provider: Zheng Xin of Fuzhou)

bare, and some deeply under the lacquer; he finally polishes bright, which makes the picture full of artistic profundities like a fish swimming in green water (Fig. 12.20). Zheng Yikun's technology was affected by the "重ね研切莳绘" of Japan, but not copied at all. With the wisdom of a Chinese, he makes the polished painting with cheap metal foil powder far better than Japan's polished painting with expensive gold powder in artistic effect. Before the birth of artificial grinding stone and water sand paper of rich models, Chinese lacquer artists were extremely difficult in creating such a technology of great difficulty that could reuse foil powder and repeatedly cover bright, and outshone others by its flexible polishing.

12.10 Returning to the Green, Returning to Civilian Use: Future Development of the Decoration Technology with Lacquer

There were many kinds of lacquer ware in the Warring States period and the Qin and Han Dynasties, which were used in all aspects of life, but strictly speaking, the lacquer ware made in the periods was mainly to serve the aristocrats. In the Song Dynasty, civil lacquer ware got unprecedented development, but the lacquer ware with mother-of-pearl inlay and the carved lacquer ware that are excellent in techniques and cost much were reserved for the upper class. In the late Ming Dynasty and the Qing Dynasty, as the civil culture developed unprecedentedly, the decoration technology with lacquer began to be used in the folk on a large scale. A big wood decorated with lacquer had to be made "in the method with fifteen steps altogether…" [27] (Fig. 12.21) As for small wood decorated with lacquer, in addition to lacquered *qin*, lacquered furniture, and lacquered Buddha statues, there were also court lacquer ware, literati lacquer ware, citizen lacquer ware, religious utensils,

Fig. 12.21 A big wood-carved component decorated with lacquer produced in Chaozhou of the Qing Dynasty (Photographed in Guangdong Museum by the author)

minority's lacquer ware. Each has its own characteristics. The decoration technology with natural lacquer had been widely used throughout China until the end of the twentieth century.

In the 1950s, the products made in the factories of industrial arts in China began to be exported. The abandonment of domestic market and the error in aesthetic orientation, however, led to the artistic deviation of lacquer ware, so that the decoration technology with lacquer was misunderstood as a diabolic trick and wicked craft, or a flamboyant practice. When export orders of lacquer ware were not distributed by the government any longer, and meanwhile the domestic market had not yet opened, the factories of lacquer ware had to be closed in succession. After it came to the industrial society, chemical coatings have been widely used, which has caused both serious pollution to the living environment of mankind and great harm to the workers. As a result, the lacquer ware decorated with natural lacquer has gradually become rare, or been collected by the rich, or made just for certain exhibition. The decoration technology with natural lacquer has come to be marginalized. In view of this, the author has put forward the viewpoint of "green lacquer art," which calls on Chinese people to return to the manual craft itself by using natural wood, hemp, and raw lacquer in order to cherish life and protect the environment. To "return" is not to return to the ancients at all, but to start a renewal of the decoration technology with lacquer under the changed situation for human beings from industrial civilization to ecological civilization.

But in the vast rural areas of China which is less affected by the commercial atmosphere, the tradition of making folk tools by the decoration technology with raw lacquer has not been annihilated. The wood-carved furniture decorated with golden lacquer has been widely used in the folk of Zhejiang, Anhui, Jiangxi, Hunan, and Guangdong. Its surface is decorated with red lacquer, its wood-carved parts with golden lacquer, which looks vigorous in carving and well-stacked in pattern, and caters to the folk preference of joyous and bustling practice. In Yinzhou of Ningbo, Zhejiang Province, there was set up an art museum of the decorated woodcarving with golden or red lacquer. The decorated wood-carved furniture with golden or red lacquer as well as the powdered gold-coated lacquer-decorated buckets, cases, baskets, and others is still necessary supplies for folk marriage in the area of Ningbo. Both Zhejiang Provincial Museum and the "Shili Hongzhuang Museum" in Ninghai have a large collection of the decorated furniture and utensils with red lacquer used in the folk weddings of this region. In the provinces of Jiangxi, Yunnan, Zhejiang and Taiwan, the lacquer ware such as decorated basket with lacquer and decorated tea plate with lacquer (Figs. 12.22, 12.23) which use wooden or pottery preformed elements is still made and used. Those modern lacquer artists in Taiwan even decorate Chinese knots with lacquer to make them waterproof, damp proof, and durable, so that used as a decoration on interior walls (Fig. 12.24). When the Chinese people re-understand the value of the decoration technology with raw lacquer and consciously stay away from those chemical "lacquer ware," the Chinese decoration technology with lacquer, which has had a long history of 8,000 years, will be born again.

After forming film, natural lacquer will also enjoy such high performances as resistance to high temperature, to impact, to certain lighting, to many kinds of solvents

Fig. 12.22 A modern folk hair-combing bucket decorated with red lacquer and traced in gold in Zhejiang Province (Photographed by the author in Zhejiang Provincial Museum)

Fig. 12.23 Modern folk wood-carved baskets with bamboo handle decorated with lacquer in Zhejiang Province (Photographed by the author in Zhejiang Provincial Museum)

Fig. 12.24 A modern lacquer Chinese knot made in Taiwan (Selected from *The Craftsmanship in Taiwan*)

and to corrosion of weak acid-base, whose rust resistance, insulation performance, anti-radiation of protons, and anti-adhesion of marine organism far exceed those of chemical coatings. Nowadays, the decoration technology with natural lacquer is used in national defense industry, petrochemical industry, transportation and water conservancy, textile printing and dyeing, aerospace, and other fields. Its potential values will continue to be excavated and discovered by modern people.

References

1. [The Han Dynasty] Zheng Xuan (Annotated). [The Tang Dynasty] Jia Gongyan (Commented). *Annotated and Commented Zhou Li*. The edition of *Si Ku Quan Shu* (*The Imperial Collection of Four*) by Pavilion of Literary Profundity. Taipei: Commercial Press, 1986, (90): p. 504.
2. [The Han Dynasty] Ban Gu. *History of the Former Han Dynasty*. Yan Shigu (Annotated): "It is called decoration with lacquer to paint articles with the lacquer." Beijing: Zhonghua Book Company, 1962, (12): p. 3989.

3. Kehe, Zang. 1999. *Characters Revision and Explanation of Shang Shu*, 103. Shanghai: Shanghai Education Press.
4. Jusun, Lan. 1982. *Current Translation of the Book of Songs: Guo Feng*, 327. Chengdu: Sichuan People's Publishing House.
5. [The Han Dynasty] Sima Qian. *Historical Records*. Beijing: Zhonghua Book Company, 1982, (No.7): p. 2143. Lacquer Garden as a place name refers to three different regions in the *Ancient and Modern Geographic Names Dictionary*. Zhuang Zhou came from Meng of the State of Song. The "Lacquer Garden" here should be located in the northeast of Shangqiu, Henan.
6. [The Han Dynasty] Sima Qian. *Historical Records*. Beijing: Zhonghua Book Company, 1982, (No.10): p. 3253, 3272, 3274.
7. [The Jin Dynasty] Cui Bao. *Gu Jin Zhu* (*Notes to the Ancient and Modern Times*). Beijing: Commercial Press, 1956: p. 19.
8. [The Song Dynasty] Li Fang. *Tai Ping Yu Lan* (Vol.766). Shijiazhuang: Hebei Education Press, 1994, (No.7): p. 176.
9. *Han Feizi: Shi Guo*: "King Yao abdicated and handed over the crown to King Shun, and Shun accepted it. Shun cut off mountain wood to make dinnerware. He cut and sawed, and finally decorated them with lacquer… Shun abdicated and handed over the crown to Yu. Yu tried to make sacrificial utensils, and dyed them with lacquer outside, while painted them in red inside." *See* Department of Aesthetics, Department of Philosophy, Peking University (Ed.). *Selected Materials of Chinese Aesthetic History*. Beijing: Zhonghua Book Company, 1980: p. 73.
10. In *Zhou Li (Vol. 6): Chun Guan Zong Bo (Part 2)*, "Wang Zhi Wu Lu" (the five kinds of vehicles of the emperor) was recorded. Zheng Xuan annotated "Ge Lu" as "fasten leather on", and "Mu Lu" as "not fasten leather on, but decorate it with lacquer".
11. [The Han Dynasty] Zheng Xuan (Annotated). [The Tang Dynasty] Jia Gongyan (Commented). *Annotated and Commented Zhou Li*. The edition of *Si Ku Quan Shu* (*The Imperial Collection of Four*) by Pavilion of Literary Profundity. Taipei: Commercial Press, 1986, (No.90): p. 499, 500.
12. [The Han Dynasty] Ban Gu. *History of the Former Han Dynasty*. Beijing: Zhonghua Book Company, 1962, (No.12): p. 3989.
13. As for the division of labors in making lacquer ware in the inscription on the lacquer eared cup of the Western Han Dynasty unearthed in Pingba, Qingzhen, Guizhou Province, public opinions are divergent, which can be seen in detail in *Journal of Hunan Provincial Museum* (2007, No.4). The explanation of it can refer to the author: *A Record of Decoration with Lacquer and East Asian Lacquer Art: Research on the Traditional Decoration Technology with Lacquer*. Beijing: People's Fine Arts Publishing House, 2014: p. 45.
14. [The Han Dynasty] Huan Kuan. *Yan Tie Lun*. The edition of *Si Ku Quan Shu* (*The Imperial Collection of Four*) by Pavilion of Literary Profundity. Taipei: Commercial Press, 1986, (No.695): p. 577, 582.
15. [The Han Dynasty] Ban Gu. *History of the Former Han Dynasty*. Beijing: Zhonghua Book Company, 1962, (No.10): p. 3070.
16. Tan Zuowen (Translated and annotated). *Yan Shi Jia Xun* (*The Yan's Traditional Injunctions*) (Vol.5)*: Sheng Shi Pian (No.12)*. Beijing: Zhonghua Book Company, 2007: p. 181.
17. Lithargite: lead oxide, commonly known as "yellow lead".
18. Zheng Shixu. *Qi Qi Kao* (*Textual Research of Lacquer Ware*). Yangzhou: Guangling Ancient Books Printing Press, Jiangsu, 1991: p 18.
19. [The Jin Dynasty] Wang Xizhi. *Bi Jing*. *A General View of the Novels of Five Dynasties* published in 1926 collected by Nanjing Library. Shanghai: Shanghai Literature & Art Publishing House, 1991: p. 298.
20. [The Qing Dynasty] Xie Kun. *Jin Yu Suo Sui*: "Beforehand antlers were burnt into ash, and covered with golden lacquer to fade in color; in the course of time, it would look like snake belly-print, of classic beauty and in elegant taste." *See* Huang Binhong & Deng Shi (Eds.). *Fine Arts Series*. Nanjing: Jiangsu Ancient Books Publishing House, 1986, (Vol.2): p. 1820.
21. [The Ming Dynasty] Huang Cheng. *A Record of Decoration with Lacquer*. Yang Ming (Annotated). *See* Chang Bei. *A Record of Decoration with Lacquer and East Asian Lacquer Art: Research on the Traditional Decoration Technology with Lacquer*. Beijing: People's Fine Arts Publishing House, 2014: p. 155.

22. The saying of "the emperor's riding carriage… was decorated with the patterns of dragons traced in gold foils…" in *History of the Later Han Dynasty* written by Sima Biao of the Jin Dynasty refers to the fact that the emperor's riding carriage in ancient China was completely decorated with the patterns of crisscrossed dragons traced in gold foils, rather than the patterns traced in flying gold. Beijing: Zhonghua Book Company, 1965 (Vol.12): p. 3644.
23. [The Ming Dynasty] Chen Ting. *Liang Shan Mo Tan*. The renewed edition of *Si Ku Quan Shu* (*The Imperial Collection of Four*). Shanghai: Shanghai Ancient Books Publishing House, 1995, (Vol. 1143): p.354.
24. [The Ming Dynasty] Gao Lian. *Zun Sheng Ba Jian*. *See* Chang Bei. *Notes and Research of the Books of Ancient Chinese Art*. Tianjin: Tianjin People's Publishing House, 2008: p.322.
25. "Original Makie": as the painting which is produced by adding gold bolus powder and then finely ground is original to Japanese Makie, Japanese lacquerers thus call the kind of Makie which covers the five categories of "ground Makie," "flat Makie," "high Makie," "meat Makie" and "wood Makie" "Original Makie." In addition, as the "color powder Makie," "powder-eliminated Makie" and "Pingji Makie" can not be ground, they do not belong to "Original Makie,"
26. Sitting screen: also known as ground screen, whose plate is inserted into a wooden block, the left and right of which one holding pillar is installed on to clamp the screen plate, and the holding pillar is clamped by a standing tooth-like thing in the front and rear respectively; the wooden blocks, standing tooth-like things and holding pillars are often carved. 山-shaped sitting screen: it refers to the three-piece sitting screen whose middle piece is made much higher than the two side pieces. Surrounding screen: connected by double-numbered pieces, with two pieces being one fold, so six-fold screen refers to the surrounding screen with 12 pieces. Stove screen: refers to the small-scale surrounding screen that is used to encircle a stove. Table screen: also known as ink-stone screen, placed on a table for the owner to closely appreciate. Hanging screen: covering several different types such as strip screen, sheets screen, square screen, and horizontal scroll screen, of single piece, double pieces, or four pieces, etc.
27. [The Qing Dynasty] Li Dou. *A Record of the Gaily-painted Pleasure-boats in Yangzhou (Vol.17): Record of Constructed Workshop Sections*. Wang Beiping, Tu Yugong (Reviewed and revised). Beijing: Zhonghua Book Company, 1960: p. 415.

Bei Chang Born in Yangzhou, Jiangsu in 1944, Professor of School of Art in Southeast University, the librarian of Jiangsu Provincial Literature and History Research Center, Vice President of Chinese Traditional Arts and Crafts Research Association. After the age of 60, she gave herself a pen name "Changbei".

Research interests: Paint art; Arts and crafts; Study on the History of Ancient Art.

Major works:*She has published more than 20 books and nearly 400 papers and reviews,. A national project: Painting Record and East Asian Paint Art: A Study of Traditional Painting System.*

Awards:The Outline of Chinese Art History won the second prize for outstanding research achievements in humanities and social sciences from 2006-2009 in universities across the country; the first prize of outstanding achievements in Nanjing Philosophy and Social Sciences; the first prize of outstanding scientific research works of the Chinese Higher Education Aesthetic Education Research Association.

Chapter 13
Papermaking

Jixing Pan

As an important invention in ancient China, the technology of papermaking has been recognized by the world for its remarkable contribution in promoting the development of human civilization.

13.1 The Origin and Development of Papermaking

Before discussing when the technology of papermaking was invented in China, we have to make it clear the definition of paper. The so-called paper refers to the schistose fiber with certain strength resulted from the plant fibrous raw material that is first mechanically (being cut up, pounded, and mashed) and chemically (stewed in alkaline liquor) treated, then pulped, manufactured with pulp, and dried. Ancient paper could be good or poor in quality, but both used in packaging and writing. As for its microstructure, the fiber should be pure and dispersed, interwoven in different directions, and appears short and brooming.

There are generally two different views on the origin of papermaking. One is Cai Lun (about 62 AD–121 AD, Fig. 13.1), a eunuch of the Eastern Han Dynasty (25 AD–220 AD), invented paper in 105 AD, which was concluded mainly based on current literature. This had been a popular theory for a thousand years until new archaeological discoveries. The other is that paper had appeared in the Western Han Dynasty (206 BC–25 AD) before Cai Lun, which was concluded mainly based on modern archaeological discoveries. This view was followed from the fact that archaeologist Huang Wenbi (1893–1966) unearthed the hemp paper (49 BC) made in the Western Han Dynasty at the beacon fire and smoke relics built in the Han Dynasty in Lop Nor, Xinjiang in 1933 and put forward the papermaking theory of the Western

J. Pan (✉)
The Institute for History of Natural Sciences, Chinese Academy of Sciences, Beijing, China
e-mail: baojie@mfa.gov.cn

J. Hua and L. Feng (eds.), *Thirty Great Inventions of China*,
https://doi.org/10.1007/978-981-15-6525-0_13

Fig. 13.1 Cai Lun

Han Dynasty. Since 1949, each of the two views has had its say. Besides, archaeologists discovered several other batches of such hemp paper at different locations during 1957–1992 (Fig. 13.2). With the tests by scientists, the papermaking theory of the Western Han Dynasty has had enough evidence to refute the theory of Cai Lun inventing paper.

Cai Lun was not the actual inventor of paper at all. His historical contribution lies in summing up the technical experience of the forefathers in making hemp paper since the Western Han Dynasty, organizing the production of a large amount of high-quality hemp paper to be offered to the court, and taking advantage of the government to improve and popularize it. He advocated the development of Broussonetia papyrifera paper and achieved the technological breakthrough of papermaking with the phloem fiber of woody plant. This invention expanded the source of raw materials and promoted the development of papermaking. In a word, Cai Lun was an innovator of the papermaking technology who linked the past with the future.

After examining the paper of the Han Dynasty and making a comparison of ancient paper, the author went to the rural areas in Sichuan, Shaanxi, and other regions where paper was produced in the autumn of 1965 to study and investigate the folk local method of making hemp paper. Analyzed in combination with the 16-step process of hemp papermaking in Zhifang Village, Fengxiang County, Shaanxi Province, the original papermaking process should be as follows: (1) soaking wet, (2) cutting up,

Fig. 13.2 Maquanwan paper (Collected in the Museum of Chinese History)

(3) washing, (4) pounding and mashing, (5) pulping, (6) manufacturing with pulp, (7) drying paper, and (8) peeling paper (Fig. 13.3). To evolve from the 8-step process to the 16-step process adopted in Fengxiang County, there should have experienced a series of technical ladders. There were most likely 11 kinds of such ladders. The author had taken waste hemp head and hemp cloth as raw materials to simulate the experimental schemes concerning the 11 kinds of ladders with ancient equipments in a handmade paper workshop. In the process, all important experiments were repeated at least 2–3 times to exclude any accidental factor. As a result, the fourth simulated experiment scheme, which covered the procedures of (1) soaking wet, (2) cutting up, (3) washing, (4) dipping in lime water, (5) pounding and mashing, (6) washing, (7) pulping, (8) manufacturing with pulp, (9) drying paper, and (10) peeling paper, was considered to produce the paper with the same features of that in the Western Han Dynasty. By identification and comparison with that in the Western Han Dynasty under magnifying glass, it is concluded that the paper produced with this scheme belonged to the same technical ladder as the Baqiao paper made in Western Han Dynasty.

The Western Han Dynasty saw the budding stage of hemp paper, whose output was smaller, whose place of production was limited, and whose quality was not high, so not enough to replace the bamboo slips and silk. During the reign of Emperor Hedi of the Eastern Han Dynasty, Cai Lun, who then held a post in the court, supervised to produce a batch of high-quality hemp paper by exploiting sufficient manpower and material resources. He presented the paper to the court in 105 AD, and the paper was thus promoted and "got wide use in the whole country."

At the end of the Eastern Han Dynasty, the most famous papermaker was certainly Zuo Bo who was from Laiyang, Shandong Province. The hemp paper he produced

1. washing materials 2. cutting up materials 3. washing materials 4. burning plant ash water 5. stewing 6. pounding and mashing materials 7. stirring the pulp in vat 8. manufacturing with pulp 9. drying and peeling paper

Fig. 13.3 The flow chart of papermaking in the Han Dynasty (Designed by Pan Jixing, painted by Zhang Xiaoyou, 1979)

was absolutely "excellent" and had been famous for a time. At that time, Zuo Bo paper, Zhang Zhi writing brush, and Wei Dan ink enjoyed equal popularity.

13.1.1 Papermaking in the Wei and Jin and the Southern and Northern Dynasties

In the competition with bamboo slips and silk, paper won many victories in succession and became one of the main writing materials in the Jin Dynasty (the fourth century). Hand-copied books on paper had been increasing rapidly, which promoted the development of culture, education, science and technology, and religion.

Compared with the Han Dynasty, the Wei and Jin and the Southern and Northern Dynasties had improved the papermaking in terms of output, quality, and processing. The raw materials used at that time had been constantly expanded, the equipment updated, new techniques emerged, the place of production broadened, and the spread of paper widened. Moreover, famous papermakers came out in succession.

Judged from the entities remained, the paper produced in the Wei and Jin and the Southern and Northern Dynasties had made obvious progress compared with that in the Han Dynasty, which was mainly demonstrated in the increased whiteness, the smoother surface, the more compact structure, and the fine texture with obvious screener's lines. Besides, there were fewer fiber bundles in the paper, and the pulping degree of fiber used to make some paper of the Jin Dynasty reached 70%, close to machine-made paper.

The progress in the technology of papermaking and in the quality of paper in the Wei and Jin and the Southern and Northern Dynasties could also be seen from the poems by the literati who made the chant of paper at that time. Xiao Yi of the Liang Dynasty (502–557) in the Southern Dynasties (420–589) once wrote a poem *Ode to Paper*: "White as frost and snow, you shape like the square layout of chess. On you we can express or narrate, who can realize you as fishing nets?" If the writing material in the Han Dynasty was mainly bamboo slips and silk, which could not be completely replaced by paper as a new material, it had changed fundamentally in the Jin Dynasty. As a large number of white, smooth, and square sheets of paper could be produced then, people did not need to use the expensive and bulky bamboo slips and silk, and even came to be accustomed to writing with paper, which eventually eliminated the bamboo slips and silk. In the last years of the Eastern Jin Dynasty, Huan Xuan abandoned Emperor Andi and called himself Emperor, and then ordered, "In ancient times, there was no paper, so bamboo slips and silk were used. Now the bamboo slips and silk should be replaced by yellow paper to use." Archaeological excavations show that the documents unearthed from the tombs or sites since the Eastern Jin Dynasty were no longer made from bamboo slips and silk, but paper.

The widespread use of paper in China had greatly promoted both the rapid increase of book-type documents and the dissemination of science and culture. On the other hand, the development of science and culture and books promoted the production of more and better paper. At the beginning of the Jin Dynasty, the collection of official books was calculated by thousands of volumes, which caused the popularity of copying books and promoted both the development of calligraphy art and the evolution of the fonts of Chinese characters. After the Jin Dynasty, the calligraphy of Chinese characters transited from the official script current in the Han Dynasty to the regular script-type official script, which finally developed into the regular script popular nowadays. The cursive script also developed during this period. It can be said that the evolution of Chinese characters had been caused by the use of paper, on which one could write with a brush as one pleased. The emergence of the outstanding calligraphers like Wang Xizhi and Wang Xianzhi in the Jin Dynasty was largely due to the widespread use of paper (Fig. 13.4). As such, it could produce better artistic results to draw on paper. At that time, the copying of Buddhist and Taoist scriptures consumed a large amount of paper, probably more than that of copying non-religious works (Fig. 13.5).

At this time, many official and private papermaking workshops had been built in the north and south to use local materials for papermaking. With Chang'an (today's Xi'an), Luoyang, Shanxi, Shandong, Hebei, and other regions as the center in the north, hemp paper, Broussonetia papyrifera paper, and mulberry (bark) paper were produced. Since the Eastern Jin Dynasty, current Shaoxing, southern Anhui, Nanjing, Yangzhou, Guangzhou, and other places had become the papermaking centers in the

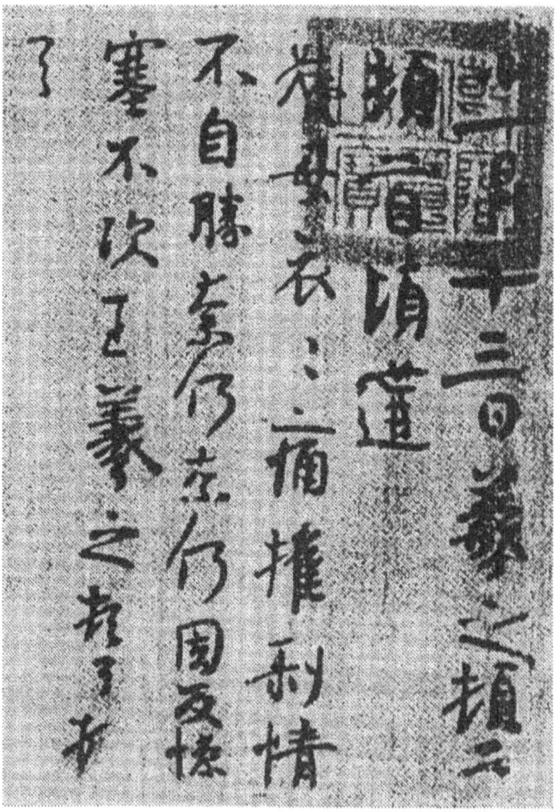

Fig. 13.4 The calligraphy by Wang Xizhi of the Eastern Jin Dynasty (Taken from Cultural Relics, No. 11, 1965)

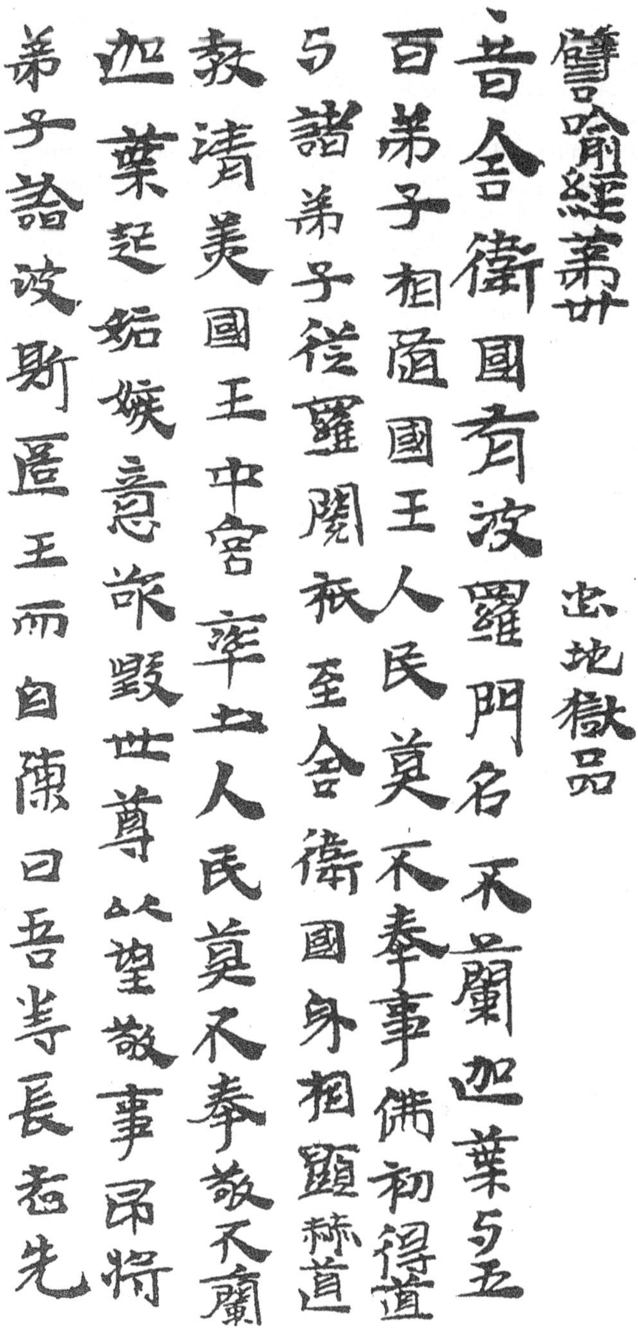
譬喻經第廿　出地獄品
昔舍衛國有婆羅門名不蘭迦葉与五
百弟子相隨國王人民莫不奉事佛初得道
与諸弟子從羅閱祇至舍衛國身相顯赫道
教清美國王中宮羣臣人民莫不奉敬不蘭
迦葉起妬嫉意欲毀世尊以望敬事即將
弟子詣波斯匿王而自陳曰吾等長者先

Fig. 13.5 The hand-copied book on paper *Pi Yu Jing* (*Buddhist Epigrammatic Parables*) written in 359 unearthed in Dunhuang (Collected in the Calligraphy Museum, Tokyo)

south, where the paper varieties produced were basically the same as those in the north. Besides, the coastal area along the Yanxi River, Shengzhou, Zhejiang, became the manufacturing center of rattan paper.

The paper made by Zhang Yong of the Southern Dynasties was superior to the imperial paper employed by the emperor. In addition to original color paper, he also produced all kinds of colored paper. Besides single raw material, he also used mixed bark fiber and hemp fiber to produce pulp to make paper.

Judging from the unearthed ancient paper, the papermaking machine with movable bed of screener had been widely used in the Wei and Jin Dynasties at the latest. The shape and size of the papermaking machine was the same as that of the paper made. This kind of papermaking machine had been used all over the world for more than 1,000 years since then. After testing dozens of such paper patterns in kind, the author found that the papermaking machine as well as the paper produced in the Wei and Jin and the Southern and Northern Dynasties was mostly rectangular, 30–55 cm in length and 23–27 cm in width. The machine was easy for one person to operate, and fewer large-size sheets of paper could be seen in later generations. As the size of the paper was small, it was mainly used for writing, while large-size silk was still used for painting.

In the Wei and Jin and the Southern and Northern Dynasties, the processing technologies of paper such as gluing, surface coating, and dyeing had also developed.

In order to improve the performance of paper, the Jin Dynasty had seen the technology of gluing. The early glue used was made by vegetable starch. The kind of glue was brushed on the surface of paper, or mixed into paper pulp, in order to increase the paper's resistance to the permeability of water by shutting off the capillaries between fibers, or to improve the suspension property of paper pulp. According to the test by the author, the hand-copied book on paper *Lv Zang Chu Fen* that was written in 416 had been treated with the technology of gluing. So far, the earliest glued paper that we have found is the unearthed paper of list of the buried objects with the dead in 384, the surface of which the starch was coated and then polished.

The surface coating technology is to evenly brush the white mineral powder on the surface of paper with adhesives, and then burnish it with stone, which can not only increase the whiteness and smoothness of the surface of paper, but also reduce the transmittance, making the surface of paper more compact and the absorption of ink better. The author found by tests that both the document paper used in 348 and the hand-copied book on paper *The Records of Three Kingdoms: Biography of Sun Quan* written in the Eastern Jin Dynasty were coated on the surface, more than 1,400 years earlier than those in Europe.

Another processing technology of paper is dyeing. In addition to increasing the appearance of paper, dyeing could sometimes improve its performance. Yellow paper was the most commonly used color paper, which had been used as early as in the Han Dynasty. In the Wei and Jin Dynasties, the technology of dyeing got further developed, as the technology of either writing first and then dyeing, or dyeing first and then writing, had been achieved. Yellow paper was widely used in public and private occasions, especially in copying Buddhist and Taoist scriptures. The dye used in dyeing was phellodendron bark of deciduous tree which belongs to Rutaceae. In addition to dyeing, this dye can also repel insects and prevent moths. Wang Xizhi and Wang Xianzhi both loved to write on yellow paper. Besides yellow paper, there were also green, red, light blue, green, peach blossom paper, etc.

13.1.2 Papermaking in the Sui, Tang, and the Five Dynasties

The Sui, Tang, and the Five Dynasties witnessed an important stage of development in the history of papermaking in China. During the period, the varieties of raw materials for papermaking were further expanded, the paper products were widely used in the daily life of the people, the areas of making paper were extended all over the north and the south, and some progress had been made in improving the properties of paper pulp and in the innovation of papermaking equipments so that larger-size fine paper could be produced (Fig. 13.6), which satisfied the special requirements of the art of calligraphy and painting. And the processing of paper became more fastidious, with some famous and precious processed paper going down in history and followed in later generations. Due to the development of both the transportation and the scientific and cultural exchanges between China and foreign countries, the technology of papermaking invented in China began to spread to foreign countries in different directions. Moreover, the invention of woodblock printing stimulated the further development of paper industry.

The expansion of raw materials for papermaking has been a symbol of the technological progress in papermaking. In the Sui and Tang Dynasties, the raw materials used included the categories of hemp, Broussonetia papyrifera, mulberry (bark), rattan (bark), Daphne odora (bark), Hibiscus mutabilis L. (bark), bamboo, etc. (Fig. 13.7). Bamboo paper as a major invention in the history of papermaking began to show up prominently in Guangdong and Zhejiang in the ninth century. Although the hemp paper was mainly made at this time, the output of the paper made of other raw materials had greatly increased compared with that in the Wei and Jin and the Southern and Northern Dynasties, and there had appeared more kinds of paper made of mixed raw materials. In the Tang Dynasty, the papermaking also used the raw fiber of wild hemp, which should be retted and unglued, trickier than rags, but the raw material was rich and cost lower. Also in the dynasty, the production of rattan paper reached its peak, with the place of production beyond Zhejiang. The green, yellow, and white rattan paper enjoyed different uses. In Luozhou of Guangdong (now Lianjiang), the barks of agalloch eaglewood were used for papermaking, named fragrant-skin paper. Xue Tao (?–832), a woman poet of the Tang Dynasty, used the barks of cottonrose hibiscus as raw material to make paper in Sichuan. She mixed the juice of cottonrose hibiscus into the paper pulp and produced the Xue Tao letter paper of pink and other colors that had been a celebrity for a time.

With the expansion of the materials and the spread of the technology of papermaking, the areas where paper was produced in the Sui, Tang, and the Five Dynasties had covered the whole country (Fig. 13.8). According to ancient books, in the Tang Dynasty, there were actually the following 11 cities or capitals which paid paper as

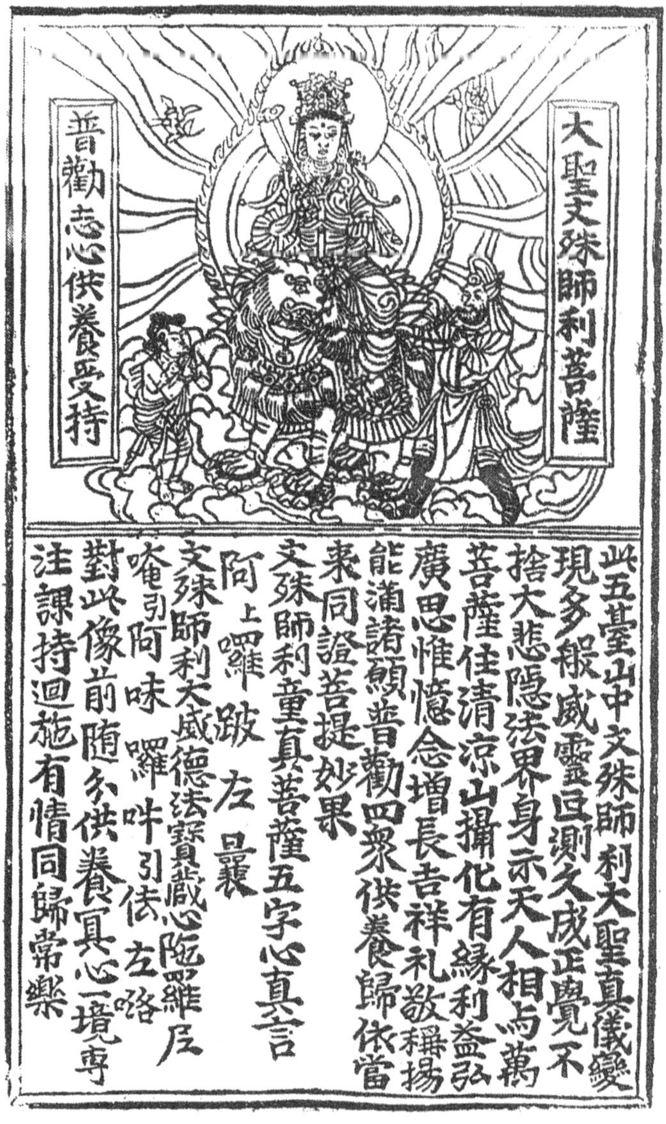

Fig. 13.6 The single-sheet *The Statue of Manjusri Bodhisattva* (31 cm × 20 cm, collected in the National Library of China) printed in the five dynasties (about in 950) unearthed in the Dunhuang Caves

tribute to the court: Changzhou, Hangzhou, Yuezhou, Wuzhou, Quzhou, Xuanzhou, Shezhou, Chizhou, Jiangzhou, Xinzhou, and Hengzhou. A large amount of paper was produced at that time, and the consumption of paper was also considerable. Take the Jixian Academy set up in the imperial storehouse of the Tang Dynasty as an example. The academy used up more than 10,000 sheets of Shu (Sichuan) paper

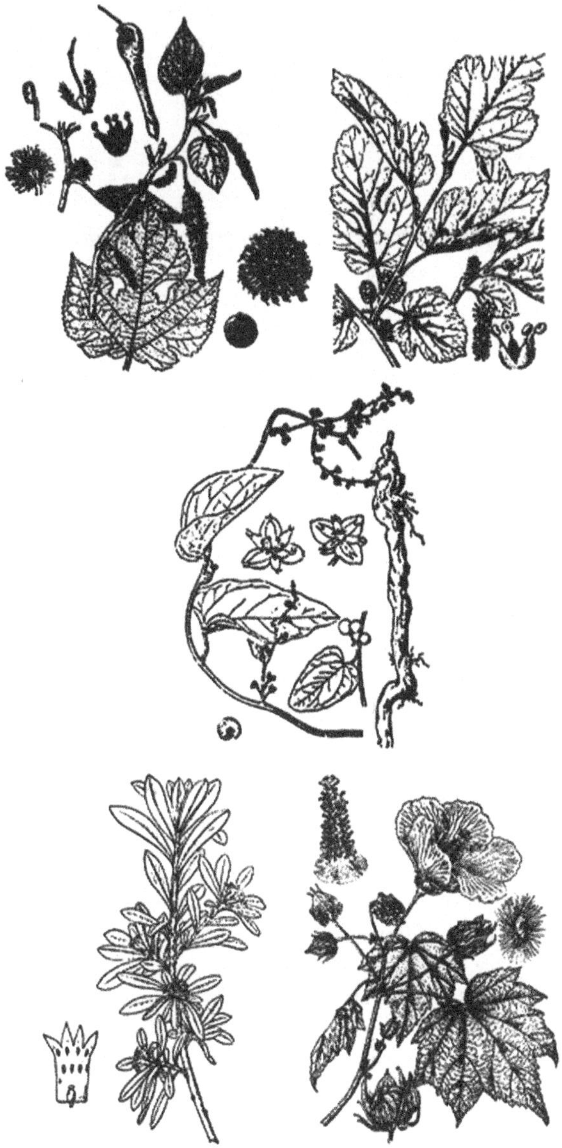

Fig. 13.7 Raw materials for papermaking. Broussonetia papyrifera, mulberry, green rattan, yellow Daphne odora, Hibiscus mutabilis L

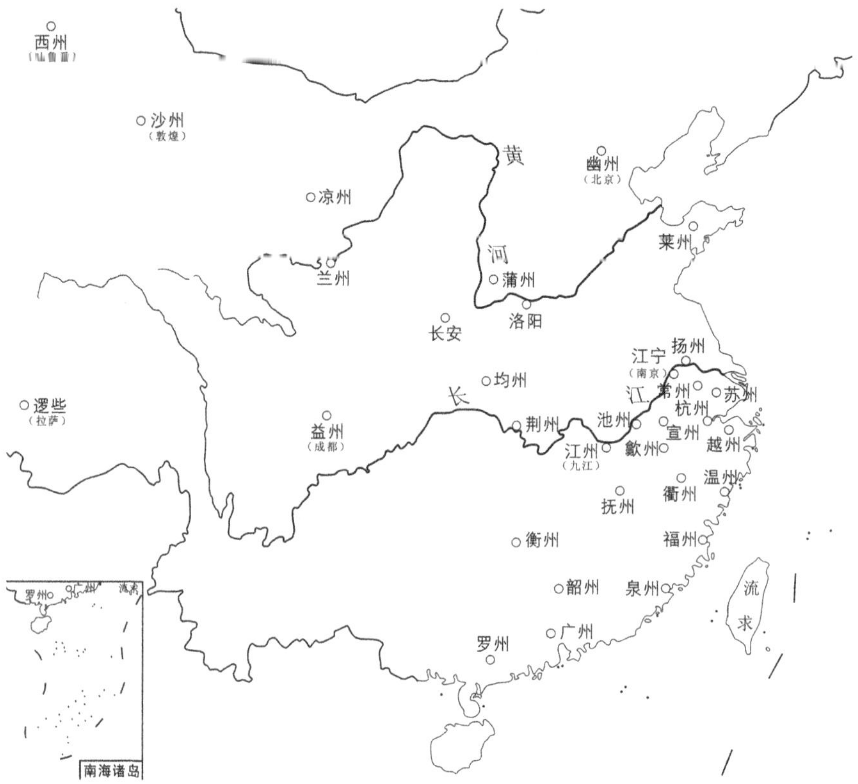

Fig. 13.8 The distribution map of papermaking centers in the Sui, Tang, and the five dynasties

and copied 365 volumes of books only in the year of 849. Paper became the material backing for the prosperity of the social science and culture then.

In the Tang Dynasty, as larger-size paper could be produced, paper paintings had sharply increased from this period, including the colored *Picture of Five Oxen* by Han Huang that was handed down, and the unearthed paintings of flowers and birds, and figures. In addition to the fields of culture and documents, many daily necessities in the Sui, Tang, and the Five Dynasties were also made of paper rather than other expensive materials. For example, paper was used to make lantern or paste window; general paper was oiled on the surface into waterproof paper, which could be used as a substitute for silk. Paper clothes, paper cap, paper quilt, paper mosquito net, paper armor, paper flower, paper cutting and wrapping paper, etc., all could replace the past textiles. "Flying currency" invented in the Tang Dynasty, which was the pioneer of current paper money, was used to replace the metal currency. In addition, paper was also used to make paper people and paper money that would be burned for the funeral of the deceased. The use of paper was becoming wider, and China had truly entered the era of paper (Figs. 13.9, 13.10).

Fig. 13.9 Paper screen (Taken from *Kan Shu Tu* by Wang Qihan in the Five Dynasties)

Fig. 13.10 Paper crown of the Tang Dynasty unearthed in Turpan (Taken from *Cultural Relics*, No.10, 1973)

The paper made in the Sui, Tang, and the Five Dynasties exceeded that in the previous generations in both quality and processing technology, and all kinds of famous and precious paper made then have always been on the lips of later generations. In order to meet the needs of writing and painting, paper was divided into raw paper and ripe paper in the Tang Dynasty. Raw paper refers to the one that was dried but not processed, while ripe paper to the processed raw paper. When to paint colored fine brushwork or to write small type, we usually use ripe paper, while to paint wash drawing or to write large type, we should use raw paper. From the technical point of view, in order to keep the brush from staining due to a momentary lapse of ink, it is necessary to shut off the capillaries between fibers to achieve expected artistic effect. Effective measures involved include burnishing, pulping, powder filling, waxing, gluing, etc. In the Tang Dynasty, there were special makers of ripe paper, who were engaged in the processing of raw paper into ripe paper with special quality, color, and appearance required for writing. Besides starch, animal gelatin and vegetable gelatin were also used for gluing in the Tang Dynasty. In order to effectively disperse the particles of animal gelatin, alum was added as precipitant at the same time. Or the gelatin and alum were brushed on the surface of paper, or mixed into paper pulp. If someone wants to write with Chinese ink which actually contains gelatin, the more glued paper tends to make his or her writing or painting dull and not vivid. Some people thus prefer to use raw paper. For general document, draft, or folk cultural affairs, pulp-pounded paper or raw paper will still be used. It was not until 1337 in Europe that animal gelatin was used for gluing in papermaking, more than 600 years later than in China.

Among the yellow paper made in the Tang Dynasty, there was a special genre which was processed with wax, named hard yellow paper. The paper was the most famous and precious, and used to write on only on those important or solemn occasions, such as the Taoist scripture *Wu Shang Mi Yao* (written in 718), the Buddhist sutra *The Lotus Sutra* (Fig. 13.11), etc. found in the Dunhuang Caves. This paper was first dyed with phellodendron bark, and then coated with yellow wax on the surface, so looks decorous and shiny of timeless appeal. It was not until 1866 in Europe that wax-coated paper appeared, more than 1,000 years later than in China.

In the Tang Dynasty, there was also produced wax-coated letter paper which was filled with white minerals, or named powder-wax letter paper. White mineral fine powder was first calendered onto the surface of paper, and then wax was calendered. This paper combined the features of both powder letter paper and wax paper, absolutely an innovation of combining the powder filling technology of the Wei and Jin and the Southern and Northern Dynasties with the wax coating technology of the Tang Dynasty. Paper workers of the Tang Dynasty also adopted some decorative techniques of lacquer men and silk workers, and invented the processing technology by coating gold and silver sheets or powder on the surface of paper, named gold-pattern paper, silver-pattern paper or sprayed gold and silver paper, cold gold paper, etc. In order to make the precious metals dazzling, a variety of colored paper was used. According to *Han Lin Zhi* by Li Zhao of the Tang Dynasty, all the letters of appointment issued by the court to military and political leaders had to adopt the five-color damask silk paper with gold patterns.

Besides, there were papers with screener's lines and pattern-calendered paper in the Tang Dynasty. These genres of paper can show bright line or pattern on the surface when examined at the light, which adds a sense of potential beauty of paper. The process of making paper with screener's lines is to weave certain line or pattern with threads on the papermaking bamboo screener, and make them protrude on the surface of the screener, where the pulp is thinned when making paper, so the line will be shiny and presented on paper. The process of making pattern-calendered paper is to

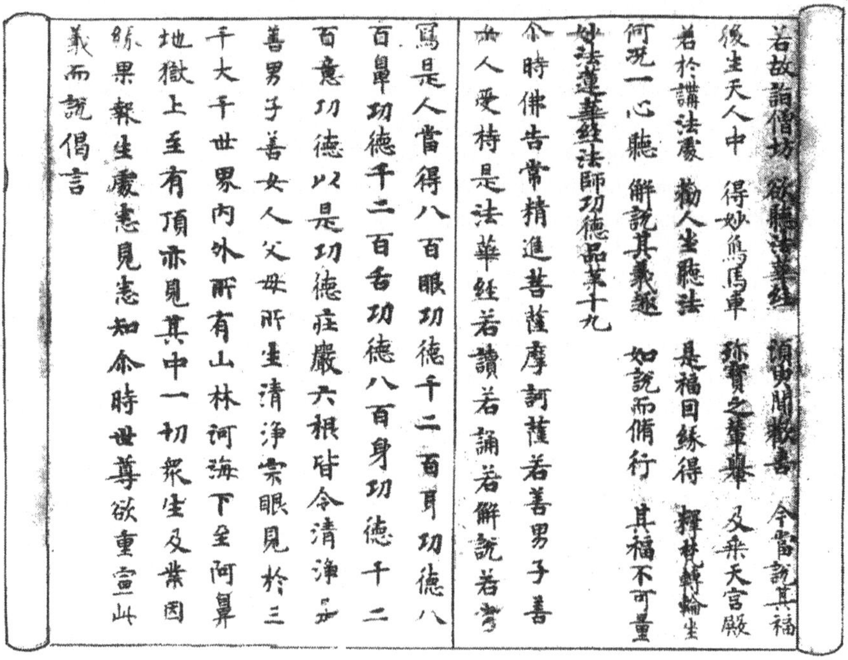
若故詣僧坊 欲聽法華經 須臾聞歡喜 今當說其福
後生天人中 得妙象馬車 珍寶之輦輿 及乘天宮殿
若於講法處 勸人坐聽法 是福因緣得 釋梵轉輪坐
何況一心聽 解說其義趣 如說而脩行 其福不可量
妙法蓮華經法師功德品第十九
尒時佛告常精進菩薩摩訶薩若善男子善
女人受持是法華經若讀若誦若解說若書
寫是人當得八百眼功德千二百耳功德八
百鼻功德千二百舌功德八百身功德千二
百意功德以是功德莊嚴六根皆令清淨是
善男子善女人父母所生清淨肉眼見於三
千大千世界內外所有山林河海下至阿鼻
地獄上至有頂亦見其中一切衆生及業因
緣果報生處悉見悉知尒時世尊欲重宣此
義而說偈言

Fig. 13.11 The hand-copied book on paper *The Lotus Sutra: Merits and Virtues of Masters* made in the late Sui Dynasty (early seventh century) found in the Dunhuang Caves (Made with mulberry (bark) paper, 26.7 cm × 43.5 cm per page, dyed yellowish, and collected by Pan Jixing)

forcibly press a wood block carved with certain line or pattern on the surface of paper, so that the line or pattern will be presented on the surface of paper. The commonly used paper for making securities, currency, certain documents and letters in later generations are made on the same principles. Yang Shen (1488–1559), a scholar of the Ming Dynasty, wrote in his *Dan Qian Zong Lu* (*General Records of Danqian*): "In the Tang Dynasty, there was produced white paper, also named diffractive wave letter paper, possibly because its lines look like water wave." Obviously, the diffractive wave letter paper made in the Tang Dynasty could be said to be water wave paper.

The technology of papermaking in the Five Dynasties directly inherited that in the Sui and Tang Dynasties, but due to the separation of the North and South and the social instability at that time, the development of technology was very uneven in various regions. Most of the hemp paper made in northwest China unearthed in the Dunhuang Caves is not excellent at all, and the calligraphy written on it is far from perfect, which reflects the social and economic depression at that time. But the "Cheng Xin Tang paper" the rulers then used had ranked first for a time. Su Yijian wrote in the *Four Manuals of the Study: Manual of Paper*: "There was produced Cheng Xin Tang paper in the Southern Tang Dynasty, which looks fine and smooth, ranking first for the time being."

13.1.3 Papermaking in the Song and Yuan Dynasties

The Song and Yuan Dynasties saw the mature technology of papermaking in China. During the period, raw materials for papermaking were further expanded, and the development of bamboo paper and rice and wheat-straw paper marked a new era in the history of papermaking. The areas, varieties, and processing technologies of papermaking developed more and more extensively and profoundly, and paper had once again been popularized in all aspects of society. In the two dynasties, most of paper was used for printing. As bamboo paper rose, large-size high-quality bark paper appeared, which was one different feature of the period compared with previous generations. In addition, the works of special study on paper were written, which was also absent in previous generations.

After the Northern Song Dynasty, bamboo paper began to develop actually. The samples of bamboo paper we can see today were just made after the Northern Song Dynasty. Bamboo paper was made in Europe in 1875, nearly a thousand years later than in China. There have been abundant bamboo resources in the Yangtze River basin, the provinces within the regions south of the Yangtze River, and even the south of the Yellow River basin. According to incomplete statistics, there are more than 50 genres of bamboo suitable for papermaking, whose output is absolutely large and distribution wide. Su Shi mentioned it in his *Dong Po Zhi Lin* (*Essays by Su Shi*) (Vol. 9): "Nowadays bamboo is made into paper, such as never previously existed." Clearly, bamboo paper was still new in the eyes of people in the Northern Song Dynasty. Zhou Mi of the Southern Song Dynasty made such a note in his *Gui Xin Za Shi* (*Guixin Notes of Historical Materials*): *Frontal Anthology*: "In the late period of Emperor Xiaozong of the Northern Song Dynasty, bamboo paper began to be used, which was a few *cun* high, more than one *chi* wide." In the *Four Manuals of the Study: Manual of Paper*, Su Yijian wrote: "Nowadays, tender bamboo is made into paper in the regions of Jiangsu and Zhejiang. If the paper is used to write a private letter, no one dares to open it, as it is easy to be torn apart, without any chance to be pasted on." This is to say that the bamboo paper made in the Northern Song Dynasty was not strong in tension, easily to be broken apart, so used to write private letters. Its quality was improved in the Southern Song Dynasty. In *Kuai Ji Zhi* (*Annals of Kuaiji*) (Vol. 1: Part Two), Shi Su concluded five advantages of bamboo paper: smooth surface, better receptivity of ink, easy to write, unchanged ink color and stronger anti-moth, without any mention of its cheapness and easy obtaining. According to the author's test, *Shan Hu Tie* (Fig. 13.12) by Mi Fu of the Northern Song Dynasty used the bamboo paper produced in Kuaiji, which was calendered and looks pale yellow and smooth on the surface. The copies of the Song Dynasty of *Yu Hou Tie* by Wang Xizhi and *Zhong Qiu Tie* by Wang Xianzhi both used bamboo paper.

In the Song and Yuan Dynasties, bamboo paper was also used to print a large number of books. The block-printed books of the Song and Yuan Dynasties were produced centering on Hangzhou of Zhejiang, Jianyang of Fujian, Chengdu of Sichuan, Jizhou of Jiangxi, etc., among which the block-printed editions in Fujian were printed mostly with bamboo paper and widely spread (Fig. 13.13).

Fig. 13.12 Bamboo paper book *Shan Hu Tie* by Mi Fu of the Northern Song Dynasty (collected in the palace museum, Beijing)

Su Yijian also mentioned in the *Four Manuals of the Study: Manual of Paper* that "people of Zhejiang thought that wheat stems or rice stalks were too brittle to be used for papermaking, while the wheat straws and the celastraceous were the best." It can be seen that the wheat stems and rice stalks had been used for papermaking in the tenth century, while in Europe the straws were used to make paper in 1857–1860. However, early straw paper was brittle and thin. As straws belong to the raw materials of short fiber, this kind of paper was later used to make "fire paper" in sacrifice, toilet paper, wrapping paper, etc.

Due to the shortage of raw materials, the rattan paper that had prevailed in the Jin and Tang Dynasties gradually disappeared from the historical stage in the Song and Yuan Dynasties. The hemp paper was produced only in a limited area, and its dominant position was given to bamboo paper and bark paper, which as a trend remained until the late Qing Dynasty at the end of the nineteenth century. In the Song and Yuan Dynasties, the bark paper was mainly used in calligraphy and painting, engraving books and public and private documents. Its output and quality both exceeded those made in the Sui, Tang, and the Five Dynasties, which also made calligraphers and painters prefer to create on bark paper.

The block-printed books of the Song and Yuan Dynasties mostly adopted bark paper. For example, the sutra *Kai Bao Zang* (engraved in 973, printed in 1108) of the Northern Song Dynasty that is collected in Beijing Library just used high-grade mulberry bark paper.

During the two dynasties, paper of mixed raw materials was also made, which is one more great achievement involved. For example, *Gong Yi Tie* and *Xin En Tie* by Mi Fu of the Northern Song Dynasty were made of mixed bamboo and hemp, *Han Guang Tie* of mixed bamboo and Broussonetia papyrifera, and *Gao Shi San*

Fig. 13.13 Illustration in *Shi Lin Guang Ji*, a block-printed book of bamboo paper in the Yuan Dynasty (Collected in the National Library of China)

Tu Shi of mixed hemp and Broussonetia papyrifera. These books are all collected in Beijing Library now. The manufacture of the paper of mixed raw materials is of great technical and economic significance. The genre of paper absorbs the advantages of different raw materials, ranking a unique technology of papermaking of China.

During this time, in order to expand the source of raw materials for papermaking and reduce the production cost, used paper was recycled by mixing it into new paper pulp to make recycled paper, called "revived paper". Song Yingxing of the Ming Dynasty recorded it in his *Tian Gong Kai Wu*: "Waste paper is washed away the ink and dirtiness, and soaked into the pulp vat to remake, which can save the original stewing and soaking... called revived paper." This achieved the purposes of making good use of waste materials and reducing production costs.

In the same period, there had also appeared a variety of processed paper and famous genres of paper.

According to documents, in the Song Dynasty, hemp paper made in Sichuan had such brands as Yuxie, Xiegu, and others; Qingjiang rattan paper, Xin'an jade-plate letter paper, and Cheng Xin Tang paper copied after the Song Dynasty made in Jiangxi were particularly loved by literati; Chinese alpine rush paper made in Shezhou looked smooth and white. Bamboo paper made in Zhejiang was the finest, and the letter paper made in the Kingdom of Wu also competed with Shu (Sichuan) paper in bamboo material. In the North, mulberry bark paper was solid and tough, and there were also such genres of processed paper as Biyun letter paper, Chunshu letter paper, Tuanhua letter paper, and Jinhua letter paper. Another great achievement of papermaking in the Song Dynasty was the ability to produce piece paper which was three to five *zhang* long, said to be the largest-size paper in the world at that time. The works involved that were handed down from ancient times include *The Thousand-Character Classic* written by Emperor Huizong of the Song Dynasty in rapid cursive style of writing that is collected in the Liaoning Provincial Museum. Moreover, there were Guanyin paper with screener's lines, swan-white paper, colored powder letter paper, and so on. In the Yuan Dynasty, there were colored powder letter paper, wax letter paper, yellow letter paper, printed letter paper, letter paper with ribs, etc. in Shaoxing, while in Jiangxi there were white account paper, Guanyin paper, Qingjiang paper as well as calendered letter paper, diffractive wave letter paper, and so on. The name of the "ice-like wing paper" made in the Song Dynasty just indicates that the genre of paper looks dense, thin, and white, which required the hand of a master to make. The major areas of papermaking in the Song and Yuan Dynasties mainly covered Jiangsu, Zhejiang, Sichuan, Anhui, Jiangxi, and Hebei Provinces. In the artistic processing of paper, the colored letter paper with golden paint that was invented in the Tang Dynasty had been further developed in the Song Dynasty. *History of the Song Dynasty: Annals of Popular Clothing* refers to the fact that the official documents of royal court were written on damask silk paper with golden paint and silver cloud. This genre of colored letter paper with gold and silver followed silk products in shape and structure, costing too much. As for colored powder letter paper and wax letter paper, they were mainly used to write on and then mounted into scrolls to be vertically hung within rooms.

In history, Xie Gong letter paper of Sichuan invented in the Song Dynasty was as famous as Xue Tao letter paper in the Tang Dynasty. Xie Gong refers to Xie Jingchu, who had made ten-color letter paper. The most famous paper of the Song Dynasty should be sutra paper of Jinsushan Mountain, shortened as Jinsu letter paper. This genre of paper had been used to write tripitaka which were inscribed in the Song Dynasty and is stored in the Jinsushan Temple in the southwest of Haiyan, Zhejiang Province. Each sheet of them was printed with the words of "Sutra Paper of Jinsushan Mountain" in small red seal, waxed on both sides, without texture. From the technological point of view, Jinsu letter paper is obviously the follower of hard yellow paper in the Tang Dynasty, much thicker, each made of two layers of thin paper, thus could be torn apart into two sheets.

Ming Ren Dian paper and Duan Ben Tang paper made in the Yuan Dynasty were processed art paper for the use of emperors in imperial storehouse. *Chuo Geng Lu* (*Stopping Farming Records*) by Tao Zongyi of the Yuan Dynasty has such a record: "Ming Ren Dian paper is slightly the same as Duan Ben Tang paper, with the three-character seal 'Ming Ren Dian' with golden paint in official script printed on." Paper with ribs and Lian Shi paper of the Ming and Qing Dynasties were both developed on the basis of the Song and Yuan Dynasties. According to *Shu Jian Pu* (*Manual of Sichuan Paper*) of the Yuan Dynasty, "All paper can be divided into the kinds of Lian 2 (conjoint two sheets), Lian 3 (conjoint three sheets), and Lian 4 (conjoint four sheets)." Lian 4 is followed by Lian Shi, which is soft and white bamboo paper. Paper with ribs has fine textures on that intersect vertically and horizontally. In the Yuan Dynasty, there was also Gusu paper, which was colored powder paper, sometimes printed with golden and silver flowers on.

Chinese painting mostly adopted silk as material in the Han, Jin, and Tang Dynasties. Since the Tang Dynasty, paper was gradually used; in the Song and Yuan Dynasties, bark paper had been mostly used. The size of paper became larger and larger, but the screener's lines on paper became finer. The painting paper used in the Tang Dynasty was about 650 square centimeters on average, about 2,412 square centimeters in the Song Dynasty, and about 2,937 square centimeters in the Yuan Dynasty. The epigraphy became prosperous in the Song Dynasty, which required making rubbings from inscriptions on stone tablets or bronze vessels with ink. The paper used here should be thin and firm, pliable and tough, and acceptable of ink, rather difficult to make. Another important use of paper in the Song and Yuan Dynasties was the printing of currency (Fig. 13.14), which saved a lot of metal and was easy to carry, certainly a revolution in the field of economic circulation.

The use of paper products in the Song and Yuan Dynasties was also much wider than that in the Tang and the Five Dynasties. At that time, paper products ranged from paper mosquito net, paper clothes, paper umbrella to paper quilt, paper-cut, paper flowers, etc., as well as paper game cards which were popular among the people. There was also paper shadow drama performed in the folk. Paper was also widely used in industrial production. Powder flasks, powder bags, and lead wires used in the manufacture of fireworks and firearms in the Song Dynasty were all made of paper. It's on mulberry bark paper that female moths laid eggs in silkworm breeding (Fig. 13.15).

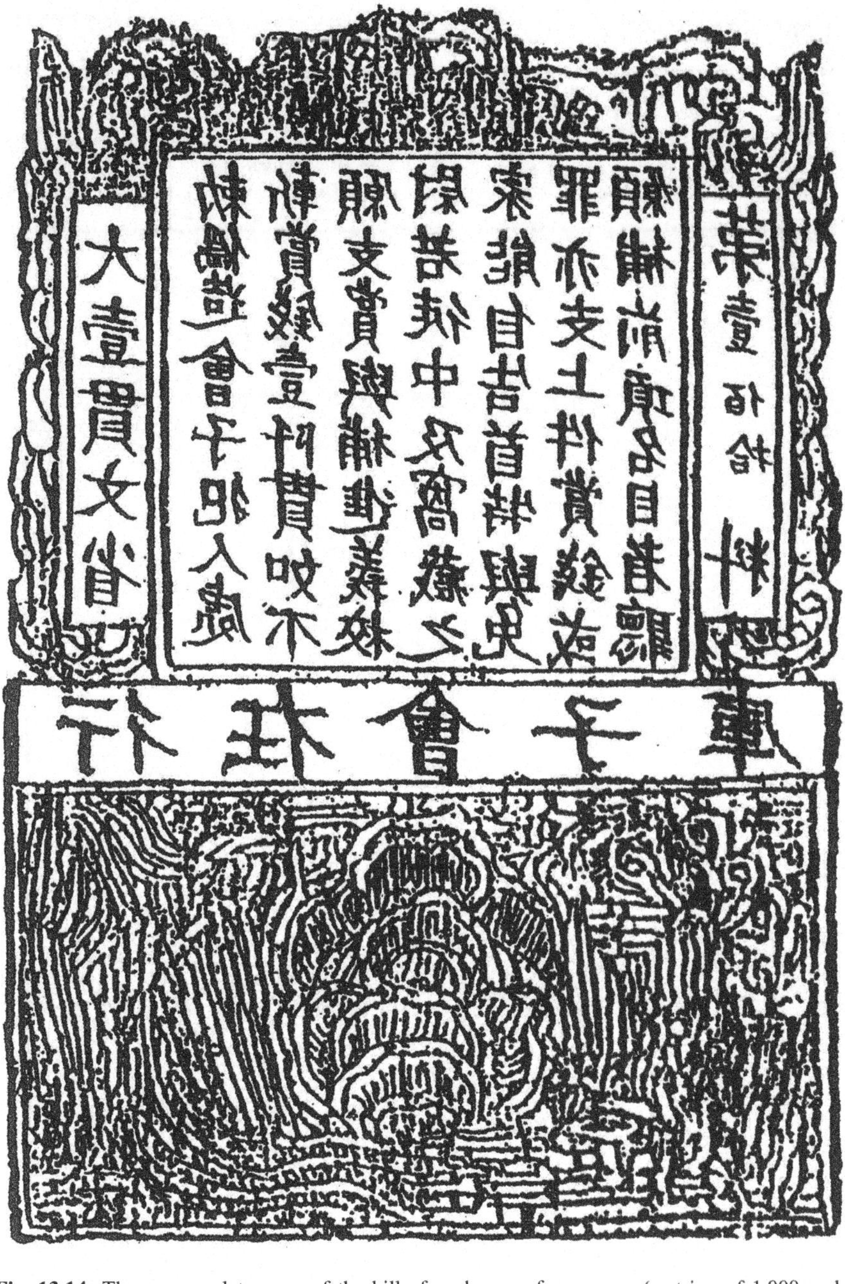

Fig. 13.14 The copper-plate copy of the bill of exchange of one *guan* (a string of 1,000 cash) cast in Hangzhou in 1161–1168 of the Southern Song Dynasty (17.8 cm × 12 cm, collected in the National Museum of China)

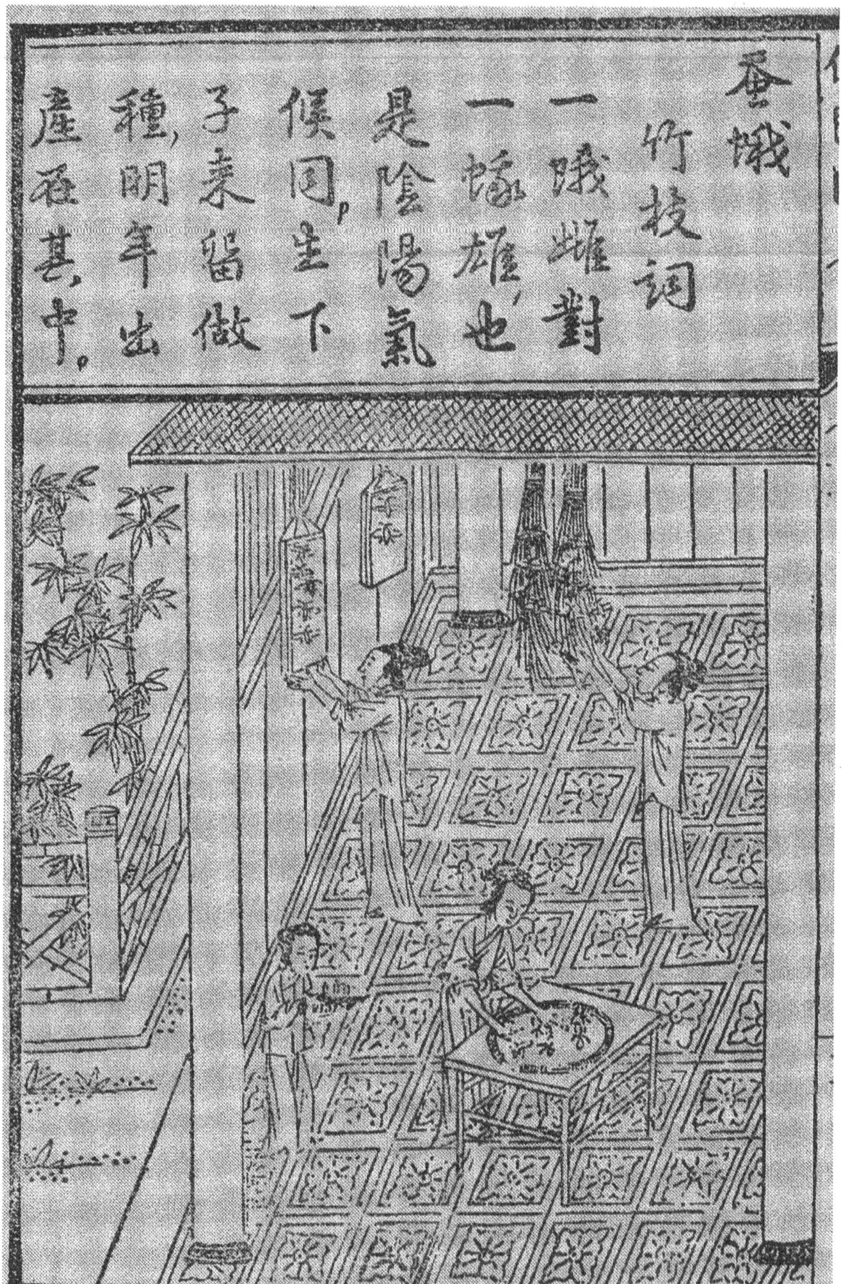

Fig. 13.15 Using thick mulberry bark paper as silkworm breeding paper (Taken from *Bian Min Tu Zuan* by Kuang Fan, 1494)

In the Song and Yuan Dynasties, the water power trip-hammer for husking rice was sometimes used to pound and mash paper materials instead of manpower (Fig. 13.16), which reduced labor intensity and increased efficiency. The water power trip-hammer for husking rice had existed in the Han Dynasty, never used in papermaking, but began to be used for pounding and mashing paper materials after the Tang and Song Dynasties. In order to improve the suspension properties of fiber, plant mucus was added into paper pulp as suspending agent, which was also treated as a technological innovation. In 1901, when chemically examining the paper used by the documents of the Tang Dynasty in China, Wiesner, an Austrian scientist, found that there was lichen in the paper pulp whose water infusion was stick slip, which indicates that this technological innovation had a long history. *Gui Xin Za Shi* (*Guixin Notes of Historical Materials*) by Zhou Mi of the Southern Song Dynasty even records different kinds of plant mucus, including the water infusion of the stem leaves of Hibiscus manihot

Fig. 13.16 The picture of conjoint water power trip-hammers for husking rice in *Nong Shu* written by Wang Zhen of the Yuan Dynasty (Block-printed edition in 1530, collected in the Beijing Library)

L., vines of carambola, leaves of hibiscus syriacus, and wild grapes, which is also mentioned in other documents, and referred to as "slide water", "paper medicine", or "water of paper medicine". The improvement of paper quality in the Song and Yuan Dynasties was obviously related to the widespread use of paper medicine.

13.1.4 Papermaking in the Ming and Qing Dynasties

The Ming and Qing Dynasties were an agglomeration of papermaking in material, technology, equipment and processing in the history, when the output, quality, use, and origin of paper were also in a higher stage of development than ever before. Besides, there appeared illustrated monograph on the technology of papermaking, without parallel in history. With the close communication between China and foreign countries, the elaborately made paper, paper products, and papermaking technology of China were incessantly spread abroad. At the end of the Qing Dynasty, China introduced the technology of papermaking with machine from the west, which was turning a new page of papermaking.

Most of the papermaking workshops in the Ming and Qing Dynasties were distributed in the southern provinces of Jiangxi, Fujian, Zhejiang, Anhui, Guangdong, and Sichuan, as well as the northern provinces mainly of Shaanxi, Shanxi, and Hebei. Raw materials included bamboo, hemp, bark, and straw, among which the output of bamboo paper was the first, especially in the southern mountain areas. Bark paper was often used in calligraphy and painting or printing books, and the output of hemp paper gradually reduced.

In the Ming and Qing Dynasties, Xuan paper specially made in Jingxian County, Anhui Province, was regarded as the best. It's named after Xuanzhou Prefecture to which Jingxian County belonged in the past. It uses mainly the barks of wingceltis as raw material, which belongs to deciduous tree only planted in China. Its branches and phloem are used to make paper. Among the genres of bamboo paper, Lian Shi paper and Mao Bian paper made in Jiangxi and Fujian, respectively, were the most common, mostly used for printing books. Hemp paper was mainly produced in the north, with the output not large. Bark paper was made both in the north and south, with the output the second. Rice and wheat-straw paper was used to make secondary paper, wrapping paper, "touch paper" or cardboard.

The papermaking of the Ming Dynasty was often mentioned in the works written in the Ming Dynasty. For example, according to *Letter Paper* in the second volume of *Kao Pan Yu Shi* by Tu Long, a state agency was set up in Xishan near Nanchang, Jiangxi Province to make paper during the reign of Emperor Yongle (1403–1424) of the Ming Dynasty. "The thickest and finest paper is named Lian 7, and Guanyin paper. The paper used to write memorial to the throne is made in Qianshan of Jiangxi, while the paper used to write list of successful candidates in imperial examinations is made in Changshan of Zhejiang, and Yingshan of Luzhou, Zhili (today's Hebei Province). There is small letter paper which is made in Linchuan of Jiangxi, and there is large letter paper which is made in Shangyu of Zhejiang…".

The *Book of Broussonetia Papyrifera* in *Great Annals of Jiangxi* written and revised by Wang Zongmu of the Ming Dynasty describes the actual situation of making Broussonetia papyrifera paper by the state agency of Jiangxi in the early Ming Dynasty, including the source of raw materials, working procedure, and so on. Among them, there were 22 processes of papermaking, including three times of stewing, two times of natural bleaching, and three times of washing. Song Yingxing of the Ming Dynasty made a detailed record of bamboo paper in *Tian Gong Kai Wu: The Technology of Drying* (Fig. 13.17), reflecting the folk technology of bamboo papermaking in the south of the Yangtze River, whose process was concise and cost a little. In the Qing Dynasty, there were also quite a lot of records of papermaking, especially of bamboo papermaking. In the volume of *Household Utensils Made of Mountain Products* of *Review of the Frontiers of Three Provinces*, Yan Ruyu described in detail the bamboo papermaking in Dingyuan and Xixiang, Shaanxi, whose technology was similar to that described in *Tian Gong Kai Wu*, although the latter was much more advanced.

Xuan paper was very famous in the Ming and Qing Dynasties. According to *Cao's Genealogy* of the Qing Dynasty, Cao Dasan moved from Xuancheng to Xiaoling, Xixiang of Jingxian County at the time of the Song and Yuan Dynasties. He found that the mountains there were rich in wingceltis, so began to making paper as a living, which was thus handed down from generation to generation. Xuan paper is characterized by its whiteness, good flexibility, smooth surface and better receptivity of ink, gradually developing into a famous genre of paper, and had attracted the attention of literati in the Ming Dynasty. The excellence of Xuan paper lies in its meticulous production. Later, the deforestation of wingceltis trees led to the insufficiency of raw materials in making Xuan paper. As a result, Broussonetia papyrifera bark or straw was often mixed into the paper pulp. During the Ming and Qing Dynasties, superior Xuan paper was used only in imperial palace, or for official documents, calligraphy, and painting. It was sometimes used to write the over one-*zhang* list of successful candidates in imperial examinations.

The Ming and Qing Dynasties were an agglomeration of the processing technologies of paper, when all genres of famous paper in history were restored to production, and meanwhile some new genres were invented. Among the processed paper of the Ming Dynasty, Xuande tribute paper was the most famous, which was made in the reign of Emperor Xuande (1426–1435) and had such varieties as five-color powder paper, five-color gold paper, porcelain blue paper, etc. Since then, the paper had been kept on producing and used in imperial palace; later, it was spread out of the imperial palace and came to be praised highly in the world. At the beginning of the Qing Dynasty, Xuande tribute paper and Cheng Xin Tang paper of the Southern Tang Dynasty had been called rare famous paper. We guessed that Xuande paper should be made by the state agency of Jiangxi, not the Xuan paper made in Jingxian County of Anhui. According to *Xi Qing Bi Ji* by Shen Chu, the "goat brain letter paper" made of Xuande porcelain blue paper was also a famous product. Goat brain and top ink were stored in basement for a long time, then coated on the surface of paper, and calendered into letter paper, which looks black as lacquer, but clear as a mirror. The scriptures written with this genre of paper could be used for a long time, not easy

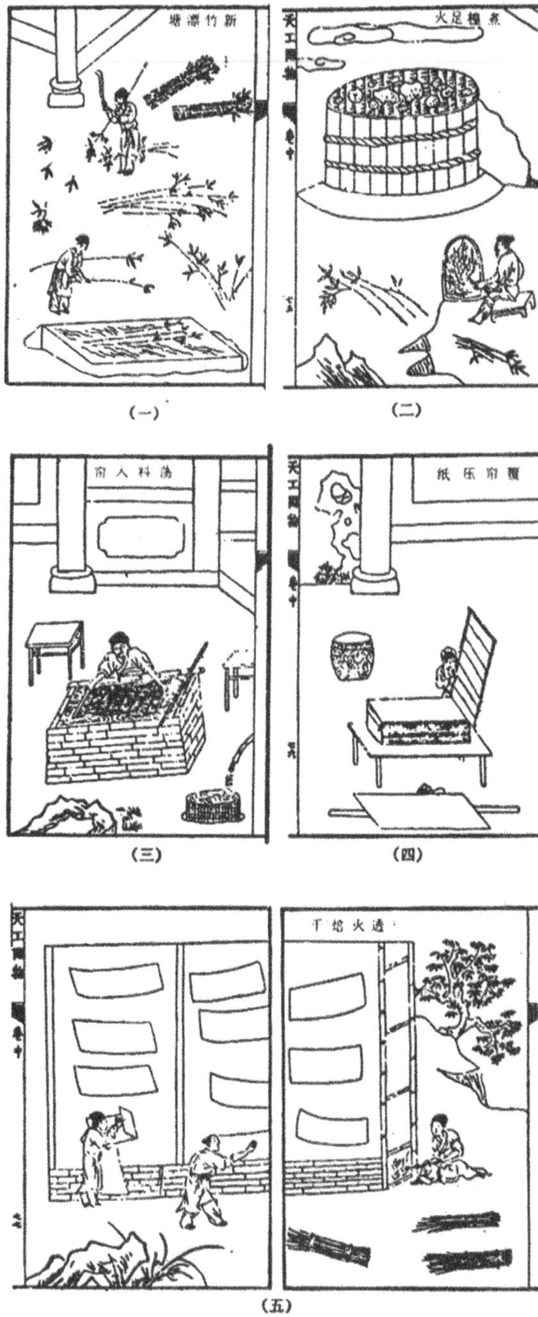

Fig. 13.17 The process chart of bamboo papermaking included in *Tian Gong Kai Wu*

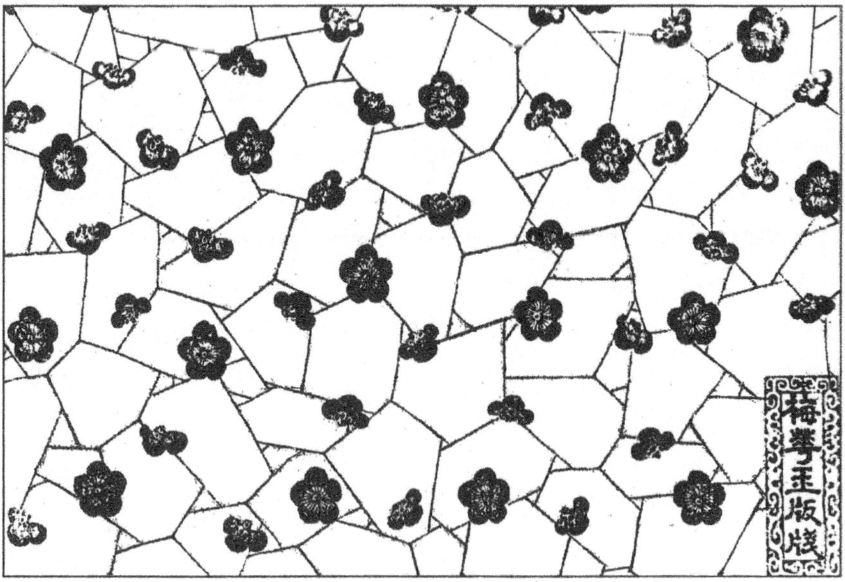

Fig. 13.18 Part of the jade-plate letter paper with plum blossoms made in the reign of Emperor Qianlong of the Qing Dynasty (Collected in the Palace Museum of Beijing)

to be torn and damaged by worms. Xuande porcelain blue paper is made by dyeing with indigo dye, whose color is the same as blue porcelain, hence the name.

In the Ming Dynasty, the sprayed gold letter paper of Suzhou and the Tan letter paper of Songjiang were also famous for a time. In the Qing Dynasty, there appeared many genres of processed paper, such as jade-plate paper with plum blossoms (Fig. 13.18), powder-wax paper with cloud-dragon in gold, paper with ribs, hair paper, mica paper, and so on. In addition, there were copied Cheng Xin Tang paper (Fig. 13.19), Jinsu letter paper of the Song Dynasty, Xue Tao letter paper, Ming Ren Dian paper of the Yuan Dynasty (Fig. 13.20), and so on.

During the Ming and Qing Dynasties, the technologies of how to make various genres of processed paper were also recorded. The two books of *Kao Pan Yu Shi* by Tu Long and *Kuai Xue Tang Man Lu* by Feng Mengzhen (1548–1605) made more detailed discussion on the processing techniques of various genres of paper, including the dyeing method of letter paper of the Song Dynasty, the printing method of letter paper with gold and silver, the beetling method of white paper, and the painting method by dyeing paper without glue, which have important reference value to our current restoring of traditional papermaking.

Fig. 13.19 The colored wax letter paper with landscape paintings by golden paint that was processed by coping the Cheng Xin Tang paper made in the reign of Emperor Qianlong of the Qing Dynasty (Collected in the Palace Museum of Beijing)

13.2 The Spread of the Technology of Papermaking of China

13.2.1 The Papermaking of China in East, South, and Southeast Asia

China is located in the eastern Asian continent, so the technology of papermaking of China had been obviously spread to its neighboring Asian countries first. China is adjacent to Vietnam on land, to which the paper and books made in China had been spread in the second century. Due to the social turmoil happened at the end of the Han Dynasty, a large number of Chinese took refuge in Vietnam and brought the Central Plains culture and production technologies. So in Vietnam, paper could be

Fig. 13.20 The "Ming Ren Dian paper" of the Yuan Dynasty that was copied in the reign of Emperor Qianlong of the Qing Dynasty (Thick bark paper dyed yellow, painted with ruyi (an S-shaped ornamental object) lines with golden paint, double-sided wax, calendered, powdered on the back, sprayed gold sheet, 121.5 cm × 53 cm, collected in the Palace Museum of Beijing)

made in the third century at the latest, including Broussonetia papyrifera paper and agallochum bark paper. In 284, East Roman offered 30,000 pieces of agallochum bark paper to Emperor Wudi of the Jin Dynasty.

Chinese paper was introduced into the Korean Peninsula from the late Han Dynasty to the Wei and Jin and the Southern and Northern Dynasties (Fig. 13.21). In the peninsula, the time that paper was made could be traced back to the fourth–fifth century, when hemp paper was mainly produced, but most of the paper used was still obtained from China. After the Koryo Dynasty (936–1392) unified the Korean Peninsula, the dynasty stayed in touch with the Song Dynasty of China. In this period, its paper industry developed, and bark paper became a new genre of paper. The paper made in this dynasty was characterized by its massiness, strength, and tenacity as well as whiteness, deeply loved by the scholar-bureaucrats of the Song Dynasty (Fig. 13.22). Xian Yushu of the Yuan Dynasty labeled in his *Zhi Jian Pu* (*Manual of Letter Paper*) the paper made in Koryo as one of the famous genres of paper.

Daiwa paper made in Japan has been popular among artists in calligraphy and painting, also widely used in daily life. According to *Ri Ben Shu Ji* (*The Official History of Japan*), people of Baiji in Korea brought Chinese books to Japan in 285; in 610, King of Koryo sent an eminent monk named Tan Zheng (579–631) to Japan.

Fig. 13.21 Two Korean papermakers swinging the screener to manufacture paper with pulp (Taken from *Papermaking: The History and Technology of an Ancient Craft* by Dard Hunter, 1947)

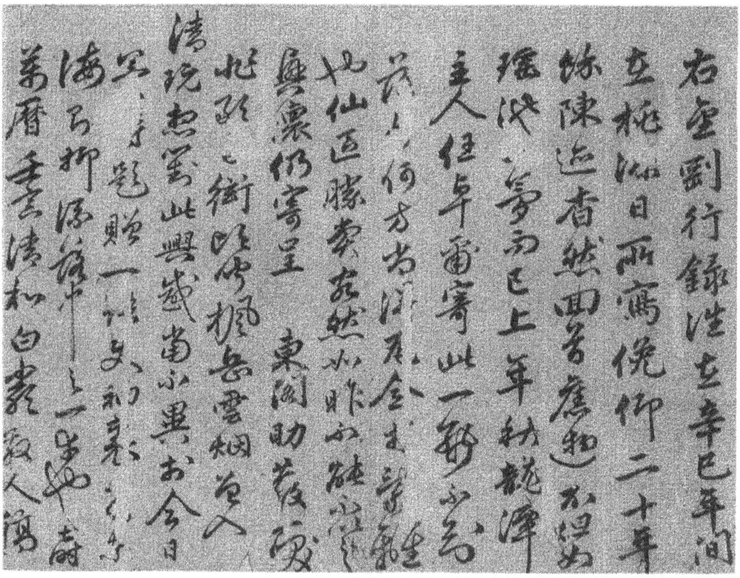

Fig. 13.22 *Jin Gang Xing Lu Zhu Ba* written by Wu Yun (1543–1620), Korean Tongzheng Minister (a senior official in charge of memorials to the throne and appeals) and calligraphist, on Broussonetia papyrifera paper in 1602 (Collected in the Qixing paper History Museum of Seoul, South Korea)

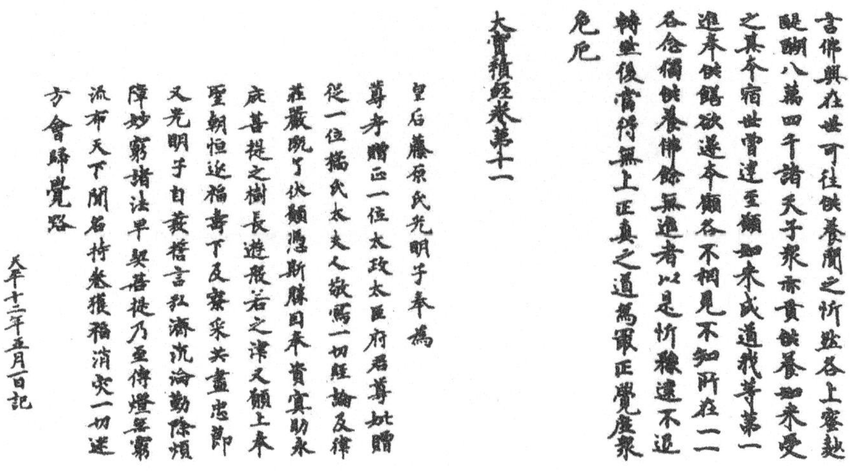

Fig. 13.23 *Da Bao Ji Jing* written on hemp paper in 740 of the Nara Dynasty (815 cm in length, with 18 sheets of paper altogether, 26.4 cm × 45.8 cm per page, scroll bound. Collected in the National Museum of Nara)

The monk had read the *Five Classics* written in China and was able to make paper and ink, so the historians used to believe that Japan began to make paper in 610. In fact, China had frequent direct exchanges with Japan during the Southern and Northern Dynasties, in which Chinese papermaking had been possibly introduced to Japan. In Japan, the making of Yanpi (bark of Yan tree) paper began in the Nara era, and the technologies of both hemp paper and Broussonetia papyrifera papermaking were consistent with those of the Tang Dynasty (Figs. 13.23, 13.24 and 13.25).

The paper and its making technology of China were spread southward to today's India (Figs. 13.26, 13.27), Pakistan, Bangladesh, Nepal, Thailand, Cambodia, Myanmar, the Philippines, Indonesia, and others. All these countries were later in papermaking than Vietnam, Korea, and Japan.

13.2.2 The Papermaking of China in Central Asia, West Asia, and North Africa

The countries in Central Asia and Western Asia were commonly known as countries of the Western Regions in ancient books. As early as in the Western Han Dynasty, these countries had contact with China, from which paper was thus imported. In 1907, Stein found in Dunhuang nine letters that were written by businessman Nanai Vandak of the Central Asia in 311–313 to his friend Samarkand in the language of Sogdian of the Central Asia. The nine letters were all written on hemp paper (Fig. 13.28), which indicates that Chinese paper had been used there as early as in the Eastern Jin Dynasty, and meanwhile was transported to Central Asia.

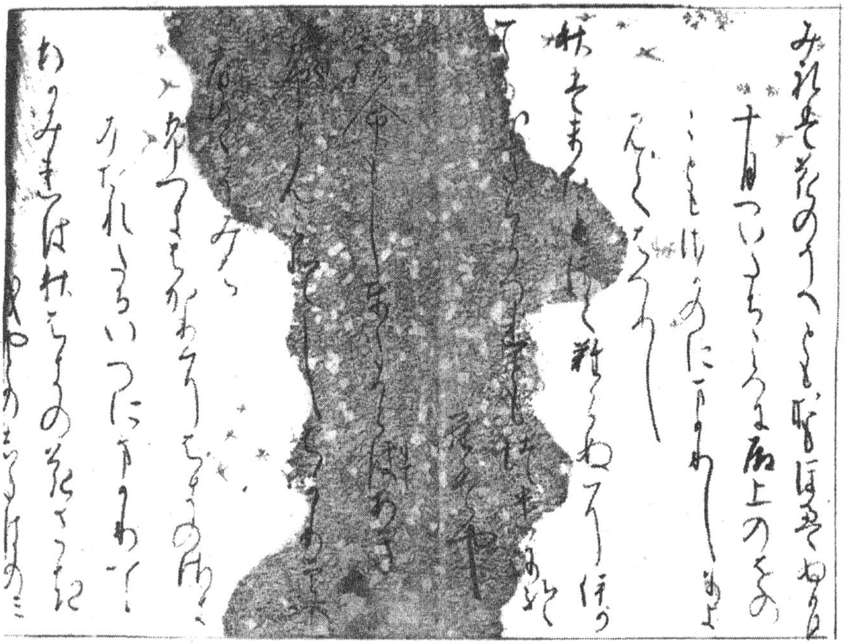

Fig. 13.24 *Collection of Yuanfu* in *36 Ren Jia Ji*, a paper collection of waka published in 1112 (Calendered with colored micas on Yanpi paper. Stored in the original Nishi Hongan-ji Temple)

In 751, the Empire of the Tang Dynasty was at war with the Arab Empire in Talas (present Kazakhstan) of Central Asia. During the fighting, some of the Tang soldiers were captured, including papermakers. Afterwards, the paper mill in Samarkand instructed by Chinese began to produce hemp paper (Fig. 13.29), which was exported to Europe later. As the Arabian forces extended to Africa, the technology of papermaking was introduced there. In 641, the Umayyad Empire conquered Egypt, and then the first paper mill of Africa was established in Cairo in about 900. In about 1100, Egypt established its second papermaking base in Africa in Fez within Morocco. Thus six paper mills had been built successively in Central Asia, West Asia, and North Africa, all of which produced hemp paper according to Chinese technologies, and paper began to be widely used in the Arab world (Fig. 13.30).

13.2.3 The Papermaking of China in Europe, America, and Oceania

The technology of Chinese papermaking was introduced to Europe via Arab, and Spain, France, and Italy appeared to be the earliest transit points for the papermaking to be spread in Europe. After the Umayyad Empire of Arab was overthrown, the

1. Cutting up Broussonetia papyrifera 2. Stewing in clear water 3. Peeling barks 4. Soaking in water 5. Removing external green layer of skin 6. Washing materials after stewing in plant ash water 7. Pounding and mashing materials 8. Manufacturing with pulp 9. Drying paper

Fig. 13.25 The process charts of making Broussonetia papyrifera paper collected in *Zhi Lu Chong Bao Ji* painted by Dan Yu Tao Xi in 1788

Fig. 13.26 The screener and wooden frames for manufacturing with pulp used in Punjab, India (Taken from *Papermaking: The History and Technique of an Ancient Craft* by Dard Hunter, 1947)

Fig. 13.27 The Buddhist sutra that was written in Sanskrit of East India in 1500 (Collected by the British Library)

former prince royal Abd al-Rahman took a group of people to take refuge in North Africa and then to Spain, where he established a regime in 756 and came to rule Spain. This made it the first European country to make paper. The earliest paper mill of Spain was built in Xátiva in 1150.

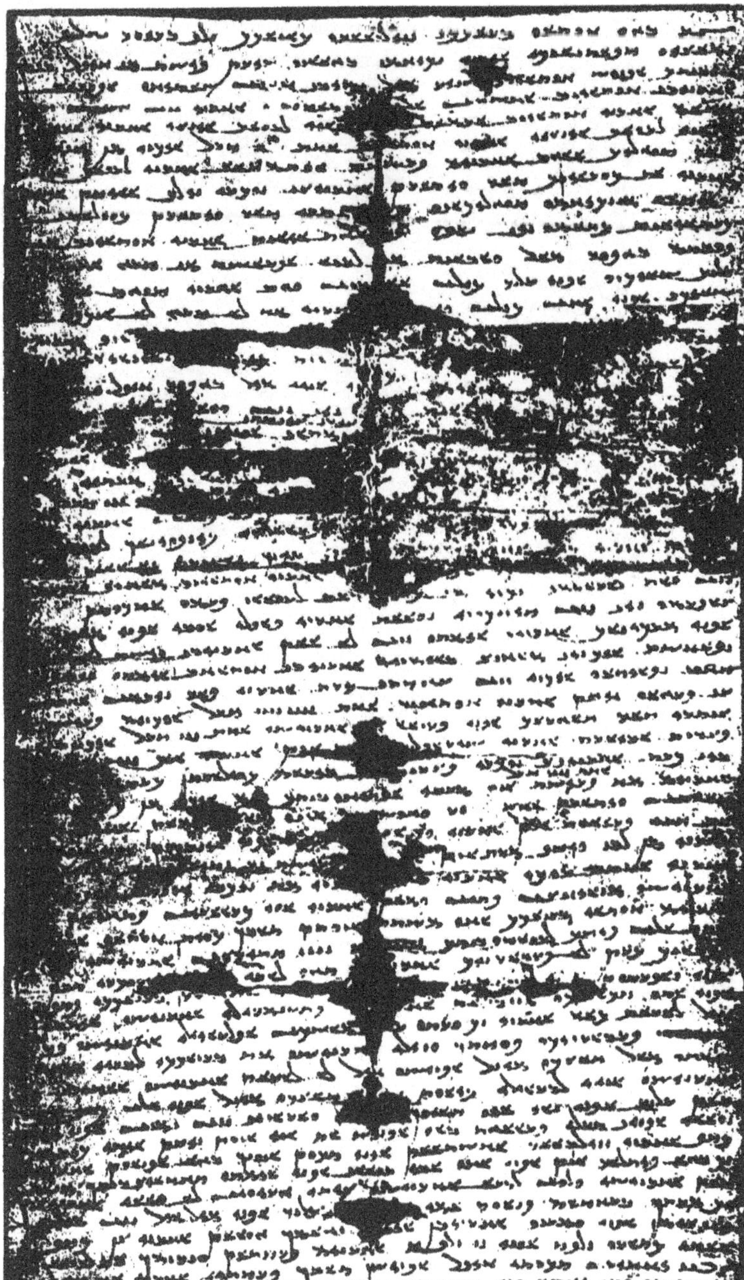

Fig. 13.28 The paper letter written in Sogdian of Central Asia that was discovered at the beacon fire and smoke relics of the Great Wall near Dunhuang in 1907 (Written in 313, collected by the British Museum)

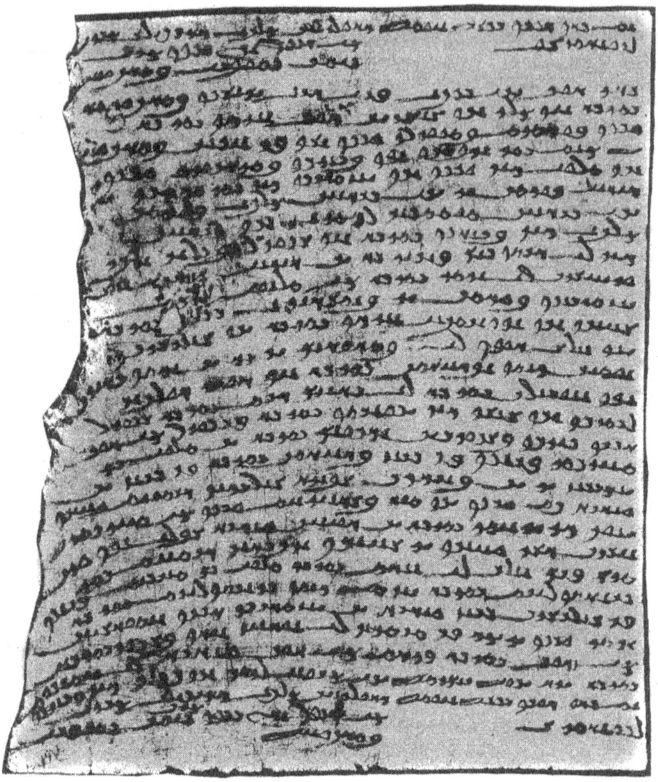

Fig. 13.29 The paper letter written in Sogdian at the beginning of the eighth century stored in the State Archives owned by Divastick, king of the Kangju Kingdom, that was found in Mount Mug near Samarkand in the 1930s (29.5 cm × 25.5 cm, stored in the Institute of Oriental Studies, Petersburg)

Adjacent to Spain, France might introduce the technology of papermaking from Spain. In France, the first paper mill was built in Hérault in the south in 1189, but as its output was very small, a large amount of paper had to be imported from Damascus and Spain (Fig. 13.31). Undoubtedly, hemp paper was mainly produced in Spain and France.

During the eleventh and twelfth centuries, Arab paper was also transported to Italy from Egypt and Morocco in North Africa via the Mediterranean Sea, and then transferred to countries on the European continent, by which the technology of papermaking was introduced into Italy. In 1276, the first paper mill was established in Montefano to produce hemp paper. Later, Italy rapidly developed the paper industry, and had become a supplier of paper to Europe in the fourteenth century, producing more paper than Spain and Damascus (Fig. 13.32).

The consumption of paper in Germany also increased in the late fourteenth century. In 1390, the first paper mill (Figs. 13.33, 13.34) was established in Nuremberg, the

Fig. 13.30 The paper document written in Greek in 1151 that was found in Egypt (Collected by the Freer Gallery of Art of Washington, USA)

printing center, and meanwhile the other paper mill was built in Cologne. Since Germany developed its paper industry, it came to be a new center that spread the technology of papermaking to Europe. In the Netherlands that borders Germany and France, an early paper mill was built in 1428, and then a larger paper mill was built in Dordrecht in 1586. In Switzerland where German is current, the first paper mill was established in the printing center Basel in 1433, while in Austria that is situated to the south of Germany, the first paper mill was built in Vienna in 1498. In Poland that borders Germany, the first paper mill was built in Krakow in 1491. Paper was available in Russia much earlier, but it was not until 1576 that the earliest paper mill was built in Moscow, and German technicians were hired (Fig. 13.35).

Fig. 13.31 The copper-plate etching made in 1791 that depicts the interior of a paper mill in France (Collected by the National Library of Paris. The production processes like manufacturing with pulp, mold unloading to pile up wet paper, pressing and squeezing, etc. can be seen in the picture.)

Fig. 13.32 The interior of the paper mill in Fabriano, Italy (Taken from *Novum Italiae Theatrum*, 1724. The production processes like manufacturing with pulp, pressing and squeezing, pounding and mashing materials, etc. can be seen in the picture.)

Separated from the European continent by a sea, Britain had imported paper from Spain in the fourteenth century, but it was much later in making paper. Its earliest paper mill was built in Hertfordshire in 1494 (Fig. 13.36). Northern European countries also made paper later due to their remote location. Denmark began making

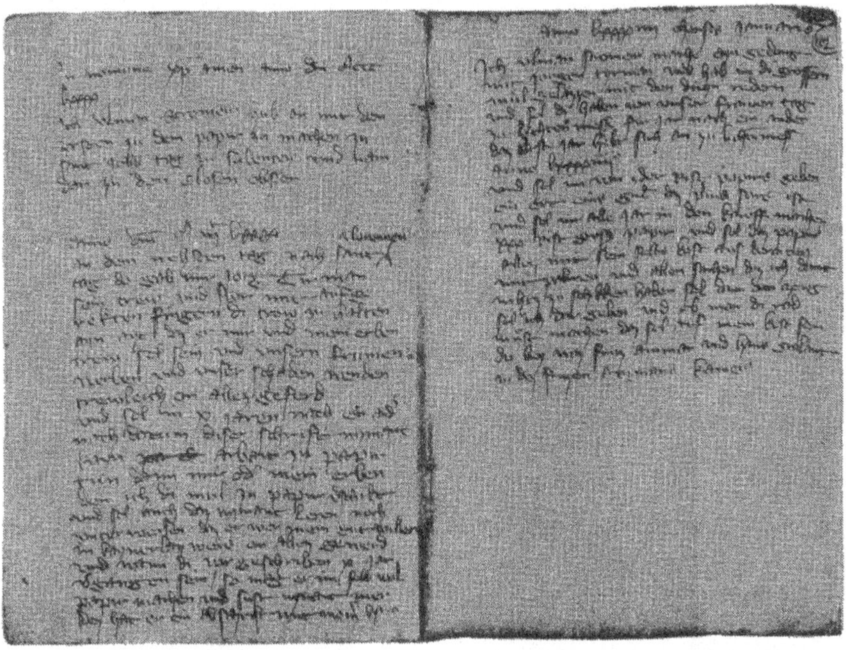

Fig. 13.33 The diary written in German on paper by Strohm, founder of the first paper mill built in Germany, in the fourteenth century (Collected in the National Museum, Nuremberg)

paper in 1635, while Norway in 1690. By the seventeenth century, however, almost all European countries had established paper industry.

It was not until 1575 that the first paper mill was built in the Americas by Spaniards after they had immigrated to Mexico. Previously, ancient materials such as sheepskins were still used in the Americas, and sometimes paper had to be imported from Europe. The earliest paper mill in the United States was founded in Philadelphia in 1690, while it was in Saint Andrews in 1803 that the first paper mill was built in Canada of North America.

As for Oceania, the first paper mill was built in Melbourne, Australia, in 1868.

By the second half of the nineteenth century, the technology of papermaking in China had completed its world-wide journey, spread over the world (Fig. 13.37).

Fig. 13.34 The paper mill built by Strohm on the outskirts of Nuremberg in 1390 (lower right corner) (Taken from Hartmann Schedel: *Liber Chronicarum*, 1493)

Fig. 13.35 The earliest picture of papermaking in Europe (Taken from the German version *Das Stāndebuch*, 1568)

Fig. 13.36 The worker manufacturing with pulp in the paper mill in Maidstone, UK (Taken from the book *Papermaking: The History and Technique of an Ancient Craft* by Dard Hunter, 1947)

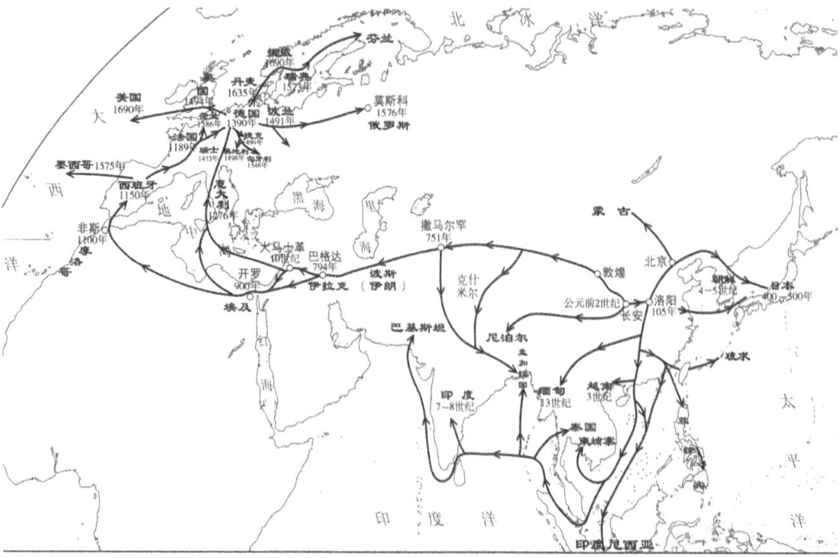

Fig. 13.37 The outward spread of the technology of papermaking of China

Jixing Pan Born in Beining, Liaoning in 1931. He graduated from the School of Chemical Engineering of Dalian University of Technology in 1954 and then worked as a researcher in the Institute for the History of Natural Sciences, Chinese Academy of Sciences.

1981-1982, visiting professor in University of Pennsylvania;

1982, visiting researcher in Robinson College, Cambridge University;

1983, communication academician of the International Academy of the History of Science (Paris);

1987, visiting professor in Institute of Human Sciences, Kyoto University, Japan;

1992, Member of the National Ancient Books Consolidation Group.

Research interests: History of Science and Technology; the History of Science and Technology Exchanges between China and Foreign Countries.

Major works:

The History of Chinese Papermaking technology, Carl Schorlemmer, The Critical Biography of SONG Yingxing, The Annotate and Study on Exploitation of the Works of Nature (Tian Gong Kai Wu), The History of Chinese Metal Movable-Type Printing Technology, Science Exchanges between Chinese and Foreign Countries, The Four Great Inventions of China, The History of Chinese gunpowder, The History of science and technology exchanges between China and foreign Countries.

He has published 210 papers in total.

Awards:

4 first prizes and 4 second prizes for excellent books.

Chapter 14
Traditional Chinese Medicine Diagnosis and Treatment

Yahua Niu

Traditional Chinese medicine enjoys a long history. In the long-range struggle with diseases, our ancestors achieved rich medical experience, and formed a unique medical system by creating effective methods of disease diagnosis as well as various treatments, making great contribution to a healthy and prosperous Chinese nation.

14.1 The Theory of Traditional Chinese Medicine

The theory of traditional Chinese medicine (TCM) began in Spring and Autumn and the Warring States Periods. *Yellow Emperor's Canon of Medicine* (Figs. 14.1, 14.2), including *Su Wen* (*Plain Conversation*) and *Ling Shu* (*Spiritual Pivot*), comprehensively summarized the medical achievements and experience of the Pre-Qin Period, laying the theoretical foundation of TCM. The medical achievements and experience were later interpreted and developed by the diagnosticians of later dynasties, and gradually formed a theoretical system in modern times that mainly covers yin-yang and wuxing (five elements), zang-xiang (visceral manifestation), qi-blood-body fluid, channels and vessels, etiology and morbidity, pathogenesis, principles of diagnosis and prevention and treatment, etc.

Y. Niu (✉)
Institute of Information on Traditional Chinese Medicine, China Academy of Chinese Medical Sciences, Beijing, China
e-mail: niuyahua@hotmail.com

J. Hua and L. Feng (eds.), *Thirty Great Inventions of China*,
https://doi.org/10.1007/978-981-15-6525-0_14

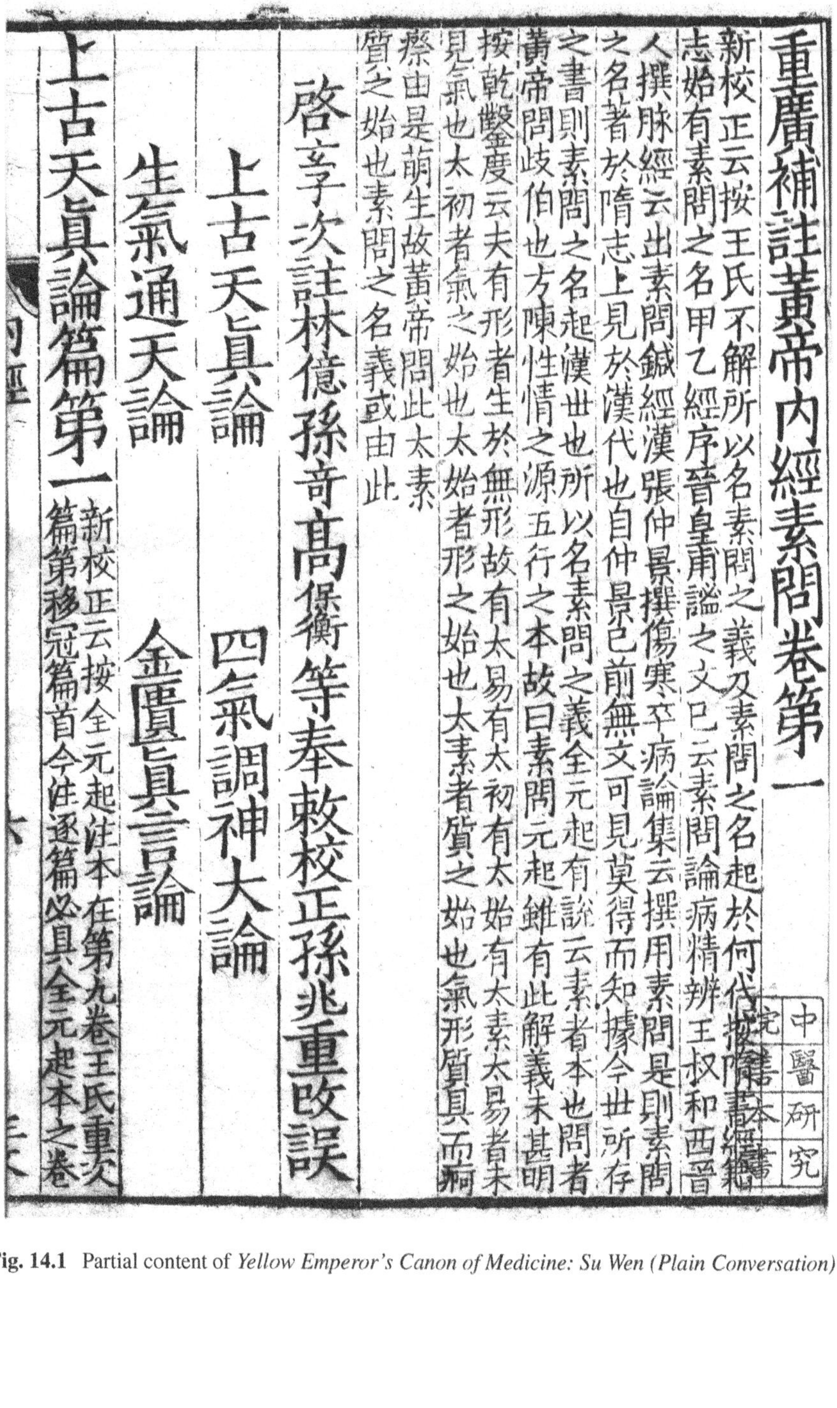
重廣補註黃帝內經素問卷第一
新校正云按王氏不解所以名素問之義及素問之名起於何代按隋書經志始有素問之名甲乙經序晉皇甫謐之文已云素問論病精辨王叔和西晉人撰脉經云出素問鍼經漢張仲景撰傷寒卒病論集云撰用素問是則素問之名著於隋志上見於漢代也自仲景已前無文可見莫得而知據今世所存之書則素問之名起漢世也所以名素問之義全元起有說云素者本也問者黃帝問岐伯也方陳性情之源五行之本故曰素問元起雖有此解義未甚明按乾鑿度云夫有形者生於無形故有太易有太初有太始有太素太易者未見氣也太初者氣之始也太始者形之始也太素者質之始也氣形質具而痾瘵由是萌生故黃帝問此太素質之始也素問之名義或由此
啓玄子次註林億孫奇高保衡等奉敕校正孫兆重改誤
上古天真論　四氣調神大論
生氣通天論　金匱真言論
上古天真論篇第一　新校正云按全元起注本在第九卷王氏重次篇第移冠篇首今注逐篇必具全元起本之卷

Fig. 14.1 Partial content of *Yellow Emperor's Canon of Medicine: Su Wen (Plain Conversation)*

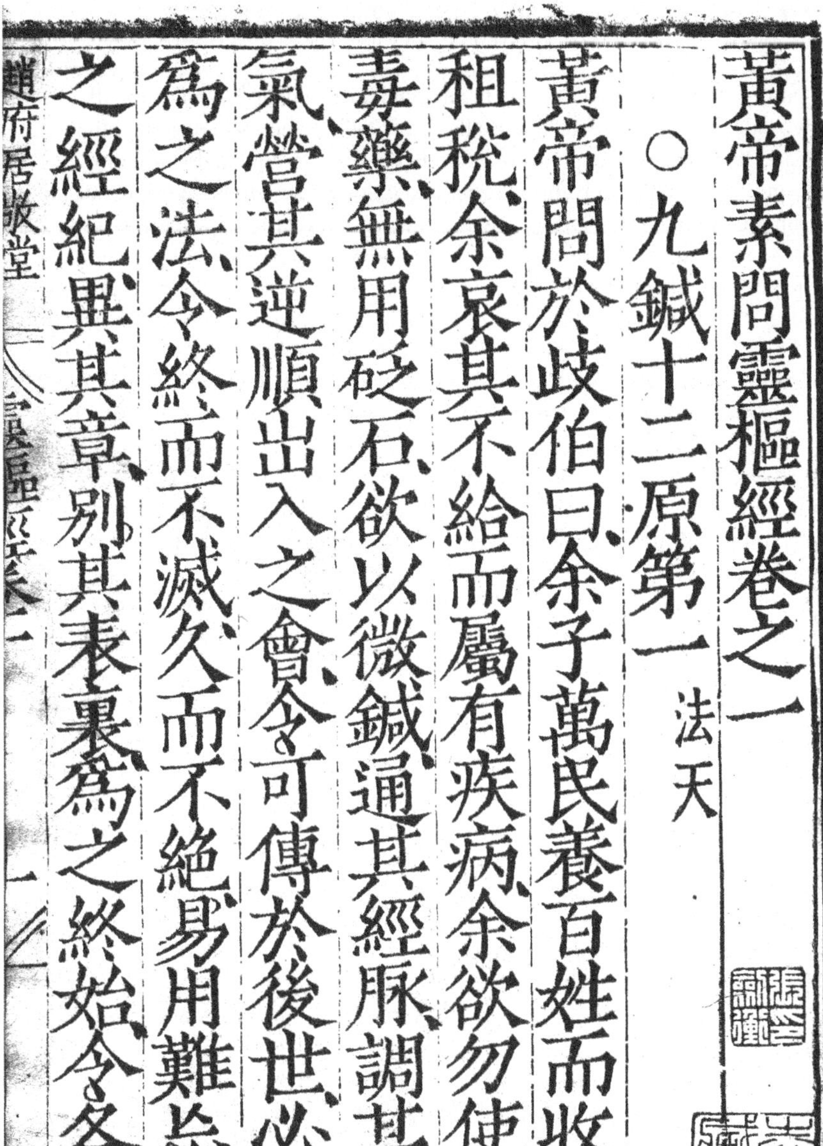

黃帝素問靈樞經卷之一

○九鍼十二原第一 法天

黃帝問於歧伯曰余子萬民養百姓而收其租稅余哀其不給而屬有疾病余欲勿使被毒藥無用砭石欲以微鍼通其經脈調其血氣營其逆順出入之會令可傳於後世必明爲之法令終而不滅久而不絕易用難忘爲之經紀異其章別其表裏爲之終始令各有

趙府居敬堂 靈樞經卷一 一 孝

Fig. 14.2 Partial content of *Yellow Emperor's Canon of Medicine: Ling Shu (Spiritual Pivot)*

14.1.1 The Yin-Yang and the Five Elements Theories

As a pair of philosophical concepts in ancient China, yin and yang originally refer to the two sides of the sun. The side to the sun is called yang whereas the other side is yin. Later the paired concepts were extended to refer to cold and warm climates, the upper and lower, left and right, internal and external positions, the states of motion and tranquility, and so on. Eventually the pair was gradually abstracted as a summary of the two opposite sides of some interrelated things and phenomena in natural world.

The yin-yang theory had become the guiding principle of the theory of TCM in the period of *Yellow Emperor's Canon of Medicine*. According to *Plain Conversation: Yingyang Yingxiang Dalun* (*Major Discussion on the Theory of Yin and Yang and the Corresponding Relationships among all the Things in Nature*), "Yin and Yang serves as the Dao (law) of the heaven and earth, the fundamental principle of all things, the origin of all changes, the root of life and death, and the storehouse of spirit. As a theoretical tool of TCM, the yin-yang theory runs through all aspects of the theoretical system of TCM, and is used to explain the structure of human body, the physiological functions, the occurrence and development of diseases, as well as to guide the diagnosis and prevention and treatment of diseases.

TCM holds that the structure of human body is an organic whole, within which there are the opposites and unity between yin and yang everywhere. According to *Plain Conversation: Baoming Quanxing Lun* (*Discussion on Preserving Health and Protecting Life*), "Human is born with a physical body, which cannot exist without yin and yang." All structures and functions of human body can be understood from the two perspectives of yin and yang. Just as *Plain Conversation: Jingui Zhenyan Lun* (*Discussion on Important Ideas in the Golden Chamber*) writes, "For human, the external is yang and the internal is yin. For human body, the back is yang and the abdomen is yin. For the zang-fu organs, the zang organs are yin and the fu organs are yang. So liver, heart, spleen, lung and the five internal organs are all yin, while gallbladder, stomach, large intestine, small intestine, urinary bladder, san-jiao (triple energizer), and the six fu organs are all yang."

The orderly performed life activities of human are the result of the interaction between yin and yang. Yin and yang are mutually interdependent and waxing and waning. The balance of yin and yang can guard the health, whereas the violation of the rules of yin and yang leads to diseases. Therefore, the diagnosis of diseases should start with the differentiation of yin and yang. It is recorded in *Plain Conversation: Yinyang Yingxiang Dalun* (*Major Discussion on the Theory of Yin and Yang and the Corresponding Relationships among all the Things in Nature*) that "a skilled diagnostician always examines the countenance and feels the pulse to differentiate yin and yang first." The general principle of treatment is to achieve and maintain the dynamic balance between yin and yang. *Plain Conversation: Zhizhenyao Dalun* (*Major Discussion on the Most Important and Abstruse Theory*) clearly points out that "careful observation of yin and yang should be made in order to regulate and achieve a balance between them." As for the significance of yin and yang, *Jingyue Quanshu: Chuanzhong Lu* (*Jingyue's Encyclopedia: the Chronicle of Faithfulness*) makes a good summary: "Every diagnostician must first differentiate yin and yang, which is the law of treatment. If there is no false differentiation of yin and yang, how can there be poor treatment? Although the diagnosis and treatment are complicated in TCM, it is believed that yin and yang are the only concepts that can be said in a word. Therefore, there are yin and yang in symptom, in pulse as well as in medicine… If yin and yang can be clearly understood and distinguished, the medical theory, although mysterious to some extent, is considered to be half way comprehended."

The five elements theory as a concept also belongs to ancient Chinese philosophy and holds that all things in the world are evolved from the transportation and transformation of the five basic substances including wood, fire, earth, metal, and water. The theory adopts the method of image comparison to argue that wood has the characteristics of growing and generating, so it belongs to the east; fire has the hot, upward characteristics, so belongs to the south; earth has the cultivating and breeding characteristics, so belongs to the central; metal, with the characteristics of quietness and ending, belongs to the west; and water, with cold, downward characteristics, belongs to the north. Borrowing the method of analogy as well as reasoning and deduction adopted by the five elements theory, TCM links the five directions, the five seasons, the five qi, the five tastes and the five colors of nature with the five internal organs, the six fu organs, the five body constituents and the five sense organs to explain the organization and structure of human body, physiological function, and pathological changes, so as to guide the diagnosis and prevention and treatment of diseases.

As a reasoning tool of TCM, the five elements theory runs through all aspects of the theoretical system of TCM. When explaining the physiological function of the five internal organs, the theory links the five internal organs to wood, fire, earth, metal, and water, respectively, and argues that the liver governs coursing and discharge, and has the physiological function of smoothing and regulating the qi flow in the entire body; wood can bend or straighten, grow and generate branches and leaves, thus enjoying the characteristics of generation, so the liver belongs to wood. Likewise, heart yang refers to the yang qi of the heart and is active, with the nature of warm-heat, so the heart belongs to fire. The spleen has the special function of promoting the transportation and transformation of food, water, and essence, thus nourishing the five internal organs, the six fu organs as well as the four limbs and the bones, so it is the source of qi, blood, essence, and other body fluids; since earth is thick in nature and has the characteristics of generating and transforming all things, the spleen belongs to earth. The lung inhales fresh air, transports body fluids, qi and blood downward, and depurates. The nature of metal is clear and converging, so the lung belongs to gold. The main function of the kidney is to store essence and govern water, and water shows the characteristics of moistening and storing essence, so the kidney belongs to water. Therefore, it is believed that the five internal organs have the same functions of generation, restriction, inhibition, and transformation as the five elements. For example, wood engenders fire, the liver belongs to wood, and the heart belongs to fire, so the liver is believed to engender the heart. Since fire engenders earth, the heart engenders the spleen. The liver belongs to water, so the heart is restricted by the kidney. Therefore, the generation and restriction among the five internal organs can follow this analogy.

The five elements theory explains pathological changes, including "disorder of mother-organ involving its child-organ" and "illness of child-organ affecting the motherone." The former refers to the transmission of diseases from mother-organ to child-organ. For example, the kidney belongs to water and the liver belongs to wood; as water can grow wood, the kidney belongs to mother-organ, and the liver belongs to child-organ, so the kidney disease can be transmitted to the liver, namely, disorder of mother-organ involving its child-organ. The latter refers to the transmission of diseases from child-organ to mother-organ. For example, the liver belongs to wood, and the heart belongs to fire; as wood can make fire, the liver belongs to mother-organ and the heart belongs to child-organ, so the heart disease can be transmitted to the liver, namely, illness of child-organ affecting the motherone. In diagnosis, TCM argues that human body is an organic whole of internal and external coordination. So, the pathological changes of the internal organs are often reflected externally, which can be detected through the examination of the patient's change in color, sound, appearance, pulse, and other aspects, combined with the attachment of the five colors, the five notes, and the five tastes to the five internal organs. According to *Canon of Difficult Issues: the Sixty-first Difficult Issue*, "He who knows what he sees can know the patient's disease by watching his five colors. He who knows what he hears can differentiate the patient's disease by hearing his five notes. He who knows what he asks can know the cause and location of the patient's disease by asking his desired five tastes. He who knows by feeling the pulse can know what and where the patient's disease is by feeling the cunkou pulse to see its deficiency and excess."

In the treatment of diseases, we should also consider the law of generation and restriction of the five elements among the zang-fu organs; in addition to treating the sick organs, we should also adjust the balance among all the organs. *Canon of Difficult Issues: the Seventy-seventh Difficult Issue* writes: "If a disease is in the liver, you should know that the disease has been transmitted to the spleen. Therefore, the spleen qi must be enriched first of all."

14.1.2 The Visceral Manifestation Theory

Zang-xiang (visceral manifestation) is also called internal organs manifestation. Zang refers to the internal organs hidden in the body, including heart, lung, spleen, liver, and kidney which are collectively called the five Zang organs; gallbladder, stomach, small intestine, large intestine, urinary bladder, and triple energizer which are collectively called the six fu organs; and brain, pulp, bone, pulse, gall, and uterus which are collectively called the qi-heng house. Xiang is generally believed to have two meanings: one refers to the manifested structure and appearance of the zang-fu organs, and the other refers to the visible physiological and pathological phenomena. The visceral manifestation theory is the theory which studies the physiological function and pathological change of each of the zang-fu organs in human body and their interrelations by observing and analyzing the physiological and pathological phenomena of human body.

In TCM, zang refers to physically existed, meaning "storage" of essence; fu refers to a cavity, a space for storing food and water; the qi-heng house is a cavity in form, with the function of storing vital essence like dirty, thus the name qi-heng house: qi means "strange" while heng "constant."

The visceral manifestation theory links zang and fu to ying and yang, respectively: zang is yin and fu is yang, connecting with each other. For example, heart connects with small intestine, lung with large intestine, spleen with stomach, liver with gallbladder, kidney with urinary bladder, and pericardium with triple energizer. The zang-fu organs are connected with each other through meridians, and closely cooperate to form a whole. The five internal organs have special connection with body orifices. For example, heart is reflected in face, charged by blood vessels, with the aperture in tongue; lung is reflected in hair, charged by skin, with the aperture in nose; spleen is reflected in lips and the acupuncture point of sibai, charged by flesh, with the aperture in mouth; liver is reflected in hands, charged by veins, with the aperture in eyes; and kidney is reflected in hair, charged by bones, with the aperture in ears and two lower orifices. Face, blood vessels, and tongue reflect the condition of the heart; hair, skin, and nose reflect the condition of the lung; and so on, each reflected externally. Using this theory, we can examine the zang-fu organs through external organs.

The theory also links the five internal organs to the five seasons: liver, heart, spleen, lung, and kidney are associated with spring, summer, long summer, autumn, and winter, respectively. Human body is thus linked to natural environment. The five functional systems centered on the five internal organs are affected by yin and yang in four seasons in physiological function and pathological changes.

The formation of the visceral manifestation theory has a certain anatomical basis. According to *Ling Shu* (*Spiritual Pivot*)*: Jingshui*, "A man of eight *chi* has flesh and skin, which can be measured externally and followed; if he dies, the body can be dissected and examined. The rockiness of each zang organ, the size of each fu organ, the amount of food, the length of pulse, the state of blood… all can be understood." The description of the structure of the zang-fu organs in human body in *Yellow Emperor's Canon of Medicine* and *Canon of Difficult Issues* coincides with the truth better. Wang Mang, one of the emperors of the Eastern Han Dynasty, once performed human anatomy. In addition, through long-term observation and medical practice, the ancients found that if the skin was affected by cold, a man would suffer from aversion to cold, nasal stuffiness, rhinorrhea, cough, shortness of breath, etc. They accordingly recognized the close relationship between the lung and the hair and the nose, and deduced such arguments as "lung connects with hair," "lung governs breathing," "cough is the voice of lung," and so on. The use of some kidney-tonifying drugs can promote the healing of fractures, as it is believed that the essential qi within kidney has the effect of promoting bone growth, thus the theory of "kidney governs bone." Likewise, the application of liver treatments can promote the recovery of eye disease, as "the aperture of liver is in eyes." The knowledge derived from practice combined with the thought and method of "the inside must be reflected in the outside," and formed a relatively complete system of knowledge about the structure, physiology, and pathology of the zang-fu organs in human body under the guidance of the yin-yang and the five elements theories (Figs. 14.3, 14.4).

It should be noted that although the names of the internal organs in TCM are the same as those in Western medicine, they are not exactly the same in implication. In the visceral manifestation theory of TCM, the zang-fu organs are not merely an anatomical concept, but a concept that covers the physiology and pathophysiology of a certain system of human body.

14.1.3 The Qi-Blood-Body Fluid Theory

In ancient philosophy, qi was an abstract concept as the primary substance of all things. Blood is a red liquid substance that runs in meridians. Jin-ye (body fluid) is a collective name of the normal water within human body; the clear and thin body fluid that runs superficially is called jin, while the thick body fluid that runs within the internal organs and the bones is called ye. According to TCM, qi, blood and body fluid are not only the product of the physiological activities of zang-fu organs, but also the material basis of the functional activities of zang-fu organs. Therefore, the formation and operation of qi, blood and body fluid are closely related to the functional activities of the zang-fu organs.

The ancients believed that all things in nature were formed by the aggregation of qi, and man was also the product of the intersection of heaven and earth as an integral part of all things in the universe. According to *Plain Conversation: Tianyuanji Dalun* (*Major Discussion on the Law of Motions and Changes in Nature*), "The mutual sense of the qi in heaven and the form on earth brings all things into being." *Plain Conversation: Baoming Quanxing Lun* (*Discussion on Preserving Health and Protecting Life*) writes, "Man is born out of the qi of heaven and earth, out of the law of the four seasons," "the combined qi of heaven and earth is said to be human." In other words, qi is the most primary substance of human body.

Qi is also the most primary substance for sustaining life activities. The growth, development, and life activities of human body require the exchange of matter and energy with the surrounding environment. *Plain Conversation: Liujie Zangxiang Lun* (*Discussion on Six-plus-Six System and the Visceral Manifestations*) says, "Heave supplies man with five qi, and earth supplies man with five tastes. The five qi go into the nose and are stored in the heart and lung, so that the five colors can be embellished and voice and sound can be displayed. The five tastes go into the mouse and are stored in the intestines and stomach, so that the five qi can be nourished. When the qi is existed in harmony, supplemented by body fluid, spirit is naturally produced in human body." Man needs to absorb nutrients from the "qi of heaven and earth" to nourish the five internal organs, in order to maintain the physiological activities of the body.

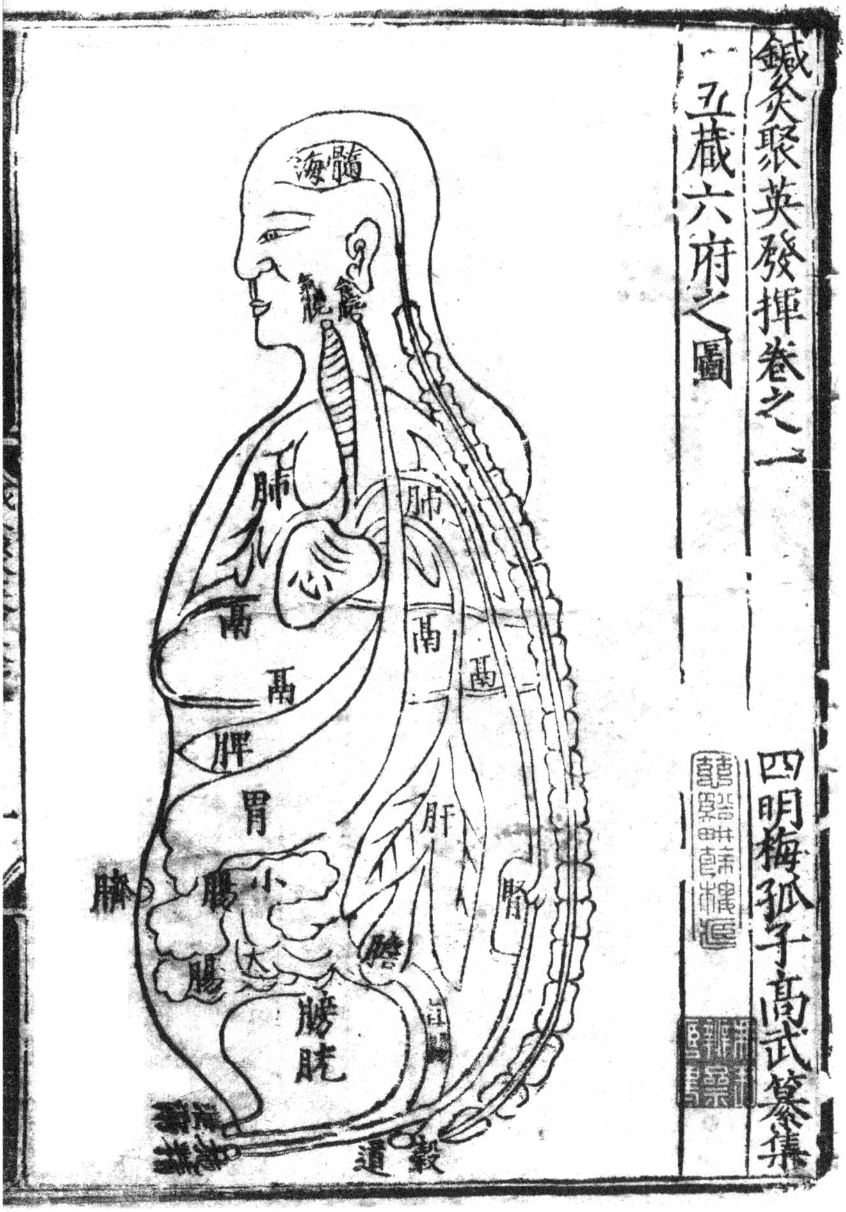

Fig. 14.3 The inside page of *Zhenjiu Juying*

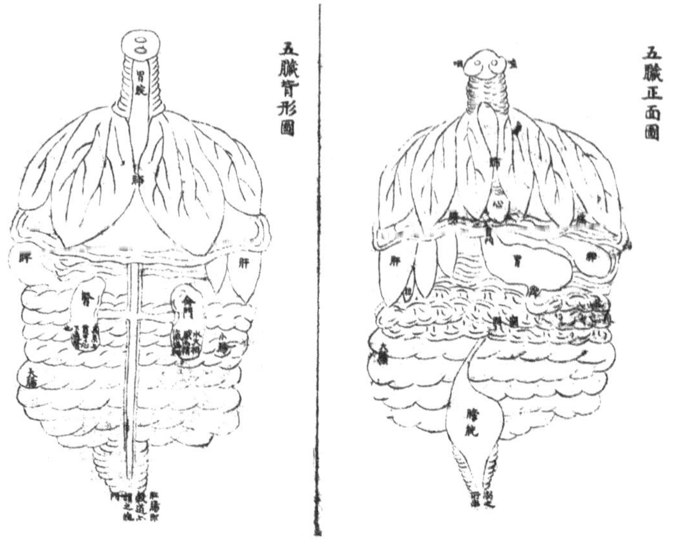

The reverse side of the five internal organs

The front side of the five internal organs

Fig. 14.4 View of the zang-fu organs in *Xunjing Kaoxue Bian*

The source of the qi in human body mainly covers the clear qi in nature, the food essence generated from food and water, and the essential qi fine generated in kidney. TCM holds that kidney stores the innate essential qi from parents, so the essential qi fine generated in kidney is also called innate qi; the food essence generated from food and water and the clear qi inhaled from nature by the lung are called acquired qi. If the innate qi interacts with the food essence in kidney, the produced qi is called original qi; if the clear qi in nature and the food essence are combined in the lung, the formed qi is called pectoral qi. The food essence circulating in blood vessels is called nutrient qi and the food essence circulating outside of meridians is called defense qi.

Original qi roots in kidney and circulates in the whole body through triple energizer. Its main function is to promote the growth and development of human body, warm and stimulate the physiological activities of all zang-fu organs and channels and vessels. If original qi is sufficient, the function of channels and vessels of all zang-fu organs is normal, so the body will be strong and healthy; if original qi is inadequate or wasted too much, all kinds of pathological changes will occur. Pectoral qi mainly governs the operation of qi-blood, breath, vision, listening, speaking, movement, and other functional activities. It can regulate many kinds of physiological activities of human body such as movement and sensation. According to *Random Notes while Reading about Medicine*, "Pectoral qi is motorial qi. The strength of all motorial breathing, speaking, sounding and the movement of limbs is dependent on pectoral qi." The main physiological function of nutrient qi covers two sides: nourishing the whole body and transforming and generating blood. The essence in food and water is the main component of nutrient qi, the essential nutrient for the physiological activities of all zang-fu organs and channels and vessels, and also a part of blood. According to *Ling Shu* (*Spiritual Pivot*), "Nutrient qi produces body fluid, which is then poured into pulse and changed into blood to enrich four limbs, and internally poured into the five internal organs and the six fu organs." Defense qi can protect the surface of skin; defend external evil invasion; warm and nourish the zang-fu organs, muscle, and hair; regulate and control the excretion of sweat; maintain a constant temperature, etc. Nutrient qi and defense qi belong to yin and yang, respectively, and their coordination can maintain a normal physiological function.

Qi functions by means of its circulation. The circulation of qi is called "qi movement," covering upward transmission, downward transmission, exhalation, and inhalation. The zang-fu organs of human body are the places where qi moves. The lung governs breathing: the inhalation of the clear qi in nature means inhalation, the discharge of the turbid qi means exhalation, the dispersion of qi means upward transmission, and the depuration and descent of qi mean downward transmission. The spleen and stomach mainly govern digestion and absorption: the spleen's absorption and transfer are upward transmission, while the stomach's transmission and transformation of food and water are downward transmission. The supersession of kidney in water can make clear qi go upperward and turbid qi go downward to urinary bladder. Each of the qi movements has different strength in each of the zang-fu organs. In the case of the liver and the lung, liver qi governs the ascent of the clear and the coursing and discharge of the turbid, and lung qi governs the dispersion and the depuration and descent of qi. If the upward transmission and the downward transmission of lung qi can assort with each other and keep balance, an orderly operation of qi and a normal operation and transmission of body fluid are maintained. For the spleen and the stomach, spleen qi governs upward transmission and stomach qi governs downward transmission. The coordination of the two will make the clear qi ascend and the turbid qi descend, so that food can be digested, food essence be transmitted, and the dregs be excreted. If the movements of qi assort with each other and keep balance, they are smooth qi movement, which means that the physiological activities of human body are normal. If the movements of qi are abnormal, they are said to be imbalance of qi movement; once qi movement is stopped, it means the end of the life activity.

Blood nourishes and moisturizes the whole body. Just due to the blood's nourishment of the internal organs, human body can carry out all kinds of activities. According to *Plain Conversation: Wuzang Shengcheng Pian* (*Discussion on Various Relationships Concerning the Wuzang (Five Internal Organs)*), "Liver which receives blood can see, feet which receive blood can walk, palms which receive blood can hold, and fingers which receive blood can take." *Jingyue Books: Blood Syndromes* says: "Therefore, blood can make the seven apertures in human head nimble, the four limbs useful, the bones and muscles soft, and the muscles abundant. It can even nourish the zang-fu organs, soothe ethereal soul, moisten the appearance, enrich nutrient qi and defense qi, and make body fluid accessible and the two lower orifices smoothly adjusted, contributing to everything that has form and quality." Blood runs in meridians and circulates endlessly, delivering nutrients to all the zang-fu organs in human body.

According to TCM, blood comes from the essence of food and water. Food is transformed into food essence through spleen and stomach, which is then converted into body fluid by nutrient qi. The body fluid is poured into veins, hence the blood. The essence of kidney can also be converted into blood, and liver has the generating function of blood, too. Blood runs to the zang-fu organs to function.

The movement of blood depends on the pulsatility of heart and the promotion of qi. Liver regulates the movement of qi, thus regulating the operation and the amount of blood. The containment of spleen qi can prevent blood from overflowing the arteries and veins. The blood of the whole body gathers in the lung through meridians and carries on the exchange of the clear and turbid qi. The warming of kidney yang also plays an important role in the normal operation of blood.

Abundant blood can make muscles well developed and sturdy, skin and hair sleek and glossy, mind clear, senses keen, and movement easy and free. Insufficient blood or excessive waste of blood can cause pathological changes of blood deficiency, such as dizziness, dim complexion, dry hair and skin, numbness of limbs, trance of mind, and so on.

Body fluid consists of the internal humors of the zang-fu organs and their normal secretions, such as gastric juice, intestinal fluid as well as nasal discharge, saliva, sweat, urine, tears, joint fluid, etc. Generally speaking, the clear and thin body fluid with better flowability that mainly spreads in the surface of skin, in muscles and in orifices, and can soak into blood veins and moisturize is known as jin; the thick body fluid with less flowability that can be perfused into bone nodes, the zang-fu organs, brain, pulp, and other tissues and nourish is called ye. *Ling Shu: Wulong Jinye Bie* says, "Each body fluid follows its own passage, so the triple energizers discharge qi to warm muscles and moisturize skin, namely jin; the body fluid that closedly flows is ye." *Ling Shu: Jueqi* points out that "the body fluid that vents in the grain of skin and the texture of the subcutaneous flesh and sweats is jin"; "the body fluid that is born out of food and rich qi, and poured into the bending and stretching bones to replenish and restore brain marrow and moisturize skin is called ye."

Body fluid is derived from food and water and is produced through the digestion and absorption of spleen, stomach, small intestine, and large intestine. Under the combined function of lung, spleen, kidney, liver, and triple energizers, body fluid is transferred to and dispersed in human body to moisturize and nourish the whole body, and transform and generate blood. The metabolic body fluid is excreted mainly in the sweat and urine, and a small amount of it is excreted in the breaths and faeces.

There is a close relationship between qi, blood and body fluid. Qi and blood have mutual rooting of yin and yang, so they are interdependent and mutually promoted. According to *Comprehensive Medicine by Doctor Zhang Lu*, "Qi and blood are mutually attached. The qi that exists without blood is scattered and abnormal; the blood that exists without qi is frozen and still." So is the relationship between qi and body fluid. Qi is invisible and governs movement, belonging to yang; blood and body fluid are visible and governs stillness, belonging to yin. Qi can stimulate and promote the production, operation and excretion of blood and body fluid, while blood and body fluid can carry qi, as the invisible qi has to attach itself to the visible blood and body fluid to function. If the qi is abundant, its functions of transforming and generating blood as well as circulating blood and body fluid will be stronger; if the qi is deficient, its above functions will be weaker, even leading to blood or body fluid deficiency as well as blood stasis, phlegm and retained fluid, water-dampness, and other diseases.

14.1.4 The Main and Collateral Channels Theory

The main and collateral channels theory is an important part of the theoretical system of TCM. Together with the visceral manifestation theory and the qi-blood-body fluid theory, it forms the theoretical framework of TCM, which is used to explain physiology and guide diagnosis and treatment.

Main and collateral channels are a collective name of the meridians and the collaterals. The meridians are mainly longitudinal, most of which are distributed among the deep fleshes, and have a close connection with the zang-fu organs. The meridians refer to the branches of channels, which are crisscrossed and distributed shallowly, having no direct connection with the zang-fu organs. According to *Ling Shu: Maidu*, "The meridians are in the inner, and the branches of channels are the collaterals. The branches of branch collaterals are minute collaterals." *Ling Shu: Jingmai* says: "There are twelve meridians, invisibly distributed among the deep fleshes. The visible one is the meridian of Foot-Taiyin, which is on the external malleolus, beyond being seen. All those which are visible are collaterals." Main and collateral channels are regarded as a network of passages, through which qi and blood circulate in the whole body, the zang-fu organs and limbs as well as nods are contacted, and along which the acupuncture points are distributed.

Main and collateral channels are composed of meridians, collaterals, muscles along the regular meridians, and skin divisions.

The meridians, including 12 meridians, 8 extra channels, and branches of 12 meridians, are the backbone of the main and collateral channel systems. Twelve meridians, also known as 12 regular meridians, are the main passages of qi and blood movement, holding certain circulating positions, moving and connecting laws, distribution law, circulating order, owner-member, and exterior-interior relationships among the zang-fu organs. Twelve meridians include three parts: hand or foot, yin or yang, zang or fu. Those which run in the upper limbs are hand channels; those which run in the lower limbs are foot channels; those which run inside of the limbs are yin channels, belonging to zang; and those which run outer side of the limbs are yang channels, belonging to fu. Three meridians are distributed on each side of the limbs, with three yin meridians on the inside, namely, Taiyin, Jueyin, and Shaoyin and three yang meridians on the outside, namely, Yangming, Shaoyang, and Taiyang, respectively. Their names are Taiyin Lung Meridian of Hand, Yangming Large Intestine Meridian of Hand, Yangming Stomach Meridian of Foot, Taiyin

Spleen Meridian of Foot, Shaoyin Heart Meridian of Hand, Taiyang Small Intestine Meridian of Hand, Taiyang Urinary Bladder Meridian of Foot, Shaoyin Kidney Meridian of Foot, Jueyin Pericardium Meridian of Hand, Shaoyang Triple Energizer Meridian of Hand, Shaoyang Gallbladder Meridian of Foot, and Jueyin Liver Meridian of Foot, respectively. As the main passages of qi and blood movement, these 12 meridians are symmetrically distributed around the head, trunk, and limbs.

Eight extra channels are the collective name of governor meridian, Ren channel, Chong channel, Dai channel, yin-qiao channel, yang-qiao channel, yin-wei channel, and yang-wei channel. They contact the 12 meridians, and regulate the qi and blood of 12 meridians by storage and infiltrating irrigation.

Branches of 12 meridians start from the limbs, run in the deep body cavities and the zang-fu organs, and end in the neck. Their function is to strengthen the internal and external connections of 12 meridians, as well as the connections among the zang-fu organs in the deep body cavities.

Branch collaterals, also branches of the meridians, are mostly distributed in the body surface. Each of 12 meridians has branch collateral, which, together with the collaterals of governor meridian and Ren channel and the large collaterals of spleen, amounts to 15. Branch collaterals are the main part of the collaterals, playing a leading role in the countless collaterals of the whole body. Branch collaterals of 12 meridians mainly link the exterior and the interior meridians, and the 3 collaterals in the trunk play the role of infiltrating and irrigating qi and blood. The fine collaterals separated from branch collaterals are called "Sunluo (minute collaterals)," just as "the branches of branch collaterals are minute collaterals" written in *Ling Shu: Maidu*. The collaterals distributed on the surface of skin are called "Fuluo (floating collaterals)."

Muscles along the regular meridians refer to the system of 12 meridians connecting muscles and flesh, a system of 12 meridians "knotting, gathering, dispersing, and winding" muscles and flesh. With the function of contacting the four limbs and all bones and restraining bones, they mainly govern joint movement (Figs. 14.5, 14.6).

Skin divisions refer to the distributed body surfaces of skin according to the meridians. Twelve meridians and their branch collaterals have a certain range of distribution in the body surface, and similarly the skin of the whole body is also divided into 12 sections, called 12 skin divisions. Different skin division has a corresponding relationship with different meridian and different zang-fu organ. The skin division theory is significant in guiding diagnosis and treatment. The observation of the changes of the color and pattern of skin in different locations is helpful to diagnose the pathological changes of some zang-fu organs and the meridians. And the treatments on certain skin divisions by application, warm moxibustion, heat shocking, and others can treat the corresponding internal organs.

The meridians are crisscrossed in human body and naturally link all the zang-fu organs, forming an organic whole of internal and external coordination and unity. Qi and blood circulate within the meridians, nourish the zang-fu organs, and regulate the balance of their functions by transmission and interaction.

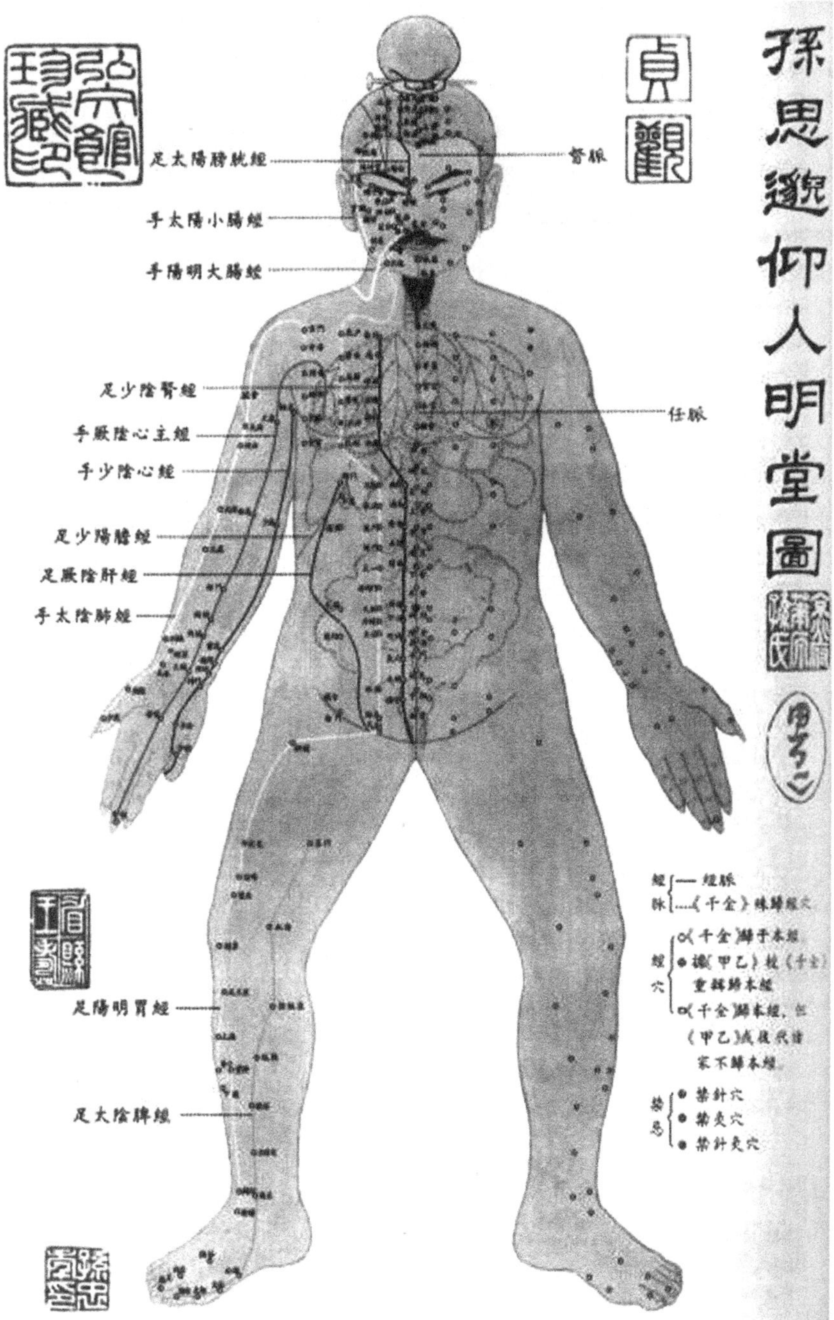

Fig. 14.5 The restored *Ming Tang Tu* by Sun Simiao

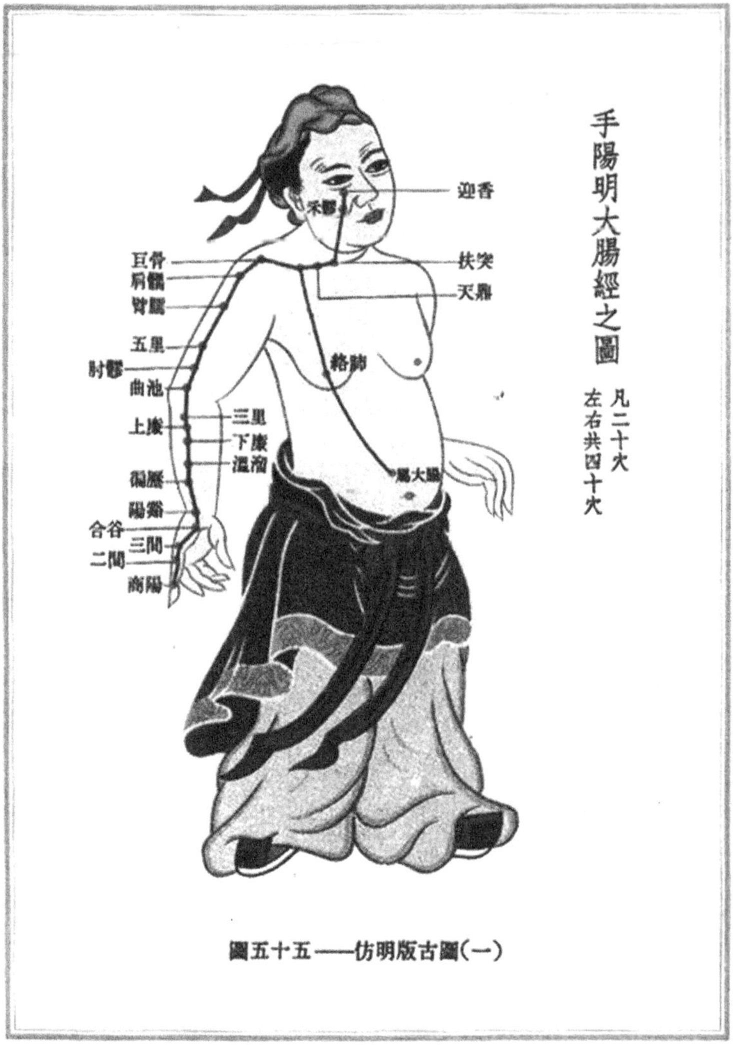

Fig. 14.6 The Diagram of Yangming Large Intestine Meridian of Hand in *New Acupuncture and Moxibustion* by Zhu Lian

14.2 Diagnostic Technology of Traditional Chinese Medicine

The diagnostic technology of TCM includes two parts: diagnostic method and syndrome differentiation. The diagnostic method refers to the techniques of diagnosis based on observation, auscultation and smelling, interrogation, and pulse feeling and palpation, the four ways of diagnosis. The syndrome differentiation, on the basis of the four ways of diagnosis, is to comprehensively analyze and sum up the series of symptoms and signs which have internal connection developed in the course of a disease, in order to explore the cause, location, nature of disease, and so on, so as to determine a scientific treatment plan.

14.2.1 Development of the Diagnostic Technology of TCM

With a long history, the diagnostic technology of TCM is diversified in ways, and has gradually formed the "four ways of diagnosis," namely, observation, auscultation and smelling, interrogation, and pulse feeling and palpation.

As early as in *Zhou Li*, a classical book in ancient China, there had been a record of the diagnosis and treatment by "physician": "Apply the five tastes, the five cereals and the five medicines to make the patient recover from a disease; conclude the severity of a disease by observing the five qi, the five voices and the five colors of the patient; analyze the disease by examining the change of the patient's nine orifices; consult the movement of the nine zang organs." At that time, the physicians could judge whether a patient could be saved or not by smelling, listening to and observing him or her. They observed the changes of the so-called "nine orifices" as eyes, ears, nose, mouth, and two lower orifices to examine the situation of the patient's zang-fu organs. Here the ways of auscultation and smelling, observation, pulse feeling and palpation, and so on were actually involved.

Shi Ji (*Historical Records*) records: "Bian Que (a famous doctor in ancient China) could prescribe without pulse feeling and palpation, observation, auscultation and smelling, and external observation." Here the "external observation" refers to the observation of the overall state of the patient. This record reveals a fact that Bian Que as a physician was expert in medical skills, and could cure diseases without applying these routine ways of diagnosis. In fact, when Bian Que passed through the State of Guo, he heard that the prince died violently. After he had asked the cause and time of his death, he judged that the prince could be dead syncope, speaking of "hearing his tinnitus and observing his dyspnea and flaring of nares; when examining his two legs even the private parts, I actually felt the warmth of his body." Here the ways of interrogation, auscultation and smelling, observation, and pulse feeling and palpation were actually involved, with the pulse feeling and palpation only in the legs. Moreover, Bian Que also judged by a comprehensive observation of the prince's face and manner and concluded that there had been a gradual change of the prince's disease from the grain of skin and the texture of the subcutaneous flesh, blood vessels, intestines, and stomach to bone marrow. There are also more records of the pulse feeling and palpation in the medical books written by Chunyu Yi, a famous doctor of the Western Han Dynasty of China. All of these indicate that the four ways of diagnosis had been widely applied in the Pre-Qin Period.

Yin-Yang Mai Si Hou and *Mai Fa* written on bamboo slips unearthed in the tomb of the Han Dynasty in Zhangjiashan and *Yin-Yang Mai Si Hou* and *Mai Fa* written on silk unearthed in the tomb of the Han Dynasty in Mawangdui are the earliest books about pulse feeling and palpation that exist nowadays. These books mostly talk about pulse feeling and palpation. According to the research of experts, these bamboo and silk books concern two kinds of pulse-taking locations: one is in the arterial position which is five cun above the malleolus medialis and in the arterial position which is immediately above the medial malleolus and the other is in the three arterial positions of foot-Shaoyin in the lower limbs, and hand-Taiyin and hand-Shaoyin in the upper limbs. The pulse manifestations concerned only cover six kinds of three pairs as the waxing and the waning, the slippery and the astringent, and the static and the dynamic. In addition, the pulse manifestations such as three-yin pulse death sign and others which concern prognosis judgement are also involved. Compared with those recorded in *Yellow Emperor's Canon of Medicine*, these ways of pulse feeling are rather simple, belonging to the ancient ones created before *Yellow Emperor's Canon of Medicine.*

Yellow Emperor's Canon of Medicine records very rich diagnostic ways. It also mentions such diagnostic ways as "watching the spirit," "observing the countenance," "hearing the voice," "asking for illness," "feeling the pulse," "examining the skin," and so on. Abdominal examination is also concerned. According to *Plain Conversation: Yinyang Yingxiang Dalun* (*Major Discussion on the Theory of Yin and Yang and the Corresponding Relationships among all the Things in Nature*), "A good physician is always to observe the patient's countenance and feel the pulse in order to first differentiate yin or yang that the possible disease belongs to and carefully observe the clear and the turbid qi, so as to partially know the disease; he watches the breath of patient and listens to his or her voice to know what the patient is suffering; he judges and weighs the medical laws to know what the disease is governed by; he knows the cause of the disease by taking the pulse on Chi Cun and examining the current state of the patient. Without correct diagnosis, without correct treatment." It is clearly pointed out that only the series of examination process including such diagnostic ways as "observing the countenance," "feeling the pulse," "watching the breath," "hearing the voice," and "taking the pulse on Chi Cun" can find the exact cause of illness so as to achieve a correct treatment.

Yellow Emperor's Canon of Medicine has established a theoretical basis for the "four ways of diagnosis." *Ling Shu: Wuyue Wushi* explains the correspondence of the five internal organs to the five sense organs: "The five sense organs are the manifestation of the five zang organs… The nose is the manifestation of the lung; the eyes are the manifestation of the liver; the lips of the mouth are the manifestation of the spleen; the tongue is the manifestation of the heart; the ears are the manifestation of the kidney." This establishes a theoretical basis for the disease diagnosis of the zang-fu organs by observing the five sense organs. According to *Ling Shu: Wuyue Wushi*, "… The one who suffers from lung disease suffers from dyspnea and flaring of nares; the one who suffers from liver disease suffers from black angles of eyes; the one who suffers from spleen disease suffers from yellow lips; the one who suffers from heart disease suffers from tortile and short tongue and red cheekbones; the one who suffers from kidney disease suffers from black cheekbones." This passage generally describes the correspondence of the colorful pulse manifestations to the five seasons and the five internal organs, and the exact manifestations in appearance when there are pathological changes in the five internal organs.

As for observation, *Yellow Emperor's Canon of Medicine* established the relationship between facial color and the five internal organs and diseases. The five colors match the five internal organs: "Green links to liver, red to heart, white to lung, yellow to spleen, and black to kidney." The five colors reflect different diseases: "Blue-black links to pain, yellow–red to hot, white to cold." The changes of facial color correspond to heat diseases: "The one who suffers from hepatopyretic disease will first turn red on the left cheek; the one who suffers from cardiopyretic disease will first turn red on the face; the one who suffers from splenopyretic disease will first turn red on the nose; the one who suffers from pyretic pulmonary disease will first turn red on the right cheek; the one who suffers from pyretic nephropathy will first turn red on the cheeks." The colors of different facial parts reflect different diseases: "If the eyebrows turn lightly moist, the disease lies in cortex; if the lips turn green-yellow, red-white or black, the disease lies in muscle; if the nutrient qi turns moistened, the disease lies in qi and blood; if the eyes turn green-yellow, red-white or black, the disease lies in sinew; if the ears turn scorched with dust and dirt, the disease lies in bone."

As for auscultation and smelling, *Plain Conversation: Yinyang Yingxiang Dalun* (*Major Discussion on the Theory of Yin and Yang and the Corresponding Relationships among all the Things in Nature*) proposed the theory of the five notes and the five voices corresponding to the five internal organs: as for liver, "its note is jiao, and its voice is shout"; as for heart, "its note is zhi, and its voice is laugh"; as for spleen, "its note is gong, and its voice is song"; as for lung, "its note is shang, and its voice is cry"; as for kidney, "its note is yu, and its voice is groan"; and "listening to the patient can infer what he or she is suffering from". According to *Plain Conversation: Xuanming Wuqi* (*Discussion on the Elucidation of Five-qi*), "As for the correspondence of the five qi to diseases, the disease in heart leads to belching, the disease in lung to cough, the disease in liver to abnormal language state, the disease in spleen to deglutition, the disease in kidney to sternutation, and the disease in stomach to adverseness, vomite and fear." It is pointed out that when qi of the five internal organs gets sick, there will be different changes in voice. As for specific diagnosis of diseases, *Yellow Emperor's Canon of Medicine* requires listening to the patients, which involves not only such unusual notes and voices as breath, cough, sternutation, groan, eructation, adverseness, and so on, but also such abnormal language states as light words, wild language, severe exhaustion of qi, anaudia, voice abnormalities, etc., as well as such notes and voices as laughter, cry, and sigh produced when the emotion changes.

The topic of smelling has also been discussed. For example, *Plain Conversation: Jingui Zhenyan Lun* (*Discussion on Important Ideas in the Golden Chamber*) links liver, heart, spleen, lung, and kidney to shame, anxiety, fragrance, fishy smell, and rancidity. "All kinds of sour vomits" recorded in *Plain Conversation: Zhizhenyao Dalun* (*Major Discussion on the Most Important and Abstruse Theory*) refer to the vomits with sour odor. "The one who gets sick will send out odor first" recorded in *Plain Conversation: Fuzhong Lun* (*Discussion on the Abdominal Disorders*) indicates that diseases would also be diagnosed by smelling at that time.

Yellow Emperor's Canon of Medicine also values the diagnostic way of interrogation, which covers widely from the patient's history of disease and treatment to the history of life such as economic status, social status, emotion, habits of living and eating, and so on. *Plain Conversation: Sanbu Jiuhou Lun* (*Discussion on the Three Regions and Nine Divisions*) writes: "It is necessary to question about the patient's original diseases as well as current disease, then feel his or her pulses, and try to judge according to the condition of his or her main and collateral channels as well as adverseness." It is proposed that the examination of a patient should start from questioning about the cause of the attack, and then conduct pulse feeling and palpation as well as observation. According to *Plain Conversation: Zhengsishi Lun* (*Discussion on the Four Therapeutic Errors*), "If a doctor does not first ask the cause of the attack, whether the patient's diet loses its balance or not, whether the patient's sleep is excessive or not, and whether the patient is hurt by any poison, and does not feel the patient's pulse on the wrist, how can he give a correct treatment?" *Plain Conversation: Shuwuguo Lun* (*Discussion on Five Errors Frequently Made in Diagnosis*) requires doctors that "for those who are not diagnosed, we have to ask whether he or she has experienced ups and downs of life. The one who lives nobly and then humbly may suffer from ying exhaustion as the disease is born from the inside; the one who experiences the rich and then the poor will suffer from waste of essence… For those who are going to be diagnosed, we have to ask about his or her diet and living. The one who eats and drinks too much, or enjoys too much or suffers too much, or enjoys and then suffers will suffer from waste of essential qi. The exhaustion of essential qi will make one dispirited and sick. Rage will consume yin, and ecstasy consume yang, both which make chaotic qi go up and create confusion to pulses."

Yellow Emperor's Canon of Medicine records the most abundant content of pulse feeling and palpation, covering many chapters that introduce the theories of sphygmology and the method of pulse feeling, including the ways, time, and parts of pulse feeling as well as the physiological and pathological changes of pulses, etc. The method of pulse feeling includes such ways as "the feeling of twelve meridians," "the feeling of Sanbu Jiuhou (the three regions and nine divisions)," "the feeling of renying pulse and cunkou pulse," "the feeling of cunkou pulse," "the feeling of measurement," and so on. The "feeling of twelve meridians" refers to the way of examining the patient's all diseases by feeling one of the arteries of each meridian that is representative or convenient for diagnosis. *The Internal Canon of Medicine* argues that if the meridian(s) of twelve meridians that should beat feebly turn(s) to beat fast and violently, it is a sign of illness. For example, "in twelve meridians, hand-Taiyin, foot-Shaoyin and Yangming beat for ever" (*Ling Shu: Dongshu*); "Yangming often beats, and Juyang and Shaoyang almost do not beat. If they turn to beat violently, it is a sign of serious disease" (*Plain Conversation: Bingneng Lun* (*Discussion on the Manifestations of Diseases*)). The "feeling of the three regions and nine divisions" involves the regions of the upper (head), the middle (hand), and the lower (foot), each of which has the three divisions of heaven, earth, and man, hence a total of nine divisions. The upper region: division of heaven, feeling the artery of the two foreheads; division of man, feeling the artery of preauricula; and division of earth, feeling the artery of the two cheeks. The middle region: division of heaven, feeling the hand-Taiyin meridian to serve lung; division of man, feeling the hand-Shaoyin meridian to serve heart; and division of earth, feeling the hand-Yangming meridian to serve the qi in stomach. The lower region: division of heaven, feeling the foot-Jueyin meridian to serve liver; division of man, feeling the foot-Taiyin meridian to serve spleen and stomach; and division of earth, feeling the foot-Shaoyin meridian to serve kidney. This is a whole-body feeling of pulse, also known as "all-around diagnosis." The "feeling of renying pulse and cunkou pulse" refers to the way of pulse feeling that renying pulse and cunkou pulse are cross-referenced and analyzed. Renying pulse (cephalic artery) mainly reflects the situation of body surface, while cunkou pulse mainly reflects the situation of the internal organs. This diagnosis comes from *Ling Shu: Zhongshi*: "Feeling renying pulse and cunkou pulse to know whether yin and yang are enough or deficient, calm or not calm." This way belongs to local-body feeling of pulse, simpler and easier to operate than whole-body feeling of pulse. *Yellow Emperor's Canon of Medicine* records more than 50 pulse names such as floating, sinking, large, small, slippery, uneven, thready, rapid, retarded, slow and intermittent, hook, abundant, irritable, pant, number, string, moistening, soft, weak, light, virtual, long, solid, strong, micro, weak, urgent, scattered, hair, firm, nutrient, stone, beat, static, tight, knotted, movement, short, slow, absolute, cross, thin, plateau, few, normal, hide, drum, leather, abrupt, strength, bounding, full, sparse pulse, etc.

Yellow Emperor's Canon of Medicine also applies the diagnosis of body palpation such as chi-skin examination, abdominal examination, and so on. "Chi-skin examination" is to determine diseases by examining the moistness, roughness, and heat of the skin from below the elbow joint (the chize acupoint) to cunkou, combined with systemic symptoms, pulse signs, and others. "Abdominal examination" is a method of examining diseases by pressing the patient's abdomen. *Yellow Emperor's Canon of Medicine* records many experiences of "abdominal examination." For example, *Plain Conversation: Fuzhong Lun* (*Discussion on the Abdominal Disorders*) says, "The disease of rigidity and fullness of lower abdomen roots in the left and right sides of the upper and lower sides… also called Fu Liang (mass located in the upper or lower abdomen)." According to *Ling Shu: Shuizhang Pian*, "Water is the cause of abdominal diseases… When pressing the patient's abdomen with hand, if the abdomen rises with the hand as wrapped water, it is the symptom," "when pressing the abdomen of the one with abdominal distension, if the abdomen is sunken and does not rise and its color is unchanged, it is the symptom," "as for mass in the abdomen… when it begins to grow, it is as big as a hen's egg, and will grow bigger until the patient looks as if he or she is pregnant. The mass will exist even for several years. If pressing it, it feels hard; if pushing it, it will move." *Ling Shu: Xieqi Zangfu Bingxing* says, "If someone suffers from urinary bladder disease, his or her lower abdomen will be swollen and painful; if to press it with hand, the patient will falsely want to urinate." At that time, the doctors had distinguished different tumors by examining the hardness of the abdomen. For example, "an intestinal tumor will grow for a few years, during which if to press it with hand, it feels soft. For a full-blown tumor, it feels hard." In *Ling Shu: Baibing Shisheng*, there is a record of "if the tumor is notable in bilateral shallow tendon of spine behind the intestines, it can be touched in hunger; otherwise, it can't be touched. No pressing." It is suggested that abdominal examination should be made when in hunger in order to avoid a false diagnosis of disease.

The diagnostic contents of *Yellow Emperor's Canon of Medicine* comprehensively summarized the achievements in diagnostics made from the Pre-Qin Period to the Han Dynasty, laying a solid foundation for the development of later diagnostics.

Canon of Difficult Issues (Fig. 14.7) is one more classic medical work that was produced later than *Yellow Emperor's Canon of Medicine. The Sixty-first Difficult Issue* covered in the book summarizes for the first time the function of the four ways of diagnosis: "The one who knows by observing is a god, the one who knows by smelling is a sageg, the one who knows by questioning is an expert, and the one who knows by feeling pulse is an artificer." The book discusses more about pulse diagnosis, including the basic knowledge and basic theories of pulse diagnosis as

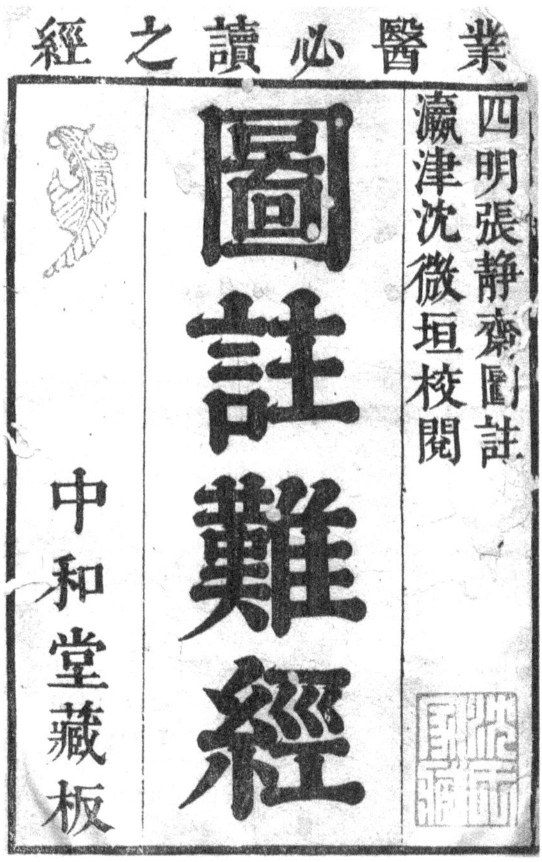

Fig. 14.7 Partial content of *Canon of Difficult Issues*

well as normal and abnormal pulse conditions, etc. In particular, it put forward the pulse diagnosis of "feeling only cunkou," which had become the development trend of pulse diagnosis for more than 1,000 years.

Treatise on Cold Pathogenic and Miscellaneous Diseases (Fig. 14.8) by Zhang Zhongjing applies the previous achievements of diagnosis and treatment to clinical differentiation of symptoms and signs, creating a diagnosis model of combining diseases and syndromes. It achieved further development in the four ways of diagnosis. It is especially worth mentioning that the book applies more of abdominal examination and divides abdominal examination into the parts as chest, flank, heart, and abdomen, which includes such syndromes as lump in the abdomen, fullness, hardness, tightness, softness, creeping, sounding, and so on. For example, "when a patient's abdomen distends, if he or she feels no pain when pressed, it is deficiency; otherwise, it is excess," "hardness of the upper abdomen when pressed," "hardness of the upper abdomen as large as a plate," and "lingering of the upper abdomen when pressed," all these symptoms are only perceived when the abdomen is touched and pressed.

During the Jin and Tang Dynasties, diagnostic technology continued to develop. By learning widely, *The Pulse Classic* (Fig. 14.9) written by Wang Shuhe reduces pulse manifestation into 24 kinds and expounds the pulse theory in detail, so that it can be applied more widely in clinical practice.

Treatise on the Causes and Manifestations of Various Diseases (Fig. 14.10) edited by the physicians of the Sui Dynasty as Chao Yuanfang, et al. has a rich diagnostic content. As for observation and pulse-taking and palpation, it applies language, note and voice, cough, the upper qi, adverseness, and so on in the diagnosis of many internal diseases such as deficiency, stroke, qi disease, cough, and so on, so as to judge the exterior and interior chills and fever, prognostic good or ill luck, and pathological changes of diseases. The book records much more about "abdominal examination," concluding eight ways of pressing of abdomen: restrained pressing, rising pressing, hidden feeling, pushing moving, feeling pressing, moving shaking, turning leaning, and holding. It also describes in detail the location, size, shape, pain and hardness, edge, sound, abdominal muscle tension, etc. of abdominal masses caused by abdominal pain, abdominal dropsy, intestinal carbuncle, etc. [1]. For example, in "sheshou disease," it says that "the disease is in the abdomen, and the mass feels like a snake when hiddenly felt." In "yupi disease," it says that "the mass feels like a fish when carried." In the diagnosis of "hypochondrium fluid retention," it says "when pressing below the costal regions, there is produced water sound"; in the diagnosis of "chronic hypochondium mass," it says "when pressing below the costal regions, there is produced much water sound"; in the diagnosis of "washmarking," it says "when restrainedly pressing heart and abdomen, there is produced water sound." It should be said that in the Sui and Tang Dynasties, the diagnostic technology of China had reached a higher level. In the Tang Dynasty, Sun Simiao also invented a "penetrated pleura examination" for the diagnosis of thoracic-dorsal abscess. That is, thin paper or bamboo intima is applied to levelly paste on a patient's open thoracic-dorsal sore, and then the patient's breath is observed in bright place. If the paper still keeps motionless, it means that the pleura have not been penetrated yet. If the paper is

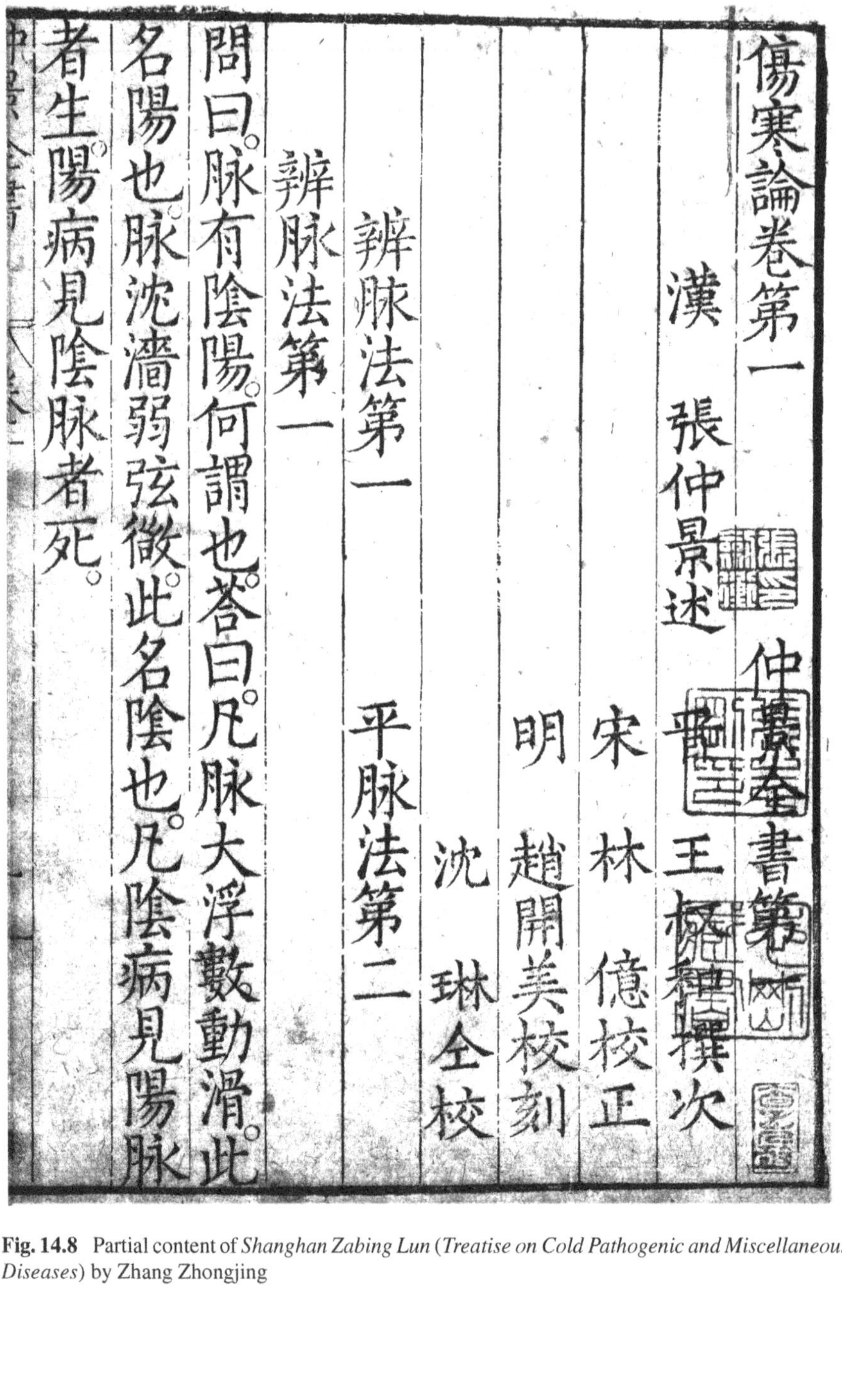
傷寒論卷第一　仲景全書第一

漢　張仲景述　晉　王叔和撰次

宋　林億校正

明　趙開美校刻

沈　琳仝校

辨脉法第一　平脉法第二

辨脉法第一

問曰。脉有陰陽。何謂也。荅曰。凡脉大浮數動滑。此名陽也。脉沈濇弱弦微。此名陰也。凡陰病見陽脉者生。陽病見陰脉者死。

Fig. 14.8 Partial content of *Shanghan Zabing Lun* (*Treatise on Cold Pathogenic and Miscellaneous Diseases*) by Zhang Zhongjing

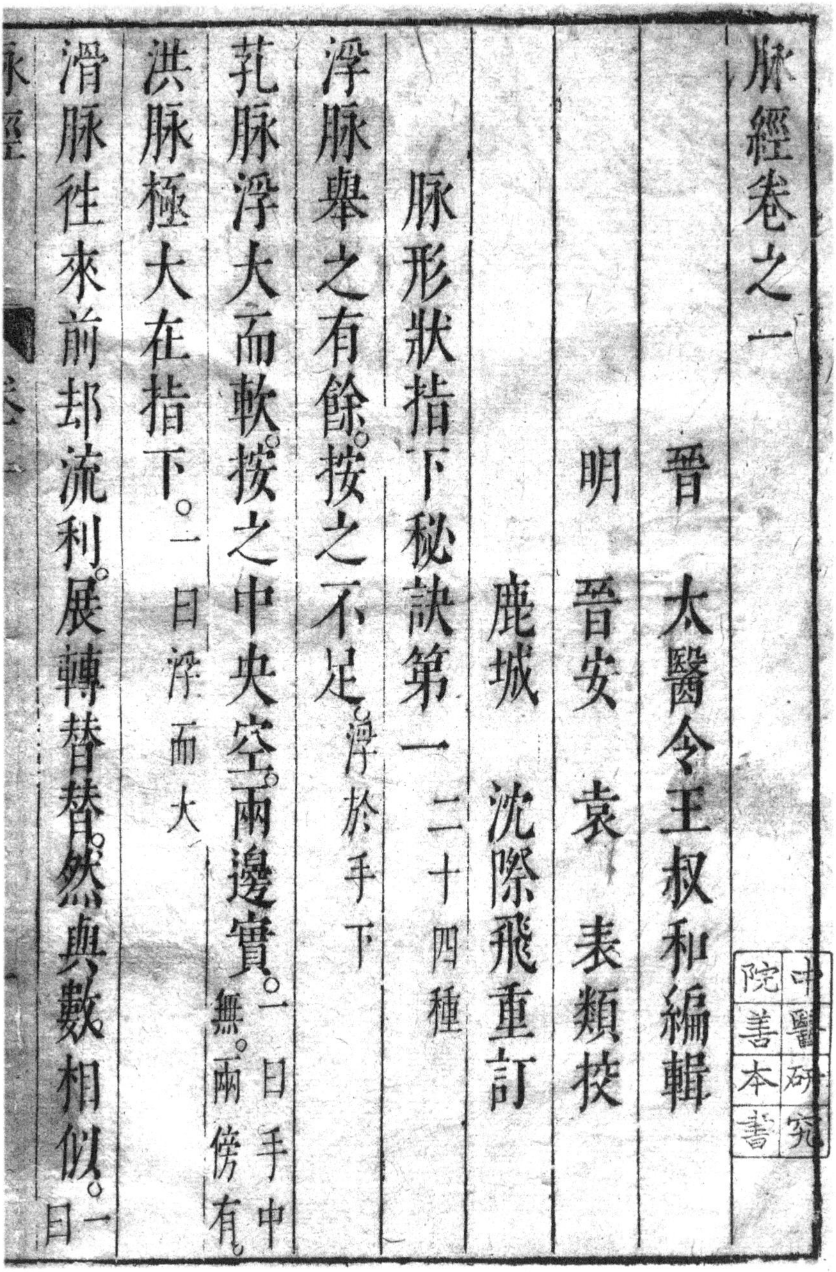

脉經卷之一

晉 太醫令王叔和編輯

明 晉安 袁 表類校

鹿城 沈際飛重訂

脉形狀指下秘訣第一 二十四種

浮脉舉之有餘。按之不足。浮於手下

芤脉浮大而軟。按之中央空。兩邊實。一曰手中無。兩傍有。

洪脉極大在指下。一曰浮而大

滑脉往來前却流利。展轉替替然。與數相似。一曰

中醫研究院善本書

Fig. 14.9 Partial content of *The Pulse Classic* by Wang Shuhe

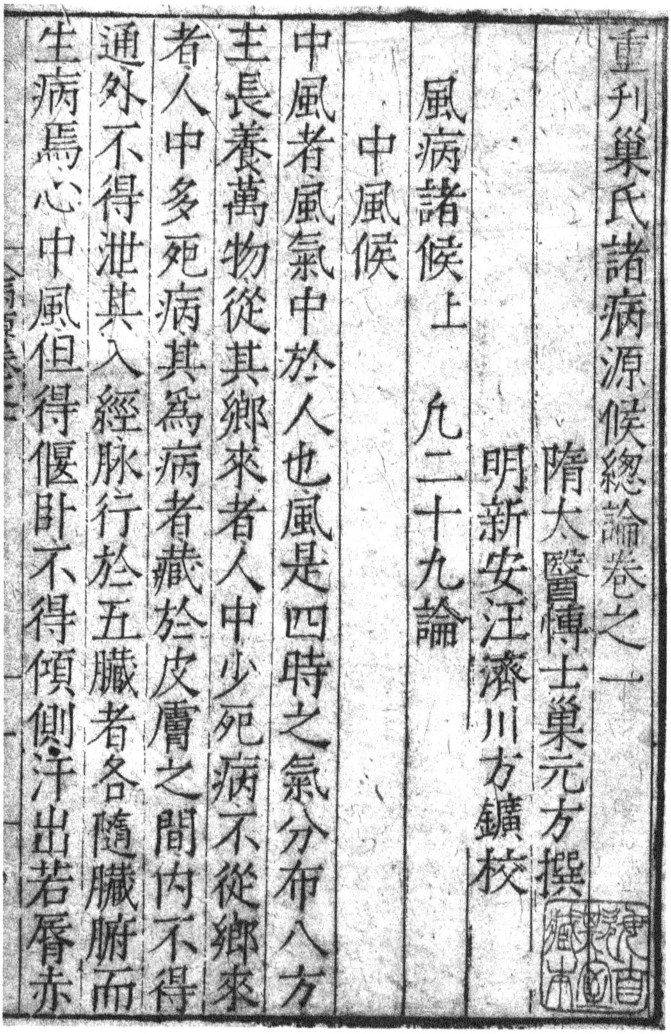
重刊巢氏諸病源候總論卷之一
隋太醫博士巢元方撰
明新安汪濟川方鑛校
風病諸候上　凡二十九論
中風候
中風者風氣中於人也風是四時之氣分布八方
主長養萬物從其鄉來者人中少死病不從鄉來
者人中多死病其爲病者藏於皮膚之間内不得
通外不得泄其入經脉行於五臟者各隨臟腑而
生病焉心中風但得偃卧不得傾側汗出若脣赤

Fig. 14.10 Partial content of *Treatise on the Causes and Manifestations of Various Diseases* by Chao Yuanfang

sunken or evaginated with the deep breath of the patient, it means a confirmation of an empyema. This examination is undoubtedly reliable [2].

During the Song and Yuan Dynasties, a hundred schools of thought emerged in the field of Chinese medicine, and many monographs on diagnostic methods were produced one by one, such as *A Guide to Diagnosis of Diseases* by Shi Fa, *Zhenjia Shuyao* (Fig. 14.11) by Hua Shou, and *Pulse Formulas* by Cui Jiayan. *Aoshi Shanghai Jinjing Lu* is the first existing monograph of tongue examination with both texts and pictures in China. The book summarizes the previous achievements made in tongue

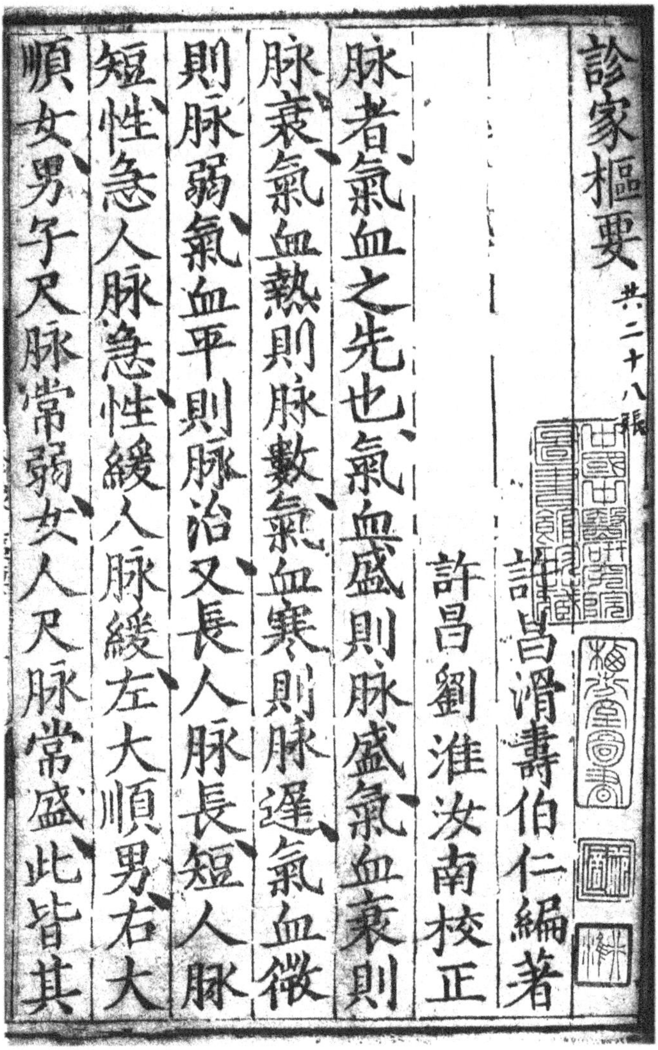

診家樞要 共二十八張
許昌滑壽伯仁編著
許昌劉淮汝南校正
脉者氣血之先也氣血盛則脉盛氣血衰則
脉衰氣血熱則脉數氣血寒則脉遲氣血微
則脉弱氣血平則脉治又長人脉長短人脉
短性急人脉急性緩人脉緩左大順男右大
順女男子尺脉常弱女人尺脉常盛此皆其

Fig. 14.11 Partial content of *Zhenjia Shuyao* by Hua Shou

diagnosis and draws common clinical tongue conditions into 36 types of maps, with a text attached to each map below. As for each pathological tongue condition, this book elaborates the etiology and pathology, treatment and prognostic judgment of the main syndrome combined with pulse conditions. In this period, fingerprint examination was also put forward. *A New Book of Pediatrics* that is a medical work of pediatrics written by Liu Fang of the Song Dynasty advocates that children under 3 years old should be diagnosed in fingerprint examination instead of pulse feeling and palpation. This kind of examination of collateral marks has been applied till now.

After the Song and Yuan Dynasties, diagnostic technology gradually got standardized, and the early rich diagnostic techniques were summarized as the current four ways of diagnosis. The "whole-body examination" of pulse-taking gradually faded out; abdominal diagnosis, although often involved in some medical works, had been less applied in reality. Zhang Gao pointed out in his *Yi Shuo*: "Nowadays, people from a noble family lives in gorgeous houses in purdah, covered with silk on arms. A doctor can neither amazingly examine his or her countenance, nor artfully feel his or her pulse." So, the patients at that time were not willing to bare their bodies, so doctors were not able to carry out a general examination of them. As a result, feeling cunkou pulse became a major way of pulse-taking. When explaining *Zhou Li*, Xiang Anshi of the Song Dynasty argued that "the examination of the change of nine orifices is current way to find the proof of diseases, assisted with the examination of the arteries of nine zang organs, which is just today's pulse-taking" [3]. It is obviously an interpretation of the ancient, showing that the four ways of diagnosis had become popular at that time.

In the Ming and Qing Dynasties, TCM had also made great progress, and diagnostic monographs emerged in large numbers, such as *Sizhen Juewei* by Lin Zhihan and *Golden Mirror of the Medical Tradition: the Essence of the Mental Cultivation Methods of the Four Ways of Diagnosis* by Wu Qian et al. which summarized the gist of the four ways of diagnosis. As for the way of observation, *Wangzhen Zunjing* by Wang Hong, learning widely from others' strong points, is unique in both theory and method. The development of the science of epidemic febrile disease of TCM accumulated a lot of valuable and unique experience for observation, especially in tongue examination, teeth examination, macula differentiation, white carbuncle differentiation, and so on. Zhang Jingyue, a famous medical practitioner, summed up the content of observation and created a "Ten-question Song," which not only made a high generalization of the content of observation, but also brought convenience to clinical inquiry, thus widely sung till now. Besides, the pulse monographs, such as *Binhu Maixue* by Li Shizhen, *Sanzhi Chan* by Zhou Xueting, *Maili Qiuzhen* by Huang Gongxiu, *Mai Yu* by Wu Kun, and *Xingse Waizhen Jianmo* by Zhou Xuehai, have had their practical value up to now. Moreover, *Precepts for Physicians* and *Yuyi Cao* by Yu Jiayan not only discussed the content and method of the four ways of diagnoses, but also created the writing rules of the clinical records of TCM on the basis of *Case Record* that was first created by Chunyu Yi, a famous doctor of the Western Han Dynasty.

14.2.2 The Content and Method of the Four Ways of Diagnosis

14.2.2.1 Observation

The so-called observation is to observe the patient's spirit, color, shape, posture, facial features, tongue picture as well as discharge, excreta, etc., in order to understand the patient's condition.

Observation includes the following aspects: observing the spirit, observing the color, observing the shape, observing the five senses, observing the skin, observing the front and back private parts, observing the main and collateral channels, observing the excrement, and so on. Observing the spirit mainly observes the eyes, facial expressions, and mental activities of the patient; observing the color observes the facial changes in color and lustre of the patient, mainly including the changes of the blue, red, yellow, white and black and their locations; observing the shape observes the patient's body, including muscles, bones, skin, size as well as posture and mobility, etc.; observing the five senses observes the changes in look and shape of the patient's eyes, ears, nose, mouth, and tongue; observing the skin observes the color and lustre and pattern of the patient's skin; observing the front and back private parts observes the patient's genital organ, urethra and anus; observing the main and collateral channels observes the changes in shape and color of the yuji of hands, index fingers, and nail collateral channels to examine diseases, mainly used in children less than 3 years old; observing the excrement observes the shape, color, quality and quantity of the patient's vomit, sputum, salivation, snot, saliva, shit and emiction, emmenia, pus, etc.

TCM holds that the external body of human is closely related to the five internal organs and the six fu organs; the changes in the functions of the zang-fu organs must be reflected in the external spirit, color, shape, and posture of human body. The five internal organs and the six fu organs link up with the body surface by 12 meridians and match the 5 body constituents (tendon, vessel, muscle, hair and skin, bone), respectively, and the lung governs the hair and skin, the liver governs the tendon, the spleen governs the muscle, the heart governs the blood and vessels, and the kidney governs the bone. The five senses are also related to the five internal organs: the eyes are the orifices of the liver, the ears are the orifices of the kidney, the nose is the orifice of the lung, the mouth is the orifice of the spleen, and the tongue is the orifice of the heart. The changes of the internal organs can be inferred by observing the changes of body surface and facial features, namely, "governing exterior to infer interior." In clinical practice, the focus of observation lies in the spirit, color of face, and the tongue (Fig. 14.12).

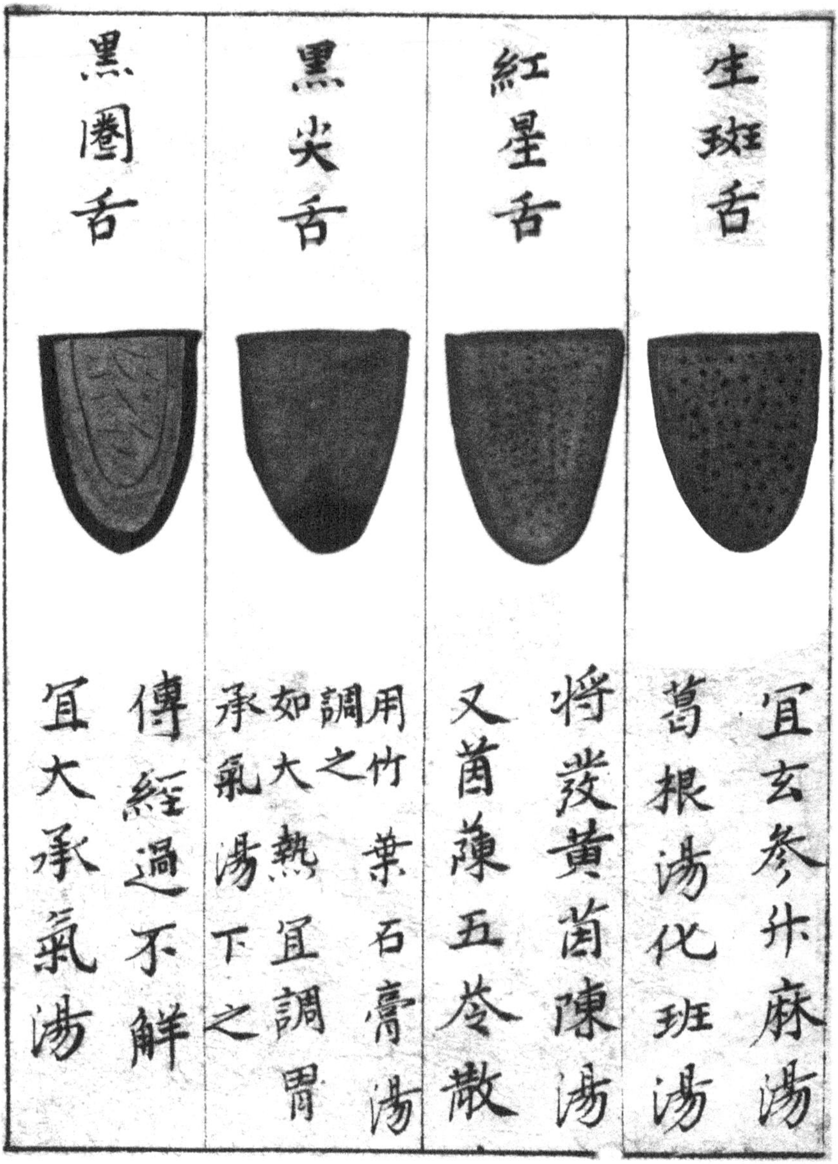

Fig. 14.12 The diagram of tongue pictures in *Shanghan Diandianjin*

14.2.2.2 Auscultation and Smelling

The way of auscultation and smelling is to identify diseases by the patient's language, breath as well as the smell of the discharge. The patient's chills and fever as well as deficiency and excess can be differentiated by listening to his or her words, breath, cough, hiccup, eructation, and so on; the patient's disease can be determined by smell his or her breath, secretion and excrement as well as certain abnormal smell of the ward. Generally, loud, gruff, and turbid voice belongs to excess syndrome; otherwise, it belongs to deficiency syndrome. Stale and foul smell mostly belongs to heat syndrome and odorless smell or slight fishy smell mostly belongs to cold syndrome. The odor of the ward is emitted by the patient's body and his or her excreta. For example, the ward of the patient who suffers from plague smells of moldy and stale odor, while the ward of the patient who suffers from hemorrhage smells of blood.

14.2.2.3 Interrogation

Interrogation is to find out the patient's normal health condition, dietary bias, cause of the disease, self-perceived symptoms, disease condition as well as treatment, medical history and family medical history, and so on by the inquiry of the patient and the insider(s). Zhang Jingyue, a physician of the Ming Dynasty, created the "Ten-question Song," which was later adapted by Chen Xiuyuan as follows: question one is about cold and heat, question two is about sweat, question three is about head and body, question four is about shit or piss, question five is about diet, question six about breast and abdomen, question seven is about deafness, question eight is about thirst, question nine is about past diseases, and question ten is about causes, assisted by taking medicine and referenced by chance variation; for women, the menstrual period is especially to be asked about, while for children, the diseases as smallpox and measles should be examined. Later, according to the notice by the Department of Traditional Chinese Medicine of the Ministry of Health, the song was adapted as follows: interrogation should start from the general (information), and then ask about the relevant. Question one is about cold and heat, question two is about sweat, question three is about head and body, question four is about shit or piss, question five is about diet, question six is about breast, question seven is about deafness, question eight is about thirst, question nine is about past diseases, and question ten is about causes; then make a diagnosis and give treatment referenced by personal and family medical histories; women should be asked about the conditions of menstrual disease and fetal production, while children should be asked about the conditions of infectious disease, inoculation, acute disease, cowpox, convulsion, malnutrition, and dietary bias.

14.2.2.4 Pulse Feeling and Palpation

Pulse feeling and palpation is to detect the pathological changes of zang-fu organs by examining the patient's pulse and other parts of body. Just as its name implies, the way of diagnosis should include pulse feeling and palpation. However, palpation is rarely used in China; in Japan, the palpations such as abdominal diagnosis are mainly adopted. As described earlier, pulse feeling in ancient times had a variety of methods, but nowadays only cunkou pulse is basically felt. When feeling pulse, the doctor would ask the patient to sit down or lie on his or her back, with the forearms set almost on the same level as the heart and the palms set upward. Lay the forearms on their side so as to make blood flow smoothly. If feeling pulse of adults, use three fingers to locate and press the three regions of cun, guan, and chi, respectively, whose pulse is called cun pulse, guan pulse, and chi pulse, respectively. The pulse of different regions corresponds to the condition of a specific zang-fu organ.

The doctors of all dynasties have different views on the length of each of the three regions of cun, guan, and chi. According to *The Pulse Classic*, "Yuji acupoint is one cun far from gaogu acupoint, and in the middle of them lies cunkou acupoint, of one chi far from which lies chize, hence chicun. The region behind cun and before chi is called guan, which is the boundary of yang and yin." It is now generally accepted that the region of styloid process of radius is guan, before which (point of wrist) is cun, and behind which (point of elbow) is chi (Fig. 14.13). As for the correspondence of the three regions of cun, guan, and chi to zang-fu organs, there have been many arguments in past dynasties. Now the commonly used clinical classification is: the cun pulse of left hand corresponds to heart, guan pulse to liver and gallbladder, and chi pulse to kidney; the cun pulse of right hand corresponds to lung, guan pulse to spleen and stomach, and chi pulse to life gate. One more principle is: cun pulse corresponds to the upper part of human body and chi pulse corresponds to the lower part of human body.

The method of pulse feeling also needs careful study. Doctors usually touch and press pulse with finger pulp and begin to apply light force on the skin, which is a kind of floating feeling, called "lifting"; then apply moderate force to touch and press the muscles, which is a kind of moderate feeling, called "seeking"; and finally apply more force to touch and press the muscles and bones, which is a kind of deep feeling, called "pressing." According to the requirement of clinical practice, a doctor can repeatedly touch and press in the order of lifting, seeking and pressing or in reverse order, or press with only one finger.

Each of the three regions of cun, guan, and chi has floating, moderate, and deep divisions for pulse feeling, called three regions and nine divisions (not ancient three regions and nine divisions). It is generally believed that flat pulse belongs to normal pulse, while floating pulse, deep pulse, slow pulse, rapid pulse, feeble pulse, forceful pulse, smooth pulse, full pulse, thready pulse, string-like pulse, etc. belong to abnormal pulse.

Although pulse feeling ranks the last one of the "four ways of diagnosis," it has been the most characteristic diagnostic method of TCM. Clinically, TCM values the comprehensive application of the "four ways of diagnosis," so as to fully and

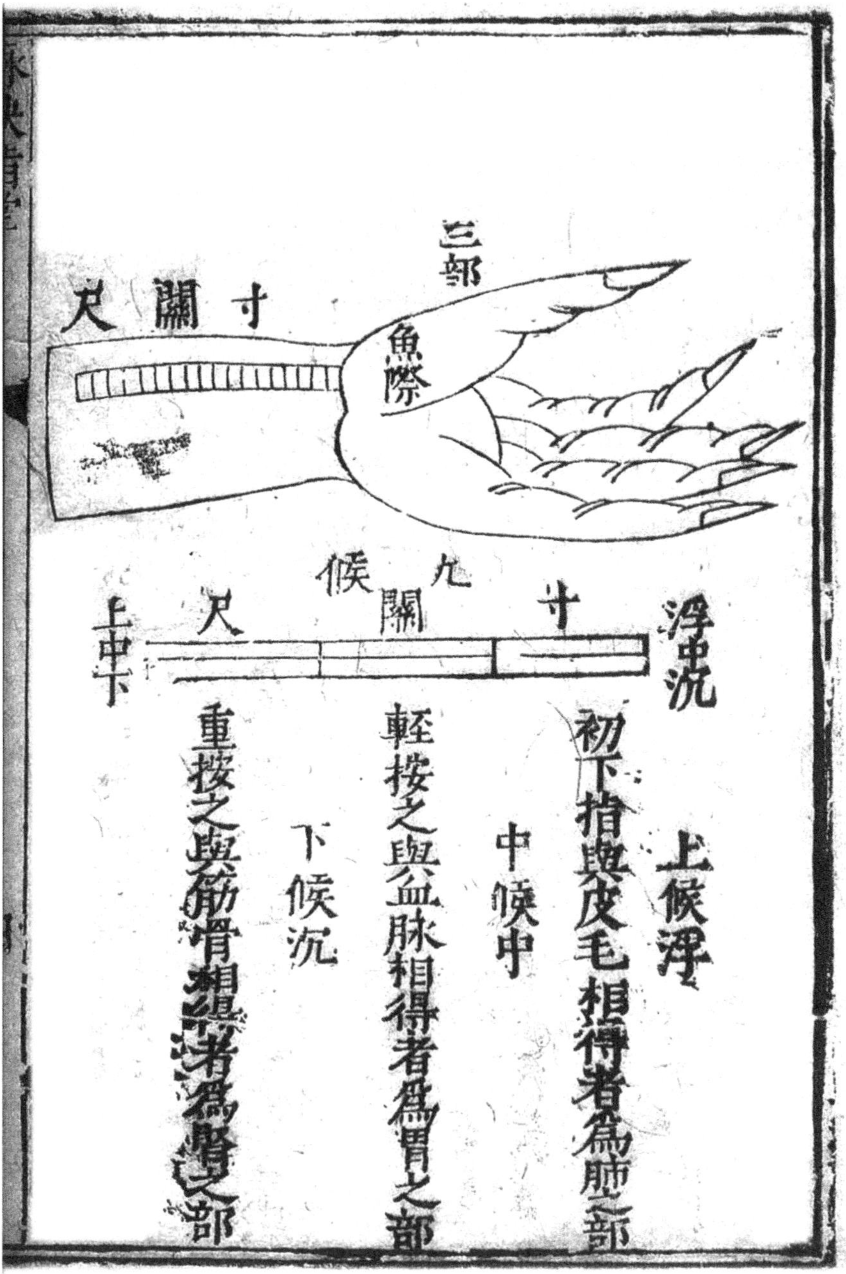

Fig. 14.13 The diagram of pulses in *Maijue Zhizhang*

systematically understand a patient's condition and make correct diagnosis. This is also known as "cooperative application of the four ways of diagnosis."

14.2.3 Syndrome Differentiation

In diagnosis, Western medicine has to clarify what "disease" the patient is suffering from, and then determine the treatment, while TCM is to first find out the "syndrome" of how to treat it. The so-called "syndrome" does not simply refer to the symptoms such as headache, vomiting, fever, etc., but a summary of the pathological conditions of one patient's cause of disease, location, pathogenic and vital contention, prosperity and decline of yin and yang, etc. by observing all the symptoms appeared.

Yellow Emperor's Canon of Medicine has recorded the diagnosis of a large number of diseases as well as the discussion related to syndrome differentiation. Zhang Zhongjing of the Eastern Han Dynasty made a clear distinction between disease and syndrome for the first time, and also juxtaposed "disease, pulse and syndrome," which pioneered the diagnostic combination of differentiation of disease and syndrome differentiation in TCM. His *Treatise on Cold Pathogenic and Miscellaneous Diseases* established the basic category of syndrome differentiation in terms of yin and yang, exterior and interior, chills and fever, and deficiency and excess, laying the foundation of syndrome differentiation, and setting up the first complete system of syndrome differentiation in diagnostics of TCM: syndrome differentiation method of six meridians of typhoid fever.

In the Sui and Tang Dynasties, syndrome differentiation got further development. As the first monograph on the diagnosis of the causes and syndromes of diseases in China, *Treatise on the Causes and Manifestations of Various Diseases* by Chaoyuan Fang is disease-oriented and covers 1,720 divisions of 67 categories. *Qianjin Yaofang* (*Valuable Prescriptions for Emergencies*) by Sun Simiao discusses the comprehensive syndromes of typhoid fever according to diseases, as well as the respective syndromes of miscellaneous diseases according to zang-fu organs, contributing to the development of syndrome differentiation of zang-fu organs.

During the Song, Jin, and Yuan Dynasties, many medical works discussed the topic of syndromes. For example, Shanghan Zongbing Lun (*A General Treatise on Typhoid Fever*) by Pang Anshi discusses the respective syndromes of six meridians of typhoid fever; *Shanghan Mingli Lun* (*Concise Exposition on Cold Pathogenic Diseases*) by Cheng wuji differentiates 50 kinds of syndromes of typhoid fever; *Zangfu Biaoben Hanre Xushi Yongyaoshi* (*A Practical Pharmacy for the Chills and Fever and the Deficiency and Excess of the Tip and Root of Zang-fu Organs*) by Zhang Yuansu combines the syndromes of the chills and fever and the deficiency and excess of zang-fu organs with the invigorant and lustramentum for the chills and fever; and *Sanyinji Yibingzheng Fanglun* (*Discussion on Pathology based on Triple Etiology Doctrine*) by Chen Yan is a complete medical work in etiological differentiation of symptoms.

During the Ming and Qing Dynasties, many comprehensive medical works were named with the Chinese character "zheng (diagnosis)," such as *Zhengzhi Yaojue* (*Key to the Diagnosis and Treatment*) by Dai Yuanli, *Zhengzhi Zhunsheng* (*Standards for the Diagnosis and Treatment*) by Wang Kentang, *Zhengzhi Huibu* (*Recollection of the Diagnosis and Treatment*) by Li Yongzi, *Bianzheng Lu* (*Records on Syndrome Differentiation*) by Chen Shiduo, etc., showing the importance of syndrome differentiation. *Linzheng Zhinan Yi'an* (*A Guide to Clinical Practice with Medical Record*) by Ye Tianshi of the Qing Dynasty even points out that "the art of healing cares about syndrome recognition, treatment determination and prescription establishment, which are the three key points," "but among the three, syndrome recognition is especially important." Up to now, syndrome differentiation and treatment has become the major feature of TCM and produced several methods of syndrome differentiation including eight principal syndromes differentiation, qi-blood-body fluid syndrome differentiation, zang-fu organs syndrome differentiation, six meridians syndrome differentiation, defense qi and ying blood syndrome differentiation, triple energizer syndrome differentiation, main and collateral channels syndrome differentiation, and so on. With their own characteristics and emphasis, these methods are interrelated and complementary to each other in clinical practice.

14.2.3.1 Eight Principal Syndromes Differentiation

"Eight principal syndromes" refer to the eight principal syndromes as yin, yang, exterior, interior, cold, heat, deficiency, and excess, exactly the four pairs of guiding syndromes as exterior and interior, cold and heat, deficiency and excess, and yin and yang. As the most basic method of differentiation of symptoms, eight principal syndromes differentiation started from *Yellow Emperor's Canon of Medicine*, achieved a great development made by Zhang Zhongjing, initially took shape in the Song and Yuan Dynasties, and got perfect and stereotyped in the Ming and Qing Dynasties.

Within the eight principal syndromes differentiation, the pair of exterior and interior is the guiding principle for differentiating the location, depth, and tendency of pathological changes. Any syndrome caused by exogenous evil, with the pathological changes in skin and much superficial part of muscles, belongs to exterior one and any syndrome caused by the seven human emotions, diet, fatigue, and other factors, with the zang-fu organs damaged, belongs to interior one. The pair of cold and heat explains the nature of diseases. Any disease caused by yin evil or deficiency of yang qi in human body belongs to cold syndrome of which yin qi is excessive and any disease caused by yang evil or deficiency of yin essence in human body belongs to heat syndrome of which yang qi is excessive. The pair of deficiency and excess is the guiding principle for differentiating evils and putting right deficiency and excess. Deficiency syndrome is the symptom for deficiency of healthy qi and excess syndrome is the symptom of evil qi, including the common types of deficiency-cold, deficiency-heat, excess-cold, and excess-heat. The pair of yin and yang is the general guiding principle in the eight principal syndromes differentiation. All the three types of syndrome of exterior, excess, and heat belong to yang, while the three types of syndrome of interior, deficiency, and cold belong to yin.

14.2.3.2 Qi-Blood-Body Fluid Syndrome Differentiation

TCM holds that qi, blood and body fluids form the material basis of human life activities and maintain normal physiological activities of zang-fu organs. Once they experience pathological changes, the normal life activities of human body will be affected. Qi-blood-body fluid syndrome differentiation can be divided into qi disease syndrome differentiation, blood disease syndrome differentiation, and body fluid disease syndrome differentiation. The qi disease syndrome differentiation is generally summarized as qi deficiency, qi collapse, qi stagnation, and qi adverseness. The common syndromes of blood disease can be summarized as blood deficiency syndrome, blood stasis syndrome, and blood heat syndrome. The body fluid disease refers to the dysbolism of body fluid or the loss of body fluid produced by various causes. The pathological changes of body fluid can be summarized as deficiency of body fluid and retention of body fluid.

14.2.3.3 Six Meridians Syndrome Differentiation

Six meridians refer to the six meridians of Taiyang, Yangming, Shaoyang, Taiyin, Jueyin, and Shaoyin. According to *Plain Conversation: Re Lun* (*Discussion on Heat*), "The suffering from exogenous febrile disease for one day will injure Juyang… for two days Yangming… for three days Shaoyang… for four days Taiyin… for five days Shaoyin… and for six days Jueyin …" On this basis, Zhang Zhongjing of the Eastern Han Dynasty created the method of six meridians syndrome differentiation. According to different natures of syndrome, this method is divided into six types: Taiyang disease, Yangming disease, Shaoyang disease, Taiyin disease, Shaoyin disease, and Jueyin disease, generally three types of yang disease and three types of yin disease. Generally speaking, the three types of yang disease have strong resistance and are serious, while the three types of yin disease have weak resistance and are not serious at all. Six meridians syndrome differentiation is mainly applied in the differentiation of the symptoms of exogenous diseases.

Several other methods of syndrome differentiation are complementary to the methods mentioned above. In clinical practice, appropriate method should be adopted according to specific condition, or several methods be combined.

14.2.3.4 Defense Qi and Ying Blood Syndrome Differentiation

This is a kind of syndrome differentiation established by Ye Tianshi of the Qing Dynasty to differentiate the symptoms of exogenous warm-heat diseases. It makes up for the deficiency of six meridians syndrome differentiation and enriches the syndrome differentiation of exogenous diseases. Defense qi and ying blood syndrome differentiation represents the syndrome classification of the four different stages of pathogenesis in the development of warm diseases. Warm disease is initially superficial, then appears the syndrome of defense tier, further appears the syndrome of qi tier, a little further appears the syndrome of ying tier, and finally appears the syndrome of blood tier.

14.2.3.5 Triple Energizer Syndrome Differentiation

The understanding of the so-called sanjiao (triple energizer) was different in different dynasties, but mostly the division of the upper, the middle, and the lower energizers is adopted to divide human body into the upper, the middle, and the lower parts, namely, the chest above the diaphragm belongs to the upper energizer, including the two organs of heart and lung; the epigastric abdomen below the diaphragm and above the navel refers to the middle energizer, including the spleen and the stomach; the abdomen below the navel refers to the lower energizer, including small intestine, large intestine, liver, kidney, urinary bladder, etc. On the basis of the concept of triple energizers, Wu Jutong of the Qing Dynasty combined with the general law of the occurrence and development of warm diseases as well as the different manifestations of the zang-fu organs to which the triple energizers belong caused by pathological changes, and established the triple energizer syndrome differentiation, the guiding principle of warm disease syndrome differentiation. The syndrome differentiation takes the upper, the middle and the lower energizers as well as the names of warm diseases as main thread, running through six meridians syndrome differentiation, zang-fu organs syndrome differentiation and defense qi and ying blood syndrome differentiation, to discuss the pathogenesis changes of the zang-fu organs which correspond to triple energizers in the process of warm diseases, and accordingly generalize the types of syndrome, and position, diagnose, and treat according to zang-fu organs.

14.3 Treatment Technology of Traditional Chinese Medicine

The treatment technology of TCM includes two aspects: principle of treatment and method of treatment.

14.3.1 *Principle of Treatment*

14.3.1.1 Preventive Treatment of Diseases

TCM attaches great importance to the prevention of diseases, just as *Plain Conversation: Siqi Tiaoshen Dalun* (*Major Discussion of Regulation of Spirit according to the Changes of the Four Seasons*) says: "A saint does not cure diseases, but makes preventive treatment of diseases; he does not cure disorder, but makes preventive treatment of disorder." The so-called "preventive treatment of diseases" includes two aspects: prevention before disease and prevention of pathological change after disease. Prevention before disease is to enhance the prevention before the attack of any disease. The main methods involved are to take good care of one's spirit and emotion, adapt to the natural environment, adjust the diet and daily life, strengthen physical exercise, and so on. *Plain Conversation: Shanggu Tianzhen Lun* (*Ancient Ideas on How to Preserve Natural Healthy Energy*) points out that "one's spirit is internally concentrated, which can prevent diseases," and advocates "idle ambition and less desire," which just mean to deal with the relationship between spirit and health. According to *Plain Conversation: Siqi Tiaoshen Dalun* (*Major Discussion of Regulation of Spirit according to the Changes of the Four Seasons*), "Yin and yang and the four seasons delimit the beginning and end of all things, and are also the root of life and death. If running counter to them, we will suffer from certain disease; likewise, we will stay healthy." An initiative adaptation to natural environment can reduce the occurrence of certain diseases. *Plain Conversation: Shanggu Tianzhen Lun* (*Ancient Ideas on How to Preserve Natural Healthy Energy*) points out that "temperate diet, regular daily life and moderate work" can "keep the body and the spirit healthy, and die old"; if someone "often becomes drunk and violent to use up his essence and exhaust his vital qi… runs counter to yin and yang and the four seasons, and lives an irregular daily life," he surely "ends in premature senility before fifty years of age." This is valuing the importance of diet, daily life, and alternate work with rest to health. In the Spring and Autumn and Warring States Period, Chinese ancestors had applied "daoyin" (namely, health exercise) and "intaking" (namely, respiring gymnastics) to prevent diseases. And later, the sports methods such as "five-animal boxing" (a gymnastics which mimicks the action of tiger, deer, ape, bear and bird), "shadow boxing," and "eight-section brocade boxing" were created to enhance one's physical fitness and improve the disease resistance.

The ancients believed that most diseases invaded into human body externally. According to *Plain Conversation: Yinyang Yingxiang Dalun* (*Major Discussion on the Theory of Yin and Yang and the Corresponding Relationships among all the Things in Nature*), "Exogenous pathogen comes quickly, so a good doctor first treats the hair, then the skin, then the muscle and vessel, then six fu organs, and finally the five internal organs. He whose five internal organs are treated meets an unfavorable prognosis with a half possibility of death." Pathogen may go from the surface to the inside, step by step, so the disease should be treated earlier to control its growth, change, and spread. If "the disease is in the liver, as liver disease is known to spread to the spleen, the spleen has to be strengthened first." According to the order of the restriction among the five elements, liver disease will lead to spleen disease, so in the treatment of liver disease, the spleen and stomach have to be strengthened simultaneously in order to control the further progress of the disease.

14.3.1.2 Searching for the Primary Cause of Disease in Treatment

Plain Conversation: Yinyang Yingxiang Dalun (*Major Discussion on the Theory of Yin and Yang and the Corresponding Relationships among all the Things in Nature*) says: "The treatment of disease has to search for its primary cause." The so-called "primary cause" refers to the essence of disease. For example, headache can be divided into exogenous one and internal one. Exogenous headache can be caused by wind-cold, wind-heat, or wind-damp and internal headache has different types of qi deficiency, blood stasis, blood stagnation, phlegmatic hygrosis, and liver yang. Accordingly, when treating headache, a doctor should analyze its specific clinical manifestations and carry out syndrome differentiation to search for its primary cause and find out its nature, so as to adopt different treatments such as relieving exterior syndrome, tonifying qi, nourishing the blood, activating blood circulation to remove blood stasis, removing dampness to reduce phlegm, and calming the liver and suppressing yang.

14.3.1.3 Regulation of Yin and Yang

TCM holds that any disease is caused by the destruction of the relative balance between yin and yang in human body, namely, imbalance of yin and yang. As for its treatment, *Plain Conversation: Zhizhenyao Dalun* (*Major Discussion on the Most Important and Abstruse Theory*) says: "Carefully examine the state of yin and yang, and adjust them until they are in balance." It is one of the fundamental treatment principles of TCM to regulate yin and yang and restore the relative balance between them by repairing or correcting defects. In concrete application, there are the following two principles: "consuming the excessive" and "polishing the deficiency."

14.3.1.4 Strengthening Body Resistance and Eliminating Evil

TCM holds that in the course of a disease's occurrence, growth, and change, there is always a state of struggle between healthy qi and evil qi. If the evil wins the healthy, the disease grows further; likewise, the disease disappears. Strengthening body resistance is to improve the healthy qi and invigorate body constitution, so as to increase the body's ability to resist diseases. The main ways involved include tonifying qi, enriching the blood, tonifying yin, and tonifying yin-yang with medicine, or by acupuncture, practicing qigong, adjusting spirit, regulating diet, etc. Eliminating evil is to get ride of pathogenic diseases and restore health. The ways involved cover attacking, clearing heat, dispelling cold, helping digestion and removing accumulation, clearing damp, activating blood circulation and removing blood stasis, promoting the circulation of qi, and so on.

14.3.1.5 Adaptation to Local Condition

TCM believes that all the factors of climate, region, social environment, and individual difference will affect the physiology and pathology of human body, so when treating diseases, a doctor has to take fully into account the onset season, climate change, and geographical environment, as well as the individual factors of the patient in terms of gender, age, physical fitness, hobby, economic condition, personality, etc. That is, the treatment of TCM is often tailored to the condition of the individual.

14.3.2 Method of Treatment

TCM has various unique methods of treatment, including such health-preserving treatments as drug treatment, acupuncture, surgical treatment, daoyin, and diet treatment as well as such external treatments as massage, cupping, and scraping, some of which had led the world.

14.3.2.1 Drug Treatment

Drug treatment is the most basic method of treatment of TCM. In the long course of life and production, our ancestors in ingesting plants and hunting were able to understand the effects of certain animals and plants as well as minerals on human body when getting in touch with them, and gradually created and accumulated some knowledge of drugs. As early as the Western Zhou Dynasty, there had been professional doctors who "gathered poisons for medical treatment." *The Classic of Mountains and Rivers* records more than 100 kinds of animal and plant drugs, many of which have been applied so far. The silk book *52 Bing Fang* (*Prescriptions for Fifty-two Diseases*) unearthed in Mawangdui records about 300 prescriptions, involving more than 240 kinds of drugs. *Sheng Nong's Herbal Classic* published at the end of the Eastern Han Dynasty contains 365 kinds of drugs and briefly describes the basic theories of pharmacy, such as of four qi and five tastes, toxicity and non-toxicity, law of compatibility of medicines, method of taking medicine as well as the drug forms of pill, powder, liquid, etc. The monograph laid a foundation work to TCM. Later, Tao Hongjing of the Liang Dynasty annotated *Sheng Nong's Herbal Classic* and supplemented it with the pharmacological achievements made since the Eastern Han Dynasty, hence *Variorum Sheng Nong's Herbal Classic*. Containing 730 kinds of drugs, the book made a detailed discussion on the origin of drugs, method of gathering and processing, identification of authenticity and falsehood, and so on, and created a classification method of drugs according to their natural attributes. By the Tang Dynasty, pharmacology had made great progress. *Newly Revised Canon of Materia Medica* issued by the government then records 844 kinds of drugs. The book contains drug atlases and corresponding written explanation, setting a precedent for the pharmaceutical works with pictures and texts.

The Song Dynasty saw the peak of the development of Chinese medicine and pharmacy, with the canons of materia medica published one after another. The government organized the compilation and revision of *Kaibao Bencao* (*Kaibao Canon of Materia Medica*) and *Jiayou Buzhu Bencao* (*Jiayou Supplementary Canon of Materia Medica*). Based on *Jiayou Buzhu Bencao* and *Bencao Tujing* (*Commentaries on the Illustrations*), Tang Shenwei compiled *Jingshi Zhenglei Beiji Bencao* (*Classified Materia Medica from Historical Classics for Emergency*) (abbreviated as *Classified Materia Medica*), which contains a total of 1,746 kinds of drugs, with an addition of 660 kinds of drugs. After that, the government revised the book three times, and printed and published it as an official book by adding the reign titles of "Daguan," "Zhenghe," and "Shaoxing". Just on the basis of it, Li Shizhen in the Ming Dynasty "spent 30 years in textually researching more than 800 medical books, altered his own medical manuscript three times," and finally compiled the epoch-making *Compendium of Materia Medica* (Fig. 14.14). This book covers 1,892 kinds of drugs, 1,160 illustrations, and more than 11,000 prescriptions. Each drug is parsed to interpret its name, to describe its place of origin, to correct its errors in past literature, and to introduce its processing method as well as its odor, main treatment, and invention. The book also contains a collection of prescriptions that had been popular in the folk. Later, Zhao Xuemin edited *A Supplement to the Compendium of Materia Medica*, which included 716 kinds of drugs that were not collected by *Compendium of Materia Medica.* By the middle Qing Dynasty, more than 2,000 kinds of drugs had been recorded in the literature.

In the long-term practice, ancient doctors found out the unique therapeutic effects of many drugs, such as the analgesic effect of Rhizoma corydalis, Ligusticum wallichii and Aconitum, the antimalarial effect of Artemisia apiacea and Antipyretic dichroa, the anti-asthmatic effect of Ephdra vulgaris, the menstruation regulating effect of Chinese angelica, the dysentery treatment effect of Rhizoma coptidis, and the parasite expelling effect of fructus meliae. The Tang Dynasty had begun to apply animal tissues and organs as well as hormone preparation. *Tang Materia Medica* records the experience of treating night blindness and improving eyesight with lamb liver; *A Supplement to Materia Medica* has a record of the application of human afterbirth as invigorator; and *Valuable Prescriptions for Emergencies* (Fig. 14.15) records the treatment of thyropathy with the thyroid gland of sheep and deer. Yeast preparations had been widely applied in medicine in the Tang Dynasty. For example, *Valuable Prescriptions for Emergencies* and *Theory of Drug Properties* by Zhen Quan both make a clear description of the nature and function of medicated leaven. All these pharmaceutical experiences have been confirmed by modern scientific experiments.

Since the twentieth century, many effective compounds have been extracted from the traditional Chinese medicine according to the pharmaceutical experiences of TCM, thus opening up a new field of modern pharmacology. In 1924, Chen Kehui and

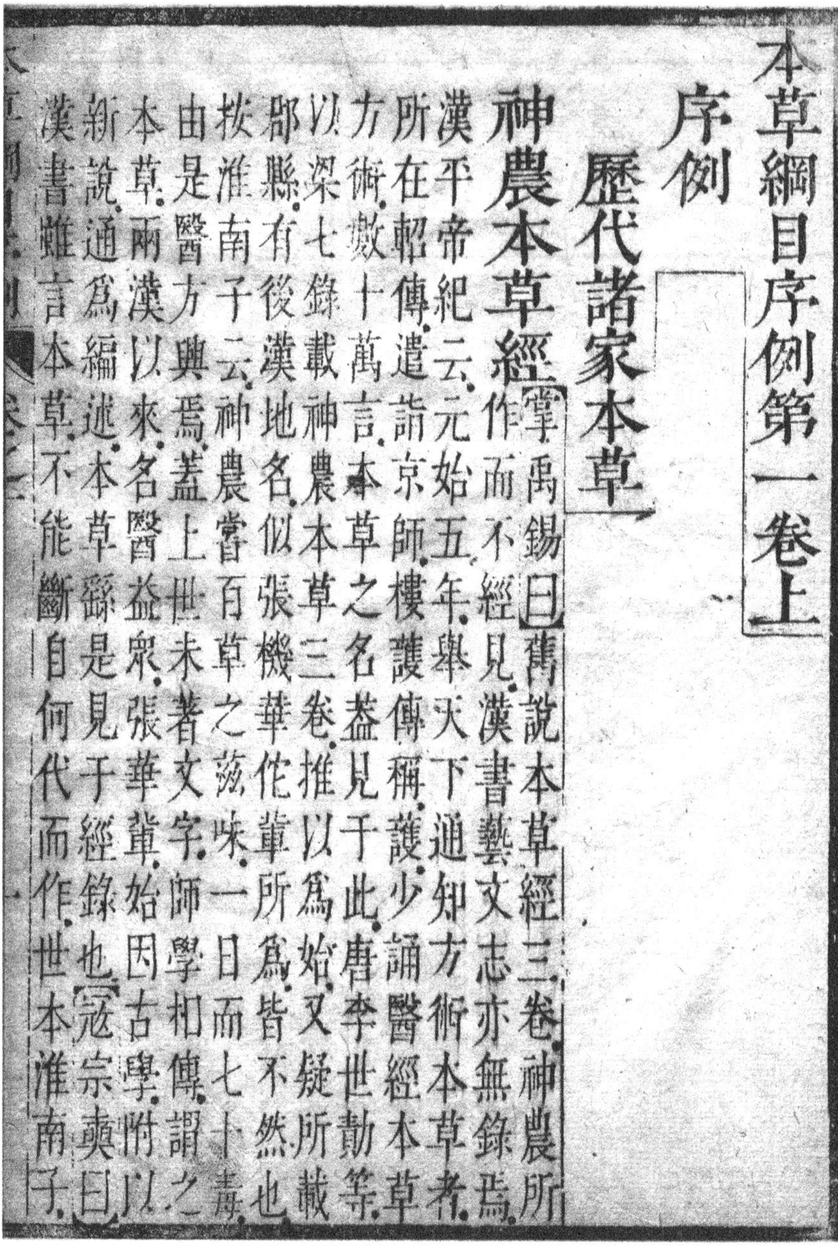

本草綱目序例第一卷上

序例

歷代諸家本草

神農本草經[掌禹錫曰]舊說本草經三卷神農所作而不經見漢書藝文志亦無錄焉漢平帝紀云元始五年舉天下通知方術本草者所在軺傳遣詣京師樓護傳稱護少誦醫經本草方術數十萬言本草之名蓋見于此唐李世勣等以梁七錄載神農本草三卷推以爲始又疑所載郡縣有後漢地名似張機華佗輩所爲皆不然也按淮南子云神農嘗百草之滋味一日而七十毒由是醫方興焉蓋上世未著文字師學相傳謂之本草兩漢以來名醫益衆張華輩始因古學附以新說通爲編述本草繇是見于經錄也[寇宗奭曰]漢書雖言本草不能斷自何代而作世本淮南子

Fig. 14.14 Partial content of *Compendium of Materia Medica* by Li Shizhen

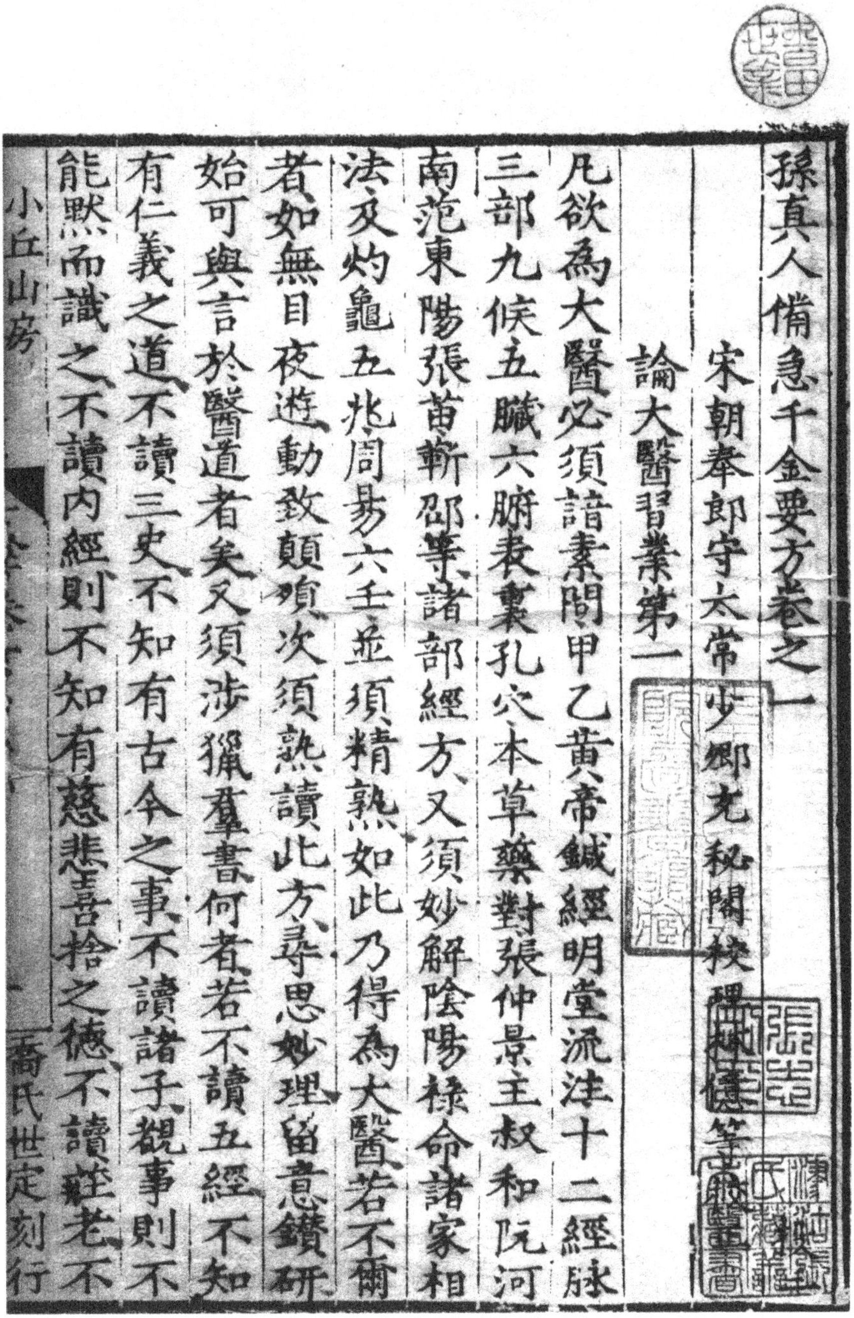

孫真人備急千金要方卷之一

宋朝奉郎守太常少卿充秘閣校理林億等校正

論大醫習業第一

凡欲為大醫必須諳素問甲乙黃帝鍼經明堂流注十二經脉三部九候五臟六腑表裏孔穴本草藥對張仲景王叔和阮河南范東陽張苗靳邵等諸部經方又須妙解陰陽祿命諸家相法及灼龜五兆周易六壬並須精熟如此乃得為大醫若不爾者如無目夜遊動致顛殞次須熟讀此方尋思妙理留意鑽研始可與言於醫道者矣又須涉獵羣書何者若不讀五經不知有仁義之道不讀三史不知有古今之事不讀諸子覩事則不能默而識之不讀內經則不知有慈悲喜捨之德不讀莊老不

小丘山房

喬氏世定刻行

Fig. 14.15 Partial content of *Valuable Prescriptions for Emergencies* by Sun Simiao

others isolated ephedrine levorotatory from Ephdra vulgaris to treat bronchial asthma, hay fever, and other allergic diseases. In the 1970s, artemisinine was isolated from Artemisia apiacea. It is an effective antimalarial monomer and has become a specific medicine for treating malarial fewer, which has a 100% protozoon inhibition ratio against rat malaria and monkey malaria. Due to her contribution to the development of artemisinine and dihydroartemisinine, Tu Youyou of China won the Lasker Award in September 2011 and the Nobel Prize for Medicine in 2015. It has become one of the important ways of modern pharmacology to carry out new drug research based on the effective components of Chinese herbal medicine.

In clinical practice, TCM mainly uses prescription to treat diseases. The so-called prescription refers to the combined medicine of several single drugs which is produced according to the principle of compatibility of medicines. The prescription enjoys a long history. *Yellow Emperor's Canon of Medicine* records 13 prescriptions. *Prescriptions for Fifty-two Diseases* as the earliest book of prescription in existence contains 286 prescriptions. In the Eastern Han Dynasty, the development of prescriptions reached a peak. *Treatise on Cold Damage* by Zhang Zhongjing recorded 113 prescriptions; *Essential Prescriptions of the Golden Cabinet* covered 262 prescriptions. These prescriptions are rightful in combination of drugs, precise and appropriate in drug selection, exact in drug ratio, and effective in treatment, most of which have been applied so far and regarded as classical prescription by later generations. Since then, the number of prescriptions had increased rapidly. *Handbook of Prescriptions for Emergencies* by Ge Hong of the Jin Dynasty recorded a large number of effective prescriptions which had been tested and were cheap and convenient to apply. In the Tang Dynasty, *Valuable Prescriptions for Emergencies* by Sun Simiao recorded 5,300 prescriptions, and *Waitai Miyao* (*Arcane Essentials from the Imperial Library*) by Wang Tao recorded more than 6,000 prescriptions. In the Song Dynasty, *Taiping Shenghui Fang* (*Taiping Holy Prescriptions for Universal Relief*) which was organized to compile by the government included 16,834 prescriptions and *Sheng Ji Zonglu* (*General Records of Holy Universal Relief*) recorded more than 20,000 prescriptions. Besides, the government then had also compiled *Heji Ju Fang* (*Prescriptions of the Pharmacy Bureau*) which recorded the prescriptions of producing and selling patent medicine and the standard of preparation. Including 297 prescriptions, this book was the first codex of patent medicines which was issued by the imperial court. *Formulas for Universal Relief* of the Ming Dynasty recorded more than 60,000 prescriptions. At present, the total number of prescriptions recorded in the literature is more than 100,000.

In a prescription, all the drugs combined can be divided into the drug of monarch, official, assistant, and envoy according to the role each drug plays. Monarch drug plays the major role of treatment; official drug assists monarch drug in the treatment to improve curative effect; assistant drug assists monarch drug to treat accompanying symptoms or secondary symptoms, or to inhibit the toxicity and severity of both monarch drug and official drug; and envoy drug leads the drugs in the prescription directly to the syndrome of disease, or reconcile all the drugs in the prescription. For example, ephedra decoction included in *Treatise on Cold Damage* is composed of the four drugs of ephedra, cassia twig, almond, and liquorice, mainly treating the wind-cold exterior excess syndromes such as aversion to cold with fever, headache and body pain, adiaphoretic asthma, thin and white-coated tongue, floating and tense pulse, etc. Among them, ephedra that is an acrid-warm herb relieving superficies, freeing lung and relieving asthma treats the main syndrome, so belongs to monarch drug; cassia twig that is an acrid-warm herb relieving superficies assists the ephedra to induce perspiration, so belongs to official drug; almond that clears the lung and downbears qi assists the ephedra to relieve asthma, so belongs to assistant drug; and liquorice reconciles the severity of ephedra and cassia twig in inducing perspiration, so belongs to envoy drug.

The treatment of disease with prescription mainly has eight methods: sweating, vomiting, purgating, reconciling, warming, clearing, replenishing, and eliminating. Sweating, also known as relieving exterior syndrome, is the application of drug that can induce perspiration to open up striae and interstitial space to remove evil out. Vomiting, also known as emetic method, is the application of emetic drug to guide disease or toxic substances out of the mouth. Purgating is to drive out the old food and settlings in the intestines and stomach by emptying the bowels. Reconciling is the application of drug that can reconcile and discharge, and regulate zang-fu organs to coordinate the functions of human body. Warming, also called cold dispelling method, is the application of warm-heat prescription to remove cold and evil and tonify yang qi. Clearing, also known as heat clearing method, is the application of severe cold prescription to clear away heat evil by clearing heat, reducing fire, cooling blood, and so on. Replenishing, also called tonifying method, is the application of the prescription with tonifying effect to eliminate weak syndrome. Eliminating, also known as dissipation method or indigestion curing method, is the application of drug that can eliminate and cure the stagnation of food, resolve hard lump, promote the circulation of qi, reduce phlegm, resolve indigestion, etc. to dissipate the excess evil kept in human body. Accordingly, prescription is also divided into sweating preparation, purgating preparation, reconciling preparation, warming preparation, replenishing preparation, and so on. In clinical practice, doctors also have to make appropriate additions and subtractions in accordance with the stage of a disease; its importance and emergency; and the patient's age, gender, and occupation as well as climate and geographical environment, so as to meet the condition of a disease and improve curative effect.

Herbs must be processed with certain methods and then turn to be traditional Chinese medicine. The purpose of drug processing is to reduce or eliminate the toxicity or side effects of drugs, to change or gentle their properties, so as to enhance their curative effect, guide them to meridians, and to reconcile the final preparations. For example, the toxicity of aconite can be reduced after soaked and boiled; the toxicity of Pinellia ternata can be reduced after rinsed and processed with alum and ginger; and the roots of some animals and hard plants need to be processed to be easy to use. Processing methods include calcining, frying, roasting, drying by heat, immersing, soaking, steaming, cooking, etc.; auxiliary materials cover wine, vinegar, honey, ginger juice, bean juice, liquorice juice, rice water, salt water, rice, alum, talcum powder, and so on. The painted book *A Supplement to Brief Introduction to Drug Processing by Lei Gong* of the Ming Dynasty (Fig. 14.16) shows the scenes

Fig. 14.16 A map of Yellow Emperor's processing of drugs collected in *A Supplement to Brief Introduction to Drug Processing by Lei Gong*

of drug processing at that time.

A prescription written by a doctor has to be made into drugs for the patient to take. There are many kinds of drugs. Among them, decoction (of herbal medicine) is the most commonly used. Decoction is said to be invented by Yi Yin of the Yin and Shang Dynasties. Actually, the *History of the Former Han Dynasty* records Yi Yin's writing of *Treatise on the Essentials of Decoction. Prescriptions for Fifty-two Diseases* contains such forms of drugs as cake, yeast, oil, pulp, pill, dust, ointment, pellet, fumigation, glue, and so on. *Yellow Emperor's Canon of Medicine* contains such forms of drugs as soup, pill, powder, ointment, alcohol, etc., which shows that the forms of drugs had been abundant in the Pre-Qin Period. The forms of drugs used in *Treatise on Cold Damage* included decoction, pill, powder, suppository, ointment, wine, vinegar, clysma, lotion, balneum, fumigant, ear drops, nasal irrigations, blow-in powder for nose, and so on, beyond 10 kinds. After hundreds of years of development, *Compendium of Materia Medica* by Li Shizhen in the Ming Dynasty had recorded more than 60 forms of drugs. In addition to the abovementioned forms of drugs, there were also produced linimentum, pastille, film agent, eye ointment, eye drops, and so on. Most of these forms of drugs are still widely applied, and some of them have been improved, exerting a far-reaching influence on the design of the forms of drugs of modern Chinese medicine [4].

14.3.2.2 Acupuncture and Massage

Acupuncture is the general term of acupuncture therapy and moxibustion therapy. Acupuncture therapy is to insert a special metal needle into certain acupoint in the body of the patient, so as to achieve the purpose of treating disease through manipulation. Moxibustion therapy is to warm the skin surface of an acupoint with burning moxa wool, so as to cure disease by means of thermal stimulus.

As early as the Stone Age, it had been known that sharp stones could be used to stab parts of human body or make it bleed to relieve pain. Gradually, stone tools were specially made for medical purpose, known as stone needle, and used to carry out surgical incision and drainage. Stone needle was once discovered in archeology in China. The sharp stone needles are called needle stone or beryllium stone. "There are stones which look like jade and can be made into needles" written in the *Classic of Mountains and Rivers* is an early record of stone needle. It can be said that stone needle is the foundation and predecessor of the tools of knife and needle in the later generations.

The occurrence of moxibustion therapy is related to the use of fire. In the process of using fire, it was found that the pain on certain part of human body could be relieved or removed by the burning or baking of fire. Later, people learnt to wrap burned stones or sand with fells or barks, and then use them to warm parts of the body. Gradually, tree branches and hays were lighted to warm parts of the body to treat diseases. After a long period of exploration, the folium artemisiae argyi that is flammable and can warm meridians was selected as the main material for moxibustion therapy, which is used to warm and stimulate locally on body surface. Moxibustion therapy thus began to be an important method of preventing and curing diseases, just like acupuncture therapy. Folium artemisiae argyi is easy to burn, full of fragrance, rich in resources, and easy to process and store, accordingly becoming the most important raw material for moxibustion therapy.

Many nationalities in the world have a history of using stone needles or needle stones, warming, etc. to treat diseases. However, none of them can be called acupuncture. Only when an acupuncture therapy is combined with main and collateral channels as well as acupoints, and guided by the theory of TCM, it can be called acupuncture.

According to the theory of TCM, main and collateral channels are regarded as a network of passages, through which qi and blood circulate and zang-fu organs are connected with body surface and all parts of whole body. As the control system of bodily functions, main and collateral channels include the meridians and the collaterals: the passages that travel through the whole body are called meridians and the tiny branches that connect the meridians are called collaterals. Main and collateral channels crisscross and are widely distributed throughout the body, linking the exterior and interior body, zang-fu organs, and limbs into an organic whole.

Zubi Shiyimai Jiujing and *Yinyang Shiyimai Jiujing* unearthed in the Han Dynasty tombs in Mawangdui are the earliest existing literature on the theory of meridians, discussing the circulation of 11 meridians and related diseases. In the later *Ling Shu: Jingmai*, 11 meridians had developed into 12, 6 yang meridians and 6 yin meridians. The book clearly records the names of 12 meridians and their circulation, and the zang-fu organs that link to subsidiary channels as well as the main diseases involved, also discussing 8 extra channels. Since then, books on acupuncture continued to come out, such as *Zhenjiu Jiayi Jing* (*A-B Classic of Acupuncture and Moxibustion*),

Tongren Shuxue Zhenjiu Tujing, *Shisijing Fahui* (*Elucidation of Fourteen Channels*), and *Qijing Bamai Kao*, so that the theory on the meridian system was gradually improved. At present, it is believed that there are 12 regular channels, 8 extra channels, 12 branches of the regular channels, and 15 collaterals in human body.

Acupoint has different other names. For example, *Yellow Emperor's Canon of Medicine* calls it "jie (knot)," "hui (meeting)," "qixue (point of qi)," "qifu (house of qi)," etc.; *A-B Classic of Acupuncture and Moxibustion* calls it "kongxue (foveola)"; *Taiping Shenghui Fang* (*Taiping Holy Prescriptions for Universal Relief*) calls it "xuedao"; *Tongren Shuxue Zhenjiu Tujing* calls it "shuxue"; and *Shenjiu Jinglun* calls it "xuewei." According to *Classified Classics: Ren Zhi Sihai*, "'Shu (infusion)', 'shu' and 'yu', all are commonly used in the book." As a result, acupoint is also known as "shuxue (point of infusion)" and "yuxue."

What is shuxue or acupoint? According to *Plain Conversation: Qifu Lun* (*Discussion on Acupoints*), "Shuxue is the point where qi of the channels arises." *Ling Shu: Jiuzhen Shier Yuan* says: "(Shuxue is the part) where the spirit and qi runs and comes in and goes out, not skin, muscles or bones at all," that is, it refers to the special part which zang-fu organs, main and collateral channels, and qi and blood are infused into and come in and go out of. Acupoints have the miraculous effects of "quick recovery by pressing it" and "dispelling diseases quickly," so become the operated parts of acupuncture, massage, qigong, and so on. To this day, however, such questions as "what is the essence of acupoints," "whether are acupoints special structure of human body," etc. still remain open.

Yellow Emperor's Canon of Medicine points out that "acupoint refers to the parts where qi of the channels arise, and each part has its own name," and records 160 names of acupoints. *A-B Classic of Acupuncture and Moxibustion* by Huanghu Mi of the Jin Dynasty records the names, other names, locations, and main treatments of 340 acupoints by combining the main and collateral channels with acupoints. Wang Weiyi of the Song Dynasty was ordered to redefine the acupoints, correct the possible errors, and cast two bronze acupuncture figures with exact acupoints marked out on the surface for the teaching and examination of acupuncture and moxibustion. The two bronze acupuncture figures were placed in the Institute of Medical Officials [5], called Heavenly Sage Bronze Figure in history (Fig. 14.17). *Qidong Yeyu* (*Speech Used by the Illiterate*) by Zhou Mi recorded what his uncle once said: "I was once given a chance to examine a bronze acupuncture figure. It was made of fine bronze with all zang-fu organs, with exact acupoints marked out on the surface, whose names were written beside with inlaid gold. The combination of the front side and the back side will form a whole human body. In the past, it was used to examine a medical worker. The location of each acupoint was coated with beeswax outside, with mercury inside, so that medical workers could try to insert the needle according to acupoints. If a needle is inserted onto the right acupoint, the mercury inside it will come out. Likewise, even a needle could not be inserted into any acupoint. It's really an ingenious instrument." [6]. Zhou Mi also wrote *Tongren Shuxue Zhenjiu Tujing* of 3 volumes (1026), which records 354 acupoints and draws a map linking 12 meridians with acupoints, marked with the name of each acupoint, actually an annotation to the bronze acupuncture figure. Later, the book was illustrated, and Xia

Fig. 14.17 The bronze acupuncture figure made in the Ming Dynasty stored in St. Petersburg

Song from the Imperial Academy was ordered to write a preface for it; then the book was printed and issued to the whole country.

There are many kinds of needles used for acupuncture and moxibustion. In ancient times, nine needles were used, namely, trowel needle, round needle, sharp needle, ensiform needle, stiletto needle, round-sharp needle, filiform needle, long needle, and large needle that are recorded in *Ling Shu: Jiuzhen Shier Yuan* and *Jiuzhen Lun*. The nine needles are different in shape and application (Fig. 14.18).

Trowel needle is mainly used for shallow needling, so as to eliminate evil, without any damage of healthy qi; round needle is mainly used to massage acupoints, mainly to rush down the evil qi among muscles and tendons, without any hurt of muscles; sharp needle is used to press to eliminate evil; ensiform needle is used for bloodletting to expel toxin and treat ulcer; stiletto needle is used for apocenosis of ulcer; round-sharp needle is used in quick piercing to treat pain or numbness, arthromyodynia, and violent qi; filiform needle is mainly used to treat ulcer caused by the evil in meridians; long needle is used for deep piercing into plump muscles to treat deep evil qi and day-long arthromyodynia; large needle is mainly used to treat diarrhea, the stagnancy of water and qi, and the diseases in joints.

Modern needles, even filiform needle, are 16.5, 3.3, 4.95, 6.6, 8.25, 9.9 cm, or 13.2 cm long and 26#, 28#, or 30# in diameter. They are selected to use according to the depth of the needled part and the condition of disease.

Massage, also known as tuina, or "anqiao" and "anwu" in ancient times, is also one of the ancient medical treatments. There are written records of "massage" in the inscriptions on bones or tortoise shells unearthed in the ruins of the Shang Dynasty. *Shi Ji* (*Historical Records*) says: "In ancient times, there was a doctor named Yu Fu.

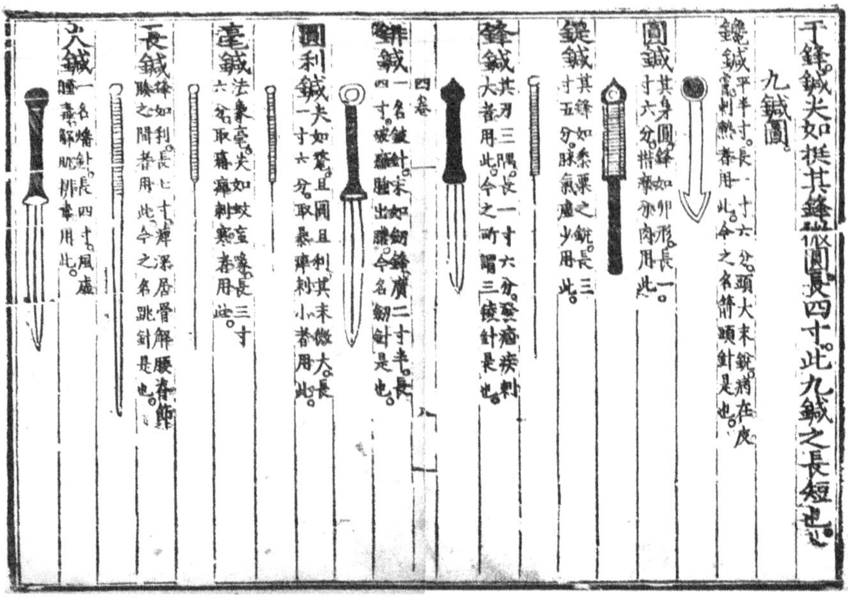

Fig. 14.18 The diagram of nine needles included in *Zhenjiu Dacheng* (*Compendium of Acupuncture and Moxibustion*) by Yang Jizhou

When he treated diseases, he did not use decoction or wine at all, but troweled stones to do qiaoyin, or do anwu…" According to researches, the term "qiaoyin" here refers to self-massage, and "anwu" to massage for others [7].

Such works as *Lao-tzu*, *Xunzi*, *Mo-tse*, and *Zhuangzi* also mention the methods of exercise and self-massage. *Zhou Lishu* records the medical case of Bian Que's cure of the prince's dead syncope of the State of Guo, and refers to the fact that "his disciple Zi You did massage for the prince of the State of Guo," which shows the role of massage in clinical application.

Yellow Emperor's Canon of Medicine talks about massage many times. According to *Plain Conversation: Xueqi Xingzhi Pian* (*Discusssion on Blood, Qi, Body and Emotions*), "if main and collateral channels as well as qi and blood are blocked, they can be treated by massage or with mellow wine medicine." It points out that the treatment of massage can dredge main and collateral channels as well as qi and blood so as to achieve therapeutic effect. *Yellow Emperor's Canon of Medicine* has the records of massage tools. The round needle recorded in *Jiuzhen* (*Nine Needles*) is used not only for acupuncture and moxibustion, but also for massage, often used together. During Qin and Han Dynasties, the treatment of massage had become one of the main medical methods.

During the Three Kingdoms Period, massage and guidance began to be used in conjunction with external drugs, resulting in the methods of ointment rubbing and fire moxibustion. When massage is carried out on the surface of human body, ointment made of traditional Chinese medicine can be applied to prevent damage to the skin of patient, and to make the medicine and the massage complement each other. As a famous ancient doctor of China, Hua Tuo once said, "When typhoid fever begins to attack, it only attacks the skin, which can be healed by ointment rubbing or fire moxibustion." In the period of Wei, Jin, Sui, and Tang Dynasties, the department of massage had been set up. *Sui Shu: Wuguan Zhi* has a record of two erudites for massage. At the end of each volume of *Treatise on the Causes and Manifestations of Various Diseases*, the methods of daoyin and massage are always offered. *Jiutang Shu: Zhiguan Zhi* has a record of 1 erudite for massage, 4 healthcare massagists, 16 massage technicians, and 15 massage apprentices. With the assistance of healthcare massagists and massage technicians, the erudite for massage instructed massage apprentices to learn the methods of massage and daoyin. Massage had become a kind of official education at that time. *Valuable Prescriptions for Emergencies* points out that "although the children are not ill, they are usually given ointment rubbing on their fontanels and archs of the feet to dispel wind-cold, when they get up in the morning." *Tang Liudian* also says, "Massage can treat eight diseases caused by wind, cold, heat, dampness, hunger, satiation, exhaustion and ease respectively." During this period, a systematic massage therapy had been basically formed. During the Sui and Tang Dynasties, the ways of ointment rubbing got enriched, such as sikimi ointment rubbing, root of red-rooted salvia ointment rubbing, aconite ointment rubbing, toxicodendron ointment rubbing, chenyuan ointment rubbing, cocculus trilobus ointment rubbing, etc. which should be selected and applied according to different conditions. As a famous medical scientist, Sun Simiao praised massage and daoyin very much. In *Valuable Prescriptions for Emergencies*, he mentioned: "Do massage three times a day, all diseases will be treated after one month, walking like running horse. This is the method of nourishing oneself."

Bao Puzi Neipian by Ge Hong of the Jin Dynasty mentions *Anmo Daoyin Jing Shijuan*, which has been regrettably lost. *Yangxing Yanming Lu* by Tao Hongjing, however, quotes parts of the scripture about daoyin: "Calm down and rub the palms to heat up, and then to warm the eyes three times. Then press the four angles of eyes to make the eyes clear.... One more method: rub the hands to heat up, and then rub the face up and down to eliminate evil qi, so as to make the face glossy. One more method: rub the hands to heat up, and then rub the body up and down, which is called dry bath, and can cure all diseases." [8]. The abovementioned contents about daoyin had been praised and cited by many books of its kind.

During the Song, Jin, and Yuan Dynasties, massage therapy was further developed. In the Ming Dynasty, Tai Yi Yuan (the Imperial Academy of Medicine) listed massage as one of the 13 departments of medical administration. In the Ming and Qing Dynasties, pediatric massage developed rapidly, and many monographs on pediatric massage were published one after another, such as *Classic of Pediatric Massage*, *The Secret of Pediatric Massage*, *Pediatric Massage of Broad Sense*, *Simple Record of Pediatric Massage*, *Method of Baochi Massage*, etc. It is worth mentioning that *Golden Mirror of the Medical Tradition* of the Qing Dynasty carried out a systematic summary of massage to the treatment of fractures and wounds, and listed "touching, setting, carrying, lifting, pressing, rubbing, pushing and holding" as the eight methods of treating fractures and wounds. To this day, massage still plays a therapeutic and healthcare role.

14.3.2.3 Surgical Treatment

It is generally believed that TCM is expert in internal medicine, while Western medicine in surgery. In fact, the surgery of ancient Chinese medicine was also developed. *Zhou Li: Tianguan Pia*n divided the doctors of that time into four kinds: physician, royal surgeon, dietician, and veterinary surgeon. Among them, royal surgeon just referred to general surgeon, who mainly treated the painful swelling on the body surface, ulcers, incised wounds, and fractures.

As early as the fifth century B.C., there had been records of the application of anesthetics in China. According to *Liezi: Tang Wen*, Gong Hu from the State of Lu and Qi Ying from the State of Zhao both suffered from disease and asked Bian Que to treat them. After the cure, Bian Que continued to carry out an operation to exchange their hearts. The story appeared to be absurd, but it is worth noting that "Bian Que made them drink poisonous liquor, so they sank into a deep coma for three days. Then Bian Que cut each of their chests, and took out and exchanged their hearts… After they woke up, they appeared as healthy as ever." It shows that at that time, people had known to use the liquor made of "poisons" as an anesthetic to conduct major surgeries.

Prescriptions for Fifty-two Diseases unearthed in Mawangdui recorded many surgical diseases such as infection, planed wound, pernio, insect bites, anal fistula, tumors, skin diseases, and so on. In the section of "Anal fistula," it recorded cutting therapy. For example, "Kill a dog and take its urinary bladder. Then use a bamboo tube to poke the urinary bladder into the rectum of the patient and blow, so as to draw forth the anal fistula. Remove its root with a knife, and apply cooked scutellaria baicalensis to it many times." There are also records of the application of elephantopus scaber and other drugs to anal fistula, and the use of smooth "collar" as a probe to examine and treat anal fistula. Besides, there are also records of the use of liquor as anesthetic to relieve pain: "To relieve the pain of incised wounds… A glass of liquor is poured to the medicine, and the patient is asked to drink it. If the patient has no tolerance for liquor, he or she is allowed to drink half. After drinking, he or she will feel painless for some time. If the pain comes again, the patient has to continue to drink the medicinal liquor." This indicates that the surgical technique at that time had reached a certain level.

According to the *Records of Three Kingdoms*, Hua Tuo carried out a laparotomy with powder for anesthesia as general anesthesia: "If the disease suffered by a patient is internally bound, beyond the reach of acupuncture and Chinese medicine, the patient has to be cut apart. He or she will thus drink the powder for anesthesia. When he or she seemingly turns to be drunk to death for a moment, the operation can be done. If the disease is in intestines, the part of intestine with disease will be cut off, then washed and sewn up, and then done ointment rubbing of the abdomen. Four or five days later, if the patient feels no pain, and does not always stay awake, he or she will recover in a month." [9]. Here, the indication of surgery, anesthetic technique, suturing technique, and after healing involved were all accurately documented, which should have certain factual basis. It is also possible for Hua Tuo to improve and produce the powder for anesthesia on the basis of his predecessors' medicinal wine. Although the ingredients of the powder for anesthesia have not been passed down, *Bian Que Xinshu* by Dou Cai records the anesthetic of "shui sheng san." In 1337, when Wei Yilin carried out a bone surgery, he emphasized that "if a patient is suffering from an injury or a pain on flesh and blood, he can't be operated immediately, but takes anesthetic first and stays until feels no pain" [10]. *Compendium of Materia Medica* also includes anesthetic prescriptions, which was clearly inspired by powder for anesthesia. By using the anesthetic mainly made of flos daturae, Japanese surgical specialist Hanaoka Seishu has performed many operations of breast cancer, gangrene of digit, cheiloschisis repairing, and so on, believed to be the first in the history of anesthetic surgery in the world. His ideas on anesthetic surgery are derived from China, so his anesthetic prescriptions are similar to those included in *Bian Que Xinshu* and *Shiyi Dexiao Fang*. In the 1930s, American scholar Lawall pointed out in his discussion of the history of anesthesia: "Some Arabian authorities referred to inhalation anasthesia, which could develop from China. Because Hua Tuo, Hippocrates of China, is said to have used a mixture made of aconite, flos daturae and other herbs to operate for this purpose." It shows that the anesthetic technique of ancient China occupies a certain position in the history of world surgery.

Treatise on the Causes and Manifestations of Various Diseases written by Chao Yuanfang of the Sui Dynasty covers much surgical information. For example, the section of "Jinchuang Changduan Hou" has a record of intestinal suture: "In the case of the wound of intestines, the chance of being saved is half according to the depth of the wound. If there appears only one end of the intestines, it cannot be linked up… If there appear two ends of the intestines, they can be sutured as soon as possible. First suture the two ends of the intestines by sewing with needle and thread, and then coat the linking with chicken blood to prevent it from leaking." It also puts forward specific requirements for nursing: "Eat only rice porridge at first; more than 20 days later, eat some broom corn millet; 100 days later, eat cooked rice. Satiation will make intestines painful and even rupture or leak." [11]. These nursing principles and requirements are still the key points for surgeons to carry out such kinds of operation. The first operation of the same kind recorded in Europe were completed by Italians in the twelfth and thirteenth centuries, about 500 years later than that recorded by Chao Yuanfang [12]. The section of "Jinchuang Cheng Yongzhong" in *Treatise on the Causes and Manifestations of Various Diseases* creatively describes the scientific method and theoretical principle of suture: "All sutures of the sores and wounds should be done vertically and horizontally. Any suture should also follow the laws made based on yin and yang, get smooth the adverseness up and down, operate gently, make yang attached to yin and yin to yang, and take care of all the grain of skin and the texture of the subcutaneous flesh as well as veins, causing everything return to normal." [11]. To this day, this method and theoretical principle of suture must be followed in traumatic suture. In addition, *Treatise on the Causes and Manifestations of Various Diseases* created the method of ligating blood vessels and established the principle of removing foreign bodies in trauma [12], which is also leading the history of medicine in the world.

Valuable Prescriptions for Emergencies by Sun Simiao of the Tang Dynasty records the use of scallion for urethral catheterization: "When urine is not in but blocked up by urinary bladder so as to block up body fluid, use one scallion and cut off its sharp tip, then insert it deeply into the penile orifice for three *cun* and blow it. And the urinary bladder will swell up, and body fluid begins to run, thus recovered" [13]. This method is not ideal enough, but can be regarded as the pioneer of urethral catheterization [14]. The book also records the suture of laceration of scrotum or prolapse of testicular, emostasis by burning, removal of foreign bodies in throat, enema by retention, enema by pressure, and so on, all reaching the highest level in history [15].

Since the Song and Yuan Dynasties, the science of surgery had experienced certain development, and more monographs on it had been published. Among them, *Weiji Baoshu* focuses on ulcer and records many medical devices such as moxibustion plate, refined knife, bamboo knife, small hook, etc. *Sheng Ji Zonglu* (*General Records of Holy Universal Relief*) records something of nasal feeding: "When suffering from acute wind stroke, the patient's jaw is tightly locked... If the teeth cannot be opened, the medicine can be infused into from the nose." The method of infusion is: "If the patient is suffering from lockjaw, use a scallion to infuse the medicine into from the nose, and the mouth will open immediately." Zhang Zihe of the Jin Dynasty made certain improvement to the method, which turned to be much closer to modern level. A stroke patient whose teeth were tightly closed was unable to be given medicine, so a doctor "took out a long clam that was polished off the blade, wrapped its tip in paper, and inserted it into the right nasal orifice of the patient. Then the patient could produce a sound of swallowing... revived immediately." *Taiping Guangji* records something of craniotomy: "The patient was put into a dark room and made drink several *sheng* of libanotus liquor, he then lost consciousness. His head was then cut apart with a sharp knife, and a worm-like lump which was two *cun* long was picked out. Then the wound was coated with plaster; meanwhile, the patient took some medicine and was asked to be moderate in eating. More than ten days later, the wound was completely healed; one month later, the patient's eyebrows grew up again, and the wound grew as smooth as before." Here the "worm-like lump" picked out possibly refers to a tumor that grew in brain.

Shiyi Dexiao Fang by Wei Yilin of the Yuan Dynasty described in detail the suturing method as well as suturing materials of abdominal dehiscence, pointing out that the root bark of white mulberry or hemp was often used as suturing thread. This suggests that the surgical repair of injuries was not uncommon at that time. In addition, some progress had been made in orthopedic reduction, and suspension reduction had been created.

More than 50 surgical works had been written in the Ming Dynasty, but surgical ideas then tended to be conservative. For example, Wang Ji argued that "the surgery has to base on the interior, and only by knowing the interior to treat the exterior, can the surgery be carried out easily," "the treatment of the exterior but disregarding the interior is the so-called bartering the trunk for the branches." [16]. Under this trend of thought, the tendency of internal treatment of surgical diseases at that time was becoming clearer well, while large surgical cases were rarely recorded. Even so, there was still a lot of progress made in those small technical fields.

Standards for the Diagnosis and Treatment by Wang Kentang described in detail the repair of cleft lip as well as the suture of trachea and esophagus fracture. In particular, it described anal atresia (also known as anal endomixis) in a more comprehensive way: "If anal intransigence happens, use an object to dredge it. Gold hairpin is the preferred, jade hairpin next. The object should be inserted deeply into the anal for two *cun*, and then a suhexiang pill is inserted into the hole until excrement comes out quickly."

Waike Zhengzong (*Orthodox Manual of External Diseases*) by Chen Shigong systematically discusses more than 150 surgical diseases and reaches a high level in the understanding of appendicitis, cancer, suppurative infection, bone and joint tuberculosis, and other diseases. It also describes a variety of surgical treatments and designs certain instruments and surgical methods for the removal of nasal polyps: "The secret method of removing rhinopolypus: first blow herba lysimachiae powder into the nasal cavity of the patient twice, and then take two fine copper chopsticks whose tips are drilled a small hole on and then threaded; the two chopsticks are set about five *fen* apart, and the tips of them are inserted straight into the root of the rhinopolypus; then tighten up the two threads and pull them down, and the rhinopolypus is naturally pulled out." [17]. This process is consistent with the principle of modern surgery. Chen Shigong advocated that when removing a mature abscess, its lower part is suitable for an incision which has to be big enough; if the rotten flesh won't fall off, cut it off; if the granulation has grown too long, cut it off. These effective methods have been used so far. In addition, the suture of trachea and esophagus fracture, the abdominal puncture to discharge pus, the amputation of phalangeal joints, etc. introduced in the book are all of practical value.

Yangyi Daquan by Gu Shicheng of the Qing Dynasty makes a more accurate record of anal atresia surgery, advocating a cutting with knife. It also discusses the surgery of vaginal atresia: "If a girl's vaginal orifice does not function, use a lead stick and push it into day by day. It will come to function for some time." This vaginal dilatation is also instructive for the treatment of other stenosing diseases [18].

It is particularly worth mentioning that the orthopedics and traumatology of TCM still plays an important role nowadays.

As early as the Jin Dynasty, *Zhouhou Jiuzu Fang* by Ge Hong had recorded the use of clamping plates (bamboo slips) to fix fractures, pointing out that after fixed, the affected limbs were not allowed to move around so as to avoid their redisplacement, and at the same time the clamping should be kept moderate. It also records the restitution of dislocation of temporomandibular joint: "A person holds the parotid sides of the patient with hands and slowly pushes…" [19]. This is the earliest restitution of dislocation of temporomandibular joint in the world, still widely used until now [20]. *Treatise on the Causes and Manifestations of Various Diseases* by Chao Yuanfang of the Sui Dynasty makes an in-depth discussion on the origin and syndrome of fracture trauma and its complications, puts forward a lot of reasonable treatments for fracture, and also points out that soft tissue fracture and open injury of joints must be sutured immediately, and broken tendons can also be sutured and fixed with thread, which is the earliest record of internal fixation for fractures. The orthopedic technology of the Tang Dynasty also reached a very high level. *Xianshou Lishang Xuduan Fang* by Lin Daoren is the earliest existing orthopedic monograph written in China. The book discussed in detail the traction, reduction, fixation, functional exercise and drug treatment of fracture and dislocation, and pointed out that before reduction, the wound should be touched with hands so as to identify the displacement of the fracture, and the fracture should be fixed with clamping plates if traction technique or manipulative correction was adopted. It put forward the treatment principle of the equal importance to both muscles and bones and the association of activity and inertia. For open fracture, the book suggested that the wound should be cleaned with boiled water, and expanded with quick knife; after the reposition of broken bones, they should be "wrapped with clean silk pieces," "never exposed in wind or touched by water." It also divided the dislocation of the hip joint into two types for the first time: anterior dislocation and posterior dislocation, and proposed that the posterior dislocation of the hip joint could be treated by the pulling with hand and kicking off with foot, and the dislocation of shoulder could be restituted by the "resetting of chair back." The book also introduced the experiences of internal and external pharmacy in fracture and injury, among which the three successful experiences in the treatment of complex fractures are still of guiding significance in today's clinical practice [21].

In the Song and Yuan Dynasties, great progress had been made in the restitution of bone injuries. Zhang Gao introduced in his medical work *Yi Shuo* the exercises of stamping with feet on a revolving shaft and rolling and relaxing tendons with a bamboo tube in order to promote the functional recovery of knee joints and ankle joints after fracture and injury, as well as the open reduction used to treat multiple tibial fractures. The Mongolians of the Yuan Dynasty were good at orthopedic treatment as well as horse riding and arrow shooting. They set up a department of bone orthopedics in the Imperial Academy of Medicine. *Shiyi Dexiao Fang* by Wei Yilin systematically sorted out the achievements of bone orthopedics made before the Yuan Dynasty, and made a lot of innovations and developments, making more perfect the treatment principle and method of fracture and joint dislocation. The treatment of spinal fracture by suspension reduction covered in the book is the first in the world.

F*ormulas for Universal Relief: Zheshang Men* written in the Ming Dynasty emphasized the importance of manual restitution, and introduced the "traction technique and clamping and jostling" used to restitute the double fractures of forearm. The method of pressing fracture and fixation with clamping plates to cure forearm fracture adopted in Chinese medicine is still in use [22]. *Jinchuang Michuan Jinfang* recorded the use of "bony crepitus" to examine fracture, and proposed to cut off the contaminated end of fracture that has pierced through the skin in order to prevent any infection in the treatment of open fracture. Besides, the book introduced the treatments of all kinds of fracture.

In the Qing Dynasty, the achievements of bone orthopedics made in previous generations were systematically summarized. *Golden Mirror of the Medical Tradition: the Essentials of Mental Cultivation Methods of Bonesetting* summed up the eight methods of bonesetting as "touching, setting, carrying, lifting, pressing, rubbing, pushing and holding," and introduced the use of the manipulation to treat such fractures as pain in waist and lower extremities, as well as the application of climbing rope, piling bricks, and bolstering behind waist and back to restitute and fix the fracture or dislocation of thoracic vertebra and lumbar vertebra. In terms of fixation, the book advocated that "the syndrome in the upper and lower bodies or on the front side should be set with produced devices, as auxiliary manipulation is not enough at all" [23] (Fig. 14.19), and created and reformed a variety of fixation devices. For example, the injury on the middle segment of the spine should be fixed with wooden supporter, the injury on the lower back with lumbar splint, the fracture of the long mainstay of the four limbs with bamboo curtain splint or cedar tube, and the fracture of patella with tool for knee fixation. *Shangke Buyao* by Qian Xiuchang also has a record of poplar splint, which was an attempt to treat bone defect by implanting artificial prosthesis into the body instead of bones.

At present, the orthopedics and traumatology of traditional Chinese medicine adopts the combinations of TCM with Western medicine, operation with manipulation, internal drugs with external drugs, traditional Chinese medicine with Western medicine, internal fixation with external fixation, small clamping plates with oulopholite, making itself achieve great progress [24].

14.3.2.4 Health-Preserving Treatment

Health-preserving was originally adopted by Taoists to live an easy life in old age and prolong life by performing various activities. Our ancestors combined the rational Taoist ideas and methods with the daily life of the people to invigorate health and prevent diseases by carrying out such methods and measures as eating and living, emotional conditioning, sports, and so on. That is, the idea of "curing no disease, but making preventive treatment of diseases" was advocated, and accordingly a unique health-preserving treatment of TCM was formed.

(1) Daily Living in the Way of Nature

TCM believes that people's daily life such as living and eating should conform to the change of the law of nature. That is, "only following the law of yin and yang, eating and drinking abstemiously, living regularly, and never overworking, one can keep the body and the spirit healthy, and die old" (*Plain Conversation: Shanggu Tianzhen Lun* (*Ancient Ideas on How to Preserve Natural Healthy Energy*)). If you violate the law of nature and live and eat incontinently, you will be ill and suffer from premature senility.

Plain Conversation: Siqi Tiaoshen Dalun (*Major Discussion of Regulation of Spirit according to the Changes of the Four Seasons*) gives some advice on the living

Fig. 14.19 Diagram of fracture reduction included in *Golden Mirror of the Medical Tradition*

behaviors throughout the year: "In the three months of spring, everything wakes up. One has to get up earlier and take a long walk in the courtyard, so as to promote sentiment. Don't kill living things, be willing to give but not to gain, and try to reward but not to punish, which conforms to the qi of spring, the way of health-preserving"; "in the three months of summer, the qi of heaven and earth interacts with each other, and all things begin to bear fruits. One has to get up earlier and live actively, so as to get rid of anger and let off evil qi, which conforms to the qi of summer, the way of dying old"; "in the three months of autumn, the qi of heaven turns to be urgent and the qi of earth to be clear. One has to lie down earlier and get up earlier just as chicken, so as to calm one's mind and reduce the bleakness of autumn. Gather the spirit and qi, calm down the qi of autumn, concentrate one's sentiment, and make the qi of lung clear, which conforms to the qi of autumn, the way of health-harvesting"; "in the three months of winter, it is the time for the storage of essence. The water is frozen and the earth cracks, without disturbance to yang. One has to lie down earlier and get up late, waiting for the sunshine, so as to make one's sentiment go down or hide. If someone has private intention or gains, he or she has to leave the cold and approach the warm, so as not to hurt the skin and let off qi, which conforms to the qi of winter, the way of essence-preserving." If so, we can achieve the effect of "curing no disease, but making preventive treatment of diseases."

(2) Adjust the Spirit and Sentiment

TCM believes that people's spirit and sentiment have a great impact on health, and emotional fluctuation has an important impact on the five internal organs. That is, "anger hurts the liver, ecstasy hurts the heart, much thinking hurts the spleen, worry hurts the lung, and fear hurts the kidney." Therefore, TCM especially emphasizes keeping oneself peaceful and relaxed, without desire or demand, living a simple life, not envious of others, which are important factors for longevity and health. According to *Plain Conversation: Shanggu Tianzhen Lunpian* (*Ancient Ideas on How to Preserve Natural Healthy Energy*), "If the spirit is kept internally, there will be no sickness. Thus, a person should live quietly with less desire, peacefully without fear, working in a moderate way, keeping qi smooth, making everything agreeable to their wishes. Therefore, one has to make food delicious, to dress naturally, to enjoy the common, and to take things as they come, which are said to be with a simple life. Accordingly, the indulgence without hurting the eyes, the obscenity without being captivated, and the free mentality without fearing of things, all conform to the law of nature. So, those who are able to be 100 years old and not failing in health enjoy complete virtues."

With the change of seasons, people's thoughts and mood will inevitably have fluctuations. TCM argues that in the spring, we should make our sentiment entirely free from worry with the rising qi; in the summer, we should keep us free from any anger; in the autumn, we should stay calm and patient; in the winter, we should make our sentiment go down or hide, so as to feel steady and sure. In this way, the vital qi in our body will be deeply hidden and favorable, and the spirit will remain firm and concentrated.

(3) Diet Treatment

Ancient Chinese doctors recognized that dietary balance helps to resist the attack of diseases, which shows the importance of daily diet to health. *Plain Conversation: Zangqi Fashi Lun* (*Discussion on the Association of the Zang-qi with the Four Seasons*) points out, "Poisons can attack evil, the five cereals can nourish, the five fruits can provide auxiliary nutrients, the five kinds of livestock are beneficial to our body, and the five vegetables are useful supplements. When their qi and tastes are compatible, eat them to replenish essence and invigorate qi." Here, the "five cereals" refer to such cereals or pulse crops as corn, sorghum, bean, wheat, rice, and so on, which are the principal food for raising human body; the "five fruits" refer to such fruits or nuts as jujube, plum, apricot, chestnut, peach, and so on; the "five kinds of livestock" refer to the meat of such livestock as cattle, sheep, pig, chicken, dog, and others; and the "five vegetables" refer to all vegetables in general. This means that when human body is attacked by disease, drugs are needed to treat, and the daily diet requires a balance of various nutrients. Once a person gets ill, he or she must be treated with diets as well as drugs.

As we all know, cereals are rich in carbohydrates that provide the necessary calories for human body; fruits and vegetables are rich in vitamins, cellulose, sugars, and organic acids; and animal foods are mostly high in protein, fat and calories, and contain essential amino acids. These are all the important nutrients that keep a normal physiological metabolism and enhance the immunity of human body. The recognition of the Chinese ancients is obviously in line with the viewpoint of modern nutrition.

In China, there is a popular saying, "all drugs have thirty percent toxicity," reminding us to use drugs cautiously. Ancient doctors also advocated that diet therapy should precede drug treatment. Sun Simiao, a famous doctor of the Tang Dynasty, said in his *Valuable Prescriptions for Emergencies*: "As for treatment, diet therapy should be provided first; if diet therapy fails, drugs can be used to treat." All medical formularies of past dynasties include a considerable number of prescriptions for diet treatment. By the Yuan Dynasty, there came out a monograph on diet therapy *Yin Shan Zhengyao* (*Principles of Correct Diet*), which was edited by Hu Sihui, an imperial doctor, and considered to be the first monograph on nutrition in the world. The book inherited those health-preserving ideas covered in ancient medical works such as *Yellow Emperor's Canon of Medicine*. It introduces the efficacy of various kinds of food and collects a large number of special prescriptions for diet therapy, as well as special discussions on the diets of such special groups as pregnant women, children, and the elderly. Many of the experiences and methods about diet therapy included in it came from ethnic minorities such as the Mongolian nationality and the Central Asian region, enjoying certain social influence in the history of nutriology of the world. In particular, the book points out, "Although the five tastes are satisfying us, our eating and drinking should not be excessive. Excessive eating and drinking will cause disease, while a moderate one is beneficial. Anyone who keeps moderate in eating delicious food is the wisest." *Weisheng Baojian* also says, "What is important to us is not to eat excessively, but to eat moderately."

TCM believes that food also has “four qi” and “five tastes”: the former refers to cold, heat, warm, and cool, while the latter to sour, sweet, bitter, pungency, and saltiness. The former was divided according to the response of the food after being eaten, and the latter was divided mainly according to the original taste of the food. “Each of the five tastes has its own benefit, to dispel, to gather, to be slow, to be urgent, to be strong, or to be soft; the four seasons conform to the five internal organs, and diseases are appropriate for the five tastes.” The basis of the diet therapy of TCM is to eat certain kinds of food according to the nature of the food as well as local conditions.

According to the theory of TCM, medicinal diet therapy should follow the law of nature and adapt to the rule of climate change. Corresponding food should thus be eaten in different seasons. For example, according to *Principles of Correct Diet*, “the qi in spring is warm, so it’s suitable to eat wheat, so as to cool qi in human body; no eating warm diet, and no wearing heavy clothes”; “the qi in summer is hot, so it’s suitable to eat beans, so as to cool qi in human body; no eating warm diet, no eating excessively, no living in wet place, and no wearing wet clothes”; “the qi in autumn is dry, so it’s suitable to eat spicy food, so as to moisten the dryness; no eating cold diet, and no wearing thin clothes”; “the qi in winter is chilly, so it’s suitable to eat millets, so as to treat the chilly with heat; no eating hot diet, no wearing heavy clothes.” Many recipes for the four seasons were summarized according to local conditions, such as mung bean soup in summer to clear away heat, mutton in winter to defend against the cold, and so on.

For those who have different physiques and those who suffer from different diseases, TCM attaches great importance to their diets. For example, *Plain Conversation: Xuanming Wuqi* (*Discusssion on the Elucidation of Five-qi*) says, “Pungency links to qi, so fewer diseases of qi, more pungent food; saltiness links to blood, so fewer blood diseases, more salty food; bitter links to bones, so fewer bone diseases, more bitter food; sweet links to fleshes, so fewer flesh diseases, more sweet food; sour links to tendons, so fewer tendon diseases, more sour food, which are said to be the five dietary taboos.” *Principles of Correct Diet* also discussed the dietary taboos of all kinds of diseases: “Liver disease abstains from pungent food, suitable to eat rice, beef, okra and so on. Heart disease abstains from salty food, suitable to eat red bean, dog meat, plum, leek and so on. Spleen disease abstains from sour food, suitable to eat soybean, pork, chestnut, leaves of pulse plants and so on. Lung disease abstains from bitter food, suitable to eat wheat, mutton, apricot and so on. Kidney disease abstains from sweet food, suitable to eat millets, chicken, peach, onion and so on.” If “eating more sour food, liver qi will be damp and spleen qi exhausted; as a result, flesh will turn to be hard and lips to cleave.” Although these taboos are not necessarily correct, the ideas they reflect are still valuable.

In a word, it is undoubtedly conducive to physical health or to the recovery of diseases to intentionally apply the features of food to the treatment or auxiliary treatment of certain diseases, and they are basically free of toxic side effect. As Zhang Xichun, a modern medical practitioner, pointed out in the book *Yixue Zhongzhong Can Xi Lu*, food “eaten by a patient can not only cure his disease, but also satisfy his hunger; not only satisfy his hunger, but also cater to his taste. If the food is symptomatic, the disease will come to be healed; if it is not symptomatic, it is harmless." "Treating disease or strengthening the body" is the most significant feature of dietary therapy.

(4) Daoyin and Qigong

Daoyin (guidance) is a kind of ancient Chinese health-preserving exercise that combines respiratory movement (dao) with limb movement (yin). As one of the dynamic exercises in qigong, it is just similar to modern hygienic gymnastics.

As early as the Spring and Autumn and Warring States period, *Zhuang Zi: Keyi* recorded a kind of daoyin as "blowing and howling-type breathing, exhaling the old and inhaling the new, and walking as a bear and expressing as a bird" performed to live a long life. One of the silk paintings unearthed in the tombs of the Han Dynasty at Mawangdui of Changsha (Fig. 14.20) was painted with a diagram of daoyin of 44 figures. Beside the diagram, the names of the moves were marked, some of which can be recognized. Among them, there are also a large number of daoyin moves that simulate animals' postures, involving snipe, crane, ape, monkey, Chinese dragon, bear, and so on. Some of the basic moves involved in today's gymnastics can also be seen in *Daoyin Tu* on the whole, which can be said to be a drawing of ancient gymnastics. It shows that before the Western Han Dynasty, daoyin had become a physical exercise among the people. During the Three Kingdoms period, Hua Tuo summarized the exercises of daoyin into five ways, and called them "five-animal exercises," namely, tiger-like exercise, deer-like exercise, bear-like exercise, ape-like exercise, and bird-like exercise, summarizing the characteristics of daoyin therapy in a more comprehensive way. The five-animal exercises created by Hua Tuo, however, have been lost. *Yangxing Yanmin Lu* by Tao Hongjing of the Southern Dynasty recorded the set of exercises, a set of daoyin programs made by imitating the normal exercises of such five birds or animals as tiger, bear, deer, ape, and bird. *Taishang Laojun Yangsheng Jue* included in *Zhengtong Daocang* also recorded the

Fig. 14.20 The restored diagram of *Daoyin Tu* unearthed at Mawangdui

"five-animal exercises," with the signature of "Hua Tuo awarding Guangling Wupu." Zhou Lvjing of the Ming Dynasty improved it by reducing the difficulty of all moves, and combined it with the circulation of qi. He also plotted out a schematized map of it as well as the written descriptions. People of the Qing Dynasty added the exercise of "osprey-like looking around" and the exercise of "lion-like dancing" to it, hence "seven-animal exercises."

In addition, there were many other daoyin exercises in ancient China, such as "Chisongzi Daoyin Fa (the daoyin of red pine nut)," "Ning Fengzi Daoyin Fa (the daoyin of Ning Fengzi)," "Hama Xingqi Fa (the promotion of qi of toad)" (combining qi with daoyin), "Pengzu Wo Yin Fa (the yin of Peng Zu lying down)," "Wang Ziqiao Daoyin Fa (the daoyin of Wang Ziqiao)," "Daolin Daoyin Yaozhi (the gist of daoyin of Daolin)," and so on. "Baduan Jin (eight-section brocade)" that appeared at the end of the Northern Song Dynasty had also been widely spread at that time. And some of them were still popular in modern times of China, greatly promoting the medical treatment and health care of future generations.

TCM is rich in treatments, such as scrapping (a popular treatment for sunstroke by scraping the patient's neck, chest, or back), cupping, sanfu (the three periods of the hot season) stickers, and some others which are still widely applied at present. The ethnic minorities such as Tibetan, Mongolian, Uyghur, Hui, and so on also have their own unique medical theory as well as diagnosis and treatment, which will not be introduced here.

TCM covers widely as a system of knowledge and experience gradually accumulated by the Chinese nation in the process of exploring the prevention and treatment of diseases. It not only contains the medical theory and diagnosis and treatment mentioned above, but also the exploration of the mystery of life. *Yellow Emperor's Canon of Medicine* systematically expounds the general law of the formation and development of life, the relationship between life activities and natural environment and social environment, and so on. Many theories of TCM came from life experience; its medicinal materials also come from life experience and even folk customs, thus enjoying a deep folk foundation. For thousands of years, the Chinese nation has followed the theories and methods of TCM to understand life, maintain health, and multiply.

TCM as a treasure of human science and culture was spread to South Korea, Japan, Vietnam, and other neighboring countries of China in history, and helped produce Korean medicine and Japanese Chinese-prescription medicine, greatly benefiting Asian people. The use of ephedrine, the treatment of malaria by artemisia annua, and the treatment of leukemia by arsenic trioxide in modern medicine all come from the theories of the clinical medication of TCM. The value of acupuncture in the treatment of diseases has been paid more and more attention by the international community, and it has become a common method to study pharmacology by using the effective components of Chinese herbal medicine. It is believed that TCM will make a greater contribution to the health of human beings.

References

1. Zhongnian, Sun. 1988. A summary of the development of abdominal diagnosis of TCM. *Shaanxi Traditional Chinese Medicine* 7: 331–332.
2. Jingwei, Li, and Sun Simiao. 1993. *Biographies of Ancient Chinese Scientists* (Part I), ed. Du Shiran, 331–342. Beijing: Science Press.
3. [The Song Dynasty] Anshi, Xiang. *Xiangshi Jiashuo*, vol. 5. Wuyingdian Juzhen Edition of Series Books of the Qing Dynasty.
4. Ailan, Zhu. 2000. A summary of the development of the forms of traditional Chinese medicine. *Journal of Anhui College of Traditional Chinese Medicine* 5: 48–50.
5. [The Song Dynasty] Tao, Li. *A Sequel to History as a Mirror*, vol. 105. *Si Ku Quan Shu* (*The Imperial Collection of Four*) by Pavilion of Literary Profundity.
6. [The Song Dynasty] Mi, Zhou. 1983. *Qidong Yeyu*, vol. 14, 251–252. Beijing: Zhonghua Book Company.
7. Binghua, Fan, and Li, Xu (eds.). 2012. *Health Science by Massage*, 2. Hangzhou: Zhejiang Science and Technology Press.
8. [The Qi and Liang Dynastics of the Southern Dynasties] Hongjing, Tao. 1990. *Yangxing Yanming Lu*. Classic Chinese Qigong: from the Pre-Qin Period to the Southern and Northern Dynasties, vol. 2, ed. Lv Guangrong, 241–242. Beijing: People's Sports Press.
9. [The Jin Dynasty] Shou, Chen. *The Book of Wei. Records of Three Kingdoms*, vol. 29. The edition of Baina Benjing in the reign of Emperor Shaoxi of the Song Dynasty.
10. Yilin, Wei. 1991. *Shiyi Dexiao Fang*, 904. Shanghai: Shanghai Science and Technology Press.
11. [The Sui Dynasty] Yuanfang, Chao. *Treatise on the Causes and Manifestations of Various Diseases*, vol. 36. *Si Ku Quan Shu* (*The Imperial Collection of Four*) by Pavilion of Literary Profundity.
12. Jingwei, Li. 1978. The contribution of *Treatise on the Causes and Manifestations of Various Diseases* to medical science and technology. *Journal of New Medicine* 8: 31–35.
13. [The Tang Dynasty] Simiao, Sun, and Pangguangfu Fang. *Valuable Prescriptions for Emergencies*, vol. 61. *Si Ku Quan Shu* (*The Imperial Collection of Four*) by Pavilion of Literary Profundity.
14. Jingwei, Li. 1981. Historical achievements in the first aid technology of traditional Chinese medicine. *Henan Traditional Chinese Medicine* 6: 17–20.
15. Jingwei, Li. 1993. *Essential Recipes for Emergent Use Worth a Thousand Gold*: The earliest clinical encyclopedia of China. In *The China 100 Series: 100 Books that Influenced China*, 375–380. Nanning: Guangxi People's Publishing House.
16. [The Qing Dynasty] Ji, Wang. Preface to *Waike Lili*. *Si Ku Quan Shu* (*The Imperial Collection of Four*) by Pavilion of Literary Profundity.
17. [The Ming Dynasty] Shigong, Chen. 'Zachuang Dumen'. *Waike Zhengzong*, vol. 4. Engraved edition in the reign of Wanli of the Ming Dynasty.
18. Jingwei, Li. 1963. Achievements in the surgery of ancient China. *Journal of the History of Science* 5: 1–11.
19. Zhijun, Shang. 1996. *Buji Zhouhou Fang (Revised edition)*, 308. Hefei: Anhui Science and Technology Press.
20. Bailing, Liu (ed.). 1998. *Science of Bone Fracture in Traditional Chinese Medicine*, 4. Beijing: People's Health Press.
21. Yizong, Wei. 1983. *History of Orthopaedic Technology in China*, 115–116. Shanghai: Shanghai Science and Technology Literature Publishing House.
22. Tianyu, Shang. 1979. The treatment of double fractures on radioulnar mainstay of forearm. *Mingtong Medicine* 12: 14–16.
23. [The Qing Dynasty] Qian, Wu. 2011. *Golden Mirror of the Medical Tradition*, Zhang Nianshun, Zhang Chi (Proofread), 960. Beijing: China Medical Science and Technology Press.
24. Yueguang, Fan (ed.). 2008. *Science of Bone Fracture in Traditional Chinese Medicine*, 7. Beijing: Higher Education Press.

Yahua Niu Born in Hohhot, Inner Mongolia in 1961, graduated from Nanjing Pharmaceutical College in 1983 with a Bachelor degree of Science; in 1989 she graduated from Inner Mongolia Normal University with a master's degree in science; in 2005, she graduated from Northwestern University with a doctorate in science and technology history. He once served as director of the editorial department of the Journal of Inner Mongolia Medical College. She is currently a researcher and doctoral supervisor of the Institute of Traditional Chinese Medicine Information, Chinese Academy of Traditional Chinese Medicine. From 1994 to 1995, she has been studied at Japan Iwate Medical University.

Research interests: Medical history; Chinese medical literature

Major works:

Papers: A Study on Two Earliest Translations of Western Anatomy; Study on Shen San Zi Fang

Biography of Peng Sixun;

Notes on *Two kinds of Chinese and Western Medical Books; Western Medicine introduced in the Ming Dynasty;*

Chief editor of *The Collection of Traditional Chinese Medicine Classics in Qifen Shi.*

Chapter 15
Porcelain

Yongshan Yang

Porcelain is a great invention in China.

Chinese researchers of ceramics and archaeologists have systematically examined and studied the ancient ceramics so far excavated for many years. It is confirmed that as early as the late Eastern Han Dynasty (before 220 A.D.), there was already mature porcelain in the region of the now Zhejiang Province, marking the successful invention of Chinese porcelain. The invention of porcelain not only means that wares with good quality could be produced and a new material with great potential was developed, but more importantly, the innovation of such technology in practice had far-reaching effects.

The invention of porcelain is not accidental, but the results of repeated practice, gradual deepening understanding, and technological transcendence and innovation on the basis of long-term accumulation of the operations in ceramic technology. Traced back, it can be seen that the development of porcelain and pottery has been following the same line. There is no doubt that without the invention of pottery, there would be no invention of porcelain.

Based on pottery making technology, the ancient ceramic craftsmen pursued reasonable use of raw materials and technologies after long-term exploration. The physicochemical properties of the ceramics were constantly improved, and the transition from pottery to porcelain has been realized. Thus, porcelain came into being.

Y. Yang (✉)
Tsinghua University, Beijing, China
e-mail: syrdytt@163.com

J. Hua and L. Feng (eds.), *Thirty Great Inventions of China*,
https://doi.org/10.1007/978-981-15-6525-0_15

15.1 The Invention of Ceramic Technology

Pottery was made from clay by firing in the early stage of human civilization, which changed the nature of clay and created the earliest artificial new material.

In their daily life, the ancestors of the Neolithic age realized that clay could be shaped into a container after being mixed with water. When dried and hardened, it had certain strength and could hold solid substances. Occasionally, after being fired, the adobe container was not only hardened, but also sintered together. Being soaked in water, the fired container could still be maintained, so the earliest pottery was produced.

The invention of pottery is a creative activity of human beings. Many kinds of potteries were fired for daily use. With dual cultural characteristics, it was the creation of both material products and spiritual civilization.

The pottery of the Neolithic Age was closely related to people's life and production. It was the most basic tool for living at that time, providing a container for storing seeds and food for farming. Hence, it played an essential role in social life and production.

The making of pottery is not only a creating activity, but also a modeling activity. The manufacture of the practical utensils and their early shapes lay a foundation for the later invention and development of porcelain.

China, a country that invented porcelain, is also one of the first countries to invent pottery. The physical materials unearthed prove that more than 10,000 years ago, the ancestors began to make pottery on the land of China. They thrived on the vast land, and through their perception and comprehension in life, they realized the truth that "Clay can be made into pottery through the interaction of water and fire," and thus made the earliest pottery.

"Pottery was created and developed independently in the centers of ancient civilizations around the world" [1]. Although the time of development differs and the pattern and the style are varied, the essence of its creative thought is the same. "Pottery is shared by all mankind, while porcelain is uniquely created in China" [2].

"The pieces of pottery or pottery fragments discovered in China before 7000 B.C. are mainly found in these sites: Zhuannian (Huairou District, Beijing), Nanzhuangtou (Xushui District, Hebei), Yujiagou (Yangyuan County, Hebei), Xianrendong and Diaotonghuan (Wannian County, Jiangxi), Yuchanyan (Dao County, Hunan), Qingtang and Niulandong (Yingde, Guangdong), Zengpiyan and Miaoyan (Guilin, Guangxi), Dayan (Lingui District, Guangxi), Dingshi Mountain (Yongning District, Guangxi). Among the pottery fragments unearthed from these ten sites, those from Miaoyan (Guilin) were measured to be the earliest, i.e., around 13,000 B.C." [3] (Figs. 15.1, 15.2 and 15.3).

In the history of human civilization, pottery technology is an important invention and creation of the early generations as well as the beginning of human science, technology, culture, and art. The actual needs of life make ceramic technology an independent craft, and the repeated operations in practice bring into being the initial process of ceramic manufacturing, inspiring the wisdom of creation. The invention

Fig. 15.1 Ancient pottery (Restored from the pieces unearthed from the Wannian Xianren Cave in Jiangxi Province)

Fig. 15.2 Ancient pottery (Restored from the pieces unearthed from the Zanpi rock in Guilin, Guangxi)

Fig. 15.3 Ancient pottery (Restored from the pieces unearthed from the Yuchan rock, Dao County, Hunan Province)

of pottery marks the beginning of the Neolithic Age, and has improved the way of life of human beings and brought profound changes to the ancient society. Pottery was the most widely used household appliances and production tools at that time. From its earliest appearance to the general application, it had never been abandoned in the long human history. In each phase of its development, there were always new creations to meet the people's practical and esthetical purposes.

Pottery making is not only the technical implementation of creation, but also the expression of formal beauty of shape, reflecting the plain design and rich imagination of our ancestors. It creates a new art, manifests the simple primitive aesthetic consciousness, and contains the basic idea of ceramic design and production (Fig. 15.4).

15.2 Early Pottery Manufacturing

The manufacture of pottery went through a transition from the initial low-level stage to the advanced stage. To make pottery, people should first choose proper materials. Based on the understanding of the property of clay and the experience of making pottery, people selected clays with few impurities, strong plasticity, and easy sintering as raw materials. By analyzing the physical data of a large number of unearthed potteries, it can be inferred that the choice of clay used at that time was gradually refined.

Starting from the invention of ceramic technology, all raw materials were locally selected–mostly sedimentary soil. The coarse sand particles and impurities were removed by panning, and then aged to obtain good plasticity so as to make it easy for the soil to be shaped. An appropriate amount of fine sand, carbon particles, and chaff were incorporated to prevent the cracking of the formed body. The material selection,

Fig. 15.4 Red pottery binaural tripod pot in the Peiligang culture period

panning, and aging process formed by the initial ceramic operation made the raw materials meet the conditions of pottery making and fit the molding operations, which is the experience of ceramics accumulated in the early Neolithic era.

Forms of pottery molding were initially small and simple, easy to be kneaded in the hand and mostly in semi-spherical, spherical, or cylindrical shapes. The molding was carried out by means of kneading or tapping to compact the clay body whose shape was fixed after being dried. The clay generally had good plasticity and was suitable for a variety of molding methods, from the initial kneading, panning, mud sheeting, and sticking, to slow wheel forming and fast wheel forming. The ceramic techniques were various, strongly adaptable, and expressive. In order to make the surface of the pottery dense and smooth, when the clay body was still in semi-dry state, clam shells or pebbles were used as a tool to hone its surface so as to make it flat and bright, which enhanced the sense of form, reducing water permeability while indicating the fun-making and the interest of handicraft.

In the process of pottery making, potters had their own understanding of the forms of wares and different aesthetic orientations, which led to the formation of various styles of pottery in different periods and in different regions. When they were making household wares initially, due to differences in raw materials, technology, natural conditions, and operation methods, they might fire variety of utensils, such

Fig. 15.5 Diamond shape pottery kettle of the Banpo type in Yangshao culture

as the natural ones. Because of this, their rich and diverse creations provided a broad platform for imagination and creativity, offering possibilities for further expansion and innovation from different perspectives (Fig. 15.5).

The primitive life of the ancestors was closely related to the fire. When they realized the impact of fire on clay, they began to make effective use of it. The earliest pottery was burned on the ground in the open air, with one layer of firewood as fuel being laid on the ground, the dried clay body placed over it, and then completely covered with the other layer of fuel. Later, the firewood was burned from the ground surface, and fuel could be added at any time. Although this method was easy to grasp, the firing temperature was not high enough and the heat was uneven, making it difficult for the pottery to have a unified color. A layer of mud as a shell, similar to the prototype of the kiln, could also be applied on the firewood, with a hole at the bottom and one on the top to ventilate and exhaust smoke. This method of firing was better at heat preservation and the temperature was increased (Fig. 15.6).

Since the middle of the Neolithic Period, horizontal cave kiln or vertical cave kiln had been used to fire pottery. These two types of caves, found in Banpo Site, one of Yangshao culture sites (5000 B.C.–3000 B.C.), were built in suitable terrains. The kilns there had relatively simple structure and small firing chamber that was roughly round with a diameter of about 1 m. Firewood as fuel was put at the bottom of the kiln and the flame from the fire path entered the kiln room. Without chimney, the temperature distribution was relatively even. Since heat loss in kilns was much smaller than that of kiln-free pottery firing, firing temperature in kilns could reach 1000 °C, slightly higher than that required to burn the red pottery from Peiligang Culture (7000 B.C.–5000 B.C.) and Cishan Culture (6500 B.C.–5000 B.C.). "From the perspective of thermal engineering, although the kiln structure is still primitive, due to small heat loss and concentrated fire entering the kiln when the fuel is burned, the temperature is easy to rise and the clay body is easy to be sintered, which is conducive to improving the quality of pottery. Therefore, the development from the kiln-free pottery burning to pottery burning in kiln is a major breakthrough in technology" [4].

Fig. 15.6 Open-air pottery burning (The Dai nationality in Xishuangbanna, Yunnan still retains the primitive method of pottery burning)

The improvement of the kiln technology, an indicator of the ability to utilize and control fire, is the key to improving the firing temperature and ensuring the quality of the product. Ancient Chinese ceramic kilns developed from the cave ones of the Neolithic Age to the ones with chimneys built in the late Western Zhou Dynasty or earlier periods, or various kilns with improved structures. The improvement of kilns raised the firing temperature from less than 1000 °C to more than 1300 °C, which created the necessary equipment conditions for China to advance from pottery making to the invention of porcelain.

15.3 The Firing of Various Potteries

After the invention of ceramic technology, the history of pottery firing began. With different raw materials and techniques, various types of potteries were fired in diverse regions. The typical potteries were red pottery, painted pottery, black pottery, white pottery, gray pottery, stamped hard pottery, and so on. The pottery types, components of raw material, molding methods, and firing processes were the most basic technical elements of the traditional ceramic technology.

The most primitive potteries were fired on the ground, with the temperature and atmosphere uncontrolled, so most of them were variegated potteries of gray red, gray brown, and other colors (Fig. 15.7). Even in the cave kiln invented later, the atmosphere could not be controlled. The clay that contained iron compounds turned red in the oxygenated atmosphere, so the pottery fired was red. There were two

Fig. 15.7 Variegated pottery tripod bowl in Dadiwan culture

kinds of red pottery: sand-filled red pottery mainly used for drinking and delicate pure-clayed red pottery mainly used for eating.

Painted pottery, a typical representative of the Yangshao culture, is mostly made by adding color to the pure-clayed red pottery. The selection and processing of raw materials were strict and the molding reached a high level. After being molded with the slow wheel, the body was trimmed and polished so as to have a round and smooth shape. The painted pottery was the outstanding creation of the ancestors of the Neolithic Age and the product of early science and technology, together with the culture and art. On the basis of molding, it opened the field of decorative arts on utensils that fully demonstrated the crafts and artistic talents of the potters, opening a glorious page in the history of Chinese ceramic culture and starting the ceramic painting art (Fig. 15.8).

Black pottery was burned in the lower reaches of the Yellow River in late Neolithic Period. The argillaceous clay locally deposited was used as the raw material. Its

Fig. 15.8 Round vortex sawtooth pattern binaural pot of painted pottery in Yangshao culture

good plasticity brought the superb molding skills into full play, raising the ceramic technology to an unprecedented level. Black pottery not only required skilled hand-molding on fast wheels, but also good mastery of spinning technique to produce thin and sleek pottery. The eggshell black pottery, with black and bright surface and eggshell-haped sidewall, was the most difficult. It was fired in a vertical kiln with temperature at about 1000 °C, and before the end of the firing, wet fuel is put into the kiln and carburized by smoke to achieve a unique black and bright effect (Fig. 15.9).

The production of white pottery began relatively late, approximately in the late Neolithic period, and developed in the Shang Dynasty. Found in the Yellow River Basin and the Yangtze River Basin, it had beautiful shapes and decorations. Because the content ratio of iron oxide in the raw material of the white pottery body was relatively low and that of alumina was high, the pottery turned white inside and outside after being fired.

According to the study of the white pottery unearthed in Chengziya Longshan Culture Site, Shandong and Yinxu Site in Anyang, Henan, the chemical composition of white pottery was very close to that of china clay and kaolin. "In the electron microscopic observation of the white pottery of Yinxu in Anyang, it was found that the main minerals included were very similar to those of kaolin. This showed

Fig. 15.9 Black pottery in Longshan culture

that our ancestors began to use china clay and kaolin at least during the Longshan culture period and the Xia Dynasty and Shang Dynasty, making China the first country in the world to use kaolin to fire utensils" [5]. China clay and kaolin are the raw materials for making porcelain, which can be burnt under high-temperature conditions. Although it was made of china clay or kaolin, white pottery could not be sintered into porcelain because its firing temperature was around 1000 °C and less than 1200 °C. Though white pottery still belonged to the category of pottery, it opened up the idea for porcelain invention. Because the raw materials of white pottery contained the possibility of "making them porcelain," the making of white pottery offered a way to raise the firing temperature and transform the materials from pottery to porcelain (Fig. 15.10).

As the most commonly used and abundant utensil, the gray pottery dominated the late Neolithic Period and developed continuously for a long time, from the Xia Dynasty, Shang Dynasty to the Zhou Dynasty. Therefore, the skills of burning gray pottery were also relatively mature. With a variety of modeling forms, gray pottery was widely used for cooking, drinking, eating, and containing. The decoration techniques included printing, carving, and pasting to express lines and geometric patterns, which created an unpretentious artistic effect (Fig. 15.11).

In addition to the composition of the clay, the formation of the gray pottery color was also inseparable from the firing atmosphere. The clay contained a certain compound of iron, which had a fluxing effect and could lower the firing temperature, and made the pottery appear gray in a weak reducing atmosphere. Compared with the early pottery firing, the techniques of firing gray pottery were relatively advanced. With good functions and various molding techniques, ceramics of gray pottery were

Fig. 15.10 White pottery geometric inscription Bu in Shang Dynasty

Fig. 5.11 Gray pottery vortex pattern pot with double handles in Zhou Dynasty

comparatively systematical and played a role of connecting the past and the future in the development of the ceramic technology.

Stamped pottery refers to the pottery whose surface is stamped with the geometric patterns. It is divided into stamped soft pottery and stamped hard pottery according to the different raw materials and firing temperature. The raw material of the stamped soft pottery is the ordinary clay with a relatively high content ratio of iron oxide. It is fusible clay and its firing temperature is below 1000 °C. However, most of the raw materials used in stamped hard pottery were the porcelain-stone clays with more impurities. The content ratio of the iron oxide was relatively low, and the firing temperature could reach 1200 °C. It was because of the composition of the stamped hard pottery that the firing temperature was increased and the hardness of the body was enhanced, approaching the standard of porcelain; and then the prototype of porcelains appeared.

The stamped hard pottery appeared later than the printed soft pottery, approximately in the late Neolithic age, and developed in Shang Dynasty and Zhou Dynasty. As a special kind of pottery, stamped hard pottery was said to represent the highest stage of the development of ancient pottery. With a variety of practical shapes, good functions, and dense and solid body, it was generally accepted and produced in the north and south of China.

Stamped hard potteries are clear in outline, line and angle, and the colors on the surface of the body are mostly purple-brown, reddish-brown, taupe, and tawny. The purple-brown has the highest firing temperature, which could reach the sintering degree. On the surface of body, it is also possible to see the luster formed by melting

Fig. 15.12 Printed ceramic cup with bow string pattern in Western Zhou Dynasty

the plant ash in the fuel and some of the materials during the high-temperature firing process (Fig. 15.12).

The raw materials used in the stamped hard pottery are very similar to those in proto-porcelain, but with a high ratio of the iron oxide. "According to archaeological excavations, the stamped hard potteries of the Shang Dynasty and Zhou Dynasty were often unearthed together with the proto-porcelain of the same period, and their ornamentation is similar or identical. Especially in the kiln sites of Shaoxing and Xiaoshan in Zhejiang, which dated back to the Spring and Autumn Period and the Warring States Period, it was also found that the stamped hard pottery and the proto-porcelain were fired in the same kiln, indicating that the relationship between the stamped hard pottery and the proto-porcelain in Shang Dynasty and Zhou Dynasty was quite close" [6]. The accumulation of the technology of firing stamped hard pottery was accompanied by the appearance of the proto-porcelain, which was the result of further development and improvement of the stamped hard pottery.

15.4 The Appearance of Proto-Porcelain

The physical data of the ceramic archaeological excavation confirmed that China first fired the green-glazed porcelain in the middle of Shang Dynasty (about 1600 B.C.–1100 B.C.), which was called proto-porcelain because it was in the initial stage, and had the properties and characteristics of porcelain. The first green-glazed drinking utensil was unearthed at the Cultural Site of the Shang Dynasty in Erligang, Zhengzhou in Henan. It was also discovered later in the sites of such provinces as Shandong, Jiangxi, Hunan, Hubei, Jiangsu, and Zhejiang. Analyzing the scientific detection, it is found that the green-glazed porcelain was made of kaolin at a temperature of more than 1200 °C there in the body and the glaze was fired simultaneously. It was in the transition stage from pottery to porcelain because its internal quality could just meet the basic standards of the porcelain inspection.

In the early stage of the development of Chinese ceramics followed by the practice of firing white pottery and printed hard pottery, the selection and treatment of raw materials were improved with the firing temperature being raised and the glaze invented. Under these basic conditions of firing porcelain, proto-porcelain was fired. It is also named primitive celadon, because its glaze was mostly turquoise, bluish yellow, or pea green. From the late Shang Dynasty to the Western Zhou Dynasty and the Eastern Zhou Dynasty, proto-porcelain developed rapidly in South China and North China, opening the prelude to the invention of mature porcelain.

Proto-porcelain, referred as "glazed pottery" when it was first discovered, was confirmed after strict scientific testing and evidencing after 1980. "There is a stage of development and improvement from pottery to porcelain. The green-glazed utensils from the Shang Dynasty to Eastern Han Dynasty should be called 'proto-porcelain' instead of glazed pottery as they were similar to porcelain in terms of chemical composition and physical properties" [7]. Proto-porcelain which began in Shang Dynasty underwent a long and important period of improvement before matured porcelain was burned in the late Eastern Han Dynasty.

In the early stage of the invention of porcelain, the molding process and decoration method of proto-porcelain were influenced by the stamped hard pottery and they were similar in process technology and firing temperature. As the process technology continued to improve, its firing temperature could reach above 1200 °C. Although proto-porcelain had the basic characteristics of porcelain and fell into the scope of porcelain, it did not meet all the conditions of the mature porcelain. Due to the low firing temperature, the raw material of the body was not completely sintered, which resulted in high water absorption rate. The glaze was not even enough and the combination of the body and the glaze was not tight enough. This primitiveness was the reason why it was called proto-porcelain or celadon (Fig. 15.13).

Raw materials are the most basic condition for the firing of ceramics, and the firing temperature is also an important factor affecting the degree of the raw material conversion. The two complement each other and are indispensable. To increase the firing temperature, improvement must be made in the firing method. In Shang Dynasty, in the southern part of China, small dragon kilns were built along the slopes. At the same time, chamber-shaped kilns with a chimney were built to improve the firing conditions, raising the firing temperature to 1200 °C, so proto-porcelain was fired. On the surface of the body of proto-porcelain, there was a layer of primitive glass-like glaze, the colors of which include blue-gray, blue-yellow, tawny or yellow-gray, and dark maroon. The invention of glaze originated from the use of wood and grass as fuel, which left a layer of glassy substance on the surface of the body during the firing process. Inspired by this, potters mixed the plant ash with other raw material to produce primitive glaze, the first high-temperature glaze in China, which was also one of the necessary conditions for making mature porcelain.

The appearance of the proto-porcelain had important technological value. The initial use of fusible clay as raw materials was changed to the choice of porcelain-stone clay and kaolin clay to make the white pottery and printed hard pottery; meanwhile, the method of firing was improved and the sintering temperature was raised.

Fig. 15.13 Original green brown green glaze celadon zun in Shang Dynasty

As a result, the proto-porcelain was fired, opening up the future for burning mature porcelain.

15.5 The Maturity of Porcelain Invention

First made in the middle of the Shang Dynasty, proto-porcelain began to develop in the Western Zhou Dynasty and entered the development stage of the early celadon in the Spring and Autumn Period and the Warring States Period. The technology was passed on from one generation to another and the product quality was improved continuously. In the late Eastern Han Dynasty, potters could produce green-glazed porcelain of high-quality, the technology of which reached a relatively high level. During this period, the preparation of raw materials was relatively standardized. The body was light gray and its thickness was moderate. Smooth and tightly combined, the glaze was mostly blue-green, blue-gray, or blue-yellow. As the molding process was very strict, the shape was regular and the firing technology improved significantly. There has also been a fundamental change in the whole, which led to the production of mature porcelain (Fig. 15.14).

"Porcelain should meet the following three requirements: first, the body should be white through the selection and processing of raw materials to increase the ratio of Al_2O_3 and reduce the ratio of Fe_2O_3; second, the body should be dense, non-absorbent and makes a crisp golden-stone-like sound when clicked by raising firing temperature above 1200 °C; third, on the surface, there is glaze fired at high temperature, which is firmly sintered with the body and is evenly distributed" [8]. The first two are the

Fig. 15.14 Green glaze double-system tank with water ripple pattern in Eastern Han Dynasty

key conditions determining the attributes and nature of the porcelain, which are also the basis for its existence and development.

Scholars of ancient ceramics first discovered that in the kiln yard of the late Eastern Han Dynasty in Shangyu District, Shaoxin in Zhejiang, mature blue-glaze porcelain was successfully produced, and it met the testing standards of porcelain. It thus realized the transition from proto-porcelain to mature porcelain, making China an inventor of porcelain. The specimens unearthed were tested by experts from the Chinese Academy of Sciences. The results showed that "the firing temperature was 1310 ± 20 °C, the water absorption rate was 0.28%, and the ratio of iron oxide was 1.64%. These data reached the general standards of modern daily-use porcelain. Therefore, Shangyu became the worldwide recognized birthplace of porcelain" [9].

The test and analysis of the porcelain fragments in the kiln site of the late Eastern Han Dynasty and the china stone near the kiln site in the Xiaoxiantan, Shangyu District shows that their chemical composition is very similar, indicating that the locally selected and processed raw materials made the green-glazed porcelain fired there in a good porcelain state. The porcelain had a low porosity and good light transmission. What is more, its body was dense and the water absorption rate was reduced. According to the test, the porcelain was fired at a high temperature between 1260 and 1310 °C. The surface of the porcelain was completely glazed, with light and elegant color; the glaze was bright, tightly combined with the body, and its thickness was increased compared with that of the proto-porcelain. Tests proved that the green-glazed porcelain in the late Eastern Han Dynasty fully reached the porcelain standard, confirming that China had already produced mature green-glazed porcelain in the late Eastern Han Dynasty at the latest.

Zhejiang is the main birthplace of ancient Chinese porcelain, with kilns in many parts of the province and a one-thousand-year history of porcelain production that began in the Han Dynasty and gradually declined in the Song Dynasty. It was here that mature porcelain was first fired. The celadon kiln sites of the Eastern Han Dynasty were mainly discovered in these cities: Ningbo, Jinhua, Deqing, and Quzhou, were parts of Yuezhou, an old region known for its outstanding achievements in porcelain-making. So the kilns excavated there were also called Yue Kiln. The mature porcelain fired there was inseparable from the local conditions: the locally sourced porcelain stones, strictly processed raw materials, sufficient water resources, excellent production technology, and improved high-temperature kilns. These conditions gave birth to the invention of porcelain.

Porcelain-making technology reached a relatively high stage in the late Han Dynasty when mature porcelain was made. It further developed and advanced in the Three Kingdoms Period, Jin Dynasty and the Northern, and Southern Dynasties, an important period of 300 years for the development and improvement of the porcelain industry when the firing skills matured day by day and a large number of excellent products were created. As the raw materials for the body and the glaze were refined and the porcelain was fully sintered, the texture of the body was solid and dense, and the glaze color was implicit and rich. A variety of products in many new styles were created, such as kettle (Fig. 15.15), the chicken-head-shaped pot, the

Fig. 15.15 Celadon plate pot in Three Kingdoms Period (220–280)

frog-shaped water bowl, the lion-shaped talisman, etc.; decorations on the utensils were molded, and their styles were elegant, simple, and interesting.

Although celadon had single form, craftsmen fired a lot of high-quality celadon with pure, clear, and smooth color and luster by giving full play to their creativity, skillfully using the characteristics of materials and effectively controlling the firing atmosphere, which laid a solid technological foundation for the brilliant achievements in porcelain-making in the Tang Dynasty and Song Dynasty.

15.6 The Development of Porcelain Technology

After the successful firing of mature porcelain in the late Eastern Han Dynasty, the porcelain craftsmanship kept improving, with standardized molding and strict processing. The quality of the green-glazed porcelain was therefore improved greatly. Its body was fine and pure, and the texture was solid and compact. As beautiful as green mountains, the glaze was rich and subtle in color. Porcelain was practically and artistically molded to suit people's daily needs. The production of the green-glazed porcelain expanded from Zhejiang to Jiangsu, Anhui, Jiangxi, Hunan, Sichuan, Fujian, Guangdong, and some other provinces.

The green-glazed porcelain was made later in the north than the south. "In 1948, a batch of celadon was unearthed at the Tombs of the Feng Family in Jing County of Hebei. According to the test, it was made from the northern porcelain clay and was burned in the northern porcelain kiln. The shapes and decorations of the utensils were magnificent and bold, quite different from the dexterous and delicate styles of the southern porcelain. Celadon kiln sites of the Northern Dynasty were also discovered in Neiqiu and Lincheng of Hebei, and Zibo, Qufu and Tai'an of Shandong" [10].

The porcelain in the early days of China belonged to the green glaze system. The raw materials of the porcelain contained a certain amount of iron. Because the amount of iron in the green glaze was different and the temperature and atmosphere of the reduction flame were not the same, the color shade of the green glaze porcelain also varied. After the long-term production operations, the craftsmen gradually mastered the relationship between the content of iron and color, so they improved the quality of the green-glazed porcelain (Fig. 15.16). Based on burning green-glazed porcelain, potters of the northern kilns successfully controlled the amount of iron in the body and glaze, and fired the early white porcelain in the Northern Dynasty. The change from green-glazed porcelain to white porcelain was the result of the deepening of porcelain-making technology, which was also a creation. Different from celadon in color, white porcelain was improved and developed on the basis of celadon. The fundamental difference between celadon and white porcelain lies in their iron content, which was basically the same in production processes. In a word, the successful firing of white porcelain was the result of the rapid improvement in the firing technology of the green-glazed porcelain, fine preparation of raw materials, significant decrease in the content of iron in the glaze and body, and the mastery of the relationship between the raw material composition and color through repeated practice.

Fig. 15.16 Lotus buddha with green glaze in Northern Dynasty

The early white porcelain clearly showed its own characteristics. The body and the glaze were milky white, the body was more delicate, and the glaze layer was clear. There was green color in the white and the green was more obvious in the thicker part, which clearly indicated that white porcelain developed from celadon. Therefore, white–green was the characteristic of the early white porcelain, and the glaze color of the white porcelain of the Sui Dynasty was still looming. "According to white-glazed porcelain excavated from tombs and ruins, as well as documents recorded in the literature, it is generally believed that the burning of the white-glazed porcelain in China should be in the Northern Dynasty. A batch of white-glazed porcelain was unearthed at the tomb of Fan Cui (a high-ranking general of the Northern Dynasty) dated 575 A.D. in Anyang, Henan. It was the earliest white-glazed porcelain found in the archaeological excavations and was characterized by white body, white and green glaze and cyan in thick glaze" [11] (Fig. 15.17).

White porcelain began to be burned in Northern Dynasty and entered a mature period in the Sui Dynasty and Tang Dynasty. The Xing Kiln in Hebei first burned celadon and then turned to white porcelain, becoming the main production area of

Fig. 15.17 White porcelain green color long neck bottle in North Dynasty

the white porcelain in Tang Dynasty, so the so-called "celadon in South and white in North" pattern in the history of Chinese ceramics was formed. The burning of white porcelain in the north represents a new pioneering technology in porcelain manufacturing, making China the first country in the world to have white porcelain, which also contributed to the development of the ceramic science and technology in the world (Fig. 15.18).

In Tang Dynasty, the white porcelain of Xing Kiln of Hebei Province was the most famous. White porcelain sprang up when the porcelain industry was dominated by green-glazed porcelain. It not only showed clear and elegant texture with its pure white, but also contained the harmonious integration in the transformation process, promoting the expansion of creative consciousness (Fig. 15.19).

The Song Dynasty was a unique period for the development of celadon. The elegant social and cultural atmosphere, the transcendental aesthetic pursuit and the exquisite skills created the historical peak of the celadon products. Ru Kiln, Guan Kiln, Jun Kiln, and Longquan Kiln were among the five famous kilns in Song Dynasty. With its own unique style, each kiln produced elegant artworks with aesthetic connotation that reached the peak of the celadon art. (Figs. 15.20, 15.21, 15.22 and 15.23).

Fig. 15.18 White shaft tank in Sui Dynasty

Fig. 15.19 White porcelain pot in Tang Dynasty

Fig. 15.20 Three-legged casket of Ru Kiln in Song Dynasty

Fig. 15.21 Guaner bottle of Guan Kiln in Song Dynasty

Fig. 15.22 Indigo glaze zun of Jun Kiln in Song Dynasty

Fig. 15.23 Furnace of dragon Spring Kiln in Song Dynasty

In Song Dynasty, the most famous white porcelain was made in Ding Kiln, one of China's five renowned kilns that produced white porcelain only. With its unique style, elegant and delicate shape, and decoration, and sophisticated production technology, porcelain from Ding Kiln opened a new chapter in the development of porcelain-making in North China. It also triggered the vigorous growth of the white porcelain in South China (Fig. 15.24).

Fig. 15.24 Ding Kiln white porcelain pot with cover in Song Dynasty

The green-white porcelain of Jingdezhen, which began in the Five Dynasties, was a typical representative of the white porcelain in South China. With pure white body and white-green glaze, it then developed into white porcelain. With rich varieties and outstanding achievements in technology and art, white porcelain of Jingdezhen rose quickly and opened up a new field of Chinese porcelain (Fig. 15.25).

The porcelain art of Jingdezhen in Yuan Dynasty became increasingly mature, and porcelain production developed rapidly and began to enter a new historical period of development. A large amount of the white porcelain products were mainly various food utensils (Fig. 15.26). Meanwhile, high-quality white porcelain also provided a new carrier for the artistic expression of painting. The famous under-glaze paintings of the Yuan Dynasty were decorated with the blue and white glazed red, and many excellent works were thus created (Figs. 15.27, 15.28).

White porcelain of Dehua in Fujian was first fired in Song Dynasty, and prospered in the Ming Dynasty. The local place was endowed with unique porcelain materials. The porcelain fired there was white and elegant, and its texture was pure like the ivory and white jade, creating a unique feature of Dehua white porcelain. The porcelain was mainly used for display and practical purposes (Fig. 15.29). Dehua porcelain sculpture became an outstanding achievement of art with unique line-based expression technique. Guanyin and Dharma were the most representative figures. The shaping techniques were exquisite, bringing the characteristics of porcelain into full play. (Fig. 15.30).

Jingdezhen became the center of Chinese porcelain production during the Ming Dynasty and Qing Dynasty. The traditional porcelain-making techniques were more standardized. With the development of the under-glaze painting and decoration, color decoration on the glaze was also enriched by innovation. The five-color decoration of the Ming Dynasty was bright and infectious (Fig. 15.31); decoration of clashing

Fig. 15.25 Celadon plum-flower pattern vase from Jingdezhen in Song Dynasty

Fig. 15.26 Egg White glaze lamp holder from Jingdezhen in Yuan Dynasty

colors was colorful and natural (Fig. 15.32). Antique-color decoration of the Qing Dynasty was very vigorous, beautiful, vivid, and general (Fig. 15.33); the famille rose decoration was soft and the picture was realistic (Fig. 15.34); the enamel decoration was rich and elegant, and refined (Fig. 15.35).

The porcelain, the beauty of the glaze was fully expressed, and excellent works of various color glazes were created (Figs. 15.36, 15.37).

In the historical development of world ceramics, China not only invented porcelain, but also developed the technology that helped to create shapes and decorations

Fig. 15.27 Blue and white lotus pond pattern bowl from Jingdezhen in Yuan Dynasty

Fig. 15.28 Under-glaze red pine jade pot with bamboo plum-flower pattern from Jingdezhen in Yuan Dynasty

Fig. 15.29 Dehua Kiln with white porcelain magnolia pattern in Ming Dynasty

in different styles. While spreading the porcelain-making technology, China also influenced and enriched the world with its profound ceramic culture.

Archaeological data confirmed that after the invention of Chinese porcelain, it was exported to North Korea, Japan, Vietnam, Malaysia, the Philippines, Indonesia, Thailand, India, Iran, Iraq, Egypt, and East Africa through the land and water ways in the Tang Dynasty. The sale of Chinese porcelain to Europe began in the early sixteenth century through the resale by Portuguese and Spanish merchants, followed by the Dutch monopoly in the seventeenth and early eighteenth centuries. Then merchants of European countries came directly to China to purchase Chinese porcelain, which was shipped to Europe in large quantities.

With the export of Chinese porcelain, porcelain technology was also spread to the world. Koreans first learned porcelain technology from China, and they began to construct kilns to fire porcelain at home in the tenth century. Japan introduced Chinese kiln technology in the eighth century and sent artisans to Fujianto learn porcelain technology in the thirteenth century, and they manufactured porcelain in Seto after returning home. Egypt succeeded in replicating Chinese porcelain in the twelfth century. In 1470, Italians learned Chinese porcelain technology and burned porcelain in Europe for the first time. In 1709, Germans mastered the Chinese porcelain-making technology and burned the first batch of high-quality porcelains. Later, France learned the Chinese porcelain technology in 1738, the United Kingdom in 1745,

Fig. 15.30 Dehua Kiln with white dharma porcelain statue in Ming Dynasty

the Netherlands in 1764, and the United States in 1890. Inspired by the shape and decoration of the Chinese porcelain, they produced many types of porcelain with their own styles. The invention of porcelain-making technology not only brought excellent utensils for human daily life, but also created works of art with exquisite quality, making great contributions to the development of material civilization in the world.

Chinese ceramics have gained new development in the modern era. Modern ceramic factories and traditional handicraft workshops spread among all the provinces in the country. Among them, there are many famous ceramic production

Fig. 15.31 Algae Jar with Wucai (colorful) fish pattern in Ming Dynasty

Fig. 15.32 Pot with flower pattern in Ming Dynasty

Fig. 15.33 Zun with ancient colorful lotus pond and phoenix tail pattern in Qing Dynasty

Fig. 15.34 Famille rose sphere bottle with peach grain pattern in Qing dynasty

Fig. 15.35 Garlic-shape bottle with enamel color and twirling branches in Qing Dynasty

Fig. 15.36 Glaze zun with Shiliu-shape in the Klin of Qing Dynasty

Fig. 15.37 Zun with blue and radish glaze in Qing Dynasty

areas including Jingdezhen in Jiangxi, Liling in Hunan, Yixing in Jiangsu, Longquan in Zhejiang, Dehua in Fujian, Zibo in Shandong, Shiwan and Chaozhou in Guangdong, Tangshan, and Handan in Hebei. Modern and new ceramics are created in these areas based on the inheritance of traditional Chinese ceramics.

As an important traditional craft of the Chinese nation, porcelain invention is not only a precious cultural heritage, but also a carrier of the technical genes. What's more, it is the embodiment of creative thinking in the practice of craftsmanship. The application range of porcelain firing has also expanded from the fields of daily-use porcelain and building materials to special ceramics, winning the same status as metals and organic polymers. The achievements of the ceramic science and technology are widely used in high-tech fields such as information, energy, biomedicine, environment, national defense, and space technology. The application value and academic significance of the invention of porcelain are far-reaching and extremely important to the modern society.

References

1. Jiazhi, Li. 2001. The emergence of early Chinese pottery and its contribution to Chinese civilization. *Journal of Ceramics* 2: 78–83.
2. The Editorial Board of Complete Works of Chinese Ceramics (ed.). 2000. *Complete Works of Chinese Ceramics*, vol. 1, 14. Shanghai: Shanghai Fine Arts Publishing House.
3. The State Administration of Cultural Heritage (ed.). 2009. *60 Years of Chinese Archaeology*, 20. Cultural Relics Publishing House.
4. Guozhen, Li, and Guo Yanyi. 1988. *An Introduction to Chinese Famous Ceramics*, 50. Shanghai Science and Technology Press.
5. China Institute of Silicate (ed.). 1982. *History of Chinese Ceramics*, 73. Beijing: Cultural Relics Publishing House.
6. China Institute of Silicate (ed.). 1982. *History of Chinese Ceramics*, 74. Beijing: Cultural Relics Publishing House.
7. China Institute of Silicate (ed.). 1982. Preface to *Collected Papers on Ancient Chinese Ceramics*. Beijing: Cultural Relics Publishing House.
8. China Institute of Silicate (ed.). 1982. *History of Chinese Ceramics*, 76. Beijing: Cultural Relics Publishing House.
9. Xiaozhong, Cheng. 1997. *Celadon*, 17. Jinan: Shandong Science and Technology Press.
10. Zhiyan, Li. 1998. *Chinese Ancient Ceramics*, 88. Beijing: The Commercial Press.
11. Jiazhi, Li. (ed.) *History of Chinese Science and Technology: Ceramics*. Beijing: Science Press, 1998:144.

Yongshan Yang Born in Laizhou, Shandong in 1938, the Professor and doctoral supervisor of the Academy of Arts and Design, Tsinghua University, the former vice president of the Central Academy of Arts and Crafts (later called the Academy of Arts and Design, Tsinghua University) and former member of the Arts Education Committee of the Ministry of Education.

Research interests: Ceramic Art Design, Basic course of Ceramic modeling, Chinese History of Ceramic Art.

Major works:

Basic Course of Ceramic Modeling; Ceramic Modeling Design; Ceramic Modeling Art; About Ceramics and Art; Chinese Ceramics; Folk Ceramics; Collections of Chinese Traditional Crafts: Ceramic; Collections of Chinese Traditional Crafts: Ceramics (Continued); Integration with Nature; Collections of YANY Yongshan's Works

Awards:

"Xieyu tea set" won the first prize of National Ceramic Design in 1986;

"Rainbow tea set" won the first prize of National Ceramic Design in 1990;

Ceramic Modeling Design won the 2nd National University Excellent Teaching Materials Award in 1992.

Collections of Chinese Traditional Crafts: Ceramic (Chief Editor) won the 2006 China Outstanding Publication Award;

Ceramic Modeling Design won the Special Award of 2008 Tsinghua University Excellent Teaching Materials.

Chapter 16
The Technology of Chinese Wooden Structure Architecture

Daiheng Guo and Peijun An

Chinese traditional architecture is dominated by wooden structure. In primitive society, "nesting" and "cave dwelling" coexisted, but in the following thousands of years of development, the wooden structures were preserved, developed, and further improved. This wood-based structure makes Chinese traditional architecture different from western architecture. In monolithic buildings, limited by the length of timber, it is inconvenient to build houses with stones like the western buildings, which can be large or small. However, it has accumulated rich experiences and superb skills in space organization of group buildings, creating a unique space form in which Chinese architecture emphasizes courtyard space and the blending of indoor and outdoor space. This feature has been blended with Chinese traditional thought in thousands of years' change and accumulation, which has materialized the ideas of "harmony between man and nature" and "harmonious coexistence." In turn, these ideas have been promoted and explained by Chinese traditional architecture. Materialized ideas and sublimated buildings have been inseparable from each other, composing the colorful movement of Chinese civilization together, and exerting a far-reaching impact on the East Asian cultural circle.

The construction of wooden structures cannot be separated from superb construction technology. The ancient Chinese wooden construction technology includes many types of work, as different dynasties have different classification standards. According to the Qing Dynasty method, it can be roughly divided into eight categories, known as the "Eight Great Works." They are wood, tile, zha, stone, soil, paint, color painting, and paste. Among them, "wood" refers to woodwork, including large and small woodwork; "tile" refers to bricks and tiles; "Zha" refers to the construction of scaffolding; "stone" refers to stone carving and masonry; "soil" refers to

D. Guo (✉)
Tsinghua University, Beijing, China
e-mail: guodh@mail.tsinghua.edu.cn

P. An
North China University of Technology, Beijing, China

J. Hua and L. Feng (eds.), *Thirty Great Inventions of China*,
https://doi.org/10.1007/978-981-15-6525-0_16

the technology of tamping soil at the foundation, walls, and other places; "paint" refers to the protection and decoration of the surface of structures and decoration components; various decorative patterns painted on the surface of some high-grade houses, known as "Color painting"; "Paste" refers to the construction technology of surface stickers on roofs and walls. Woodwork is the most important work among the eight types of work since it contains the main structure of the building and the final decoration details. The Chinese traditional wooden construction technology has now been recognized as an important part of China's intangible cultural heritage.

16.1 The Characteristics of Ancient Chinese Wooden Structure Architecture

16.1.1 Beam-Frame Structure Type

The main structural form of the ancient Chinese wooden architecture is the beam-frame structure system. The earliest is the "column rafter system," which refers to the method of using round cave wall as the wall, and the small wooden stick (rafter) around it on the central column during the semi-cave dwelling era. (Fig. 16.1).

Later, as the cave dwelling was replaced by architecture built on the ground, the column rafter system evolved into the "column purlin rafter system." At the same time, the columns were lengthened along the vertical direction of the house, with thick wooden sticks attached to each other and rafters put on them, thus forming a girder structure. In order to meet the need of the roof drainage, a beam-frame structure with a sloping roof was invented (Fig. 16.2).

There are four main types of ancient Chinese wooden structure buildings: beam-lifting, bucket-piercing, drywell, and dry railing. Among them, the beam-lifting type and bucket-piercing type are the most widely distributed. Dry railings are found only in the area of humid climate. The type of dry well is characterized by stacking logs and making walls and roofs, which are suitable for timber-rich areas.

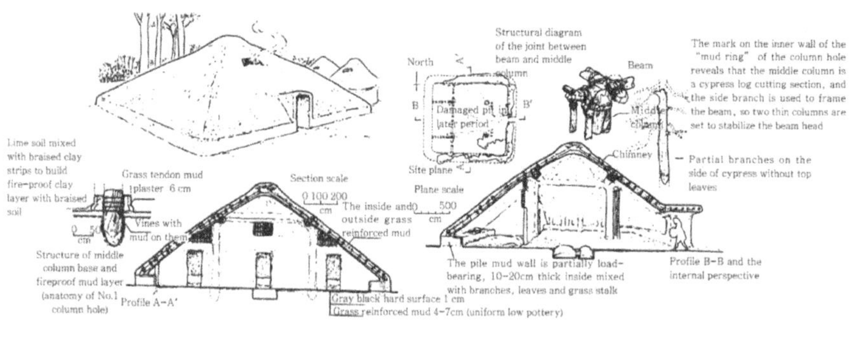

Fig. 16.1 Restoration of F1 house at Banpo site in Xi'an

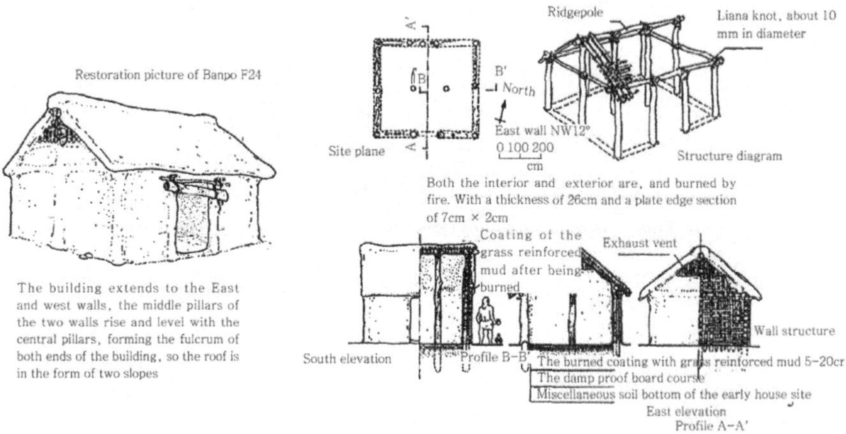

Fig. 16.2 F24 restoration map of Banpo (from the anthology of architecture and archaeology)

16.1.1.1 Beam-Lifting Girder Frame

As the name implies, the structure of a beam-lifting girder frame is to erect a layer of beams (Fig. 16.3). The first step is to support a beam with two columns; then use two short columns on the beam to support a slightly shorter beam on which two short columns are erected to support an even shorter beam. As such, with the rest done in the same manner, the beams are shortened layer by layer to form a girder frame. At both ends of each beam, a long wooden strip—purlin, is placed vertically between the two beams, which forms the rudiment of a sloping roof.

The advantage of the beam-lifting type is that it is spacious with few indoor columns, which is suitable for various functions. The disadvantage is that the columns and beams both need the wood from the grown-up trees, which indicates a certain economic strength.

16.1.1.2 Bucket-Piercing Girder Frame

Bucket-piercing is different from beam-lifting. Its origin is not the one like the semi-cave dwelling, but the nest dwelling. Therefore, the framework of purlin supported directly by columns has been developed (Fig. 16.4). Each purlin is supported by columns, which are connected by wooden loquats to form a beam frame. In order to ensure its stability, a large cross-sectional component—a horizontal tablet—is used to connect the outer eaves and columns of the two beam frames. Because of the different height of the columns, purlin is placed in different positions, forming the rudiment of the sloping roof. The advantages and disadvantages of the bucket-piercing type are opposite to those of the beam-lifting type. The advantage of the bucket-piercing type is that the cross section of the column is smaller, saving a lot of

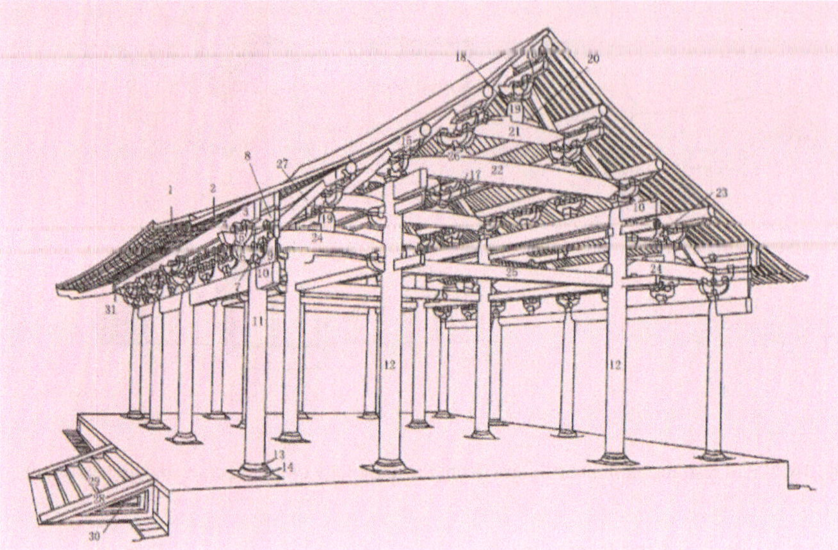

1. Flying rafters 2. Eaves rafters 3. Eaves rafters 4. Dou 5. Gong 6. Huagong 7. Ludou 8. Stigmas 9.Eye panels 10. Apertures 11. Eaves 12. Inner columns 13. Columns 14. Column bases 15. Flat frames 16. Ridge frames 17. Grips 18. Hua Ding wipes chin frames 19. Shu Zhu 20. Heta 21. Flat Beam 22. girder (four rafters) 23. tie 24. Rufu 25. String 26. Hump 27. Supporting Foot 28. Fuzi 29. Tread 30. Elephant eye 31. Pillow wood

Fig. 16.3 Song style hall beam-lifting girder frame

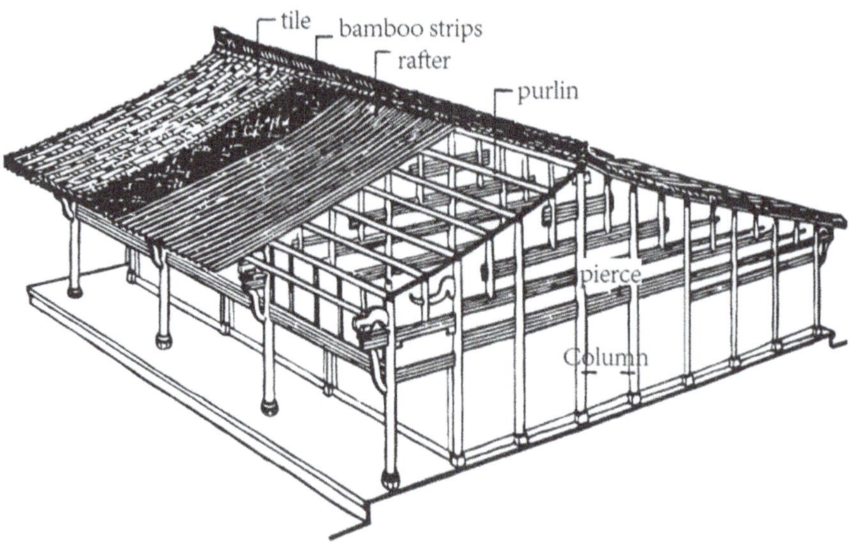

Fig. 16.4 Bucket-piercing girder frame

materials. The disadvantage is that it needs many columns so that the indoor space layout is not flexible enough.

16.1.2 Beam Front Characteristics of Frames

Most of the buildings with the beam-lifting structure are in the north of China, while those with the bucket-piercing structure are common in the South. However, due to the complementarity of the beam-lifting and bucket-piercing systems, they are sometimes used together. The bucket-piercing structure can be used in gables and other places, and the beam-lifting structure can be used in places where larger space is needed.

A building needs many beams and frames to form the structural system of using space. That being said, in order to meet the functional requirements of shelter from wind and rain, it also needs to set up walls, doors, and windows as the enclosure structure independent of the wooden frame bearing the roof load. Therefore, there has always been a saying in China that "walls collapse but roofs do not collapse."

In ancient China, the wooden structure gradually changed from the original binding joints, and the mortise and tenon joints were invented. The joints of Chinese wooden structure with tendon and mortise as the main connection mode can easily resist various external forces from different directions with good seismic performance. At the same time, the ancient Chinese craftsmen also made special treatment to the structure of a building, which is to make pillars "rise" and "side foot." "Rising" is to make the outer column of the house into different heights, while the middle ones are low, and the ones at corner are high. The columns gradually rise from the middle to the corner, so that the connecting beam on the column head forms a relaxed curve. "Side foot" is the building's outer column slightly inclined inward. Through such treatment, it will produce a force to squeeze into the middle, and strengthen the integrity of the whole framework.

In order to protect the timber structure and the wall of the house, the upper part of the girder frame is covered with a roof with far-reaching eaves. For different types of roof forms, due to the influence of ethical culture, the concept of hierarchy is artificially endowed. The highest is the top of the Ludian Hall, also known as the Four A Tops. The rest are Xieshan Top (also known as the Nine Ridges Top), Suspended Hill, Hard Hill, Single Slope, Cunjian Top, and so on. Heavy eaves are senior to single eaves. There is also another kind of cup top, which is relatively rare.

No matter what kind of roof, most of them have a unique feature which is different from that of other civilizations. In other words, the profile of sloping roof is not a straight line, but a soft curve. Why? There are different opinions on this issue. The mainstream view is that (1) it is conducive to drainage, which will get the so-called "spitting away" effect in ancient books; (2) it is convenient for day lighting, i.e., the so-called "anti-universe sunshine." It was not until the Ming Dynasty (700 years ago) that brick walls were widely used in China. Previously, most of the walls were built with rammed earth, which required high waterproofing. The effective way to improve the waterproofing of the walls is to increase the depth of the eaves. If the selected eaves are kept at the same slope as the roof, the indoor lighting condition will be poor. If the eaves are slightly raised to form a concave surface, the problem can be solved perfectly. After a period of development, various types of roofs have been formed as seen today (Fig. 16.5).

For ordinary buildings, the eaves are chosen by the rafters, but if the building grade is higher or the scale is larger, the eaves need to be further picked out. The eaves cannot meet the requirements of the rafters, so another component is needed to support the weight of the eaves. Such a component part is the rafters of ancient Chinese architecture.

16.1.3 Mortise and Tenon Joint Structure

The tenon–mortise joint structure is a way to connect two components, the basic form of which is that one component is inserted into another. The protruding part is called "tenon" and the concave part is called "mortise." Since wood has a certain degree of toughness and is easily processed for producing slight deformation with the change of humidity, mortise is the most suitable structure for wood structure. The earliest known tenon–mortise structure was found at the Hemudu site in Yuyao, Zhejiang more than 7,000 years ago. There are many kinds of mortise and tenon structures in the wooden structure relics unearthed from the site. They are mainly used in the dry-fence buildings, such as convex square tenons, circular tenons, double-layer convex tenons, dovetail tenons, and enterprise-mouth tenons (Fig. 16.6). Especially astonishing is that most of these exquisite tenons and mortises were processed with the stone tools.

The tenon–mortise structure is not the only way to connect the wooden components, rope binding is also common in early cultural relics. Banpo Site in Xi'an belongs to the Yangshao Culture, which is later than the Hemudu Culture. The unearthed buildings are mainly tied up. The binding connection is more convenient, easy to construct, and the requirement for tools is relatively low. The binding method is difficult to meet the engineering requirements of high accuracy and its durability is poor. Therefore, in the process of cultural exchange in different regions, the way of binding was eventually replaced by the tenon–mortise structure, and eventually developed a variety of forms of tenon–mortise. Although no remains of the wooden structures of the Western Zhou Dynasty have been found, the level of wooden tenons and mortises can be seen from the coffins unearthed from the tombs of the Western Zhou Dynasty (Fig. 16.7). At this time, the mortise and tenon was not only a simple way of "mortise" and "mortise" interpolation, but also a targeted design for different parts, different nodes, and different stress states. Seemingly complex, but in fact concise in the most critical parts with the most effective way, the simplest structure and the most economical material can help achieve the construction objectives, achieving the "four or two sets of kilograms" effect. The tenon–mortise structure can be roughly divided into three types: the first one is mainly the combination of

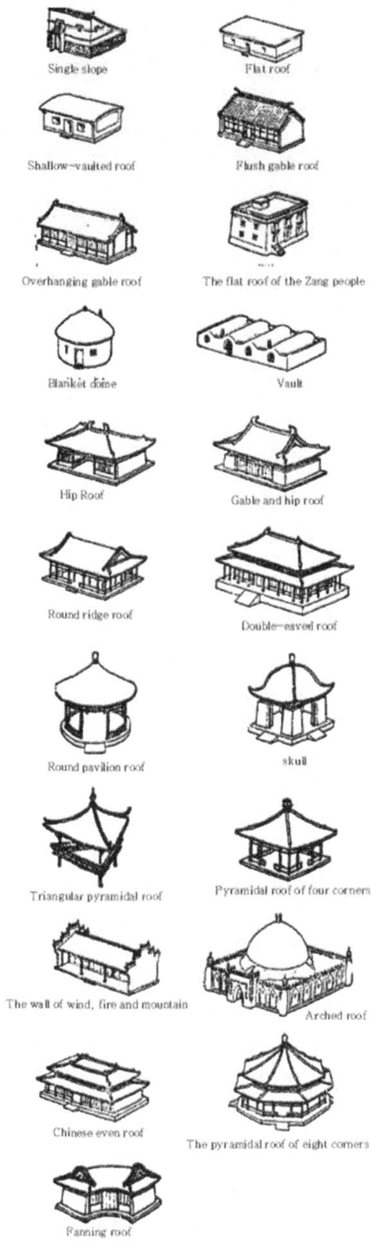

Fig. 16.5 Roof style of single building in ancient China

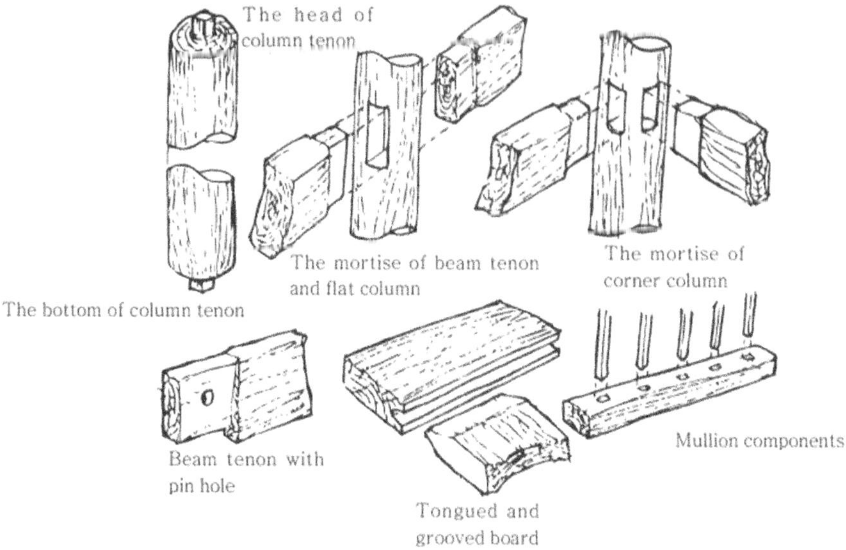

Fig. 16.6 Mortise and tenon of wood structure at Hemudu site, Yuyao, Zhejiang (*Journal of archaeology*, 1978, issue 1)

face and face, or the combination of two sides, or the combination of face and side, such as "groove mortise," "enterprise mortise," "dovetail mortise," "belt-piercing mortise," "tenon-tie," and so on. The second one is a "point" structural method, which is mainly used for T-joint, angular joint, cross joint, and extension joint of straight and arc materials such as "lattice shoulder tenon," "double tenon," "double clamp tenon," "hook tenon," "nail tenon," "half tenon," "tongue-in-tenon," and so on. The third one combines together three components that are connected with each other. This method is more complex and special than using some above-mentioned mortise–tenon joint structures such as the common "supporting tenon," "long and short tenon," "shoulder-holding tenon," "zongjiao tenon," and so on.

The combination of the tenons and mortises can make the components seamlessly linked together, contributing to an organic formation of the whole structure. It can be said that the use of the mortise and tenon structure in ancient China had reached a fascinating level.

16.1.4 The Era Characteristics of Dovetail in Beam-Lifting Frame

With regard to the bracket set, it was written in the Song Dynasty document *Ying Zao Fa Shi* as "枓"(dougong), which is recorded in the Qing Dynasty Ministry of Documentation and Works *Engineering Practice*. The bracket set has played an important

① The structural sketch and pin tenon attachment of the painted dowel unearthed from No.5 Chu tomb caojiagang, Dangyang Country, Hubei Province (Journal of archaeology, No.4, 1988).

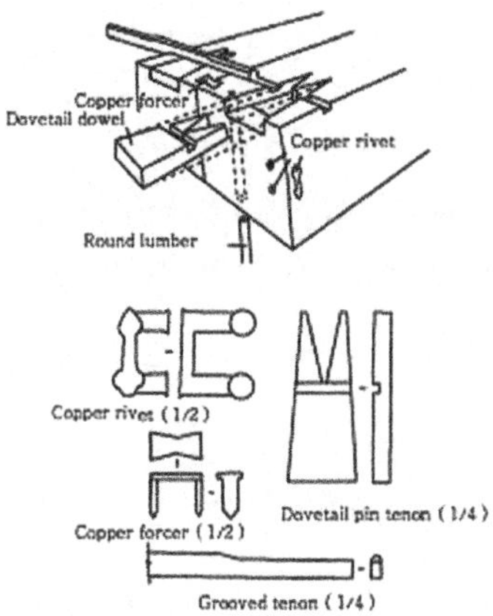

② Wooden coffins and tenons of Zhou Dynasty unearthed in Hunan and Henan.

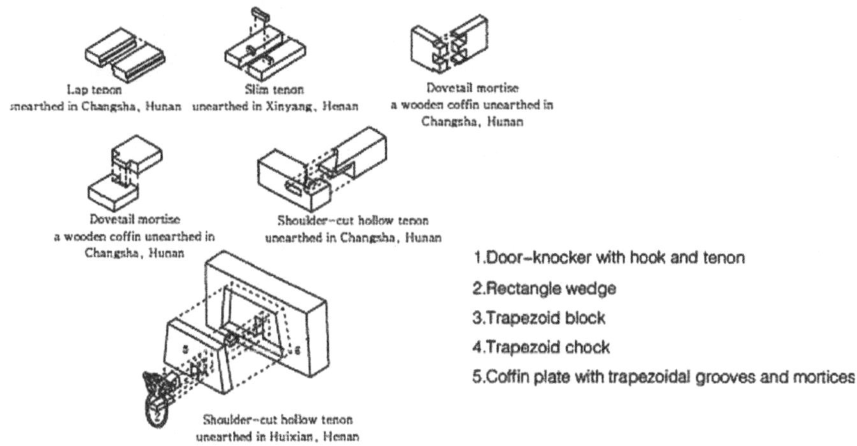

Fig. 16.7 Mortise and tenon structure of coffin in Western Zhou Dynasty

role in the development of Chinese wooden frame architecture. Its evolution can be regarded as an important symbol of the evolution of the Chinese traditional wooden frame architecture and an important basis for identifying the age of Chinese ancient wooden frame architecture. The evolution of the bracket set can be divided into three stages.

16.1.4.1 The Initial Stage (From the Western Zhou Dynasty to the Han Dynasty)

The earliest bracket set can be seen from bronze wares, such as the short wood on the pillars of mulberry picking and hunting pans, the image of a big battle on the

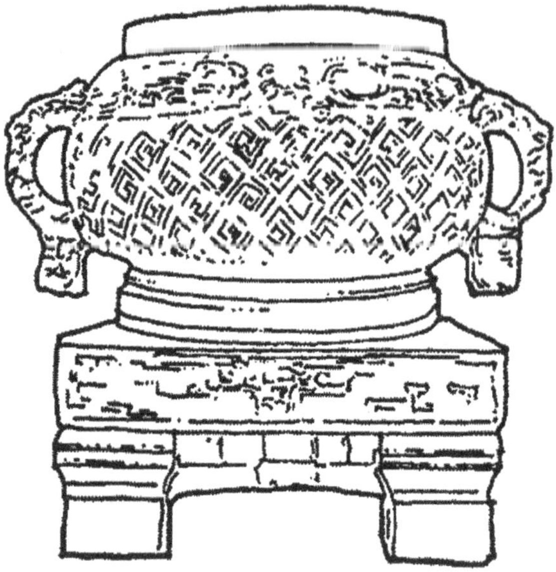

Fig. 16.8 Bronze wares of the Western Zhou Dynasty

bronze wares of the Western Zhou Dynasty (Fig. 16.8), the bronze scheme unearthed from the Zhongshan Tomb of the Warring States Period (Fig. 16.9), and the images of battles and 45 degree oblique bundles on the stone wares, Ming wares, portrait stones, and portrait bricks of the Han Dynasty. It can be seen from the Hanming pottery buildings unearthed from Hangao Yique, Muma Mountain in Sichuan, and Gaotang in Shandong that the top of the column has the bracket set supporting purlins, beams or floor floors, and the outer end of the cantilever supports the bracket set purlins, which are not connected with each other (Fig. 16.10).

Since the Han Dynasty, the bracket set has been added to the forehead of the pillars. In addition to using the same cords as those on the pillars, there are also herringbone cords, with a bucket on top, supporting the purlin cords, such as the North

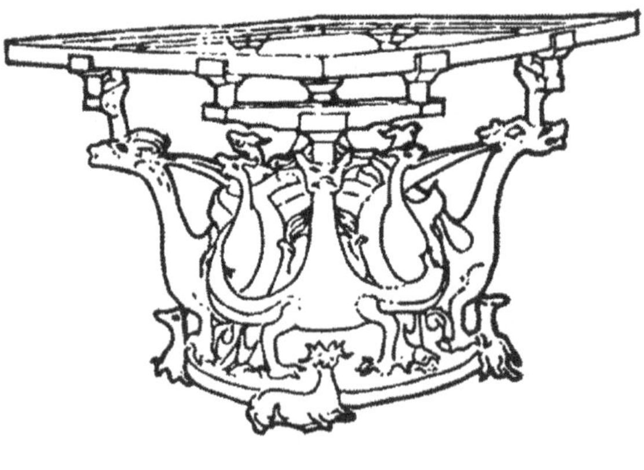

Fig. 16.9 Warring States bronze scheme

Fig. 16.10 The bracket set on the Hangao Yique

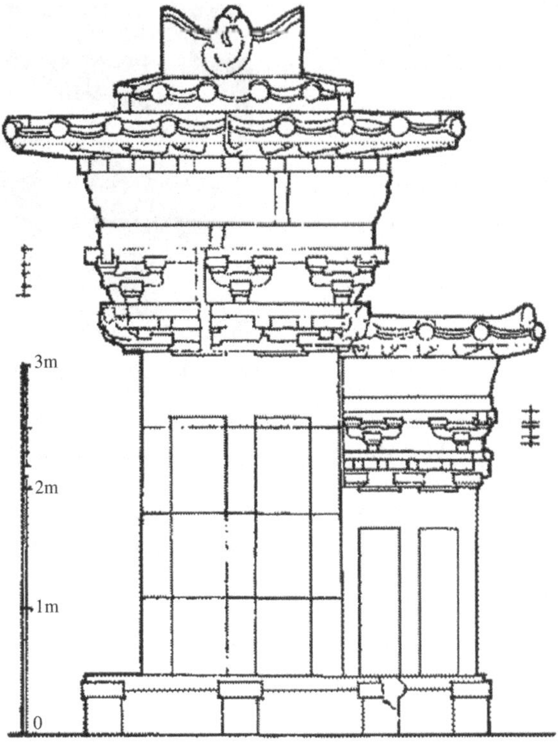

Wei Buddhist Pagoda of the Yungang Grottoes in Datong and the Beiqi Grottoes in Tianlong Mountain in Taiyuan (Figs. 16.11 and 16.12).

Fig. 16.11 The North Wei Buddhist Pagoda of the Yungang Grottoes in Datong

Fig. 16.12 the Beiqi Grottoes in Tianlong Mountain in Taiyuan

16.1.4.2 The Mature Stage (The Tang, Song, and Yuan Dynasties)

The existing wooden objects can be seen in the Tang Dynasty architecture. During this period, the beams in the house and the bracket set on the pillars have been combined into one. The bracket set along the left and right sides of the house is also intersected with the trunk-like crowns (Fig. 16.13). A circle of stigmas and beams at right angles and sides of the building are interwoven into a horizontal frame divided into several well-shaped lattices, and the bracket set becomes the reinforcement joints at each intersection. At this time, the bracket set is no longer a supporting beam or the eaves component, but an inseparable part of the horizontal frame. This horizontal frame, now known as the "paving layer," is used on the palace-shaped frame column network, playing a key role in maintaining the integrity of the wooden frame. This practice began in Tang Dynasty and gradually improved in Song Dynasty, and lasted to Yuan Dynasty (Fig. 16.14).

Fig. 16.13 East Hall of Foguang temple, Wutai Mountain

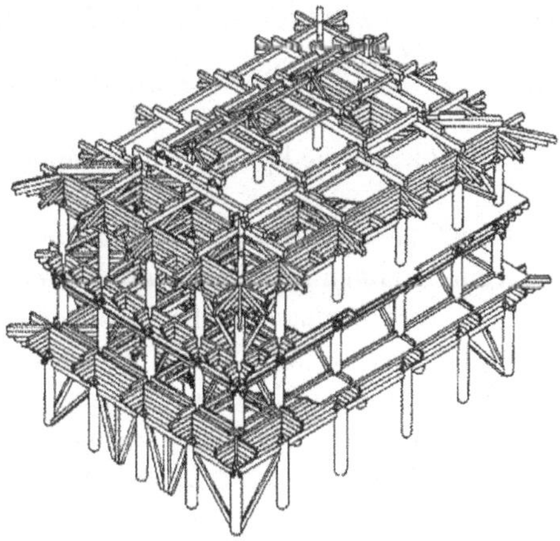

Fig. 16.14 Paving layer of Guanyin Pavilion in dule temple, Jizhou District, Tianjin

16.1.4.3 The Transitional Period of the Dougong Bracket Function (The Ming and Qing Dynasties)

Since the Ming Dynasty, the large and small foreheads and follower beams have been used between stigmas. The size of the bracket set has been shrinking and the spacing of the bracket set has been encrypted. The relationship between the bracket set and the beams in Qing architecture is no longer as close as that in Song architecture. Therefore, the bracket set has no paving layer since the Ming and Qing Dynasties. Its material and size are much smaller than those of the Song Dynasty, but it still has the function of supporting the lifting eaves, which is becoming increasingly prominent as a symbol of the level of architecture (Fig. 16.15).

Fig. 16.15 The bracket set of the building of the Forbidden city in Beijing

16.1.5 Construction Method of the Lifting Beam-Frame Bracket Set

The structure of the bracket set is simply divided into three parts: Dou, Gong, and Ang.

"Dou" is a wooden loquat that directly bears the weight of the crosswood, loquat, or beam. Dou can be divided into several types. Take the Song architecture as an example, the part placed at the bottom of a group of the bracket sets is called Ludou, which is generally used on the upper side of the middle line of the column. The transverse and vertical Gong are placed on the cross opens on Ludou. The transverse Gong revealed is called Huagong and the vertical Gong is called Nidao Gong. The part at the joint of Huagong, Gong, and Ang is called cross Dou. Gongs are supported and fixed by Dou. Qixin Dou is located at the center of Gong. The Dous placed at both ends of the transverse Gong is called San Dou. The bracket set is joined by tenon and mortise. The Ang bolt in the slant Ang penetrates into the lower Gong to fix it. Despite their different names, these buckets are almost identical in shape, except for their size and grooves.

The name of the bracket set in *Engineering Practice* of the Ministry of the Qing Dynasty is different from that in *Ying Zao Fa Shi* published in the Song Dynasty, but the main change is that the components of the bracket set become smaller and its role in structure is greatly weakened. In Song Dynasty, it was called "paving," which was divided into "pillar paving," "filling paving," "corner paving," and "corner paving". In *Ying Zao Fa Shi*, different appellations are given according to the complexity of the twigs: "One jump is called four pavements, two jumps are called five pavements, three jumps are called six pavements, four jumps are called seven pavements, and five jumps are called eight pavements." The counting unit of a spar is "paving." The smallest spar should be composed of four layers of components. The four-layer component of a spar that jumps out consists of a spar, a Chinese spar (or high), a juggler, and a square lining head. The four layers are arranged, as shown in the profile of a spar (Fig. 16.16). If there is a crossbar (parallel to the facade of the building), it is called "Jixin Making." If the crossbar is not directly supported by a layer of tower or high, it is called "mind-stealing making." The one that only uses one layer of crossbar is called "single Gong building," and two layers of crossbar are called "double Gong building" (Fig. 16.17).

The Qing style bracket sets are named differently according to the different locations. A group of bracket set is called "Yicun," "Zhu Touke" on the pillars, "Flat Family" on the forehead and "Angular Family" on the horn posts. Specifically, the number of tramples is generally referred to as trampling. The number of tramples refers to the number of tramples picked out from a trample. The expressive method of the Qing style bracket set is as follows: a few steps, several warps, and several elevations (Fig. 16.18). The simplest bucket is a bucket of three liters or a bucket of two liters of alternating hemp leaves which are not trampled on. The maximum is eleven buckets of the five-bracket sets. In the Qing bucket, there is also a kind of "gold bucket" used for the flat body section of the lower eaves of the out-reaching

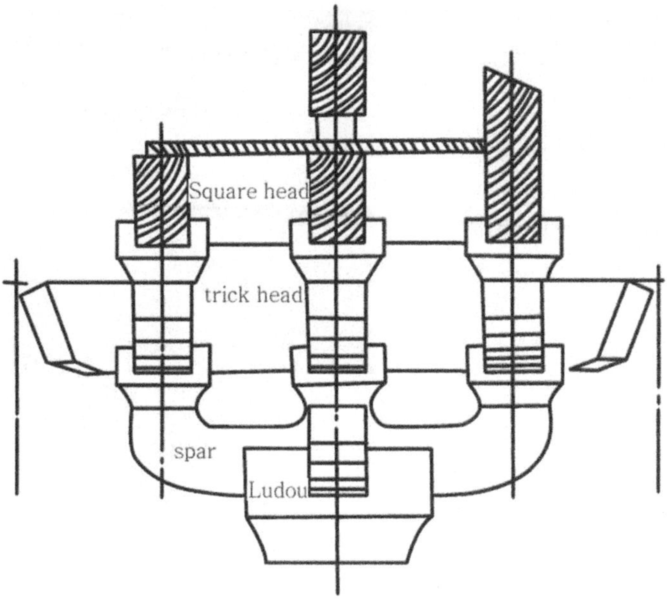

Fig. 16.16 The cross section of the four pavements of the Song style

eaves or heavy eaves. It has a high front and a long slanting one-step stand at the back. It is evolved from the Song style lower-lift, i.e., the weighing pole is raised, but the outer eaves are changed to flat. The structure of the whole structure is not as reasonable as that of the Song style.

16.2 The Originality Achievements of Ancient Chinese Wooden Architecture

When the wooden construction technology developed to the tenth–eleventh century, it was already quite mature. Li Jie of Song Dynasty compiled *Ying Zao Fa Shi*, which reveals a lot of innovative achievements about the architectural techniques, which is summarized as follows.

16.2.1 Establishment of a Modular System of Wooden Structure

This modular system chooses the largest number of components in the structure as the modulus. "Material" refers to the section of "gong" and "fang" in the building.

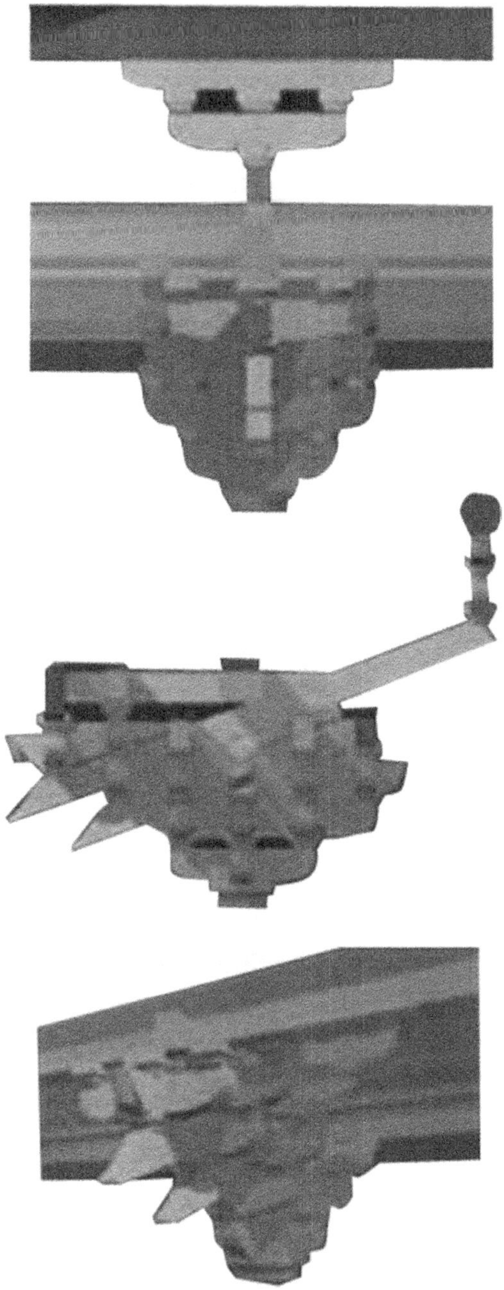

Fig. 16.17 The model of the six pavements of the Song style

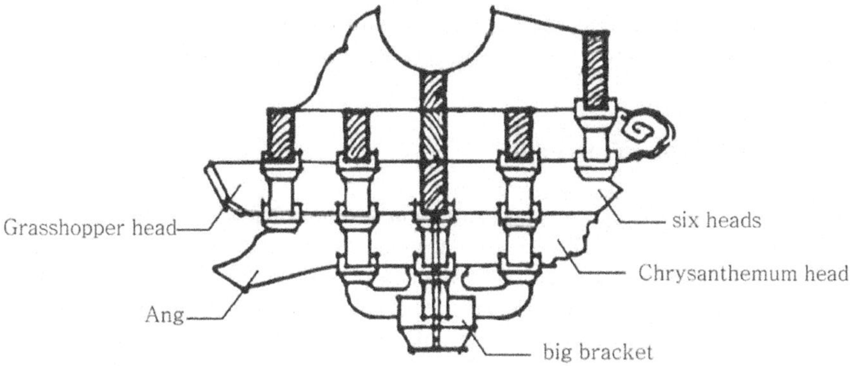

Fig. 16.18 Single step, single elevation, five warps of the Qing style

In the whole material system, there are eight grades of material, of which the largest is 9 in. by 6 in., and the smallest is 4.5 in. by 3 in. The length and straightness of the beams, columns, trusses, rafters, foreheads, and various components on the tendons of wooden structures all can be measured by material, which can be divided into "foot material" and "single material." The single material is 15 copies tall and 10 copies wide. One of them is called "one-point degree" in *Ying Zao Fa Shi* (pronounced as "fèn"). The height and width of the foot are 21 and 10 points, respectively. The difference between the single wood and foot wood is called "Cotoneaster," which is 6 points high and 4 points wide. For a building, first of all, we should choose its material grade, as is recorded in *Ying Zao Fa Shi* that "all the building systems are based on materials of eight grades, so the size of the house is used...... The height and depth of every building, the length of its name, the tendency of its straight and straight development, and the appropriateness of its rules and regulations are all based on the division of the materials used" (Fig. 16.19).

What are the more profound concepts of this modulus with a cross section of a component-material than the simple digital modulus? From the application of the

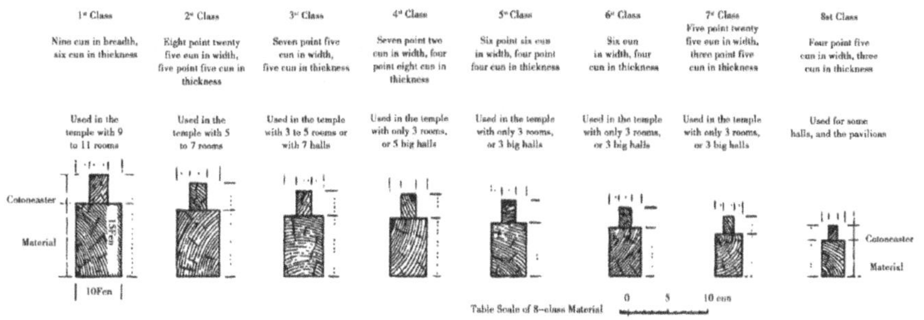

Fig. 16.19 The analysis chart of Song Dynasty's wood structure modulus system contained in *Ying Zao Fa Shi*

modular system of "material and fèn," we can see that it contains concepts of strength, scale, and structure, which are not typical of other modular systems.

16.2.1.1 Concepts of Strength

In the big wood system, the main structural components, such as the beams and appendices, which are measured by the modulus of "material and fèn," all have more scientific cross-sectional forms, which are generally recognized by architectural historians. The section form of the beam will be discussed in detail in the next part. At the same time, we can also see that in the material system advocated by the *Ying Zao Fa Shi*, the "foot material" appeared. The components using foot material as a modular unit in the big wood system are mainly cantilevered components such as Chinese Gong and Dingtou Gong. In the buildings at that time, the weight of a canopy supporting the overhanging eaves was an important factor bearing the structural stress. A canopy can be regarded as a short cantilever beam, which was much larger than other transverse forces in the pavement. Therefore, the section needs to be enlarged, but the craftsmen did not generally enlarge the height and width of the section. They only increased the height and width of the section to make it wider. The ratio is 21:10 to improve the bending resistance of the cantilever beam. The use of foot materials more clearly confirms the intention of the ancient craftsmen to incorporate the concept of strength into the modular system of "material and fèn."

16.2.1.2 The Concepts of Construction

It is precisely because of the use of the uniform "material" as the cross section of the Chinese wooden building with the system of tendon that every building can be guaranteed to have the standardized structural joints for joining the tendon so that dozens of different shapes can be laid one by one. Therefore, it is self-evident that the modular system of "material and fèn" contains the concept of structure. The structural mode of the combination of material and cottage becomes the basic pattern for the construction of joints everywhere. In the French system, when it comes to several materials, if the size of the beam height or column diameter is not specified in particular, it means the construction method of several layers of lint or lint overlapping with each other. The so-called "several materials and several cottage" is a synonym of a certain construction mode, forming a standardized construction mode.

16.2.1.3 The Concept of Scale

There is also a close relationship between the modular system of "material and fèn" and architectural art. In a group of buildings, the use of different grades of "material" can tell the difference of buildings in different scales, thus distinguishing the main, secondary, high, and low buildings. Such a layout of buildings is precisely due to the impact of ethical culture on the Chinese architecture. For example, the Shanhua Temple in Datong, Shanxi, where the materials of the main hall and the three temples are between the second- and third-grade materials, the gate is made of the fourth-grade materials and the Puxian Pavilion is made of the fifth-grade materials.

The problem of the building scale that can be considered by the modular system of "material and fèn" is still limited to a certain range. It mainly controls the scale of large timber frame. Some components, such as the height of window sill, railing, and the detail size of doors and windows, also have a great influence on the building scale. These components are not big wooden work, so they are not controlled by "material and fèn." *Ying Zao Fa Shi* gives a clear response to the scale problems reflected by them in corresponding chapters.

The production of this modular system is closely related to the productivity and production relations at that time because the official buildings at that time were constructed by the construction team of the official handicraft industry. In the process of construction, the craftsmen adopted a specialized division of labor. The craftsmen who made girders took charge of the processing and installation of all such components in the whole building group. The craftsmen who make tendons were responsible for the processing and installation of tendons in all houses of different sizes in the building group. When craftsmen accepted the construction task, they had no chance to see such detailed construction drawings as today but relied only on the oral explanation of the master craftsmen who presided over the project. Of course, it was impossible to speak about all respects. They could roughly account for the overall control range, intervals, number of spars, number of pavers, and so on. According to a set of rules passed down from generation to generation, the craftsmen determined the grade of the building materials, carried out the component processing, and finally assembled them into a building. The modular system of "material and fèn" not only ensures that the components processed by them have standardized nodes, so that they can be assembled accurately and correctly, but also ensures that the components have sufficient strength and each building in the complex has appropriate scale.

The vitality of the modular system of "material and fèn" also lies in the simplification of complex dimensions in construction. For the same type of components, their material and dimension are the same. When they are used in different grades of buildings, only their material and dimension need to be remembered, not their actual dimension. The craftsmen only need to use the eighth-grade materials to make different materials and sub-grade benchmarks, and then they can lay out and process, so as to reduce construction errors.

Unfortunately, this analog system has not been passed down. In the *Engineering Practice* of Qing Dynasty, the Ministry of Industry redesigned the digital analog system based on the size of the big bucket mouth.

16.2.2 Specification of Types of Buildings

The buildings of that time, according to *Ying Zao Fa Shi*, are divided into three categories: palaces, halls, and spare houses. And the types of their structures are also divided into three categories. There is another kind of material relic, which can be called the pavilion structure, which is grouped into the palace type.

16.2.2.1 The Palace Structure

The palace structure is used for high-grade buildings, with three characteristics:

(1) Two sets of structures (truss and grass truss) are used.
(2) The inner and outer columns are of the same height. The appendix and floor truss are placed between the columns to form a column frame layer.
(3) There are clear layers.

The structure of each palace is composed of roof, paving layer, and column frame layer by layer. In addition, those with sub-steps need to insert sub-steps frame around the hall body, i.e., half-slope roof, paving layer, and sub-steps column frame layer. A pavilion structure is created by modifying the palace structure. At that time, there were waist eaves and seats between the floors in the pavilion. So there appeared a dark layer of structure, namely, the flat floor supporting the floor in the interior, forming a podium in the exterior of the room, on which one can look far into the distance.

Existing examples are the Guanyin Pavilion of Dule Temple and the Yingxian wooden pagoda.

16.2.2.2 The Hall-Style Framework

The inner and outer columns are different in height. The inner column rises to the head of the girder or the section under the girder, on which it is subjected. Its characteristics are as follows:

(1) The beams and trusses are thoroughly made without straw trusses, and the tail of the beams is inserted into the inner column body.
(2) Paving is relatively simple. At most six beds are used, as generally four beds are enough. Because of the rise of the inner column, the back end of the beam truss can be directly inserted into the inner column body without paving. As no paving layer has been formed, the outer eaves are the main paving layer.

The structure of the hall changes with the depth of the house and the number of the inner columns.

16.2.2.3 The Remaining Roof Frame

Also known as "pillar and beam work," it belongs to a simple structure supported only by pillars and beams, and it is used in some minor buildings.

16.2.3 The Proposed Scientific Proportion of the Stressed Member-Beam Section

In *Ying Zao Fa Shi* the lengths of beams and the size of sections are stipulated. What is mentioned in the book mainly refers to the size of cross-sectional materials used for main beams in the frame. The section height is mostly 1/10–1/13 of the length of the beam, and it is required that the ratio of height to width of the section should be the ratio of height to width of the "material," i.e., the ratio of height to width is 3:2. For large-scale high-grade buildings, the length of the beams increases, and so does the section (Fig. 16.20).

Da Vinci once made a conclusion that the ratio of cross section of a beam is a scientific problem of a stressed member.

Leonardo Da Vinci put forward the principle which was considered to be of universal significance at that time: "Any supported object that can bend freely, if its cross section and material are uniform, is farthest from the fulcrum and its bending is the largest [1]." Through experiments, he concluded that the strength of the beams supported at both ends is inversely proportional to their length, but directly proportional to their width. In other words, the longer the beam with the same cross section is,

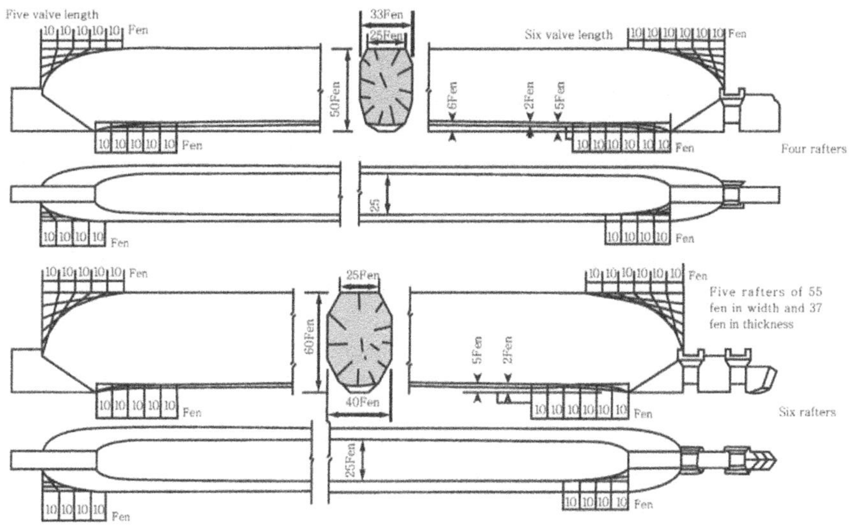

Fig. 16.20 The beams of beam-lifting girder frame contained in *Ying Zao Fa Shi*

the smaller the strength will be; the larger the width of the beam with the same length is, the greater the strength will be. Comparing this conclusion with the summary of *Ying Zao Fa Shi*, we can see that the provisions on the slenderness ratio of the beams in this book contain the meaning that the length is inversely proportional to the strength. Leonardo Da Vinci's conclusion on the relationship between strength and width of the beams is not a bit closer to the essence of the problem than the stipulation of the height–width ratio of the beams in *Ying Zao Fa Shi*. The height dimension of the beam is 1.5 times the width stipulated in *Ying Zao Fa Shi*, which shows that it was recognized that the height dimension of the beam was more important than the width dimension of the beam in the force, but da Vinci did not realize this.

It was not until in the seventeenth century than Leonardo's conclusion on this point was broken through by Galileo. Galileo put forward in *Two New Sciences*: "If the width of a wooden ruler or a thick rod is larger than its thickness, and when it is erected on a wide edge, its ability to resist fracture is greater than when it is laid flat, and its proportion is the ratio of thickness to width [2]." Here, Galileo has proved that the key factor affecting the force of the member is the height of the cross section. The bearing capacity of the member is good when it is placed vertically, which shows that the strength is closely related to the height of the cross section. It is concluded that the ratio of strength between erection and flattening is the ratio of thickness to width, which indicates that the change of width has little effect on strength. However, the most appropriate ratio of height to width of the bar section is not given. Therefore, Galileo's conclusion on this issue has not reached the depth of the conclusion that the ratio of height to width of the beam section is exactly 3:2 in the *Ying Zao Fa Shi*.

Thereafter, in a report on the bending of beams, a mathematician and physicist, Parent (1666–1716), who lived from the second half of the seventeenth century to the beginning of the eighteenth century, summed up a scientific method when he talked about how to intercept the rectangular beam with the greatest strength from a log, namely, the product of AB and AD on both sides of the rectangular beam must be the maximum. At this time, the diagonal DB of the rectangular beam is the diameter of the log, which is divided into three equal parts by the vertical lines made from A and C (Fig. 16.21) [3]. According to this conclusion, the ratio of the long side to the short side of a rectangular beam can be obtained, when the short side is 2, the long side is 2.8. This is close to the ratio of height to width of the beam 3:2 stipulated in "*Ying Zao Fa Shi.*"

From the end of the eighteenth century to the beginning of the nineteenth century, Thomas Young (1773–1829), a British scientist, also confirmed Palente's conclusion, and further found that when a rectangular beam is taken from a known cylinder, the bearing capacity of the beam varies with the height and width of the cross section:

When the ratio of height to width is $\sqrt{3}$:1, the rigidity is the largest.

When the ratio of height to width is $\sqrt{2}$:1, the strength is the largest.

When the ratio of height to width is 1:1, it is the most elastic.

The ratio of height to width of beam section is 3:2 stipulated in the *Ying Zao Fa Shi*, which can be regarded as the median value of both, taking into account both stiffness and strength.

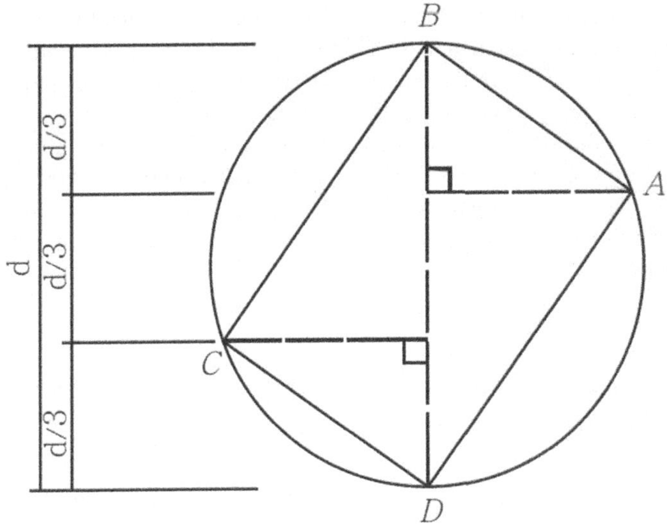

Fig. 16.21 Reasonable section diagram

Fig. 16.22 Guanyin Pavilion, Dule temple in Jizhou District, Tianjin

Maybe some people will think that *Ying Zao Fa Shi* is just a summary of the practical experience at that time, which may not necessarily be a consideration like this. However, from the retained objects at that time, it cannot be explained by a simple *Ying Zao Fa Shi*. From 95 parameters of the beam section size in 24 existing buildings, it can be seen that about 50% of the ratio of height to width of the beam section is within the range of $\sqrt{2}$:1 to $\sqrt{3}$:1. The ratio of height to width of the beam section is precisely defined as 3:2 in *Ying Zao Fa Shi*. It should be admitted that this is the conclusion drawn by the ancient craftsmen after a careful study of the strength and stiffness of the beam. Therefore, this conclusion should be regarded as a theoretical improvement. It should also be pointed out that the section size of the beams stipulated in *Ying Zao Fa Shi* is generally slightly larger than that of the existing ancient buildings in Tang and Song Dynasties. This may be due to the fact that the rules and regulations stipulated in the book *Ying Zao Fa Shi* are specially applicable to the construction of important national construction projects, thus paying special attention to the quality and safety of the projects.

At the beginning of the twelfth century (1103), it was surprising that the book *Ying Zao Fa Shi* was able to draw such a valuable conclusion, about 600 years earlier than the conclusion of Western scientist Palente. It has been scientifically confirmed by the experiments and practices of many scientists in later generations, which is really laudable.

16.2.4 Creation of a Structural System with Superior Seismic Performance

The ancient Chinese wooden structures have excellent seismic performance, not only because they adopt the tenon–mortise joints so that the overall structure can stand above and move on the ground, and absorb seismic energy, but also because they have a scientific structural system. Here are some examples.

16.2.4.1 The Guanyin Pavilion, Dule Temple, Jizhou District, Tianjin (Fig. 16.22)

The Guanyin Pavilion of Dule Temple in Jizhou District of Tianjin, built in the late tenth century, is a wooden pavilion with five rooms (20.2 m) wide, four rooms (14.2 m) deep, and three floors (two bright floors, one dark floor, 23 m high).

A pavilion-type frame with diagonal braces between columns is adopted (Fig. 16.14). The frame method is superimposed by three-storey frames, and the top of which is supported by a beam-lifting structure (Fig. 16.23). Around the three-storey frame, there are short beams—breast truss between the inner and outer columns. One end of the breast truss extends into the outer eaves and the other end extends into the inner eaves. The breast truss is joined with the bucket, spar, and eave in the paving by tenon and mortise. The vertical components such as stigma, groove, and Luohan fang intersect vertically with the breast truss to form a grid, which is distributed around the frame with three up-and-down paths, and the ceiling or floor is laid in the middle of the grid. In this way, three rigid rings are formed, which, like the ring beams used in modern buildings, play a reinforcing role in resisting horizontal loads and improve the overall stiffness of the structure. Secondly, several groups of diagonal components are arranged in the frame, which plays an important role in improving the stress condition of the frame. In the structure of the Guanyin Pavilion, inclined components are used in many parts. In the horizontal direction of the frame, there are two kinds of inclined beams: the first is four groups of 45-degree short beams at the four corners of the building. Each group has four upper and lower beams. The second

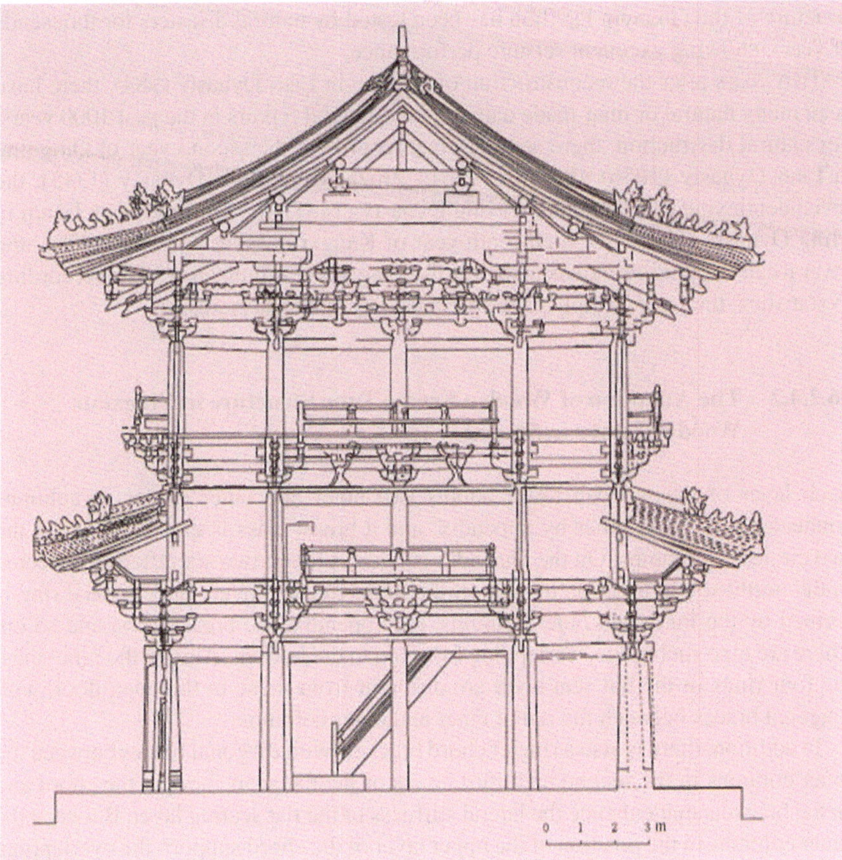

Fig. 16.23 The cross section of the Guanyin Pavilion, Dule Temple,

is the middle part of the dark layer. Four short beams are used between the inner columns to make the middle waist of the three-storey frame look like a hexagonal frame so as to place the statue of Guanyin through the three stories. In addition, in the vertical direction of the frame, there are oblique members as diagonal braces in the wall between the outer eaves and the inner columns in the hidden layer. These inclined parts are very helpful to improve the stiffness of the frame. Especially in the four corners of the pavilion, four groups of rigid supports are formed, which are similar to the space grid structure.

In his book *An Introduction to Architecture*, the Japanese scholar Naoto Fukushima once mentioned that it was only after the Anzheng earthquake in 1855 that the method of using diagonal braces between columns appeared in Japan. And this was the first appearance of Japan's seismic structural method, which is still widely used in seismic structures. However, the existing Guanyin Pavilion built in 1984 in Jizhou District of Tianjin has adopted the method of diagonal bracing between pillars. As a result, the structure of the Guanyin Pavilion has been tested by natural disasters for thousands of years, showing excellent seismic performance.

Two years after the reconstruction of Tonghe in Liao Dynasty (984), there have been many natural or man-made damages and several repairs in the past 1000 years. For natural destruction, there are historical records of the second year of Qingning in Liao Dynasty (1056), the fifth year of Zhizheng in Yuan Dynasty (1345), the seventeenth year of Chenghua in Ming Dynasty (1481), the fourth year of Tianqi in Ming Dynasty (1624), the eighteenth year of Kangxi in Qing Dynasty (1679), and several catastrophic earthquakes such as the Tangshan earthquake in 1976. According to statistics, the Dule Temple has experienced up to 28 earthquakes.

16.2.4.2 The Adoption of Wooden Frame Tube Structure in Yingxian Wooden Tower in Shanxi

Each layer of the wooden tower adopts two inner and outer circles of columns connected with each other by appendix, and a breast truss is erected between the two circles of columns. On the appendix between pillars, two six rafters are erected in the north–south direction to support the floor. In each layer, an octagonal ring is formed by the inner and outer columns, the appendix, the breast truss, and so on. There are nine such rings overlapping from top to the bottom. Among the nine rings, the four rings in the flat seat layer are different from those in the open floor, with diagonal braces between the radial inner and outer columns.

In addition, there is also a ring of chord brackets with diagonal braces between the inner columns in the second and third layers of the flat seats there in the chord and radial brackets that enhance the lateral stiffness of the flat seating layer. Between the inner columns in the position of the upper layer of the chord support, the overlapping multi-layer wooden loquats are arranged to form a closed ring beam along a slit in the middle line of the octagonal column. At each corner, the position of each side corresponding to the outer eaves column and the middle part of each side are provided with radial wooden fang connected with the outer eaves flat seat bucket, integrating the inner and outer grooves. In addition to the wooden loquat with the flat seat in the inner groove, there are also many wooden loquats in the part with the bucket loquat, which also constitute several closed octagonal ring beams. Moreover, among the eaves and columns of the wooden tower, except for the four facades, all the oblique sides are plaster walls. The original diagonal braces in the walls were removed only in 1936, and all of them were replaced by lattice doors.

The use of chord and radial braces between columns enhances the plane stiffness between columns. The closed wooden frame composed of stigma and loft acts as a ring beam, constituting the structure system of the wooden tower with an octagonal cylinder of strong rigidity. Therefore, it can be classified as the "ancient high-rise wooden frame cylinder structure." It has many similarities with the cylinder structure in modern buildings. The modern cylindrical structures are considered to have the best seismic performance, and the fact that Yingxian wooden tower is still standing after many earthquakes and shelling confirms the excellent seismic performance of ancient wooden frame cylindrical structures (Fig. 16.24).

The structural system of the wooden pagodas is the most scientific one in ancient Chinese wooden structures. It goes beyond the unstable state of parallelogram between columns and beams in the system of the lifting beams and piercing buckets, applying the diagonal braces between columns and beams and matching with the closed wooden frames of several circle beams above and below. In this way, it constitutes a kind of space structure similar to the modern one as a whole, which has left the most brilliant chapter in the history of Chinese wood structure development.

16.2.5 Proposal of Some New Technologies in Material Processing

16.2.5.1 Manufacture of the Ring Beam

When it comes to constructing a circular building by using the traditional wooden frame with the beam-lifting structure, and if they are processed with a straight piece of wood, they need to dig inside and stick outside to make an arc for circular beams, which takes both wood and labor. The craftsmen used a "wet bending method" to soak the straight wood in water and then bend it under pressure. A typical case of this practice is the Temple of Heaven in Beijing, which was rebuilt in the Ming Dynasty and rebuilt in the Qing Dynasty (Figs. 16.25 and 16.26).

16.2.5.2 Packed Pillar and Pillar Manufacturing

For hundreds of years, with the consumption of timber, bulky timber is increasingly scarce, and small timber is easy to obtain, so there are "composite columns" and "composite beams." The earliest examples of this kind of construction, such as the Baoguo Temple Hall in Ningbo, Zhejiang, built in 2013, have the most typical inner columns: Assemble four logs together, then make up the logs with the wooden strips, and make the surface into a cucurbit prism to weaken the mark of the combination. In the Ming and Qing Dynasties, the "wrapped column" appeared, with small diameter of wood as the core, using a curved cross section of the wooden bar wrapped on a layer so that the diameter of the column increased.

The building of Pavilion in Song Dynasty is to stack up the wooden frame layer by layer with each layer of pillars inserted into the lower column beam, which is known as "plunger construction." This method will produce residual deformation when encountering external forces such as earthquakes. The columns in the upper and lower layers incline irregularly to each other, and the serious ones show an S-shaped trend. After the Ming and Qing Dynasties, this situation changed. By using the method of the wrapping columns, a section of columns was lengthened, and the

Fig. 16.24 The cross section of Yingxian wooden tower

Fig. 16.25 The Temple of Heaven in Beijing

Fig. 16.26 The Hall of Prayer for Good Harvest

joints of the outsourced wooden strips and columns were interlaced with each other, so that a long column of the upper and lower integration could be formed and used on the pavilion, which is called "through-column construction."

16.2.6 Colored Painting Techniques for Protecting Against the Rise of the Times

China had the technology of drawing color paintings on wooden structures as early as the Han Dynasty 2000 years ago. This technology uses natural minerals or plant pigments to draw patterns on the surface of the wood components. It can not only protect wood from corrosion and insects but also achieve the artistic effect of beautifying the buildings. Moreover, it has adopted different patterns and paints at different times, and has become an excellent document for recording the architectural age. It can also express the aesthetic ideal of users and distinguish the rank of buildings, which is a rare technology that serves multiple purposes.

16.3 Conclusion

We have so far briefly introduced some main contents of the Chinese traditional wooden structure architecture, and have seen the ingenious tenon–mortise structure, the wooden beam frame of the beam timber, the "hook-heart-fighting angle" tenon, as well as the unique modular system, the structural system with excellent seismic performance, etc. Every item is enough to make people realize the superb and exquisite technology of ancient Chinese wooden construction. Why can we call "ancient Chinese wood construction technology" a great invention in ancient China? As an integrated system, the wood construction technology is the sum of technology, skills, knowledge, and experience of various complex scientific principles, and can perfectly meet the needs of ancient Chinese society, so it has been continuously developed and improved in thousands of years. Although it is impossible to determine the specific inventor and the time of invention, it is precisely because of the perfect relationship between technology and society that we can leave examples of the wooden structures with a life span of more than 1,000 years, which can make them stand out in the world's architectural forest and spread to Japan, the Korean Peninsula and other countries in South Asia. It is for this reason that Chinese wooden architecture is regarded as an important invention in ancient China, which is well-deserved.

References

1. [US] Tiemusheng, S. P. 1961. *History of Material Mechanics (Chinese Version)*, 5. Chang Zhenmei [trans.] Shanghai: Shanghai Science and Technology Publishing House.
2. [US] Tiemusheng, S. P. 1961. *History of Material Mechanics (Chinese Version)*, 12. Chang Zhenmei [trans.] Shanghai: Shanghai Science and Technology Publishing House.
3. [US] Tiemusheng, S. P. 1961. *History of Material Mechanics (Chinese Version)*, 38. Chang Zhenmei [trans.] Shanghai: Shanghai Science and Technology Publishing House.

Daiheng Guo Born in Beijing in 1936, graduated from Tsinghua University in 1960 and was supervised by Professor LIANG Sicheng. He is currently a professor at Tsinghua University and a national registered architect. She is the former doctoral supervisor and director of the Department of Architecture and History of Tsinghua University.

Research interests: History of Ancient Chinese architecture; urban development and architectural heritage protection.

Major works: History of *Ancient Chinese architecture; The Memory Heritage of the Old Summer Palace-Drawing files of Yangshifang.*

She has completed more than 10 academic monographs and published more than 70 academic papers in domestic and international magazines.

Awards:

The first prize of the National Natural Science and the first prize of the State Education Commission for scientific and technological progress in 1987;

the Architectural Creation Award.

Peijun An Born in Beijing in 1971, Ph.D., national registered architect, lecturer at the North China University of Technology.

Research interests: Architectural History and Theory

Major works: *The Influence of Eight Flags System in Qing Dynasty on the Shop Forms in Beijing; Study on Jianruiying Diaolou in Qing Dynasty; "The History of Architecture" or "the Architecture of History": A brief discussion on two historical views on architecture.*

Chapter 17
Chinese Cooking

Pangtong Qiu

Chinese cooking is a great invention of the Chinese ancestors. For a long time, it has played a vital role in safeguarding the health and reproduction of the Chinese nation. Sun Yat-sen said in *The Strategy for the Establishment of a Nation*: "Cooking is born in civilization. If the cooking is not rooted from the culture, you are not good at distinguishing taste. If the taste discrimination is not oblivious, then cooking is not good. The excellence of Chinese cooking also shows the depth of civilization evolution [1]." This judgment is in line with historical facts. Under the influence of traditional Chinese culture, Chinese cooking is promoted, developed, and shaped by China's natural environment and products as well as the Chinese people's mode of production, lifestyle, and wisdom. It is unique in the world's food culture and is one of the main schools of cooking.

Chinese cooking does not only refer to food preparation and seasoning methods but also refers to the whole process of selecting food raw materials, processing, and eating. It has distinct characteristics and unique connotation, mainly as follows: the diet structure with grains, fruits, and vegetables as the main food and meat as the auxiliary; The daily diet mechanism is of two kinds: staple food and non-staple food. The production concept of "Cleanliness is the most important, health and should be combined, five tastes are appropriately applied for good health." There are the philosophic thoughts and etiquette folk customs of "harmony" and "harmony between man and nature" in our diet culture. In particular, it also involves the selection of the raw materials, primary processing, knife cutting and matching, temperature and cooking methods, seasoning methods, banquet combinations, production of reconstituted raw materials, and the flavor and genre of dishes.

P. Qiu (✉)
Expert Guidance Committee of China Cuisine Association, Beijing, China
e-mail: qiupangtong@sina.com

J. Hua and L. Feng (eds.), *Thirty Great Inventions of China*,
https://doi.org/10.1007/978-981-15-6525-0_17

17.1 The History of Chinese Cooking

17.1.1 Pre-historic Age (about 4000 years Ago)

In pre-historic times (about 4,000 years ago), since the ancestors broke away from the state of "drinking blood from shavings," the basic conditions for the formation of cooking in our country have been met by the invention of the artificial fire-taking technology, the emergence of early agriculture, breeding, and aquaculture, the emergence of pottery cooking utensils and cooking pits, and the success of salt cooking.

Archaeological results show that the natives of Beijing learned to make fire manually and cooked food with fire about 500,000 years ago. During the Neolithic period, China already had cereals such as grain, broomcorn millet, millet, rice, wheat (barley, wheat), and water shield; vegetables such as cabbage, mustard, and gourd; livestock such as cattle, sheep, pigs, dogs, and chickens; aquatic products such as fish, turtle, and clam; salt, and wild honey. Because of the pottery stove, pottery caldron, caldron, tripod, caldron and caldron (Fig. 17.1), together with fire and salt, the boiled and steamed rice and porridge, the flavored soup has been produced. From the unearthed pottery pan 6,000 years ago (Fig. 17.2), it is speculated that pancakes may have been found at that time. Qinghai unearthed millet noodles 4,000 years ago (Fig. 17.3).

Fig. 17.1 Pottery

Fig. 17.2 Pottery pan in the Neolithic Period

17.1.2 The Xia, Shang, Zhou Dynasty (2100 B.C.–221 B.C.)

During this period, Chinese cooking saw critical development characterized by the improved stoves, sharpened cutting tools, and the advent of bronze cookware (Figs. 17.4, 17.5 and 17.6).

Bronze cauldrons, pots, woks, kettles, and plates came with the cooking ways of pan-frying, deep-frying, and stir-frying. A flexible kind of cookware—chopsticks—was invented. There were a wider range of ingredients and condiments, including salt and meat sauce with salty flavor, maltose, sugarcane juice, honey with sweet flavor, plums with sour flavor, fermented soybeans with bitter flavor and pepper, ginger, cinnamon, scallions, leaf mustard, shallots with spicy flavor. Alcohol was also used as seasoning. The cooking oil derived from animal grease. The selection and combination of ingredients were stringent, and many ways of cooking, like grill, deep-fry, steam, pan-fry, stew, marinate, stir-frying, and so on [2].

The staple rice, porridge had been separated from the subsidiary food and chowder. A dietary structure featuring grains, fruits, and vegetables with meat as complementary was in place. Dietary therapy started to rear its head.

Generally speaking, Chinese cooking laid its groundwork during this period.

Fig. 17.3 Noodles from the "La" historical site

Fig. 17.4 Bronze rice steamer

17.1.3 The Qin, Han, Wei, Jin, Northern and Southern Dynasty (221 B.C.–589 A.D.)

The development of Chinese cooking hit a climax.

Notable signs are the birth of iron cookers and the widespread use of rotary stone mills. The iron kettle is light and transfers heat faster, which makes the cooking method of stir-frying stand out. The rotary stone mill has a high grinding rate, and

Fig. 17.5 Bronze container

Fig. 17.6 Oven tray with cloud patterns

with a sieve basket, fine powder can be sieved to promote the rapid development of noodle.

Cooking materials are more abundant. For example, there are dozens of vegetables.

Commonly used are woad (Chinese cabbage), leek, celery, gourd, water shield, radish, eggplant, agaric, and fungus. In Han Dynasty, there were also vegetables cultivated in greenhouses [3]. Zhang Qian imported cucumbers, garlic, coriander, alfalfa, etc. through the Western Regions. There are also many fruits. Pomegranate, grape, walnut, etc. are also introduced by Zhang Qian. The famous freshwater and seawater products include abalone from Bohai Sea, perch from Wu Land [4].

There were the Bing-shaped fish from the Sichuan region, and "Luo Lei Yi Fang" which is more expensive than cattle and sheep. There are also the sea salt, pond salt, rock salt, well salt, sugar cane, sesame oil, soy sauce and vinegar, and more than 20 kinds of vegetable seasonings.

Chinese people had commanded the skills of cooking (Fig. 17.7) at that time. Stir-fry, as an important invention, was textually documented, together with Soup, Chun soup (Fig. 17.8), Processed fish, etc. With salt, alcohol, cold rice, orange peel, and cornel being sealed and brewed, the processed fish has an impact on Japanese sushi. The dishes in Yangtze River and the Central Plain have exhibited their regional traits.

Pastry-making skills were greatly improved. Pastry fermentation method has become relatively mature, with cooking utensils such as steamer, oven, strainer spoon, and baking pan (Fig. 17.9). Pastry like pancakes, pancakes with soup, shuiyin, steamed buns, wontons, baked rolls, zongzi, and rice noodles were also made. Shuiyin, a great invention by Chinese people, refers to noodles. The method of making it is well documented in Chinese ancient books.

A distinctive vegetarian culture deriving from the fusion of the folk and Buddhist vegetarian diets emerged.

It is said that the King of Huainan invented tofu, a significant event in the history of Chinese and world culinary.

Fig. 17.7 Chef drawing in Han Dynasty

Fig. 17.8 Water shield and whitebait thick soup

Fig. 17.9 Lacquer tray from the Mawangdui Han Tombs

There were more than 30 cooking books written in this period, concerning dishes, sauces, alcohol brewing, eating mode, and health preserving.

A comprehensive framework of Chinese cooking has thus been established.

17.1.4 The Sui, Tang, and Song Dynasty (581 A.D.–1279 A.D.)

This is an important period for the Chinese cooking to develop fully its stereotype.

Iron cooking became more commonly used, and coals came into daily use. The vegetables cooked on a daily basis totaled more than 40 types, lettuce and spinach were introduced into China and edible fungi could be artificially cultivated. And both freshwater and seawater fishes were cooked more frequently. In terms of condiments, many kinds of soybean sauces, vinegar, sugar, and alcohol came into being. A large number of novel dishes were invented, especially medicinal dishes and vegetarian dishes, so did some new pastry, like dumplings and steamed buns (Fig. 17.10).

Tofu as an ingredient was documented in books. And red yeast was also available for cooking.

The exchange between China and the west in cooking has been quite frequent (Fig. 17.11). Many western dishes were introduced into China and Chinese dishes, and cooking methods and banquet formats went to Japan [5]. It is also believed that tofu was exported to Japan in Tang Dynasty [6]. In Song Dynasty, the Chinese cooking has gone through comprehensive development, hitting another climax. The iron woks, ovens, hot pots, movable stoves, trays, baking pans, steamers (Fig. 17.12) were widely used in cooking. Ingredients such as tofu, gluten, rice noodles, bean sprouts, Chinese chives, preserved meat, freshwater fishes, sea food, bamboo shoots,

Fig. 17.10 Pastry unearthed in Xinjiang in Tang Dynasty

Fig. 17.11 Female tomb figure made of flour unearthed in Xinjiang in the Tang Dynasty

Fig. 17.12 Food steamer on the tile carving in Song Dynasty

and fungi were increasingly used. There was a wider variety of condiments, the term "soybean sauce" was coined. The range of available cooking oils was expanded to include flax oil, cannabis oil, almond oil, safflower seed oil, and turnip seed oil.

The main cooking methods were in place, including diversified ways of stir-frying and steaming, deep-frying, boiling with sauce, grilling, stewing, and baking.

Food was seasoned in multiple ways. A conclusive statement about seasoning with a far-reaching effect was made: "food should be seasoned to cater to the taste of diners [7]."

Various types of dishes, court dishes, folk dishes, and vegetarian dishes gained all-round developments.

New flour-made varieties like steamed stuffed bun, jiaozi, and sweet dumplings, etc. form the noodles, mangtou, etc. came into existence. The food market flourished with restaurants serving different types of dishes. Northern cuisine, southern cuisine, Sichuan cuisine, and vegetarian food emerged.

As the exchange between China and the west advanced, Japanese monks came to China to learn from our tea culture and the cooking of the vegetarian food, the skills of which were brought back to their homeland [8].

17.1.5 The Yuan, Ming, and Qing Dynasty (1271 A.D.–1911 A.D.)

This is the period when Chinese cooking came into its maturity stage.

In the Yuan Dynasty, ovens for roasting duck and lamb, and the cooking utensil for helao, a kind of Chinese pastry, appeared.

Some nationalities, like Mongolia, Nvzhen, and Hui, developed their own distinctive cuisines. Dishes with soup developed quickly. There was a wider range of medicinal dishes.

Dishes were seasoned in diversified manners, like the delicious boiled wheat bran from the city of Wuxi was added with up to 11 condiments.

Flour dishes created by minority ethnic groups emerged, like Siu Mai, spring rolls, noodles, and other pastry.

Many foreign scholars hold that the Chinese noodles were introduced into Italy and the steamed buns into Japan and Korean Peninsula during Yuan Dynasty.

The cookware saw new progress in Ming Dynasty. The iron wok sets made in Foshan, Guangdong were exported overseas.

Some precious ingredients like the bird's nests, sea cucumbers were used in cooking. Vegetables like sweet potatoes, corns, tomatoes, and peppers made their way to China.

About a dozen types of vegetal oil were available for cooking, among which soybean oil and seed oil were considered as premium oil. Tea oil and olive oil were also used at this time.

A lot of new dishes were created, and more than 30 cooking methods were in place. The dishes in many areas, such as regions south of the Yangtze River, Fujian and Guangdong, Beijing, Jiangsu, and Shandong have gained swift development.

The torn noodles (Fig. 17.13) emerged, as well as the moon cakes, rice noodles, and Guang pancakes. Besides, the Suo noodles cooked with broth can be dried and stored for future cooking, which is a great convenience. Several cooking treatises were written. In Qing Dynasty, there were a wider range of cookware (Fig. 17.14) such as woks, hot pots, saucepans, stoves, and grills for meat barbecue. The western oven was introduced and the stoves were further improved (Fig. 17.15). Famous brands for cutting tools emerged.

Food utensils such as saucer, plate, bowl, box, basin appeared with more specifications. There were even utensils that can keep food warm.

There were varieties of food resources, and peppers were introduced into Chinese provinces such as Hubei, Hunan, Sichuan, Yunnan, Guizhou as either vegetable or condiment.

Menu for food were greatly developed. There were more than 40 kinds of cooking methods, including grilling, steaming (Fig. 17.16), boiling, deep-frying, pan-frying, baking, and thousands of dishes, including Peking duck (Fig. 17.17), hot pot lamb (Fig. 17.18).

Sea Cucumber Stir-fired with Scallions from Shandong, Kung Pao Chicken (Fig. 17.19) from Sichuan, Braised Shredded Chicken with Ham and sliced dry tofu

Fig. 17.13 Torn noodles

Fig. 17.14 Hele Bed in Qing Dynasty

(Fig. 17.20) from Yangzhou, Sweet and Sour Mandarin fish from Suzhou, Sweet and Sour Pork Ribs from Wuxi, Sashimi and Roast Suckling Pig from Guangdong, roast whole lamb as a Muslim dish, stewed mixed vegetables as a vegetarian dish, and Duck with Chinese Caterpillar Fungi (Fig. 17.18) as a medicinal dish.

The skills for pastry making were fully developed with the folk customs. A lot of pastry made its way to people's dining table, like yuanxiao, zongzi, mooncake, the Double Ninth Festival cake, and rice cake (Fig. 17.21).

A flourishing catering market was characterized by intensive competition among numerous diners, bars, restaurants, and eateries.

The established banquet formats included all-lamb banquet, vegetarian banquet, eight-dish banquet, which are very demanding in terms of ingredients and taste.

Many local Chinese cuisines came into being, among which the most famous are the Lu cuisine from the Yellow River basin, Huaiyang cuisine (Fig. 17.22) from the lower reaches of the Yangtze River. Yue cuisine from the Pearl River basin, Chuan cuisine from the upper reaches the Yangtze River. These four cuisines all are delicious with distinguishing features, respectively. Lu cuisine is salty; Huaiyang

Fig. 17.15 Pancake-making drawing from the Qing Dynasty

Fig. 17.16 Steamed Wuchang Fish

Fig. 17.17 Roasted Duck in Stove

Fig. 17.18 Donglai mutton hot pot and Cordyceps Sinensis Duck Soup

Fig. 17.19 Kung Pao Chicken

Fig. 17.20 Braised Shredded Chicken with Ham and sliced dry tofu

cuisine strikes a perfect balance between being salty and sweet; Yue cuisine is light in flavor; Chuan cuisine is quite spicy.

Several cooking classics at that time were written, which include *Xian Qing Ou Ji, Yang Xiao Lu, Tiao Ding Ji, Xing Yuan Lu,* etc.

Fig. 17.21 Cake and Pastries

Fig. 17.22 Boiled Minced Crab Meatball

Thanks to the growing economic and cultural exchange, western dishes made their way to China. Meanwhile, Chinese cooking had an increasingly far-reaching influence among Asia, Europe, America, and Australia.

17.2 The Dietary Structure Featuring Vegetables and Fruits, with Meat as the Subsidiary Food

The dietary structure is closely related to people's health and the prosperity of nations. As early as 2000 years ago, Chinese ancestors established a dietary structure featuring vegetables and fruits, with meat as the subsidiary food. This is proven to be the most prominent feature and advantage of Chinese cooking.

A Chinese classic of the traditional Chinese medicine called *Huang Di's Canon of Medicine* said that people should take grains, fruits, meat, and vegetables to be healthy and strong, among which grains serve as the staple, fruits, meat, and vegetables serve as the subsidiary food. In ancient China, "five grains" refer to millet, barley, bean, wheat, and rice; "five meat" refers to beef, mutton, pork, chicken, and dog meat; "five vegetables" refer to okra, leek, leaves of pulse, shallot, and scallion. Since ancient times, Chinese people have been eating much more grains, fruits, and vegetables than meat.

The dietary structure was formed for many reasons.

First, it was because of the natural environment for cultivation and the choices made by human being in the ancient agricultural society. Many of the Chinese plains like the Loess Plateau, North China Plain, and the middle and lower reaches of Yangtze River, etc. at the time were suitable for the cultivation of grains such as barley, panicum, millet, wheat, rice, etc. In the process of foraging for food, our ancestors realized the seed of some plants were more desirable as food, so they were domesticated. In this case, green bristle grass was converted into millet, wild rice into cultivated rice. At the same time, many kinds of vegetables, including cabbage, leaf mustard, okra, and leek, and meat, including beef, mutton, chicken, and pork were included in people's daily recipes. Those grains can be harvested two times a year and are high-yielding and filling, so they accounted for a large proportion of Chinese people's diet. However, meat can only play a complementary role on account of low output.

Second, from the health-preserving perspective, grains, vegetables, and fruits can be eaten in great quantity, but excessive consumption of meat on a daily basis can endanger health. It is mentioned in *Master Lv's Spring and Autumn Annals* that too much meat and alcohol may hurt the stomach. Confucius also said that one is not supposed to eat more meat than grains.

Last, it was believed that humans should eat a wide variety of food. A Chinese classic *Da Dai Li Ji* mentioned that vegetarians may not be so intelligent, carnivores may be too belligerent, and people only feeding on grains are smart and intelligent. The formation of this dietary structure is also related to the fact that animals are omnivorous.

After the Qin and the Han Dynasty, the rationale of this dietary structure was further expounded in some medical classics, such as *Thousand Golden Prescriptions* in Tang Dynasty, *Taiping Royal Prescriptions* in Song Dynasty, and *Principles of Correct Diet* in Yuan Dynasty. New types of ingredients imported from abroad like cucumbers, spinaches, peppers, and tomatoes enriched the diet of the Chinese people.

Bai Juyi was a renowned Chinese poet in the Tang Dynasty, whose diet focused on vegetables and grains since he was middle-aged. He often ate rice, pancakes, okra, parsley, asparagus, and wild vegetables. For meat, he sometimes only ate fishes. In doing so, he drastically improved his mental and physical health.

The diet in the countryside is also worth mentioning., It is mentioned in *Record of Yongxing County* in late Qing Dynasty that the countryside diet featured rice, which is also the main ingredient for wine-making. The farmers relied on corns and sweet potatoes in summer and autumn to be energetic. If those two staples had a poor harvest, then they would have the roots of the ferns. However, fish, pork, chicken, and duck were the highlight of the banquet, with vegetables and fruits playing a minor role. In the period of the Republic of China, it is mentioned in *Record of Shunyi County* that the most commonly eaten grain was corn, and other grains like wheat, rice, sorghum, flax were less eaten; vegetables like scallions, Chinese chives, spinaches, Chinese cabbage, turnips, and leaf mustards were most common. Other foods like tofu, eggs, and meat were not eaten so often. It is also mentioned in *Record of Tongguan County* of the same period that people predominantly eat wheat and other grains, complemented by vegetables, and they eat meat only on festivals.

All the above indicates that the Chinese people have been following this dietary structure for more than 2000 years. Nowadays, a new guideline for diet has been released, which is built on the traditional dietary structure. Collin Campbell, the equivalent of Einstein in nutriology says in Chinese Health Report that a meat-based diet will lead to chronic diseases; a diet based on vegetal food is the best for health, and can effectively prevent and contain the chronic diseases [9]. This statement gives a high credit to Chinese traditional diet.

17.3 A Strict Selection of Ingredients and Sophisticated Cutting Skills

17.3.1 The Selection of Ingredients

This is an integral part of Chinese cooking. During the period before Qin Dynasty, this process has been attached great importance. Here are a few principles. First, select seasonal ingredients. *The Analects of Confucius* also emphasized this. Second, select clean and sanitary food. It is said in *The Rites of Zhou* that the meat of sick animals or stinky meat cannot be consumed, and the same thing happened to meat with cysticercus, which is in line with the practice of epidemic prevention department. Third, select the locally popular food, as was suggested by *Master Lv's Spring and Autumn Annals* therein dozens of local ingredients are enumerated, which are all well-grounded.

These three principles were particularly highlighted in Han Dynasty. Some Chinese classics also made some relevant requirements. It was mentioned in *Qi Fa* by Mei Cheng that vegetables harvested during the solar term of White Dew are the best. It is also recorded in *Ode to Shu Province* that the abalone from the east of Yangtze River, as well as the beef and the mutton from the east of Gansu are the best. It was also a testimony to the fact that the local ingredients from the east of Gansu and the east of Yangtze River have made their way to Sichuan. The classic of *Synopsis of Golden Chamber* included hundreds of food taboos, a lot of which are considered scientific.

The selection of ingredients was more meticulous in the Northern and Southern Dynasties.

Another Chinese classic *Qi Ming Yao Shu* mentions the selection of Ulva pheasant harvested in April for cooking water shield porridge, and suggests choosing the carps in spring and autumn for making fried carp.

Besides, one should choose ingredients according to the cooking method. The pork for roast must be suckling pigs, the fish for making fish pancake must be kept alive to the last minute, preferably whitefish. The goose and duck for roast must be less than 60 to 70 days old so that the meat can be tender [10].

After the Sui and Tang Dynasties, ingredient selection is distinctive in different historical periods. Yuan Mei in Qing Dynasty made a systematic exploration into the selection of ingredients.

He emphasized the selection of better breeds of animals and plants in one of his books—*Menu of Sui Yuan*. Pigskin should be thin, and the big itself should not be smelly. Chicken should be young. Ingredients of different breeds can be diametrically different. He also attached great importance to the selection of seasonal food. Unseasonal turnips are hollow inside, unseasonal bamboo sprouts have a bitter flavor, and unseasonal anchovy has hard bones. Besides, it is preferable to select ingredients according to the cooking method. One had to better cook with female chickens, male ducks, the bottom of water shield, and the root of celery.

17.3.2 Cutting Skills

Cutting skills, as a crucial part of Chinese cooking, refer to the cutting of ingredients based on the nature of ingredients, the cooking hours, the seasoning and aesthetics, as well as skills such as cutting, chopping, peeling, etc. in dealing with the raw materials.

Before the Qin Dynasty, the thin-edge copper knives appeared and the development of cutting skills gained momentum. In *The Analects of Confucius,* it is said that one should not consume improperly cut food. In Zhou Dynasty, animals can be cut into 7 pieces or 21 pieces to the needs of cooking or rituals. In addition, ingredients can be diced, sliced, shredded, and chopped up. Terms like big meat loaf, small meat loaf, meat shred, etc. have appeared.

In the Northern and Southern Dynasties, Hong'an, the cutting and cooking of meat, and Bai'an, the making of flour staple were distinguished from each other, which furthered the development of the cutting skills. Meat can be cut into transparent slices as thin as cicada's wings. It was stated in *Qi Ming Yao Shu* that the mutton for making sausage should be chopped up, and fish should be cut into thin layers. Different dishes have different demands for cutting skills.

In Tang Dynasty, cold dishes with a combination of different ingredients are very demanding in terms of cutting skills. It is mentioned in *Zhuo Kuai Shu* [11] that there are many ways of cutting fish, which represent a sophisticated craftsmanship.

In Yuan Dynasty, people started to pay attention to carving the ingredients, like the carved turnip flowers in some imperial dishes. Besides, the cutting of the ingredients should be in line with people's aesthetic taste. Proper cutting of ingredients can improve the taste of dishes. Meat chowder in Wuxi cuisine is to cut meat into beautiful patterns, and cook it in scalding water. This cutting method can increase the heated surface area of the ingredients, which is convenient for the seasoning to seep into the ingredients. Besides, the flavor contained in the ingredients can be better unlocked.

In the Qing Dynasty, there are wide range of dishes with exquisite cutting skills. Some dishes were famous for the cutting skills of cooks, like "Song Shu fish" and "chrysanthemum meat."

The cutting skills and tools are closely related to Chinese cooking (Fig. 17.23). Chinese kitchen knives come in different specifications with various purposes. For example, some knives can dice and slice the ingredients with the front part and chop things up with the rear part, which are suitable for both families and restaurants. Western dishes have little variation in terms of the shape of ingredients, so the knives in western cooking are long and narrow.

Fig. 17.23 Different kitchen knives

17.4 Stir-Fry: The Unique Chinese Cooking Method

There are about 45 kinds of cooking methods in Chinese cooking, among which stir-frying was a unique one created by Chinese people.

The key to stir-frying is to heat the wok to a very high temperature, and add oil, put in diced, sliced, or shredded ingredients and stir them with spatulas until they are well cooked. You can season at any time you want. In doing so, the cooked dishes can be very palatable.

As to when this cooking method appeared, there are a lot of controversies. There are speculations that it may appear in Shang Dynasty, in Spring and Autumn Period and Warring State Period, in Han Dynasty, or in Wei and Jin Dynasty based on the unearthed bronze cookware. It is safe to say that stir-frying has existed in the Wei and Jin Dynasty because a book called *Qi Ming Yao Shu* of this period mentions this kind of cooking method. But it is inappropriate to call this period an emerging stage of the cooking method of stir-frying.

Qi Ming Yao Shu records two texts concerned with the stir-fry:

The way to cook eggs: crack and whisk them, stir the yolk and the egg white, and stir-fry them with salt, preserved soybeans, sesame oil, and scallions, which tastes very fragrant and delicious [12].

The way to cook duck: choose a young and fat duck as large as a pheasant, chop off its head and marinate it for a while to eliminate the unpleasant smell. Then wash the duck and chop it up, stir-fry it with scallions, salt, preserved soybeans, peppers, and ginger [13].

We can infer that people back then stir-fried food with bronze cookware, the two cooking methods of stir-frying and pan-frying could not be distinguished from each other at that time, and this stir-frying cooking method can date back to the period before the Qin Dynasty.

In the Song Dynasty, stir-frying was very common. The stir-fried dishes like stir-fired chicken, rabbit, crab, clams, eel, etc. were sold at the marketplace in Bianjing and Lin'an. There was also a multiple range of stir-frying ways. People were more demanding for the flavor and texture of the stir-fried dishes. The way of cooking meat was recorded in *Wu Shi Zhong Kui Lu* at that time: slice the meat into thin layers, marinate it with soybean sauce, heat the wok, and stir-fry it on big fire until it whitens. It is similar to the way modern Chinese stir-fry meat.

In the Yuan, Ming, Qing Dynasty, people had further improved this cooking method, and there were a wider range of stir-fried dishes with diverse flavors.

A culinary book in Ming Dynasty referred to stir-fired lamb tripe and kidney. Another book in Ming Dynasty mentioned stir-fried lamb, goose, and chicken with salt and oil.

In Qing Dynasty, more stir-fried dishes (Fig. 17.24) emerged and stir-frying has become more flexible. You can stir-fry a wide range of ingredients on soft fire, big fire, with little condiments or with multiple other ingredients. You can season the food before or after the stir-frying. You can choose whether to thicken the soup with starch or not, and whether to stir-fry cooked or uncooked ingredients or not.

Fig. 17.24 Stir-fried shrimp meat

Because of large fire and hot oil, the stir-fried meat is tender and the stir-fried vegetables are crispy. By adding starch or egg white, the nutrition like vitamin C, and water in the ingredients can be better preserved. This cooking method is a testimony to the uniqueness and appeal of Chinese cooking. Stir-frying doesn't exist in western cooking, and roast and pan-frying are often used instead. There is no lexical equivalent in English either. Maybe this is exactly why westerners are so fond of Chinese cuisine.

17.5 Seasoning: The Key to Successful Cooking

Seasoning is another integral part of Chinese cooking, which can determine the flavor of dishes.

Before Qin Dynasty, there were arguments that food should be seasoned differently in different seasons, sour flavor for spring, bitter flavor for summer, spicy flavor for autumn, and salty flavor for winter [14]. The notions of preserving the original taste of the ingredients and the coordination of five flavors were advocated. It was also agreed upon that people share common perception of tastes. There were unseasoned porridge, and porridge seasoned with salt and plums, and "Wu porridge" with a sour and bitter flavor, dried meat seasoned with sauce, ginger, cinnamon, and leeks, and soft-shelled turtle seasoned with sugarcane juice.

Seasoning developed rapidly in the Han, Wei, Northern, and Southern Dynasties. In addition to the five flavors of sweet, sour, bitter, spicy, and salty, various condiments like soybean sauce, salt, fish sauce, preserved soybeans, vinegar, ginger, garlic, scallions, pepper, orange peel, mustard, pomegranate juice, and saki can create many complicated flavors.

The seasoning procedures were paid great attention to. The condiments can be added before, during, or after the cooking of dishes. A traditional Chinese dish called "pan-fried fish with honey" is made by cleaning the crucian, marinating it with alcohol, honey, and salt, and pan-frying it with oil. Another dish called "bruised eggplant" is made by bruising eggplants with condiments. The food can also be accompanied by condiments when served, like the roast oyster is served with a saucer of vinegar.

Seasoning is not just about taste, but also the good smell of food. For example, the fried fish will smell of lotus, and be more mouth-watering if steamed in the lotus leaves.

In Tang Dynasty, it was proposed that everything can be eaten as long as it is properly cooked and seasoned. There were also unseasoned dishes.

In the Five Dynasty, a dish called "red lamb" was cooked by adding red yeast to lamb. This is the first-time use of red yeast as a condiment in the world.

In Song Dynasty, a person named Su Yijian proposed that food should be seasoned to the taste of one who eats it. This then became a classical argument of seasoning.

Dishes with multiple flavors flooded the market, for example, in Lingan, there are "steamed and unseasoned sea eel," "roast quill with honey," "escargot with the flavor of alcohol," "five-flavor goose," and "spicy and sour clam" ... with excellent seasoning skills.

In Ming Dynasty, chili was introduced into China, and was used as condiment in Qing Dynasty in middle south and southwest China. An increasing number of local good-quality condiments gave rise to more diversified dishes.

It was believed in the *Menu of Suiyuan* that different ingredients should be seasoned differently to accentuate the distinct features of them so that different dishes can have their unique tastes (Fig. 17.25).

17.6 "Xian" Flavor and Extracting This Flavor from Soup

Among the five flavors, "xian" flavor is absent around the world. Chinese people are the first to identify this flavor, and the first to devise ways to extract this flavor from soup.

"Xian" can be explained in various ways, like freshness and delicacy. This flavor first derives from fresh fish. *Shuo Wen Jie Zi* associates the "Xian" flavor with fresh fishes. A man named Duan Yucai in Qing Dynasty added annotation that the "xian" flavor can be produced by cooking fresh fish. It was suggested that the flavor of fresh fish is the "xian" flavor.

Fig. 17.25 Steamed ham in honey sauce

Before Qin Dynasty, *The Songs of Chu* mentioned the flavor of "xian." Afterward, this character was frequently mentioned as a flavor, especially in ancient Chinese poetry. In Tang Dynasty an ancient Chinese poet—Bai Juyi, referred to the grilled goose as "xian" in his poem. Another poet Jia Dao directly pointed out that fish has a "xian" flavor, associating "xian" with flavor for the first time. In Song Dynasty, it was very common to relate "xian" to flavor, as people back then perceived the bamboo shoots and water chestnut as "xian."

In Ming Dynasty, "xian" was explicitly classified as a flavor in a book called *Shi Wu Gan Zhu* by Huang Yizheng.

A lot of ways of creating the flavor of "xian" were devised by ancient Chinese people, like cooking soup. According to *Qi Ming Yao Shu*, soup made of broken cattle bones and lamb bones and then seasoned with salt and preserved soybean can serve as condiments for other food to add a flavor of "xian." In Tang and Song Dynasties, this soup was used in making porridge. A book called *Yi Ya Yi Yi* in the Yuan and Ming Dynasties records the way of making soup. A book for health preservation in Song Dynasty also records in a detailed manner on how to make this soup with meat or vegetables. This soup is extremely delicious and has the "xian" flavor, and was later used in many dishes and porridge. Now the skill of making this soup is quite sophisticated. In late Qing Dynasty, this kind of soup was made with pig bones, chicken, ham, dried scallops, chopped-up pork, and the soup needs to be filtered once. Then add the meat of chicken thigh, chicken breast, and chopped-up pork and filter out the dregs. In the end, season it with salt to increase the "xian" flavor.

In addition to this kind of soup, the ancient Chinese people also made bamboo shoots soup, mushroom soup, shrimp soup, or made bamboo shoot powder, mushroom powder, and crab oil as the "xian" flavor agents. In early Qing Dynasty, a person called Li Yu observed in one of his books that bamboo shoots can add a great "xian" flavor to dishes, but its juice should be used instead of its scums. This statement does make sense.

It is worth mentioning that in 1908, the Japanese scientists succeeded in decomposing glutamic acid and extracting sodium glutamate from kelp, which was later named aginomoto and hit the market the next year. In 1914, the Japanese set up the first aginomoto factory with wheat gluten as the ingredients. However, the Chinese people lagged a little behind and made it in 1922 by Wu Yunchu under the brand of "Foshou." It is a real pity in history that the Chinese people, as the first in the world to identify and create the "xian" flavor, lag behind in making aginomoto.

However, soup with the "xian" flavor is still widely used in many dishes and is irreplaceable. Some up-end dishes, "like swallows' nest," "bamboo fungus and egg soup," "cabbage in the boiling water" must be cooked with the "xian" soup, which will impart these dishes a more profound taste.

Nowadays, the "xian" flavor, as one of the basic flavors, is universally accepted in many countries, despite the fact that the receptor of this flavor hasn't been identified yet.

17.7 Tofu

Tofu, a great invention by the Chinese people, is both food and ingredients, wielding a global influence. Many historical records show that tofu was invented by the Huainan Prince, Liu An, as recorded in *Poem of Vegetarian Food* by Zhu Xi in Song Dynasty.

However, the lack of written records of tofu from the Han Dynasty to the Tang Dynasty casts doubt on this claim.

Considering that there were soybeans, revolving mill, and coagulator in Han Dynasty, it is highly possible that Liu An invented tofu by accident. But it is equally possible that one of Liu An's contemporaries invented tofu and this feat was ascribed to Liu An.

It is noteworthy that dozens of cooking books written in the Wei and Jin Dynasties are now missing. The archaeological work in the future may help unearth some of them.

There are also scholars who believe that the fermented bean curd in *The Book of Diet* by Xie Feng in Sui Dynasty serves as evidence for the existence of tofu.

The earliest written record of tofu arose in *Qing Yi Lu* by Tao He in the Five Dynasty. It was recorded in this book that an upright and honest county official in Qingyang County had no meat but tofu on a daily basis, which suggests that tofu was a commonly eaten food and appeared much earlier than expected. People in the Yi area called tofu "special mutton."

In Song Dynasty, tofu was extensively eaten. The pan-fried tofu and tofu soup were sold at many restaurants, and there were also tofu soup eateries. A famous Chinese poet Su Dongpo was fond of eating the pan-fried tofu with honey. A cooking book records a dish featuring tofu and hibiscus flowers, and another dish named "Dongpo Tofu"featuring tofu and Chinese torreya. Another recipe book records a dish of tofu strip with five flavors. A poem "Chef in the Mountain" by Lu You also mentions tofu. Tofu is also known as "Shuru" and "Ganzhi."

During this period, people were well aware of how to make tofu. One ancient book records that the raw soybean can be ground into edible tofu. Yang Wanli, a Chinese poet, expounded that one should clean and peel soybeans, put them in the mill, and cool what comes out to make it curdle to produce something white and bland.

In Yuan Dynasty, a cooking book mentions tofu, and it was recorded that tofu soup was all over the market in a Chinese-language book for Gaoli people.

In Ming Dynasty, Li Shizhen, a famous Chinese doctor, gave a very detailed account of how to make tofu in his medical treatise *Compendium of Materia Medica.* He said: "tofu can be made with soybeans, black soybeans, white beans, peas, and mung beans. Soak the beans in the water, grind and filter them. Then, curdle them with salty marinade, plaster, or vinegar. Actually, anything that is salty, bitter, sour, or spicy can help curdle the tofu. A thin layer on the surface can be taken down and dried to make the tofu skin, an edible ingredient. Up till the Ming Dynasty, many kinds of beans could be used to make tofu per se through specific procedures, and there were 4–5 types of curdling agents available. These are the experience from several dynasties.

A book named *Mo E Xiao Lu* recorded that tofu made of the mung beans and soybeans has a good texture. Another ancient book concerning physics recorded that tofu of color, like green, yellow, black, and brown, can be made with other ingredients.

There was a wider variety of food made of tofu ever since, like soybean milk, tofu pudding, dried bean curd, bean curd skin, fried been curd, fermented been curd, and so on. And tofu itself also had many variants, like northern tofu, southern tofu, old tofu, tender tofu, frozen tofu, while dried tofu, tea dried tofu, dark dried tofu, dried tofu with shrimp egg and red, white, smelly, and spicy fermented tofu.

There was a dazzling diversity of dishes made of tofu as well, like "bean curd skin roll recorded in *Song Shi Yang Sheng Shu* concerning health preservation," which is made with roll shredded meat, shredded ginger, shredded lettuce, and shredded bamboo shoots with bean curd skin. In Qing Dynasty, there were in total more than one hundred tofu dishes as mentioned in *Shi Xian Hong Mi*. It is also mentioned in *Menu of Sui Yuan* that pan-fried tofu with small shrimps and sweet alcohol is named as "Jiang Shi Lang tofu." There are also famous tofu varieties such as the "wok-cooked tofu" from Shandong, "shrimp tofu" from Suzhou City, "mapo tofu" from Sichuan.

Tofu has been an extensively popular food since the ancient time. Despite its tender texture, it has to do with its nutrition and the medicinal attribute. It was made clear in a recipe book in Qing Dynasty that tofu can clear heat, generate saliva, moisten dryness, detoxify the body, improve stomach, and eliminate waste. Meanwhile, the cold nature of tofu was also identified. Eating too much tofu will damage the kidney, but this problem can be dealt with. Scientific research suggests that tofu is high in premium protein amino acids and unsaturated fatty acid, with no cholesterol contained. Besides, carbohydrate, vitamins, and minerals contained can help prevent cardiovascular diseases. In this case, tofu is particularly well received in Japan. The Japanese scholars are convinced that tofu was introduced to Japan in the Song or Yuan Dynasty. Tofu was introduced to Europe in late Qing Dynasty. A kind of very popular hamburgers nowadays is made with tofu patty and vegetables.

As Chinese cuisines go global, it can be anticipated that tofu will gain popularity worldwide.

17.8 Medicinal Dishes

Traditional Chinese medicine theorizes that food and medication are of the same origin, so it is advisable that people treat diseases with medicinal dishes, soup, or drink. In Shang Dynasty, a person named Yi Yin cooked soup with orange peel and ginger to treat diseases for other people. This is one of the earliest medicinal dishes called "Yi Yin soup." According to *The Rites of Zhou,* in the Zhou Dynasty, the imperial court set up the position of "food doctor" to take charge of the food, drinks, meals, and condiments of the lords. It is also recorded that people can treat diseases with five grains and five flavors.

Huang di's Canon of Medicine made a lot of statements about the treatment of diseases by food. Dozens of medicinal food were documented in *Shen Nong's Herbal Classic.*

In the Han and Wei Dynasty, medicinal dishes gained further development. Many medicines in ancient medicinal books unearthed in Wuwei County, Gansuin the Western Han Dynasty were food. One-fourth of more than 200 types of medicine in *Fifty-two Prescriptions* unearthed in Mawangdui Tomb are food. *Synopsis of Golden Chamber* by Zhang Zhongjing also included medicinal diet prescriptions. It is worth mentioning that a book by Ge Hong recorded a prescription in which food containing iodine was used to treat thyromegaly, hundreds of years ahead of Europeans. This book also recorded the treatment of diabetes with pig pancreas, the first practice to treat diseases with viscera. There were many monographs on medicinal dishes during this period, which are all missing now.

In the Sui and Tang Dynasty, a few medicinal diet treaties emerged. Sun Simiao's book included the pharmacological and medical effect, as well as the eating taboos of 154 types of fruits, vegetables, grains, and meat, like lamb liver can cure night blindness and deer kidney can reinforce people's kidney qi. Its introduction points out that doctors need to identify the cause of diseases, and should only resort to medicine when food fails to cure the diseases. This book served as an instruction on the cooking of medicinal dishes. A book concerning treatment with food by Meng Xi and Zhang Ding records more than 200 food as medicine. Another treatise included more than 300 medicinal diet prescriptions, among which 211 were preserved to date. They concern internal medicine, gynecology, pediatrics, and so on. Every prescription specifies the amount of ingredients used, cooking methods, and medical effect.

During and after the Song Dynasty, medicinal dishes have made strides, and a lot more related treatises, such as *Taiping royal prescriptions* and *Sheng Ji Zong Lu* in the Song Dynasty were written as a result, most of which collected many prescriptions. A book by an imperial doctor in the Yuan Dynasty collects 61 prescriptions for medicinal dishes and records their ingredients, ways of making and taking, and effects. *Compendium of Materia Medica* by Li Shizhen also collects enormous medicinal dish recipes, among which more than 50 are about porridge. Another book called *Porridge Recipes* by Huang Yunhao in Qing Dynasty includes over 200 kinds of porridge.

In general, Chinese medicinal diet was influenced by many traditional Chinese medicine theories like the balance between ying and yang, bodily metabolism, and five flavors related to meridians, as well as by the Taoist health-preserving philosophy. It also drew on the life experience of the general public. Medicinal diet includes dishes, porridge, pastry, drinks, etc. The dishes feature soup, cold dishes, and stewed food; porridge is made of grains, vegetables, and meat; pastry mainly includes noodles, steamed buns, and wonton; drinks include soup and alcohol.

Famous medicinal dishes include "trotter soup" for stimulating lactation, "magic soup" made of sticky rice, ginger, scallions as a cure for colds, "bazhen cake" made by steaming yam, lotus seed, tuckahoe, Gorgon fruit and sticky rice for improving the health of stomach and spleen; "stewed fish head with gastrodia elata" for curing dizziness, "stewed chicken with pseudo-ginseng" for reinforcing qi and blood, "white fungus soup" for moistening lungs, "stir-fried bitter gourd" for clearing heat and "deer horn liquor" for promoting the health of kidney. These dishes are cooked even nowadays, bearing out the vitality of these ancient prescriptions.

17.9 Vegetarian Diet

Before Qin Dynasty, Chinese had a vegetarian diet during fasting to show their reverence for ancestors and gods.

In the Han, Wei, Northern, and Southern Dynasties, folk vegetarians dishes are integrated with the Buddhist vegetarian diet. Vegetarian food recipes emerged as quite unique. Before the Qin Dynasty, vegetarian food referred to all the non-meat food; in Han Dynasty, it referred only to vegetables.

Moreover, the non-vegetarian food does not refer to meat and fish before the Wei and Jin Dynasty, it referred to the smelly vegetables. Buddhists denied themselves five types of smelly vegetables including garlic, rocambole, Hingu, leeks, and scallions in order to stay uncorrupt and obedient. But how did all this turn out to be the denial of meat?

The research reveals that the denial of meat was not originally included in the Buddhist doctrines. The Indian Buddhist monks ate both meat and vegetables when begging for food. However, a Buddhist denomination called Dacheng banned the eating of meat, and this is the denomination that most Chinese people believed in. In this case, most Chinese Buddhist monks chose not to eat meat. Besides, an emperor named Xiao Yan strongly advocated for the self-denial of meat and wrote an article about this. He was convinced that eating meat amounts to killing and that vegetables are good for people's health by increasing metabolism, improving levelheadedness, and reducing fatigue, while eating meat will do the otherwise by entailing more diseases and dizzying one's mind. These two reasons were behind the trend of eating only vegetables among the Chinese monks, and behind new types of vegetarian food consisting of vegetables and traditional vegetarian dishes.

In the Wei, Jin, Northern, and Southern Dynasties, there were a wider variety of vegetarian dishes. According to historical records, there is a monk and chef who can impart multiple flavors to one kind of ingredient [15]. According to *Qi Min Yao Shu,* the most commonly eaten 87 vegetarian dishes included scallion and leek porridge, bottle gourd porridge, fermented soybeans, pan-fried nori, etc.

In Tang Dynasty, an official from Sichuan held a vegetarian banquet for his colleagues, with dishes featuring dyed pastry and konnyaku resembling the meat [16]. They were early meat-resembling vegetarian dishes. At some temples, the vegetarian dishes were exquisitely made. It was said that Bai Juyi, a famous ancient Chinese poet, often had vegetarian dishes made of rice, arrowroot, lead mustard, and fragrant ginger with a famous monk named Tao Guang.

In Song Dynasty, vegetarian dishes have gradually gone out of temples and made their way to ordinary people's dining tables. There were restaurants for vegetarian dishes in Bianjing and Lin'an. An increasing type of ingredients was used, like tofu, gluten, rice noodles, bean sprouts, bamboo shoots, and mushrooms. Some vegetarian dishes were designed to resemble meat dishes, like false chicken, fish, duck, and sausage (Fig. 17.26).

A recipe book named *Ben Xing Zai Shipu (Menu of Vegetarian Diet)* by Chen Dasou included twenty recipes for vegetarian dishes. Another cooking book named *Shan Jia Qing Hong* by Lin Hong also mentioned many vegetarian dishes featuring

Fig. 17.26 Fortune yellow crocker

flowers, including "Xue Xia porridge" made of tofu and hibiscus and "golden rice" made of chrysanthemum (Figs. 17.27, 17.28 and 17.29).

In the Yuan Ming and Qing Dynasty, the development of the vegetarian food gained momentum in more places. More and more ingredients were used in vegetarian dishes, like fungi, wild vegetables, and nuts. Vegetarian dishes were also cooked

Fig. 17.27 Chrysanthemum walnut

Fig. 17.28 Steamed vegetable dish

Fig. 17.29 Fancy cold dish

by using many cooking methods with different flavors. There were more than one hundred vegetarian dishes in the court, households, temples, and marketplace.

Nowadays, vegetarian diet has become increasingly popular. Nutritionists tend to associate some diseases with eating too much meat, like hypertension, hyperlipidemia, hyperglycemia, diabetes, and cancers. A vegetarian diet is nutritionally rich and can drastically reduce the incidence of cardiovascular diseases because it does not contain cholesterol and saturated fatty acid. With these said, Chinese traditional vegetarian food should be inherited in the interest of people's health.

17.10 Conclusion

To sum up, Chinese cooking is a great creation intricately related to Chinese people's daily lives and the future development of the Chinese nation. It was basically shaped in the Wei, Jin, Northern, and Southern Dynasties and became globally recognized upon further development.

Some scholars think that only French cooking, Turkish cooking, Muslim cooking, and Mediterranean cooking can hold a candle to Chinese cooking. French cooking mainly uses aquatic products, beef, milk, cheese, poultry, mushroom, and flour as ingredients, and the main cooking methods are pan-frying, deep-frying, roast, baking, and so on. The French dishes are also seasoned after the cooking, with liquid condiments and spices mostly used. The staple and subsidiary food are not distinguished from each other. French cooks are good at making chowder and the presentation of dishes.

The Mediterranean cuisine is also very distinctive, featuring olive oil, grains, fruits, fish, nuts, eggs, and dairy products, which can greatly reduce the risks of cardiovascular diseases. Its dietary structure is similar to that of Chinese coastal area, but the oil used is different, and more eggs and dairy products are used in Mediterranean cooking.

A wide range of ingredients are used in Chinese cooking, a large proportion of which are grains, fruits, and vegetables. There are various ways of cooking and seasoning, and a variety of dishes with different taste. An American scholar Eugene Anderson acclaims in his book *Chinese food* that the Chinese people have created the greatest cooking in the world in the Song and Yuan Dynasty. Chinese cooking attaches great importance to health-preserving and medicinal diet and vegetarian food emerged as a result. It is doubtless that the Chinese cuisine will gain a growing popularity and have a brighter prospect around the globe.

References

1. Yat-sen, Sun. 2009. Nation-building Strategy. *The Collection of Soun Yat-sen's Essay*, 211. Guangdong People's Publisher.
2. *Thirteen Classics Explanatory Notes and Commentaries*, 1468. 1980. Zhonghua Book Company.
3. Gu, Ban. 1962. *History of the Han Dynasty*, 3642–3643. Zhonghua Book Company.
4. Xuanzhi, Yang. 1963. *The Redacted Version of the Book of Jialan in Luoyang*, 133. Zhonghua Book Company.
5. Seiichi, Tanaka. 1991. *The History of Chinese Food in Japan*, 67. Heilongjiang People's Publisher.
6. Nodo, Kou. 1997. *Chinese Food in Japan* (10): 39.
7. Hong, Lin. 1986. *Homemade Food in Mountains*, 7.
8. Nobuo, Haruda. 2011. *The Sociological History of Japanese Cuisine*, 47. China: Social Sciences Academic Press.
9. Campbell. 2006. *Chinese Public Health Report*, 1. Jilin Culture and History Publisher.
10. Sixie, Jia. Art of Chicken Raising. *Qi Min Yao Shu.*
11. Rihua, Li. *Causerie in Zitao Chamber.*
12. Sixie, Jia. Art of Goose and Duck Raising. *Qi Min Yao Shu.*
13. Sixie, Jia. Cooking of Pan-frying. *Qi Min Yao Shu.*
14. *Thirteen Classics Explanatory Notes and Commentaries*, 667. 1980. Zhonghua Book Company.
15. Silian, Yao. 1973. The Biography of Hechen. *The Book of Liang*, 548. Zhonghua Book Company.
16. Guangxian, Sun. 1981. *Bei Meng Suo Yan*, 17. Shanghai Ancient Chinese Book Publisher.

Pangtong Qiu Born in Yangzhou, Jiangsu in 1944, graduated from the Chinese Department of Beijing Normal University. He is a professor at Yangzhou University and former director of the cooking and nutrition science department of the university and is currently a member of the Guiding Committee of the Chinese Cuisine Association.

Research interests: Catering history; Ancient Chinese language.

Major works:

Books: *Overview of Ancient Chinese Cuisine Books; History of Chinese pastry; History of Chinese Cuisine;*

Papers: Study on the Origin of Stir-frying Method; Taoist Food Culture and Festival Customs; Study on the Origin of Chinese Soup Cuisine.

Awards:

History of Chinese Pastry was selected into the third "Three One Hundred" original book publishing project of the General Administration of Press and Publication.

Chapter 18
Carriage Hitching Technique and Stirrup

Wei Chen

Horse is one of the most important livestock. Since wild horses mostly spread over flat and broad plains, the ethnic groups living here domesticate horses at the earliest time. In the Potai ruins with about 5,500 years located in the North of Kazakhstan, the pottery that had been filled with horse milk and the changes in the bones caused by the domestication of horses were discovered, which has so far become the earliest known relic of horse domestication [1]. Up to now, when people refer to the nomadic tribes, the image of the nomads galloping on the horseback immediately springs to their mind. Farming peoples start to feed and use horses in the later period, but a series of improvements and new advancements of carriage hitching techniques have greatly promoted the expansion of the use of horses among the settled farming peoples. Among them, the improvement of carriage hitching techniques and the invention of stirrups are the two major technologies that exert important influences on social development, and they are closely related to China.

18.1 The Carriage Hitching Technique

Besides the function for riding, the livestock plays an important role as the prime power in driving all kinds of ancient machinery. The so-called carriage hitching technique is the way of using one or several livestock to form a traction system by a set of specially designed devices, i.e., harness, which itself has many components. A set of reasonable harness should make full use of all livestock's power that must be well coordinated. Harness used in China is different from that in America at an earlier time. It can be said that the carriage hitching techniques existing in ancient China are more advanced than those in the western world for a long period of time.

W. Chen (✉)
The Institute for History of Natural Sciences, Chinese Academy of Sciences, Beijing, China
e-mail: chenwei@ihns.ac.cn

J. Hua and L. Feng (eds.), *Thirty Great Inventions of China*,
https://doi.org/10.1007/978-981-15-6525-0_18

18.1.1 The Early Western Neckband Driving Method

The development of the carriage hitching techniques is closely related to the vehicle's shape. The earliest vehicle both in the East and West is a single hitch vehicle, using a wooden pole to connect the body of the vehicle with the livestock. If more than one livestock is used, weight and yoke will be added, respectively, at the front of the wooden pole to control the livestock. Therefore, the towing point is naturally placed along the middle line of the back of the livestock. Harness used in the Western world of early period driving method usually includes the neckband (in the front of the neck) and girth (around the back of the belly and rib). The towing point is located where the neckband and girth meet (Fig. 18.1). If the girth lacks support from the front and rear, it tends to become loose at the back, so that the neckband becomes the main part for the livestock to exert force. If it is a chariot with higher mobility requirements, it often has no hitch and girth but only uses the neckband for traction. Therefore, this kind of driving method is called the neckband driving method.

The neckband driving method may originate from yoke used in cattle farming [2]. However, the anatomical features of horses and cattle make this kind of driving method suitable for cattle, but not necessarily applicable to horses. The horse's neck is more exposed and leans forward. Its trachea protrudes forward and its carotid artery is near the surface of the body. If the neckband is used as main bearing part, it will press the horse's neck. So when the horse moves forward, it will have difficulty in breathing and it will affect the blood supply to the brain. The faster the horse runs, the more serious the situation is. The result is that it is difficult to make the most of the horses' power [3].

According to the experiment made by a French scholar called R. L. des Noëttes, the efficient traction of two horses controlled by the neckband and girth is about 1/2 ton [4]. However, if modern harness is used, a horse can easily pull a total load of 1.5

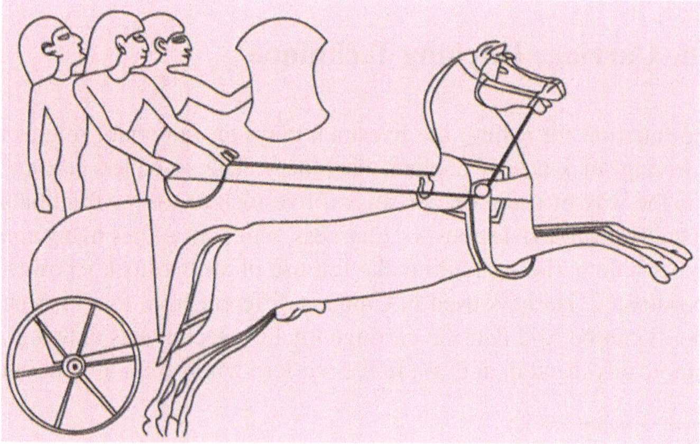

Fig. 18.1 Hittites' chariot (about 1,500 years ago)

Fig. 18.2 A chariot on earthenware in ancient Greek (in the late eighth Century B.C.)

tons. In other words, the efficiency of the neckband driving method is only equal to 1/6 of that of the modern harness [5]. So its low efficiency is evident. Compared with ancient horses, the modern horses have been greatly improved in many aspects such as breed and feed, and then the ancient carriage's load capacity in using neckband driving method is very low. In Roman *Codex Theodosianus* compiled in 438, it is stipulated that the load capacity of a two-wheeled vehicle pulled by a single horse is limited to only 154 lb (about 70 kg). Even for a mail vehicle that is pulled by several livestock, its load capacity is only less than half a ton [5].

If we consider the restrictions on load capacity of the chariot, we can know that the ancient Western chariots are rarely used for close combat, but usually used for rushing and chasing. Warriors on the chariots mainly use bows and arrows (Fig. 18.2) for the convenience of jumping off chariots to combat when they are close to enemies. The Western ancient vehicle diameter usually does not exceed 90 cm, which means that a rein rope parallel to the ground cannot connect the vehicle axle with the horse's traction point. And the force transmission relies on the hitch that is tilted upward, indicating that the horse's traction force is wasted once again. From Mesopotamia and ancient Egyptian murals, we can see that the chariot carrying one person usually needs two horses while the chariot carrying two persons needs four horses. This in turn increases the difficulty of keeping the horse in the middle and the horse on both sides well coordinated. In short, the neckband driving method has brought lots of technological restrictions on the Western ancient vehicles and horses.

18.1.2 The Early Chinese Yoke Driving Method

Different from the Western ancient chariots, vehicles in the Warring States Period have greater load capacity. In *Mo-tse,* it is recorded that craftsmen make a linchpin and carve short wood into a vehicle that can burden the weight of 50 dans (a unit of measurement in ancient China). There is a similar record in *Han Feizi* that a vehicle has the load capacity of 30 dans. 1 dan is equal to 120 jin. According to the system of weights and measures in the Warring States Period, 1 jin is equal to about 250 g.

We can see that the load capacity of the vehicle amounts to about 1 ton at that time. Obviously, its load capacity is much larger than that of the mail vehicles in Roman times. If the Chinese carts had used the neckband driving method, such load capacity would have hardly been reached.

It is worth noting that the vehicle's shape and manufacture found in antiques from the Shang Dynasty to the Warring States Period and inscriptions on ancient bronze objects have a subtle difference from that of the Western ancient vehicles, i.e., the diagonal lines linking the hitch to a part equivalent to yoke can usually be seen in the images of Chinese ancient vehicle (Fig. 18.3). The two diagonal lines are actually rein ropes attached to two yokes' inside, which can be seen in No. 2 copper cart's restoration discovered in the Mausoleum of the First Qin Emperor. The rear end of the rein ropes is attached to the ring in front of the vehicle, and such ring and axle are connected by a thick rope. Actually, the use of the rein rope moves one-hitch vehicle's original driving point (located in the center of the vehicle's balance) down to the lower end of the yoke. Considering the larger wheel diameter of the Chinese ancient vehicles (up to 1.33 m), the line from the driving point to the vehicle axle is close to a horizontal state [6]. More importantly, although there is the neckband (called as soft leather around the neck) enclosing each horse's neck below No. 2 cart in the Mausoleum of the First Qin Emperor, the instrument is only used to prevent the horse from coming off the yoke, which is not really stressed. Therefore, it cannot react against horse's neck and affect the horse's breathing, providing solutions to the biggest problem of the neckband driving method.

Naming according to the use of the main harness under force can be called "yoke driving method" [7] (Fig. 18.4). The use of the yoke driving method has brought about the popularity of chariots. Because of greater use of horse power, the chariot's volume, load capacity, and speed are remarkably favorable. If chariots spread out in horizontal formation in open plains and rush to infantry lacking of mobility, the enemies can hardly resist. During the Spring and Autumn Period, this advantage was obviously suitable for striving for hegemony with the North China Plain as its main battlefield. Therefore, the number of chariots at that time continued to rise and became the standard for measuring the military power of the vassal states. However, chariots also have a disadvantage of their inability to adapt to the complicated landscape. In rough mountains, chariots had difficulty in turning round, and could hardly fight

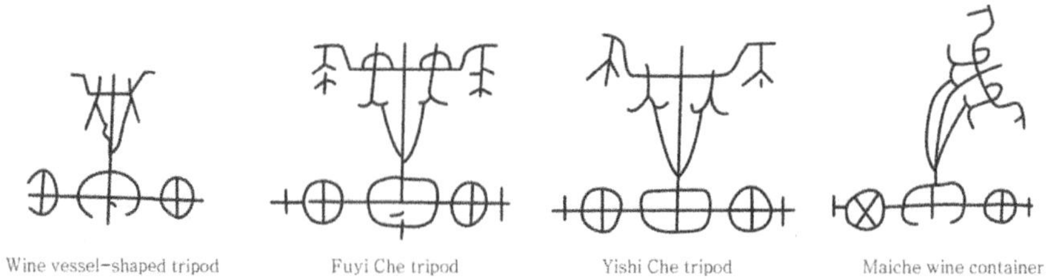

Fig. 18.3 Chinese character "Che" (vehicle in English) in inscriptions on ancient bronze objects (Image Library of Mechanical History at Tsinghua University)

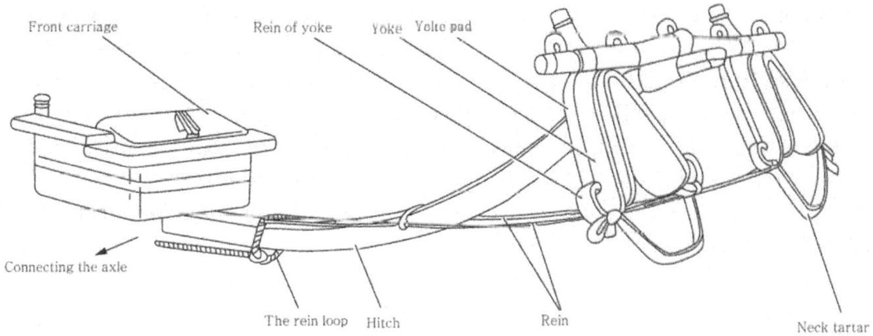

Fig. 18.4 A sketch map of the yoke driving method (quoted from Sun Ji. *Chinese Ancient Carriages and Clothing System*. Shanghai Classics Publishing House, 2013: 60)

with infantry, losing impact force and mobility. For drivers to freely use the chariot's bridle rope and regulate the rein ropes' stress size, they need a long-term training. These limitations made chariots gradually decline. After the middle of the Warring States Period, the chariots became less important with the innovation of shooting on horse in Hu dress. Double shafts vehicle, known for its speed and flexibility, started to replace one-hitch vehicle called "war vehicle" or "light vehicle."

18.1.3 The Chest Belt Driving Method

Double shaft carriages appeared earlier in West Asia. From Wuhu gold vehicle (Fig. 18.5) with four horses of the Persian Achaemenid Dynasty (550 B.C. to 330 B.C.) collected in the British Museum, it can be seen that its shape and manufacture have been relatively mature [8]. However, in terms of the driving methods, it still

Fig. 18.5 A double shaft carriage in Archimedean Period of ancient Persia

uses the neckband method, which is not advanced. The double shaft carriages in China, also appeared in the late Warring States Period at the latest, are more likely to originate from ox carts which serve to load stuff when walking on the flat ground. The earliest double shaft ox cart was discovered in the Shaanxi Fengxiang Tomb belonging to the early Warring States Period. The double shaft carriages depicted on painted wine vessel unearthed in Changsha Chu Tomb could reflect the shape of the driving method's early evolution (Fig. 18.6). The carriage hitching the carriage has a belt in front of its chest therein it may be easy to drag out a rein rope to attach to the shaft's middle. Here, the neckband only serves as a connection component between the chest belt and yoke. It can be seen from the carriage's harness that complex harness used to coordinate several horses has been exemplified. Due to the use of the double shaft, a rein rope should be placed on either side of a horse. Due to the remaining influence of old driving method, two rein ropes might be attached to either ends of the right and the left sides of the yoke at the beginning. By the time of the Western Han Dynasty, two rein ropes were not constrained by the yoke; instead, they were connected with each other to become a whole chest belt bypassing the horse chest. The yoke and affiliated soft leather around the neck were degraded as the supporting parts. This kind of driving method (that chest belt shoulders force when the horse pulls the weight) is called the "chest belt driving method." Compared with the yoke driving method, the neck belt driving method is more convenient. Its supporting point and force point shouldered by the horse's neck and chest, respectively, are separated so that partial force of the horse body is correspondingly reduced.

The chest belt driving method was also introduced to Europe through the Eurasian Steppes. Harness on either side of the horse's shoulder blades, which was discovered in the nomadic tombs of the Eastern Europe in the fifth Century, is mainly made of bone fragments, horns, and other materials. Italian regions might also get the technology from the East Goths at that time. It was not until three centuries later that the chest belt driving method gradually began to spread in other regions of Europe. Despite this, the neck belt driving method continued to be used. In the tenth century,

Fig. 18.6 A double shaft carriage depicted in painted vessel unearthed in Chu tomb in Changsha

the traditional neck belt was replaced by the neck strap which the rein ropes were directly connected to. To a great extent, it preserved the system of the neckband driving method [9].

18.1.4 The Saddle (or Neck Strap) Driving Method

The chest belt driving method is not perfect. Its defects are mainly reflected in two aspects. The first is that it retains an upward arc line with excessive curvature, which was common in the past yoke driving method. In the old system of the single hitch driving system with several horses, it is necessary to pay attention to balancing the yoke when each horse was processing. Therefore, in order to balance the yoke, the hitch definitely curves upward, which is not necessary for the double shaft. Carriage shaft curving upward has a high center of gravity, so it increases the possibility of turning over. What's more, such arc-like shaft that appears fragile and easily broken is very difficult to make with the thick and bulky wood. This also indicates the defect that the remaining elements of the old driving method cannot adapt to the new driving method, which takes a long time to improve. By the time of the Eastern Han Dynasty, people adopted a method improving safety with two reinforced poles installed between the middle of shaft and the end of yoke. However, it also makes the structure of shaft more complicated, hence not perfect either. The other defect is that the horse's chest rather than the most powerful scapularis is the main stressed part in chest belt driving method. The chest belt driving method originated from ox carts. The cattle's shoulder hunches and its scapularis get close to the yoke, so it is easy to distribute force. However, the horse's widerrist is lower than cattle's acromion, so the chest belt method cannot completely adapt to features of the horse body. The new driving method must improve the two defects, thus developing the saddle driving method (or the neck strap driving method).

In allusion to the first defect, an attempt to lower the fulcrum of shaft's end to reduce curvature of shaft was made in between the Western and the Eastern Dynasties, which is almost simultaneous with the above-mentioned installing reinforced pole. The attempt was originally made on the big carriage whose shaft is thick and slightly curved. And lowering or even flatting the shaft's curvature is also a process of improving and canceling the balance. A carriage carved on a stone relief is unearthed in Xiaotang Mountain, Feicheng from the Eastern Dynasty. In the shape of "兀," a Chinese character balanced on both lowered ends, and with the front shaft slightly raising, the shaft and balance can be connected. However, the balance whose shape is like "兀" is too complicated to make. Therefore, there appears the improved yoke-type balance on the big carriage depicted in the portrait of the Wu Clan Temple. The big carriage's shaft also needs to be raised upward while the balance starts to combine with the yoke, making the structure relatively simple and more solid. During the period from the end of the Eastern Han Dynasty to the Three Kingdoms period, the balance was basically canceled and the shaft was directly connected to yoke's end. At that time, the shaft has almost become a straight pole without any curvature;

in the meanwhile, the yoke feet have to be moved obliquely outward to accommodate the shaft's end. Such foot-moved big yoke has also become an object that needs improvement. The first defect of chest belt method is basically canceled while the second defect still exists.

After the shaft is laid flat, the upward force of the yoke component can be reduced. The horse's shoulder once again shoulders more force in pulling forward. In addition to connecting the horse body and the shaft, the yoke also serves to block up the horse's shoulder to make it similar to the structure of cattle's shoulder. However, the hard yoke is likely to slip off and easily scrapes the horse's skin. As a result, the shoulder holster or neck belt filled with soft materials starts to be around the horse's neck. The neck belt first appeared in Sichuan during the period from the late Han Dynasty to the Three Kingdoms. On some portrait bricks, around the horse's neck is obviously the thick ring (Fig. 18.7). However, due to the rise of talking about matters in a metaphysical style in the Wei-Jin period, the ox cart was more popular among the ruling class. Nevertheless, the shoulder holster or beck belt is unnecessary for the ox cart, which affects dissemination of the new driving method. The neckband method started to appear in the North in the sixth century. From a few Mogao Grottoes murals from the Northern Dynasties to the Tang Dynasty (Fig. 18.8), it can be seen that the shoulder holster filled with soft material is used on the driving-carriage horse's neck, which increases the height of the horse's widerrist to be able to better utilize the pulling force of the horse's scapularis. However, at that time the shoulder holster and yoke did not coordinate with each other until in the painting of *The Riverside*

Fig. 18.7 The carriage carved on Sichuan brick relief in the Han Dynasty (quoted from Joseph Needham. *History of Science and Technology in China*, volume 4, part 2. Science Press, 2010: 366)

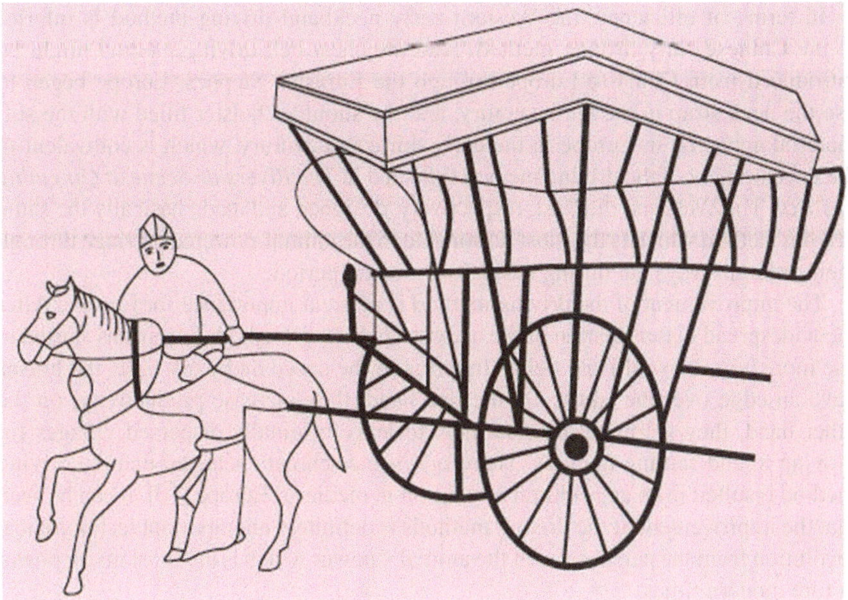

Fig. 18.8 The carriage depicted in *The Portrait of Zhang Yichao Unifying Military Travel* in the late Tang Dynasty

Scene at Qingming Festival in the Song Dynasty, there appeared a carriage pulled by four donkeys directly using the shoulder holster. Around the Southern Song Dynasty, the saddle on the horse appeared which has given the driver more convenience in terms of driving and maintaining the carriage's balance. Adopting the saddle-type driving method can help prevent the yoke from scratching the horse, lower fulcrum, lay shaft flat, and make full use of both sides of scapularis suitable for shouldering force. Since the saddle driving method can not only maintain the driving stability, but also strengthen the force of the horse-drawn carriage, the saddle driving method always has been used until today.

18.1.5 Significance of Improving the Driving Methods

As a whole, in terms of the carriage body and the driving, the ancient carriage in China is hugely different from that the ancient carriage in the Western world. And there exists an apparent difference between the yoke driving method and the Western neckband driving method, which demonstrates that the Chinese early driving technology is our own invention. A variety of the driving methods subsequently appearing in China also gradually accumulate experience in the process of incessantly raising problems and solving problems and finally achieve breakthroughs.

In terms of efficiency, the Western early neckband driving method is inferior to the Chinese early driving method. And the chest belt driving method might be introduced from China to Europe through the Eurasian Steppes. Europe began to use the neck strap in the tenth century, and the shoulder holster filled with the soft material appeared in Europe in the early thirteenth century, which is equivalent to the development of the driving method reflected in *The Riverside Scene at Qingming Festival*. The West and the East, respectively, designed and made basically the same driving method which is the most reasonable to the animal-powered carriage through their different ways but during basically the same period.

The improvement of the driving method is of great importance for Europe. After the widespread dissemination of the neck strap driving method, Europeans started to use more horses to cultivate fields. In terms of the speed and work time, the horses have an edge over the cattle. On the one hand, they increase productivity; on the other hand, they facilitate more farmers to leave originally dispersed villages for moving to and settling in cities. Therefore, some scholars believed that the driving method resulted in an agricultural revolution in medieval Europe [10]. It can be seen that the improvement of the driving methods is definitely an important technological revolution from the perspective of the animal's power which is the most major power before modern times.

18.2 Horse Stirrup

18.2.1 A Brief Introduction to Horse Stirrup

The horse stirrup is just like the shape of a semi-elliptical ring. The upper part is made of materials with high tenacity such as leather and iron; the lower edge can use wood or rattan as core; the outside part is covered with metal plate or leather to make a broader pedal. The horse stirrup often as a pair droops below the saddle. When mounting a horse, the rider can tramp on one side of the horse stirrup to step on the horse's back. When riding, the rider passes his or her foot through the horse stirrup to help stabilize the body. When galloping, the rider does not even have to sit on the saddle, but only stand on the horse stirrup with the body leaning forward, when the combination of the body and the horse presents the elegant riding posture (Fig. 18.9).

For the herdsmen on the grassland, the horse stirrup is useful but not a necessity, because they have been accompanied by the horses since an early age. Even if there is no horse stirrup and horse saddle, they also can pull the horse rein with only one hand and squeeze the horse's back with the other hand to skillfully jump on the horse. When riding, they are more accustomed to clamping the horse's body with legs to maintain stability. However, for farming people who are less familiar with the horses, the horse stirrup is almost an indispensable tool when riding. Logically speaking, according to different demands for the horse stirrup, stimulation function

Fig. 18.9 The galloping rider

caused by the invention and use of the horse stirrup, changes a lot. As the use of the horses became more and more extensive in the settled farming people, it further stimulated the invention and dissemination of the horse stirrup, which should be taken into consideration when exploring the origin of the horse stirrup.

18.2.2 The Origin of the Horse Stirrup

Some earliest physical objects or images of the horse stirrup were discovered in China. They are all single stirrups tied to one side of a horse. The earliest might be an iron stirrup that was unearthed in Wuwei, Gansu, but it became an unrepaired piece [11]. On the pottery figurine excavated in the tomb (in 302) of the Jin Dynasty in Jinpengling, Changsha, an image is depicted that there is a horse stirrup on the horse's left side, but the rider does not tramp on the stirrup [12]. What's more, a piece of the gold-filled horse stirrup (Fig. 18.10) was discovered in No. 154 tomb located in Xiaomin village, Anyang. Dating to the early of the fourth century and the middle of the fourth century, its stirrup is also placed on the horse's left side. The tomb's owner might be Han people influenced by Xianbei culture [13]. In the latter two discoveries, the only one stirrup is placed on the horse's left side, which means that the rider uses the left feet to tramp on the horse stirrup and the right feet to step over the horse's back. Such a horse-mounting way is fit for the condition that the right body of most people is more suitable for distributing force.

Why did the single stirrup helping mount horse appear at that time? And why did the early discoveries appear more in the Central Plains even in the Yangtze River downstream area? On the one hand, the reason is that after the Han Dynasty, a bridle bridge (the part like bridge caused by the arcing of the saddle's front and rear) is gradually elevated, and the rear saddle bridge tends to be higher than the front saddle bridge, which increases difficulty for mounting a horse [14]. On the other hand, during the period of the Eastern Han Dynasty and the Three Kingdoms, originally settling in the Northeastern region, Wuhuan nationality that is good at saddling horses and shooting arrows moved south. Its cavalries once even reached Jingchu and Jiaoguang regions [15]. Horses they domesticated undoubtedly entered the Central Plains in large numbers, which made residents here contact with the horse more extensively, thus increasing the possibility of the horse stirrup's appearance.

Fig. 18.10 The horse stirrup unearthed in Xiaomin Village, Anyang

The structure of the horse stirrup is simple and it is easy to make and be accepted by people. It was possible that soon after using single stirrup, people found that tramping on the stirrup was able to greatly increase stability when riding. Under such situations, horse stirrup was placed on the other side of the horse. Since then, single stirrup developed to be the double stirrups, and the function of the single auxiliary horse-mounting developed to be the function of enhancing stability when riding, which was a rapidly finished process. The horse figurines equipped with the double stirrups were discovered in No. 7 tomb of the Wang family cemetery in Nanjing (Wang Yi, the tomb owner died in 322), not far away from the above-mentioned discoveries in terms of time. The development from the single stirrup to the double stirrups conforms to the technology's evolution law from being simple to complicate.

18.2.3 Dissemination of the Horse Stirrup

The horse stirrup quickly spread to the region equivalent to the Northeastern region in modern China. In the tombs of the early fourth century such as Shiertai village in Chaoyang, Liaoning, only one piece of the horse stirrup was discovered in the whole set of the horse harnesses [16]. In the tombs of later period, in the same region, the horse stirrups in pairs (Fig. 18.11) were discovered [17], which showed that the Western Liaoning also experienced the complete evolution from the single stirrup to the double stirrups. Chinese mainland was the origin of the technological invention, the horse stirrup, and then the horse stirrup in depth entered the region in which Xianbei nationality who lived social life in the Central Plains Dynasty were situated since the end of the Eastern Han Dynasty. The process is the starting point of the horse stirrup's outward dissemination. Since then, the horse stirrup has been continuing to spread from the Northwest to the North. The region around modern Ji'an of Jilin in, Gaogouli lacked the stage of using the single stirrup, where the earliest used double

Fig. 18.11 The horse stirrups unearthed in Feng Sufu tomb in Beiyan, Beipiao, Liaoning

stirrups appeared in the middle of the fourth century. And the long-handled stirrup which was common in Xianbei region also became the typical type of horse stirrup later in the Korean Peninsula and Japan [18].

The spread of the horse stirrup to the Northwest could be roughly divided into two stages. First of all, as Xianbei nationality moved to the Northwest, the horse stirrup also gradually began to be brought to the Central Asian. The early horse stirrup found here appeared in the shape of number 8, an Arabic numeral caused by the connection between the stirrup ring and the stirrup handle, i.e., the stirrup ring is a bent metal strip. Such a horse stirrup with simple style was not found in Chinese mainland. It was possible that grasslands nationality adjusting their measures to local conditions adopted relatively simple manufacturing technology to develop it. Turks prompted the horse stirrup's dissemination into the second stage. Since Turks were known for being adept at smelting iron, the horse stirrup undoubtedly contributed a lot to their predomination over the whole Eurasia. The horse stirrup used by Turks adopted a new-type Guishou. Such horse stirrup type almost simultaneously appeared in the sixth century from Chinese mainland to the Central Asia. Such dissemination might happen as Turks expanded to the West (Fig. 18.12) [18].

The grasslands nationality using the horse stirrup forged new military power outwards, and the horse stirrup thereupon spread to other farming regions. According to knight images on the murals and coins, it can be seen that Persians used the horse stirrup around since the sixth century. Moreover, when settled in the Southern Russian Steppes, Avars organized cavalries to march toward Byzantium and the Eastern Europe after having learned to use the horse stirrup. In order to resist invasion, the Byzantine emperor Morris (in 582–602) commanded the Imperial Cavalry be equipped with the saddle and two horse stirrups, so that the horse stirrup was introduced to the Eastern Mediterranean coastal area [19]. The earliest physical object of the horse stirrup was discovered in Hungary, no later than the seventh century [20]. Later in the eighth century, the horse stirrups began to be extensively used in Europe.

Fig. 18.12 The stirrup under the saddle of the terracotta figurine (a collection from the Tajikistan archaeological museum, from the sixth to the eighth century)

18.2.4 The Meaning of the Horse Stirrup

In ancient times, the cavalries' fighting capacity in wild operations was much better than that of the infantries, which was mainly reflected in its extraordinary maneuverability and stunning effects of the frontal assault. The invention of the horse stirrup marked the completeness of the riding harnesses and marked the development of the cavalry enter a new stage, and there out, it exerted an influence on the military affairs and the development of the whole society.

Before the occurrence of the horse stirrup, if the horse abruptly rushed forward, the rider was easily cast down from the horse's back due to inertia. Therefore, at that time, the cavalry of grasslands nationality such as that of Xiongnu usually used flanking maneuvers to shoot enemies with bows and arrows, and avoided directly fighting with the opponents (Fig. 18.13). At the same time, since the lower limbs of the rider were mainly used to maintain stability, his riding posture was greatly constrained. If using a long spear, the rider could only use power of the shoulder and the upper arm to assassinate forward. With the use of horse stirrups, it was easier for the rider to control and tame the horse. His torso got relaxed; meanwhile, the riding became more comfortable and labor-saving (Fig. 18.14). The close combination of the horse and the person forms a combat unit. The cavalry can not only stand on the horse stirrup while marching to shoot arrows in a wider field and huger power, but also swing the sword right and left to slash. More importantly, the horse stirrup makes cavalry charge forward toward the enemies. Making use of its speed and weight to destroy the infantry squares, the heavy cavalry equipped with armor in the Middle Ages is almost equivalent to modern tank in function. Some scholars even believed that the horse stirrup had facilitated the appearance of the European knights, thus resulting in the birth of the European feudalism [21]. Such views, of course, deserve further discussing. However, for riding, the horse stirrup originating from China is absolutely a remarkable invention.

Fig. 18.13 Stone relief of shooting deer in the Western Han Dynasty unearthed in Luoyang, Henan

Fig. 18.14 A mural of Hu people playing polo in Liyong tomb in the Tang Dynasty

References

1. Levine, M.A. 1999. Botai and the origins of horse domestication. *Journal of Anthropological Archaeology* 29–78.
2 [English] Needham, Joseph. 2010. *The History of Science and Technology in China*, 335. Science Press.
3. The viewpoint was proposed by a French military officer called R. L. des Noëttes at the beginning of the 20th Century that such neckband driving method is easy to hold horse's trachea and causes horse to suffocate when moving forward and reduces efficiency and the load capacity of animal force. Such view once was broadly cited. Joseph Needham also mainly cited the view in his book called The History of Science and Technology in China in volume 4, part 2, and made Chinese academic community convince that the yoke method is better than the neckband method. However, such view was already outdated. After studying the horse breed, roads, images and literatures, the Western scholars believed that Noëttes' mistake was to place the driving point on the horse's back on the shoulder joint which made horse the most painful. Nevertheless, in fact, the popular driving point was below the shoulder in Roman age (distributed along the Mediterranean coastal areas). This kind of driving method is called the dorsal yoke. From the Seine to the north of the Rhine, the driving point was above shoulder. It was called neck yoke. The two methods would not affect horse's breathing. Whether it prevented breathing more than the earlier driving method in Greek-Rome age, it needed further studying. It could refer to Judith A. Weller. Roman Traction Systems. http://www.Humanist.De/rome/rts/index.html. Considering comprehensive summary about the driving method in classic age,

it could refer to G. Raepsaet. Riding, Harness, and Vehicles. *Handbook of Engineering and Technology in the Classical World*. Oxford: Oxford University Press, 2008: 580–605.
4. The experiment made by J. Spruytte in 1977 demonstrated that the efficiency of the neckband method was far higher than what Noëttes expected that the cart's load capacity was similar to that of modern driving method (J. Spruytte, Early Harness Systems: Experimental Studies, a contribution to the history of the horse, J.A. Allen, 1983).
5. des Noëttes, R.L. 1931. L'attelage et le cheval de sellea travers lesâges, 5. Paris.
6. Ji, Sun. 2013. *A Collection of Chinese Ancient Carriages and Clothing* (Revised Edition). Shanghai Classics Publishing House, 60.
7. Ji, Sun. 1988. Three kinds of driving methods of carriages in ancient China. *History of Nature Sciences*, Part 2, 1984: 169–176. According to No. 2 bonze horse-drawn carriage unearthed in Qin's Mausoleum in Shaanxi, the article demonstrated in detail the shape of the yoke method. However, his article followed the analysis of the Chinese character "carriage" form appearing in Oracle and inscriptions on the ancient bronze objects from the earlier articles. He also regarded the analysis as evidence that the yoke method already existed and believed that the yoke method was unique in the ancient world and a unique tradition of Chinese driving technology. Later Sun Ji revised some views in the article written in 1984, but he believed that the yoke method as a Chinese original idea was still retained in the latest revision of *A Collection of Chinese Ancient Carriages and Clothing* (Shanghai Classics Publishing House, 2013). Nevertheless, such idea might not be accurate because images similar to the ancient writing glyphs could be seen in plenty of the rock paintings in a large region of the Eastern Eurasia from the outer Caucasus to the Inner Mongolia. Such images could still be seen in the bronze sculptures in the Warring States (Sun Ji's *A Collection of Chinese Ancient Carriages and Clothing*, p. 13). According to approximate spoke's numbers and axle's positions in these images, Edward L. Shaughnessy believed that there were the carriages of the same type in the Eastern Eurasia, and the appearance, development and decline of the carriages in the outer Caucasus were in average earlier than those of the Eastern Asia (Edward L. Shaughnessy. Historical perspectives on the introduction of the chariot into China. *Harvard Journal of Asiatic Studies*, 1988, (1): 189–237). Although there was no definite evidence of the driving methods in the Eastern Asia like No. 2 bonze horse-drawn carriage unearthed in Qin's Mausoleum, it was inadvisable to regard the yoke method as Chinese important technological achievement. What's more, Zhong Shaoyi and others doubted whether a rope could conduct traction or not, and believed that the yoke method was still the neckband method in nature (Zhong Shaoyi and others. On driving methods of the single shaft carriage in the Shang and Zhou Dynasties. *History of Mechanical Technology-Collected Papers of the First Session China-Japan International Academic Conference on Mechanical Technological History*, 233–234).
8. Mongiatti, A., N. Meeks, and J. Simpson. 2010. A gold four-horse model chariot from the oxus treasure: a fine illustration of Achaemend gold work. *The British Museum Technical Research Bulletin* 4: 27–38.
9. From the images such as the relief coins in Roman times (the earliest appeared in the early 2nd Century), it seemed that there was the neck strap driving method of the present sense at that time (the main differentiating evidence is whether the belly belt and yoke exist or not).
10. Gimpel, Jean. 1976. *The Medieval Machine: The Industrial Revolution of the Middle Ages*, 29–58. New York: Penguin Books.
11. Wuwei Regional Museum. 1987. A tomb of the Wei-Jin Period, Nantan, Wuwei, Gansu. *Cultural Relics* 9: 87–93.
12. Hunan Provincial Museum. 1959. The excavation report for the Wei, Jin, the Southern, Sui Dynasties' Tombs in Changsha. *Acta Archaeologica Sinica* 3: 75–105.
13. Nanjing City Museum. 1972. A briefing of cleaning No. 5, 6, 7 tombs in Xiangshan, Nanjing. *Cultural Relics* (11): 23–41.
14. Ji, Sun. 2013. *A Collection of Chinese Ancient Carriages and Clothing* (Revised Edition), 97. Shanghai Classics Publishing House.
15. Ye, Fan. 1972. *Book of Later Han*, 1286. Zhonghua Book Company.

16. Liaoning Provincial Institute of Cultural Relics and Archaeology, Chaoyang City Museum. 1997. The 88 M1 excavation report for brick factory in Shiertai Village, Chaoyang. *Cultural Relics* (11): 19–32.
17. Cultural Relics Team of Liaoning Provincial Museum. 1984. Mural tombs of the Western Jin Dynasty in Yuantaizi. *Chaoyang Cultural Relics* 6: 29–45.
18. Ling, Cheng. 2009. The horse stirrups' origin and its new dissemination discussion in the middle ancient times. *European and Asian Journal* 9: 204.
19. ShahÎd, Irfan. 1995. *Byzantium and the Arabs in the Sixth Century*, 575. Harvard, Mass: Dumbarton Oaks.
20. Curta, Florin. 2007. *The Other Europe in the Middle Ages: Avars, Bulgars, Khazars and Cumans, Kononklijke,* 316. Brill NV.
21. White, Lynn. 1966. *Medieval Technology and Social Change*, 28–38. Oxford: Oxford University Press.

Wei Chen Born in Changyuan, Henan in 1985, Ph.D., assistant researcher at the Institute of Natural Science History, Chinese Academy of Sciences.

Research interests: History of technology; History of mathematics.

Major works: *The Transformation of Chinese Scissors Form and Its Possible External Influence in the 11–13th Century; Distribution and Technical Dissemination of Hoof Iron in Ancient China; Field Area Algorithm and Land System in the Middle Ages Computation: An Investigation of the "Tiancao" Volume; Popular Science in the Perspective of Cultural History-Reading R. Daenduns "Hypnotism and the End of the French Enlightenment"*.

Chapter 19
Printing

Xiumin Zhang and Qi Han

Most modern scholars believe that the invention of block printing began in the early Tang Dynasty (seventh century). Its history is roughly divided into four periods: (1)The Tang Dynasty till the Five Dynasties was the invention and epidemic period of block printing; (2) The Song Dynasty was its heyday. It continued to develop in the Liao, Jin, and Xixia Dynasties; (3) Printing in the Yuan Dynasty; (4) Printing in the Ming and Qing Dynasties. Movable-type printing was invented during the reign of Qingli (1041–1048) in the Song Dynasty. It was developed into tin, wood, copper, lead, and other movable-type printing in the Yuan, Ming, and Qing Dynasties, among which wood type was used most. Over the past 1,300 years from the early Tang Dynasty to the end of the Qing Dynasty, China had been focusing on block printing. After the War of the Opium, the traditional block printings and the old movable type were gradually replaced by the Western lithographs printing and letterpress printing.

19.1 Block Printing

Block printing was the process of carving characters on a whole piece of wood or other materials to make a wood-block printing plate and painting ink on this whole plate, which was also called whole-page printing.

X. Zhang · Q. Han (✉)
The Institute for History of Natural Sciences, Chinese Academy of Sciences, Beijing, China

J. Hua and L. Feng (eds.), *Thirty Great Inventions of China*,
https://doi.org/10.1007/978-981-15-6525-0_19

19.1.1 The Invention and Popularization of Block Printing (from About 636 in the Tang Dynasty to 960 in the Five Dynasties)

There was a long-standing debate in history over when the block printing began. People in the Qing Dynasty tended to think that it began in the Five Dynasties. However, some people said it began in the Sui Dynasty, or at the turn of the Sui and Tang Dynasties, or the Han Dynasty, or the Eastern Jin Dynasty, or the Northern Song Dynasty. But most scholars believed that it began in the Tang Dynasty. As for the accurate year of the beginning of the block printing in the Tang Dynasty, there were also different opinions. Scholars tended to think that it began in the seventh century when emperor Zhenguan had *Nv'ze* written by Queen Zhangsun printed in 636.

The earliest extant printed copies of the Tang Dynasty include the *Clean and Bright Toloni Sutra* (Fig. 19.1) in Chinese characters of Empress Wu, which was found in the State of Han in 1966. In Turpan, Xinjiang, the sacred volume of the *Lotus Sutra* printed in Chinese characters of Empress Wu collected by the Japanese was found. *Vajra Cchedikā Prajñā Paramita* (Fig. 19.2), which was discovered in Dunhuang in the ninth year of Xiantong, was particularly famous. It was recorded that the Bian family in Long Chifang in Chengdu County sold books of mantra in Chinese and Sanskrit. They began to print Calendar in Shangdu (Xi'an at present) in the Tang Dynasty. During the reign of Dezong (783–785), a voucher for a businessman's tax payment on the market was called "Yin Zhi." After the ninth century, the printing of the private calendar prevailed in Sichuan, Huainan, and Jiangdong. Luoyang re-carved the Buddhist scriptures, and Jiangxi carved the Taoist book *Biography of Liu Hong*. In the third year of Zhonghe (883), Liu Pin saw the block printing versions of *Yin Yang Miscellaneous Notes*, *Oneiromancy*, and other books.

The Five Dynasties and the Ten Kingdoms (907–960) were a period of fragmentation in history, but the book-printing business was still quite popular. The most influential book to the later generations was the classics of Confucianism initiated by

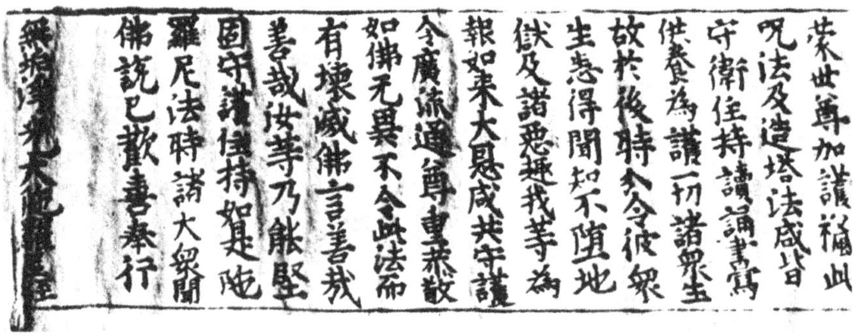

蒙世尊加護福此
呪法及造塔法咸皆
守衛住持讀誦書寫
供養為護一切諸衆生
故於後時亦令彼衆
生悉得聞知不墮地
獄及諸惡趣我等為
報如來大恩咸共守護
令廣流通尊重恭敬
如佛无異不令此法而
有壞滅佛言善哉
善哉汝等乃能堅
固守護住持如是陁
羅尼法時諸大衆聞
佛說已歡喜奉行
無垢淨光大陁羅尼經

Fig. 19.1 Clean and Bright Toloni Sutra

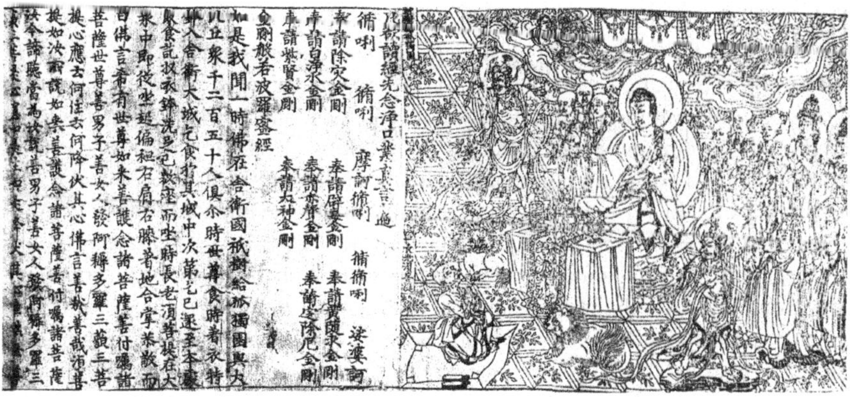

Fig. 19.2 Vajra Cchedikā Prajñā Paramita

Prime Minister Feng Dao in the Imperial College. From the third year of Changxing in the Tang Dynasty to the third year of Zhou Guangshun (932–953), *Nine Sutra* was created. Since the second year (935) of Mingde, Wu Zhaoyi had employed craftsmen to print the works of *Selected Works, The Primary Anthology* and *Nine Sutra,* etc. which was the beginning of printing of Official Edition. In the state of Wu and Yue, Qian Hongchu, king of the state of Yue, made 3 copies of *Sarva Tathagata Hridaya Dhatu.* The monk named Yanshou printed more books of mantra, which laid the foundation for the Hang carvings in the Song Dynasty.

19.1.2 The Heyday of the Block Printing (the Song Dynasty, 960–1279)

19.1.2.1 Printing in the Song Dynasty (960–1279)

The characteristics of block printing in this period were (1) the central government and local government paid much attention to printing books; (2) the block printing was popular in many places; (3) the content is rich; (4) the fonts, ink, and decoration are delicate. After the Zhao family unified China, education and culture flourished. The capital city, Kaifeng, had Taixue (the top educational institute) and martial arts, law, mathematics, medicine, painting, and other colleges. Township school, family school, and bookstores scattered all over the Lane in Hangzhou. The school was always full of students and half of the urban residents were educated in Fuzhou. Due to large number of students, textbooks and various books were naturally needed.

The key areas for block printing in the Song Dynasty were as follows:

Kaifeng. It was the capital city, the political, economic, and cultural center of the Northern Song Dynasty. The publishing institutions include imperial court,

Chongwen court, secretary of the imperial court, the national history court, the imperial court, the ministry of punishment, the Da Li temple, and the imperial court.

Hangzhou. Yemeng Deyun, a collector in the Song Dynasty said that, "Hangzhou is ranked at the top in printing, followed by Sichuan and Fujian." Since the block printings in Hangzhou were excellent, many books of the Imperial College were printed there. After the founding of the Southern Song Dynasty, many books were published by the Imperial College, the Deshou Temple, the Imperial Hospital, Lin An House, and Fu Xue. There were also many private bookshops that published books.

Chengdu. Like Hangzhou, it had a good foundation in the printing industry. In the early Song Dynasty, the first Buddhist scriptures, the *Kaibao Tripitaka* with 5,048 volumes (also known as *Song Kaibao, Sakamoto Tripitaka, Kai Bao Sutra*), was published in Song Dynasty, which was the mother edition of various sutras. It began in the fourth year of Kaibao (971) and was carved into 130,000 pieces in the eighth year of Taiping Xingguo (983). There were only about ten volumes in existence both at home and abroad.

Fuzhou. The Fuzhou officials and monks in the Northern Song Dynasty carved a large version of the sutra in the Dong Chan Temple, named *Chongning Sutra*. The Kai Yuan Buddhist Temple has also carved the *Bilu Sutra*, with more than 6,000 volumes in each of the two sutras. There were only a few copies that remained. In the fourth year of Zhenghe (1114), Huang Chang also applied to building the *Flying Sutra* in Guang County, a total of 540 letters, 5,481 volumes, entitled *Zhenghe Wanshou Dao Sutra*. In Fuzhou, two *Buddhist Sutras* and one *Dao Sutra* were printed. Hence, Fuzhou became the classic publishing center of Buddhism.

Huzhou. Wang Yongcong and his brothers donated to print 5,480 volumes of the *Great Buddhist Scriptures* in the Sixi Yuanju Temple in Huzhou. It was almost finished in the second year of Shaoxing in the Southern Song Dynasty (1132). The version was moved to the Zi Fu Temple after the Chunyou period. These volumes were also named *Tang Shu* and *History of Five Generations*. The beginning month or year of the carving was unknown, but it was finished in the second year of the Southern Song Dynasty (1175), 5,490 volumes in total. In the third year of Jiaxi (1239), Huzhou Sixi Fabao Zifu Temple published 5,940 volumes of Buddhist sutras. In the late Qing Dynasty, Yang Shoujing purchased one from Japan, which is now kept in the National Library of China.

Pingjiang Prefecture (now Suzhou). Known as the "golden piggy bank" in the Song Dynasty, Pingjiang was home to the famous collections of poems written by Li bai, Du fu, Bai Juyi, Wei Yingwu, and other poets of the Tang Dynasty. Since the fourth year of Shaoding (1231), the devout men and women, monk's and laymen collected money to set up a workshop of sutra in Yansheng Court and created the *Qi Sha Sutra*.

Jianning. Situated in the northern part of Fujian, Jianning was one of the publishing centers from the Song Dynasty to the end of the Ming Dynasty. The two workshops (the Masha and Chonghua) affiliated to Jianyang County were called the "Book House" and their publications were sold around, far away to Koryo and Japan.

In addition, there were about 80 kinds of block printing in Yanzhou (now Jiande), which were called the "Yanzhou version." There are 66 kinds in Jiankangfu House (now Nanjing), with about 20,000 printed books. Next was the Qingyuanfu House (now Ningbo), which printed the "Mingzhou version." Books printed by the Shaoxing House were called the Yuezhou version. Quzhou and Wuzhou (now Jinhua) were also famous for printing books, especially good at pirating books. In the Southern Song Dynasty, books were printed almost everywhere, and even the distant Qiongzhou in Hainan Island printed medical books.

The academic culture of the Song Dynasty flourished, providing many manuscripts for the printing industry, and the publication of many printed books promoted the advancement of culture and technology. The printing of the Song Dynasty was rich in content, including ancient history books, and hundreds of kinds of geography books. In addition to the ancient scholars, one division of our ancient books called Zi Bu also published and printed the *Ten Calculating Method Books*, ancient books on agriculture, etc. The government attached great importance to the medical books, repeatedly revised and published Chinese medical books, and promulgated the Emperor Taizong's and Emperor Huizong's medical prescription. Scholar-officials, such as Su Shi, Shen Kuo, etc., preferred to publish up to 50 kinds of their family's prescriptions. Books about pediatrics, obstetrics and gynecology, acupuncture and moxibustion had also been published. There were many versions of the collections of poets such as Han Yu and Liu Zongyuan, and 23 printed editions of Su Dongpo. There were other printed materials such as the banknotes called "Jiaozi," "Meizi," and "Guanzi," and many vouchers of transportation and marketing transactions, handprint, woodcut, etc.

19.1.2.2 Printing of the Liao, Jin, Xixia and Dali Dynasties

From 916 to 1125, in the Liao Dynasty, the rulers of the Qidans believed in Buddhism, built temples and pagodas in Wujing, and rebuilt the vast Buddhist scriptures in Fangshan County. They also carved two, *The Tripitaka*, one of which was the large scroll with 5,048 volumes, and 12 volumes of them were found in Shanxi in 1974; the other one was a small paper book with less than a thousand volumes, which can only be found in the records of the Koryo. The Liao Dynasty Publishing Center was in Nanjing (or Yanjing, now Beijing), which had a printing house. Temples such as Da Min Zhong Temple (now Beijing Fa Yuan Temple), Da Hao Tian Temple (now destroyed), and the bookstore also printed Buddhist scriptures. The general books included the book *Long Yao Hand Mirror*, the medical book *The Elbow Rear*, Su Dongpo's *Da Su Xiao Ji*, and the newly discovered children's reading material *Mounting* printed version by Li Han in the Tang Dynasty.

In the Jin Dynasty, from 1115 to 1234, the Jin military broke into Kaifeng, the capital of the Northern Song Dynasty, and brought the Hui, Qin two emperors, and the countless people to the heel. They also took away the books of *Imperial College*, *Mige* and *Sanguan* the *Guozijian* printed version, as well as the *Tao Sutra*. When the *Six Classics* and *Seventeen Histories* were printed by the imperial college of the Jin Dynasty, the previous Song edition obtained from the Song Dynasty was printed again and *The Proposal of Su Dongpo* was updated. In addition, Nanjing (now Kaifeng), Pingyang, Taiyuan, Ningjin, and other bookstores printed books. The southeastern Jindai River Road (southwest of the present mountain) Buddhist believers donated money, scorpions, pear trees, carving knives, etc., and the *Golden Sutra* was carved by the monks, commonly known as *Zhao Cheng Sutra.* At that time, many Taoist works were published. Zhong Du Tian Chang Guan (now Beijing Bai Yun Guan) created the *Dajin Xuandu Treasure* 6,545 volume based on the Song Dynasty *Tao Sutra* edition, and another thousand volumes of remained sutras. In the 18th year of the Yuan Dynasty (1281), Baoding, Zhending, Taiyuan, Pingyang, and Hezhongfu (Puzhou Town, Yongji County, Shanxi Province), Wang Zu Shi Shantou, Kansai, etc., all had Tao Sutra scriptures. Except for the Pingyang edition published in the Mongolian period, the rest were secretly preserved by the Taoist priests in 1258 when the Mongolian sect burned. There are up to six or seven editions of the *Dao Sutra* inside and outside of the Jin Dynasty capital, which was a grand event in the history of Chinese Taoism. Jin set up the Nvzhen Guozi School and Nvzhen Fuxue School, with 3000 students. Nvzhen's *Five Classics, Han Shu,* and *New Tang Shu* were translated and promulgated successively, and Nuzhen's *Filial Piety Classic* was distributed to the guards.

The Xixia Dynasty was from 1038 to 1227. In Ningxia area, the minority regime was established by the tangutqiang. The rulers also believed in the Buddha. At the beginning of the founding of the empire, they purchased *The Book of the Tripitaka* with horses from the Northern Song Dynasty for six times, and asked Song for a book printed by the imperial college. Emperor Li Renxiao of the western Xia Dynasty distributed 50,000 or 100,000 Buddhist sutras at a time. In the western Xia Dynasty, many books of war such as *Sun tzu's Art of War, Liu Tao, Strategies of Huang Shi Gong, the Analects of Confucius,* and *Er Ya* were translated into Chinese. And today we still have the *Pan Han He Shi Zhang Zhong Zhu*, the 1190 version in the Renzhong period of Xixia.

The Dali Dynasty (937–1253), the Dali National Journal, composed of the Bai people in Yunnan, had *Buddhism, Long Life,* and Other sutras.

19.1.3 Printing in the Yuan Dynasty (1271–1368)

The most important places for block printing were Dadu (now Beijing), Hangzhou, Jianning, and Turpan.

Da Du. It was the political, economic, and cultural center of the Yuan Dynasty. Emperor Taizong built the Editing Institute in Yanjing. Emperor Shizu sent his emissaries to Hangzhou and other places to collect the printed official books to Beijing, and these old Song editions became the Yuan version. The Guangcheng Bureau was one of the central book publishing institutions. Xingwen Department was also set up to print many history books with the *Zi Zhi Tong Jian* (*Comprehensive Mirror for Aid in Government)* as the beginning. The imperial college printed the fine version of *The Treatise on Febrile Diseases*, the Tai Hospital Journal *Shengji General Record*, and the operas from the Yuan dynasty.

Pingyang. Emperor Taizong took the advice of Yelv Chucai to set up an editing institute in Pingyang, and compiled the classics of history. Thus, Pingyang became the publishing center of the Jin and Yuan dynasties.

Hangzhou. The Yuan government also liked to send the official books to Hangzhou for printing. For example, the new manuscript of *The Song History* was sent to Hangzhou by horses, and the selected books would print 100 copies. The *West Lake Academy* in Hangzhou, formerly the imperial college of the Song Dynasty, had a collection of over 200,000 old books and had 92 skilled workers repairing over 7000 pieces of the damaged books in the reign of Zheng from 21st to the 22nd year (1361–1362).

Jianning. Much more books were printed by the Jian'an and Jianyang Bookstore.

Turpan. It has found that the classics were printed in six languages.

The great Sutra classics include *Puning Sutra* in Hangzhou road, *Moraine Sand Sutra* after the Song Dynasty, and two sutras of Fuzhou in the Song Dynasty. Kublai Khan sent a special envoy to Koryo to repair *Koryo Sutra.* There were also *Collections of the Mongolian Language, Collections of the Sutra Language, Collections of the Xixia Language.* The official master monk Guan Zhuba of the Songjiang Prefecture had printed more than 50 Sutras, and about 30 books of the Hexi dialects. The Taoist priest Song Defang and his disciple Qin Zhi'an used more than 500 workers to print more than 7,800 volumes of the *Xuandu Treasure* in 1244 in Xuandu, Pingyang. The Mongolian translations of *Filial Piety, Zhenguan Dignitaries,* and *Interpretation of Great Learning* were printed, along with many salt tea citations and paper money.

19.1.4 Printing in the Ming Dynasty (1368–1644)

In the first year of Emperor Hongwu (1368) in the Ming Dynasty, the tax book was exempted in August, which made the book industry flourish. After Emperor Jiajing, more than 1,200 people were engaged in publishing and printing. The number of official and private block printing books was large, and there were many varieties, surpassing the amount in the Song and Yuan Dynasties. At that time, the two Jing and thirteen provinces all printed, with emphasis on Nanjing, Beijing, Hangzhou, Huzhou, Suzhou, Huizhou, and Jianning.

Nanjing. In the early Ming Dynasty, Nanjing imperial college received more than 100 kinds of old versions of the imperial college of the Southern Song Dynasty stored in the west lake academy of Hangzhou in the Yuan Dynasty, and took about 300 books from the local edition and the new edition of the imperial college. The various departments of Nanjing and Ying Tianfu also printed books, and there were more than a hundred bookstores.

Beijing. In the 19th year of the Yongle period (1421) in the Ming Dynasty, the capital was moved to Beijing. The publishing industry was under the supervision of the eunuch department, and has printed about 200 kinds of scriptures, which were called the *Factory book.* Most books printed by Beijing imperial college were based on south college version, which were about 90 ones. The important ones were *Twenty-One Historical Books*. Every three years, the Ministry of Rites published *Deng Kelu* and *Trial Record*, and the Military Department's *Record of Military Tests*. The court had also published a book, and Qin Tianjian had printed and produced the *Large Calendar Day* every year, which had been issued at home and abroad.

Hangzhou. The Chief Secretary, the Inspectorate, and the Hangzhou Government had many books and many bookstores. In the last years of the Yongle period, a Buddhist collection of *Wulin Sutra* was printed.

Huzhou. This city was well known for its imprint. The Ling and Yan families had a total of 144 kinds of imprint.

Suzhou. Collecting books and the block printing books were quite popular among many people in the area. Before Wanli (1573–1620), there were 177 kinds of books, which ranked the highest in the country. There were also many book shops. Mao Jin, a famous collector and publisher of Changshu County, had published more than 100,000 copies of about 600 kinds.

Huizhou. Since the Song Dynasty, it has been the origin of paper and ink. In the Ming Dynasty, many books were by private publishing houses and bookstores.

Jianning. In the Song and Yuan Dynasties, the bookstores were mostly located in Jian'an County. In the Ming Dynasty, the bookstores were mostly in two places in Jianyang—Masha and Chonghua with more than 100 bookstores, and the total number of books published ranked the first in the country. Chonghua had a book market which opened on the first and sixth day every month. During those days, merchants flooded to the book market, while it is uncommon in other provinces.

In the Ming Dynasty, there were 39 independent Imperial Palace publishing houses, with 400 kinds of inscriptions. The inscriptions were well mounted, which was called "Fanfu Ben." The books printed by the prince Ning's palace of Nanchang and its descendants Prince Yiyang ranked the first, followed by Prince Shu. Eunuchs also had their own contributions to print books, which have not been found in the past. In addition to the block printing of ancient books, many works of the people of the Dynasty have been published. In the early Ming Dynasty, there were so-called "book-making" (or "letter of decree"), which was compiled by Emperor Taizu himself mostly, and the most important one was the three parts of *Da Gao*. In order to satisfy the needs of imperial examinations, he also published many Baguwen (Stereotyped Writing). Emperor Yongle attached great importance to local chronicles and issued two revisions to them. There were about 1,500 kinds of local chronicles in the Ming Dynasty, and only half of them exist now. In the Ming Dynasty, those whoever passed the royal examinations must have a printed manuscript of about 6,000 kinds of poems and essays. Popular literature *Romance of the Three Kingdoms*, *Water Margin*, *Journey to the West*, and so on were loved by people, and have been published again and again. There were more than 70 versions of *Pipa Ji*. There were many scientific and technological books, as well as the medical books published in the Ming Dynasty. Some western scientific and technological books, such as *Euclid's Geometry Original* and Xiong Sanbao's *Taixi Water Law*, were also introduced. Nanjing had *Hongwu Nanzang*, *Yongle Nanzang*, and its supplementary sculptures, Beijing had *Yongle Beizang* and *Panzang* in Tibetan, Hangzhou Jingshan cable edition *Jiaxing Sutra* and *Orthodox Taoist Sutra*. Beijing Nantang has published the Catholic Church book.

19.1.5 Printing in the Qing Dynasty (1644–1911)

Nurhachi had been known to use printing to serve the politics, and now *Official Articles to Emperor Ming Wanli* remains as a record. After entering the northeast area, the Chinese and Manchu books were published, which were called *Neifuben* and imitated the Jing Changben of the Ming Dynasty. In the nineteenth year of Kangxi (1680), a book repair center was set up in the gallery around the Wuying Hall, which administered the publishing and decorating work of books. Emperor Qianlong carved *Notes on the Thirteen Classics* and *Twenty-Four Books on History*, so the famous Wu Ying Dian versions were called the *Wu Ying Dian Edition* for short. It was the most exquisite carved under the reign of Kangxi and Qianlong. The palace plate was made of Kaihua paper and ink, which was the best-printed version of the Qing Dynasty. There were 382 copies saved in the temples, and smaller the number after the Jiadao period. Beijing imperial college was only used as the house of Wuying Dian and Beijian in the Ming Dynasty to store the books. In the fourteenth year of the Daoguang period (1834), 64 editions were stored and printed, with nearly 150,000 pieces.

In the Qing Dynasty, the local government offices did not print as many books as those in the Song and Ming Dynasties. After the Taiping Heavenly Kingdom, the rulers put forward the slogan of "Maintaining the world and cultivating the people." In 1863, Zeng Guofan founded the Jinling Bookstore in Nanjing. Since then, more than ten provinces have followed his example establishing official bookstores in the provincial capitals. The famous ones were Zhejiang Official Bookstore, Wuchang Chongwen Bookstore, and Guangya Bookstore in Guangzhou. Each Bureau printed about 1,000 kinds, known as "official printed edition." For popularization, the prices of those books were low, so the quality of paper and ink was poor. Due to the short span of time from then on, the editions of the Zhejiang Bureau and Guangya Bureau each had 1,560,000 pieces. In the Qing Dynasty, the bookstores in Nanjing and Hangzhou had declined, mainly located in Beijing. There were more than 100 bookstores (some people said 300), mostly in the area outside Liulichang in Xuanwumen. There were about 50 bookstores in Suzhou, followed by Guangzhou and Foshan.

In addition to the block printing of ancient books, a large number of Qing poetry and essays were printed and published in the Qing Dynasty. There were as many as thousands of pamphlets in popular literature, such as children's books and folk songs. There were about 7,000 kinds of local chronicles in the Qing Dynasty, most of which were from Emperor Kangxi and Emperor Qianlong. There were nearly 3,000 series (including about 70,000 books), mostly compiled by Jiangsu and Zhejiang people, and there were many local series. The Manchu documents were mostly carved in Neifu and Beijing Bookstores, and there were 180 kinds of them. The

Buddha sutras contained 79,036 pieces of *The Long Sutra* of the double-sided pear wood version, which was the only complete version of *The Dazang Sutra* existing in China. In addition, there were many kinds of incomplete abbreviations of *The Dazang Sutra* inscribed by Jinling Sutra Inscription Office and various temples in the late Qing Dynasty. Emperor Qianlong proudly completed the publication of the Manchu *Dazang Sutra* in his later years. The Tibetan scriptures were carved and printed in the ancient edition of Naitang in the Yuan Dynasty (1312–1313); Yongle edition, Wanli edition, Tar Temple edition, Changdu edition, and Litang edition in the Ming Dynasty; Beijing edition, Zhuoni edition, Dege edition, New edition of Naitang in the Qing Dynasty; and Lhasa edition in 1933, among which the Dege version was the best.

The Mongolian Dazang Sutra has been translated and printed four times: (1) in the year of Yuan Dade (1297–1307), it was printed in Tibet; (2) in the period of the Ming Wanli (1573–1620), it was added to the publication; (3) in the year of Kangxi (1683), it was published in Ganzhu'er; (4) in the period of Qianlong (1741–1749), it was reprinted in Danzhu'er from the sixth to the fourteenth year of Qianlong. Islam, Catholicism, and Christianity also had their own books published.

In the first year of the Zhizheng period (1341), the *Vajracchedikā-prajñāpāramitā sūtra* (Fig. 19.3), inscribed by the unknown monk at the Zifu Temple on Zhongxing Road (now Jiangling), was overprinted in both Zhu (red) and Black ink. In the

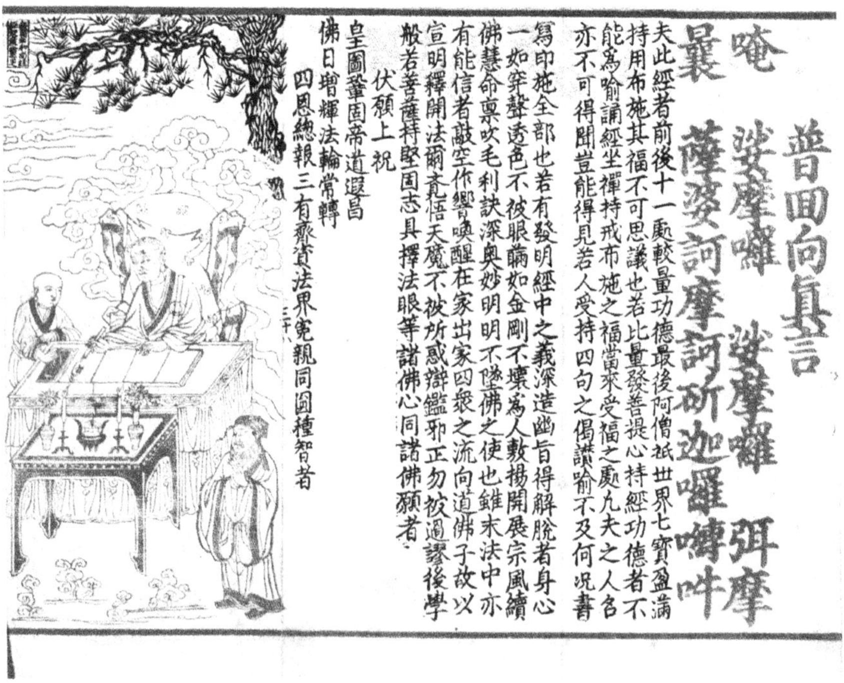

Fig. 19.3 Vajracchedikā-prajñāpāramitā sūtra

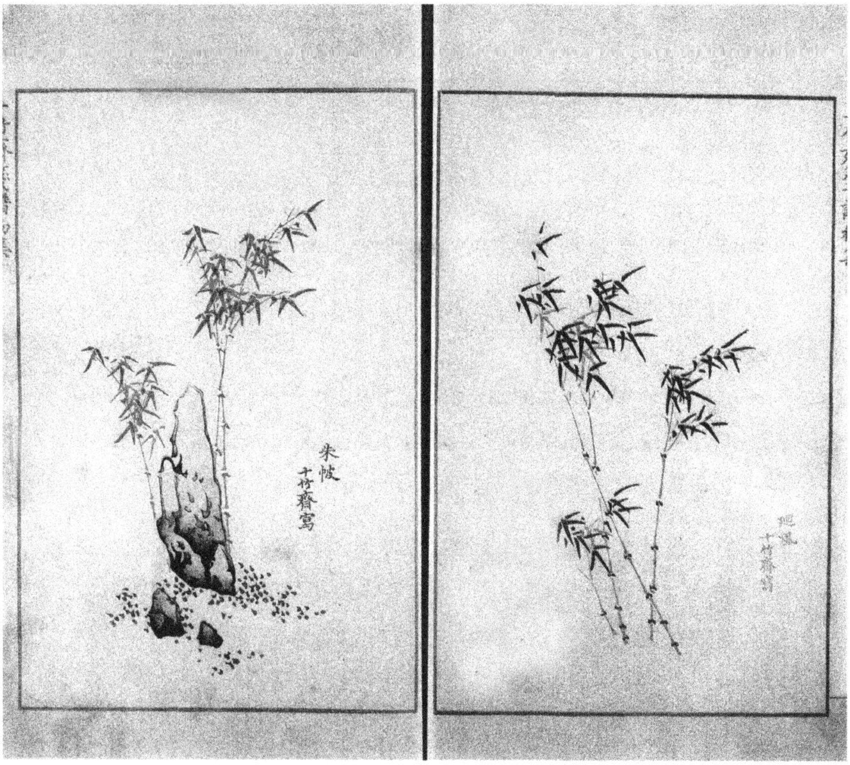

Fig. 19.4 Hu Zhengyan's Shizhuzhai painting (**a**)

history of printing, bamboo plate color printing was another great leap forward, and it was very successful in the early seventeenth century. The most prominent representative works were *Luoxuan Become Ancient Notes* carved by Wu Faxiang, a native of Jiangning, which were overprinted by bamboo plate and arch-flower method. *Luoxuan Become Ancient Notes* was printed in 1626, a year earlier than Hu Zhengyan's *Shizhuzhai Painting and Calligraphy Spectrum* (1627, Figs. 19.4 and 19.5), and 19 years earlier than *Shizhuzhai Notebook*. During the Qianlong period, Ding Liangxian and Ding Yingzong from Suzhou printed many flower-and-bird paintings (Fig. 19.6) on bamboo plate, with fine sculpture and arch-flower technique. They were also excellent products in cross-color printing.

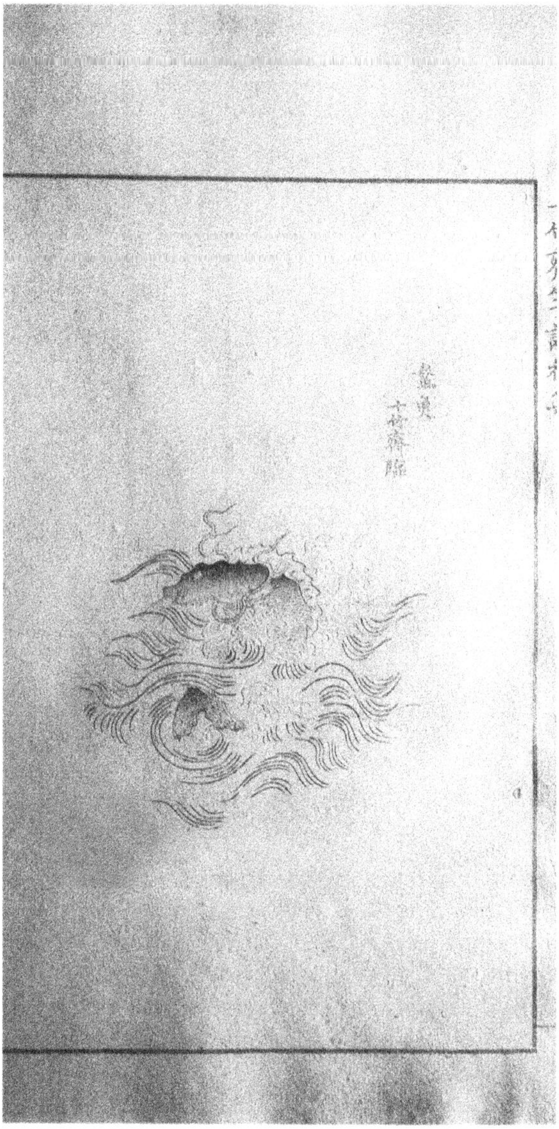

Fig. 19.5 Hu Zhengyan's Shizhuzhai painting (**b**)

19.2 Wax Printing, Tin Casting Board, and Clay Plate

In the Song Dynasty, people not only used various wooden and copper plates as printing tools, but also invented wax for printing. In the first year of Shao Sheng (1094), people in Kaifeng were eager to publish the list of the new number one scholar of the national test. They could not wait for the boards to be printed, therefore printed it with wax instead. This record could be found in Volume 2 of *Chu Zhu Records*

Fig. 19.6 Flower-and-bird paintings on bamboo plates by Ding Liangxian

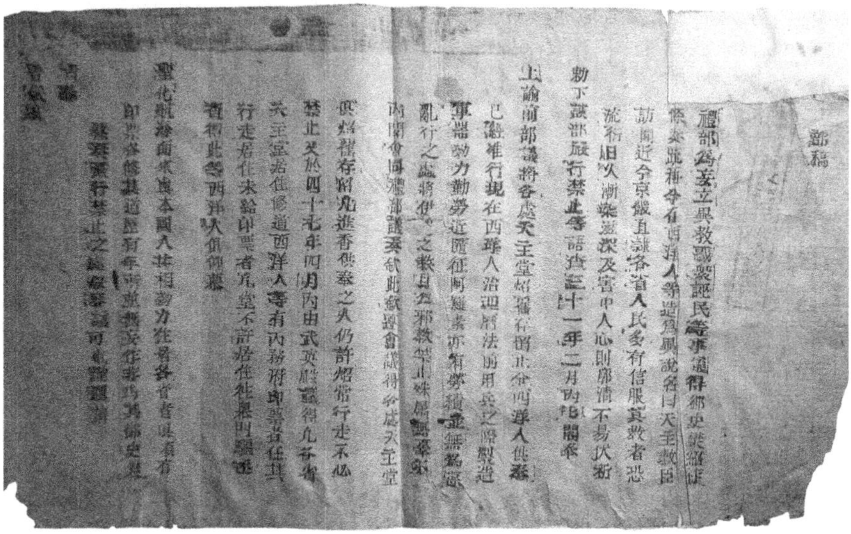
部稿
禮部為[illegible]立異教[illegible][illegible]民[illegible]事[illegible]得御史蔡維[illegible]
條奏疏稱今有西洋人等造為異說名曰天主教臣
訪聞近今京城直隸各省人民多有信服其教者恐
流布日以漸染滋深及害中人心則廓清不易伏祈
勅下該部嚴行禁止等語查[illegible]十一年二月內[illegible][illegible]
上諭前部議將各處天主堂[illegible][illegible][illegible][illegible]並令西洋人供奉
已經准行現在西洋人治理曆法兼用其人之際製造
事器效力勤勞近隨征[illegible][illegible]素亦有勞績並無為匪
亂行之處將伊等之教目為邪教禁止殊屬無[illegible]
內閣會同禮部議奏欽此欽遵會議得各處天主堂
供奉禮拜存留凡進香供奉之人仍許照常行走不必
禁止又於四十七年四月內由武英殿議得凡各省
天主堂若住修道西洋人等有內務府印票者任其
行走若住未給印票者凡堂不許居住社[illegible]四鄉[illegible]
者概此等西洋人俱撤
聖化執法而求其本國人共相勸力往各省者具有
印票各修其道並有每年前赴京城安住者均其係史[illegible]
嚴[illegible]宣行禁止之處應毋庸議可也為此通行
[illegible]

Fig. 19.7 Wax printing

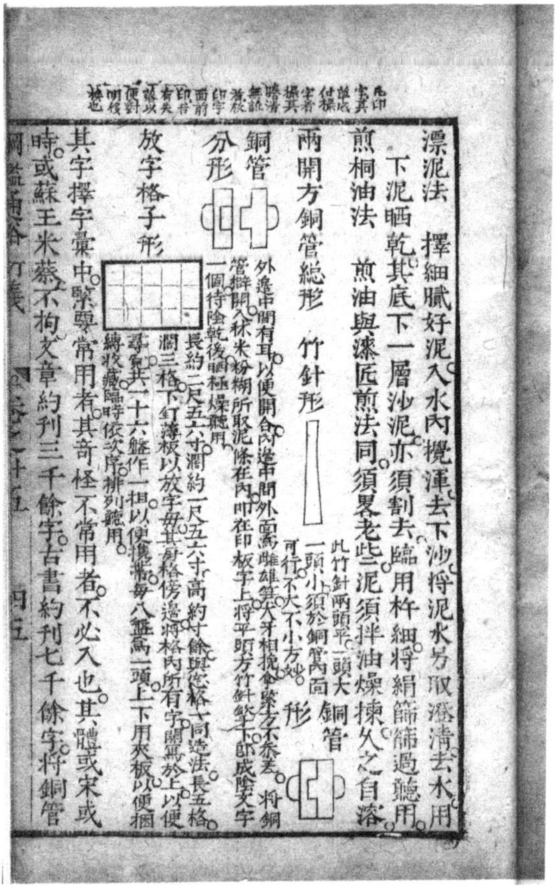

Fig. 19.8 The popular evolution of the 21st historical books

of the Song Dynasty be He Wei, Bi Jian is number one, and Zhao Shen comes the second. At the beginning of the song, people were eager to report it and print it on wax plate. Initial words are modeled, but no ink was found at the point of water. The reporter cried out to the public: "Bi beat Zhao Shen, winning the champion!" All people said it sounded ominous. Then Zhao Shen was accused of conspiracy. At that time, the new champion was Bi Jian, but because of the oiliness of the wax, the word "Jian" was not printed together with three dots before the word. This wax printing was suitable for urgent situation and time-consuming works. Whether it was used in the Yuan and Ming Dynasties has not been recorded, but it was often used in printing newspapers in the Qing Dynasty (Fig. 19.7).

In the early Ming Dynasty, some people used tin plates to print counterfeit banknotes and they had been executed for the forgery. In the 52nd year of the reign of Qianlong (1787), Chengdun, Shexian County, printed Volume 1 of the *Wadang Characters of the Qin and Han Dynasties*. It began to use jujube wood to copy. Though it had been proofread, there were still some mistakes in that copy. Later, the method of printing with sand was adopted by the Han people, and Benwa was taken as a model to melt tin into sand. Chengre printed with the molten tin can be called as the innovative printing.

The newly certified Xiucai Lv Fu (1671–1742) made clay plates with self-made clay characters in the first year of the reign of Qianlong (1736) and printed 26 volumes of his book, *The Popular Evolution of the 21st Historical Books* (Fig. 19.8), in which the printing method was introduced in detail. As a writer and a printer, it was rare in the history of the Chinese printing, but Lv Fu was earlier than Zhai Jinsheng and Liang A'fa.

19.3 Movable Printing

19.3.1 Clay Movable Printing

The Movable clay type was an irregular word made of clay for typesetting and printing. According to Shen Kuo (1030–1095), a scientist in the Northern Song Dynasty, in Volume 18 of *Mengxi Bi Tan*, "Skills" (Fig. 19.9), the clay movable characters were invented by Bi Sheng, a civilian during the Qingli period (1041–1048). The method was to use glutinous mud to print letters; each word was printed, molded with fire to make the muddy movable letters. They were alternately typesetted and printed on two iron plates. Shen Kuo called this method "extremely efficient for

使是八除者增二便是但一位一因之若
位數少則頗簡捷位數多則愈繁不若乘
除之有常然算術不患多學見簡即用見
繁即變不膠一法乃爲通術也
版印書籍唐人尚未盛爲之自馮瀛王始印
五經已後典籍皆爲版本慶曆中有布衣
畢昇又爲活版其法用膠泥刻字薄如錢
脣每字爲一印火燒令堅先設一鐵版其
上以松脂臘和紙灰之類冒之欲印則以
一鐵範置鐵板上乃密布字印滿鐵範爲
一板持就火煬之藥稍鎔則以一平板按
其面則字平如砥若止印三二本未爲簡
易若印數十百千本則極爲神速常作二
鐵板一板印刷一板已自布字此印者纔
畢則第二板已具更互用之瞬息可就每
一字皆有數印如之也等字每字有二十
餘印以備一板內有重複者不用則以紙
貼之每韻爲一貼木格貯之有奇字素無
備者旋刻之以草火燒瞬息可成不以木
爲之者木理有疎密沾水則高下不平兼

Fig. 19.9 Mengxi Bi Tan of Yuan edition

printing hundreds of thousands of copies." Bi Sheng's invention of the clay movable-type printing marked the birth of the movable-type printing, about 400 years earlier than the Gutenberg movable-type printing in Germany.

In addition to Bi Sheng, Shen Kuo and his nephews, a poem in *Deng Su's Collection of Works* by Mr. Chu Yu (1091–1132) in the late Northern Song Dynasty and the early Southern Song Dynasty (no later than 1132 years ago) also recorded the relevant historical facts of Bi Sheng's "two-plate iron" method, which showed that people still had a certain degree of understanding of the movable-type printing calligraphy at that time.

Again, according to Volume 198 of *The Collection of Zhou Yiguo*, when Zhou was in Changsha in the fourth year of the Shaoxi's reign in the Southern Song Dynasty (1193), he said that the book *Yu Tang Za Ji* written by himself was presented to friends, which "used Shen Cunzhong's method—clay copper plate." According to Volume 15 of the *Collection of Mu'an* by Yao Shu, Yang Gu (Yao Shu's student), nephew of Yao Shu in the Yuan Dynasty, used "Shen's movable board" to print in Zhu Xi's *Xiao Xue, Jinsi Lu,* and Lu Zuqian's *Dong Lai Jing Shi Lun Shuo* from 1241 to 1251. The so-called " Shen Cunzhong's method" and "Shen's movable board" referred to the movable-type version of Bi Sheng clay described by Shen Kuo, which proved that the Bi Sheng method has been mastered and applied in the Song and Yuan Dynasties.

In the tenth year of the Daoguang's reign (1830), Li Yao of Suzhou employed a dozen of workers when he lived in Hangzhou. He imitated the printing method of the Song Dynasty's clay plate and printed the *Historical Survey of Southern Xinjiang*. In the 24th year of reign of Daoguang (1844), Zhai Jinsheng, a teacher in Jingxian County, Anhui, made more than 100,000 mud characters, divided into large, medium, small, sub-small to which the smallest is in number 5 (Fig. 19.10). They were printed in his own poetry anthology, *The First Edition of Mud Plate Trial Printing* (Fig. 19.11), all in Song style, and were printed clearly; later, it was used

Fig. 19.10 Mud characters made by Zhai Jinsheng

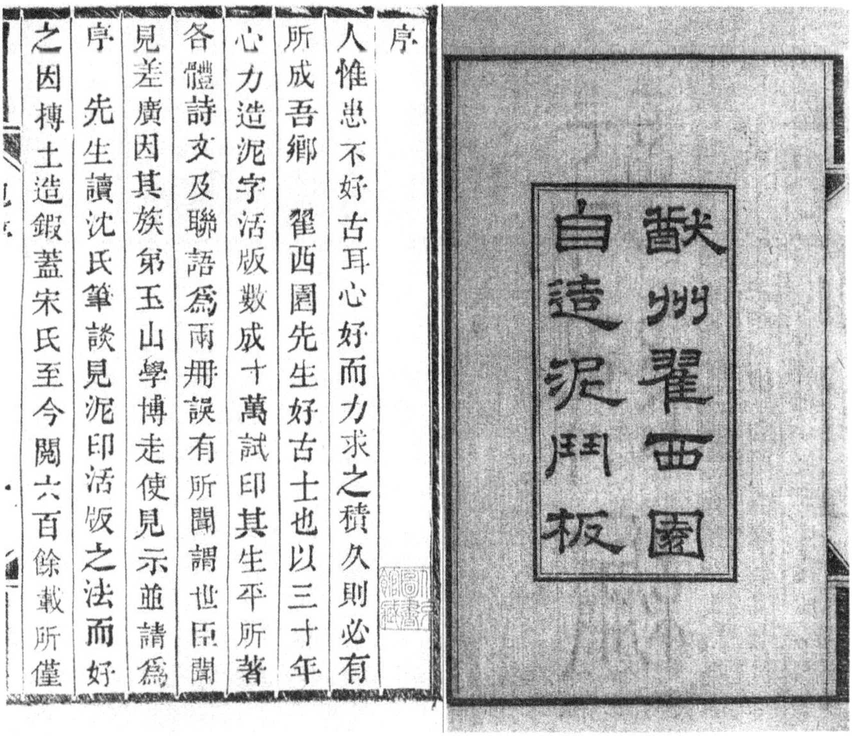
序
人惟患不好古耳心好而力求之積久則必有
所成吾鄉　翟西園先生好古士也以三十年
心力造泥字活版數成十萬試印其生平所著
各體詩文及聯語爲兩册誤有所聞謂世臣闡
見差廣因其族弟玉山學博走使見示並請爲
序　先生讀沈氏筆談見泥印活版之法而好
之因摶土造鍛蓋宋氏至今閱六百餘載所僅

猷州翟西園
自造泥鬥板

Fig. 19.11 The first edition of mud plate trial printing

in printing by his friend Huang Juezi's *Xian Ping Shui Wu Chu Ji* and his younger brother's *Shui Dong Zhai Shi Zong Pu.*

In 1987, in Wuwei, Gansu, the West Xia Buddhist Sutra *Vimalakirti Sutra* (Fig. 19.12) was found. It was identified as a print in Renzong's reign after research. According to the phenomena of the air-hole, deformations and trace in strokes, scholars regarded it as an example of the clay movable-type printing.

19.3.2 The Wood Movable Printing

In the early Yuan Dynasty, the agronomist Wang Zhenyu created more than 30,000 wood movable characters in the second year of Dade's reign (1298) after Bi Sheng's experiment with the clay movable-type printing in the Northern Song Dynasty, and successfully edited his own Dade County Chronicle, which was the first local record of the wood movable type in China. He invited the craftsmen to make the wood movable type and worked out the method of typesetting and printing. In 1298, he printed Dade County Chronicle, and compiled the whole set into The Calligraphy

Fig. 19.12 Vimalakirti Sutra

of Making Movable-Type Printing, which was attached to the volume of Agricultural Books for publication. The main method was to paste the paper pattern on the board, carved the words as before, sawed them into words, and then trimmed them with a knife to unify the size and height. Then the typesetting lines, spaced with bamboo, filled with a plate of frame, with small bamboo filling and tightening after the ink-coated paper, brown brush straight along the border brush. At the same time, he also created a runner typesetting frame (Fig. 19.13). The runner was placed in wooden characters according to rhyme, which promoted the runner to use words as human beings and made it easy to retrieve and return words (see Agricultural Books Movable Character Printing Calligraphy). From the Yuan Dynasty to the second year of Zhizhi's reign (1322), Ma Shengde of Zhizhou in Fenghua, Zhejiang, printed 100,000 characters in movable-type edition and printed such books as Daxue Yanyi. In Dunhuang Thousand Buddha Cave, it was once found the Yuan Dynasty Uighur characters were made of hardwood. In the Wanli period of the Ming Dynasty, the wood movable characters were popular. Some Fanwang offices, academies, and private houses were mostly used for printing books. In the eleventh year of Chongzhen's reign (1638), the wooden movable printed pattern "Di Bao" was used until the end of the Qing Dynasty.

In 1774, the thirty-ninth year of Qianlong's reign in Qing Dynasty, Jin Jian was the Vice President of the Four Treasuries, who managed the carving of the Wuying Palace. According to the statistics of Peiwen Poetry Rhyme, six thousand and hundreds of words were collected, which were stored in wooden cabinets of the eight-storey drawers according to the rhyme entries and carved into 253,500 jujube-wood movable characters and printed in 138 kinds of "Wu Ying Dian Juzhen Edition Series" (including 4 kinds of block printing) and several other single-line editions

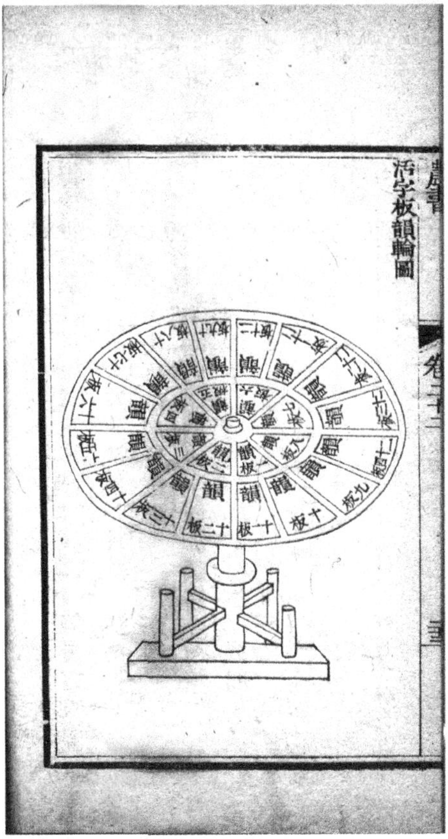

Fig. 19.13 A runner typesetting frame in agricultural books movable character printing calligraphy

in more than 2,300 volumes. Emperor Qianlong changed the name "Movable Character Edition" into "Juzhen Edition." In 1776, Jin Jianju wrote the book The Program for the Collection of Precious Editions in Wuying Dian (Fig. 19.14) describing the printing process with detailed contents. After that, 14 provincial wood movable-type books were imitated, and most of them were poems and essays. Other places such as Shaoxing, Changzhou, and Huizhou had thousands of genealogies. Seventy-eight of them is wood movable scripts. The wood movable-type printing was prevalent in ancient China, second to block printing, which had the names of "sub-board" and "plywood."

Although the movable-type block printing saves labor and time, it has not been vigorously promoted for 800 years. Besides the political and economic reasons, it also had its own technical defects. Generally, individuals or bookstores were limited to capital and had only thousands of movable characters. Due to the limitation of the number of characters, they had to take the method of one-sided typesetting, disassemble the one-sided plate, and arrange it again. The size of the same movable type cannot be uniform. The pad was uneven. The font was skewed. The ink color

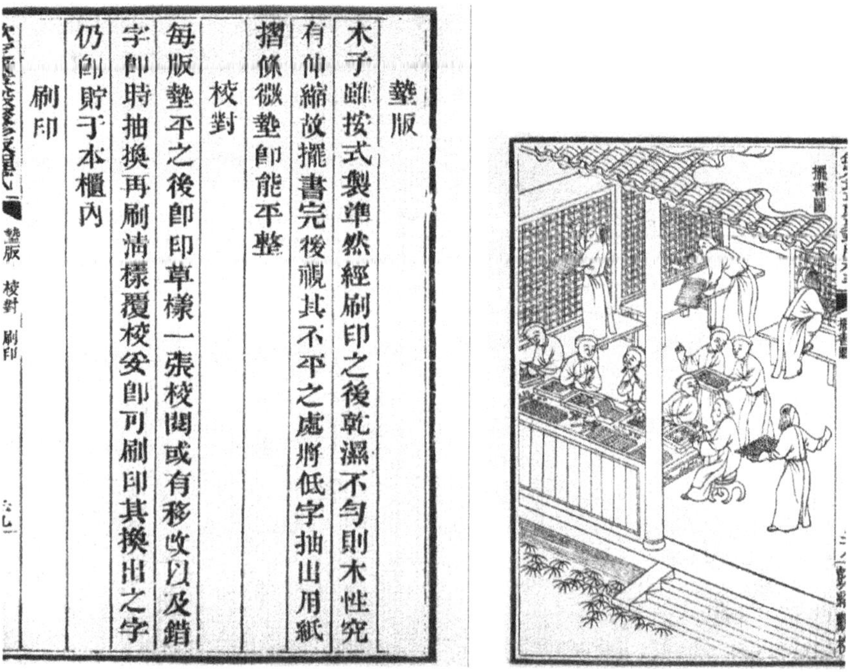

Fig. 19.14 The program for the collection of precious editions in Wuying Dian

was uneven, and there were many wrongly used words due to careless proofreading, so the movable typesetting was not taken seriously. Some people thought that the movable-type edition was only a temporary measure. Only when it was carved into a full edition can it be regarded as an official publication. Because of the old paperless type, if a book was to be reprinted, it must be re-typesetted. On the contrary, block printing can be reprinted more economically and conveniently, so letterpress printing cannot replace block printing in China.

In the Xixia Dynasty, the wood movable-type printing also appeared, such as the remnants of *Dafang Guangfo Huayan Sutra,* and the sentence "vow to carve broken characters" appeared in the vow-making article. In addition, there were parts of the Western Xia Buddhist Sutra *Auspicious Everywhere to the Mouth and the Sequel* (Late Xixia, Fig. 19.15), which was the oldest living wooden character book.

19.3.3 The Copper Movable Printing

The copper movable printing was an irregular word cast in copper for typesetting and printing. Chinese copper movable type was popular in the south from the end of the fifteenth century to the beginning of the sixteenth century. The earliest was the third

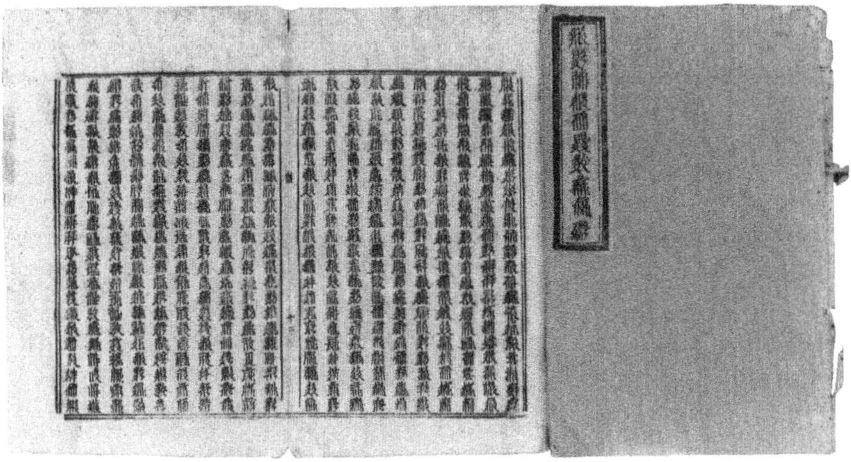

Fig. 19.15 Western Xia Buddhist Sutra auspicious everywhere to the mouth and the sequel

year of Hongzhi's reign in the Ming Dynasty (1490) in Wuxi, Jiangsu (1439–1513), which was printed in bronze movable type in 50 volumes of the *Annotations of the Song Dynasty's Ministers in the Imprinting Hall of Huitong* (Fig. 19.16). Later, he printed many books, such as *The Splendid Millennium Valley* and *Baichuan Xuehai.* Among them, 8 kinds of books published before 13th year of Hongzhi (1500) were equivalent to the cradle books in Europe, which were very precious. Hua Sui's uncle, Hua Sui, made a "very precise letterpress" which could be published in a few days and printed *Lu Fang Weng Ji.* Hua Xie's nephew Hua Jian's Lanxuetang had a collection of poems and essays by Cai Yong of the Han Dynasty, Bai Juyi of the Tang Dynasty and Yuan Zhen of the Tang Dynasty. The rich and powerful person—AN Guo (1481–1534) "Cast movable-type copper plate, printed on the secretaries," of which Zhengde *Dongguang County Chronicle* was the only copper movable-type local chronicle in China. In addition, there were copper plates in Changzhou, Suzhou, and Nanjing. The Qingyuan Copper Plate of Zhejiang Province printed *Zhuge Kong Ming Xinshu.* Chicheng (now Jia'ou, Fujian) had Jiajing Copper Plate Blueprint *Mozi.* Jianyang bookseller You Rong and Rao Shiren jointly used "Tong (Copper) Plate Movable Character" to print more than 100 pieces of *Taiping Royal View* in Jiangsu.

In the late Kangxi period of the Qing Dynasty, the government manufactured millions of bronze movable characters and printed astronomical and mathematical books such as *Qinruo Almanac, Mathematical Essence,* and *Lv Lv Zheng Yi.* In the sixth year of Yongzheng's reign (1728), with large and small two-size bronze characters, Chen Menglei's *Ancient and Modern Book Integration* in 10,000 volumes, 64series with 5020 pieces each were printed. The folk copperplate had a copy of Quinoa Pavilion in Jiangsu, which was earlier than the Copper Character in Neifu. In the twelfth year of Jiaqing's reign (1807), Taiwanese Army General Wulong carved bronze characters and printed the *Oracle Guangxun.* Lin Chunqi, Fuzhou, made the most copper characters. He merged more than 200,000 silver. In 21 years within

會通館印正本諸臣奏議卷第一
龍圖閣直學士開國伯趙汝愚輯
君道門
君道一
論人君之大德有三 司馬光

Fig. 19.16 Annotations of the Song Dynasty's ministers in the imprinting hall of Huitong

the 26 years in Daoguang's reign (1846), he employed more than 400,000 bronze characters of large and small, with exquisite regular scripts. They printed *Yin Lun, Shi Ben Yin,* and *Four Books Bian Meng* by Gu Yanwu of the Qing Dynasty, known as *Fu Tian Shu Hai* (Fig. 19.17). There were also copperplate collections of poems and essays in Hangzhou, with almost the same font as those in Futian.

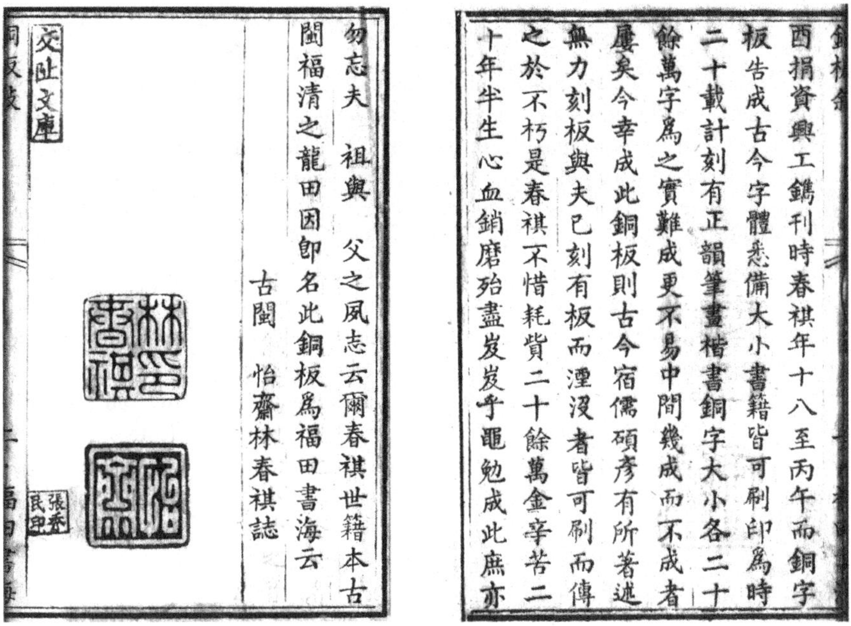
西捐資興工鐫刊時春祺年十八至丙午而銅字
板告成古今字體悉備大小書籍皆可刷印為時
二十載計刻有正韻筆畫楷書銅字大小各二十
餘萬字為之實難成更不易中間幾成而不成者
屢矣今幸成此銅板則古今宿儒碩彥有所著述
無力刻板與夫已刻有板而湮沒者皆可刷而傳
之於不朽是春祺不惜耗貲二十餘萬金辛苦二
十年半生心血銷磨殆盡岌岌乎黽勉成此庶亦

勿忘夫 祖與 父之夙志云爾春祺世籍本古
閩福清之龍田因即名此銅板為福田書海云
古閩 怡齋林春祺誌

Fig. 19.17 Fu Tian Shu Hai

19.3.4 The Tin Movable Printing

Tin movable type was an irregular word cast from tin for typesetting and printing. In the early Yuan Dynasty, Wang Zheng wrote in *Agricultural Book: Creating Movable Character Printing Calligraphy*: "In modern times, we cast characters with tin. We use iron bars to run through it. They are embedded in the helmet and printed by the boundary line." But the tin character "is difficult to make ink, and the rate of printing is worse, so it can not last long." It can be seen that there were tin movable-type printing books before Wang Zhen's book. Because of the poor hydrophilicity of tin, the performance of ink was not good. This was the earliest record of the tin movable printing in historical documents, one or two hundred years earlier than the creation of the European metal movable characters. By the mid-Ming Dynasty, the famous publisher Hua Sui "Fan Tong Ban Tin Character" showed that he had made tin characters besides copper characters, but lacked detailed records. In addition, in the early Ming Dynasty, there were also descriptions of printing or forging paper money with tin plate, which was also a supplementary proof of using tin to make printing plates.

Lottery gambling was popular in Foshan Town of Guangdong in the Qing Dynasty. According to Wei Sanwei (an American, a printer and bookseller), an unknown Mr. Deng printed two kinds of lottery tickets, thus started casting tin movable printing in the 30th year of Daoguang's reign (1850) at a cost of more than 10,000 yuan, resulting in three pairs of movable characters: one in flat letters, others in long letters, and small letters, totaling more than 200,000. In the first year of Xianfeng (1851), Yuan Ma Duanlin's *General Examination of Literature* was printed in 120 volumes. It was said that the characters were pleasant, the paper was white and the ink was bright. Wechsler also introduced Deng's method of casting tin characters and printing: using carved wood characters to make clay patterns on the mud, pouring molten tin into the mud patterns, crushing the clay patterns after solidification, and taking out the movable characters. After it was finished, the layout was tightened in a smooth and solid pear tray, bounded by brass, with 10 lines in half page and the center of the plate, such as block printing pattern. Then ink was used to brush printing. Deng Shixi's movable-type copies were still handed down from generation to generation, such as *Santong* (*Tongdian, Tongzhi, General Examination of Documents*), *Collection of Chen Tongfu, Spring and Autumn Period of Sixteen Kingdoms.*

Printing in China has played an important role in inheriting culture and extending Chinese history for thousands of years. Printing was also spread to Korean, Vietnam, Ryukyu, Japan, and other Chinese cultural circles and spread to Europe through trade and missionary channels. It inspired the invention of movable-type printing in Europe.

Xiumin Zhang Born in Shengzhou, Zhejiang, graduated from Xiamen University in 1981, and now works as the associate research fellow in National Library of China.

Research interests: Printing History of China; Textual Bibliography; History of China-Vietnam Relations.

Major works: *The Invention of Chinese Printing and its influence, A Collection of Papers of Zhang Xiumin on the History of Printing, Printing History of China, The History of Chinese movable-type printing (Co-authored with HAN Qi), A Collection of Papers on the History of China-Vietnam Relations, Data Directory of Taiping Heavenly Kingdom.*

Awards: The revised edition of Printing History of China won the first national "Three One Hundred" original books publishing project award, the First Prize of Outstanding Ancient Books Award in 2006, Special Prize of the 16th Zhejiang Shuren Publishing Award, the Honor Award of Tenth Outstanding Ancient Books Award of East China.

Qi Han Born in Shengzhou, Zhejiang in 1963, PhD, researcher at the Institute of Natural Science History, Chinese Academy of Sciences, chief editor of *Research on the History of Natural Science*, editor of the *Archive for History of Exact Sciences* and *Annals of Science*, Member of the International Advisory Committee of *Historia Scientiarum*. He has been invited to visit the United States, Japan, France, Britain, Germany, Portugal, Italy and other countries.

Research interests: History of Chinese science; History of Chinese and Western cultural exchanges; Catholic History in Ming and Qing Dynasties.

Major works: *The Westward Spread of Chinese Science and Technology and Its Influence (1582 –1793); Chinese printing history; Collection of Catholic Literature during the Reign of Emperor Yongzheng and Qianlong Preserved in Europe; China and Europe: Printing and the History of Book.*

He has published nearly 100 papers in academic journals at home and abroad.

Chapter 20
Tea Planting and Making

Jiahua Zhou

As one of the three largest non-alcoholic beverages in the world, tea is planted and produced in more than 50 countries. Currently, there are nearly 3 billion tea lovers in more than 160 countries and regions worldwide. Studies have shown that regular tea drinking has a variety of health values. For example, tea polyphenols have much stronger anti-aging effect than vitamin E. A cup of 300 ml green tea has the same antioxidant function as a bottle of claret. EGCG, the main component of tea polyphenols, is the nemesis of all cancers. Daily tea drinking helps lose weight, so tea ranks No. 1 in the United States weight-losing products. In addition, the amino acid in tea promotes the massive secretion of dopamine, which helps to generate a sense of pleasure. Various health values of tea have further promoted tea drinking.

20.1 China—Hometown of Tea

Based on plant taxonomy, tea is a kind of Camellia from the angiosperm family, belonging to the dicotyledonous, primitive genus subclass. At present, a large number of artificially cultivated tea trees are generally called Camellia sinensis. In 1950, Chinese botanist Qian Chongshu determined the scientific name of tea tree as Camellia sinensis (L.) O. Kuntze according to the international naming and researches on the characteristics of tea. Although the wild tea plants are distributed in many places of the world, it is the Chinese ancestors who first discovered and put tea plant in use in the world. The Swedish scientist Carl von Linne (1707–1778), who founded the scientific botanical classification system, named the tea tree. "The sinensis" in the book *Plant Species* published in 1753, and then revised it as Camellia sinensis L. The Latin meaning of "sinensis" is China. This shows that Linne and the

J. Zhou (✉)
The Institute for History of Natural Sciences, Chinese Academy of Sciences, Beijing, China
e-mail: zhoujiahua0129@sina.cn

J. Hua and L. Feng (eds.), *Thirty Great Inventions of China*,
https://doi.org/10.1007/978-981-15-6525-0_20

scientific community at that time recognized China as the first country to know and utilize the tea plants, so the naming of the tea plant is related to China.

In the ancient Chinese literature, the names of tea are various. The character "茶" was first seen in *the Book of Songs*. A Chinese poem says "who says the tea is bitter, it is as sweet as fruit". In another poem, the line "采荼薪樗, 食我农夫" presents farmers' tea-drinking life at that time [1], although "荼" here may refer to a kind of bitter vegetable. Then, in the earliest book in ancient times, Er-ya (in the second century BC between the Qin and Han Dynasties), there is an article mentioning tea as "檟". Guo Pu of the Eastern Jin Dynasty commented: the morning tea is called "荼", and the evening tea "茗" or "荈", while it has another name of "苦荼" in Sichuan. Yang Xiong in the late Western Han Dynasty said in the book *Dialects*, Sichuan people called tea as "蔎". In addition, there are more than 10 tea names such as "诧", "姹", and "茗". Given the variety of dialects, it is normal for tea to have different names in different places, which shows that the tea plants are grown in many parts of China. In the Tang Dynasty, Lu Yu pointed out in the *Cha Jing* (*The Classic of Tea*): "The names of tea are various such as "荼, 檟, 蔎, 茗, 荈". Among them, "荼" is the most frequently used. In order to unify the glyphs, Lu Yu rewrites "荼" into "茶". With the spread of the Classic of Tea, 茶(tea) was gradually fixed. However, in the vast Chinese territory, there are also differences in the pronunciation of "茶". For example, in Cantonese "茶" is pronounced as "cha", in Fuzhou as "ta", in Xiamen, in Shantou as "te", and the Yangtze River Basin and North China as "chai", "zhou", or "cha"; Yunnan Dai people pronounced "茶" as "la", Guizhou Miao people pronounced "chu ta", and so on. Since tea is a bulk commodity of the Sino-Democratic flow, tea is delivered to many parts of the world, so are the tea names. The pronunciations of "茶" in different countries can be divided into two categories. One is that the countries that import tea by sea, including the Western European countries such as Britain, France, and Germany, and the Indian Ocean countries such as India and Sri Lanka, whose pronunciations of "茶" is similar to "cha", "ta", and "te" of the Fujian and Guangdong coastal areas. Another category includes the countries that import tea along the route of the Silk Road to the north and the west, including Russia, Iran, Turkey, and other countries, whose pronunciation of "茶" is "chai", close to the dialects of North China (Table 20.1). The above-mentioned information, to some extent, confirms that China is the origin of tea and its various names.

At present, there are more than 380 species of Camellia in 23 genera in the world, of which more than 260 species in 15 genera exist in China, which are distributed mostly in Yunnan, Guizhou, and Sichuan. According to field surveys and literature, China is the first country to be found with largest quantities of wild tea plants. The article "Tongjun Lu" in the book *Wupu Bencao* written in the Three Kingdoms Period, records that big tea trees were found in south China, and its bitter soup can lead to staying up all night. Lu Yu also describes the condition of wild big tea tree in *Cha Jing*: "In the Ba Mountain Valley, there is a huge melon-shaped tree with thin and long leaves, white rosebush flowers, palm-like fruits, lilac stalks, and walnut-like roots. There are a lot of similar historical materials indicating the fact that our ancestors were familiar with and used tea trees as early as 1,700 years ago.

Table 20.1 Tea pronunciations in the world (Taken from *Hui Shi Tian Gong, The Collection of Cultural Relics Made in China*, edited by the State Administration of Cultural Heritage)

Land transmission with "CHA"	Sea Route transmission with "TAY"
Korean (차)	Dutch (thee)
Japanese (茶)	English (tea)
Vietnamese (trà)	German (Tee)
Thai (ชา)	Hungarian (tea)
Nepali (चिया)	Yiddish (טיי)
Hindi (चाय)	Hebrew (תה)
Urdu (چاے)	French (thé)
Bengalese (চা)	Spanish (té)
Persian (چای)	Italian (tè)
Turkish (çay)	Latin (tea)
Arabi (شاي)	Danish (te)
Kiswahili (chai)	Norwegian (te)
Greek (τσάι)	Swedish (te)
Bulgarian (чай)	Finnish (tee)
Romanian (ceai)	Estonian (tee)
Serbian (чај)	Latvian (tēja)
Croatian (čaj)	Icelandic (te)
Albanian (çaj)	Armenian (թեյ)
Czech (čaj)	Bahasa Indonesia (teh)
Slovak (čaj)	Malay (teh)
Ukrainian (чай)	Tamil (தேயிலை)
Russian (чай)	
Portuguese (chá)	

According to incomplete statistics, the wild tea trees have been found in 198 places in 10 Chinese provinces. The so-called wild big tea trees usually refer to trees that are not artificially cultivated and their leaves are rarely picked. Actually, before the human beings knew how to use and cultivate the tea trees, they were wild and later cultivated and modified to the nowadays modern tea plants (Figs. 20.1 and 20.2). It can be seen that there is no absolute limit between the wild tea and the cultivated one. The majority of the current Camellia plants are distributed in Yunnan, Guizhou, and Sichuan whose terrain is varied with stacked mountains and crossed valleys. The small landforms and microclimate areas there with such high-altitude and changing climate provide an easy bed for intra-specific variation in tea planting. The varied

Fig. 20.1 Cultivated wild big tea tree (Xishuangbanna Region, Yunnan)

Fig. 20.2 Contemporary tea garden

ecological environments give birth to different tropical and subtropical large leaf species, middle leaf tea trees, temperate mid-leaf species, and lobular tea plants. This is one of the reasons why most varieties of the tea trees are found in Yunnan, Guizhou, and Sichuan. The concentrated distribution of the wild big tea trees and the variety of tea tree species all prove that China is the birthplace of tea.

20.2 The Development of Tea Drinking

Although the presence of wild large tea trees is an important evidence for China to be the original place of the tea trees, the place where the wild tea trees are found is not necessarily the origin of tea because the tea drinking custom is not necessarily developed there. After all, there is a process of understanding and utilizing the leaves of the tea trees. In ancient China, there are folk legends of Emperor Yan who found and used the tea trees (Fig. 20.3). The earliest Chinese herbal book, *Shen Nong's Herbal Records* says: "Emperor Yan tasted hundreds of herbals, got poisoned and was detoxified by tea." The original *Shen Nong's Herbal Records* were lost, and there have been quite a lot of copies since the Ming and Qing Dynasties, but the above story is not recorded. However, its impact still goes with the later generations. Lu Yu in the Tang Dynasty said in *Cha Jing* (Fig. 20.4) that "tea becomes a kind of drink due to Emperor Yan's discovery". In fact, Emperor Yan was conceived by the

Fig. 20.3 Emperor Yan collects herbs (Liao, Anonymous)

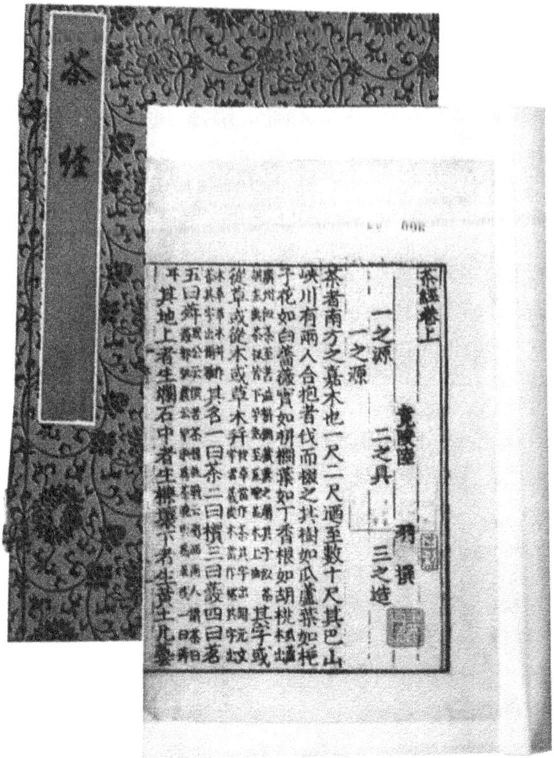

Fig. 20.4 Lu Yu's *Cha Jing* book shadow

ordinary people. Legend has it that Emperor Yan came at the right time and right place to teach the people to make tools and to cultivate (from *White Tiger Stories* by Ban Gu in the Han Dynasty). In this way, Emperor Yan is the collective embodiment of the ancestors who pioneered Chinese agricultural making more than 5,000 years ago. It is the Chinese ancestors who discovered in the labor life that the tea leaves are non-toxic and edible with certain medicinal effects. Therefore, the tea is gradually cultivated as a cash crop and develops into an industry.

There is a practical process for the planting of tea trees and the understanding of tea efficacy. Tea processing techniques and drinking methods also have undergone a gradual process. At first, people know that the tea leaves are non-toxic and beneficial. "The ancient people chew grass for its juice", that is to say, tea leaves were first chewed and tasted bitter but harmless, and gave people a feeling of fragrance, cleansing, and refreshment. Especially when people were poisoned by eating certain plants, chewing tea seemed to be able to resolve the pain of poisoning. Over time, the chewing of tea leaves became a hobby and spread. At the beginning, the tea was cherished as a sacrifice. When the output increased, the tea gradually entered many households as a normal drink. Yan Ying said in "Yanzi Chunqiu": In the Spring and Autumn Period, when Yan Ying was the prime minister of Qi Jinggong, he ate millet-free rice, two or three roasted wild birds, pickled vegetables, and drinks tea."

This record indicates that tea was used as a soup or drink for the public. After all, the taste of chewing fresh tea leaves directly is not good. People found that tea soup tasted better when they cooked tea leaves with water, especially after adding salt, garlic, pepper, and other ingredients. So, people gradually changed from chewing tea leaves to cooking tea soup, which is not only a simple change in the way of drinking tea, but the advancement of people's knowledge of tea.

After the fresh leaves of the tea tree are picked, they usually dry or rot in a period of time. Therefore, chewing fresh tea leaves is strictly limited to only a short period of time in a year by place and time. If the tea leaves are cooked, fresh leaves as well as dry leaves can be used as long as they are properly stored, so that people can enjoy the good taste of tea at any place and time. The method of cooking tea is also acceptable after adding other ingredients, such as cassia, ginger, salt, and some spices, to further strengthen the color and aroma of the drink. The easiest way to store fresh leaves is to dry them in the sun, which is probably the most primitive method of tea processing. Dry tea leaves, different from fresh tea leaves, are more difficult to swallow after chewing, so the edible method of tea leaves can only be changed to cooking or soaking with hot water.

The reason for chewing tea is detoxification. Some bitter vegetables also have the same effect. People couldn't distinguish tea from bitter vegetables at first, so they regarded tea as an edible dish (Fig. 20.5). The Dai, Hani, and Jingpo people who have been living in Yunnan have the habit of eating "bamboo tea". The so-called bamboo tea is a kind of tea dish by compacting the fresh tea leaves into a bamboo tube after being sun-dried or steamed. After tea juice is filtered out, seal the tube mouth, and let tea leaves in the tube slowly ferment. Two or three months later, open the bamboo tube. Cook the tea once more and put it into a jar. Add the sesame oil to make a dish.

There are many ways to drink tea. Zhang Yi, a Wei Person in the Three Kingdoms Period, recorded the method of making and drinking tea at that time—People in the Jin and Ba Regions pick tea leaves to make tea rice paste or cook the tea into red-colored drink into porcelain vessels, mixed with onion, ginger, and orange peel. The drink is sobering and refreshing. *Jin Shu* says: "Wu people pick tea to cook porridge". Tea cooking was very popular at that time. People made tea leaves into caked tea, baked it then crushed it into powder-like pieces before the tea is brewed with some onion, ginger, and orange peel. This method of cooking tea has been passed down till today, such as Mongolian salty milk tea, Tibetan butter tea (Fig. 20.6), oil tea of Miao and Dong nationalities, Tujia tea, Naxi salt tea, Dai bamboo tube tea and so on. Baked tea is one of the old methods of tea processing by picking the tender shoots of the tea tree, directly baking it on the fire to a burnt yellow color, then putting it into the teapot for cooking. Later, in order to make the tea for long storage, people smashed the fresh leaves with wooden sticks, made the pieces into cake shape or ball shape and placed them in a damp-proof container after drying. Before drinking, people kneaded a small piece, crushed it, and put it into a pot, added boiling water or boiled. The above method of making and drinking tea is also described in *Guang Ya*.

The southwestern part of China, especially the Ba and Shu area, was once the earliest tea making region in ancient China and the center of tea making and drinking.

Fig. 20.5 Jinuo's cold tea dish

After the unification in the Qin Dynasty, due to economic and cultural exchanges, tea-related activities began to spread to east and southeast China in which the climate and environment were suitable for growing tea. At that time, tea drinking was a fashion among scholars, officials, and temples, so the tea making and drinking were greatly developed. In the Sui Dynasty, economic and cultural exchanges between the North and the South China were more active, and the fashion of tea drinking began to spread to north China. With the prosperity of the Tang Dynasty, tea drinking was gradually popularized in the northern regions after the mid-Tang Dynasty.

The making of tea industry has developed greatly. Due to the differences in natural environment, types of tea trees and processing techniques, various famous tea products came into being. At that time, four tea categories were formed according to different processing methods: crude tea, loose-leaf tea, crushed tea, and caked tea. To make the crude tea, the buds, leaves, and stalks of tea plants are picked, chopped up and then dried. Loose-leaf tea is made by steaming the fresh tea leaf, drying, and then sealing for future cooking. The crushed tea is the crushed tea leaves after

Fig. 20.6 Tibetan butter tea

stoving or frying. Caked tea is a cake-shaped tea that is made through the process of steaming, crushing, flattening, and then drying, which is convenient for packaging.

In the Tang Dynasty, tea drinking prevailed in the ruling class and the public as an important part of life. "Tea is a kind of food as important as rice and salt". "Serving a cup of tea for guests" has become an important etiquette and traditional social lifestyle. As a result, more attention is paid to the quality of tea, the way of drinking tea and the tools of drinking tea. Especially when the elements of culture and arts are added in the fashion of drinking tea, the function of tea drinking is not only limited to thirst relief and health care, but also a demonstration of cultural display and spiritual enjoyment. When tea-drinking techniques are emphasized, "tea art" came to the stage; "tea etiquette" was formed then; and finally "tea ceremony" developed (Fig. 20.7). Therefore, it is more demanding in the fine processing of tea, the color, shape, and flavor of tea. The level and quality control of tea products and

Fig. 20.7 Brick with tea cooking portrait in the Northern Song Dynasty

Fig. 20.8 Tea competition painting by Liu Songnian in the Southern Song Dynasty

drinking techniques are formed, and tea-drinking not only meets personal hobbies and enjoyment but also becomes the social and commercial "tea contest" which started to prevail in the Song Dynasty. The essence of tea contest is the evaluation of tea and the display of tea art (Fig. 20.8). This kind of artistic activity that shows elegance and self-expression promotes the further popularization of tea and tea-making skills. Tea drinking is integrated into cultural activities to form tea culture which in turn contributes to the further prosperity of tea industry. In the practice of tea contest and appreciation, it has been found that brewing loose-leaf tea, more convenient than cooking tea, tastes better. Hence, brewing tea has spread quickly and even replaced caked tea cooking in some areas and ethnic groups, which has led to a faster development of the technology of the green loose-leaf tea, the source of modern tea-making traditions.

20.3 Evolution of Tea Processing

The picking of fresh tea leaves is seasonal, so the storage technique is a must to keep them from decaying for the purpose of continuous supply and transportation. The main processing method for fresh leaves is drying, by sun or by heating, more than 90% water in the tea leaves to achieve preservation purpose by slowing the enzymatic activity of leaf cells. Otherwise, under the influence of certain temperature, humidity, oxygen, and light, the fresh tea leaves will undergo oxidation, degradation, and transformation of various chemical components that bring about qualitative change and eventually decay, which makes tea preservation difficult in ancient times. According to *Guang Ya* written by Zhang Yi, during the Three Kingdoms Period, people kept the tea leaves that had been boiled, steamed, and then pressed into cake shapes for drying. The skill to steam fresh leaves is called steam-green; slight boil skill called fry-green. After fixation, tea leaves are pressed into cake shapes, then dried, stored, or transported. This method of processing has gradually spread, and become perfect in the Tang Dynasty.

The improvement of tea processing technology in the Song Dynasty was first reflected in the making of the tribute tea. As a kind of tribute tea, Longfeng tea was decorated with dragon and phoenix paintings on its caked shape. The manufacturing process is described in *Beiwan Bielu* by Zhao Ruli in the Northern Song Dynasty (the year of 1186), which records the processing steps as tea steaming, pressing, grinding, making, yellowing, baking, etc., namely, select the tea buds, wash and steam them, then rinse them with cold water, squeeze out the water and tea juice, and place them in the pots with water to grind before putting them into the dragon and phoenix mold, and finally dry the caked tea for storage and transportation. The Xiaolongfeng tea at that time is made of selected fine leaf buds, which is of course extremely expensive. Thus, the important step of fine processing is tea-buds selection.

In the Song Dynasty, the popular loose-leaf tea making is simpler in green-removing technique than the making of the dragon and phoenix caked tea. As for tea quality, the loose-leaf tea is not worse than the caked one, for the former can be brewed with the boiling water, much easier than tea cooking. Therefore, the steam-green or fry-green skills become increasingly popular. Zhu Yuanzhang, the founding emperor of the Ming Dynasty and also a farmer's son, gave an order in the year of 1391 that the tribute tea could be the loose-leaf tea buds to save labor [2], which makes steam-green or fry-green a main stream of tea processing. Many historical famous teas came into being in this period.

The steam-green and fry-green techniques differ in orientations of tea processing. The former uses the high temperature of steam to remove green; the latter uses the high temperature of the frying pan. Both methods are to inactivate most of the enzymes in fresh tea leaves at high temperatures, and remove green to facilitate the preservation, but the effects are different. The boiling point of the green component in the fresh leaves is around 160 °C. The temperature of the boiling water vapor is only about 100 °C and cannot fully volatilize the green component, so the tea aroma is hardly to form. The temperature of the fry-green can reach above

160 °C, most of the green components in the fresh leaves can be volatilized within a few minutes, and the tea aroma is formed under high temperature. In addition, the fried baking can also promote the formation and volatilization of many aroma substances in fresh leaves, adding flavor greatly. Quick stir in fry-green can reduce the decay of fresh leaves during the making process, so it can ensure the freshness and taste of the tea. Therefore, in practice, the technology of the fry-green, which is said to appear in the Tang Dynasty, is gradually promoted. In the processing period of the Tang, Song and Yuan Dynasties, people realized that the fried green tea not only saved time and effort, but also improved the tea quality. This is the reason why it has been rapidly developed in the Ming Dynasty.

The fry-green methods are various in different places with many unique processing skills to produce different teas with special qualities. However, the basic processing techniques are similar, and the main processes passed down to the present as a tradition include high-temperature green-removing, simmering, re-frying, baking, and packaging.

The yellow tea, black tea, oolong tea, and white tea are all related to the green tea frying process and can be regarded as the innovation of the green tea making.

In the frying of green tea, when the temperature is too low, the time is too long or the cooling is not timely after the green-removing, or the tea leaves are not dried in time after the simmering or smashing, the color of the tea soup will turn yellow. This is the source of yellow tea whose taste is also very good with a special flavor. Therefore, the above-mentioned improperness in the tea making process is especially highlighted, with the means of yellow braise, to form the unique making skill of the yellow tea.

Black tea first appeared in the sixteenth century in the Wuyi Mountain area in Fujian where was once the main producing area of the tribute tea in the Tang and Song Dynasties, according to the record of Duonengbishi. Legend has it that at the time of the turmoil in the late Ming Dynasty, an army entered Fujian from Jiangxi, passing through Renyi Town of Chong'an County, and occupied a tea farm. The tea to be processed could not be dried in time and thus turned black. The tea farmers, in order to recover the loss, used flammable pine to dry and accidentally made the black tea with pine aroma and dried longan taste, also called Lapsang Souchong black tea. In the seventh century, black tea was transported by Shanxi merchants to Wuhan, Tianjin, and Datong, and exported to Europe by the Dutch and East Indian Company. It was popular for the European upper class and the royal family to make tea instead of drinking wine. At the beginning of the eighteenth century, a group of missionaries came to China. They came to the Wuyi Mountain area to observe making of black tea. Then the tea tree seeds and tea-making techniques were then brought to the British colony in India, the source of Indian Assam black tea and Darjeeling black tea. At the same time, Fujian's black tea making skills gradually spread to 12 provinces. In 1875, Yu Ganchen, who had been a magistrate of Chong'an County, resigned to his hometown in Anhui, and set up a black tea house in Dongzhi County and Qimen County where gradually changed from green tea making base to black tea base. Qimen Black Tea has also become a model for Kongfu Black Tea and black tea has become the largest export tea in China. Lapsang Souchong black

Fig. 20.9 Tea soup of white tea, green tea, black tea, and oolong tea

tea, the pride of Chinese black tea, is deeply loved in some European countries. It is called "bohea", which is the homonym of Wuyi.

Oolong tea, also called green tea, is represented by Wuyi Dahongpao and Anxi Tie Guanyin, which combines the aroma of green tea and flowers, and the mellow flavor of black tea. Yet its making process is slightly more complicated than the green tea and black tea. Its processing not only absorbs the mechanism of black tea fermentation, but also combines the means of controlling fermentation in green tea making. The procedure does not completely destroy the whole leaf tissue and slightly scratches the leaf edge tissue. The cellular components in the leaves are not completely changed, and some of them are oxidized. It is this complex process that creates the unique color and aroma of oolong tea. The green tea and black tea making technology are integrated in this process (Fig. 20.9).

The early high-quality white tea is made of the white tea leaves with sun-heat drying instead of frying. Because the white tea leaves are covered with white tiny hair, its tea soup is green with light flavor and sweet taste, and the white tea tastes special and is highly appreciated by some scholars.

Dark tea appeared no later than the middle of the Ming Dynasty. In 1524, Chen Jiangshu recorded the making of dark tea: "dark tea, the middle-class tea with less fake goods, is collected in its limited making sites wherein dark tea is produced and packaged in bamboo baskets with printed commodity signs of the governmental tea divisions, half for border trade (sell tea to buy horses) and half sold in Markets". It is recorded that the dark tea making amount was not much at that time and its making area was limited, the middle-quality dark tea then was made by sun-heating after steaming.

The dark tea may be created in the making process of green tea. The tea leaves may become dark when the amount of leaves is too much or the fire temperature is too low, or the time is too long, or fermentation is caused by high temperature and humidity in the piled tea leaves. These tea leaves have lost the original color and aroma of green tea, but they have another mellow flavor. So, the dark tea manufacturing comes into shape by adding the piling process after the process of steaming and smashing. The so-called piling is to pile the tea leaves in the clean and darkroom to let it ferment for about 10 h at room temperature above 25 °C, and the relative humidity of about 85%. The piling process is the key part for black tea processing. The dark tea taste is closer to that of the caked tea in the Tang and Song Dynasties, and has been loved by many ethnic minorities who have not adapted themselves to green tea. The dark tea produced in the Mainland is transported to the side sill, so it is necessary to press it into square bricks or round cakes for easy transportation and storage. Therefore, the pressing process is added in the making of the dark tea. Most of the dark tea making areas also make green tea. Most tea farmers use the green tea from the early spring buds. Later, they pick the thicker leaf branches (4–6 leaves of one bud) as raw materials to make the pressed dark tea, by the processes of green-removing, smashing, piling, re-smashing, baking, steaming, pressing, shaping, and drying.

Flower tea belongs to the reprocessed tea with both tea and floral aromas. To make diversified tea taste, people in the Tang Dynasty often added their favorite ingredients in the tea soup such as ginger, onions, salt, orange peel, and dates. In the Song Dynasty, the flower tea was recorded in texts. When brewing tea, people added some spices or flowers. In the Ming Dynasty, flower tea making has been developed quite a lot, and much experience has been accumulated in selection and proportion of the flowers matched with the loose-leaf tea. After the Qing Dynasty, Fuzhou and Suzhou became centers of the cellar making flower tea in which the processed fresh flowers are mixed with the dried tea leaves to combine floral fragrance and tea aroma. The flower tea is also called the cellared flower tea, smoked flower tea, and scented tea due to its different ways to absorb floral fragrance. Common flower tea includes jasmine tea, osmanthus green tea, honeysuckle tea, white orchid tea, and rose black tea.

As the hometown of tea, China has accumulated rich experience in the selection and planting of tea trees after thousands of years, and has cultivated hundreds of excellent varieties of tea trees. The making technologies of tea making are also various to make the green tea, white tea, yellow tea, Oolong tea, black tea, and dark tea, picking, green-removing, simmering and baking, etc. In China, tea can be described as the "national drink", and tea for guests has become the most basic etiquette. Daily tea drinking is also a common practice. Drinking tea is a good habit for health, an etiquette as well as an art, which becomes an important carrier of the traditional Chinese culture. In addition to the enjoyment of its unique color, aroma, taste, shape, and benefits, drinking tea can also relieve spirit, cleanse the soul, and calm people down with strong appeal and vitality. In short, tea is an ideal drink, so it can be widely spread around the world.

20.4 International Communication of Tea and the Tea-Making Technology

Since the Tang Dynasty, tea, together with silk, has been exported to foreign countries as a popular specialty through the Silk Road and Sea Silk Route. According to Japanese literature, Japanese tea drinking began around the early days of Nara era (year of 710 794) when Sino-Japanese cultural exchanges were very frequent. In 729, when hundreds of monks were convened by the Japanese government to teach the Mahaprajnaparamita Sutra, they were granted tea to drink. Zuicheng Chan master of Japan, the founder of the Tiantai sect, returned to China in 805, and his companion, Konghai Monk, returned in 806. They brought back a lot of tea seeds to plant in Japan, and taught Japanese the tea making techniques. From then on, Japan developed its own tea industry, absorbed and digested the Chinese tea-making techniques, tea cooking methods and tea culture, and integrated the Japanese traditional culture, especially the understanding of Zen Tao, forming the unique Japanese tea ceremony represented by Mocha Ceremony and Fried Tea Ceremony. The neighboring Korean Peninsula also developed its tea ceremony with special Korean nationality features by imitating tea cooking methods in the Tang Dynasty Around the sixth century during the Silla reunification period, tea planting was recorded, so the introduction of the Chinese tea-making techniques and tea-drinking practices to Korean was earlier than Japan. Subsequently, the custom of drinking tea was gradually promoted.

Tea was exported to the west through two routes: one is the Silk Road from Shaanxi, Gansu, Xinjiang to West Asia and Europe; the other sets off from Xiamen or Guangzhou, passing through the South China Sea, the Bay of Bengal, the Indian Ocean to Southeast Asia, South Asia, West Asia, coastal countries in East Africa, and some European countries (Fig. 20.10).

Fig. 20.10 Tea making craft in Jiangnan district (on nineteenth century British copperplate)

The exchanges along the silk road, including the Western Expedition of Genghis Khan, have made many ethnic groups in Central Asia aware of tea and developed the practice of drinking tea. However, the tea trade along the marine route was more influential. At first, the Arabian merchants monopolized the trade between the East and the West. When the Portuguese Da Gama discovered the real India at the end of the fifteenth century, the Portuguese seized control of the trade between the East and the West. In the sixteenth century, the Dutch who had formed an alliance with Portugal also came to Guangzhou through this route. Together with the Portuguese merchants, they bought tea, the beverage that can help digestion, and shipped them back to Europe. In this way, tea was gradually exported into Portugal, Spain, France, and the United Kingdom. The aristocrats of European countries have a strong interest in tea. Especially when the Lapsang Sauchong's black tea appeared at the aristocratic banquet, the color and aroma of the tea soup almost attracted all guests. Since then, in British eyes, tea has been "healthy liquid and soul drink". In the eighteenth century, tea became the most popular drink in UK. Tea replaced beer in breakfast and dinner and replaced gin in the rest of the day.

Tea consumption increased from 30.3 tons in 1701 to 2229.6 tons in 1781 and reached 6847.8 tons in 1791 (Fig. 20.11).

Today, according to the tea consumption annual per capita, the UK is the third country after Turkey and Ireland, but the world's largest tea brand is an English company that enjoys a market share of 23 billion USD every year. In other European countries, although the popularity of drinking tea is not as much as that of the United Kingdom, it has become an indispensable part of life. For example, the French regard tea as "the gentlest, most romantic, and most poetic drink". Some countries in Russia

Fig. 20.11 Tea trade in Xiamen (nineteenth century British copperplate print)

and Eastern Europe popularized tea drinking in all walks of life in the eighteenth century, although it was known in the sixteenth century. The Russian upper class drinks tea in a distinctive manner as the European aristocrats, from tea set to tea drinking, from tea custom to tea ceremony.

In the Tang Dynasty of China, the Silk Road spread tea to Central Asia and India. In fact, there were also wild tea trees in India, and its natural environment is also suitable for growing tea, but tea planting was not developed until the eighteenth century when it became a British colony. Tea was processed into black tea then and shipped to UK, and resold by the British merchants to Europe and even the Americas. Since the mid-nineteenth century, India has become the world's largest tea producer. Tea is India's invaluable asset that has changed the way of life of Indians, which activated the Indian industrial economy. Sri Lanka, located on the east and west traffic backbone, is an ideal garden for tea trees with abundant rainfall to the hills and the direct sunlight. Merchant ships also gather there. Before the eighteenth century, Sri Lanka was only a transit point for Chinese tea exports. When the British understood that it was very difficult to use opium instead of silver to exchange for Chinese tea, they started to cultivate tea trees in Sri Lanka. Soon tea produced in Sri Lanka and India replaced the status of Chinese tea, and Sri Lanka became a major tea producer.

By the nineteenth century, most people around the world recognized that tea was an excellent drink. In 1886, China's tea exports reached 2.68 million dans (Chinese counting unit). The tea-drinking fashion around the world is directly or indirectly related to Chinese tea. The Chinese people have created tea-making techniques and tea-drinking methods, and have made immortal contributions to the civilization and healthy life of all nations of the world.

References

1. Jin Qihua. 1984. *The Translation of The Book of Songs*. Jiangsu Ancient Books Publishing House.
2. Zhang Tingyu et al. 1974. *Food Log, the History of the Ming Dynasty*, 1955. Zhonghua Book Company.

Jiahua Zhou Born in Ruian, Zhejiang, in 1942, graduated from the Department of Chemistry Department of Guangxi University and is now a researcher of the Institute of Natural Science, Chinese Academy of Sciences.

Research interests: History of Chemical; Processing of traditional agricultural, livestock and mineral products; history of wine and so on.

Major works: *History of Chemical Thoughts, Biography of Chemists, History of Ancient Chinese Chemistry, History of Chinese Science and Technology: Chemistry, History of World Chemistry, Yongli and the Yellow Sea, Complete Works of Chinese Traditional Crafts: Brewing.*

Awards: *A brief history of science and technology in the 20th century (co-authored)* won the Second Prize of Natural Science in Chinese Academy of Sciences in 1989;

The Five Milestones of Human Understanding of the World (co-authored) won the Second Prize of the 2006 National Science and Technology Progress Award;

Principle—The Masterpiece of Times (co-authored) won 1988 National Outstanding Book Award;

Chinese Cultural Annals: Records of Science, Technology and Chemistry won the 2000 Book Honor Award;

History of Chinese Chemistry won the 14th China Book Award.

Chapter 21
The Invention and Use of Celestial Equatorial Measurements

Yunli Shi

From the existing literature, the observation, research, and forecast of astronomical phenomena have become an important issue of concern to the Chinese Dynasties since the Xia Dynasty. It is the responsibility of special officials and institutions, and forms several thousand years of tradition of continuous development, making astronomy one of the most developed scientific fields in ancient China. There are a large number of important inventions and creations in almanac compilation, calculation of celestial calendar of five phases, astronomical observation and instrument research and development, etc., which once took the lead in the world. Among all these inventions, the most fundamental one is the invention of equatorial astrometry measuring instruments, based on the celestial equatorial coordinate system.

Since ancient times, astronomers have used three main coordinate systems to record and describe the position of celestial bodies (Fig. 21.1): the horizontal coordinate system based on the horizon circle and zenith, representing the positions of the horizon azimuth Az[1] and the horizon altitude Alt(h)[2]; the ecliptic coordinate system based on the ecliptic circle and the ecliptic Arctic, showing the position of celestial bodies represented by the ecliptic longitude λ[3] and the latitude β[4]; the equatorial coordinate system of celestial body position represented by equatorial longitude α[5] and equatorial latitude δ[6] on the basis of equatorial circle and north celestial pole. On

[1]The angle between the arc of the celestial body and the zenith to the north point.

[2]The opening angle of the celestial body passing the zenith circle to the horizon circle.

[3]The opening angle of the arc of the celestial body and the ecliptic to the vernal equinox.

[4]The opening angle of the celestial body passing along the ecliptic Arctic Circle to the horn of the ecliptic.

[5]The opening angle of passing the horn of the celestial body and the northern celestial pole to the point of the vernal equinox.

[6]The opening angle of the celestial body passing the northern celestial circle to the ecliptic sphere.

Y. Shi (✉)
University of Science and Technology of China, Hefei, China
e-mail: ylshi@ustc.edu.cn

J. Hua and L. Feng (eds.), *Thirty Great Inventions of China*,
https://doi.org/10.1007/978-981-15-6525-0_21

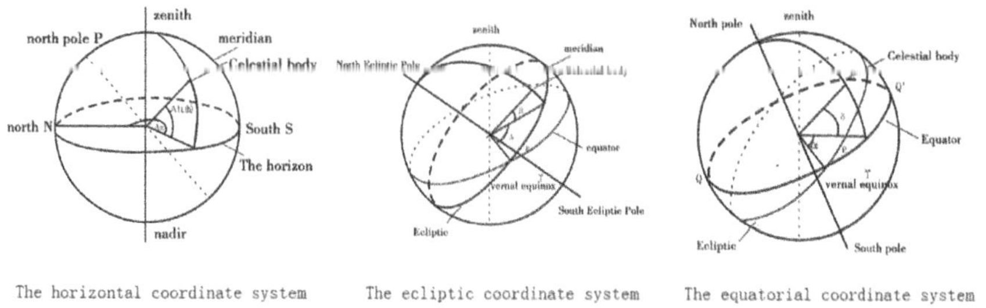

Fig. 21.1 Three main celestial coordinate systems

the one hand, the horizon coordinates of a celestial body vary with each other because of different geographical latitudes. On the other hand, the horizon coordinates of a celestial body will also change with the rotation of the Earth at all times, so various star catalogues generally do not use the horizon coordinate system. Although the ecliptic coordinates of the celestial body do not change like the horizontal coordinates, the orientation of the ecliptic will change at any time due to the rotation of the earth. Therefore, when the system is used in the observation instrument, it will cause inconvenience in aiming, and the structure of the instrument will become more complicated and cumbersome. Only the equatorial coordinate system can maintain the fixation of the celestial coordinates, which is easy to aim during the observation. The structure of the corresponding observation instrument is, therefore, relatively simple. Especially in the automatic tracking and observation of celestial bodies, equatorial instruments are also well adapted to the daily operation of the celestial sphere from east to west, so they have irreplaceable advantages. Because of this, Equatorial Mount has been widely used in modern and contemporary astronomical telescopes.

In ancient China and Europe, the above three coordinate systems were used in different forms. However, at least from Hipparchus (about 190 B.C.–125 B.C.) and Claudius Ptolemaeus (about 90 A.D.–168 A.D.), the ecliptic coordinate system dominated the development of European astronomy theory, observation, and instruments. It was until the end of the seventeenth century that Tycho Brahe (1546–1601), a Danish astronomer, truly realized the advantages of the equatorial instruments. From 1581 to 1583, he designed and manufactured three equatorial astrometry measuring instruments[7], which gradually popularized equatorial instruments in Europe. China took this step more than 1,700 years earlier than Europe. The equatorial astrometry instrument appeared in the Qin and Han Dynasties and became the leading astrometry instrument. These instruments not only greatly promoted the development of the

[7]Namely, three-rings equatorial armillary sphere (composed of a meridian ring, an equatorial ring, and a right ascension ring), four-ring equatorial armillary sphere (composed of a meridian ring, a prime vertical ring, an equatorial ring, and a right ascension ring) and a great equatorial armillary sphere with half equatorial ring. The diameter of the former two instruments is 1.6 m, and the latter is 3 m.

ancient Chinese astronomy, but also had a direct impact on Japan, Korea, and other surrounding countries.

21.1 The Origin of the Circular Sphere and Equatorial Astrometric Measuring Instrument in China

Ancient China experienced a process of development from being coarse to being fine in the research and utilization of the celestial movement. At first, people may determine the arrival of the New Year and a certain season based on the direction of sunrise and sunset and the stars reaching the middle sky at dusk and dawn. And the change of the month is determined by the change of the phase of the moon. It is in this way that people obtain the approximate period of the movement of the sun and the moon, and adjust the possible plate connection between the month and the season according to the actual situation to meet the needs of making the calendar. However, since the Spring and Autumn Period, the accuracy of the calendar has been considered as a key factor in national future, and the maintenance of the precise calendar has become a symbol of the Monarch power and legitimacy. From the Warring States Period on, more complex astrological systems have also needed to be based on more sophisticated calendar astronomy, which includes not only the study of the movement of the sun and the moon, but also the movement of the five planets.

The silk manuscripts *Five Stars Divination* [1] (Fig. 21.2), which was unearthed in the Tomb of Mawangdui in Changsha in 1973, is the most important evidence for this development. These silk manuscripts were written in 168 B.C. The main content of the manuscripts was used for divination according to the pattern of planets. As the astronomical basis of such divination, the manuscripts also give various data concerning the cycle and dynamics of the planetary movement, including a quantitative description of the speed of planets. For example, the First Emperor of the Qin Dynasty (259 B.C.–210 B.C.) in the first month of the year, the Jupiter rotated twelve points per day, one degree per twelve days. Eventually, it rotated thirty degrees and one hundred five points. In the lunar January of the first year during the reign of the First Emperor of Qin, the Saturn was located at the planet Yingshi. It rotated eight points per day, one degree per thirty days. Eventually, it rotated twelve degrees and forty points. This shows that the corresponding astrometric measuring instrument and the corresponding celestial coordinate system appeared at that time. Some scholars even speculated that China should already have had the so-called "pre-Qin armillary sphere" [2] before that. Unfortunately, these speculations are not based on literature and physical evidence.

In 1977, a "twenty-eight lunar mansions" was unearthed from the tomb of Ruyinhou Xia Houzao (?–165 B.C.) in Fuyang, Anhui, and provided a persuasive material for the solution of this problem. Twenty-eight lunar mansions [3] are divided into two sets, the upper set and the down set, and each has a small hole in the heart. We can use a thin round stick to pierce the two sets together.

Fig. 21.2 The No. 3 Han tomb unearthed in Mawangdui, Changsha, *Five Stars*

The upper plate (the right side of Figs. 21.3 and 21.4) is 1.7 cm thick and 23 cm in diameter; the center of the plate is marked with a crosswire shaped like " + " and decorated with Big Dipper; the circumference of the plate has a beveled edge with a width of 4 cm with 365 pinholes, which corresponds to the degree of the universe. The lower plate (the left side of Figs. 21.3 and 21.4) is 0.8 cm thick and

Fig. 21.3 The unearthed state of the twenty-eight lunar mansions

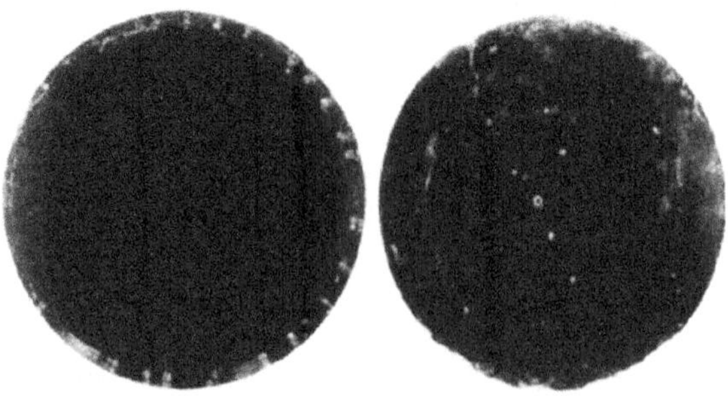

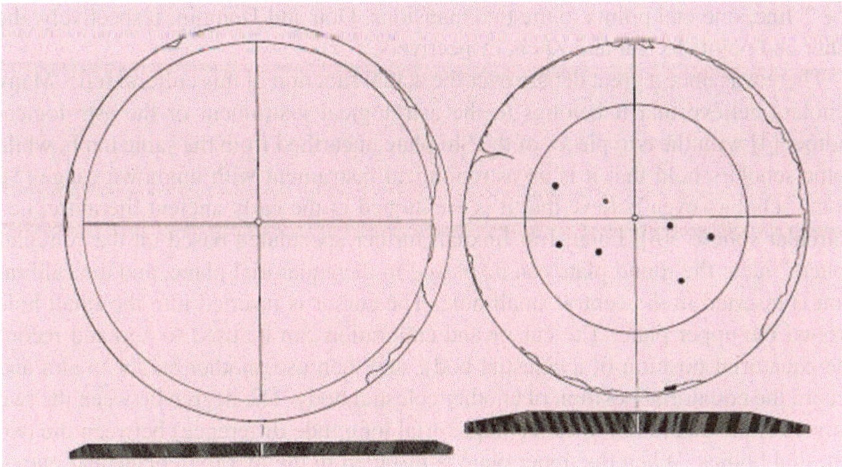

Fig. 21.4 The copy of the twenty-eight lunar mansions

25.6 cm in diameter; the periphery has a beveled edge with a width of 1.1 cm. The top is engraved with the name of each of the twenty-eight lunar mansions and the distance between the sets, which is the degree of each constellation (Fig. 21.5); the mansions distribution on each set is not the same, they are obviously arranged according to the actual distance; the center of the plate is also marked with the

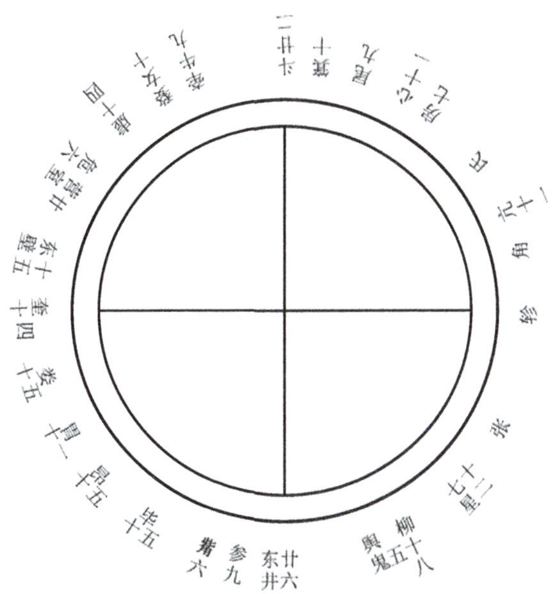

Fig. 21.5 The names and distance distribution of the twenty-eight lunar mansions

" + " line, one end points to the two mansions, Dou and Dongjin, respectively, the other end points to Kui and Zhen, respectively.

There was once a great debate over the actual function of this cultural relic. Many scholars believe that it belongs to the astrological instrument or the astrological aidtool [4] with the two pieces of the Shi plate unearthed from the same tomb, while some scholars hold that it is an astronomical instrument with unknown usage [5]. Some scholars even believe that it is mentioned in the early ancient literature, i.e., "circular sphere" [6]. Later, Liu Jinxuan further speculated based on the "circular sphere" idea: the round plate can be placed in the equatorial plane, and the calibration is inserted in the central small hole. The cursor is inserted into the small hole around the upper plate. The cursor and calibration can be used to aim and record the equatorial position of a celestial body, and then use another cursor to aim and record the equatorial position of another celestial body. The degree between the two cursors is the equatorial distance (equatorial longitude difference) between the two celestial bodies. When the upper plate is installed in the meridian plane and one of the " + " lines of the plate faces the North Pole, the equator Quji degree (equivalent to equatorial latitude) [7] can be measured by using the calibration and the cursor in the celestial body. Along this line of thought, Pan Nai further pointed out that a "shi plate frame" (also referred to as "lacquer box" in the archaeological briefing) on the side of the twenty-eight lunar mansions when unearthed may be part of the instrument [8].

Recently, Shi Yunli and other people have analyzed the detailed materials of the Shi plate frame (Figs. 21.6 and 21.7) in the excavation archives of the Ruyinhou tomb. Based on the measured data of its structure and size, it is found to be used to support the twenty-eighth lunar mansions (Fig. 21.8). Through its support, Liu Jinyu envisioned that the elevation angle of the calibration of the twenty-eight lunar mansions is 32°63', which is exactly the same as the geographical latitude (32°90') of the Fuyang area. The twenty-eight round mansions perpendicular to the calibration is located in the equatorial plane of the Fuyang area (Fig. 21.9).

Through the method speculated by Liu Jinxi, the equator longitude can be measured by using the upper set. The twenty-eight lunar mansions on the chassis just provide an equatorial reference frame of mansion degree for this instrument, which

Fig. 21.6 The old photo of the unearthed (Fuyang Museum) "plate frame" in the tomb of Xia Houzao

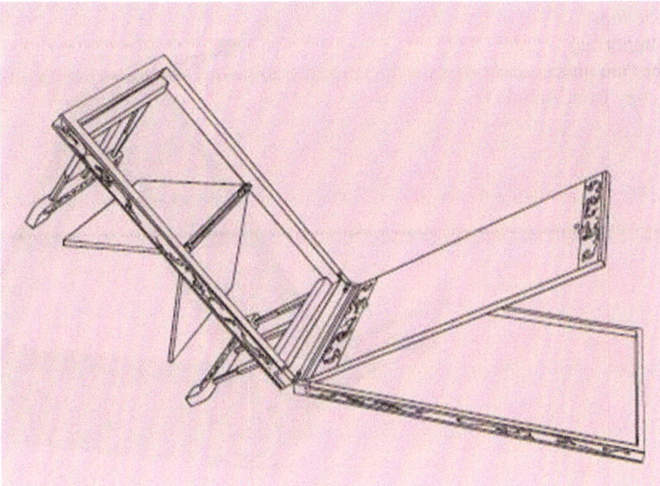

Fig. 21.7 The main structure of "shi plate frame" (drawing according to the original one from Fuyang Museum)

can be used for simple celestial position calculation: first, aim the first cursor of the upper plate with a known celestial body; then use another cursor to aim a celestial body with unknown position (such as a certain planet); put the frame flat, and align the first cursor with the mansion degree of the first celestial body on the chassis, then the second cursor toward the SU degree is the equatorial extent of the second celestial body. Moreover, this method can determine the equatorial mansion degree of multiple celestial bodies at one time. In other words, the twenty-eight lunar mansions of the tomb of the Ruyinhou is indeed an instrument for measuring the position of the celestial equator. Its appearance indicates that the corresponding celestial equatorial coordinate system has been initially formed in the Ruyinhou time [9].

In fact, such instruments using a calibration and a cursor to perform astronomical observations along a circumference were not exceptional in the early Western Han Dynasty. The two existing "sundial" of the Western Han Dynasty have differences in the graduation and the upper set of the twenty-eight lunar mansions (the tick mark engraved on the "sundial"), but its main structure has big similarities with the latter—the center has a large round hole, which is obviously left for the central calibration; the intersect between each time-scaled tick mark and the circumference has a small round hole, which is obviously reserved for inserting the cursor. In addition, the diameter of their circumference is basically around 23.4 cm, which is also equivalent to the diameter of 23.6 cm on the upper set of the twenty-eight lunar mansions, corresponding to the measurement method of the Han Dynasty (Fig. 21.10).

There is still controversy about the use of this instrument. Some people think that it is an equatorial sundial [10] while others think that it is a relatively horizontal type sundial [11]. It is also considered that it is mainly a "sundial instrument" [12] that measures the sunrise and sunset orientation. Chen Meidong believes that it is not

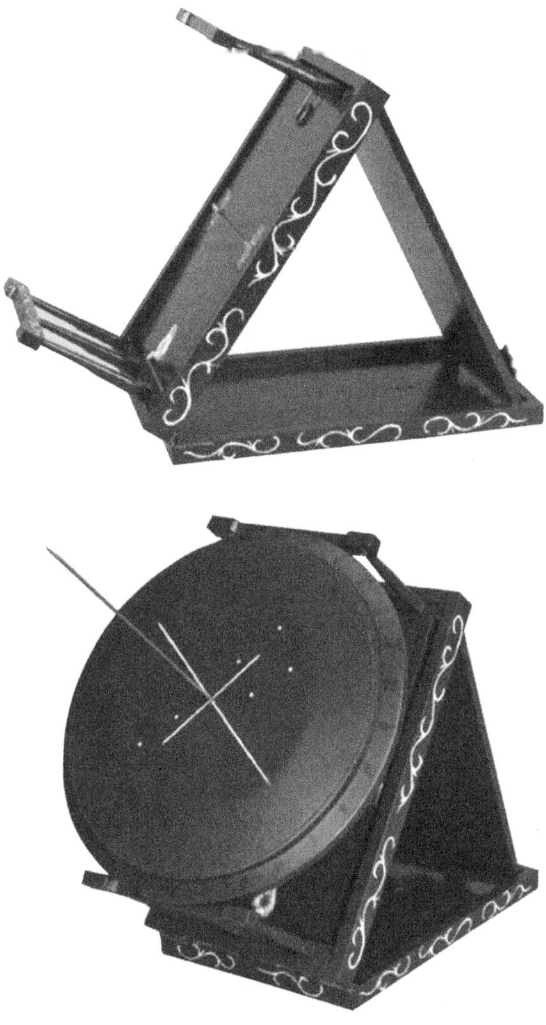

Fig. 21.8 The most reasonable supporting method of the "shi plate frame" (the copy from Anhui Museum)

only an equatorial sundial, but also can be used for the measurement of the celestial equatorial distance. The measurement method is consistent with the usage of the twenty-eight lunar mansions that Liu Jinzhen estimated [13]. The above study on the equatorial observation function of the twenty-eight lunar mansions of the Xiahouzao tomb shows that the method of the equator installation has not been known until the early Western Han Dynasty at the latest. In this case, it is natural to use such an instrument to observe the sunshade in the equatorial plane to determine the time, and to apply the same principle to the observation of the celestial equatorial distance. Therefore, Chen Meidong's speculation is completely reasonable.

The research conducted by Li Zhichao suggests that the "Yi" in Chinese used by the ancient people when referring to the instrument is the meaning of the "benchmark"

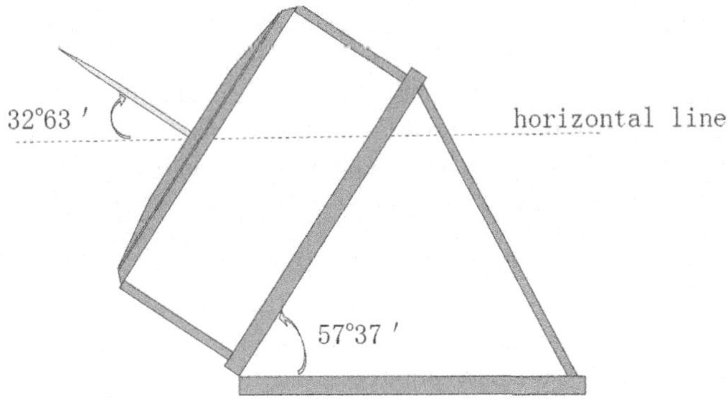

Fig. 21.9 The installment effect of equator of twenty-eight lunar mansions

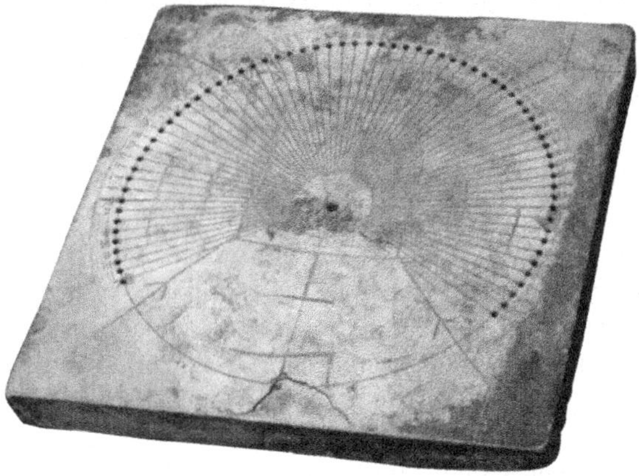

Fig. 21.10 The "sundial" unearthed from Tokto, Inner Mongolia in 1897 (National Museum of China)

(such as "*Er Ya Shi Gu*" reads: "Yi means standard".), whose extended meaning is the calibration and cursor for aiming [14]. In this way, the twenty-eight lunar mansions and the Western Han Dynasty "Sundial" are in line with this kind of structure, i.e., a circle plus a calibration and a cursor. Therefore, they should be the "circular sphere" mentioned in the early literature. The invention of circular sphere marks the emergence of the equatorial system and the equatorial astrometry measuring instrument in China, and has become the direct ancestor of the armillary sphere and the most important astrometry measuring instrument in ancient China.

21.2 The Invention of Armillary Sphere

According to Zhang Heng's statement and Liu Hui's opinions, "turning circular sphere into armillary sphere" and "turning circular sphere into Wan", Liu Jinyi believes that the birth of ancient Chinese armillary sphere has had a development course from "circular sphere to armillary sphere". Although the best observation area of the armillary sphere is the North of the equator, the calibration and the cursor are parallel to each other. When it is used for aiming, a sphere will be shaped instead of a line. So if the calibration is long enough, it is also possible to observe celestial bodies within a certain range in the south of the equator (for example, planets and moons moving to the equatorial south). The concrete methods are as follows: first, raise the eye from the surface of the plate, and make the calibration cut to the upper plate and coincide with the celestial bodies to be observed in the equatorial south. Second, insert the cursor on the upper plate between the calibration and the celestial body, and keep a state of the three points and one line, then record the position of the celestial body on the equator. Of course, observations cannot be made for the southern constellation that is too far from the equator. It may be for this reason that the ancient people thought of replacing the plate with a ring to completely eliminate the occlusion. In order to fix the ring and the corresponding aiming device, it is necessary to set a plurality of rings, so that there is a transition from a flat circle to a vertical circle, and an instrument called "armillary sphere" named by Zhang Heng. However, although a plate became a ring, and a circle became an armillary, the core structure has remained unchanged for nearly 20 centuries, which was the equatorial observing system.

From the historical data, the invention of armillary sphere can be traced back to the period when Emperor Wu of Han changed the calendar in around 104 B.C. The inventor was Luo Xiahong, an astronomer in the Western Han Dynasty of China and the main founder of the Taichu calendar recruited by the imperial court from Ba County (now Langzhong County, Sichuan) to participate in changing the calendar. There are two most important literature bases. One is Yang Xiong (53 B.C. to 18 B.C.). The first volume of the *Fa Yan*: "Some people turn to armillary sphere for help, which said: Luo Xiahong built it, Xuanyu Wangren measured it, and Geng Zhongcheng imitated it". The other is Chen Shou (233–297) *Yi Bu Qi Jiu Zhuan*: "Hong, also called Changgong, knew astronomy well and resided in the Luoxia, the Emperor of the Wu ordered Taishi to change the '*Zhuanxu Calendar*' into '*Taichu Calendar*'".

According to the characters of Chen Shou, "armillary sphere" (Hun Tian), is obviously an instrument that can be used to perform measurement operations ("rotating") at selected standard locations ("the center of places"), also called "浑仪" (Hun Yi) by later generations. However, in Yang Xiong's writings, "armillary sphere" is more like talking about theory of sphere-heavens, i.e., understanding the universe as a spherical concept wrapped outside the sphere. Yang Xiong's seems to indicate that the theory of sphere-heavens was put forward by Luo Xiahong, and Xianyu Wangren (living around 78 B.C.), an astronomer in the Western Han Dynasty, matched the degree for the model. Geng Shouchang ("Geng Zhongcheng", living around 91 B.C.–48 B.C.) made an instrument, that is "celestial globe" called by later generation, which is used to demonstrate this cosmological model. Because this kind of theory better reflects the actual situation of the celestial movement than the theory of canopy-heavens (the theory holds that the sky is above and the ground below), so Yang Xiong said:

"It can't be violated". Since the armillary sphere should appear with the theory of sphere-heavens at the same time, the two usages of the word "armillary" are not contradictory. However, the structure of the armillary sphere of Luo Xiahong is not clearly recorded in historical books.

Judging from the records in the *Hou Han Shu* · *Lv Li Zhi* in 92 A.D., Jia Kui's (30–101) speech during the presiding of the calendar discussion has shown that the celestial equatorial coordinate system has been improved greatly from the late Western Han Dynasty to the early Eastern Han Dynasty. First, the two coordinate system elements of the "Equator" and "Pole" (i.e., the South and North Pole) have been very clear; second, the concepts of the equatorial longitude and latitude have been established—the former uses the equatorial longitude difference between the celestial body and the sign star ("distance star") of the twenty-eight lunar mansions to represent the "Rusu" degree, while the latter uses the arc distance from the celestial body to the northern celestial pole ("depolarization" degree), or from the celestial body to the equator's arc distance to represent the "out of the equator" degree; again, the difference among the ecliptic (the large circle of the sun doing the annual viewing movement), the equator and its coordinate functions was now well understood by astronomers. Therefore, the armillary sphere should have two basic elements at the latest at that time—the right ascension circle for celestial body targeting and equatorial latitude measurement, and the equatorial circle for the equatorial longitude measurement.

However, in the Jia Kui era, astronomers also realized that it is more direct to use the ecliptic to measure the movement of the sun and the moon. So Jia Kui suggested making the corresponding instruments. In July, 103 A.D., the emperor ordered to make a "Taishi ecliptic bronze instrument". Obviously, this is an armillary sphere that uses the ecliptic to replace the equator as the axis. However, due to the Sunday's movement of the celestial sphere, the position of the ecliptic on the celestial sphere will change at any time, so it is difficult to target. As a result, the astronomers will not always "follow the change" because of the "difficulties of waiting for the good time".

At the end of the Eastern Han Dynasty, the astronomer Zhang Heng (78–139) made an armillary sphere (also known as celestial globe) when he was an imperial astronomer, and placed it on the Lingtai, a building of observing the celestial phenomena. In addition, Zhang Heng also made an armillary sphere driven and controlled by an hourglass to demonstrate the celestial body's Sunday movement or operation and achieved a high integrating degree. It is a pity that the specific records of these instruments have not been found in the existing historical materials. The first instrument in the history book with a clear description of the structure is the "Kong Ting armillary sphere" (Fig. 21.11). According to the *Sui Shu* · *Astronomical Treatise*, it is a copper instrument composed of internal and external parts. The outer layer is made up of a flat single ring, an equatorial single ring, and a meridian double ring. They not only function as skeletons but also have scales, which is called component of the six cardinal points by later generations. The inner layer consists of a right ascension double ring with a scale and a diameter of about 171 cm. The ring is equipped with a radial shaft which is installed in a circular hole representing

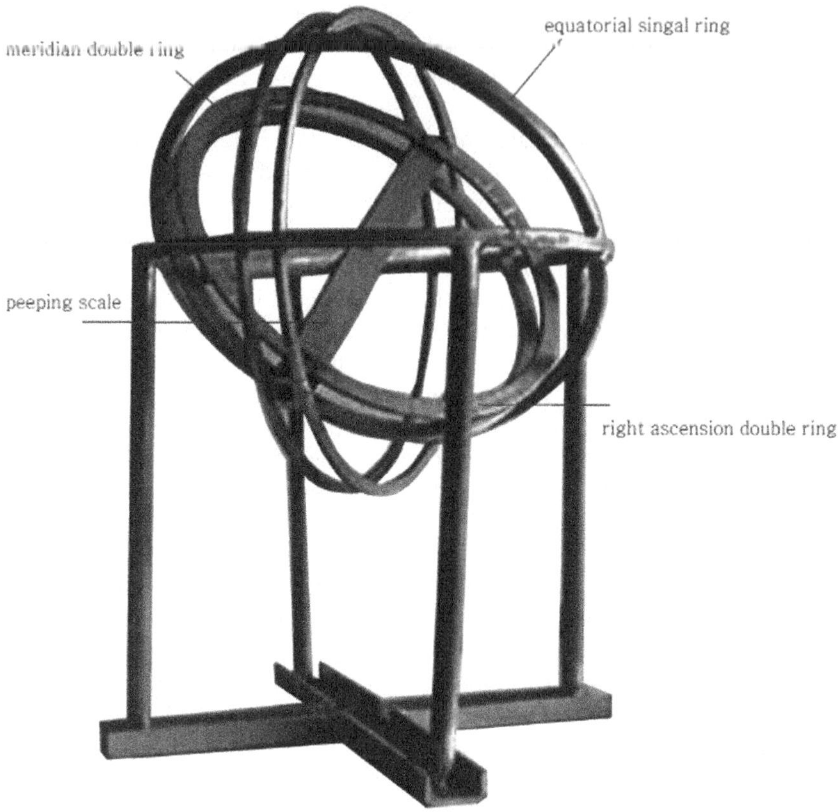

Fig. 21.11 Kong Ting armillary sphere restoration model (Li Zhichao)

the North Pole and the South Pole on the meridian double ring, enabling the right ascension double ring to rotate around the polar axis. Therefore, the later generations call it the Four-You-ring or Four-You-instrument. The ring equipped with a peeping tube for aiming can rotate around the ring center in the ring surface. By adjusting the positions of the Four-You and a peeping tube, any orientation on the celestial sphere can be targeted for measurement. The peeping tube indicates depolarization on the right ascension double ring, and the right ascension double ring on the equatorial ring indicates right ascension.

In the early Tang Dynasty, the Astronomer Royal Li Chunfeng (602–670) submitted a written statement and pointed out that the "Kong Ting armillary sphere" only had the equator but no ecliptic. So, he suggested rebuilding the instrument. Upon approval, Li made an "Armillary ecliptic sphere" in 633 A.D. This instrument is divided into three layers, establishing the shape of the triple structure of the Chinese armillary sphere. The outermost "component of the six cardinal points" and the innermost "Four-You-instrument" are the same as those on the "Kong Ting armillary sphere", but a "component of the three arrangers of time" is added between that two. The component of the three arrangers of time consists of the equator, the ecliptic, and the white channel single ring and the right ascension ring. It, 245.6 cm in diameter, can rotate around the polar axis. For the first time on the component of the three arrangers of time, the white road ring was installed, with the aim of making accurate observations of the moon movement. In order to adapt to the change of intersections of the Huangbaidao, i.e., moving one degree every 249 draconitic months along the Ecliptic, Li Chunfeng also played 249 holes on the ecliptic ring, and moved the position of moon's path by one hole every time after a draconitic month.

In 721, Zhang Sui, a famous monk of the Tang Dynasty, accepted an edict to change the calendar, and suggested the creation of a new instrument. Two years later, the copper instrument was made, called "Taishi ecliptic You-instrument" (Fig. 21.12). This instrument is also triple-like as Li Chunfeng's armillary sphere, but it has significant changes in both the outer and middle layers: the first change was getting rid of the equatorial ring in Li Chunfeng's component of the six cardinal points and replacing it with the unitary ring in the east and west. The second one was making a hole per degree on the equatorial ring, so that the position of the ecliptic ring

Fig. 21.12 Taishi ecliptic you-instrument duplicating model (Li Zhichao)

can imitate the precession phenomenon understood by the ancients, and continue to retreat along the equatorial ring. That's why it was called "ecliptic you-instrument". The third one was making a hole per degree on the ecliptic ring to simulate the retreat of the intersection of ecliptic plane and orbit plane rather than 249 holes.

The Northern Song Dynasty was one of the dynasties that produced the most armillary spheres in numbers in ancient China. From 995 to 1124, seven big armillary spheres were built. These armillary spheres basically maintained the triple structure of the Tang Dynasty's armillary sphere, but there were many improvements. For example, the practice of conducting both Baike and Fenke in Chinese, methods of calculating time in ancient time on the horizon ring before canceling the Huangyou new armillary sphere made during the third year of Huangyou (1051). And putting Baike and Fenke on a fixed equatorial ring has played the role of the time plate. It can be used directly for the local sun and star measurements. Similarly, in order to reduce the aiming error, Shen Kuo (1031–1095) carried out a reform on the Xi Ning armillary sphere made in 1074 that the diameter of the lower hole of the peeping tube was scaled down to one-fifth of the upper bore for accurate targeting. In Southern Song Dynasty, Su Song (1020–1101) installed an armillary sphere instrument on top of water carriage image table which he took part in manufacturing (Fig. 21.13). Under the drive and control of the timing machine, the Four-You-instrument can automatically track a given celestial body. This function is not possible to achieve with instruments using other coordinate systems, which fully demonstrates the advantages of the equatorial device.

The astronomical instruments of the Northern Song Dynasty were finally swept to Yanjing in the Jin Dynasty, and many were preserved in the Ming Dynasty and shipped to Nanjing. In 1439, The Imperial Board of Astronomy copied a large number of instruments of the former Dynasties, one of which was imitation of the armillary sphere from the Song Dynasty. In fact, up to now, it is still preserved in Zijin Mountain in Nanjing.

21.3 The Invention of Abridged Armillary

The structure of Chinese armillary sphere has undergone a development process from the simple one to the complex one, from two main structures to three main ones, from the development of the equatorial ring to the addition of the ecliptic and the white ring and other rings. Each improvement marks the progress made by people in understanding the laws of celestial movement. However, such development has also brought some drawbacks. First, the structure of the multi-ring overlapping has brought difficulties to precision manufacturing, which easily causes the misalignment of the center of each overlapping ring, thus affecting the improvement of the observation accuracy. Second, if there are more rings, the acre of the sky area will be more covered, and it is not easy to finish observations of certain celestial bodies. Third, the instrument structure is complex and difficult to operate.

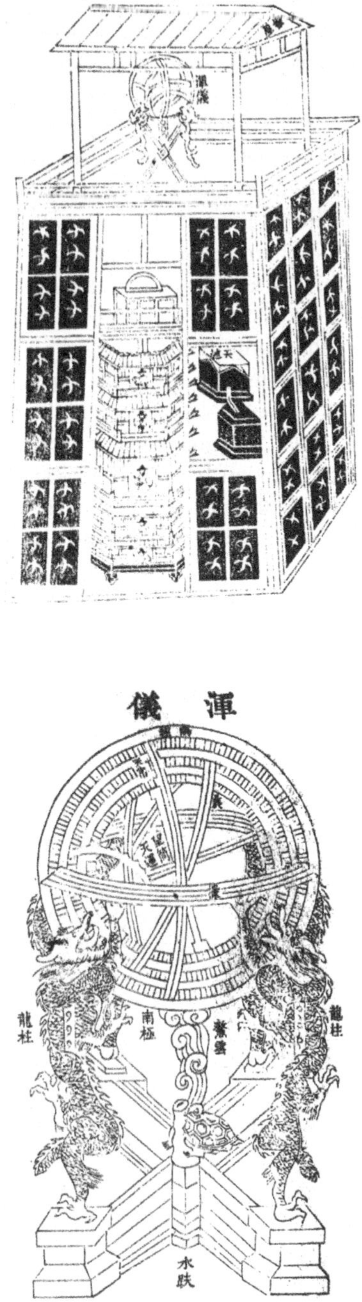

Fig. 21.13 The armillary sphere on Su Song Waterborne instrument and its enlarged picture

During the Northern Song Dynasty, astronomers were aware of these problems. Shen Kuo simplified the structure of his new armillary sphere. The first is to eliminate the white ring, and at the same time reduced the cross-sectional area of some parts to make it light and easy to use; the second is toad just the positions of the ecliptic, equator and horizon ring and make them into flat round rings to decrease their occlusion of the line of sight. In the Yuan Dynasty, Guo Shoujing (1231–1316) completely simplified and reconstructed the structure of armillary sphere, and invented the abridged armillary between 1276 and 1279. In 1439, The Imperial Board of Astronomy copied an abridged armillary in Beijing (Fig. 21.14). The preservation of this instrument gives us a glimpse of the greatness of this invention. The abridged armillary is mainly composed of an equatorial armillary sphere and an astronomical theodolite. The equatorial armillary sphere is the core of the instrument. It only keeps the three rings of the Four-You, equator, and Baike on armillary sphere. The latter two rings overlap each other and are installed together at the southern end of the Four-You-ring, so that the sky area above is almost unobstructed. There are four small cylinders between Baike and the equatorial ring, which act as "roller bearings" to reduce the resistance when the equatorial ring rotates. The peephole on the Four-You-ring is also reduced to be a copper strip with a transverse ear at both ends, known as the "peeping scale". There are small holes in the center of the two transverse ears, and each of the small holes is equipped with a thin line for aiming to improve the precision. In addition, there are two radial copper bars on the equatorial ring, which is called "boundary balance". The two ends of the boundary balance are connected with the North Pole by thin copper bars to form two right ascension planes. When observing, we should let these two right ascension planes target the two celestial bodies. The difference in the right ascension between the celestial bodies can be read

Fig. 21.14 The imitated abridged armillary during years of the Zhengtong Era in the Ming Dynasty (Nanjing Zijin Mountain Observatory)

on the equatorial circle. The north end of the Four-You-ring is also equipped with a "Houji ring" to determine the direction of the North Pole.

The astronomical theodolite consists of a "horizon ring" and a "vertical revolving ring" perpendicular to it, and is therefore separately referred to as the "vertical revolving instrument". The structure of the vertical revolving ring is similar to the Four-You-ring and also has an aiming device. During the observation, the height of the ground plane can be read from the vertical revolving ring, and the squared angle of the ground plane can be read from the ground level ring.

21.4 The Promotion of the Development of Ancient Chinese Astronomy by the Equatorial Astrometry Instrument

Astronomy is an observation-based quantitative science therein instruments and measurements play an important and basic role in its development. In addition to the observations and divining of the so-called specific celestial phenomena such as meteors, comets, new stars, sunspots, and celestial light color changes, the research objects of ancient Chinese astronomy are mainly movements of five stars and fixed stars. For the sun movement, the main observation instrument in ancient China was gnomon. According to the change of the length of the noon shadow, the ancient people can measure the arrival of the solar terms and the length of the tropical year. For other celestial bodies, the equatorial measuring instrument is the most important observation tool. They can be used to measure the position of these celestial bodies at a given time and then analyze their motion patterns.

In the measurement of celestial position, stellar coordinate measurement has a fundamental role, because the stellar coordinates provide a reference frame for the observation and measurement of other celestial phenomena. The accuracy of the observation of the stellar position (especially the position of the twenty-eight lunar mansions) directly determines the measurement of the positions of the five stars and the accuracy of the calculation. China is one of the earliest nations in the world to conduct accurate measurement for the position of the Zhoutian fixed stars. The Shi's "*Star Sutra*" from the Warring States to the Han Dynasty now contains the world's oldest star list, listing the "Rusu degree of flag star(equivalent to right ascension)" of the 120 constellations (In ancient times, China called them "asterism") and "degrees of depolarization(equivalent to declination)". This achievement is clearly inseparable from the invention and use of circular sphere and armillary sphere, so it appears with the invention of these two important instruments at the same time. Because of such precision measuring instruments, it was the *Taichu Calendar* compiled in 104 B.C. that the first calendar technique for quantitative calculation of the five stars movement of the sun and the moon appeared, and the equatorial coordinate system was used in these calendars.

After the Tang Dynasty, the development of the armillary sphere continued to promote the progress of the stellar observation and calendar astronomy. For example,

Seng Yixing used the "ecliptic you instrument" to re-observe the twenty-eight lunar mansions, and found the ancient and modern differences of the stellar coordinates, thus improved the accuracy of the winter solstice measurement and provided an important guarantee for the improvement of the calendar level of the past ages. Over the past 100 years of the Northern Song Dynasty, seven armillary spheres were made, and seven large-scale all-day star position measurements were also carried out accordingly.

Correspondingly, the overall standard of the calendar in the Song Dynasty also continues to improve. In the Yuan dynasty, Guo Shoujing invented the abridged armillary. Through the improvement of the aiming device and the calibration mode, its accuracy of reading improves about five times compared with instruments of past ages, which makes star's measuring accuracy at that time to reach a new high. In the meanwhile, Guo Shuojing and others compiled *The Timing Calendar*, which has also become a calendar with the highest measuring accuracy in ancient China reaching a high peak of the development of Chinese calendar astronomy. Its achievements definitely owe to the abridged armillary.

In addition to astrometry, the equatorial astrometry instruments also generate an incidental achievement of cosmology: the armillary sphere originating from circular sphere undoubtedly helped people understand that the universe is round, the theory of sphere-heavens is proposed, which explains and displays its specific structure through the armillary sphere. This is similar to what is said in *An Instruction Book of the Armillary Sphere* written by Zhang Heng in the Eastern Han Dynasty. The book briefly summarizes the basic thoughts of the theory of sphere-heavens as that the universe is round, with the earth in it just like the egg yolks in an egg and the universe is bigger than the earth. It also clearly describes basic circles and points of the armillary sphere model, including equatorial circle, zodiacal circle, horizontal circle, the North Pole, the South Pole, circle of perpetual apparition, circle of perpetual occultation and other basic elements, and the position relation among them. All circles and points basically respond to the rings on the armillary sphere. It can be seen that the leap from the popular theory of the canopy-heavens to the theory of sphere-heavens in the Han Dynasty is to a great extent relative to the appearance and development of equatorial instruments. Although the later celestial demonstration instruments like the celestial globe (a celestial sphere model) appeared, the armillary sphere itself was continuously regarded as an intuitive expression of the armillary sphere structure by astronomers. The reason why they successively add zodiacal circle and white circle on the armillary sphere and even also try to reflect the two's position changes relative to the equator (Li Chunfeng and Seng Yixing) is that their increasing knowledge of the celestial sphere model has constantly deepened, apart from the observation purposes.

Since the sixth century, the Chinese calendar and astronomical system started to be worshiped and imitated by Japan and North Korea. Since then the astronomy of the two countries has been in line with China. Although the knowledge about the equatorial astrometric instruments was already introduced to the two countries, imitations of these instruments were relatively late according to the written records. In 1433, astronomers under the rule of King Sejong in the Korean Li Dynasty made a set of

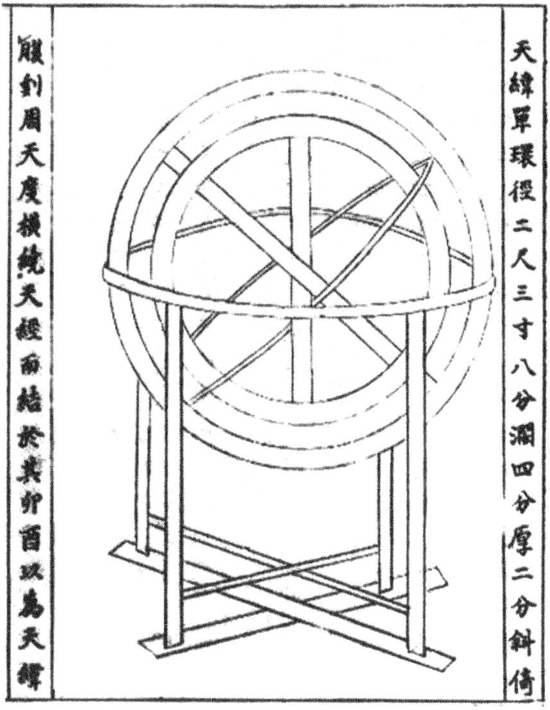

Fig. 21.15 The Chinese armillary sphere in *Tian Wen Qiong Tong* (1698) written by Shibukawa Shunkai

the astronomical instruments, including the abridged armillary, the small abridged armillary, the armillary sphere, respectively. In addition, according to the principles of the Chinese equatorial instruments, astronomers in the Korean Li Dynasty successively made the hanging sundial, the horizontal sundial, the direct south sundial, the sun-and-stars timing apparatus, the small sun-and-stars timing apparatus, and other instruments to measure time, which makes great contributions [15] to improving the equipment level of the astronomical instruments in Korea. In Japan, Shibukawa Shunkai (1639–1715) helped the Japanese Shogunate finish reforming Zheng Heng calendar in 1983 and then was appointed as an Astronomer Royal. In order to equip the observatory of the Japanese Shogunate, he designed and made a series of astronomical instruments, including the China-type armillary sphere (Fig. 21.15).

References

1. Jiyu, Ren. 1993. *The Silk Book Sorting Group of Mawangdui Han Tomb: Five stars Transcriptions: Chinese Science and Technology Classics Collection Astronomy*, vol. 1, 85–94. Elephant Press.
2. Zhentao, Xu. 1976. From the perspective of the silk book (Five Stars Transcriptions): the innovations of "pre-Qin armillary sphere". *Archaeology* 84: 89–94.

3. Anhui Provincial Work Team. 1978. Briefing on the excavation of the Fuyang Shuanggudui tomb of the high official Ruyinhou of the Western Han Dynasty. *Heritage* 12–31.
4. Harper, Donald J. 1978. The Han cosmic board (Shih). *Early China* 1–10.
5. Difei, Yin. 1978. Divining and astronomical instruments unearthed from the high official Ruyin tomb of the Western Han Dynasty. *Archeology* 338–343.
6. Dunjie, Yan. 1978. On the style plate and divining in the Early Western Han Dynasty. *Archeology* 334–337.
7. Jinyi, Liu. 1984. *From Circular Sphere to Armillary Sphere—Inspiration from the Twenty-eight Lunar Mansions in the Early Han Dynasty*, 205–213. Science Press.
8. Ding, Pan. 2015. *Colorful Literature: The History of Ancient Astronomical Instruments of China*, 49. Shanxi Education Press.
9. Yunli, Shi, Lin Fang and Chao Han. 2012. A new exploration of astronomical instruments unearthed from Xia Houzao's tomb in the Western Han Dynasty. *Studies in the History of Natural Science* 1–13.
10. Chinese Astronomical History Research Group. 1981. *Chinese Astronomical History*, 180–182. Science Press.
11. Shengchi, Guo. 1984. *On the Western Han Dynasty*, 214–236. Science Press.
12. Jiancheng, Li. 1989. *The Sundial—The Oldest Surviving Astronomical Instrument in China*, 145–153. Cultural Relics Publishing House.
13. Meidong, Chen. 2003. *Science and Technology History of China, Astronomical Volume*, 129–130. Science Press.
14. Zhichao, Li. 1995. *Testing the Instrument*, 173–179. Henan Education Press.
15. Needham J., Lu Gwei-Djen, J.H. Combridge and J.S. Major. 1986. *The Hall of Heavenly Records: Korean Astronomical Instruments and Clocks*, 1380–1780. Cambridge: Cambridge University Press.

Yunli Shi Born in Susong, Anhui in 1964, Doctor of science, Professor and Executive Director of the Department of History and Archaeology of Science and Technology of the University of Science and Technology of China, former vice chairman of the East Asian Society for the History of Science, Chairman of the Working Group on Historical Instruments of the International Astronomical Union, and member of the Chinese Society for the History of Science.

Research interests: History of astronomy

Major works: *History of Ancient Chinese Science and Technology: Astronomy, Collection of Overseas Rare Chinese Science and Technology Classics*

Awards: Second Prize for Outstanding Achievements of Philosophy and Social Science in Anhui Province in 2007.

Chapter 22
Bulkhead

Longfei Xi

The bulkhead of a ship is constructed on the basis of the ship's keel and the bottom shell plating and clings to the side plate and the vertical flat plate of the deck. If a ship is supported by a large number of bulkheads, it must have sufficient strength and stiffness. If the gaps between the bulkhead strakes and the gaps between the bulkhead and the shell plate and the deck are filled with a mixture of hemp thread, tung oil, and lime called "tung oil ash," it will possess sufficient watertightness. When a tank is flooded, it will not affect the adjacent tanks, which will enable the ship to improve its capability to resist sinking. In other words, it will possess unsinkability.

As early as the early fifth century in the Yixi Period of the Jin Dynasty, China built ships with watertight bulkheads. Later generations had developed and improved watertight bulkheads, making ancient Chinese ships equipped with high strength and high resistance to sinking, and featured by its watertight bulkheads.

22.1 The Invention of Watertight Bulkhead

The earliest ship with watertight bulkheads in China was called the "eight-cabin warship." It was built by Lu Xun, who followed Sun En, a famous rebel leader, to revolt in the Jin Dynasty. The eight-cabin warship is characterized by the use of watertight bulkheads to divide the hull into eight cabins. In this way, even if there is a leak on a certain cabin, the warship can still be safe and will not sink.

In October of 399 A.D., Sun En led hundreds of thousand people to uprise from an island, killed the county magistrate of Shangyu, and captured Kuaiji (now Shaoxing, Zhejiang) [1], and then quickly occupied the other seven counties within ten days, calling himself General of Eastward Expedition [2]. In December, Sun En was

L. Xi (✉)
China Maritime History Studies Association, Beijing, China
e-mail: Xilongfei428@163.com

J. Hua and L. Feng (eds.), *Thirty Great Inventions of China*,
https://doi.org/10.1007/978-981-15-6525-0_22

defeated by the official army and fled back to the island. In 402 A.D., Sun En led his followers to attack the coastal areas of Zhejiang and was defeated by the official army and his followers almost died. Then, Sun En drowned himself in the sea, and thousands of people thus re-elected Lu Xun, Sun En's brother-in-law, as the leader [3].

In 403 A.D., Lu Xun led his army to attack Dongyang, Zhejiang and assaulted Yongjia in August, but failed. In 404 A.D., he sailed south in October and conquered Panyu (now Guangzhou). Lu Xun governed those places and claimed himself as General Pacifying the South [4]. In 410 A.D., Lu Xun's army went northward from Guangzhou to occupy Yuzhang (now Nanchang, Jiangxi) and other counties, and then went along the Yangtze River to approach Jiankang (now Nanjing). The biography of Lu Xun in *The Book of Jin* has the following records: "Lu Xun led his sizeable army that boasted a hundred thousand people and about a thousand ships to go along the Yangtze River, defeated the general LIU Yi in Sangluozhou (now Susong, Anhui), and arrived in Jiangning (now Nanjing)." [5]

In 411 A.D., as defeated several times, Lu Xun failed to conquer Guangzhou for the second time. Then he went south to Longbian of Jiaozhou (now Hanoi, Vietnam). Given that the prefectural governor of Longbian commanded soldiers to "throw the tail torch to burn Lu Xun's fleet," Lu Xun's army was defeated and he drowned himself.

There are many records of Sun En and Lu Xun's revolts and their bad endings of throwing themselves into the sea due to their failure against the official army in the biography of Sun En and the biography of Lu Xun in *the Book of Jin* and the relevant ten years' record in *History as a Mirror*. Sun En and Lu Xun are deemed as bandits in the historical documents. Many documents do not record the eight-cabin warship built by Lu Xun or praise their achievements. However, the record of words and deeds of Emperor Yixi and the biography of Emperor Wu of the Song Dynasty have, to some extent, revealed the fact that Lu Xun invented the eight-cabin warship.

Yi Wen Lei Ju quoted from *Yi Xi Qi Ju Zhu*: "Lu Xun made nine eight-cabin ships, with four floors and a height of more than ten zhangs (about 110 feet)." [6]

The biography of the Emperor Wu in *the Book of Song* records about Liu Yu's suppression on the navy of Lu Xun: "Lu Xun led his army to set out for Baling (now Yueyang, Hunan) on that day, with nine four-floor eight-cabin ships." [7]

The eight-cabin ship built by Lu Xun is considered to be a ship that divides its hull into eight compartments with watertight bulkheads. The watertight bulkhead was an invention of China in 410 A.D., whose inventor was Lu Xun, one of the insurgent leaders of the Jin Dynasty.

Watertight bulkheads can date back to the Shang Dynasty when "舟" (boat) was inscribed on the oracle bones in several shapes. Oracle bones inscriptions are pictographic so it can be seen that the word "舟" looks like a boat, which is composed of vertical and horizontal parts. What does the horizontal line in the character boat mean? It looks like a rib and also resembles a bulkhead. Either or both of them might be the real meaning of the horizontal line.

Western scholars believe that it is natural for the ancient Chinese people to invent the watertight bulkheads by imitating the diaphragms of bamboo. Robert K. G.

Fig. 22.1 The Tang Dynasty wooden boat discovered in Rugao, Jiangsu

Temple, an American historian of science and technology, said: "The Chinese people got the inspiration of building the bulkhead of ship from observing the structure of the bamboo—the diaphragms of the bamboo divide the bamboo into many empty bamboo tubes. Considering there was no bamboo in Europe, so Europeans could not acquire such an inspiration." [8]

In his article "the Ratio of Bamboo: Research on the Structure of Ancient Ships," the French ship archaeologist Jane Burut has a similar view, saying: "Bamboo, because of its integral solid and thick stem (like hull) and the transverse bamboo joints (like the supporting partition plate) inside, is often compared to the structure of ancient Chinese wooden boats by some scholars such as GRG Worcester and J. Needham." [9]

The invention of watertight bulkhead in China is not only proved by historical records, but also the unearthed ancient ships. Although the bulkheads invented before the Jin Dynasty or in the Jin Dynasty have not been discovered so far, two ancient ships built in the Tang Dynasty were found to be equipped with the watertight bulkheads.

One is the Tang Dynasty wooden boat discovered in June, 1973 in Rugao, Jiangsu (Fig. 22.1) [10]. The ship is about 18 m long and is divided into 9 compartments with the watertight bulkheads between every two compartments. The longest compartment is 2.86 m long and the shortest is 0.96 m.

The other is the Tang Dynasty wooden boat discovered in Shiqiao Town of Yangzhou, Jiangsu in March 1960 (Fig. 22.2) [11]. The renovated ship is about 24 m in length and is divided into 5 large compartments. The boat has a solid structure, and the connection between the wooden boards is combined with the tenons and iron nails. And their joints are filled with putty, making good watertightness possible.

The eight-cabin ship's navigation area was from the coast of Zhejiang to the coast of Guangdong, and then to the present Beibu Gulf and today's coast of Vietnam. The restoration study suggests that the eight-cabin ship was designed as a sea boat with a sharp bottom and two upturned ends. A model of this type of sea boat was exhibited at the Jiaxing Ship Culture Museum.

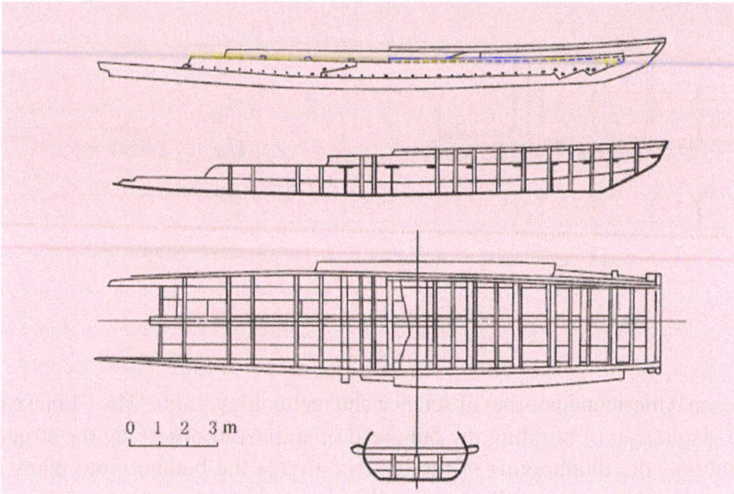

Fig. 22.2 Ancient wooden boat in Shiqiao, Yangzhou

22.2 Improvement and Wide Application of the Watertight Bulkhead Technology

From the Tang Dynasty to the Song Dynasty, the watertight bulkhead technology became more and more mature. The Song Dynasty sea boat (Fig. 22.3) unearthed in Quanzhou Bay in 1974 is equipped with 12 watertight bulkheads which divide the ship into 13 cargo holds. The bulkhead plate, about 100–120 mm thick, is usually made of fir wood and the last row of the bulkhead plate is made of the camphor wood. The Chinese unearthed ships of the Song and Yuan Dynasties all have very mature watertight bulkheads.

Figure 22.4 shows the map of the ship wreckage of the Song Dynasty found in the Quanzhou Bay. As shown in Fig. C of the figure, the intersection of the bulkhead and the hull is surrounded by surrounding frames. This not only improves the watertightness of the bulkhead, but also strictly limits the displacement of the bulkhead in the ship's head and tail. To avoid this problem, as shown in Fig. 22.4A, B, before the middle part of the ship, frames are placed behind the bulkhead to limit the rearward displacement of the bulkhead; since the shell plate before the bulkhead is narrower, the forward displacement is even impossible. Similarly, behind the middle part of the ship, frames are placed in front of the bulkhead so that the bulkhead cannot be displaced forward. Since the shell plate behind the bulkhead is narrower, the bulkhead is less likely to move backward.

Figure 22.5 shows the wreckage of the Yuan Dynasty sea ship salvaged in Sinan-gun,South Korea. The ship has a total of 7 bulkheads that divide the hull into 8 cabins. The surrounding frames can be seen behind the previous three bulkheads in the ship. There are not such peripheral frames behind the four bulkheads after the

Fig. 22.3 The site of the unearthed ship of the Song Dynasty in Quanzhou Bay

middle part of the ship. The reason is that after the middle part of the ship, frames are placed in front of the bulkhead, which is not visible in the picture. The maturity of the bulkhead surrounding frame technology is self-evident.

The bulkhead mapping of the seagoing vessel found in Quanzhou is shown in Fig. C. of the Fig. 22.4. There is a circular hole at the lowest point of the bulkhead, namely, the upper plane of the keel, called the drain hole. It is used to drain sewage when the cabin is washed. When the unloading is completed, the cabin needs to be washed. At this time, the cork of the drain hole should be removed. When the ship is slightly tailed, the sewage in each cabin will automatically flow to the tail cabin, and then the sailor can remove the sewage in the tail cabin. After that, plugging the drain holes with a cork will restore the watertightness of each bulkhead. Figure 22.6 shows the drain holes of the bulkhead on the ancient ship found in Sinan-gun, South Korea. Given that there were not watertight bulkheads in ancient Korean ships, Korean scholars are not familiar with the watertight bulkhead technology and their drain holes. When they saw the holes at the lower end of each bulkhead of the ship found in Sinan-gun, they were very surprised and further "strongly doubt about Joseph Needham's opinion that the Chinese ancient ships usually have several completely watertight transverse bulkheads." [12]

Figure 22.7 shows the mapping of the No. 3 and No. 5 watertight bulkheads of the No. 1 Ming Dynasty ancient ships found in Penglai. It can be seen from the figure that compared with the previous generations, the watertight bulkhead technology in the Ming Dynasty has improved. On the one hand, the adjacent two rows of bulkheads plates are not simply flat butt joints, but are butted by concave and convex grooved

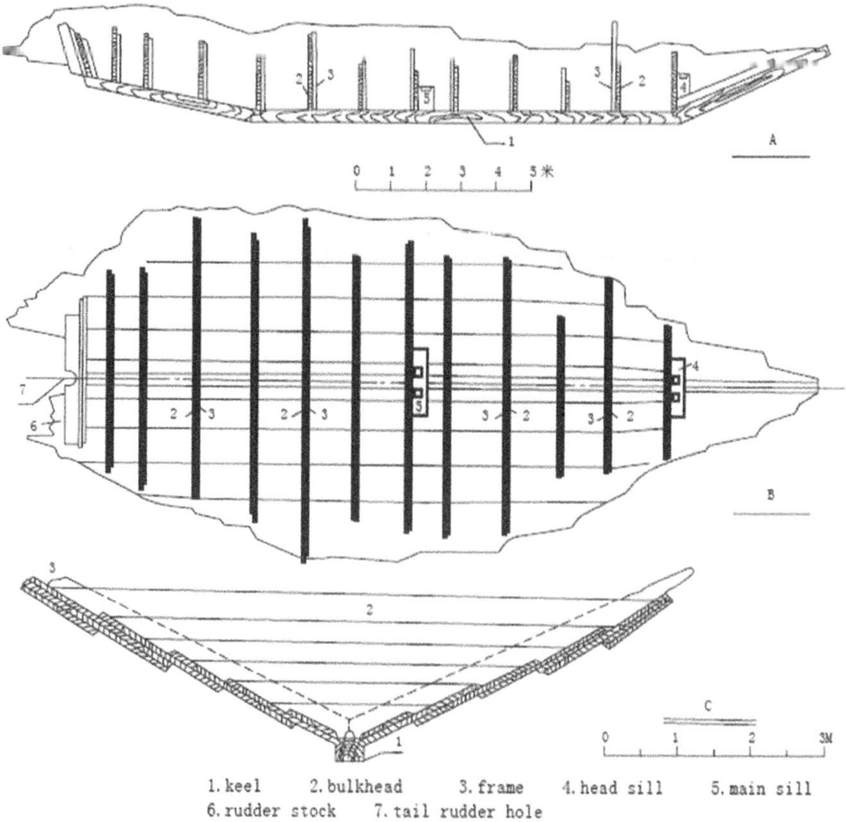

Fig. 22.4 Mapping of the wreckage of the ship in the Song Dynasty in Quanzhou Bay

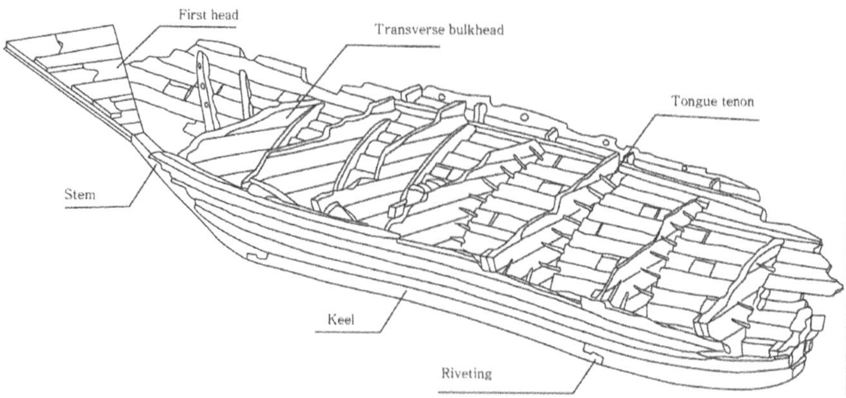

Fig. 22.5 The shipwreck of the Yuan Dynasty in Sinan-gun, South Korea

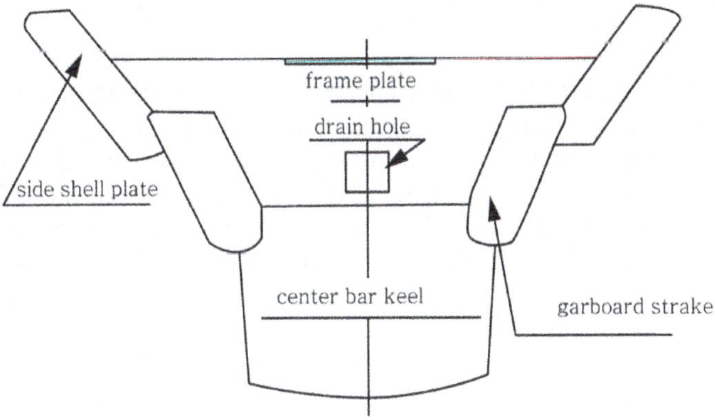

Fig. 22.6 The drain hole at the bottom of the Yuan Dynasty ship's bulkhead discovered in Sinan-gun, South Korea

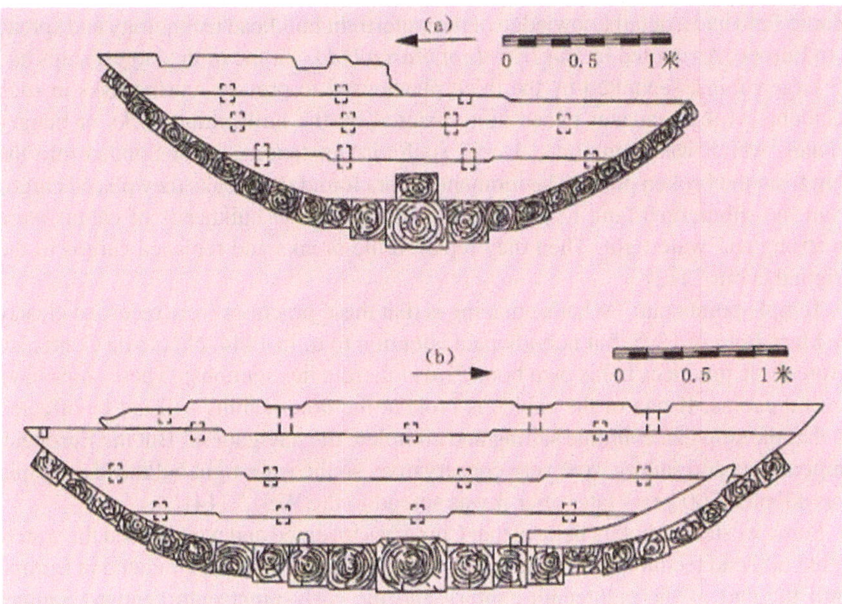

Fig. 22.7 The bulkhead mapping of No. 3 (top) and No. 5 (bottom) of Penglai No. 1 ancient ship

plates. In this way, it is guaranteed that there will be no displacement and deformation of the adjacent two rows of bulkhead plates. On the other hand, four square tenons are embedded between two adjacent rows of bulkheads so that the bulkhead can be prevented from being displaced and deformed. In Fig. 22.7, there are drain holes above the second-row plates on the left and right sides of the No. 5 bulkhead keel.

Up to now, the bulkheads in all unearthed Chinese ancient ships have the drain holes. As long as the cork is plugged into the drain hole, the bulkhead will be watertight. The watertight bulkhead technology invented in China has three important functions: (1) even if a ship strikes a reef, it can also prevent water flooding from reaching the adjacent tank, thus ensuring that the ship will not sink; (2) the shell plate and deck are supported by many bulkheads, which increase the stiffness and strength of the hull; (3) the bulkhead provides a sturdy transverse structure for the hull so that the mast can be closely attached to bulkheads and tightly connected with the hull, making it possible to use many masts and sails in ancient Chinese sailing boats.

22.3 The Spread and Modern Development of the Watertight Bulkhead Technology

Marco Polo had a detail knowledge of the watertight bulkhead technology and spread it to Europe. As quoted in *The Travels of Marco Polo*, "Some of the largest ships had 13 large cabins, separated by the thick plates used to prevent marine risks in such accidents as the seawater penetration resulted by the hull hitting rocks or hungry whales, which was common… Under such circumstances, water flooded into the ship from the broken place. The moment sailors found the breaks they placed cargos from the submerged tank to the adjacent cabin, for the bulkheads of cabins were so strong and watertight. Then they repaired the breaks and replaced cargos to the original cabin." [13]

Temple points out: "What's amazing is that these practices were recorded clearly by Mark Polo in 1295, but nobody paid attention to it. In 1444, Nicolo de Conti also wrote these practices in his own book *Travel* therein he mentions: 'These ships have several cabins. If one of the cabins is broken, the other cabins will not be affected so that the ship can continue sailing and complete the navigation.' But the European shipbuilders and sailors were very conservative, so the watertight bulkhead principle was adopted 500 years later after it was spread to the West." [14]

Some western scholars believe that China has invented and widely used the watertight bulkhead technology for thousands of years, and it was not emulated in Europe until the end of the eighteenth century and the early nineteenth century. Samuel Bentham (1757–1831), a British navy chief engineer, was the first one to design the watertight cabin of a ship in Europe. He was ordered by the First Lord of the Admiralty to design and build six nautical ships with a new structure, "As today's Chinese practice, the divided cabin is used to reinforce the structure of the ship and prevent the ship from sinking." [15]

Concerning "separating cargo holds with transverse bulkheads," Joseph Needham wrote: "We know that in the early 19th Century, the European shipbuilding industry took advantage of this kind of watertight bulkhead due to full awareness of the pioneering practice in China." [16]

Fig. 22.8 Titanic hit an iceberg (1912)

On April 10, 1912, Titanic, the new British luxury liner, after four days of her maiden voyage, collided with an iceberg near Newfoundland Island (Fig. 22.8). The cabins of the bow were broken and water flooded into them. More than 2 h later, the liner sank and 1,320 of the more than 2,500 passengers on board lost their lives. Although they radioed for help, none of the other ships' radio operators responded for they were sleeping. This serious shipwreck has shocked the world shipping industry.

Under the direct influence of this incident, the representatives of the major maritime countries gathered in London in 1914 and signed the International Convention for Safety of Life at Sea on January 24. But it was not put into practice because of the outbreak of the First World War. Later, in 1929, 1948, and 1960, the second, third, and fourth international conferences on safety of life at sea were, respectively, held and the International Convention for the Safety of Life at Sea1 was signed and amended. The Convention provides the basic requirements for ships' life-saving equipment, radio communication equipment (including radio operators who should be on duty 24 h a day), and navigation aids for ships sailing on the high sea. It also specifies the requirements for the ship's anti-sinking capability [17].

In order to guarantee and improve the ship's anti-sinking property, the ship's molded depth or freeboard should be increased at a certain load. When the load and molded depth are constant, it is necessary to increase the number of watertight bulkheads and reduce the length of the cabin for the purpose of ensuring the resistance to sinking. Watertight bulkhead technology is now gaining much attention. *The Rules for the Anti-Sinking Ability of Seagoer* in all countries complies with the current International Convention for the Safety of Life at Sea, and requires the minimum number of the watertight bulkheads for all types of the seagoing vessels.

The watertight bulkhead technology is a crucial invention in China's history of shipbuilding. At present, the construction of the wooden boats is gradually decreasing, and the watertight bulkhead shipbuilding technology has become China's

Fig. 22.9 Liu Xixiu (right) and Liu Sanji (middle) teaching watertight bulkhead technology to apprentices. Photo taken on December 8, 2010

"intangible cultural heritage in urgent need of protection." On November 15, 2010, at the fifth meeting of the UNESCO Intergovernmental Committee for the Protection of Intangible Cultural Heritage in Nairobi, Kenya, the watertight bulkhead technology project was reviewed and included in the 2010 Intangible Cultural Heritage List in Urgent Need of Protection. Figure 22.9 is the scene where Liu Xixiu and Liu Sanji, the intangible cultural heritage representative inheritors of the Ningde watertight bulkhead shipbuilding technology project, are teaching the watertight bulkhead construction skills to the apprentices in the shipbuilding.

References

1. Guang, Sima. 1956. *History as a Mirror*, 3497. Shanghai: Zhonghua Book Company.
2. Xuanling, Fang, et al. 1974. *Book of Jin: Biography of Lu Xun*, 2632. Shanghai: Zhonghua Book Company.
3. Guang, Sima. 1956. *History as a Mirror*, 3451. Shanghai: Zhonghua Book Company.
4. Xuanling, Fang, et al. 1974. *Book of Jin: Biography of Lu Xun*, 2634. Shanghai: Zhonghua Book Company.
5. Xuanling, Fang, et al. 1974. *Book of Jin: Biography of Sun En*, 2635. Shanghai: Zhonghua Book Company.
6. Xun, Ouyang. 1982. *Yi Wen Lei Ju*, 1234. Shanghai: Shanghai Classics Publishing House.
7. Yue, Shen. 1974. *Book of Song: Records of Emperor Wu*, 18. Shanghai: Zhonghua Book Company.
8. Robert K.G. 1995. Temple. *China: The Kingdom of Invention and Discovery—China's 100 Worlds' First*, 397. 21st Century Publishing Group.
9. Jane Burut. 2005. *The Ratio of Bamboo—Research on the Structure of Ancient Ships*, 514. Social Sciences Academic Press.
10. Museum, Nanjing. 1974. The Tang Dynasty wooden boats found in Rugao. *Cultural Relics* 5: 84–90.

11. Jiangsu Provincial Cultural Relics Working Team. 1961. The Tang Dynasty Wooden Boat Discovered in Shiqiao, Yangzhou. *Cultural Relics* 52–54.
12. Li, Liu. 2011. *Watertight Bulkheads and Their Drain Holes in Chinese Ancient Ships*, 57–64. Shanghai: Shanghai Bookstore Publishing House.
13. Polo, Marco. 1937. *The Travels of Marco Polo*, 191. Beijing: The Commercial Press.
14. Robert K.G. 1995. Temple. *China: The Kingdom of Invention and Discovery—China's 100 Worlds' First*, 396. 21st Century Press.
15. Robert, K.G. 1995. Temple. *China: The Kingdom of Invention and Discovery—China's 100 Worlds' First*, 395–396. 21st Century Press.
16. Pan Jixing. 1986. *Collected Works of Joseph Needham*, 258–259. Liaoning Science and Technology Publishing House.
17. Longfei, Xi. 1991. *Introduction to Ships*, 12. Beijing: China Communications Publishing and Media Management Corporation.

Longfei Xi Born in Meihekou, Jilin province in 1930, graduated from School of Naval Architecture &Ocean Engineering, Dalian University of Technology, and then served as a professor in the Research Center of Shipbuilding History, Wuhan University of Technology. He once served as Deputy Director of the Shipbuilding Department, Director of the Ship Design Teaching and Research Office; chief editor of the university journal. He was a member of the China Shipbuilding Engineering Society since 1988;

1984–2006, vice president of the China Shipping History Research Association and has served as honorary president since 2006. He is currently a life-long advisor to the China Maritime History Research Association and a member of the International Academic Committee on the *History of Maritime Relations*.

Research interests: Ship design; history of shipbuilding in China.

Major works: *The Principle of Ship Design, Introduction to Ships, China Shipbuilding Industry, History of Shipbuilding in China, the History of Chinese Science and Technology (Transportation), Ship Culture, General History of Chinese Shipbuilding*.

Awards: The first prize in Science and Technology Progress granted by State Science and Technology Commission in 1988;

The Rise and Fall of the Medieval Chinese Shipbuilding Industry published in the British *Shipbuilding Yearbook in 2004* won the Bronze Award of the Royal Shipbuilding Engineering Society in 2005.

Chapter 23
Gunpowder

Zhanhong You

Gunpowder, also known as the black powder or smoke powder, is a chemical material that can burn rapidly and regularly when the external force has been applied appropriately, generating a larger amount of gas at high temperature. It is a mixture of saltpeter, sulfur, charcoal, and the other combustibles. Since saltpeter and sulfur had been used as medicine in ancient China, and the combination of such matters is highly inflammable, the Chinese name of the gunpowder is known as "Combustible Medicines". Gunpowder is the first kind of the explosives that our human beings ever controlled, and it's one of the Four Great Chinese Inventions, which made a significant contribution to the development of the world's civilization.

Black powder is a kind of explosive which can burn quickly, producing a great amount of gases, without additional combustion improver. It usually consists of 75% saltpeter, 10% sulfur, and 15% charcoal, which are evenly mixed together.

Saltpeter plays the role as oxidizer in the reaction, which is also called potassium nitrate in modern chemistry. It is highly combustible and easily explosive when exposed to the organics. One of the important features of the gunpowder is that it contains oxidant in itself, which allows it to burn without taking the oxygen from the air because the potassium nitrate will release the oxygen. This forms a self-combustion system and undergoes intense oxidation-reduction reaction with combustibles such as charcoal and sulfur, and releases a large amount of heat.

In the process of burning, sulfur plays the role of the reducing agent, which is an important factor for the gunpowder to explode. Sulfur mainly acts as a combustion adjuvant in black powder, that is, sulfur is oxidized to sulfur dioxide (SO_2) and sulfur trioxide (SO_3), and the heat is released. Sulfur also acts as the adhesion agent and a catalyzer to lower the ignition point of the gunpowder.

Z. You (✉)
Institute for History of Science and Technology & Ancient Texts, Tsinghua University, Beijing, China
e-mail: youzhh@lib.tsinghau.edu.cn

J. Hua and L. Feng (eds.), *Thirty Great Inventions of China*,
https://doi.org/10.1007/978-981-15-6525-0_23

Carbon also acts as a combustion adjuvant in gunpowder, which is equivalent to the fuel in black powder. The reaction of black powder is mainly about the carbon being oxidized. Therefore, the type of charcoal and the degree of carbonization have a great impact on the gunpowder's performance.

The black powder is prone to get damp. It will release smoke when it explodes, and its capacity for destruction is smaller. It is also a necessary part in ignition powder, propellant powder, and blasting fuses. Before the twentieth century, the black powder was the gunpowder and explosives for guns and canons, but it was gradually replaced by the more powerful smokeless gunpowder and explosives.

The black powder nowadays is widely used as the gelatins explosive and igniting agent for pyrotechnics, as well as propellant or ignition; a propellant for incineration bombs, incendiary bombs, flares, and shotguns' bullets and fireworks; it is also used as the explosive in mining industry.

23.1 Alchemist and Gunpowder

Ancient Chinese pharmacists and alchemists played a special role in the invention of gunpowder. At first, they tried to add the saltpeter and sulfur in the formulation of a medicine for immortality, and they gradually discovered the explosive properties of the sulfur and saltpeter in the process. Later, the original version of gunpowder, which contained nitrate, sulfur, and charcoal, was refined no later than 808 in the Tang Dynasty.

Early in the Spring and Autumn Period and the Warring States Period (770 B.C.–221 B.C.), the alchemists had already used saltpeter and sulfur to make the medicine of immortality. *Sheng Nong's Herbal Classic*, the earliest pharmacopeia in China, ranked nitrite the sixth in the list of 120 top-quality medicines (Figs. 23.1 and 23.2), and the sulfur ranked as the second in that list (Figs. 23.3 and 23.4). Ge Hong (281–341), a famous alchemist in the Eastern Jin Dynasty, recorded the experiment of refining realgar with the saltpeter, Xuantong Sausage (pig intestine) and the pine resin in the book of *Bao Buzi—Xianyao*, the Tao's classics. It has been proved by the experiment that when the amount of saltpeter is small, arsenic and elemental arsenic can be obtained with these three substances; when the amount of saltpeter is large, it will explode when heated extensively.

In the Southern Dynasties, Tao Hongjing (456–536) recorded an experiment in which the purple-blue color flame rose when the saltpeter burned in the book of *Shen Nong's Herbal Classic*.

In the later part of the Sui Dynasty and early Tang Dynasty, Sun Simiao, (581–682), the Chinese medical scientist and alchemist, wrote the book *Sun Zhenren Alchemy* (in *Methods of Supreme Alchemy*). It has documented "fohu method with sulfur" (Fig. 23.5). First, grind the saltpeter and sulfur into powder. The powder then is placed in a silver pot or a sand pot. Dig a pit on the ground, and put the pot in the pit. Later, take three soap horns, roasting their surfaces till it is burnt and the flames are extinguished. Place about three pounds of charcoal to heat the pot. Wait for the

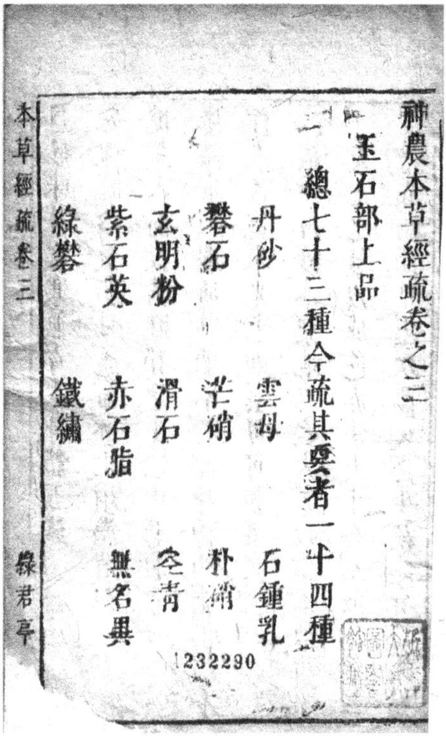
神農本草經疏卷之三
玉石部上品
總七十三種今疏其要者一十四種
丹砂 雲母 石鍾乳
礬石 芒硝 朴消
玄明粉 滑石 空青
紫石英 赤石脂 無名異
綠礬 鐵鏽
1232290
本草經疏卷三 綠君亭

Fig. 23.1 The photo of saltpeter described in *Sheng Nong's herbal classic* [1]

charcoal to burn to one-third, then immediately put out the fire until it cools down [3]. Such a method only produced a mixture of sulfur, saltpeter, and charcoal, but by adding the soap horns, the combustion rate and intensity is constrained.

In the Tang Dynasty, Qing Xuzi, an alchemist, wrote The Secret of Jindan, the Great Saint-ancestor of Tai Shang (Selected from Jiageng's Integration of Lead and Mercury), recording "fohu method with alum" (Fig. 23.6), as is described in the book "mix appropriate amount of sulfur, saltpeter and Fructus Aristolochiae and grind them into powder; the mixed powder is then put into a pot placed under the soil; heat the pot with bullet shape of fire and wait for the smoke to rise, then contain the smoke with thick layers of wet paper; after the smoke disappears and the pot has cooled down, the sulfur is successfully made" [4].

This method blended the Fructus Aristolochiae with the same amount of sulfur and saltpeter. The combination of these three substances forms a self-combustion system, which is the early prototype of gunpowder. It can be inferred that the formula of these three substances of saltpeter, sulfur, and charcoal by the ancient Chinese alchemists had been invented in Tang Dynasty [5].

After the mid-Tang Dynasty, the formula of the gunpowder has recorded in alchemists' works that the combination of saltpeter, sulfur, and charcoal would cause fire and even destroy houses. There is a brief introduction to "fohu method with saltpeter" (Fig. 23.7) in Synopsis of Zhenyuan Fine Methods: "Mix the sulfur, realgar

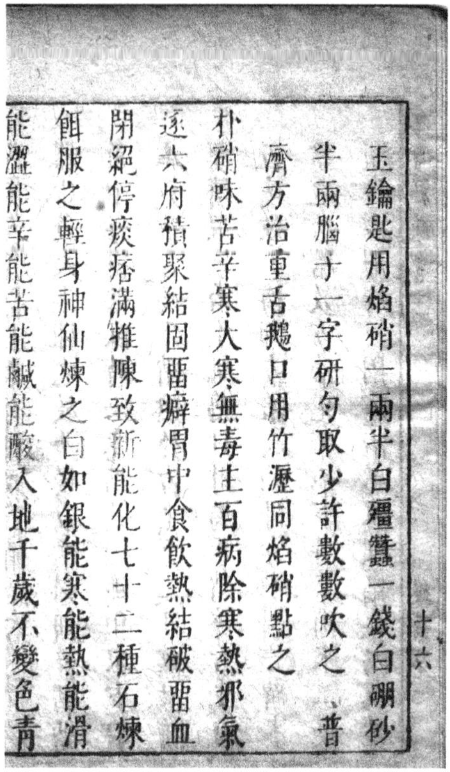
十六
玉鑰匙用焰硝一兩半白殭蠶一錢白硼砂
半兩腦子一字研勻取少許數數吹之 普
濟方治重舌鵝口用竹瀝同焰硝點之
朴硝味苦辛寒大寒無毒主百病除寒熱邪氣
逐六府積聚結固留癖胃中食飲熱結破留血
閉絕停痰痞滿推陳致新能化七十二種石煉
餌服之輕身神仙煉之白如銀能寒能熱能滑
能澀能辛能苦能鹹能酸入地千歲不變色青

Fig. 23.2 The photo of saltpeter described in the text of *Sheng Nong's herbal classic* [1]

and saltpeter together, and ignite them with honey; be careful with the rising flame, it may reach hands or heads, or even damage the house…; the appropriate amount of saltpeter can successfully produce the right medicine, but the excessive amount of saltpeter may fail; people should not ignite the combination of these three substances, otherwise it will cause a fire" [6].

Alchemists used saltpeter, sulfur, and charcoal to refine the medicine, in which they tried to prevent these three substances from exploding, but inadvertently discover the original version of the gunpowder. For military strategists in ancient China, the medicine that contained saltpeter, sulfur, and charcoal is the best material for fire attacks. The strategists knew the method of the fire attack from Sun Tzu in its 12th chapter. However, before the alchemists improved the gunpowder with better combustion performance, the only way for fire attacks was to attach flammable materials to the arrowhead and ignite it, then shooting the ignited arrows with bow in order to destroy the enemy's wooden fortifications, weapons, and supplies.

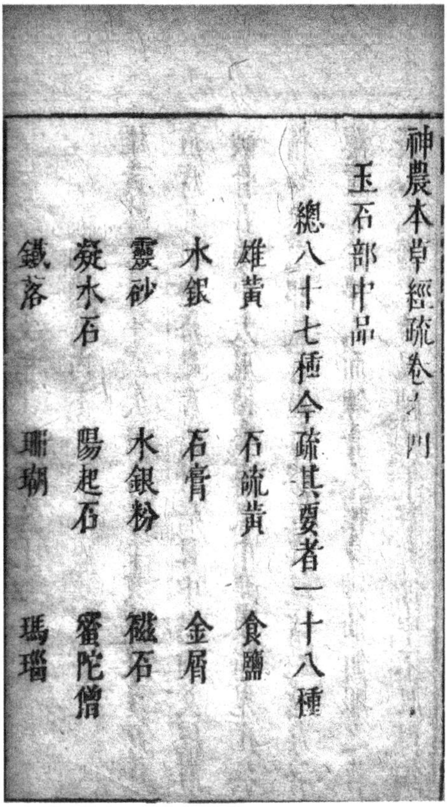
神農本草經疏卷之四
玉石部中品
總八十七種今疏其要者一十八種
雄黃 石硫黃 食鹽
水銀 石膏 金屑
靈砂 水銀粉 磁石
凝水石 陽起石 蜜陀僧
鐵落 珊瑚 瑪瑙

Fig. 23.3 The photo of sulfur described in *Sheng Nong's herbal classic* [2]

23.2 Invention of Gunpowder for Military Usage and Its Improvement

From the Northern Song Dynasty, gunpowder was used for military purposes and its formula was modified to suit the needs of the war. In 1044, led by Zeng Gongliang, the Hanlin Bachelor and also an official from the Ministry of Rites (999–1078), edited by Ding Du, the Hanlin Bachelor and an official from Ministry of Industry, the masterpiece *Collection of the most Important Military Techniques* was compiled with a preface written by the Emperor Song Renzhong. There are 40 volumes in the book, and its completion takes the effort of all the great minds of the country. The masterpiece documented the world's first three military gunpowder formula: gunpowder fireball, weaved gunpowder fireball, gunpowder fireball with toxic smoke (Figs. 23.8, 23.9 and 23.10).

The ingredients of the gunpowder fireball are as follow: Jinzhou sulfur 522.2 g, Wohuang 26.11 g, saltpeter 1492 g, rhizoma gastrodiae 37.3 g, dry hog gum 37.3 g, yellow arsenic 37.3 g, led powder 37.3 g, Caulis Bambusae in Taeniam 37.3 g, Plumbum Rubrum 37.3 g, yellow wax 18.65 g grinded in powder, colza oil 0.23 g,

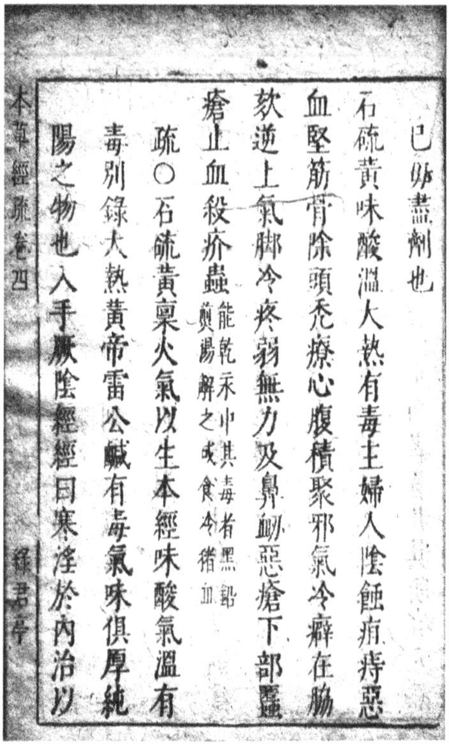
已卯盡劑也
石硫黃味酸溫大熱有毒主婦人陰蝕疽痔惡
血堅筋骨除頭禿療心腹積聚邪氣冷癖在脇
欬逆上氣脚冷疼弱無力及鼻衄惡瘡下部䘌
瘡止血殺疥蟲能乾汞中其毒者黑鉛煎湯解之或食冷豬血
疏○石硫黃稟火氣以生本經味酸氣溫有
毒別錄大熱黃帝雷公鹹有毒氣味俱厚純
陽之物也入手厥陰經經曰寒淫於內治以
本草經疏卷四 綠君亭

Fig. 23.4 The photo of sulfur described in the text of *Sheng Nong's Herbal Classic* [2]

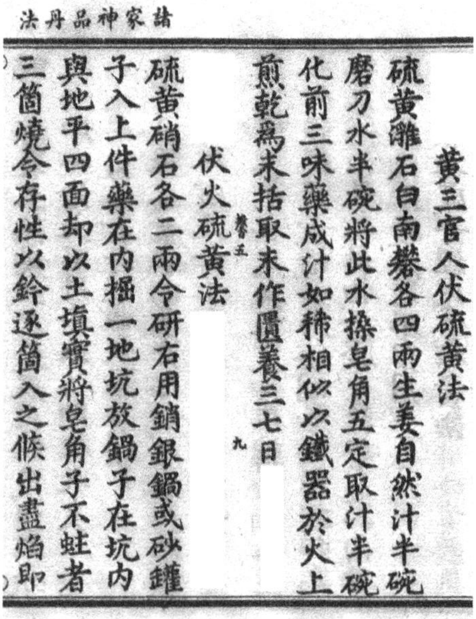
諸家神品丹法
黃三官人伏硫黃法
硫黃雌石白南礬各四兩生姜自然汁半碗
磨刀水半碗將此水搽皂角五定取汁半碗
化前三味藥成汁如稀相似以鐵器於火上
煎乾為末括取末作匱養三七日
卷五 九
伏火硫黃法
硫黃硝石各二兩令研右用銷銀鍋或砂罐
子入上件藥在內掘一地坑放鍋子在坑內
與地平四面却以土填實將皂角子不蛀者
三箇燒令存性以鈴逐箇入之候出盡焰即

Fig. 23.5 The photo of fohu method with sulfur [3]

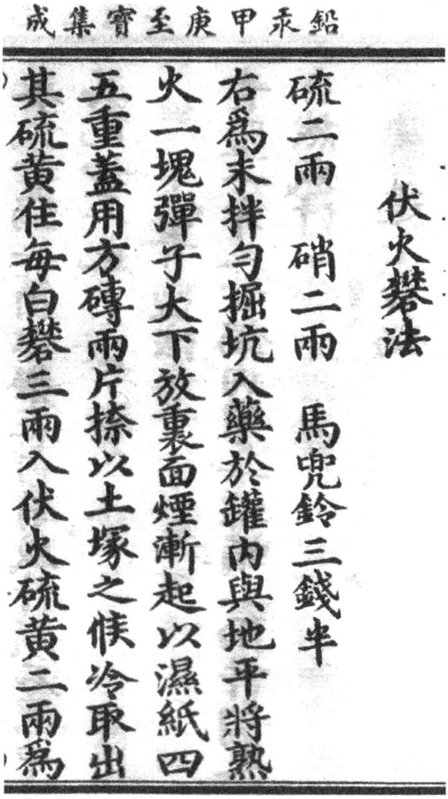
鉛汞甲庚至寶集成

伏火礬法

硫二兩　硝二兩　馬兜鈴三錢半
右爲末拌勻掘坑入藥於罐內與地平將熟
火一塊彈子大下放裏面煙漸起以濕紙四
五重蓋用方磚兩片捺以土塚之候冷取出
其硫黄住每白礬三兩入伏火硫黄二兩爲

Fig. 23.6 The photo of fohu method with sulfur [4]

turpentine 18.65 g, and thick oil 0.23 g. The ratio of the three substances of saltpeter, sulfur, and charcoal in this formula is as follow: saltpeter takes1492 g, accounting for 50.6%; sulfur, and Wohuang take 783.3 g, accounting for 26.6%; various carbonaceous materials takes 672.146 g, accounting for 22.8% [7].

The ingredients of the weaved gunpowder fireball are as follow: sulfur 746 g, saltpeter 1492 g, charcoal 186.5 g, asphalt 93.25 g in powder, dry hog gum 93.25 g in powder; bamboo mushroom and jute mushroom each 37.53 g, both of which are cut into pieces; tung oil, oil, and wax each 93.25 g, all of which are heated and mixed. The ratio of the three substances of saltpeter, sulfur, and charcoal in this formula is as follow: saltpeter takes 1492 g, accounting for 50%; sulfur takes 746 g, accounting for 25%; various carbonaceous materials take 711.31 g, accounting roughly for 25% [8].

The ingredients of the gunpowder fireball with toxic smoke are as follow: sulfur 559.5 g, Aconitum 186.5 g, saltpeter 1119 g, charcoal 186.5 g, seeds of Glabrousleaf Pittosporum 186.5 g, Gigantic Euphorbia Root 186.5 g, Tung oil and oil each 93.25 g, charcoal 186.5 g, asphalt 93.25 g, Asadin 37.3 g, bamboo mushroom and jute mushroom each 37.53 g, both of which are mixed as ball shape; the ratio of the three substances of saltpeter, sulfur, and charcoal in this formula is as follow: saltpeter

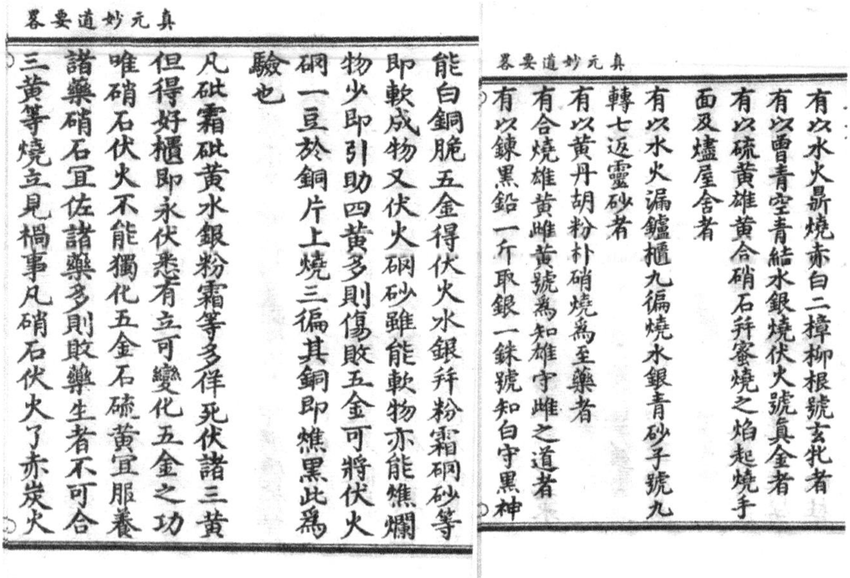

真元妙道要畧

有以水火鼎燒赤白二樟柳根號玄牝者
有以曾青空青結水銀燒伏火號真金者
有以硫黃雄黃合硝石并蜜燒之焰起燒手
面及燼屋舍者
有以水火漏鑪櫃九徧燒水銀青砂子號九
轉七返靈砂者
有以黃丹胡粉朴硝燒為至藥者
有合燒雄黃雌黃號為知雄守雌之道者
有以鍊黑鉛一斤取銀一銖號知白守黑神

真元妙道要畧

能白銅脆五金得伏火水銀幷粉霜硐砂等
即軟成物又伏火硐砂雖能軟物亦能焦爛
物少即引助四黃多則傷敗五金可將伏火
硐一豆於銅片上燒三徧其銅即焦黑此為
驗也
凡砒霜砒黃水銀粉霜等多伴死伏諸三黃
但得好櫃即永伏悉有立可變化五金之功
唯硝石伏火不能獨化五金石硫黃宜服養
諸藥硝石宜佐諸藥多則敗藥生者不可合
三黃等燒立見禍事凡硝石伏火了赤炭火

Fig. 23.7 The photo of fohu method with saltpeter [6]

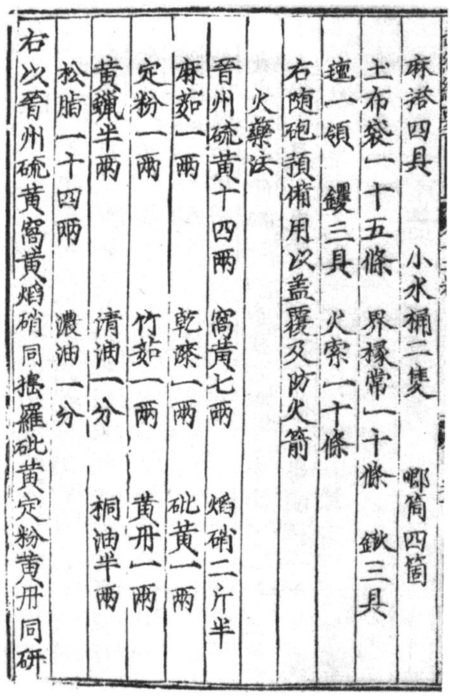

麻搭四具　小水桶二隻　唧筒四箇
土布袋一十五條　界樑常一十條　鍬三具
氈一領　钁三具　火索一十條
右隨砲預備用以蓋覆及防火箭
火藥法
晉州硫黃十四兩　窩黃七兩　焰硝二斤半
麻茹一兩　乾漆一兩　砒黃一兩
定粉一兩　竹茹一兩　黃丹一兩
黃蠟半兩　清油一分　桐油半兩
松脂一十四兩　濃油一分
右以晉州硫黃窩黃焰硝同擣羅砒黃定粉黃丹同研

Fig. 23.8 The photo of gunpowder fireball formula [9]

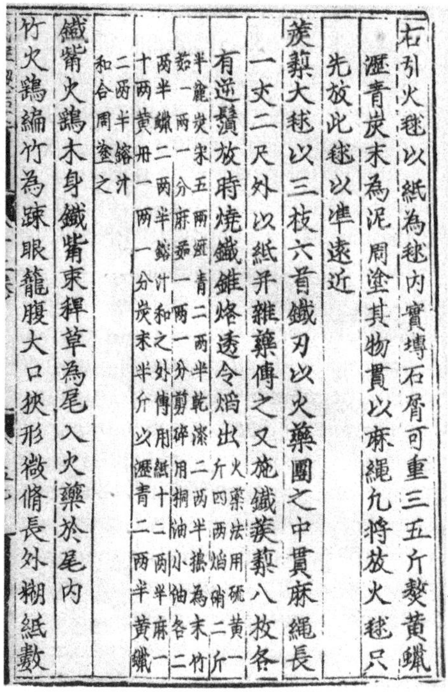

右引火毬以紙為毬內實塼石屑可重三五斤𤇆黃蠟
瀝青炭末為泥周塗其物貫以麻繩凡將放火毬只
先放此毬以準遠近
蒺藜火毬以三枝六首鐵刃以火藥團之中貫麻繩長
一丈二尺外以紙并雜藥傅之又施鐵蒺藜八枚各
有逆鬚放時燒鐵錐烙透令焰出火藥法用硫黃一斤四兩焰硝二斤半麄炭末五兩瀝青二兩半乾漆二兩半擣為末竹茹一兩一分麻茹一兩一分剪碎用桐油小油各二兩半蠟二兩半鎔汁和之外傅用紙十二兩半麻一十兩黃丹一兩一分炭末半斤以瀝青二兩半黃蠟二兩半鎔汁和合周塗之
鐵嘴火鷂木身鐵嘴束稈草為尾入火藥於尾內
竹火鷂編竹為疎眼籠腹大口狹形微脩長外糊紙數

Fig. 23.9 The photo of weaved gunpowder fireball formula [9]

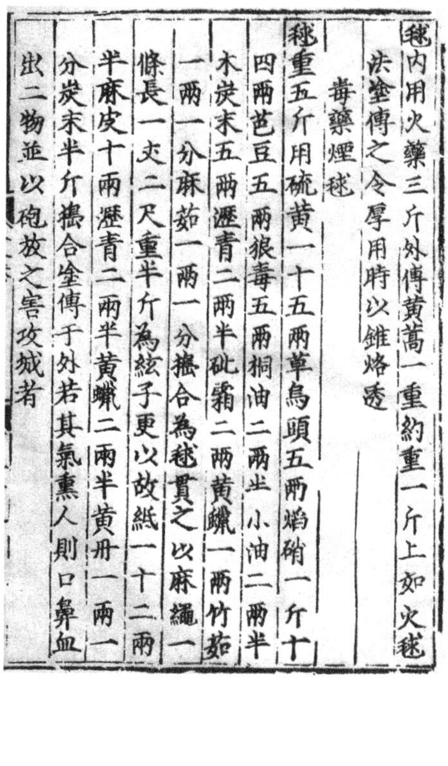

毬內用火藥三斤外傅黃蒿一重約重一斤上如火毬
法塗傅之令厚用時以錐烙透
毒藥煙毬
毬重五斤用硫黃一十五兩草烏頭五兩焰硝一斤十
四兩芭豆五兩狼毒五兩桐油二兩半小油二兩半
木炭末五兩瀝青二兩半砒霜二兩黃蠟一兩竹茹
一兩一分麻茹一兩一分擣合為毬貫之以麻繩一
條長一丈二尺重半斤為弦子更以故紙一十二兩
半麻皮十兩瀝青二兩半黃蠟二兩半黃丹一兩一
分炭末半斤擣合塗傅于外若其氣熏人則口鼻血
出二物並以砲放之害攻城者

Fig. 23.10 The photo of gunpowder fireball with toxic smoke formula [10]

takes 1119 g, accounting for 49.06%; sulfur takes 559.5 g, accounting for 24.8%; various carbonaceous materials takes 562.11 g, accounting roughly for 25.6% [8].

The above three gunpowder formula use saltpeter, sulfur, and charcoal as the basic ingredients. Although there is still a certain gap between those formula and the modern standard military black powder, they have the functions of attacking or releasing toxic smoke, even poisoning after adding other substances, which were suitable for the needs of wars. Compared with the traditional fire arrows ignited with oil, the paste-type gunpowder fireball, wrapped with flammable materials such as the linen paper, pine resin, and yellow wax, ignited by the pyrotechnic cone and projected to the enemy camp by a lithotripter, which is not easy to extinguish. In the Northern Song Dynasty, it has a "gunpowder kiln" in Kaifeng, located in Henan, specializing in the production of gunpowder.

At the beginning of the twelveth century, the granular gunpowder with high nitrite content was developed, which improved the gunpowder's capacity for launching and explosion, and could be used to produce new types of firearms. The advantages of the granular gunpowder lie in that "the three components of granular gunpowder are in a stable state, and the gaps between the particles in gunpowder are evenly distributed, so the explosion of the granular gunpowder is more consistent, almost instantaneous exploding" [11]. The invention of the granular gunpowder can be regarded as a revolution of the gunpowder technology.

The gunpowder in the Yuan Dynasty increased the content of the nitrite. *Magical Methods of Defeating Enemies*, Yuan Dynasty Firearms Monograph, recorded thirty kinds of gunpowder formulas, introduced in its volume 10—*Firearms Drugs*. The ratios of saltpeter, sulfur, and carbon in gunpowder (propellant) were 76%:5%:19%, 78%:8%:14% for gunfire (bomb), 86%:1%:13% for line gunpowder (ignition with retarded propellant) [12]. In August 1974, a bronze bracelet from the early fourteenth century was unearthed in Xi'an, and blocks of the gunpowder remained in the bracelet chamber. Through scientific analysis and detection, the ratio of saltpeter, sulfur, and carbon in gunpowder is 60%:20%:20% [13]. Compared with the gunpowder in the Song Dynasty, the improved gunpowder in Yuan Dynasty increases the content of the salt peter, so does its purity. It thus becomes a better type of the granular gunpowder.

In addition, according to Zhou Mi's *Pre-collection of Guixin Miscellaneous Knowledge and Gunfire* in the Southern Song Dynasty (in 1280), the Yangzhou gunfire depot exploded massively. As the accident is described: "All the gunfire exploded as loud as the landslides and tsunamis… the power shook the roofs and tiles a hundred miles away… all the guards in the depot were crushed dead after the explosion, and their weapons were blown over ten miles away. The flat ground was smashed into pits, with some of them as deep as valleys. Over 200 families and its residents disappeared after the disaster" [14]. The reason for this accident started with the processing of sulfur. The sparks occurred in the process, which brought up the fire. Then the fire quickly spread to the gunpowder depot, causing a violent explosion. This accident proved that the performance of the gunpowder in the early Yuan Dynasty was much more powerful than that in the Northern Song Dynasty.

In 1384, the feudal office set up a workshop specifically for the production of gunpowder for military use; in 1395, a weapons bureau was set up, specializing in the production of gunpowder and various firearms. The formula of gunpowder at the early stage of Ming Dynasty apparently increased the content of saltpeter as is recorded in the book Fiery Dragon Classic, written by Jiao Yu (completed in 1412). Its formulas of various kinds of gunpowder (Fig. 23.11) contained saltpeter for more than 90%, such as the formula of gunpowder for attacking at night: saltpeter 149.2 g (76.9%), sulfur 7.46 g (3.9%), and fir ash 37.3 g (19.2%); the explosive gunpowder formula: saltpeter 74.6 g (91.3%), sulfur 11.19 g (6.8%), and fir ash 1.84 g (1.9%); and the flaming gunpowder formula: saltpeter 74.6 (57.1%), sulfur 14.92 g (5.7%), and the other materials as fir ash, bamboo ash (broad leaf bamboo) all together 48.49 g (37.2%). These three formulas proved that the content of saltpeter and the purity of the gunpowder increased, and the speed and the intensity of combustion were also improved.

After the mid-Ming Dynasty, the ratios of saltpeter, sulfur, and charcoal in gunpowder formulas gradually coincided with those in other countries (75% of saltpeter, 10% of sulfur, and 15% of charcoal). They were all recorded in such military books in Ming Dynasty as *Wubian* (1558), *Jixiao Xinshu* (1560), *Shenqi Pu* (1598), *Military Records* (1606), *Wubeizhi* (1621). In the book *Jixiao Xinshu of Chapter Fifteen "Weapons"* by Qi Jiguang, a military expert, had a detailed introduction to the making of gunpowder for ancient firearms (Fig. 23.12): saltpeter 37.3 g, sulfur 4.56 g, and charcoal 1.84 g; the standard formula: saltpeter 1492 g, sulfur 205.15 g, and charcoal 268.56 g. The ratio of nitrate, sulfur, and charcoal is 75.75%, 10.6%, and 13.65% [16], which basically matches the formula of the black powder under zero oxygen condition (74.84% nitrate, 11.84% sulfur, and 13.32% carbon) [17], as well as the ratio of the black powder used in modern countries (75% nitrate, 10% sulfur, and 15% carbon). The gunpowder by this formula can be fully and quickly

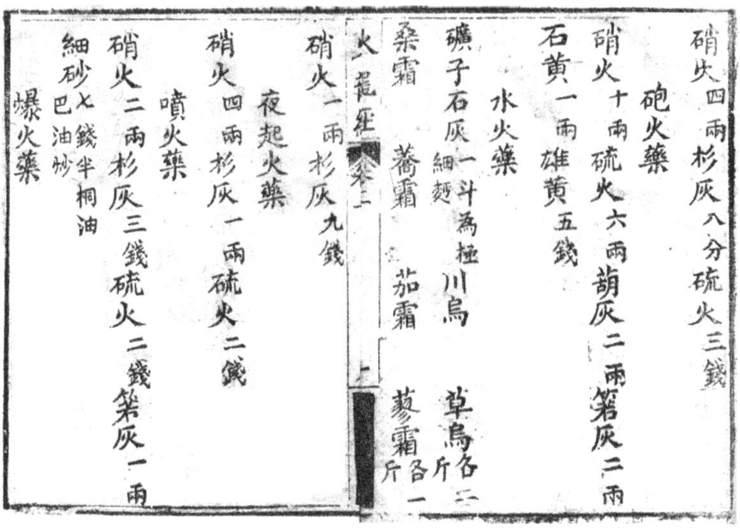
硝火四兩杉灰八分硫火三錢
砲火藥
硝火十兩硫火六兩葫灰二兩箬灰二兩
石黃一兩雄黃五錢
水火藥
礦子石灰一斗為極細麪 川烏 草烏各三斤
桑霜 蕎霜 茄霜 蓼霜各一斤
火龍經 卷上
硝火一兩杉灰九錢
夜起火藥
硝火四兩杉灰一兩硫火二錢
噴火藥
硝火二兩杉灰三錢硫火二錢箬灰一兩
細砂七錢半 桐油 巴油炒
爆火藥

Fig. 23.11 The photo of the gunpowder formula in the Ming Dynasty [15]

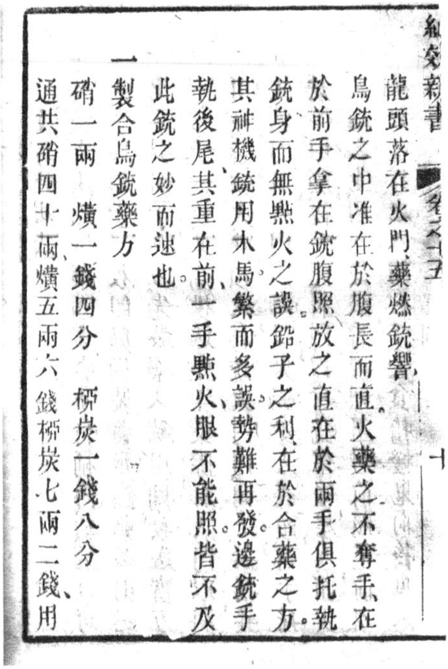
紀效新書　卷之十五　十

龍頭落在火門、藥燃銃響、
鳥銃之中准在於腹長而直、火藥之不奪手、在
於前手拿在銃腹照放之直在於兩手俱托執
銃身而無點火之誤。鉛子之利、在於合藥之方。執
其神機銃用木馬、繁而多誤、勢難再發。邊銃手
執後尾、其重在前、一手點火、眼不能照。皆不及
此銃之妙而速也。
一製合鳥銃藥方
硝一兩　熿一錢四分　柳炭一錢八分
通共硝四十兩、熿五兩六錢柳炭七兩二錢、用

Fig. 23.12 The photo of the gunpowder formula for firearms [18]

burned, as is described in the book, "with 7.46 g gunpowder placed on hand, burn it down but it would not scorch the hand, then it was ready for firearms" [18].

At the middle or later period of the Ming dynasty, the refined granular gunpowder was produced, which further improved its combustion and explosion performance. Zhao Shizhen, one of the gunpowder inventor, wrote the book *Shenqi Pu* (Magic Artifact Spectrum) therein it emphasized that the granules of the gunpowder should be evenly made, "if the granule is too pasty, it should not be used; if the granule is too delicate, it should not be used; only the granule with the size of millet is qualified" [19].

From the Song Dynasty to the Ming Dynasty, the content of the saltpeter in the gunpowder formula increased gradually from 50.6% to 75.8%. Its burning speed and power improved while the residue decreased. Because of the crystal transformation, the sulfur after being heated absorbs a large amount of heat so that the initial ignition stage of the black powder is difficult to proceed. Therefore, the content of sulfur in the gunpowder gradually decreased over time, and the combustion performance is improved.

23.3 The Gunpowder Technology Spread to the West

In the Northern Song Dynasty, *Wu Jing Zong Yao (Collection of the Most Important Military Techniques)* (completed in 1044) documented that three military gunpowder formula and Mongolian cavalry launched the third Western Expedition (1252–1260). The development and dissemination of the gunpowder and firearms lasted for more than 200 years. First, it was invented in the Northern Song Dynasty, spread to Khitan, and then to the Jin Dynasty. The gunpowder was further spread to Mongolia, then from Mongolia to Arab and European regions, and finally to the whole world. In addition to international trade and personnel exchanges, war itself had played a very important role in promoting the dissemination of the gunpowder. During the war, munitions, including the gunpowder and firearms, were often captured by the enemies, along with the factories and craftsmen who made the gunpowder and firearms, as well as the officers and soldiers who were good at using firearms to fight. Thus the technology of the gunpowder and firearms was known by the other side. In order to compete for advantages in technology, all parties had continuously imitated and improved the formula of the gunpowder and the firearms, further accelerating its dissemination and its improvement.

In the early stage of the Southern Song Dynasty, Arabs had acquired knowledge about saltpeter in Chinese gunpowder through trade, which was called "Chinese Snow" at that time. Abu-Muhâmmâd Abdullah ibn-Ahmadibnal-Baytar (1197–1248), completed the book *The Compilation of Medicines in Arabic* around 1248, which first mentioned saltpeter from China as Bārūd in Arabic [20].

With the large-scale Western Expedition of Mongolian cavalry, the technology of the gunpowder and firearms invented by China was further introduced into the Arab region. Around 1280, the Syrian Al-Hassanal-Rammâh Najmâl-Dinal-Ahdab (1256–1295) wrote *The Collection of Equestrian and War Strategies in Arabic*, which recorded the formula of the gunpowder such as "flying fire", "Chinese arrow", "Chinese spark wheel", and "Khitan spark". The "Flying Fire" formula contained saltpeter, sulfur, charcoal, Chinese iron, and other substances, of which the ratio of nitrate, sulfur, and carbon is 10:1:3; the ratio of saltpeter, sulfur, and carbon of "Chinese Arrow" is 10:1.125:2.625 [21], which was closer to the traditional Chinese gunpowder formula.

Another Arab military book, *Methods of Fire Attacking the Enemies*, published later than *The Collection of Equestrian and War Strategies*, recorded 35 fire attacking methods, of which the 13th and 33rd methods mentioned the making of the gunpowder with saltpeter, sulfur, and charcoal. For example, as in the 13th Method, the second kind of "flying fire" was made by the following method: "take 1 lb of sulfur, 2 lb of charcoal, and 6 lb of saltpeter, and grind them on the marble in powder, then put it into the cartridges as the "flying fire" and "thundering dynamite" require. Attention must be paid that the cartridges should be thin and long, and the powder should be pressed firm when making the "flying fire", but the cartridges of "thundering dynamite" should be short and thick, and only half of the powder in "flying fire" should be loaded in the cartridges. Finally tie the cartridge's two ends properly with wire" [22]. This kind of gunpowder formula was not different from the traditional Chinese gunpowder.

The Europeans first learned the power of the gunpowder and firearms during the Western Expedition of the Mongolian cavalry. In 1238, the Mongolian army attacked Moscow, Rostov and Vramir with iron artillery loaded with the gunpowder. In the winter of 1240, Badu, the Mongolian military expert, led his army to attack and occupy Qiwa (now Kiev) with the same weapon.

In 1241, during the Battle of Lèignitz, the Mongolian army defeated the Polish army under the command of Grand Duke Rick II of Krakow (1238–1241) with firing arrows by the gunpowder and fireballs with toxic smoke. The Poles did not know it was the gunpowder and believed that the Mongolian army used black magic. In churches near Lèignitz, there are murals of performing the trick of "flying dragons" by the Mongolian army, which in fact were the firing arrows [23].

Since then, the Pope and the King have sent messengers to Karakorum, the palace of Mongolia. King Louis IX of France (1214–1270) sent French Franciscans missionary Guillaume de Rubrouck (1215–1270) to Mongolia in 1252, accompanied by the Italian associate Bartolomeo de Cremona. They arrived in Helin at the end of 1253, then summoned by Montgomery of Yuan Xianzong at the beginning of 1254. After months of learning, they returned to France in 1255. Rubrouck wrote *The Journey of William of Rubruk to the Eastern Parts* in 1253. In his *Journey to the East*, Rubrouck said that he met Germans, Russians, French, British, and Hungarians who were working there during his visit to Helin, Mongolia [24].

In addition to sending missionary working as craftsmen for the Mongolian Khan, Europeans also learned about the Chinese gunpowder and firearms by translating Arab technological works, including military books, into Latin.

The first European scholar to introduce gunpowder was Roger Bacon (about 1214–1292), a famous British philosopher and pioneer of modern experimental science. In 1267, he first mentioned the loud noise caused by igniting the saltpeter in *Opus Majus*. In *Opus Tertium*, he also described the power of explosives wrapped in paper: "a special kind of salt called saltpeter, mixed with sulfur and charcoal, and sealed in a lambskin carton, when it exploded, it made such a scaring sound louder than thunder, and it flashed stronger than the brightest lightning" [25]. Bacon's knowledge of gunpowder and artillery came from two sources: one is from the European missionaries who have been sent to the Mongolian Khan Court. He had met Rubrouck in Paris and read through his books. *Opus Majus* mentioned his names. The other is from Arabic technological works. He was familiar with Arabic language as well as the culture, thus he could understand technological works in Arabic. Hermann Alemann, a German translator who translated Arabic works into Latin, had a close relationship with him [24]. Bacon even got samples of the Chinese paper cannons from missionaries returning from China and carry out experimental research on gunpowder at that time.

Another European who introduced the Chinese gunpowder was Saint Albertus Magnus from Germany (1200–1280). Being as famous as Bacon at that time, Albertus was considered to be one of the most knowledgeable men in the medieval Europe. The gunpowder and the "flying fire" formula were introduced in his work *De Mirabilibus Mundi*: take 1 lb of sulfur, 2 lb of charcoal, and 6 lb of saltpeter, and grind them on the marble in powder, then put it into the cartridges as the "flying fire" and "thundering dynamite" require. Attention must be paid that the cartridges should be thin and long, and the powder should be pressed firm when making the "flying fire", but the cartridges of "thundering dynamite" should be short and thick, and only half of the powder in "flying fire" should be loaded in the cartridges. Finally, the cartridge's two ends are tied properly with wire [26]. This method is almost the same as the formula in the 13th Method of the Arabian military book *Methods of Fire Attacking the Enemies*. Albertus did not understand Arabic, and his description of gunpowder in De Mirabilibus Mundi was directly quoted from the Latin-translated Methods of Fire Attacking the Enemies.

In the fourteenth century, the gunpowder and firearms were applied in war by European countries. In the fifteenth century, the standard formula of the granular gunpowder appeared in Europe, which had triggered the revolution in technology for military usage and the scientific revolution, even the Industrial Revolution in Britain.

References

1. [The Ming Dynasty] Xiyong Miao. 1625. Volume 3 of *Shen Nong's Herbal Classic*. A block-printed edition.
2. [The Ming Dynasty] Xiyong Miao. 1625. Volume 4 of *Shen Nong's Herbal Classic*. A block-printed edition.
3. Daozang Dong Shenbu Zhushu. *Volume 5 of Zhujia Shenpindan Law*, vol. 594, 11. Shanghai.
4. Daozang Dongshenbu Zhushu Category. 1924. *Volume 2, Lead and Mercury Jiageng Zhibao Integration*, 13 years, vol. 595. Shanghai.
5. Zhaochun Wang. 1991. *History of Firearms in China*, 9. Military Science Press.
6. Daozang, Dongshen Department, Zhushu Category. 1924. *A Brief Introduction to Zhenyuan and Miaodao*, 13 years, vol. 596, 9. Shanghai.
7. Zhaochun Wang. 1991. *History of Firearms in China*, 10–11. Military Science Press.
8. Zhaochun Wang. 1991. *History of Firearms in China*, 11. Military Science Press.
9. Gongliang Zeng and Ding Du. 1959. *Volume 12 of the Preface Collection of General Essentials of Wujing*.
10. Gongliang Zeng and Ding Du. 1959. *Volume 11 of the Preface Collection of General Essentials of Wujing*.
11. [US] T.N. Dupuy. 1985. *The Evolution of Weapons and War*. Yan Ruichi and Li Zhixing (trans.), 119. Military Science Press.
12. Jixing Pan. 2016. *History of Chinese Gunpowder* (Collection of Illustrations), 76. Shanghai Far East Publishing House.
13. Zhaochun Wang. 1991. *History of Firearms in China*, 49. Military Science Press.
14. [The Southern Song Dynasty] Meticulous: Pre-collection of Kuxin Miscellaneous Knowledge. 1986. *Cannon Disaster, Four Books of Wenyuan Pavilion*, vol. 1040, 10. Taiwan Commercial Press.
15. [The Ming Dynasty] Jiaoyu. Volume 1 of *the Fire Dragon Classic*. Ten Years (1412) of Yongle, the Ming Dynasty.
16. Zhaochun Wang. 1991. *History of Firearms in China*, 196. Military Science Press.
17. Zhaochun Wang. 1991. *History of Firearms in China*, 204. Military Science Press.

18. [The Ming Dynasty] Jiguang Qi. 1829. *Ji Xiao Xinshu Volume 15, Qing Daoguang Nine Years.*
19. [The Ming Dynasty] Zhao Shizhen. *Artifacts Spectrum, Xuanlangtang Series*, vol. 85, 29.
20. Jixing Pan. 2002. *Four Great Inventions in Ancient China: Origin, Outreach and World Impact*, 454, 447–448, 462, 464–465. China University of Science and Technology Press.
21. Hong Yang, Cheng Dong and Zhong Shaoyai. 1991. *Chinese Military Encyclopedia, Ancient Weapons Subvolume*, 117. Military Science Press.
22. Jixing Pan. 2002. *Four Great Inventions in Ancient China: Origin, Outreach and World Impact*, 454. China University of Science and Technology Press.
23. Zhaochun Wang. 2007. *History of World Firearms*, 47–48. Military Science Press.
24. Jixing Pan. 2002. *Four Great Inventions in Ancient China: Origin, Outreach and World Impact*, 462. China University of Science and Technology Press.
25. Hong Yang, Cheng Dong and Zhong Shaoyai. 1991. *Chinese Military Encyclopedia, Ancient Weapons Subvolume*, 118. Military Science Press.
26. Jixing Pan. 2002. *Four Great Inventions in Ancient China: Origin, Outreach and World Impact*, 464–465. China University of Science and Technology Press.

Zhanhong You Born in Xinhua, Hunan in 1965, Doctorate of science at the Chinese Academy of Sciences, associate researcher at the Museum of Cartography at Tsinghua University, visiting researcher at the Chinese Character Museum.

Research interests: History of Military science and technology

Major works: Participating in the writing of the second part of *History of Chinese Mechanical Engineering Invention; Chinese Military Encyclopedia: History of World War*.

Papers: President Kennedy's Scientific consultant Wiesner and American Arms Control Policy; Technical Reflection on the Failure of Beiyang Navy; On the Historical Experience of Pugwash and Its Significance

Awards: *21 Lectures on Science and Technology History* won the first prize of Tsinghua University's excellent textbooks in 2008;

In 2013, the research report *Thinking on the Development Trends of International Strategic Patterns and Countermeasures* that he co-authored won the first prize for outstanding achievements in military science of the Academy of Military Sciences.

Chapter 24
Compass

Nianzu Dai

The invention and development of compass is an integral part of societal development and human civilization. Its birth and application made possible Zheng He's voyage, European nautical expedition in sixteenth and seventeenth centuries and hence Columbus' discovery of new continent, as well as Magellen's global navigation. The New World and global sailing routes have then become the prerequisites for accumulation of capital and burgeon of capitalism during the development of Western societies, which have laid down the foundation for cultural recognitions and exchanges among world citizens.

24.1 South-Controlling Spoon

There were two ancient Chinese inventions that would bring confusion. One is south-controlling spoon, and the other is south-pointing vehicle. The former is a magnetic pointer, following principles of magnetism. While the latter is a mechanical pointer, conform to mechanics. The spoon rests on the natural pointing features of the magnetic bar, and the car, however, depends on gear pairs and automatic clutch system which requires human control. From the Han Dynasty, many books have confused the south-controlling spoon with the south-pointing car. As a result, controversies were aroused in academia.

The magnetic pointer, known as south-controlling spoon, was the archetype of a compass. It was a spoon made from a magnetic bar, which might stop in the south-north direction in its free status. Wang Chong in the Western Han Dynasty wrote in his *Lun-heng:*

N. Dai (✉)
The Institute for History of Natural Sciences, Chinese Academy of Sciences, Beijing, China
e-mail: nzdai@sina.com

J. Hua and L. Feng (eds.), *Thirty Great Inventions of China*,
https://doi.org/10.1007/978-981-15-6525-0_24

> Thus, the Quyi grass perhaps never existed and instead was a fabricated idea, or it did exist, but it was a rumor that it can help to point out an obsequious villain. Assuming that this kind of grass was in essence capable of pointing out certain people, it could be possible that it was born to move at the sight of every human. Ancient Chinese are simple-minded. At the sight of the moving grass, they claimed that it could point at people and at even brown-noses. The handle of the south-controlling spoon could also tell where the south was when put on the ground. As for maggots resided in rotten fish, they crawled northwards because this is a part of their nature. And the Quyi grass likewise had this pointing strain in their instincts. Three objects are mentioned in this paragraph, i.e., Quyi grass, south-controlling spoon, and maggots in fish. They all embrace the capacity of direction telling.

Grass shivered in front of humans and then was deemed to own the power of pointing; maggots could travel towards north; south-controlling spoon's handle would point at south. Directionality is engraved in their nature. As for the south-pointing spoon, no matter in the context of ancient China or today's common sense in science, the magnetic bar is the only object with this pointing nature. The mentioned "handle of south-controlling spoon" is named the "magnetic spoon" at present.

Wang Chong's passage reminds words in *Han Feizi-Law* written at the end of the Warring States Period:

> If a vassal attempts to invade his king's land, he would adopt the strategy to attack the nearest according to the terrain. And the king would finally lose the directions because of the constantly changing landscape. Hence, the king set up a south-controlling device to locate. Upon reading, one might realize that the device in the above passage is about the same as the spoon mentioned by Wang Chong.

Mr. Wang Zhenduo (1913–1992), an expert in cultural relics and museum studies, has stringently examined those two passages above in 1940 and 1950s [1]. In his studies, he found that the Chinese character "杓" (which was already rendered as "spoon" in English) in *Lun-heng* equaled the character "勺" (which is "spoon" in modern Chinese); and the character "柢" meant handle of the spoon; the character "投" stood for "set up"; the character "地" used by the diviners, referred to the Earth Section of the geomantic compass with a smooth panel and characters and symbols which indicated directions. According to Wang's examination, those two quoted passages were of significant value in the history of science. From then, Wang Zhenduo unveiled a new chapter of the history of compass.

Spoons found in relics are different from ones in daily use. They are dissimilar in material, size, straightness of handle, and aesthetic features (Figs. 24.1 and 24.2). When duplicating the south-controlling spoon, Wang Zhenduo used the curve handles due to his aestheticism (Fig. 24.3). The duplication has long received appreciations from audiences, which owes partly to aestheticism in it. However, demagnetization might happen during the transformation from a natural magnet to a curve magnetic spoon. Even after assembling, its magnetic induction would wane in a short time due to its curve shape. That was a fact beyond Mr. Wang's expectation back then.

For more than 20 years to date, experts have proposed various improved schemes on Wang's duplication in order to overcome the demagnetization and to retain the spoon's pointing feature. And one scholar was inspired by Wei Zhao's *Poem of Gourd Ladle* (the Tang Dynasty) and put forward an idea that "the handle of the spoon"

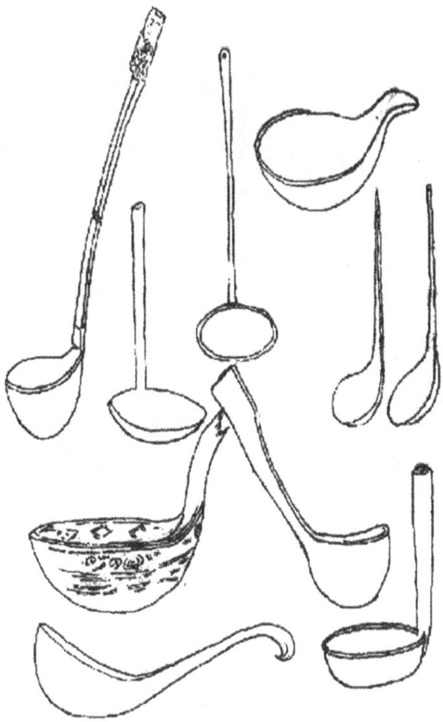

Fig. 24.1 Ancient spoons drawn by Wang Zhenduo

Fig. 24.2 A curve spoon in ancient painting (from Cultural Relic Task Force of Gansu.Excavation Report of a Tomb with Murals at Jiayuguan Pass. Cultural Relics Publishing House, 1985: Fig. 32)

stood for a gourd ladle with a lodestone inside which it could point to south when floating on a mercury tank [2].

Some scholars believed that a gourd ladle with lodestone on water or on a smooth copper plate could also function as a compass when a wooden stick was placed on

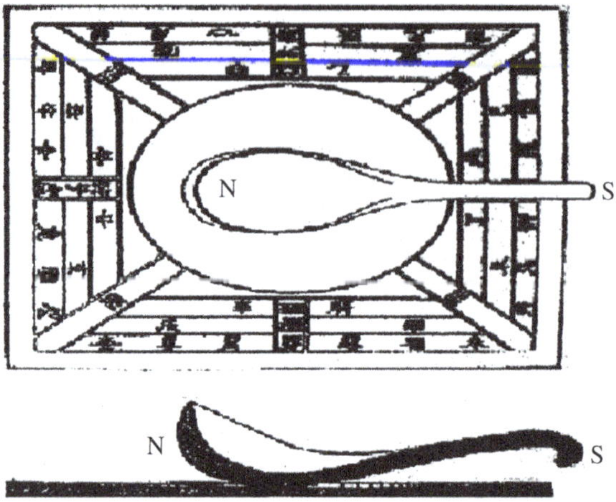

Fig. 24.3 Restored painting of the south-controlling spoon by Wang Zhenduo

the surface of the gourd ladle to show directions [3]. Other scholars argued that when making a south-controlling spoon, the magnetic spoon should be small in size and the Earth Section of the geomantic compass should be simple in design. They deemed those were basic rules for further speculation [4]. As for "handle of the spoon", it should be a small spoon with a straight handle for direction indication. In addition, the handle might be short because of the gravity issue. Also, we, the author, have pondered over the design of the spoon and thought that it just looked like a spoon, but not a real one itself. In other words, it is not necessary to crave the lodestone into a spoon, nor into Wang Zhenduo's aesthetic duplication. A slowly-carved straight spoon-shaped one (Figs. 24.4 and 24.5) is enough to maintain the magnetism and the pointing function [5]. More explorations were tried based on Wang Zhenduo's studies and his duplications.

In Fig. 24.4, ABCD is a crude magnetic bar from a natural magnet and EFG (the spoon) will be obtained after further polish. End E needs no hollow part, while Cambered Surface G should be smooth enough.

In history, there were places popular with lodestone. That is why they were called Ci Mountain (Ci is a homophone of the Chinese character which stands for magnet). In ancient China, seeking a good lodestone was no way a difficult task. In his *Herb of the Tang*, Su Gong of the Tang Dynasty had mentioned that a brand-new lodestone could attract an iron knife of 500 g. And Joseph Needham once told people that the

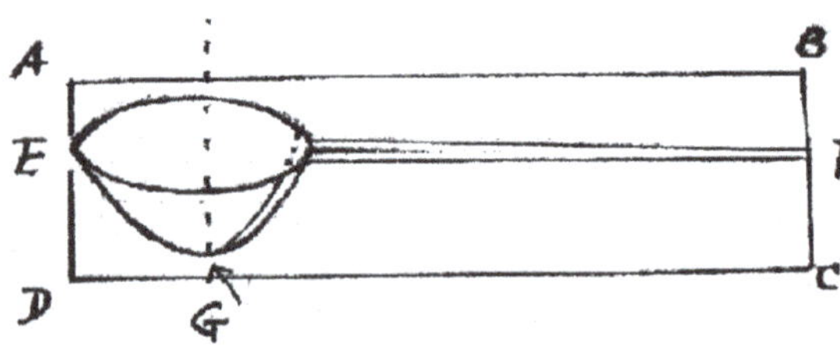

Fig. 24.4 Draft of spoon with straight handle made of natural magnet

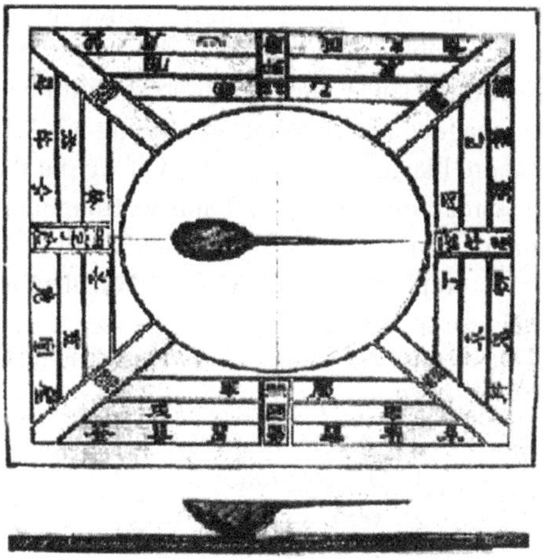

Fig. 24.5 Restoration of the south-controlling spoon with straight handle (vertical view and end view respectively)

giant magic lodestone used by J. A. Dallabella to establish the magnetic inverse-square law of 1799 in Lisbon was actually a gift to the King of Portugal from the Chinese Emperor more than a century ago [6]. A calculation of time may lead to the conclusion that the lodestone was a present from the Emperor Kangxi (reign: 1662–1722). With such a lodestone, a magnetic spoon in Fig. 24.4 could be easily obtained by processing. And of course, the extracted natural lodestone will be subject to demagnetization along with time, unless humans help to magnetize it again.

South-controlling spoons in *Han Feizi* and *Lun-Heng* were defined as a magnetic pointing spoon, which was not a groundless claim. 200 years before Wang Chong, Chinese already learnt about the magnet's ability to point direction. Zeng, a guest of Liu An (King of Huainan in Han Dynasty 179 B.C.–122 B.C.), has complied *Huainan Encyclopedia*. Even though this book was lost, *Taiping Imperial Encyclopedia* by Li Fang in the Song Dynasty had a number of quotations from it. For example:

> Lower the magnet into a well and the lost may return home.
>
> Get the girdle from the lost and use it to wrap a lodestone. Put the wrapped stone into well and the lost may find his way home. (*Taiping Encyclopedia*, volume 9: quotation 88)

Li Shizhen's *Compendium of Materia Medica*, Volume 1: *Stone* also referred to the aforementioned lines, but replaced "the lost" with "the stray". While in *Four Branches of Literature* in *Taiping Encyclopedia*, the wrapped magnet was hanged at home instead of in well. Some books in the Qing Dynasty had it as "hang in the room". "The lost" and "the stray" as quoted in these texts means those missing from home or lose their ways. Hanged free and not disturbed by the wind, a magnetic bar will stop at the north-south direction. "Hang in the well" is an attempt to simulate this condition. With this condition, the same result will be reached no matter the lodestone is hanged in room, at home or in well.

For those who could not fully understand the words in *Huainan Encyclopedia*, they tended to believe the hanged lodestone is an irregular item without edges. However, it was not the case. The necromancy in the story was to help the lost find his way home, which meant the hanged lodestone must embrace the feature of direction pointing. Hence, the lodestone should be of the bar shape. If it was nothing about the magnet's directionality, hanging a lodestone in a well would seem useless. Isn't it more like a witch's ceremony if we put the lodestone at the doorstep or the nearest crossroad from home? What is worth mentioning is that the wrap has no effect on the magnetism. Surrounding with the cloth of the lost person is no more than a deceitful tactic employed by a sorcerer. And it is also an evidence of scientific practice in disguise of necromancy.

It is said that thousands of necromancers lived in Liu An's mansion (King of Huainan) (*Book of Han, Story of King of Huainan*). The necromancers were familiar with alchemy and there were also some men of letters who were engaged in scientific research. Today's *Huainanzi* and *Huianan Encyclopedia*, also known as *Book of King of Huainan*, were said to be works of guests and necromancers in Liu's mansion. They were guests of King of Huainan while at the same time the biggest necromancer group in the Han Dynasty they happened to be, judging from the scientific knowledge embraced by their works, the largest scientific group in the Han Dynasty. Science in their "inventions" included surface tension shown in the story that a pin with dandruff in the pinprick might float on water, re-boil happened in the anecdote that the sound of thunder would be made by dropping a sealed copper urn into a well, prototype of hot-air balloon, prototype of periscope, ice lenses, electrostatic flash, attraction and repulsion of magnet, etc [7]. And the pointing feature of the magnet is one of their most important discoveries. Someone among them might go through lots of trials of hanging a magnetic bar and finally discovered its directionality, which was later wrapped with a cover of necromancy.

It is known to all that necromancy, witchcraft, and magic played the role as midwives during the birth of science in the history. The necromancers experimented with the suspended magnet and observed the result, which was exactly the cognitive basis or knowledge basis for the birth of the magnetic spoon. In this sense, it could be said that the text in *Huainan Encyclopedia* was the first evidence of the magnet's polar pointing feature [8]. With regard to the same magnet, the necromancers wrapped their discovery with alchemy, illusion, witchcraft, or magic, and he or she might be ideologically deceitful. However, they must also point out that their so-called witchcraft was an "endless extension and groundless speculation" [9]. These are not two contradictory concepts. And it is, in fact, a necessary way of thinking for the historians.

According to the Song Dynasty's remnant *Lun-heng*, the word "杓" in "司南之杓" (the handle of the south-controlling spoon) was written as "酌". Therefore, the frivolous annotation of "酌" as "行" (locomotion) finally leaded to interpreting the south-controlling device as a "south-pointing car" [10]. And the fact was, 70 years ago, scholars who studied *Lun-heng* had pointed out that "酌" mentioned was a misspell character [11]. It can be said that the "south-controlling device" in *Lun-heng* is a magnetic spoon, and so far has not been shaken by any dissent [12].

By quoting the story that Qian Linzhao was entrusted in 1952 to make a south-controlling spoon and failed, someone tried to negate the existence of south-controlling spoon. During 1980–1992, the author once worked under the Qian Linzhao's leadership and served as a secretary and a co-complier for *Encyclopedia of China-Physics*. When discussing the session of "compass" drafted by Wang Zhenduo, Qian Linzhao mentioned that he was entrusted with the reproduction of the south-controlling spoon. We not only affirmed Wang Zhenduo's important contributions to the ancient Chinese science and technology, but also discussed the possibility of future revision of reproduction of the south-controlling spoon. And Qian has never denied the claim that the south-controlling device was a magnetic spoon. In his *Encyclopedia of China-Physics*, there is a session of "South-Controlling Spoon" in the Chapter of "Compass", which is the evidence [13] we can find.

24.2 South-Pointing Needle Compass

South-pointing needle compass, in the very beginning, consisted of a magnetized steel pin or a sewing needle, whose tip might be the N pole, pointing at the south; or it could possibly be the S pole and pointed at the north. Chinese traditional culture has always favored the auspicious position of "sitting north and facing south". Therefore, regardless of pointing at the north or south, the magnetized pin was called the "south-pointing needle". And certainly, the discovery of magnetic induction must precede the invention of south-pointing needle compass.

The earlier written records of the phenomenon of magnetic induction are found in the *Special Records of Renowned Doctors* by Xiao Liangchao and Tao Hongjing (456–536):

> There is also good lodestone in the south and the suspended stone can attract needles. The lodestone which can attract 3 or 4 needles at a time is of good quality. (Li Shizhen. *Compendium of Materia Medica*, Volume 1: *Stone* from *Special Records of Renowned Doctors*).

Su Gong in Xianqing Period of Gaozong's reign in the Qing Dynasty (656–660 years) also wrote in his *Notes on Herbs in the Tang Dynasty*:

> The newly mined magnet can attract 10 needles at one time and move an iron knife of 500 g. (Tang Shenwei. *Classified Materia Medica*, Volume 4: Session of Jade from *Notes on Herbs in the Tang Dynasty*).

The steel needle is attracted by the magnet and becomes a magnet, which is called magnetic induction. Herbal pharmacologists also used magnets for medicine. Before being used for medicine, it was necessary to judge the authenticity and quality of the magnet. Hence, they finally discovered the phenomenon of magnetic induction. With this finding, the invention of the south-pointing needle compass would not be far away. Try one more step: place the magnetized steel needle on a smooth surface, or on water, or suspend it at its center of gravity and when the steel needle is stationary, it points at north or south. And now men have a south-pointing needle compass. However, it took 200 years for this step forward.

About 200 years after SU Gong, Duan Chengshi in late Tang Dynasty (about 803–863) has the words of magnet, steel needle, and floating needle in his book *Sacrificial Utensils in Youyang*. Duan Chengshi (courtesy name: Ke Gu), was from Linyi, Shandong. His ancestor, Duan Zhixuan had joined the army of Emperor Taizong of the Tang and made great contributions; his father Duan Wenchang served as the premier in Emperor Xianzong's reign at the end of Yuanhe Period (820 AD). As for Duan Chengshi himself, he was appointed as Official Collator at the Imperial Library in late Tang Dynasty and used to be the governors of Luling, Jinyun and Jiangzhou. His highest rank was the Ritual Official at the Taichang Temple. His stories were also recorded in both *New Book of the Tang* and the *Old Book of the Tang*. On the summer of the third year of the Huichang Period, Duan and his friends Zhang Xifu (courtesy name: Shan Ji) and Zheng Fu (courtesy name: Meng Fu) visited temples in Chang'an. Essays in *Trip of Temples and Pagodas* were then written by him and split into two sections, which were collected in Volume 5 and 6 of *Sequel of Sacrificial Utensils in Youyang,* respectively. The three of them invited the abbot for sightseeing and enjoyed making poems together. When they traveled to the Temple of Bodhisattva at Pingkang, they wrote "Shu Shi Lian Ju", in which two lines by Zheng Fu read:

> "Steps, touches and flags fluttering; songs, peeps and water clean."

And the abbot who styled himself "Sheng Shang Ren" followed:

> "Travel bravely with a magnetic stone [14], to avoid losing one's way in acres of dangerous vines."

During their stay at the Tian Wang Pavilion, another poem was made. In the poem, Duan's lines were:

> "We tied our horses on a pine and throw a pin into a bowl. We recall the poems of Tangshi and sang it loudly and clearly [15]."

All those lines indicate that in late Tang Dynasty, the magnets and steel needles were regarded as the travel essentials. Even if a new needle is ground with a magnet on the spot, the needle will be magnetized. With this needle placed on the surface of the water, one will know where the north/south is, which can be concluded by linking up all their lines of poem. Although there were no lines like "pointing at north" or "pointing at south" in the poem, they tied the horses beneath the pine and put a pin on a bowl of water so as to tell the direction in the wild. That is an exact description of the application of a compass in travel. "Travel bravely" might imply that the south-pointing needle compass was all but a new invention, while no trace of the positioning panel was found in those lines.

About 150 years after the demise of Duan Cheng, i.e., in the early Song Dynasty, the knowledge of magnetism became popular overnight, and many documents of the geomantic compass or south-pointing needle compass were produced almost at the same period.

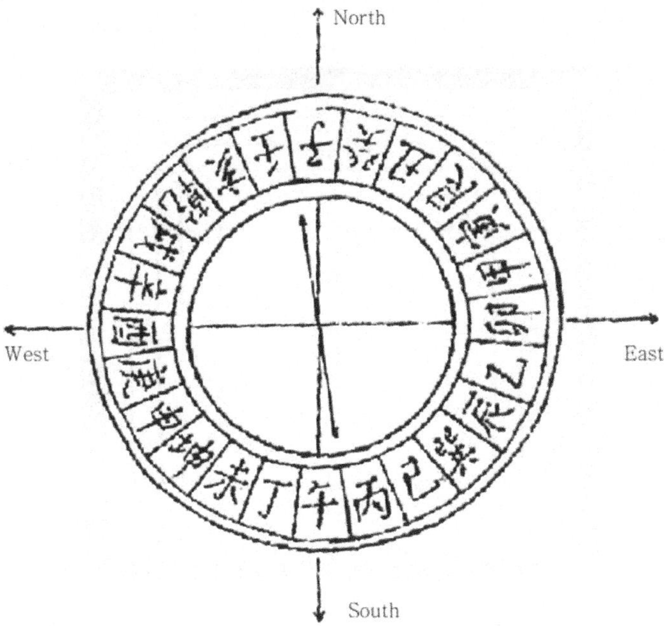

Fig. 24.6 The assumed geomantic compass used by Yang Weide

The astronomer and diviner Yang Weide (circa 987–1056) wrote *A Record of Tomb* on the first year of the Qingli Period (1041 AD), and the remnant of this book is still kept in the National Library of China. One can find in the book:

> If one wants to determine where the east, west, north, or south is, one should put the pin point at Bing Wu direction, and when the pin points at the middle between Bing and Wu, we get the exact direction of south. Hence, the middle points between Bing and Wu as well as between Ren and Zi indicate the direction of south and north respectively (*The Record of Tomb,* Volume 1: *About Main Mountain*). The aforementioned words are about the methods to differentiate the geographical direction by the geomancers (Fig. 24.6).

Bing Wu and Ren Zi are characters engraved on the positioning panel for direction telling. This panel of position might probably be the Earth Section of the geomantic compass used by the diviners for thousands of years. The combination of the magnetic needle and the positioning panel could be said to give birth to the magnetic compass or compass, in short, in the history of science and technology. Yang Weide has long been an official in the Imperial Directorate of Astronomy and Calendar and was once the observer and recorder of the famous supernova in 1054. In his later years, he was engaged in the Feng Shui telling of the tomb [16]. In almost the same Period, the geomancer Wang Ji (around 988–1058 AD) wrote the *Poem about Pin Reading*:

> Between Wei and Xu, the exact south lies at the upper 3 units of the Zhang. If one knows nothing about the Kan and Li, and huge deviations may happen, which leads to failure in Fung Shui telling.

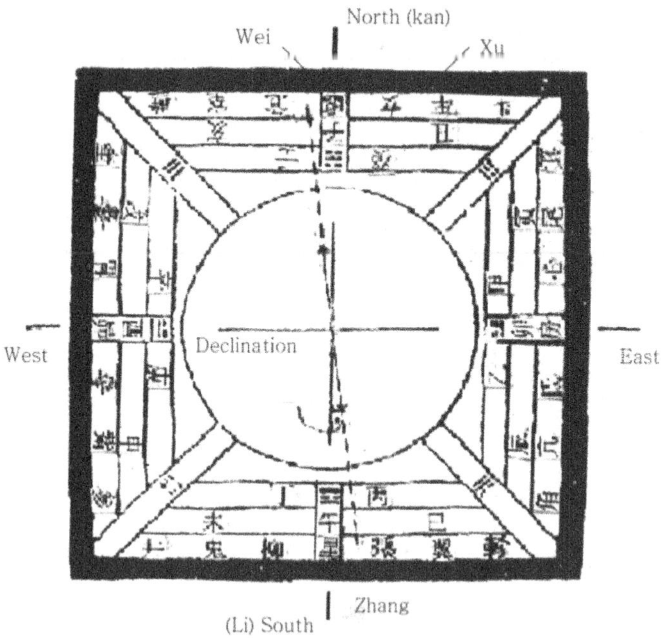

Fig. 24.7 The assumed geomantic compass used by Wang Ji

Those lines were collected in the *Ancient and Modern Books Integration—Art Session*, Volume 655: *Geomancer*, which was quoted from *GUAN's Book on Geography—Middle*. Wang Ji had annotated the book therein he wrote the precious poem. The compass mentioned in the poem and its characters on pane are shown in Fig. 24.7. Xu, Wei and Zhang are all the names of constellations and used for direction differentiation: While Xu indicates the direction of north, Wei lies in the west of Xu and Zhang is a constellation in the southeast. And Kan and Li are pointing at north and south. The magnetic needle pointing at the middle between Xu and Wei is about the same of Yang Weide's claim that aiming at the midpoint between Bing Wu and Reng Zi. That is because the magnetic declination in the capital Kaifeng was approximately 7.5°. The works of Yang Weide and Wang Ji also proved the invention of the compass and the discovery of the magnetic declination. They also showed that the compass was an indispensable observation instrument in the hands of the Feng Shui master. Bowl, water, and floating needle in Duan Chengshi's work in late Tang Dynasty could help one to presume that the compass of the early Song Dynasty might be a water compass.

One more evidence showing the popularity of the compass and compass knowledge in early Song Dynasty is the book *Session One* of *Wu Jing Zong Yao* (*The Collection of Books for Warriors)* completed by the military theorist Zeng Gongliang (999–1078) in 1040–1044, in which (Volume 15) another way to make a compass was presented. Zeng named it "south-pointing fish" in his book. Heat a piece of iron

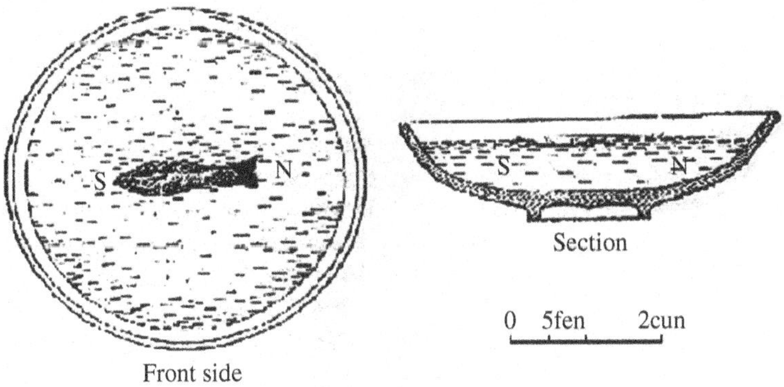

Fig. 24.8 Zeng Gongliang's south-pointing fish and the floating method by Wang Zhenduo

that has been cut into a fish shape, and then quench it in a north-south direction. After cooling, the iron piece becomes a magnetic "south-pointing fish" (Fig. 24.8). This was the second method to obtain magnetic induction since Tao Hongjing: earth induction. That is to say, in addition to using a magnet to induce a steel needle for a compass, it is also possible to use a geomagnetic field to magnetize a piece of steel and form a compass.

About 50 years after the sudden explosion of the compass knowledge, the scientist Shen Kuo (1031–1095) in the Song Dynasty summarized his understanding of the manufacture, installation of the compass and the knowledge of magnetic declination in his book *Dream Pool Essays*. He wrote:

> The geomancer grinds a pin with a lodestone and then the pin can point at south. However, it is not always the case and it usually sways toward southeast. The magnetic needle may wobble on the water, at hands or on the rim of a bowl for the reason that it moves fast. And it may slip and drop from a hard surface. The hanging method may have the best performance, i.e., when taking the best strand of newly-produced cocoon silk, applying the wax on the strand with the seeds of mustard, tying it in the middle of the needle, and hanging it in the windless place, the needle always points to the south. Some needles may point north, while others stop at south. (*Dream Pool Essays*, Volume 24: *Essays One*)

And those descriptions resemble a piece of today's scientific report. The record involved not only the pointing feature of a magnetic needle, but 4 ways to form a compass: floating on water, placing on nails, or putting it on the rim of a bowl, and hanging with a string. Among them, the floating method is "making a needle go through the flame and place it above water" (Kou Zongshuang. *Augmented Materia Medica*, Volume 5: *Lodestone*), while the hanging method was adopted by the European for instrument manufacture in eighteenth Century [17]. It should be especially pointed out that there was little written description of the non-floating compass in the historical documents, but it was also the invention of the Chinese.

The "Porcelain Figurine of Immortal Zhang" [18] excavated from the tomb of the Song Dynasty in Linchuan, Jiangxi holds a geomantic compass in his right hand. The needle, box, and scale on compass were clear and easy to identify (Figs. 24.9

Fig. 24.9 Porcelain Figurine of the Immortal ZHANG excavated from a tomb of the Song Dynasty at Linchuan, Jiangxi

and 24.10). The occupant of the tomb was Zhu Jinan, who passed away in the third year of Qingyuan Period (1197 AD) and was buried next year. It is evident that the non-floating compass in the late twelveth century was a common tool of the Feng Shui masters.

The compass manufacturing was further innovated when the non-floating compass was invented or even earlier. *The magic of the Immortal* (Chen Yuanliang. *Daily Encyclopedia*, volume I) has recorded that putting a magnetic bar into a wooden fish or turtle contributes to fashioning a new type of the south-pointing fish or turtle (Figs. 24.11 and 24.12). While the fish compass could be used as a floating compass, the turtle one could function on land.

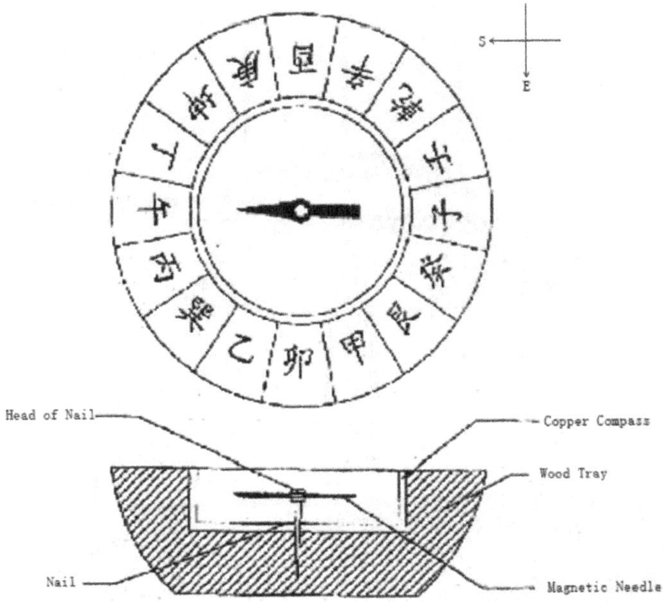

Fig. 24.10 The restoration painting of the non-floating geomantic compass of Porcelain Figurine of the Immortal Zhang (Pan Jixing. The Four Great Inventions of Ancient China, Their Origin, Development, Head of Nail Spread and Influence. World Press of University of Science and Technology of China, Nail 2002: p. 346, Fig. 232)

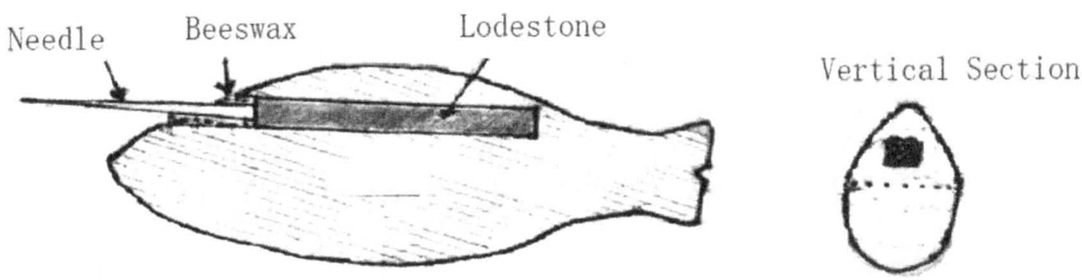

Fig. 24.11 Wooden south-pointing fish by Chen Yuanliang

24.3 Geomantic Compass and Navigation

As mentioned earlier, to form a geomantic compass is to place the south-pointing needle on a positioning panel (also known as the panel of dial), which has a developmental history of its own. The geomantic compass used by the diviners in the Qin and Han Dynasties (Figs. 24.5 and 24.7) is the same as the positioning panel. If a circular hollow for water is installed at the center of the compass, it will become a floating compass. However, the early design of the positioning panel was mostly square, an inconvenient shape for one to observe the magnetic needle. Later in around the

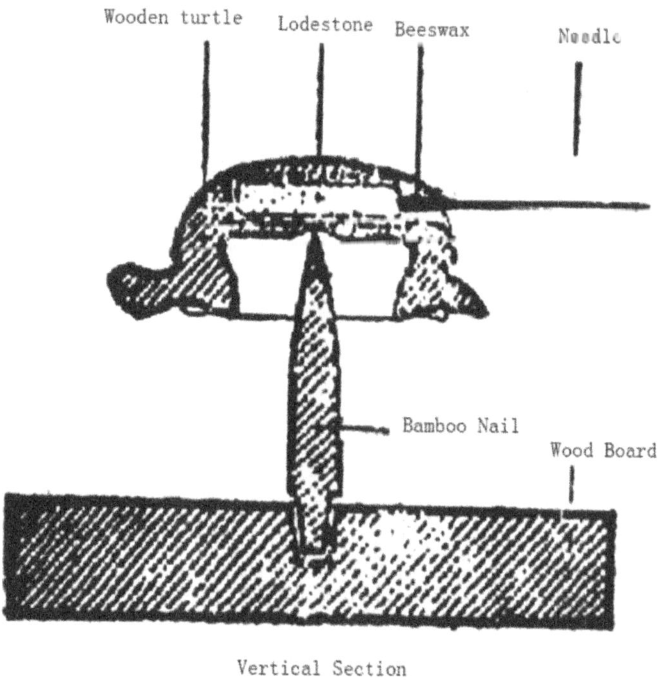

Fig. 24.12 The wooden south-pointing turtle (Vertical Section) by Chen Yuanliang

middle and late tenth century, the positioning panel was adapted to a circular design. At the beginning of the Song Dynasty, the military theorist Xu Dong (1980-1011 AD) has compiled his *Book of Huqian* in the first year of the Jingde Period (1004 AD). In his book, an article called "Augury for Birds" was illustrated with a square panel of position engraved with Eight Diagrams and with a circle of the Twelve Earth Branches for direction telling (Fig. 24.13). The "Augury and Birds" was extracted from the classic books of the Sui and Tang Dynasties. That is the reason why the circular panel of position was a product of either military strategists or geomancers from the Tang Dynasty. As shown in Fig. 24.13, the circular panel of position and the south-pointing needle can be combined into a geomantic compass, and the direction telling will become simple and clear.

The development of the twelve-direction (Earthly Branches) compass into a twenty-four one owes definitely to the geomancers' credit. The geomancers combined 12 Earthly Branches, 8 Heavenly Stems, and 4 Hexagrams (also known as 4 dimensions) to form twenty-four directions, each of which was 15° (Fig. 24.6). And the geomancers' geomantic compass complicated itself gradually. In *Book of Cape by Fairy*, there was an illustration called Panel of Floating-Needle Compass (Fig. 24.14), an augmented geomantic compass designed by the geomancers in the Song Dynasty. From the inside outlays 5 rings: the inner empty circle is actually the hollow for the

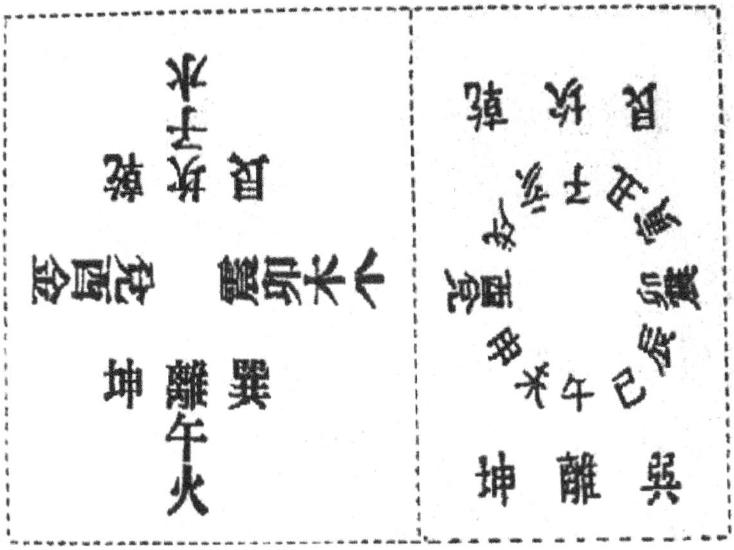

Fig. 24.13 The eight diagrams square geomantic compass (left) and the twelve earthly branches round the geomantic compass (right) drawn by Xu Dong

Fig. 24.14 Panel of the floating-needle compass

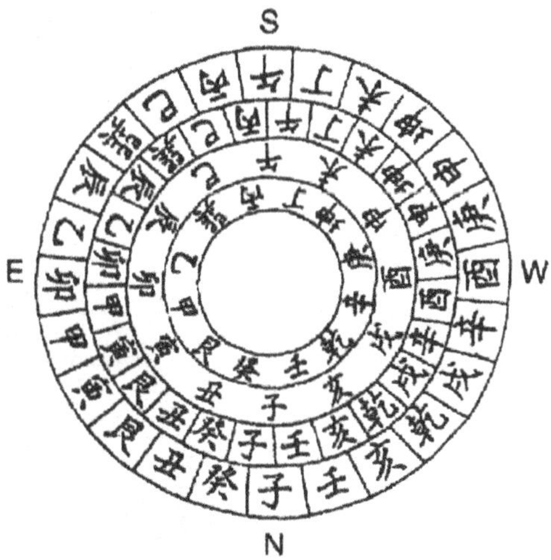

floating needle; the second ring is engraved with twelve directions, i.e., a combination of 8 Heavenly Stems and 4 dimensions; twelve ways are marked on the third ring, which is represented by the 12 Earth Branches. This layer is consistent with the earlier paint of the circular geomantic compass by Xu Dong; as for the fourth and fifth, they indicate 24 directions, a result of combining the second and the third rings. This geomantic compass was probably a portrayal of the history of early compass development. The fourth and fifth rings differ by 7.5°, which may be the testimony of the magnetic declination of the capital in the Northern Song Dynasty. Since then, the geomantic compass has become more complicated and more mysterious and has been regarded as the *Book of Heaven* in the realm of geomancy.

The compass for navigation is relatively simple and clear (Figs. 24.15 and 24.16). Interestingly, in addition to the geomantic compass which was handy in position telling, it was possible that human once used porcelain bowls and plates as part of the geomantic compass in the early days of navigation. Excavation of such objects demonstrated that the ancient Chinese might emboss the compass scale and needle placement at the bottom of the plate (Fig. 24.17) or people would paint in the bottom of the bowl a magnetic needle as well as rushes, and at the rim of the bottom there

Fig. 24.15 The floating-needle compass of voyage in the Ming Dynasty

Fig. 24.16 The non-floating-needle compass of voyage in the Ming Dynasty

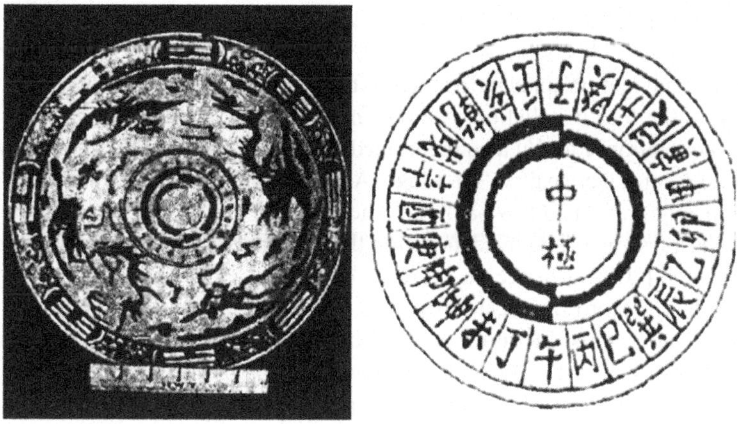

Fig. 24.17 The porcelain geomantic compass in the Ming Dynasty. On the right is the glaze-painted center part of the compass

was a character of Zhen (Fig. 24.18) [19]. The latter bowl would be connected to a cardboard or sheet of wood with directions on it to form a floating-needle geomantic compass. During navigation, there were a large number of such bowls and plates in the cabinet on a ship for immediate replacement or urgent use. And they could also be sold or gave away to foreigners in the ports of different countries.

The earliest written records of the compass for navigation can be found in Zhu Yu's (1075–1140) *A Record at Pingzhou* in the first year of the Xuanhe Period (1119 AD). There are lines in the book:

> Sailors have their ways to know about position: at night, they observe stars; during the day, they study the sun; on cloudy days, a compass will be used. (*A Record at Pingzhou*, Volume 2).

The author's father, Zhu Fu, served in the Yuanfeng Period (1078–1085 AD) as a lieutenant in Laizhou and Runzhou, and later as the commander of Guangzhou.

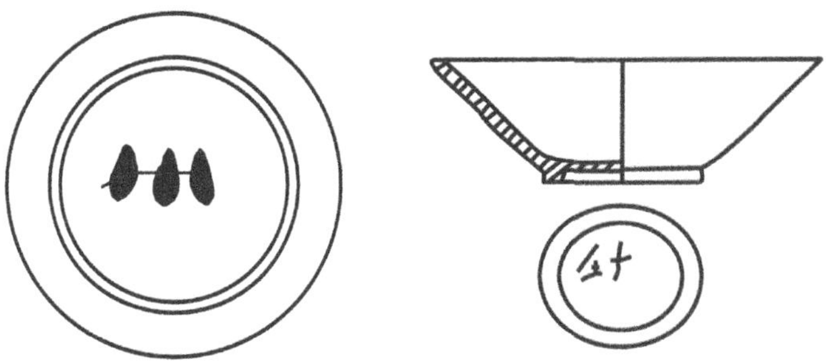

Fig. 24.18 The compass bowl in the Yuan Dynasty (on the left is the glaze painting of the magnetic needles and rushes at the bottom of the bowl; on the right is the side view of the bowl and the character of Zhen at the bottom)

The descriptions of the foreigners' residential areas as well as the docks and ships of Guangzhou in Zhu's book were mostly from his father's narration. Therefore, the compass for navigation appeared as early as the late eleventh century, and it was about the same as the four installation methods in Shen Kuo's *Dream Pool Essays.* Apart from this quotation, Zhu Yu also mentioned that the maritime regulations issued by the imperial court were called "Jia Ling". At the beginning of the twentieth century, when someone translated this paragraph, he rendered the "Jia Ling" as the name of an Arab businessman. The misconception that the Arab first used magnetic needles for navigation [20] was hence mongered.

After Zhu's record, applications of the geomantic compass in navigation were often the content of the literature. In the 5th year of the Xuanhe Period in Northern Song Dynasty (1123 AD), Xu Wei was sent on a diplomatic mission to Koryo, and used a south-pointing needle compass during his voyage (*Collections of Paints of the Mission to Koryo at Xuanhe Period*, Volume 34); it was the first year of the Baoqing Period (1225 AD) that Zhao Rushi described a navigation in the South China Sea in his essay "A Record of Foreign Countries" therein paragraphs of using a compass for direction differentiation were included. Wu Zimu has depicted the voyage of the Zhejiang merchant in his *Book of the Dream* in the 10th year of the Southern Song Dynasty (1274 AD):

> Zhejiang embraces passages connecting the river and the ocean, and there are ships of merchant in various sizes. The larger ones have a capacity to hold 5000-unit goods and around 500–600 people; the middle ones 2000–1000-unit goods and 200–300 passengers; the rest of them may have 8 or 6 oars and can accommodate a hundred sailors. ... Sailing off land, the ship enters the endless ocean and dangers lurk, the dragons and evil spirits. The sky turns dark whenever storm breaks, and people on ship can only navigate by a compass and clutch at a torch for lighting. No one can afford a mistake on reading the compass, because it is concerned with a ship of human lives. I had heard lots of detailed stories of this kind from some rich merchants. (*Book of Dream*, Volume 12: *River, Ocean, Boats and Ships*).

It is said that when ships enter the South China Sea and travels through the "Seven Oceans" (that is the eastern waters of today's Hainan Province), the weather is usually unpredictable. Rain pours suddenly, stiff wind roars and giant waves soars, which is extremely scary. Water around the reef is by no means deep, in which one often witnesses ship wreck. The only lifeline is a south-pointing needle compass. Any mis-operation may lead to ending up in the fish's belly (*Book of Dream*, Volume 12: *River, Ocean, Boats and Ships*).

The "south-pointing needle compass" mentioned here refers to the "geomantic compass". A compass is called "Di Luo" in ancient books (Zeng Sanyi. *Yin Hua Novels Collection*) therein "Luo" was identical to one of the characters in "Luo Pan" (geomantic compass) or someone named it "Luo Jing"; it was also known as "Luo Xing" or "Zhai Jing", "Zhai Jin" in mouth of geomancers [21]. Despite of the different names, they are the same as the modern compass. There were sophisticated canal system for the East China Sea and the southeast coast in Yuan Dynasty. Grains and cloth in the south were transported to the north all year round. And directions along the route were marked one by one in the nautical atlas, which led to the emergence of the nautical maps like *Maps of Nautical Passages*. Zhou Daguan (about 1275–1346 AD) was sent to Zhenla (today's Cambodia) during the first year of Yuan Zhen (1295 AD) and created *Life in Zhenla*. His voyage from Wenzhou to Zhenla was minutely narrated between covers. During the Yongle Period of the Ming Dynasty (1403–1424 AD), the navigator Zheng He (1371 or 1375–1433 or 1435 AD) had made 7 voyages. His huge fleet went deep into the South China Sea, traveled through the Straits of Malacca, sailed to the East Indian Ocean and finally reached the Lion Kingdom (today's Sri Lanka). And then he set forth for the Indian Ocean until arriving at the east coast of Africa and the entrance of the Red Sea. Until 1497 AD, the Portuguese navigator Vasco da Gama (circa 1460–1524 AD) has successfully bypassed the Cape of Good Hope in Africa and disembarked in India. The south-pointing needle compass or geomantic compass invented by Chinese has contributed to blazing the trail toward Indian Ocean and to laying a solid foundation for the European discovery of Cape of Good Hope. Half of the sea routes connecting the East and West through the southern tip of Africa were opened by the Chinese.

Knowledge about the magnetic needles and compasses was disseminated from China to Europe by the Arabs. Metaphorically, once the compass is on board, it spreads at the same speed as that of the ship. A British man named Alexander Neckam (1157–1217 AD) was the first European who mentioned the use of the magnetic needles for nautical navigation in 1190. It was a century after Zhu Fu's and Zhu Yu's narrations of the nautical compass. When it comes to the earliest designer of the European compass, the story has it that the designer was a Frenchman named Petrus Peregrinus of the thirteenth century, completed his design in 1269; but it is said that this credit should be given to Flavio Gioja of Naples, an immigrate of southern Italy and the invention date should be 1302 [22].

What needs to be pointed out is that the manufacture of the compass was quite prosperous in the Ming and Qing Dynasties. After nearly a hundred years of changes, some areas in provinces such as Jiangxi, Fujian and Anhui still keep manufacturing the traditional compasses. Among them, the production techniques and methods of the Wan'an Compass in Xiuning, Anhui (Figs. 24.19 and 24.20) were selected as items of "National Craftsmanship" in 2006 and were listed as well as protected in the category of National Intangible Cultural Heritage.

Fig. 24.19 The modern geomantic compass made in Wanan, Xiuning

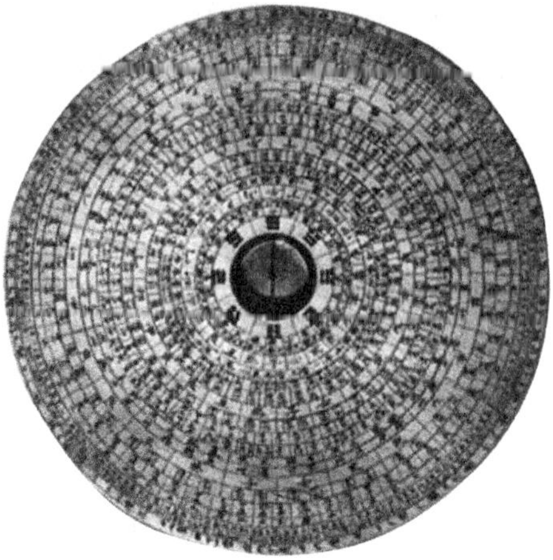

Fig. 24.20 The modern geomantic compass made in Wanan, Xiuning

References

1. Zhenduo, Wang. 1948–1951. South-controlling spoon, South-pointing compass and geomantic compass. In *Acta Archaeologica Sinica*, vol. 3, 4, 5; and Zhenduo Wang. 1989. *Archaeometry*, 50–218. Cultural Relics Publishing House.
2. Jinguang, Wang, and Wen Renjun. 1986. New research on South-controlling spoon in *Lun-heng* and the restoration. *Literature and History* 31: 25–32.
3. Zhichao, Li. 2004. Further discussion on South-controlling spoon. In *Yellow River Cultural Forum*, vol. 11, 69–77. Shanxi People's Publishing House.
4. Jixing, Pan. 2004. Source of compass. In *Yellow River Cultural Forum*, vol. 11, 16–67. Shanxi People's Publishing House.
5. Nianzu, Dai. 2004. South-controlling spoon and geomantic compass. In *Yellow River Cultural Forum*, vol. 11, 82–110. Shanxi People's Publishing House.
6. Needham, Joseph. *Science and Civilization in China*, vol. 4, Part 1: p. 234. Chinese version translated by Lu Xueshan et al. *Science and Civilization in China*, volume 4 and book 1. Science Press, 2003: p. 220. It is said by the author's colleague Mr. Han Qi that he has seen this lodestone twice in 1995 and 2014 respectively, and has affirmed that this lodestone was a gift from Emperor Qianlong.
7. Zhenhuan, Hong. 1983. Huainan encyclopedia and the physics in it. *The Chinese Journal for the History of Science and Technology* (3): 31–36; Zhichao Li. 1995. Huainan encyclopedia and its physical values; Zhichao Li. 1995. *Framework of the Chinese Technological History*, 325. Henan Education Press.
8. Nianzu, Dai. 2004. South-controlling spoon and geomantic compass. *Yellow River Cultural Forum*, vol. 11, 82–110. Shanxi People's Publishing House.
9. Nianzu, Dai (editor-in-chief). 2001. *Science and Civilization in China-Physics*, 407. Science Press.
10. Sun, Kong. 2005. About South-controlling spoon and South-controlling jade plate. *Journal of National Museum of China* 4: 4–11.
11. Hui, Huang. 1990. *Lun-heng and Endnotes (Collection Version)*, 759–760. Zhonghua Book Company.
12. Nianzu, Dai. 2014. Magnetic pointer: South-controlling spoon and discussion with Mr. Sun Ji. *Studies in the History of Natural Sciences* (4): 385–393.
13. Zhenduo, Wang and Lin Wenzhao. 1998. Compass, *Encyclopedia of China Physics*, 1232. Encyclopedia of China Publishing House.

14. Chinese characters "CI ZHEN SHI" (Lodestone) was rendered as "ZHAN ZHEN SHI" in Zhao's print in Changshu in the reign of Wanli of the Ming Dynasty. Character "Zhan" was misspelled in Zhao's version, and there was no such a thing as "ZHAN ZHEN SHI". In Mao's *Jin Dai* at the end of the Ming Dynasty and the beginning of the Qing Dynasty and in Zhang's *Xue Jin* in the reign of Jiaqing in the Qing Dynasty, character "Zhang" was corrected as "Ci".
15. The first character Lang should be Zhu (help).
16. Xin, Bai, and Wang Luoyin. 2013. Comments on Yang Weide and his scientific achievements. *Studies in the History of Natural Sciences* 2: 203–213.
17. Nianzu, Dai. 2002. *A Series of Books on History of Physics in China-A History of Static Electricity and Static Magnetism*, 180–181. Hunan Education Publishing House.
18. Dingrong, Chen and Xu Jianchang. 1988. The tomb of the song Dynasty in Linchuan, Jiangxi. *Archaeology* (4): 329–334.
19. Zhenduo, Wang. 1989. About the magnetic needle painted in the porcelain of the Yuan Dynasty. *Journal of National Museum of Chinese History* (1): 73–79; Zhenduo Wang. 1989. *Archaeometry*, 229–237. Cultural Relics Publishing House.
20. Jixing, Pan. 2012. *On the History of Scientific and Technological Exchange between China and Foreign Countries*, 133–143. China Social Sciences Press.
21. Zhenduo, Wang. 1948–1951. South-Controlling Spoon, South-Pointing Compass and Geomantic Compass. In *Acta Archaeologica Sinica*, vol. 3, 4, 5; Zhenduo Wang. 1989. *Archaeometry*, 50–218. Cultural Relics Publishing House.
22. Cajori, F. 2002. *History of Physics*, 20–21. Translated by Dai Nianzu and proofread by Fan Dainian. Guangxi Normal University Press.

Nianzu Dai Born in Changding, Fujian in 1942, graduated from the Physics Department of Xiamen University and is now a researcher in the Institute of Natural Science History, Chinese Academy of Sciences. He used to be the executive director of the Chinese Society of Science and Technology History, chairman of the Physical History Professional Committee, and director of the Institute of Natural History of the Chinese Academy of Sciences.

Research interests: History of Chinese Physics

Major works: *History of Chinese Physics, History of Chinese Phonology, History of Chinese Mechanics*

Awards: The Second Prize of Natural Science of the Chinese Academy of Sciences.

Chapter 25
The Technology of Ultra Deep Drilling

Jixing Pan and Xiaowu Guan

China has a well salt mining history of more than 2,000 years, which can be roughly divided into three stages: The first one is called the large-diameter shallow well stage, from the end of the Warring States Period to the Mid-Northern Song Dynasty (256 B.C.–1041 A.D.) featuring shallow wells, dug by labor, with a diameter of 2–9 m and a depth of 20–250 m. The second is regarded as the bamboo salt well stage or the small-diameter deep well stage, from the Mid-Northern Song Dynasty to the Mid-Qing Dynasty (1041–1815). A bamboo salt well was usually about 30 cm wide in diameter and 100–800 m deep, from which salt and natural gas as well as oil can be mined at the same time, and it is drilled in the rope percussive drilling method with an iron drill. Last but not least, the comprehensive improvement stage from the mid and late Qing Dynasty to the 1920s (1815–1920), characterizing salt wells and natural gas wells, with 30–32 cm in diameter and 800–1,200 m deep. In this stage, people were still adopting rope percussive drilling method to drill into the underground rock formation. The biggest improvement at this period was that the drilling jar invented in the Ming Dynasty was attached on drilling tools. Workers used the drilling tools with a jar to drilled more than 1,000 m deep. The drilling tools with a jar made China's drilling technology to a new height. Since the 1920s, the modern mechanical drilling stage has begun, which is beyond the scope of this article.

With a focus on the development of salt well drilling technology and its introduction to the west, this chapter describes the features, value, and influence of deep well drilling technology as a great invention in ancient China.

J. Pan (✉) · X. Guan
The Institute for History of Natural Sciences, Chinese Academy of Sciences, Beijing, China
e-mail: baojie@mfa.gov.cn

J. Hua and L. Feng (eds.), *Thirty Great Inventions of China*,
https://doi.org/10.1007/978-981-15-6525-0_25

25.1 The Early Drilling Technology for Digging Large-Diameter Shallow Wells

In 1974, a copper shaft of the Warring States Period (400 B.C.) was excavated in Tonglvshan in Daye, Hubei, with 50 m deep and a square mouth, 1.1–1.3 m wide, which has solved downhole operation problems such as ventilation, drainage, excavation and lifting, and lighting [1]. Sichuan is the birthplace of the salt well. People in Sichuan began to produce salt from wells at the end of the Warring States Period (300 B.C.), with the aid of ancient mining and shaft sinking technology.

According to *Records of Huayang Kingdom* [2], written by Chang Qu in the Jin Dynasty and *The Historical Records*, Emperor Zhaoxiang of the Qin Dynasty (306–251 B.C.) made Shu (Sichuan's ancient name) an important base wherein he controlled the floods caused by flooding of the Minjiang River (a tributary of the Yangtze River) and appointed Li Bing as the governor of Shu. During Li Bing's tenure, he presided over the construction of Dujiangyan Irrigation System, a large irrigation project, which irrigated tens of thousands of hectares of farmland and helped promote the development of agriculture. In addition, wells were sunk to get salt in Guangdu (now Shuangliu, Sichuan). All these efforts made Shu a land of abundant resources. According to the development of well drilling technology at that time and the later salt well drilling technology, the salt wells at the time of Li Bing were possibly large-diameter shallow wells.

Salt mining technology was further developed in the Han Dynasty, leading to the expansion of salt area and the increase of salt output. According to *Records of King Shu*, written by Yang Xiong in Western Han Dynasty, dozens of salt wells were mined in 3 years during the reign of Emperor Hanxuan [3]. According to *Huo Zhi biography* in *History of the Early Han* by Ban Gu, Luo Bao, a Chengdu native, ran business by using Pingling (now Xianyang, Shaanxi) people's capital and commuting between Ba and Shu during the reign of Emperor Cheng and Emperor Ai. He accumulated thousands of gold and became a millionaire in several years [4]. This shows that in the Western Han Dynasty, some people became rich by operating salt wells to increase the value of their capital.

Portrait bricks [5] (Fig. 25.1), found in the tombs of the Eastern Han Dynasty in Chengdu and Qionglai County in the 1950s, vividly show the salt wells and the working scene of salt miners. The salt well, shown on the left, has a high wooden frame at its top with a pulley attached to rotate the bucket up and down in the well. The frame has two floors, each with two people standing opposite to each other. The two people on the right side pull the empty bucket down into the well to fill brines, and the two on the left lift the bucket filled with brines. Then brine is poured into the tank on the right of the well and flows through the bamboo pipes to the five pots next to the salt boiling hearth. One uses a fan to help the fire burn stronger in front of the hearth with the chimney behind. Another two with salt bags on their back walk down the mountain and take them to the warehouse. In the picture, there are also two men shooting wild animals in the mountain, showing that the location of the salt well is in a remote mountain area. The well can hold two persons working

Fig. 25.1 The portrait brick of salt wells in 100 A.D. unearthed in Qionglai County, Sichuan (stored in Sichuan Provincial Museum)

at its bottom. They arrive there by sitting in a bamboo basket attached by the pulley above the well. Then they use cone, spades, shovels, chisels, and other iron tools to smash rocks and dig and remove the detritus out of the well. After repeated digging and removing, the well gets deeper and deeper. They won't stop digging until they reach the saline water layer. Lime, sand, loess, and glutinous rice paste are mixed to make a kind of cement called trinity mixture fill, which are then applied around the wall of the well. After that, thick boards or slabs of stone are used to prevent the wall from collapsing. The early salt wells since the times of Li Bing may probably be dug in the above-mentioned processes.

As for the depth of the ancient salt wells, according to *Tai Ping Guang Ji*, written by Li Fang in 978, the salt well in Lingzhou, Sichuan, dug by Zhang Daoling in the later Han Dynasty, was 9.2 m wide and 124.2 m deep [6]. Moreover, according to *Yuan He Jun Xian Tu Zhi* written by Li Jifu in 813, the Ling Well in Renshou County, Sichuan was 93 m in diameter and 257 m deep [7], a gigantic one that outperformed the other salt wells in both size and depth.

As time went by, the limitations of the drilling technology for the ancient large-diameter salt well became more and more prominent. First of all, though the input of labor and capital was high, the concentration of the extracted brines was low, leading to increased workload of the salt boiling workers. Second, the drilling was so slow that it often took more than 10 months or even several years to finish one project. Due to overexploitation, the gradual depletion of the shallow underground brines made it necessary for the brine producers to dig deeper to get more brines.

However, the existing well drilling technology ended up being ineffective. Coupled with unfavorable salt policy since the late Tang Dynasty and the Five Dynasties, salt production in large-diameter well began to wane since the tenth century.

25.2 The Bamboo Salt Well Drilling or the Small-Diameter Deep Well Drilling

The second stage of salt well development started in the Northern Song Dynasty, from 960 to 1126. The small-diameter deep well drilling method was invented in Si Chuan in the middle Northern Song Dynasty, from 1016 to 1071, also known as the bamboo salt well drilling technology.

The expression of the bamboo salt well was first seen in a memorial to the throne presented by Wen Tong of the Northern Song Dynasty [8]. This newly invented bamboo salt well drilling technology in Jingyan County, Lingzhou, spread rapidly during 1041–1048. The bamboo salt wells were easily covered by the salt producers who can then avoid salt tax. Moreover, they also employed refugees to work for them. During the reign of Emperor Shenzong, from 1068 to 1077, Wen Tong, as the magistrate of Lingzhou, sent a memorial to the emperor and solicited aid to strengthen supervision over salt production in Jingyan County. Su Shi, a distinguished poet, makes supplemental statements on the shape and structure of the bamboo salt wells and their drilling methods [9]. Therefore, according to their description, the bamboo salt well drilling technology first appeared during the reign of Emperor Shenzong, from 1068 to 1077, in Jingyan County to the south of Chengdufu Road. During the period from 1049 to 1077, less than three decades, this drilling technology had become well-known in Lingzhou. Then it quickly spread to neighboring Jiazhou and Zizhou. At that time, there were more than 1,000 wells and every well owner hired about 20–50 employees. They used a round-tipped iron drill to dig small-diameter salt wells with 20–30 cm in diameter and 100–300 m deep, so as to extract highly concentrated brines. The ancient large-diameter salt wells were dug by workers under the well with crude tools, while small-diameter salt wells were drilled by percussion drilling with a round-tipped iron drill. With tremendous force, percussion drilling could help drill into hard formations and reach enough depth.

After drilling a small-diameter deep well, workers made seven bamboo pipes, sometimes eight, with the bamboo nodes being cut off and with each 10 feet or so in length and 10–30 cm in diameter. Then with mortise and tenon joints, they were assembled into one long hollow bamboo pipe, 23–35 m long, and the seams were tightened with hemp ropes and consolidated with cement and lacquer. After that, the long bamboo pipe was fit right into the well, as a casing pipe to prevent the well from collapsing and avoid freshwater infiltration. Another similar long bamboo pipe, smaller in diameter, was used as a brine drawing pipe which had a valve made of cooked cowhide at its bottom. After the brine drawing pipe entered the well and reach its bottom, the brines then rushed into the valve until the pressure of the brines inside the pipe was large enough to shut the valve. Finally, the workers lifted the pipe, opened the valve, and released the brines into tanks and boiled them later.

The popularization of the bamboo wells greatly promoted the salt production in Song Dynasty. According to the book, *History the Song Dynasty and Shi Huo Zhi*, up to 1132 A.D., there were more than 4,900 salt wells in Sichuan, producing more than 3,000 kg salt annually [10]. The number of salt wells was 7 times greater than that in the early Song Dynasty, the annual salt output, 2.7 times and the annual salt tax, 5 times [11].

After the bamboo well drilling technology took its shape, it continued to develop and reached its climax in the Ming and Qing Dynasties. The book, *Jing Yan Tu Shuo*, written by Ma Ji in 1578, gave a thorough description of the salt well drilling. Ma Ji based on his field work during his tenure as the magistrate of Shehong County, Sichuan. This book was illustrated by Yue Yufang, the preface of which was written by Guo Zizhang, a local official of Sichuan. Although the book had never been published, it was copied and thus passed down by manuscript. The texts in the book were included in three books, namely, *Shu Zhong Guang Ji* [12] by Cao Xuequan in 1610, *Tian Xia Jun Guo Li Bing Shu-Shu Zhong Fang Wu Ji- Jing Fa* by Gu Yanwu during the transition of the Ming and Qing Dynasties, and *She Hong Xian Zhi-Shi Huo Zhi* by Xie Tingjun and others in 1886. However, the illustrations were lost. Fortunately, The illustrations in *Yan Jing Tu Shuo*, the second volume of *Si Chuan Yan Fa Zhi* (1882), edited by Ding Baozhen, Governor of Sichuan, could make up for the lost. *Zhi Liu Jing Feng Wu Ming Shi Shuo* [13], written by Wu Dingli and published in 1872 by the publisher—Fu Shun Si Yuan Tang, was one of the few books which monographed salt zone management and salt production technology in artesian well district, Fuyang County. Based on Wu Dingli's field work, the book, with illustrations, is informative and reliable. Also, based on field work, Li Rong wrote *Zi Liu Jing Ji* [14] in 1876, in which he described salt production and salt sources, especially in the deep well geologic horizon.

Yan Jing Tu Shuo, *Zi Liu Jing Shuo*, *Zi Liu Jing Ji*, and others have related written records about the process and key points of operation of China's rope deep well drilling technology since the Ming Dynasty, while *Si Chuan Yan Fa Zhi* gave illustrations about that.

According to *Jing Yan Tu Shuo*, "when deciding the location of the well, sinking workers preferred places on the river sides, near steep mountains or close to sand." Generally speaking, the area in the vicinity of the rivers in Shehong County has abundant fresh water. Water in the mountain areas is scarce and salty. In the high mountain area, the low and plain land is suitable to drill a well, while another ideal location is in a bulge of the low mountain area. *Zi Liu Jing Ji* recorded the sampling characteristics by drilling at different formation. When deciding the well location, the rock in the ground must be examined. The red rock layer is first drilled, followed by the limestone, ginger rock, oil, grass white rock, yellow sandstone, natural gas, blue sandstone, mung bean rock (a kind of clay rock), and finally black water (concentrated brines layer). Every layer is named according to its rock color and shape. All the rocks may not be seen during drilling, but there must be ginger rock and mung bean rock.

During the bamboo well period in the Song Dynasty, 7 or 8 bamboo pipes with a diameter slightly smaller than that of the well were combined into one long hollow bamboo pipe, sometimes 22–25 m long, and inserted into the well as a casing pipe. There is no need to add a casing pipe in the well downward because the rock formation is firm. But the soil of the upper well is so soft that the bamboo casing pipe, under considerable tension and pressure, may be easy to crack after a long time. There is some improvement in cementing the well in the Ming Dynasty. After selecting the well location and leveling the surrounding land, workers excavated a wellhead (Fig. 25.2). 30 stone rings, square outside and round inside, (Fig. 25.3) with diameter ranging from 26 to 30 cm or 36–40 cm, circumference 60 cm and thickness 0.3–0.6 m, were put into the well one after another. Then soil and gravel were filled around the stone rings to prevent the upper well from being crushed by external force.

In the Ming and Qing Dynasties, besides protecting the well with the stone rings, the hollow long wood pipe was also used as a well-cementing casing pipe. At first, it is necessary to drill a large wellhead with a large drilling tool, called Fishtail File (Fig. 25.4). Then a head frame was built at the top of the well, which was equipped with a pedal frame, an ox-drawn winch, and a transmission system. The winch is used to draw in and draw out the rope to drive the drill bit to hit the rock or by lifting and dropping the pedal. No matter which way is adopted, when every 1–2 chi (1/3–2/3 m) the workers drilled, they needed to take the drill bit out and put the brine drawing pipe into the well to extract the broken rock (cuttings) and muddy water. They repeated the above operations until they reached the red rock layer.

Large drill tools were made of iron about 3 m in length and 30–40 cm in width, with a weight of 50–100 kg and a fishtail bottom, historically known as the Fishtail

Fig. 25.2 The digging of a wellhead (from *Chuan Yan Ji Yao*, Vol. 2)

Fig. 25.3 The insertion of stone rings (from *Chuan Yan Ji Yao*)

Fig. 25.4 Cleaning of the broken rock and mud (from *Chuan Yan Ji Yao*)

File (Fig. 25.5). When rotating the drill bit, the drill pipe with an iron ring makes a vertical drill hole with a drill jar. The twisted and plaited bamboo cables connected all the drilling parts, which embodied the characteristics of the Chinese rope percussion drilling technology. In the process of drilling, workers have to pour water into the well for a depth of about 6–10 m and then the spring would come out itself, no need for pouring more water [15]. Generally speaking, a large-diameter well should be drilled about 62–93 m deep with a diameter of 26–40 cm.

In the Ming and Qing Dynasties, the pine wood and cypress were used for stronger casing pipe. According to *Yan Jing Tu Shuo*, the pine wood was vertically sawed in half and the two halves were then hollowed but could be combined into a pipe. The two ends were connected with mortise-and-tenon joints and the seams were tightened with the hemp ropes. Then they would be consolidated with the tung oil and Lime (Fig. 25.6). The inner diameter of the pine wood cylinder was slightly smaller than that of the well. The bottom of the pipe was also wrapped with rope, on which tung oil and lime wrapped up, so that it could fit right into the well as a casing pipe. The wooden casing pipe was put into the well through the pulleys on the head frame, and

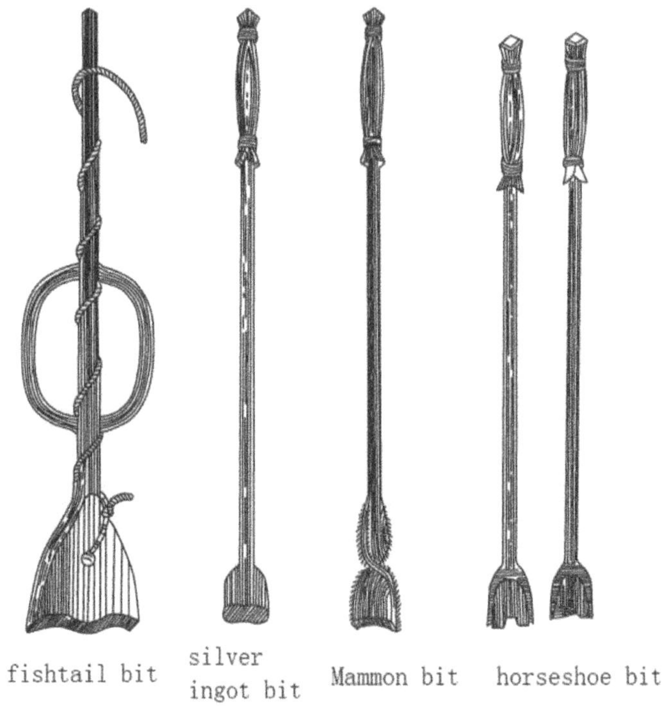

Fig. 25.5 The main well drilling tools used in the salt well district in Sichuan (taken from *Chuan Yan Ji Yao* (1916))

Fig. 25.6 The making of the wooden casing pipes (from *Chuan Yan Ji Yao*)

the uppermost pulley was connected to the transmission system and the ox-drawn winch by the twisted and plaited bamboo cables (Fig. 25.7). The total length of the wooden casing pipe could be about 45–90 m.

Fig. 25.7 The insertion of the wooden casing pipe (from *Chuan Yan Ji Yao*)

After the pine wood casing pipe was installed into the well, a small iron drill bit was used to drill a small diameter deep well, formerly known as Xiao Qiao. With a total length of 1.2 zhangs (about 3.72 m) and a weight of 40–70 kg, the drill consisted of a long handle, a drill jar and a drill bit like silver ingot, thus formerly called Silver Ingot Bit (Fig. 25.5). The drill bit was about 2–2.3 decimeters or 2.6–3 decimeters high, with an elliptical bottom and sharp left and right sides. The drill jar, also called Zhuan Cao Zi in Chinese, was driven by a winch (Fig. 25.8), which is the same way as drilling a large wellhead. This part accounted for 80–90% of the whole salt well and it was the main body and also the open hole of the well. Because of hard formations, it took a long time to drill a well, usually about 4 or 5 years or more than a decade. In terms of the well diameter, according to *Zi Liu Jing Shuo*, the previously drilled well had a diameter of 8–8.25 cm, the bigger one 11 cm…the Xiao Yan wells were 13–16 cm in diameter. As for the well depth, it varied greatly because of the geological conditions. According to *San Sheng Bian Fang Bei Lan* written by Yan Ruyu in 1822, the deepest well could be 960–1280 m in depth, which broke the world record at that time.

When colored salt water or brine was found, the drilling would be finished. Then the brine drawing pipe was sent into the well and the brines rushed into the valve until the pressure of the brines inside the pipe was heavy enough to shut the valve. The workers lifted the pipe, opened the valve with an iron hook, and released the brines into the tank, ready for boiling. On the top of the well, a considerably high head frame was built, which was supported by several timbers and equipped with a fixed upper pulley and a lower pulley, also called sky pulley and ground pulley. Beside the well, there was a thatched shack, under which was an ox-drawn winch with a perimeter of 5 m, wound with long twisted and plaited bamboo cables. The tractive force of the winch was greater than before. A guide wheel with a pivot was installed between the ground pulley and the winch to change force direction. All the parts mentioned above were connected with the twisted and plaited bamboo cables (Fig. 25.9).

Since the eleventh century, a set of tools and methods for solving drilling accidents have been developed, which became more sound and complete during the Ming and Qing Dynasties. Before coping with the accident, an experienced sinker would first place a gage stick with barb (Fig. 25.10) into the well to find out the cause of the accident, what the falling objects were and their location in the well, and then take corresponding measures. For example, in the process of drilling, the drilling tools and the twisted and plaited bamboo cables were broken and dropped into the well, and they can be taken out by the iron five-prong rake. And if the well is blocked by mud, the iron pole with teeth at its end will be used to stir the sediments so that they can be removed from the well by a bamboo pipe, formerly known as scraper pipe.

In the Qing Dynasty, 70 kinds of tools for drilling and salvaging, and dozens of tools for solving downhole accidents were invented. Another common downhole accident was that when the drill bit hit the rock formation and got stuck into it due to excessive force, the drilling would be interrupted. If the sinking workers pulled it out with excessive force, the drill and the twisted and plaited bamboo cables would break. Ancient people invented the drill jar called "Zhuang Zi Qian". Placed between

Fig. 25.8 The drilling of a small-diameter deep well (from *Chuan Yan Ji Yao*)

Fig. 25.9 Drawing brines (from *Chuan Yan Ji Yao*)

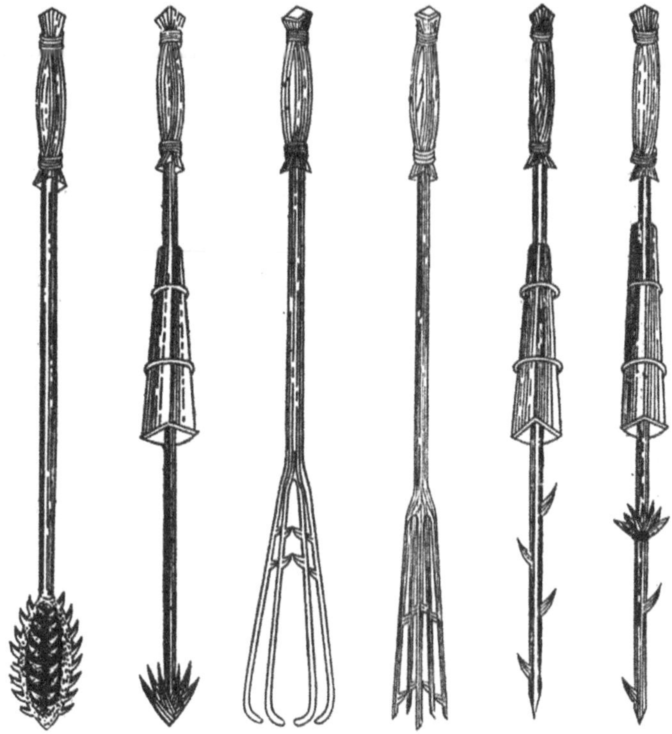

Fig. 25.10 The salvaging tools used in Sichuan salt well area (from *Chuan Yan Ji Yao*)

the drill and the twisted and plaited bamboo cables, it could produce a rebound force after the drill bit hit the rock formation, so that the drill bit would not be trapped and played a role of automatically freeing. This device was used in deep well drilling in Yancheng District, Sichuan, at the beginning of the fourteenth–fifteenth century. In the Ming and Qing Dynasties, there were different names for Zhuang Zi Qian, such as Ting Zi and Zhuan Cao Zi, and there were many kinds for different downhole operations. These devices like Zhuang Zi Qian were placed between the drilling, salvaging, and drawing tools and the twisted and plaited bamboo cables connecting the lifting system above the well, with functions such as hanging, righting, directing, shaking, protecting, and stuck freeing.

The design and structure of the drill jar or Zhuan Cao Zi was simple. It was composed of rectangular iron rod, historically called Ji Jiao Gan, a pedestal with its top wide and its bottom narrow, known as Qiu Qiu, and an iron sleeve, or Ji Dan Ke (Fig. 25.11). The drill jar had a length of 1.9 m and a weight of 20 kg. The rod has a rope hole in its upper end to tie a bamboo cable which stretches into the well. The lower part of the iron rod was square. The cross section of the pedestal is conical in shape and the iron sleeve can slide up to the top of the iron rod and down to the pedestal. The rod, the sleeve, and the pedestal of the drill jar are connected with the

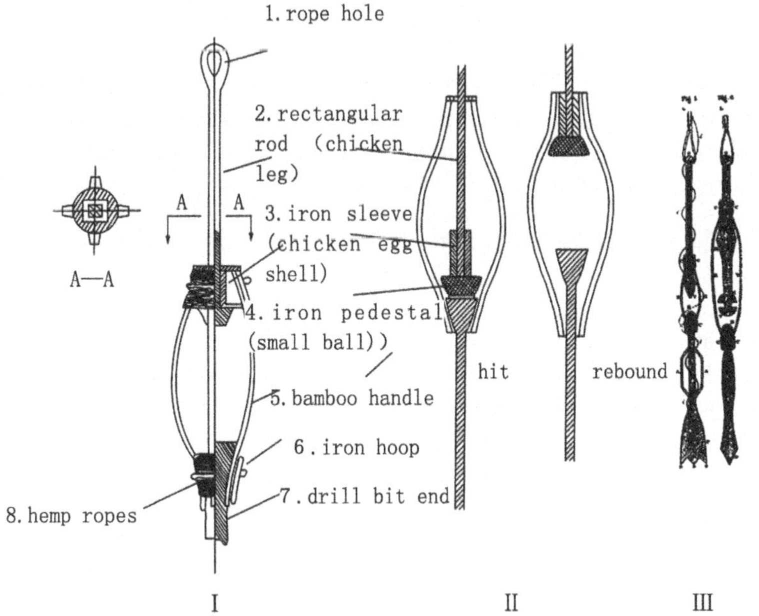

Fig. 25.11 The structure and operation sketch maps of the Chinese drill jar (Zhuan Cao Zi)

tail of the drill bit through a handle. The handle is composed of four bamboo pieces bent into an elliptical shape, evenly distributed at its front, back, left, and right, and the joint is wrapped with the hemp ropes and then tightened with an iron hoop. The rod hits up and down within the handles, for the purpose of shaking and stuck freeing.

Since the drill jar made the drilling tool heavier, the force of the drill bit to break the rock formation is enhanced. If the cable is moderately long, the drill bit will not work when it reaches the well bottom. At this time, the drill jar will descend until it knocked against the drill bit and emitted a clattering sound, indicating that the drill bit has reached the well bottom. If the cable rope is too short or too long, the sound cannot be heard outside the well, so the length of the cable rope should be adjusted in time [16].

Therefore, based on the bamboo well in the Song Dynasty, the drilling technology had become gradually mature since the Mid-Ming Dynasty and the drilling process gradually became more standard and specific than before, including identifying well location, installing stone rings, sinking shallow pits, extracting mud and broken rock, inserting the bamboo pipe and drilling the boreholes [17]. The well-cementing technology was improved. So was the well structure. The stone rings, wooden pipes, and bamboo pipes were used in the upper well to consolidate the soft formation and prevent penetration of fresh water, while the lower well kept the open hole to prevent collapse. The combined use of the stone rings, wooden pipes, and bamboo pipes as the cementing casing pipes broke the limitation of only using bamboo casing pipes in the Song Dynasty. The downhole accident treatment technology has also made new progress. Tools such as the stirring shovel, iron five-prong rake, drill jar, were used to salvage falling objects down the salt well and tools such as souzi, caoqian, scraper pipes and others were used to clean the well, which effectively handled the accidents during well drilling and salt well production. Advances in these technologies have created conditions for deep well drilling and salt extraction from the underground. Guo Zizhang, a local official of Sichuan in the Ming Dynasty, said that during 1573–1619 the salt wells in Shehong County near Chengdu were 155–186 m deep, and the deepest one had reached 310 m [18].

The prosperity of salt production in the Qing Dynasty promoted the maturity of drilling technology [19]. The drilling process was more refined than that of the Ming Dynasty including identifying the well location, sinking wellhead, installing the stone rings, sinking the shallow pits, installing the wooden tubs, drilling the boreholes, and so on. The drilling procedures were more complete and the drilling technology and tools were also improved. In the Ming Dynasty, there were only two kinds of drill bits, big ones, and small ones, while in the Qing Dynasty, various types of drill bits were invented to suit different rock formations, such as the fishtail bits, silver ingot bits, horseshoe bits, Dian Gen Zi bits, and the Mammon bits. Drilling tools such as Zhuan Cao Zi (the drill jar), handles, and long iron poles improved drilling efficiency and well quality. The application of well measure technology and the methods for correcting well inclination tended to be mature. Approaches to tackling wellbore bending and deviation, such as marking by hanging ink, well measuring and inclination correcting, were adopted so as to ensure well quality. The well repairing and treatment technology has developed greatly compared with that of the Ming Dynasty. It has formed a complete set of techniques such as filling cavities, salvaging, repairing wooden pipes, and cleaning of the well. For example, dozens of fishing tools have been used in the Mid-Qing Dynasty, so that accidents can be eliminated in time during drilling and operations. In the Qing Dynasty, the development of well drilling and treatment technology reached a new height, providing an important foundation for the salt production industry. In 1730, the salt wells in Sichuan spread to 40 prefectures and counties, with a total of 6,116 wells producing more than 92 million kilograms of salt [20]. In 1758, another 2,000 wells were added. During 1815–1820, the depth of the salt wells could reach 960–1,280 m [21].

Although China's deep well drilling technology was mainly used for the exploitation of the salt underground, it was also used to exploit the oil underground in the thirteenth–sixteenth century and natural gas in the eighteenth century. Later, it was followed by countries around the world. The deep well drilling technology played an decisive role in advancing oil exploitation and oil product processing technologies and so on in early modern world and also helped to establish corresponding engineering disciplines. The Shenhai Well (Fig. 25.12) at the Zigong Salt Field was drilled in 1835, serving as a living specimen of ancient China's deep-drilling technology. As the deepest well in the world at that time, the Shenhai Well had a depth of 1001.42 m, the part above 125 m with a diameter of 11.4 cm and the part from the well bottom to the 125 m deep part with a diameter of 10.7 cm, producing 8,500 m^3 of natural gas and 14 m^3 of black brines per day [22]. In 1985, the Shenhai Well was restored to its original layout and structure in Qing Dynasty. In January, 1988,

Fig. 25.12 The Shenhai Well site (from http://blog.sina.com.cn/s/blog_4cd33b790100wu9j.html)

it was listed as one of the important heritage sites under state protection by the State Council of China.

25.3 The Spread of Deep Well Drilling Technology to the West

From ancient times till early modern days, the table salt in Europe had mainly been the sea salt as the exploitation of salt by sinking the salt wells was not known until the sixteenth century. The drilling and salt extraction technology mastered by Europeans was introduced from China. The introduction of Chinese deep well drilling to Europe can be roughly divided into three stages. In the first stage (fourteenth–sixteenthcenturies), China's salt well drilling was initially introduced to Europe and was put into practice. In the second stage (seventeenth–eighteenthcenturies), Chinese deep well drilling technology with the help of iron drill bits was introduced to Europe and was put into practical use. The last stage occurs in 1990–1930s when China's percussion drilling technology spread to the whole Europe and North America. It was used until the late nineteenth and early twentieth centuries, laying a solid foundation for modern drilling technology in the world.

The Italian Marco Polo first reported the message of the Chinese salt well drilling to Europe. He came to China in 1275 and returned to his homeland after living in China for 17 years. In 1298, *The Travels of Marco Polo* was first recorded in ancient French, and then it was translated into Italian, Latin, and German during fourteenth and sixteenth centuries. According to his book, salt was found deep underground in the China inland areas and mined from the drilled wells [23].

Western European countries with the sea salt had no urgent need for the well salt, but Eastern Europe and the Central Europe inland countries and regions were urgently in need of new sources for salt. After learning the method of salt exploitation from the drilled wells in *the Travels of Marco Polo*, they soon began to drill salt wells. In the sixteenth Century, the first batch of salt wells was sunk and operated in Wieliczka [24] (near Kraków, an industrial and commercial city in southern Poland), Bochnia and Sóowár in Hungary. Marco Polo did not introduce specific salt well drilling approaches so the Europeans believed that the salt wells in China were common water wells or mines and used conventional methods to drill wells. In this case, the first batch of salt wells in Europe in the sixteenth Century was large-diameter wells and the drilling technology used was equivalent to that of the Han and Tang Dynasties. However, those wells brought considerable benefits.

In 1500, the annual salt output of the salt wells in Vijlicika was 3 million kilograms, and Bochnia, 5 million kilograms. The salt was extracted from the well by manpower treadmill (Fig. 25.13). In 1670, some salt wells were about 360 m deep. Workers were sent into the well and lifted by a horse-drawn winch through a trolley, and there were

Fig. 25.13 The manually dug salt well in the seventeenth century, in Wieliczka, Poland (restored by A. Dlugosza and stored in Muzeum Zup Krakowskich Wieliczka)

lights down the well. According to the visitors' records, in 1673 the salt mined by Suvor had different colors and the salt well was 190 m deep, from which salt was taken out by a trolley. In 1724, the salt wells were 256 m deep with an annual salt output of 6 million kilograms, and the horse buckets were used to take the salt out.

When Western European countries learnt that Poland had accumulated a large amount of wealth via the exploitation of salt from the salt wells, they soon followed it. During sixteenth and seventeenth centuries, salt wells were drilled in Saxony, Bavaria, and Cheshire. These salt wells were large-diameter shallow wells drilled in Vielichka's way. Although they were not deep enough to reach the rock salt, valuable brines could be drawn from them.

The large-diameter shallow wells are not deep enough, so concentrated brines and halite cannot be found, limiting the further development of the European salt well production. In this case, China's advanced drilling technology was introduced to Europe through missionaries at the turn of the seventeenth and eighteenth centuries. According to *Les Merveilles de la Science*, written by Guillaume Louis Figuier in 1870, a French scholar, on October 11th, 1704, the Bishop of Tabrska (é vêque de Tabrasca) published a book, *Lettres édifiantes*, in which he introduced China's salt wells [25]. Figuier also cited the French version *Voyage Pittoresque*, published in the Netherlands in the last few years of the seveteenth century, in which the Sichuan salt wells were also recorded. The reports on the Sichuan salt well were recorded and sent back to Europe by the missionaries or Jesuits who were preaching in Sichuan at that time.

During the reign of Emperor Kangxi in the early Qing Dynasty, all of the salt wells in the salt well district of Sichuan Province were small-diameter deep wells. At the turn of the seventeenth and eighteenth centuries, the European missionaries visited the salt wells in Sichuan and learnt that Chinese people used rope percussion drilling technology with an iron-blade drill bit and dug small-diameter deep wells, from which they successfully extracted brines and rock salt. Therefore, at that time, the drilling technology the missionaries introduced to Europe was rope percussion drilling with an iron-blade drill bit. After repeated trials, the Europeans mastered this drilling technology with an iron-blade drill bit, replaced the simple digging method with it, and made a historic breakthrough in the western drilling technology.

In 1713, J. J. Batels, a German machinist, invented a drilling machine, called Bohrmaschine, to improve the traditional method of drilling a open shaft well in the mining areas. He used rope percussion drilling with a steel-blade drill bit to successfully drill a small-diameter shaft well, 25 cm deep. The drill was tied to a lifter driven by a tandem tread wheel [26]. This was the earliest European attempt to drill a well by using Chinese rope percussion drilling technology. However, when they reached the hard rock formation, they failed to further drill the well by using this method. Later, this technology was abandoned by the Europeans, thus making very little impact on later generations.

In 1714, J. C. Lehmann, a professor of medical science at the University of Leipzig, published a book, *Beschreibung des Bergbohrers*, in which he described and drew drilling equipment, including six percussion drill bits, four components for removing debris and water out of wells and four components for salvaging broken drill stems

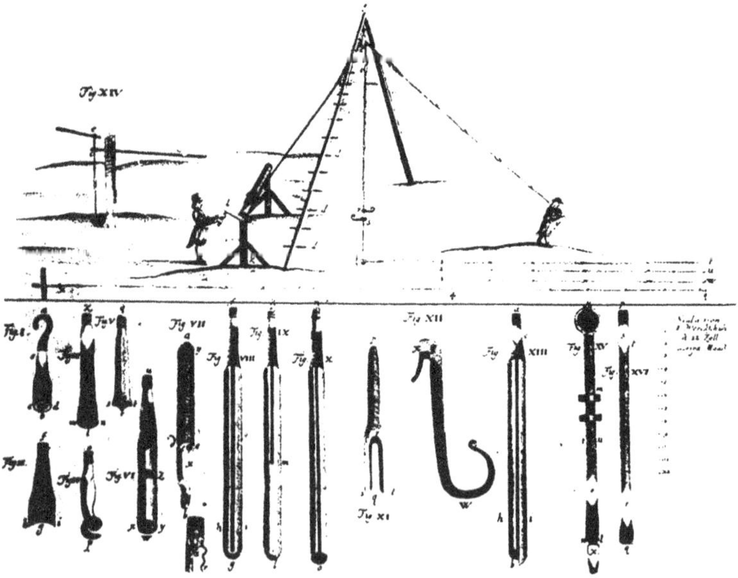

Fig. 25.14 The salt well drilled by Europe's rod percussion drilling method (top) and different drill bits used (bottom) in the early eigteenth century (from *Beschreibung des Bergbohrers*, by Lehmann)

and drill bits [27]. These components were hung on a pulley that was supported by three pillars and lifted by a pulley pulled by labor. Lehman adopted the idea of Chinese percussion drilling technology for small-diameter wells because the drill bit he described was similar to that used by the Sichuan well workers. However, he used a long hard rod to pull the drill bit instead of the bamboo strips or ropes (Fig. 25.14). He mentioned many uses of the well driller, but practice showed that its most important use was to drill wells to obtain salt, especially in inland areas where salt deposits were in deep underground.

Many years later, in order to improve the production capability of Saxony Salt Plant, J. G. Borlach, the former assistant of Lyman focused on deepening the well and combined excavation and drilling in Artern during 1725–1731, making the well as deep as 140 m. From 1744 to 1764, he managed to obtain brines by drilling in Dürrenberg [28].

The French well drilling projects multiplied in the late eighteenth century but they were usually used to extract the ground water. In 1781, a 100 m-deep well was drilled in the northern city of Lille, while in 1784, a 580 m deep well was drilled in Paris [29], a record depth held for more than a century in Europe. The lack of saline groundwater in France limited the development of the salt wells. The salt deposits were only found occasionally in Vic, near Germany, in 1819.

The American salt well drilling dated back to the early nineteenth century at the Kanawha salt works, Virginia (now West Virginia State) [30]. In 1809, the two brothers, David Ruffner and Joseph Ruffner dug a well on salt-lick, using a 1 m-diameter hollow maple wood as a reinforced well casing pipe, and one of them removed mud out of the well and dug 4 m deep to reach the saline water. In another place, they dug a second well, 14 m deep, where they used an iron-blade drill bit to drill and get the saline water. In the first well, they continued to drill to 17.5 m deep with a steel shovel drill bit and reached the saline water again [31]. We can see that the American salt welling drilling technology of the early ninteenth century followed the mining technology popular in Europe, especially in Germany after the eighteenth century, which was the product of Chinese drilling techniques introduced to Europe.

People in the surrounding area imitated the Raffner brothers' example in developing the salt wells. In 1819, 15–20 salt wells were opened with the average well depth of 116 m, and 30 boiled salt plants were produced. The annual output of salt was 18 million kilograms. In 1827, there were 61 wells with a depth of over 91.4 m in the Kanawha region, and 120 wells with an average depth of 116 m in 1830. The salt wells have also expanded to other states, especially in Ohio, where there were wells over 300 m deep in 1831. Since 1828, the Raffner brothers have driven the drill bit with a steam engine, which marked the emergence of the mechanical drilling.

During the westward spread of the Chinese well drilling technology in the seventeenth and eighteenth centuries, European and American countries took it for granted that the drill pipe was directly connected to the drill bit, making the hidden danger caused by it last more than one hundred years. At the end of the eighteenth century, Europeans noticed the defects of such drills and thus made various improvements. However, they were all palliative measures and could not get to the root of the problem.

At the beginning of the nineteenth century, Chinese well drilling technology was introduced to Europe again. The technical information was transmitted by François Imbert. After coming to China during the reign of Emperor Daoguang in the late Qing Dynasty (1825), he went to the nearby salt well district to preach during his hospital stay in Chongqing parish, Sichuan. Under the guidance and explanation of the local well owners, well technicians, foremen and well workers, he made a detailed on-site inspection of the well salt technology. In 1828, *Annales de l'Association de la Propagation de la Foi*, published in Lyon, included the letter *Consulter sur les puits de sel* [32] Imbert wrote in September 1826 to Langlois, the Dean of the French Overseas Missionary Theological Seminary. It pointed out that there are more than 1000 salt wells in Sichuan, with well depth ranging from 490 m to 585 m. They were drilled with 130–180 kg iron drills that had the upper part of the umbrella shape suspended by the rattan rope. Some people rhythmically jumped on the pedal to ensure that the drilling action was coordinated; the brine was transported in the long bamboo pipe and then was boiled with coal or natural gas emitted from the well [32]. The above-mentioned umbrella part was actually an impactor or a rotary groove, which acted as a buffer between the drill bit and the drill pipe.

At that time, some French drilling experts suspected whether the Chinese salt well could drill to a depth of 585 m [32]. To this end, Imbert went to the salt well region in Fushun County, Xuzhou District, Sichuan. On September 1, 1829, he wrote a letter to Langlois at the artesian wells in Fushun County, included in Book Four of *Annales de l'Association de la Propagation de la Foi* published in 1830. It is said in his letter that the drill was suspended by a rattan-wrapped rope and drilled in the well; the other end of the rope was connected to a drum-shaped winch that provides lifting power. The rope encompassed 50 turns and the circumference of the winch was 12 m. So, the winding rope was 600 m long and the total length was 945 m. In this case, the well depth really reached 610 m, which proved that the well depth data he had reported originally was correct [33].

Imbert's report on Sichuan well salt technology quickly attracted the attention of relevant European and American people and became the basis for improving existing drilling techniques and tools. In 1830, the German drilling specialist Karl C. F. Glenk did not know about this kind of device, so the drill he used could not effectively weaken the impact of the drill bit. But in 1839 his compatriot Heinrich von Dechen first mentioned the drill in the article Drilling work in 1811–1837 in Aten town and said in June, 1834, Karl von Oeynhausen introduced a new device called "Transducer" (Wechselschere) or "Slider" (Schieber) during his drilling for the new salt plant (Neusalzwerk bei Rehme) near Neum [34]. In 1858 a German August H. Beer provided a picture of the device developed by Karl von Oeynhausen in his book *Erdbohrkunde* (Fig. 25.15). It can be seen from the figure that the drill pipe and the drill bit were not directly connected, but a cylindrical device that can slide up and down was placed between them, which has the same structure and function as the impactor or rotary groove used in the Ming and Qing Dynasties in China. This device thus provides a new type of tool necessary for deep drilling. In 1842, C. G. Kind, Karl C. F. Glenk's assistant, claimed in his book *Anleitung zum Abteufen der Bohrlöcher Luxemburg* that he had drilled a 435 m-deep salt well. In the following year, he developed the "Freifall-Apparat", which was installed between the drill pipe and the drill, providing great significance for the success of deep drilling. In 1847, the Frenchman J. B. Jobard published the article "Über das Brunnenbohren mit dem Seilbohren" in German, claiming that only a French well worker successfully drilled 89 wells by taking advantage of the Chinese method [35]. Therefore, the successful drilling with the help of a rope drill has been confirmed by the practices of the European drillers.

However, those who tried the rope drill reported that it was difficult to maintain the verticality of the borehole with this method because the rope is stretchable and is not an effective medium for the vertical transmission of force. In fact, the Chinese never used the stretchy hemp ropes but the bamboo ropes that were twisted with the split bamboo strips. In the twentieth century, two Americans, Wallace Crawford and MT Archer, saw Chinese use the tangled flexional bamboo strips as the drilling ropes. Each rope was usually 4.6 cm [36] or 4.5 cm–15 cm in diameter and 12 m long [37]. The ropes were connected one by one. The deeper the well was, the thicker the rope was. The bamboo strip ropes were resilient enough to be encompassed on the winch to lift the drill and were rigid enough that well worker could rotate the rope counterclockwise from time to time to ensure that the borehole was vertical. The rattan rope introduced by Imbert is twisted by the stem of Calamus platyacanthoides and has the same effect as the bamboo strip rope. Even though there was no bamboo

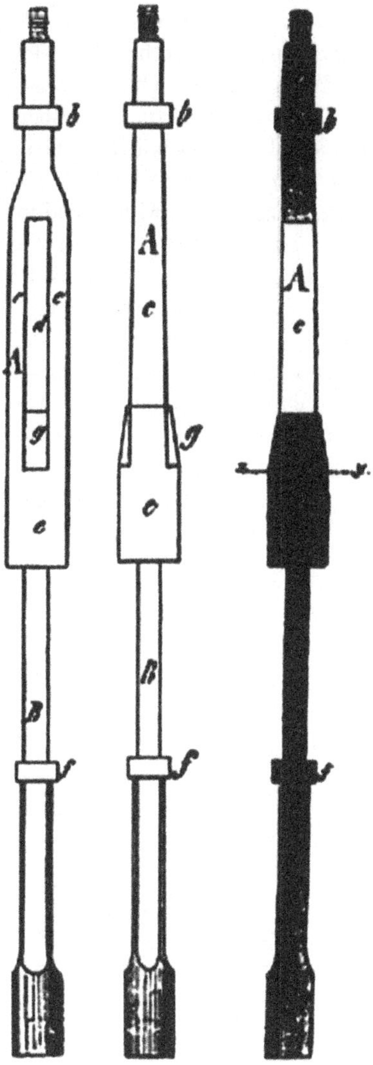

Fig. 25.15 Karl von Oeynhausen's drill jar (Wechselschere or Schieber) (taken from Erdbohrkunde)

or rattan in Europe, this kind of rope could be imported from China, or be replaced by ropes of similar properties. Americans even used wire ropes for the same purpose.

Americans used the rope drill method in 1830, which was earlier than the Europeans. In 1840, jars were introduced and drills began to be driven by the steam engines to accelerate the development of well salt drilling. Since the United States did not have bamboo, the drill was suspended by a wire rope instead, and the other end of the rope was connected to the lifting device. The wire ropes were good substitute because they had the characteristics of the bamboo strip ropes, namely, rigidity and bendability, and could be produced by machine. Its working principle and function were the same as the Chinese competitor or the rotary slot. William Morris, a technician in the Kanawha Salt Wells, Virginia, USA, was awarded the patent of "jars" (U.S Patent 2243) from the US government on September 4, 1841.

In 1876, John P. Hale, a scholar who studied the history of the salt production in Kanawha, wrote *Manufacture of Salt in Kanawha* [38] according to the memoirs of the old local residents, pointing out that Morris had invented jars as early as in 1831, but patented it ten years later. On the basis of the illustrations provided by Hal (Fig. 25.16) and the description, it is a device installed between the wire rope and the drill bit, and there is also a sleeve that can slide up and down near the lower part of the drill bit.

The jar drilling tools in Europe and the United States might be developed at the same time without the influence from each other, but they all have a common source, i.e., the inspiration of the impactor or rotary groove used in the salt well region of Sichuan, China, in the Ming and Qing Dynasties.

China had a variety of tools needed to drag for the damaged equipment in the well. They were very delicate and were introduced to Europe in time because Westerners

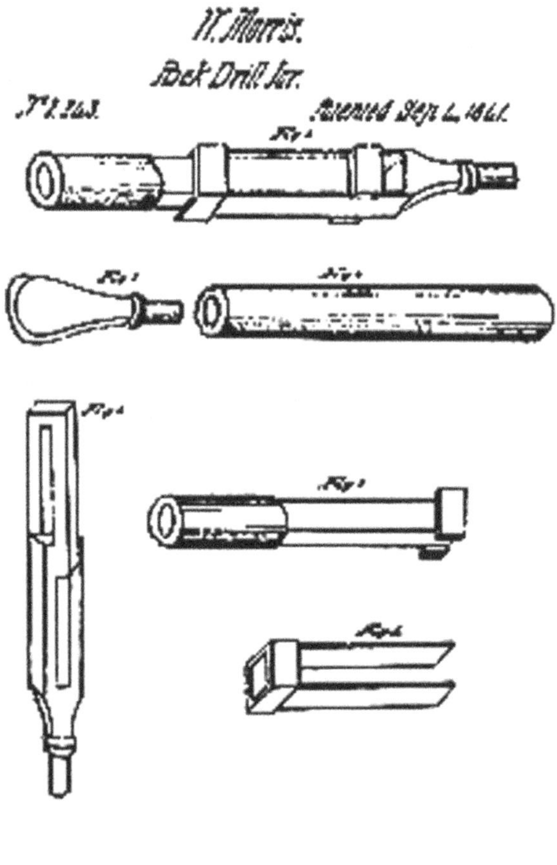

Fig. 25.16 Morris's drill jar

had noticed that the salvage tools in China and Europe were very similar, but there were more types of salvage tools in China. It is an invention that Europeans replace wood and stone casing pipes with iron pipes as well casing pipes. In 1846, Fauvelle, a Frenchman, introduced the water injection drilling method: injecting water through the hollow drill pipe and discharging the broken debris from the well [39]. In fact, since the Song and Ming dynasties in China, the water injection technology has been used with the help of the hollow bamboo pipe with a one-way cock. In 1865, the German Kicken proposed to use a centrifugal pump, and thus greatly improved the efficiency [40]. In 1860, Sch ningen, took example by the practice of the Kanawha salt plant in the United States and used the steam engine as the drilling power, instead of the human power or the marathon winch, and finally drilled a salt well of 580 m or more in depth in Sachsenmilch, Germany [41].

From pole drills in the early eighteenth century to rope drills in the 1930s, Europeans finally grasped the true meaning of the deep well drilling technologies in China, made some technological innovations and developed new tools such as jars similar to the rotary slot. In the 1880s, Europeans drew on the strengths of the United States and made the entire Western deep well drilling technology enter a modern stage, ending the westward history of the Chinese well drilling technology.

In the history of the well salt mining, China has invented the bamboo and wood sleeve water-blocking method, deep well jarring drill method and related drilling tools and salvage tools, especially the impactor or rotary groove that can be used for hanging, indicating, jarring, righting, and pipe-free agent were, as the key part of small-mouth deep well drilling. The invention and development of these deep-drilling and mining techniques have enabled the exploitation of natural brines deep in the ground. The westward spread of related drilling information and technology has greatly boosted the development of the modern drilling technology. The complete well salt mining and producing technology and process remained in Da'ying County in Zigong, Sichuan has become the living fossil of the world deep well drilling. In 2006, this technology was included in the first batch of China's national intangible cultural heritage. As the British scholar Jeseph Needham said, although the ancient Chinese well drilling was mainly for mining bittern, the deep well drilling technology is undoubtedly a pioneering invention in China and provides inspiration for oil drilling and mining in modern China and around the world.

References

1. The Tonglvshan archaeological excavation team. 1975. *Briefing on the Excavation of the Tonglvshan Ruins of the Spring and Autumn Period, Hubei Province*, vol. 2, 1–12. Cultural Relics.
2. Chang, Qu. 1919. *Hua Yang Guo Zhi*, vol. 3. Shu Zhi: The Commercial Press.
3. Xiong, Yang. 1960. *Shu Wang Ben Ji*. See *Tai Ping Yu Lan*, vol. 865, 3840. Li Fang, Zhonghua Book Company.

4. Gu, Ban. 1962. *History of the Han Dynasty*, vol. 91, *Huo Zhi Lie Zhuan*, vol. 11, 3690. Zhonghua Book Company.
5. Haoliang, Yu. 1955. *Tomb No. 1 in Yangzi Mountain, Chengdu*, in *Heritage Reference Materials*, vol. 9, 70–84; Zhiyuan Liu et al. 1983. *Portrait Bricks of the Han Dynasty and the Han Society in Sichuan*. Cultural Relics Publishing House.
6. Fang, Li. 1983. *Tai Ping Guang Ji*, vol. 399, *the Salt Wells*, the version of *Overview of Notebook Novels*, vol. 5, 136. Guangling Ancient Books Printing Press.
7. Jifu, Li. *Yuan He Jun Xian Zhi*, vol. 33. The version of Si Ku Quan Shu.
8. Tong, Weng. 1986. *Dan Yuan Ji*, vol. 34, *Zou Wei Qi Cha Jing Chao Guan Zhi Jing Yan Xian Shl*, vol. 1096, 738. Taiwan Commercial Press.
9. Shi, Su. 1983. *Dong Po Zhi Lin*, 123. East China Normal University Press.
10. Tuo, Tuo, and A Lu Tu et al. 1977. *History of Song Dynasty*, vol. 183, *Shi Huo Zhi Xia Wu*, vol. 13, 4475. Zhonghua Book Company.
11. Yuanxiong, Lin, Song Liangxi, and Zhong Changyong. 1987. *The History of Chinese Well Salt Technology*, 152. Sichuan Science and Technology Press.
12. Ji, Ma. 1986. *Yan Jing Tu Shuo*. See *Shu Zhong Guang Ji*, vol. 66. Fang Wu Yan Pu. Cao Xue, vol. 592, 111–112. Taiwan Commercial Press.
13. Dingli, Wu. 1872. *Zhi Liu Jing Feng Wu Ming Shi Shuo*, 1–16. Fu Shun Si Yuan Tang.
14. Rong, Li. 1890. *Zi Liu Jing Ji. Shi San Feng Shu Wu Quan Ji Wen Gao*, vol. 1. Xiang Long An Shu Yuan.
15. Ji, Ma. 1986. *Yan Jing Tu Shuo*. See *Shu Zhong Guang Ji*, vol. 66, Fang Wu Yan Pu. Cao Xue, vol. 592, 111. Taiwan Commercial Press.
16. Delin, Liu, and Zhou Zhizheng. 1730. *The Study of Chinese Ancient Drilling Tools*, vol. 36. 138–140, 260–264. London: Shandong Science and Technology Press.
17. Yuanxiong, Lin, Song Liangxi, and Zhong Changyong. 1987. *The History of Chinese Well Salt Technology*, 173–180. Sichuan Science and Technology Press.
18. Zizhang, Guo. 1886. *Yan Jing Tu Shuo Xu*. See *She Hong Xian Zhi*, vol. 5, Shi Huo Zhi, Xie Tingjun et al.
19. Yuanxiong, Lin, Song Liangxi, and Zhong Changyong. 1987. *The History of Chinese Well Salt Technology*, 190–243. Sichuan Science and Technology Press.
20. Baozhen, Ding. 1882. *Si Chuan Jing Fa Zhi*, vol. 7. The engraved version.
21. Ruyu, Yan. 1822. *San Sheng Bian Fang Bei Lan*, vol. 1. The engraved version.
22. Lishan, Hu. 1957. *Research on Artesian Well Structure and Natural Gas Extraction in Sichuan Basin*, 15. Petroleum Industry Press.
23. The Travels of Marco Polo, ed. 9th edn, New York: Garden City, 1932, 173.
24. Hrdina, J.N. 1842. *Geschichte der Wieliczkaer Saline*. Vienna.
25. Figuier, G.L. 1870. *Les Merveilles de la Science. Paris* 4: 530–533.
26. Calvor, Hennig. 1763. *Acta Historico-Chronologico-Mechanica Circa Metallurgiam in Hercynia Superiori*, vol. 1, 3–6. Brunswick.
27. Lehmann, J.C. 1714. *Beschreibung des Bergbohrers*. Leipzig.
28. von Dechen, Heinrich. 1838. Die Auffindung von Steinsalz bei der Saline zu Artern, Archiv für Mineralogie, Geognosie. *Berobau und Hüttenkunde (Berlin)* 2: 232–239.
29. Theodor Tecklenburg. *Handbuch der Tiefbohrkunde*, Bd.1: pp. 84–85, Leipzig, 1886.
30. Disbrow, Levi. 1831. *Disbrow's Exposé of Water Boring*. New York.
31. Ruffner, David. 1818. The Kanawha salt works. *Niles Weekly Register* 8: 135.
32. Imbert, François. 1829. Lettre de M. Imbert, missionnaire apostique, à Messieurs les directeurs (Langlois) du séminaire des Missions Etrangères (Ou-tong-kiao, Kia-ting-fou, Sept. 1826). *Consulter sur les puits de sel. Annales de l'Association de la Propagation de la Foi (Lyon)* 3: 369–381.
33. Imbert, François. 1830. Lettre de M.Imbert, missionnaire apostique,à Messieurs les directeurs (Langlois) du séminaire des Missions Etrangères (Tse-lieou-king, le 18 Sept. 1828). *Annales de l'Association de la Propagation de la Foi (Lyon)* 4: 411–415.
34. von Dechen, Heinrich. 1839. *Die Bohrarbeit zu Artern in den Jahren 1831 bis 1837*. Archiv für Mineralogie, Geognosie, Bergbau und Hüttenkunde (Berlin) 12: 39–120.

35. Jobard, J.B. 1847. Über das Brunnenbohren mit dem Seilbohren. *Polytechnisches Journal (Stuttgart)* 105: 14 24.
36. Crawford, Wallace. 1926. The salt industry of Tzeliutsing. *China Journal of Science and Art (Shanghai)* 4: 174.
37. Archer, M.T. 1961. Drilling in Tzu-Liu-Ching, China. In *History of Petroleum Engineering*, 146–152. Dallas: American Petroleum Institute.
38. Hale, John P. 1876. Manufacture of Salt in Kanawha. In *History of Kanawha County*, ed. G.W. Atkinson, 149–223. West Virginia, Charleston.
39. Bruckmann, Ernest. 1846. An Account of the Imperial Salt Works of Sóowár. Philosophical Transactions of the Royal Society of Fauvelle. *Sur un nouveau système de forage. Annales de Chimie et Physique (Paris)* 3, 18: 328–331.
40. Kicken. 1865. *Ueber das Hiederbringen von Bohrlocher und Schächte vermittelst eines durch hohle Bohrgestäng geführten Wasserstroms*, vol. 13, 177–180. Zeitschrift für das Berg-, Hütten-, und Salinenwesen im Preussischen Staate (Berlin).
41. Multhauf, Robertp. 1978. *Naptune's Gift. A History of Common Salt*, 182–183. Baltimore: John Hopkins University Press.

Jixing Pan Born in Beining, Liaoning in 1931. He graduated from the School of Chemical Engineering of Dalian University of Technology in 1954 and then worked as a researcher in the Institute for the History of Natural Sciences, Chinese Academy of Sciences.

1981–1982, visiting professor in University of Pennsylvania;

1982, visiting researcher in Robinson College, Cambridge University;

1983, communication academician of the International Academy of the History of Science (Paris);

1987, visiting professor in Institute of Human Sciences, Kyoto University, Japan;

1992, Member of the National Ancient Books Consolidation Group.

Research interests: History of Science and Technology; the History of Science and Technology Exchanges between China and Foreign Countries.

Major works: *The History of Chinese Papermaking technology, Carl Schorlemmer, The Critical Biography of SONG Yingxing, The Annotate and Study on Exploitation of the Works of Nature (Tian Gong Kai Wu), The History of Chinese Metal Movable-Type Printing Technology, Science Exchanges between Chinese and Foreign Countries, The Four Great Inventions of China, The History of Chinese gunpowder, The History of science and technology exchanges between China and foreign Countries.*

He has published 210 papers in total.

Awards: 4 first prizes and 4 second prizes for excellent books.

Xiaowu Guan Born in Feidong, Anhui in 1970, PhD, researcher at the Institute of Natural Science History, Chinese Academy of Sciences, the assistant director of the Institute, director of the Chinese Academy of Sciences Cultural Heritage Science and Technology Cognitive Research Center, deputy secretary-general of the Chinese Traditional Arts and Crafts Research Association.

Research interests: History of technology; Traditional crafts; Archaeometry, etc.

Major works: *Tracing the Source—History of Composing Bronze Clock*.

Chapter 26
Intensive Cultivation and Eco-Agriculture

Zongdian Min

The intensive cultivation and eco-agriculture is a great invention that was made by ancient Chinese and had played a prominent part in the ancient agricultural production. It, however, did not refer to a single technology at all, but a series of technical measures. When there are more people and less land, the adoption of intensive production and management could not only raise the grain yield per unit, but also maintain ecological equilibrium as much as possible so as to make sustainable use of farmland.
The intensive cultivation and eco-agriculture covers appropriate use of farmland, improvement of soil condition, cultivation of crops, command of agricultural timing, collection and application of manure, selection and breeding of seeds, pest control, and so on. In addition, all these fields interact with one another, providing such proper conditions for agricultural production in air, temperature, water, fertilizer, soil, seed, etc.

26.1 Reasonable Utilization and Improvement of Farmlands

China is a country with more mountains but less plain, with larger population but less farmlands. In order to promote agricultural production, the ancient Chinese had developed and improved many barren lands into fertile farmlands, providing necessary bases for agricultural production and accumulating rich experience in land use and soil improvement.

Z. Min (✉)
China Agricultural Museum, Beijing, China
e-mail: minzongdian@hotmail.com

J. Hua and L. Feng (eds.), *Thirty Great Inventions of China*,
https://doi.org/10.1007/978-981-15-6525-0_26

26.1.1 Reasonable Utilization of Land

Throughout the history, sterile lands such as hilly land, flood land, and coastal land have always been developed into farmlands in different ways.

The early utilization of hilly land by Chinese ancients was just the plantation of crops along the slopes, which was called Shetian, slash-and-burn cultivation. The lands adopted with such cultivation could serve 3 years in general due to severe soil and water loss. This was the original utilization form of hilly lands. In the Song Dynasty, a new utilization form appeared, that is, to build low banks of earth with the same contour line in accordance with hill slopes, and then to develop farmlands inside those low banks, which played a significant role in preventing the loss of soil and water. In this way, ancient Chinese realized both the development of hilly areas and the maintenance of soil and water. This kind of farmlands ascended like terraces, hence the name terraced field (Fig. 26.1). The name "terraced field" (Titian in Chinese) was first seen in the travel notes *Can Luan Lu* (*Register of Mounting a Simurgh*) by Fan Chengda in the Song Dynasty. One more way to open up hilly lands was to divide a hilly land into seven layers from top to bottom; the lower five layers could be planted, from the bottom layer and then upward. In this way, the fertile mud could be gathered in the fields; with ditches built alongside, rainwater could be stored for irrigation. This was indeed an ingenious utilization of hilly lands as it not only developed farmlands, but also prevented soil erosion.

Developing flood land into farmland has been an important measure adopted along the Yangtze River. The main method involved is to build dikes to enclose fields inside and block water outer side. The field developed out of this method was called diked field, also known as polder field (low-lying paddy field). The original polder field was relatively simple and later evolved into a much complex one which combined dikes, culverts, sluices, and ditches (Fig. 26.2). This kind of polder field "just functions like a big city, with ditches in the middle and sluices outer side. The sluices will open in drought to let in the water from the Yangtze River to irrigate, and shut down when overwatered to prevent any flooding disaster." Accordingly, the former wasteland could be transferred into fertile fields that "are free from drought and flood, and benefit numerous farmers" [1]. In the Northern Song Dynasty, the length of the dyke banks within Dangtu County and Wuhu County of Anhui Province reached 240 km, and in the Southern Song Dynasty; there were as many as 1,498 polder fields around the Taihu Lake. Polder fields had played a considerable role in expanding the farmlands at that time. Besides, there were also silted field, countered field, and sandy field developed, which were small in scale, but all represented a method to develop flood land suiting local conditions.

To develop coastal land, the first thing was to build a dike to block water, and then an improvement was made by "digging a ditch beside the field to gather rainwater for irrigation in drought" [2]. This belonged to the cultivating procedure of salt elimination by digging a ditch and freshwater storage for irrigation. As coastal lands contain high salinity, it was impossible to grow crops on them until a desalination process was completed. The biological desalination method adopted was: "First plant

Fig. 26.1 Terraced field

tares to absorb extra salt in the soil. When the salinity is exhausted, the land is ready for cultivation" [2]. After such kind of management, "the land can produce ten times of grain yield as much as average land." The profit was indeed striking (Fig. 26.3).

26.1.2 Soil Improvement

As for soil, ancient Chinese recognized that "the conditions of soil are quite different. Some are fertile, while others sterile." At the same time, they also figured out a series of measures to improve the different conditions of soil. "Whether a land is ideal for cultivation depends on how it is managed. Well managed, any land can be prolific"

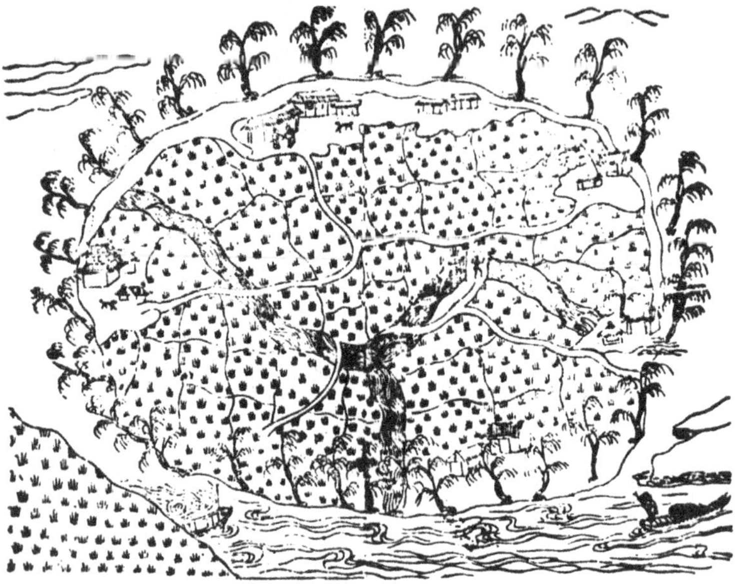

Fig. 26.2 Polder field

[3]. Hereunder, the achievements that ancient Chinese had made in soil improvement will be discussed by presenting examples of the improvement of the saline-alkali land in Northern China and the improvement of the cold waterlogged field in Southern China.

The earliest method adopted to improve saline-alkali land was to plant paddy rice, which ingeniously realized the dual effects of growing rice and improving soil condition through a combination of paddy rice planting and salt leaching. This method was adopted as early as in the Warring States Period by Ximen Bao, the then governor of Ye (now in Hebei Province), and worked so remarkably that "rice and sorghum could even grow on most of the infertile saline-alkali land." It had been adopted generation by generation.

Digging ditches to drain off salt was also an ancient method to manage saline-alkali land. In the Warring States Period, the high level of salt and alkali in soil was listed as a major problem of agricultural production in *Lv Shi Chun Qiu: Ren Di* (*The Annals of Lv Buwei: The Requirements of the Land*), which records that "can you decrease the level of salt and alkali in soil through irrigating soil by digging ditches?" This method took relatively simple procedures, so had also been adopted by later generations.

Irrigating soil with silt to force salt down in soil was also a method adopted to improve saline-alkali land in the Warring States Period. When the First Emperor Yingzheng of Qin gave the order to build the Zhengguo Canal in 246 B.C., "the silt in the Zhengguo Canal was used to irrigate saline-alkali land" [4]. In this way, a large area of saline-alkali land was improved, and as a result, the Guanzhong basin

Fig. 26.3 Coastal land

(now the central Shaanxi plain) was changed into a fertile place, reputed as "the most productive place in the world" [5]. In the reign of Emperor Shenzong of the Song Dynasty, this method was also adopted in the provinces of Henan, Hebei, Shanxi, and Shaanxi. According to records, "in the regions of Shen, Ji, Cang, and Ying, only the land which could be irrigated with silt by the Yellow River, the Hutuo River and the Zhangshui River became fertile, otherwise saline-alkali land," [6] which shows the efficiency of this method.

In addition, many other effective biological measures such as green manure application and tree planting were involved to improve saline-alkali land. Moreover, deep plowing was also a method of desalination. These methods were widely adopted during the Ming and Qing Dynasties and turned out to be effective, according to local historical records [7].

Distributed in Sothern China, cold waterlogged field was acidic in nature, low in soil temperature, short of phosphor and kalium, which largely hindered the growth and yield of rice. Accordingly, it was imperative to improve the conditions of cold waterlogged field in order to increase its rice yield. One of the methods involved was to heat up the temperature of soil through plowing and smoldering. According to *Agricultural Book* written by Chen Fu of the Song Dynasty, "Most of the swamps among mountains and rivers are low in temperature. In winter, plow them deep and drain off water; frozen up by snow and frost, the soil will become crisp. When spring comes, burn withered grass to warm up the soil. In doing so, the germination percentage of seed can be largely improved; even the icy springs are unable to affect the seedlings. If not, few seedlings could grow in such cold soil under the influence of cold springs." The second method was to plow frozen fields in winter. *Annals of Guiyang County* in the Qing Dynasty records: "The fields near mountains are relatively low in temperature…when in winter, store water in them and plow, planting no crops then. If a field is left dry in winter, its yield will probably fail in the following year." The third method was to apply lime and ashes to manage, which could neutralize the acidity of soil, supply phosphor and kalium, and loosen the soil. This method can be found in some local records [8], and its effect was easily perceptible. In addition, solarization was also a good way to improve cold waterlogged field. According to records, "cold waterlogged fields in Xinchang County and Shengxian County are too cold to plant early rice. Rice seedlings may not be transplanted until summer solstice. When rice seedlings grow stronger, there is no need to irrigate. Leave the fields bathed in blazing sunshine until they become cracked. In the end of July and the beginning of August, it begins to rain, so the fields turn to be loose and the rice seedlings are growing luxuriant. If it doesn't rain, draw water to irrigate the fields. Both want of sun exposure or early exposure to rain will cause rice plants grow thin, rarely growing into tufts" [9].

26.2 Crop Rotation and Multiple Cropping

It was an important measure in ancient agricultural production to improve land use capability and crop yield per unit through rotation and re-plantation. This measure had appeared even in the Warring States Period. The records "plant millet this year and wheat next year" in *Lv Shi Chun Qiu: Ren Di* just refers to the millet–wheat rotation. The notes in *Zhou Li: Ti Shi* (*The Rites of Zhou: The One Who Weeds*) made by Zheng Xuan in the Eastern Han Dynasty made such a description: "In today's countryside, millet is firstly planted in fields, then wheat, and finally legume." It shows that the millet–wheat rotation had further evolved into a millet–wheat–legume rotation, that is, triple cropping achieved in 2 years.

On the basis of the millet–wheat–legume rotation developed in the Han Dynasty, a legume-centered plantation system was formed during the Wei, Jin, Southern, and Northern Dynasties (220-589 A.D.). It included the legume–grain crop rotation system and the green manure crop rotation system which featured by the rotation of legume with other crops. Jia Sixie, a famous agronomist in the Northern Wei Dynasty of China, made a systematic summary of rotation methods in his book *Qi Min Yao Shu* (*Essential Ways for Living of the Common People*).

First, rotation was a better method for most crops than sequential cropping. For example, "grains should not be planted in the same field for even two consecutive years" (*Chapter of Grain Cultivation*); "hemp should be planted in good fields, not the ones planted last year" (*Chapter of Hemp Cultivation*); "it is better to rotate rice with other crops annually" (*Chapter of Rice Cultivation*). The reasons involved are because "weeds will grow in grain fields and the yield of grain subsequently decreases" if planted sequentially; "the leaves of hemp may wither prematurely"; "rice field will be invaded by tares which are rather difficult to eliminate." Jia Sixie theoretically clarified the importance of rational rotation of crops and pointed out that rotation is able to eliminate weeds, control pests, and increase yield.

Second, green manuring rotation system was a "good method to fertilize field." Jia Sixie highly affirmed the effect of this system. According to *Qi Min Yao Shu: Cultivation*, "To fertilize field, the best green manure crop is mung bean, and red bean and benne come second. Plant green manure crop in May or June, and plow the field in July or August in preparation for planting grain in next spring. In doing this, per mu (ancient Chinese unit of area) field can yield ten *dan* (ancient Chinese unit of weight, one dan equals to 50 kg) grains. The effect of green manure is similar to the effect of excrement of silkworm or other animals." The legume was consciously brought into the rotation system in China's agricultural production. The integration of land utilization with land maintenance marked the beginning of biological land maintenance in China.

During the Tang and Song Dynasties, a large number of northern people moved southward to the lower reaches of the Yangtze River, which had been sparsely populated before and became densely settled. As a result, the demand for wheat largely increased there and thus the price of wheat soared. To deal with this problem, people there began to plant wheat in the slack seasons of paddy fields. Therefore, a double-cropping system of rice–wheat rotation took shape in the south. This system first appeared in Yunnan in the Tang Dynasty and spread to the lower reaches of the Yangtze River in the Song Dynasty. According to *Local Records of Wu Prefecture*, people in the area of Suzhou (now in Jiangsu Province) "plant rice after reaping wheat, achieving double cropping a year" at that time. In the Song Dynasty, besides the rice–wheat rotation, rice–rape rotation, rice–broad bean rotation, and rice–vegetable rotation were also adopted. In the Southern Song Dynasty, Chen Fu wrote in his *Agricultural Book*: "Soon after the harvest of rice, people begin to plough the field and expose it to the sun. After heaping excrement to the field, they plant legume, wheat or vegetables in the field." It just manifests the implementation of this rotation system. The establishment of this system in the lower reaches of the Yangtze River had great significance economically and agriculturally. Firstly, the multiple-cropping index got raised and grain yield increased, which opened up a new way for alleviating the pressure of farmland shortage. Secondly, this system helped to maintain soil conditions though rotating water and land crops, and at the same time improved soil fertility. Therefore, Chen Fu summarized that this cultivation system not only "has the effect of improving soil fertility," but also achieved strong economic significance of "saving some labor to fertilize the fields in the next year and also increasing annual yield." Nowadays, this system still remains the major cultivation system in the region of the lower reaches of the Yangtze River.

During the Ming and Qing Dynasties, with the population growth, the lack of farmland became a nationwide problem. The major solution that could be adopted was still to develop multi-cropping system and improve land utilization rate. As a result, the systems of double cropping rice and triple-cropping rice were formed in the south, while the system of triple cropping in 2 years came into being in the north.

As for double-cropping rice cultivation, there were mainly two forms. One was double continuous cropping. According to *Tian Gong Kai Wu* (*Exploitation of the Works of Nature*) written by Song Yingxing in the Ming Dynasty, "People in the southern plain plant two batches of rice in the same field and harvest twice a year"; "in June, people crop the first batch of rice in maturity, cut down rice stalk, plough the field, and then plant the second batch of rice." The other one was double intercropping, that is, "plant the first batch rice seedlings in rows of generous spacing. After ten days or so, plant the second batch between those rows. When the first batch matures in the beginning of autumn, reap it and uproot its stalk. Heap those stalks around the root of the later batch as a kind of green manure so the later batch grows luxuriant and fruitful" [10]. The form was mainly prevalent in the areas of Zhejiang, Jiangsu, and Fujian.

The triple-cropping system was mainly popular in the south of the Yangtze River Basin and the coastal areas of South China, where the temperature was relatively high. The planting pattern mainly includes double-cropping rice plus one dryland crop, and the triple rice–rice–wheat cropping in a year. This system was introduced in *Fu Zhou Fu Zhi* (*Local Records of Fuzhou Prefecture*) written in the Ming Dynasty: "After reaping wheat in April, two batches of rice will be planted, so there are three crops in a year." In addition, there had developed other patterns of triple cropping, such as rice–rice–rice pattern, rice–rice–potato pattern, rice–rice–radish pattern, and wheat–rice–legume.

Northern China is not as pleasant in temperature as that in the south, so the quadruple cropping in 3 years and the triple cropping in 2 years were mainly adopted. The basic plantation form was grain-centered, added with proper crops that can improve soil, supported by oil-bearing crop and autumn grain crop. In the Qing Dynasty, Liu Guiyang wrote in his book *Shuo Jing Can Gao* (*Manuscript on Classics Explanation*) that "the high land yields three times in two years sequentially of wheat, legume and grain crop such as durra, rice, or millet. Likewise, the low land also yields three times in two years, but sanzi (a kind of grain crop) is planted in late September… and after the harvest of wheat, legume is also planted," reflecting the dryland plantation in the north.

As the multi-cropping system was further developed during the Ming and Qing Dynasties, a new ecological agriculture system was created, which made a comprehensive use of land, water, and crop resources.

This kind of agricultural management first appeared in the area of Changshu in the reign of Emperor Jiajing of the Ming Dynasty. According to *Chang Zhao*

He Zhi Gao: Anecdotes, the low-lying land in this region was often flooded, so many local people abandoned their farmland and moved to other places. There were two brothers, Tan Xiao and Tan Zhao, who were good at management. They hired villagers to reclaim these waterlogged fields, chiseling the lower land into ponds, and enclosing the higher land into fields. As a result, the annual output was three times higher than that of ordinary fields. Besides, they also raised fish in ponds and pigs and chickens in pigsty which was built over the ponds, the excrement of pigs and chickens being the fodder for fish. Also, fruit trees like plum and peach were planted in the embankment, and rice, gromwell, water chestnut, and Gordon Euryale were planted in the low-lying land. Once again, the annual income was three times higher than that of ordinary fields. In this way did the two brothers acquire wealth. The comprehensive management mode of "grain–animal–fish–fruit–vegetable" was the initial artificial ecological agriculture. Later, this mode was spread to the areas of Jiaxing and Huzhou in Zhejiang Province. The local people developed it in accordance with the local resources. They fed livestock with grain and in turn fertilized crops with the excrement of livestock; they fed silkworm with mulberry, raised fish with silkworm feces, and fertilized mulberry with fish manure. The skillful combination and cyclic utilization of land, water, animal, and plant resources achieved satisfying results: the land became productive, soil-rich, grain luxuriant, and mulberry exuberant. This mode was further adopted in the Pearl River Delta. The locals also combined the process of fish farming, mulberry planting, rice planting, and silkworm breeding. The mulberry-based pond ecosystem, hence, was formed, acquiring obvious ecological and economic benefits. According to *The Annals of Gaoming County* written in the reign of Emperor Guangxu of the Qing Dynasty, "Dig deep those waterlogged land, and make it into pond by building ridges around it with mud. Six ridges can enclose four ponds. Mulberries are built in ridges; fish are kept in ponds; silkworms are feed with mulberry; fish are raised with silkworm excrement. These procedures interact with each other and yield a tenfold increase in the annual output." In addition to mulberry-based ecosystem, there were also fruit-based ecosystem, rice-based ecosystem, sugar-based ecosystem, vegetable-based, and so on. This marked the beginning of artificial ecological agriculture in China and received high evaluation from foreign countries.

26.3 Ways for Land Cultivation and Command of Agricultural Timing

Northern China is a region with less and mal-distributed rainfall. In the planting spring, this region is drought and windy, and water in the soil evaporates quickly. Therefore, fighting against drought has been a top priority in the land cultivation of Northern China.

The cultivation methods initially adopted included "Shen Geng Ji You (plowing deep and smashing soil blocks)" (*Guoyu: the State of Lu*) and "Shen Geng Shu You (plowing deep and grinding soil blocks)" (*Zhuangzi: Ze Yang*). According to *Lv Shi Chun Qiu: Ren Di*, "The land moisture can be ensured if being plowed deeply enough." This was because deep plowing could increase the soil's ability to store water and help crops deeply root into the ground. "You" was a wooden hammer used to smash soil blocks. "Ji You" means to smash soil blocks soon after plowing, and "Shu You" means to smash soil blocks into powder so as to reduce the evaporation of soil moisture. This was the earliest method for maintenance of land moisture, namely, "Geng-You method". In the Han Dynasty, the Geng-You method was developed into "Geng-Mo method". According to *The Book of Fanshengzhi*, "The wheat field can be cultivated in both May and June, but never in July. Just flatten the land in July with Mo for plantation." "Mo" was a kind of toothless rake that was dragged by a bull to crack soil blocks and flatten the land. Its effects in smashing soil, flattening land, and reserving moisture were much better than those of "You", so that the Geng-You method was replaced by the Geng-Mo method.

During the Wei and Jin Dynasties, the Geng-Mo method was further evolved into a Geng-Pa-Mo method to resist drought and reserve moisture. The specific procedures of this new method were even manifested in the Wei-Jin murals on the walls of the Jiayuguan Pass. Jia Sixie introduced in detail the specific procedures of the method in *Qi Min Yao Shu: Cultivation*. First, the suitable period for cultivation is up to the soil moisture. According to the book, "Be it spring or autumn, it is better to plough the fields when the soil is neither too arid nor too damp." This is because the field with moderate moisture is prone to crack and easy to be plowed, saving the bull's effort. If the soil moisture goes out of control, "arid land are preferred rather than damp land" in that "the soil will disintegrate soon after rain, while the soil of damp land tends to integrate into rigid blocks, and the water in soil will evaporate, affecting agricultural production for several years."

Second, the depth of land cultivation should vary according to farming period. According to the book, "The land should be plowed deep for the first time, but shallow for the second time." After harvesting autumn crops in the Yellow River Basin, people should plow land deep to store more water in the land from rain and snow so as to freeze–thaw the weathered soil. In the spring and summer when the Yellow River Basin is in drought, the temperature will get higher and the water will evaporate faster. If the second plowing was still deep, the raw soil would be turned up, which would make the clods vary in size, and destroy the soil structure. As a result, the soil moisture was lost and the sowing was affected.

Third, use Pa and Mo to cultivate land after tillage so as to fight against drought and maintain soil moisture. According to the book, "In the windy spring, use Mo to crack soil blocks and to flatten the land. If not so, the land will be in drought as a result of soil moisture loss"; "through repeated cultivation, the soil moisture can be conserved even in drought period" [11]; "a short period of drought cannot damage the whole land's crop. Why? It is due to the repeated cultivation by Mo" [12]. For the cultivation timing, the book says: "After spring ploughing, the land should be tilled from time to time"; "while in autumn, the land should be tilled when the soil in the surface turns gray after long exposure to sun"; "after ploughing, it is better to let the land basking in the sun if one does not plan to till it" [12].

Fourth, as Jia Sixie emphasized, as for the specific operation of the tillage, "the tooth of plow should be small and the land should be hoed over and over again" [12]. The land could only be thoroughly tilled with plow of small tooth; otherwise, wipe land strip may leave uncultivated. Moreover, the land should be hoed repeatedly after tillage so that the soil moisture can be maintained even in drought.

In addition, Da, a farm tool used to compact soil, was also used after tillage. According to the book, "In spring, the land should plow deep, and Da should be applied to compact soil; in summer, the land should plow shallow, and it is not necessary to treat it with Da." Da is used in spring in order to help seeds adhere to soil, raise soil moisture, and protect the seedlings; in summer, if treated with Da, the soil would be densely compacted once rain falls.

Since then, the cultivation system for dryland in Northern China had been shaped. The Geng-Pa-Mo method realized the drought resistance and soil moisture preservation for the dryland in the north and provided solutions for the agricultural development in lack of rain. It was an agricultural technical measure that best represented the intensive cultivation in traditional agriculture of China.

Since the Northern Wei Dynasty, the importance of harrow had been especially emphasized, and frequent and fine operations with harrow were advocated. According to the agricultural book *Han Shi Zhi Shuo* written in the Jin and Yuan Dynasties, "In ancient farming, plow once and rake six times when cultivating land. Nowadays, people only focus on deep ploughing, neglecting the significance of fine raking. Without thorough operation with harrow, the soil would be rough and loose. Even though seeds can germinate after sowing, the seedlings may die of suspension, pests, drought or others, as their roots are not closely adhered to the soil. With thorough operation with harrow, on the other hand, the soil would be fine and smooth. The roots of seedlings are tightly adhered to the soil after slightly rolled, free from drought and any other diseases." Frequent and fine operations with harrow could resist drought and conserve soil moisture, ensure the germination of seeds and the healthy growth of seedlings, and reduce pests and diseases as well. The realization of these marked that the dryland cultivation in Northern China had reached a more mature level in drought resistance and soil moisture conservation.

Similar to the Geng, Pa and Mo preparation procedures before plantation, hoeing lands after plantation was also important in soil moisture conservation, and attached great significance in ancient agricultural production. *Qi Min Yao Shu: Chapter of Grain Cultivation* records: "The so-called hoeing involves much more than just weeding, and then the soil becomes fertile, the yield and quality of crops increase." Especially in spring, people always hoed land to loosen soil and conserve soil moisture. According to the book, "Hoe in spring to loosen soil, and hoe in summer to eliminate weeds." In order to ensure the quality of inter-tillage, different farm tools and corresponding cultivation techniques were thus created at that time.

This Geng-Pa-Mo-Chu method provided favorable conditions for drought resistance, soil moisture preservation, and bumper harvest in Northern China. From the perspective of the history of the agricultural science and technology in the world, this exquisite and intensive cultivation technique was indeed an amazing invention.

Agriculture is directly constrained by seasonal and climatic conditions, so it is a basic requirement for agricultural production to master and make full use of different conditions. As the ancients said, “It is said that ‘do not start to cultivate before the right timing, and do not reap after the right timing’. Therefore, it is a must in agricultural production to grasp the agricultural timing. In doing so, no failure will occur in the process of plantation, cultivation, breeding and harvesting” [13]. This reflected that the ancients had thought highly of agricultural timing.

During the Spring and Autumn and the Warring States Period, ancient Chinese had created the 24 solar terms to grasp agricultural timing according to the position of the earth in its revolution around the sun. A year is equally divided into 24 solar terms, each of which represents different weather conditions such as cold, warm, drought, or damp. The weather conditions the solar terms indicate are exactly the same as the actual conditions in the middle and lower reaches of the Yellow River, which is thus very convenient for people to conduct agricultural production according to the 24 solar terms, the unique Chinese agricultural calendar.

Due to the different natural environments in different places, the weather condition in different regions may vary even in the same solar term, so the phenological timing method was created as a supplement in order to correctly grasp agricultural timing. Phenology is a branch of science which indicates seasonal climatic changes through the phenomena of plant germination, flowering, and fruiting, bird migration, early frost, thawing, and so on. It is more specific and correct. This method had been applied in agricultural production in the Spring and Autumn and the Warring States Period. According to *Lv Shi Chun Qiu: Ren Di*, “About two months after the Winter Solstice, calamus germinates ahead of any other grass. People thus begin to cultivate.” In the text, the germination of calamus was regarded as a phonological phenomenon and marked the beginning of cultivation period.

In the Han Dynasty, a wooden pile chronometry was invented to determine cultivation period. According to *The Book by Fan Shengzhi*, “When spring comes and the frozen land thaws, strike a wooden pale into the ground, with two *cun* (about 6.67 cm) above the ground. When temperature rises in the spring, soil blocks disintegrate and bulge, burying the whole wooden pile; then it’s time to pull out old roots. Twenty days later, cultivate the land when the soil turns rigid. The effect to cultivate at this time is much better than that to cultivate later.” The agriculture timing was thus figured out by observing the length of the wooden pile above the ground.

In the Yuan Dynasty, agronomist Wang Zhen created *Shou Shi Tu* (*Graph of Agricultural Timing*) by combining the calendar, solar terms, phenology, and farming. The graph divides a year into 12 months, each of which includes two solar terms, with the phonological features and suitable farming activities for this month listed below. In this way, people could have a better command of farming.

26.4 Selection and Breeding of Seeds

Seed selection and breeding is treated as an important part of intensive farming and ecological agronomy, and also an excellent tradition in the agricultural production of China. As early as 3,000 years ago, the *Book of Songs: Daya: Civilian* recorded the concept of "Jia Zhong", which was interpreted as "the species of good yield" or fine breed by the ancients. During the Warring States Period, *Lv Shi Chun Qiu: Ren Di* put forward the standards of fine breed: "the stalk should be straight and the stem strong," "the ear should be big and the shell thin," and "the millet should be plump and the taste excellent." However, there had been no specific measures formulated for seed selection at that time.

In the Han Dynasty, great importance had been attached to seed selection, for which the earliest measures were created. According to *The Book by Fan Shengzhi*, "To select wheat seeds, pick out big wheat ears with full grains when in maturity. Cut them down, tie them into bundles and store them in the high and dry place"; "to select rice seeds, choose the taller ones, cut them down and bundle them up. Hang these bundles in a high and dry place so that the seeds will not rot away." These records indicate that there had been clear seed selection measures at that time: to select ears of different crops separately, and store them mixed in bundles at a high and dry place to prevent them from damp.

Later in the Northern Wei Dynasty, people realized that mixed varieties would be pernicious. "If the seeds of different crops are mixed, their maturity periods will vary and the yield rates decrease. Besides, the mixed grains are unfavorable in the market as they tend to be half-cooked. Therefore, special attention should be paid to the selection and storage of seeds" [14].

In order to separate different kinds of seeds and obtain fine seeds, a measure was created: "Harvest different crops separately. Choose the big ears of the same kind according to color, cut down and hang them high. When spring comes, sow different seeds into different fields to breed seeds for the next year. The breeding field should be carefully maintained. After threshing, different kinds of seeds should be stored separately in different cellars, with their own stalks covering the cellars as a mark. If not, the risk of being mixed still exists" [14]. Such elaborate measure turned out to be an effective method in selecting and breeding fine varieties, probably regarded as the earliest seed breeding developed in China.

In the Ming Dynasty, the technique of seed selection had been further developed. Geng Yinlou, an agronomist of the Ming Dynasty, gave such a description of seed breeding in his book *Guo Mai Min Tian*: "As men have their fathers, all grains, fruits, peas and vegetables would have their own seeds, for which the land is the mother. If the land is to be fertilized and the crops are to be strong, the seeds should be carefully selected. As for the methods… that is, to carefully select several *mu* of fine land to plant on. The crops to be planted, grains or beans, should be carefully selected so that those seeds with strong and smooth quality can be sorted out for breeding. The manure and the workload cost with plowing and hoeing tools in this kind of land will be several times those in other fields… As a result, their seedlings will be stronger, and the seeds they bear will be fuller than the original… Pick up the greatest ones as the seeds for next year. When the crop yields triple within three years after repeating the same steps, the grain will be as big as the millet." This kind of seed selection, combining seed selection and breeding, was much more elaborate than that mentioned in *Qi Min Yao Shu*. By this time, the technique of mixed selection and breeding of seeds developed in ancient China had turned to be quite mature.

In the Qing Dynasty, the selection and breeding of seeds in China had undergone new development, which was manifested by the emergence of the breeding method of single heteromorphic ear of grain. The breeding method was recorded in *Ji Xia Ge Wu Bian* edited by Emperor Kangxi of the Qing Dynasty: "In the Fengze Garden (now in the Sea Palaces in Beijing), there are several areas of paddy fields where a kind of crop Yutian grain is planted. In September, the crop will be harvested. One day in late June when the grain just started to form ears, I walked along the paths among the fields. Suddenly, I saw a head of rice that stood higher than others, with the rice grain already filled. I collected the seeds of the head of rice, intending to verify their pre-maturity by next year. In next June, the rice grown from those seeds ripened earlier than other plants as expected. From then on, the kind of rice grows steadily and I could harvest many of them every year."

Actually different from the original "Yuantian grain", this kind of grain was "reddish in color and elongated in size, smelling good and tasting delicious," completely a new species. Here, a single heteromorphic plant was selected and bred into a new species, which was also the earliest known record of individual plant selection and systematic breeding in China. This event attracted the attention of Charles Darwin, British naturalist, and founder of evolution. He said in his book *The Variation of Animals and Plants under Domestication*: "In the edict of Emperor Kangxi, he advised people to select seeds that are remarkable and large. Even the emperor selected seeds himself. It's said that Imperial Rice, that is, rice grown for the emperor, was noticed by Emperor Kangxi himself in a field in the past, thus preserved." It can be seen that this event enjoyed an impact abroad.

In ancient China, great achievements had been made in seed selection and breeding, which were mainly reflected in the following aspects.

Firstly, a large number of varieties of crops had been developed. Take the two grain crops, millet and rice, as examples.

Millet is an important grain crop in Northern China. According to *The Book of Songs*, there were two kinds of millet in the Western Zhou Dynasty. In the Western Jin Dynasty, according to *Guang Zhi* (*Extensive Annals on Nature*), the number of millet varieties had increased to 11. In the Northern and Southern Dynasties, according to *Qi Min Yao Shu*, there were 86 varieties of millet. In the Qing Dynasty, according to *Shou Shi Tong Kao*, there were about 500 varieties of millet (some were counted repeatedly).

As for rise, it is the main grain crop in Southern China. According to *Guan Zi: Chapter of Di Yuan*, there were 24 varieties of rice. In the Song Dynasty, according to the modern statistics on 12 local chronicles of the dynasty, there were a total of 301 varieties. Removing those repeated varieties, there were still 212 varieties of rice. In the Qing Dynasty, according to *Shou Shi Tong Kao*, the number of the varieties of rice reached about 3,429 (some were counted repeatedly).

All these above reflect that the grain crops cultivated in ancient China were of a great variety and developed pretty rapidly.

Secondly, ancient China saw very rich varieties of grain crops. Still take rice as an example, whose types include the following:

In terms of growth period, different rice could be harvested in different periods, such as "50 days", "60 days", "100 days", "urgent 30 days", and "slow 200 days".

As for plant type, it had long-stalked and short-stalked types.

In panicle type, there were such various types as long awns, short awns, long grain, short grain, round grain, and oblate grain. There was also such type of rice with sterile lemma such as Feilaifeng rice and Fenghuang rice.

In terms of rice color, different varieties varied from snow white, buff, bright red, half purple to mixed black.

In quality, there was fragrant variety such as fragrant japonica rice and Xiangzi rice, the variety that is suitable for winemaking such as Jinchai glutinous rice and Xiaoniang glutinous rice (rice whose straw are weak), and the variety that has nourishing effect such as Kaxue glutinous rice.

In terms of resistance, there was the variety that is resistant to drought such as Pieshatian rice and Zhancheng rice, the variety that is resistant to flood such as Zhangshuihong rice and Deepwater rice, the variety that is resistant to cold such as Wukou rice, the variety that is resistant to salt and alkaline such as Salty rice and Wumang rice, and the variety that is resistant to lodging such as iron-like straw rice.

These various types of rice, some of which had made contributions to the control of insect pests, animal damage, and natural disasters, some of which served as auxiliary cultivars for multiple cropping, playing an important role in easing seasonal contradiction between crop harvests, symbolized the great achievements ancient China had made in rice breeding.

26.5 Reasonable Use of Manures

As one of the important characteristics of ancient agriculture of China, emphasizing manure collection and fertilization was a vital part of the intensive cultivation and eco-agriculture. It was also the significant reason why the agriculture of China remains flourishing after a long-term cultivation for thousands of years. Therefore, there were such sayings as "cherishing manure as gold" and "using manure as medicine" in ancient times. The former means that collecting manure is as important as accumulating gold, while the latter means that fertilizer shall be applied as appropriately as medicine. In ancient times, fertilizer was called manure, and fertilization referred to applying fertilizer on fields. Fertilization had been documented in such literatures of the Warring States Period as *Han Fei Zi: Expounding Lao-Tzu's Theory*, *Meng Zi* (*The Book of Mencius*): *Part A of Duke Wen of Teng*, and *Xun Zi* (*The Book of Xun-Tzu's Theory*): *National Prosperity*. These documents indicate that China had started attaching importance to fertilizers and fertilization as early as 2,300 years ago.

Why did ancient Chinese pay so much attention to fertilizers and fertilization? It was closely related to the understanding of ancient peasants of the roles that fertilizers could play in agriculture production.

26.5.1 Improving the Productivity of Land and Fertilizing Soil

In the Spring and Autumn and the Warring States Period, people had realized it. According to *The Book of Rites: On Order*, spreading manure "can fertilize soil and improve the hardened land," that is, conducive to making hard and heavy soil fertile. Wang Chong, a thinker and literary critic of the Eastern Han Dynasty, made a further introduction to it in his book *On Balance: Nature of Soil*: "Fertility and barrenness are two different natures of land. The land with fertile soil and sufficient irrigation is of good nature, and the crops planted on it will flourish. The land that is not flat and infertile is of bad nature, but the crops grown on it can be similar to those grown on fertile land in harvest, only if its soil fertility is improved through deep ploughing, careful hoeing and manure applying by more manpower." Here, fertilization was clearly clarified to be conducive to improving land productivity and soil fertility. Chen Fu, an agronomist of the Southern Song Dynasty, further stated in his *Agricultural Book: Chapter of Applying Manure according to the Soil*: "Different types of land differ in soil and climate. Some of the land is fertile, and some barren. But whatever it is, the method to manage it shall be appropriate. The land with black and fertile soil is certainly good, but over-fertile soil can also result in a situation in which the crop seedlings flourish while their seeds are not plump… The land with poor soil is certainly barren and of bad nature, but if it is applied with manure so as to fertilize its soil and intensively cultivated, the crop seedlings on it will also flourish, and their seeds will be also plump." In addition, the agronomist noted the decisive role of fertilization in turning barren soil into fertile one. According to *Zhi Ben Ti Gang* written in the Qing Dynasty, "although the land is barren, once it is regularly spread with manure, it can turn to be fertile farmland"; "after continuous cropping, a land may lose its fertility, so nothing will grow on it. Spread manure over the land to fertilize its soil, and its productivity will be maintained." It indicates that fertilization can improve soil fertility and land productivity. This is a consistent Chinese understanding in this regard that has been gradually developed and deepened for more than 2,000 years since the Warring States Period.

26.5.2 Turning the Waste into Valuable Resources to Improve the Yield

Fertilizers in ancient China mainly came from the waste produced in family life, the parts of agricultural products that could not be consumed by human beings and animals, the sludge in rivers and sewers, and so on. These things were originally useless, but if accumulated as fertilizers, they could turn to be treasures for crops. Therefore, China has always attached great importance to the accumulation of fertilizers and fertilization and regarded it as an important way of making the waste profitable. Wang Zhen, an agronomist of the Yuan Dynasty, said in his *Agricultural Book*: "For those manures and something decayed, people think little of them and ignore them. However, when they are applied to fields, they will be timely rainfall that moistens the soil of fields. Only people who are engaged in agriculture know it. It is what is called cherishing manure as gold." According to *Zhi Ben Ti Gang*, "As for accumulating manure as much as possible, people belittle and ignore it. But if the manure is applied to fertilize the soil of fields, the fields will be productive. It is turning stinking manure into fine fertilizer, which will double the yield of grains, make vegetables and fruits grow more splendidly, and help people live an abundant life." Here, the significance of fertilization and fertilizer was recognized from the economic perspective of turning the waste into valuable resources to improve crop production.

In ancient times, China had paid special attention to the utilization of waste. Almost every waste that could be utilized was used as fertilizer. Therefore, ancient China held varieties of fertilizer, which were commonly called farm manure.

In the Warring States Period, human waste, animal manures, weeds, and plant ash had been used as fertilizer in China. In the Qin and Han Dynasties, the use of silkworm excrement, ground pupa, ground bones, beanstalks, and river mud as fertilizer had been recorded, among which the use of animal manures was particularly developed. The pigsty connected with latrine in this period had been unearthed in China. The discovery of this pigsty reflects the importance attached to manure accumulation by breeding pigs at that time, and also the long history of manure accumulation by breeding pigs in China.

In the Wei, Jin, and Southern and Northern Dynasties, besides the kinds of fertilizer mentioned above, both old-wall soil and cultivated green manure were also used as fertilizer. The cultivated green manure was of great significance in the development of Chinese fertilizer as a renewable and inexhaustible resource of fertilizer. In the Song and Yuan Dynasties, some inorganic fertilizers such as lime, plaster, and brimstone began to be applied to agriculture production. According to statistics, there were about 45 varieties of fertilizer in this period [15], among which the use of cake fertilizer and inorganic fertilizer stood as a new development of this period.

During the Ming and Qing Dynasties, as the multiple-cropping developed rapidly, the needs for fertilizers also increased at an unprecedented speed. It was an urgent requirement of this period to expand the sources of fertilizer by taking every possible step. The varieties of fertilizer were thus expanded, reaching more than 130 kinds of

12 categories [15]. It can be seen that there were various kinds of fertilizer in ancient China, in which organic fertilizers accounted for the vast majority, thereby forming the characteristic of Chinese fertilization: main application of organic fertilizers with the supplement of inorganic fertilizers.

Ancient China not only attached great importance to the accumulation of fertilizers, but also to the manufacturing and processing of fertilizers so as to improve the efficiency of fertilizers. The methods involved include the following:

(a) Retting miscellaneous fertilizers. In the Warring States Period, China already had the weeds in fields retted as fertilizer by utilizing the high temperature in summer, which was the beginning of the use of waterlogged compost. Chen Fu of the Southern Song Dynasty introduced one more method of retting miscellaneous fertilizers in his *Agricultural Book*: pile up the husks gotten through hulling and the decayed stalks and leaves into a pond, and collect the dishwater or washing water of rice to soak and decompose them; after a long time interval, they will decompose into waterlogged compost. In the Ming Dynasty, the method of retting fertilizers adopted in the region of the Taihu Lake was to evenly mix the stalks and shells of Chinese milk vetch or broad bean with river mud, and then accumulate them to soak and compost. This method of composting was called "pit plant" or "pit broad bean," [16] now called "cellar mixed with pond sludge" in Southern China.

(b) Composting animal manures. *Qi Min Yao Shu* written in the Northern Wei Dynasty records a method called manure treading: "After autumn harvest, each peasant family would collect all the stalks and falling leaves on the farmland, and store them in some place. Every day, these collected stalks and leaves would be spread on the cattle pen about three *cun* thick for the cattle to tread on. They would turn to be fertilizers overnight, which were collected and piled up next morning. In this way, one farm cattle could tread thirty carts of manure in total in a winter." This was a composting way that combined dry grass and earth spread in a pigsty with manure accumulation, actually the earliest method of manure composting created in China. In the Qing Dynasty, Sun Zhaikui described a method called "fertilizer making" in *Teaching Book of Crop-planting*, introducing in detail the accumulation and making of manures produced by cattle, sheep, horses, mules, donkeys, and pigs. The method was similar to but more elaborate than that of manure threading.

(c) Cake fertilizer fermentation. It was a method found in the Song Dynasty. According to *Agricultural Book* by Chen Fu, "maku (sesame dregs after oil extraction) is difficult to use. It should be finely pulverized with a pestle, and piled up with burnt plaster to make a cellar cover. As making wine, when the manure ferments and generates 'shumao' (one kind of monospora) after long-time fermentation, it should be spread out from the middle part. For warm manure, it should be set outer side. For cold manure, it should be placed in the middle and then covered with the cellar again. Repeat this process several times. The manure can be used until it cools down; otherwise, it will burn the seedlings." maku is an organic fertilizer with high quality and contains a large amount of nitrogen. If it is applied to the farmland directly, it will give out heat after fermenting and do harm to the growing seedlings. To avoid the result, such measures as pulverizing maku, mixing it with burnt plaster, and piling fermentation were taken at that time. Marked by the "emergence of shumao," the cake fertilizer should be treated with the previous steps for three or four times after fermentation, and then put to use until it cooled down. This indicates that ancient China had accumulated rich experience in the use of cake fertilizer, which was rare and estimable in the twelfth century.

(d) Burning lime–soil. It was a method of accumulating and making soil-fertilizer that was invented in the Song and Yuan Dynasties. Two approaches were involved. One was to "stack and burn the soil swept out, burnt-out ash, chaffs gain through winnowing, broken branches and falling leaves" [13], which was similar to the current way of burning plaster. The other was to "stack and burn the soil with vegetation; when the heated soil cools down, grind it with a stone roller" [3], which was exactly the same as the current way of smoldering soil.

(e) Making up manure pellets. Manure pellet is a mixed fertilizer with high concentration. Its main materials include human excrement, animal manure, poultry manure, sesame dregs, bean dregs, black-soybean dregs, animal carcasses, and their innards, hair, and blood. In addition, inorganic fertilizers such as black alum, arsenic, and sulfur are also included. The mixed materials were sealed in a pit or buried underground after sealed in a crock. After decomposition, the manure would be dried and crushed for use. This kind of fertilizer had extremely high fertilizer efficiency, one *dou* (ten liters) of which was equivalent to ten *dan* of human excrement in fertilizer efficiency, generally used as seed manure. This marked the beginning of the refining of compound fertilizer in China.

Ancient Chinese not only paid attention to the accumulation and making of fertilizers, but also to the maintenance of fertilizer efficiency. Two methods were covered. One was to build a manure house: "Beside each farmhouse, a manure house with low eaves should be built to prevent manure from erosion of wind and rain. If manure is exposed for several months, it will lose its fertilizer efficiency. Therefore, a deep holding pond should be dug out in the manure house, with four walls built with bricks to prevent leakage" [13]. The other was to set up a manure pit beside the field, which could maintain the fertilizer efficiency and facilitate the transportation. The way involved was described as "Brick manure pit is often built beside the field by experienced farmers in ancient Southern China. Applied with the manure fully fermented in the pit, the fields will be much fertile" [3]. The latter one had been still used in the farmland in Southern China until modern times.

Ancient Chinese had been advocating "manure accumulation for fertilizing fields" and "reasonable use of manures," which just refers to the rational fertilization in modern terms. Then what is rational fertilization? It means to select the type of fertilizer suitable to the nature of the soil and apply it in proper amount, timing, and approach. By the Song and Yuan Dynasties, China had had a clear understanding and specific measures in this respect.

Chen Fu in the Southern Song Dynasty said in his *Agricultural Book*: "It makes sense to clarify the nature of the soil and use suitable manure to fertilize the soil. As a slung goes, 'fertilization is just like medicine feeding'," which points out that the application of fertilizer depends on the soil, that is, to apply fertilizer suitable to the soil. According to *Agricultural Book* by Wang Zhen of the Yuan Dynasty, "The fields, however, shall be fertilized in a moderate manner. If the manure without being processed is applied suddenly or too much manure is applied, then crops may get hindered by the heat released out from the fermentation of such fertilizer." This means that the amount of fertilizer applied should be appropriate and that the fertilizer applied should be thoroughly decomposed.

After the Song and Yuan Dynasties, such a principle of rational fertilization had been advocated in agricultural production of China.

In the *Book of Farming Encouraging in Baodi*, Yuan Huang of the Ming Dynasty claimed, "Fertilizers made from river mud are suitable for clay and sandy soil, while fertilizers made from lime and plant ash are suitable for acid soil." *The Shen's Book on Farming* has such a record: "Sheep manure is appropriate to be applied to land, while pig manure is ideal to be spread on field. Ash should be avoided being used to heap land as it tends to remove fertility, but could be used for fertilizing field as it is soft." All of these belonged to the method of applying fertilizer according to the conditions of soil.

According to *Collection of the Anecdotes of Wuxing*, "According to an old peasant living in Huzhou (now Huzhou in Zhejiang Province), manure could not be applied in field too early, as early fertilization could result in fertilizer shortage so that when crops would fail and not flourish in autumn. When planting crops, river mud had to be used firstly as base material, whose soil fertility increased slowly but extended. When hot summer arrives, spread some fertilizers made by burnt plaster or vegetable dregs, whose soil fertility also increased slowly. After the beginning of autumn at the Limit of Heat, a large amount of fertilizers should be applied to the field, whose soil fertility would double so that the ears of grain could grow rapidly." This belonged to the method of applying fertilizer according to different timing.

As *The Shen's Book on Farming* stated, "Fertilizers shall be applied when the seeds of wheat are sown, and applied when the vegetables are flowering." According to *Qi Min Si Shu*, "As for the fertilization of the crops of wheat, wheat should be fertilized in winter, while barley in Chunshe (the day of sacrifice to the god of the land in spring). Therefore, there goes the saying, 'Fertilizers should be spread over the awn of barley, and applied to the seedlings of wheat'." This means that different crops require different methods of fertilization, that is, appropriate fertilizers should be applied according to different varieties.

Later, Yang Shen of the Qing Dynasty summarized these previous experiences into "three adaptations in fertilization" in his book *Zhi Ben Ti Gang*, namely, adaptations to a different timing, soil, and variety. According to the book, "As for adaptation to different timing, different fertilizers have different nature, and thus should be applied in different seasons. In spring, it is appropriate to apply human excrement and animal manure; in summer, fertilizers made from weeds, mud, and seedlings fertilizers are suitable; in autumn, fertilizers produced from burnt plaster are efficient; in winter, fertilizers made from sea shells and furs are adaptable. As for adaptation to soil, different soils are of different nature, some fertile and some barren, so fertilizers should be applied according to the soil, which is just like prescribing medicine according to the symptom. For example, it is appropriate to apply fertilizers made from burnt plaster on the soil which is damp and fertile, to apply fertilizers made from dregs on the soil which is yellow (acid soil), to apply fertilizers made from weeds and mud on the soil which is sandy, to apply fertilizers made from furs, hooves, horns and sea shells in paddy fields, and to apply pig manure and the like on extremely dry land. All these experiences have been tested and verified to be reasonable. But it is inappropriate to apply fertilizers on saline-alkali soil; otherwise, the land will be damaged, and no crops can grow. As for adaptation to variety, different crops differ in nature, so fertilizers should be applied according to their features. For example, fertilizers made from sea shells, hooves, horns and furs are suitable for paddy fields; fertilizer made from soybean dregs and seedling dregs are suitable for wheat and millet; human manure and fertilizers made from sesame dregs are suitable for vegetables and melons. Practices testified that if applied appropriate fertilizers, all crops can have a good harvest." The "three adaptations in fertilization" fully demonstrated the delicate and scientific fertilization technique developed in ancient China, which was surely a significant contribution to fertilizer science made by ancient Chinese.

Fertilization by diagnosing seedlings in the rice cultivation of Southern China was one of the remarkable achievements in fertilization technique. According to *The Shen's Book on Farming*, fertilization by diagnosing seedlings was a fertilization technique originated in the Ming and Qing Dynasties in the Hangzhou–Jiaxing–Huzhou Plain (now in Zhejiang Province). "As for farming, it is easy to do everything except the fertilization. When fertilizing, farmers shall wait for an appropriate time and observe the color of seedlings, which is the most critical in farming." Here, "wait for an appropriate time and observe the color of seedlings" means to observe the growth stages and nutritional status of rice. As the book says, "Subsequent fertilizers shall be applied after the Limit of Heat, the time when the rice begins to bear grains and the seedlings turn yellow. If the seedlings do not turn yellow, subsequent fertilizers must not be applied. So long as the seedlings still remain green, subsequent fertilizers cannot be spread at later stage. If the seedlings grow thick, it is probably that nutrient supply is insufficient. Then when the rice shooting out into ears, three *dou* of cake fertilizers should be spread into every single *mu* of field so as to improve the soil fertility. Never apply fertilizers into the fields when the seedlings still remain green; otherwise, it will result in exuberant plants but poor fruits." The time when the rice begins to bear grains is also the stage when the single cropping of late rice changes from vegetative growth to reproductive growth, so the color of rice seedlings will change from deep green into light green, which is just "when the seedlings turn yellow" described in the book, the mark for applying extra fertilizer. During this period, if the seedlings do not turn light green, it indicates that the rice seedlings still enjoy enough nutrition, or not experience the transition from vegetative growth to reproductive growth. Therefore, extra fertilizer cannot be applied without careful consideration; otherwise, the rice seedlings will remain green and get flattened, leading to the result that "luxuriant seedlings fail to produce rice." According to the book, "The family with few fertilizers would suffer from poor soil fertility as well as bad harvest, while the family with enough fertilizers often worried about blighted grain caused by over-fertilization. The reason for this lies in the fertilization of young seedlings." This further pointed out the importance of fertilization by observing seedlings. The method of applying fertilizer according to the color of seedlings was adopted based on a full knowledge of rice growth, so it is undoubtedly scientific, which marks an outstanding achievement in fertilization technique made in China.

26.6 Crop Protection and Pest Control

Insect pest is the main danger that agricultural production has to meet, the solution to which is thus of great significance to ensure a good harvest. As for pest control, the ancients took various measures, which, according to category, could be divided into three kinds: agricultural control, chemical control, and biological control. Among them, agricultural control is widely adopted.

Agricultural control is a method adopted to control the biological communities in fields, change the living environment of pests, and eliminate or reduce the source of pests through various agricultural operations. There are many specific ways involved, such as deep plowing. According to *Lv Shi Chun Qiu: Ren Di*, "The land should be ploughed five times before sowing seeds, and hoed five times after sowing seeds, which must be completed carefully and thoroughly. If plowed deeply enough, the land moisture can be ensured. In this way, there will be no weeds or insect pests in fields." Crop rotation could also control insect pests. According to *Nong Zheng Quan Shu* (*Complete Treatise on Agriculture*), "In the place with high terrain where both cotton and rice can be planted, the cotton shall be planted for two consecutive years and then the rice shall be planted for one year. In this way, the root of weeds will rot, the soil will be fertile, and there will be no insect pests. A crop rotation period cannot exceed three years; otherwise insect pests will breed." Timely planting can prevent crops from breeding pests. According to *Lv Shi Chun Qiu: Shen Shi* (*Annals of Lv Buwei: Examining the Seasons*), "The hemp that is timely planted will not attract locusts"; "the bean that is timely planted will not breed insect pests"; "the wheat that is timely planted will not produce maggots." According to *The Shen's Book on Farming*, "The farming doesn't seek early planting. If the soil is not fertile at all, earlier plantation of crops would lead to insect pests. If there is abundant rain for farming in that year, then it will be better to transplant rice seedlings before or after the season of Grain in Ear. If a waterwheel has to be used to irrigate fields in a year of drought, then it is acceptable to transplant rice seedlings until the Summer Solstice." This means that the transplanting of rice seedlings at an appropriate time is helpful to prevent insect pests. Selecting and planting pest-resistant seeds is one more way to prevent insect pests. *Qi Min Yao Shu: Grain Cultivation* recorded 14 varieties of seed such as Zhugu (scarlet grain) and Gaojuhuang (yellow grain grown in high places) which were free of insect pests. In addition, some cultivation measures like weeds eradication can be adopted to reduce insect pests. According to *The Shen's Book on Farming*, "All pests that damage seedlings often lay eggs in the fields or mud flats. Therefore, the root of weeds shall be eradicated in winter and new soil be added to the original soil. This is also a way of protecting seedlings by killing pests." *Complete Methods of Locust Control* also says: "In order to prevent crops from the damage of locusts, all weeds in the places where the nymphs of locusts breed should be eradicated firstly, which thus eliminates the source of locusts."

Besides agricultural control, drugs were also used to control pests in ancient China. Among the drugs used, some were vegetative, and some mineral-based. The vegetative drugs include dozens of varieties, such as plant ash, radix sophora flavescens, radix stemonae, tung oil, and so on. *Qi Min Yao Shu: Melon Planting* recorded such a method of controlling pests of melons: "Early in the morning, when the dew is not evaporated, raise the melon vines with a stick, and sprinkle ashes over the melon roots. Then one or two days later, hill up the melon roots with soil, and there will be no pests any longer." According to *Agricultural Book* by Wang Zhen, "When there are pests in vegetables, mash the radix sophorae flavescentis, mix it with limewater, and then sprinkle the mixture over the vegetables, and the pests will die." *Book on Locust Control* recorded that radix stemonae could be used to kill the nymphs of locusts: "The method is to decoct the radix stemonae (stemona japonica) into juice, and then mix it with highly concentrated lye water and mature vinegar"; puncture the holes in fields where the pest eggs were hiding with a piece of iron wire. "Then pour the mixture in the kettle into the holes," and the next day "sprinkle lime water into the holes. The remaining eggs of pests will rot and never be hatched into larvae." Vegetative drug was a kind of pesticide which was widely used in ancient China. Without residual toxicity and pollution, it had a favorable effect on agricultural ecological protection.

As for mineral-based drugs, lime, sulfide, and arsenical were usually used. Chen Fu said in his *Agricultural Book*: "When sowing time is coming, sprinkle lime into wet mud to get rid of insect pests." Wang Zhen had a similar description of it in his *Agricultural Book*: "All mulberries and fruits breed pests, which must be exterminated… One method is to make smoke by exploiting sulphur and realgar to fumigate the pests to death." According to *Tian Gong Kai Wu* by Song Yingxing, "As for wheat fields… in the Luo River basin of Shaanxi and Henan, the seeds would be mixed with arsenic to control pests"; "as for arsenic…it was always used to mix with the seeds of bean and wheat in Shanxi, which could also drive away the rats in fields; it was also dipped by the roots of seedlings to ensure a good harvest in the areas of Ningbo and Shaoxing in Zhejiang."

Biological pest control was a great invention made in ancient China. Early in the Jin Dynasty, weaver ants had been employed to control citrus pests in citrus orchards in South China, which can be seen in *Descriptions of Plants in the South* by Ji Han. Until the Qing Dynasty, the method had been applied in citrus and lemon orchards. In the Ming and Qing Dynasties, biological pest control was expanded from orchards to grain fields, and ducks were bred to control locusts. In the areas of Qingshan and Shunde in Guangdong in the Ming Dynasty, ducks were initially raised to successfully control amphibious crabs in rice fields, and later to control locusts. According to *Records of Locust Control* by Lu Shiyi in the late Ming and early Qing Dynasties, "Ducks can prey locusts that are unable to fly yet. Therefore, drive hundreds of ducks into rice fields to prey locust, which will be eaten up in a moment. This is also one of the methods to catch the larvae of locusts in the lower reaches of the Yangtze River." This biological pest control works safely, without pollution but with good ecological effect, therefore opening up a new way for pest control.

In addition to pest control, corresponding measures had also been adopted to deal with the damages caused by sudden and disastrous weather. Building a net to prevent crops from natural disasters and reduce losses, plant protection has been an integral part of the intensive cultivation and eco-agriculture in China.

For over 2,000 years, the intensive cultivation and eco-agriculture has made great contribution to the development of Chinese agriculture, which is manifested in the following:

Firstly, the yield of crops per unit had been continuously increasing. The yield of dry crop (millet) per mu in Northern China was about 42.5 kg in the Warring States Period, which was increased to about 60 kg in the Han Dynasty, and about 75.5 kg in the Qing Dynasty. As for the yield of rice per mu in Southern China, it was about 111 kg in the Song Dynasty, which was increased to about 176.5 kg in the Ming Dynasty, and about 201.5 kg in the Qing Dynasty. In 2,000 years from the Warring States Period to the Qing Dynasty, the yield of dry grain per mu was increased by 77%, and that of rice by 81% from the Song Dynasty to the Qing Dynasty [17].

Secondly, the utilization rate of land had been continuously increased. In the Warring States Period, a continuous cropping of one crop 1 year was implemented in China. After the Tang and Song Dynasties, the system of three crops in 2 years and partly two crops in 1 year was developed in the middle and lower reaches of the Yellow River, the system of two crops in 1 year was developed in the lower reaches of the Yangtze River, and the system of three crops in 1 year was developed in the Pearl River Delta. If the utilization rate of land of China was set to be 100% under the system of one crop in 1 year, it had been improved by 50%, 100%, and 200% correspondingly since the Warring States Period. There was no such a precedent in the world that the multiple-cropping index of land had been continuously increased without soil fertility decline after perennial cultivation.

Thirdly, a large population had been fed by a small amount of cultivated land. Before the reign of Emperor Kangxi in the Qing Dynasty, China's population was about 60 millions and reached 400 millions in the reign of Emperor Daoguang. If the population of 59.87 millions in 1381 in the Ming Dynasty was taken as a base, China's population had increased by 5.7 times in 453 years from 1381 to 1834. With the increase of population, the area of cultivated farmland per capita had declined sharply. The area of cultivated land per capita was 9.67 *mu* in 2 A.D. in the Western Han Dynasty, 6-7 *mu* in the Ming Dynasty, 3.56 *mu* in 1766 in the Qing Dynasty, and 2.41 *mu* in 1887. The fact reflected that in the Ming and Qing Dynasties, the contradiction between large population and scarce land had become rather sharp, and the farmland needed to feed 1 mouth was much lower than that in the Han Dynasty.

It is thus clear that the agricultural technique which had realized the breeding of a large population with a small amount of farmland was indeed a major achievement made in ancient Chinese agriculture.

Just as the well-known German chemist Liebig said, "Guided by practical experience and observation, Chinese agriculture managed to maintain the soil fertility for a long time, and continuously improve grain yield to accommodate the population growth, thus creating incomparable agricultural methods" [18].

References

1. (Northern Song Dynasty) Zhongyan, Fan. *Collected Works of Fan Zhongyan.*
2. (Yuan Dynasty) Zhen, Wang. *Nongshu (Agricultural Book).*
3. (Yuan Dynasty) Zhen, Wang. *Agricultural Book: Chapter of Applying Fertilizers to Soil.*
4. (Western Han Dynasty) Qian, Sima. *Historical Records: The Annals of Rivers and Canals.*
5. (Eastern Han Dynasty) Gu, Ban. *The History of the Former Han Dynasty: Biography of Dongfang Shuo.*
6. (Northern Song Dynasty) Kuo, Shen. *Meng Xi Bi Tan.*
7. *The Annals of Fugou County* written in the reign of Emperor Daoguang of the Qing Dynasty; *The Annals of Guancheng County* written in the reign of Emperor Daoguang of the Qing Dynasty; *The Annals of Funing County* written in the reign of Emperor Guangxu of the Qing Dynasty.
8. *The Annals of Qianyang County* written in the reign of Emperor Qianlong of the Qing Dynasty; *The Annals of Yongming County* written in the reign of Emperor Kangxi of the Qing Dynasty; *The Annals of Changning County* written in the reign of Emperor Daoguang of the Qing Dynasty.
9. (Ming Dynasty) Rong, Lu. *Shu Yuan Miscellaneous Notes.*
10. (Ming Dynasty) Zhenyi, Changgu. *Nong Tian Yu Hua.*
11. (Northern Wei Dynasty) Sixie, Jia. *Qi Min Yao Shu (Important Arts for the People's Welfare): Cultivation.*
12. (Northern Wei Dynasty) Sixie, Jia. *Qi Min Yao Shu (Important Arts for the People's Welfare): Different Versions.*
13. (Southern Song Dynasty) Fu, Chen. *Agricultural Book.*
14. (Northern Wei Dynasty) Sixie, Jia. *Qi Min Yao Shu (Important Arts for the People's Welfare): Harvest.*
15. Department of Chinese Agricultural Heritage, Chinese Academy of Sciences. 1985. *Compendium of the History of Agricultural Science and Technology in Ancient China*, 135. Jiangsu Science and Technology Press.
16. (Ming Dynasty) The Shen's. *The Shen's Book on Farming.*
17. Zongdian, Min, and Dong Kaichen. 1982. Some problems in the history of agricultural technology in China. *Agricultural Archaeology* 2: 12–22.
18. [German] von Liebig, Justus. 1981. Farming and History: Preface to *Application of Chemistry in Agriculture and Physiology* (6th). Dong Kaichen (Trans.). *Scientific Research Data Repertory* by Beijing Agricultural University (6).

Zongdian Min Born in Huzhou, Zhejiang in 1933. He graduated from the History Department of Nanjing University in 1955, and in 1957graduated from the Chinese Revolutionary History Research Course of Renmin University of China. He once served as director of the Agricultural History Research Office of the China Agricultural Museum, researcher of the China Agricultural Museum, chief editor of Ancient and Modern Agriculture magazine, and vice president of the Chinese Agricultural History Society.

Research interest: The history of agriculture in ancient China

Major works: *Chinese Agricultural History Record; History of Chinese Farming.*

The chief editor of *Illustrated Handbook of Agricultural Science and Technology History in Ancient China; History of Chinese Modern Agricultural Science and Technology. Participating in the writing of China Encyclopedia: Agriculture; Chinese Agricultural Encyclopedia: History of Agriculture; History of Chinese Agricultural Science and Technology; Pictures of Chinese Farming Instruments; Biography of Jia Sixie; The Natural Resources and Regional Development of Agriculture in China; The Land Use of Chinese agriculture.*

Awards: The second prize of the National Science and Technology Progress Award in 1991;

The second prize of outstanding achievements in humanities and social science research granted by the National Education Commission in 1995;

The first prize for Scientific Progress of Ministry of Agriculture in 1996;

The third prize of the National Science and Technology Progress Award in 1998.

Chapter 27
Traditional Chinese Calculation Method with Abacus

Lisheng Feng

Chinese abacus calculation is a computing method with abacus to make arithmetical operations such as addition, subtraction, multiplication, division, and extraction. Based on mathematical theory and algorithm, this method helps people to calculate by moving up and down the beads with fingers according to certain formulas. As the tool of abacus calculation, Chinese abacus is usually made of wood and sometimes of other materials like metal, bone, or ivory. Frame, rod, beam, and bead can be seen in an abacus. A series of vertical rods strung with beads extend from the top to the bottom of the frame and a beam divides two beads in the upper deck and five beads in the bottom on each rod. In Chinese abacus calculation, each rod represents a decimal place with high order to the left and lower order to the right. Each bead in the left rod is ten times as big as its neighbor in the right and ten should be replaced by one in neighboring place to the left. Operations can be completed by sliding the beads with thumb, index, and the middle fingers. To show a number, beads are moved against the beam. Each bead in bottom deck has a value of 1, and the upper one has a value of 5. One upper bead must be taken if all five bottom beads in the same rod are counted. With its high efficiency, abacus had become the most important calculating tool in China since the Yuan Dynasty and played an important role in mathematics and social life.

Abacus calculation is a major invention of Chinese people in long-term social practice and first recorded in the late East Han Dynasty by mathematician Xu Yue in his work *Shushu jiyi* (*Notes on Traditions of Arithmetic Methods*), which has survived with a commentary from Zhen Luan in 535. In this book, we can see 14 popular calculating methods, including abacus calculation, and early abacus with one upper bead and four bottom beads available for grouping, quinary, and decimal systems. Abacus with rods and beam first appeared in the Song Dynasty was further

L. Feng (✉)
Institute for History of Science and Technology & Ancient Texts, Tsinghua University, Tsinghua, China
e-mail: fls@tsinghua.edu.cn

J. Hua and L. Feng (eds.), *Thirty Great Inventions of China*,
https://doi.org/10.1007/978-981-15-6525-0_27

improved in the Yuan Dynasty and widely used in the Ming Dynasty, and finally took the place of the counting rod to become the major counting tool in China. In the Ming Dynasty, abacus was introduced to Japan, Korea, and other East Asian countries, and was popularized in these countries.

27.1 The Origin of Abacus Calculation and the Invention of Abacus

As one of the main computing tools in ancient China, counting rod is simple and easy to handle. However, it also has some disadvantages, such as the oversized layout in massive calculations and untraceable mistakes in quick operations, and so on. Therefore, the reform was started very early and reflected in 14 calculating methods mentioned in Xu Yue's *Shushu jiyi* (*Notes on Traditions of Arithmetic Methods*). According to Zhen Luan's commentary, abacus calculation, one of the 14 methods, is performed in abacus with 3 decks. Beads are arrayed in the upper and bottom decks, and answer is put in the middle. Five beads are bunched in each column. One in the upper deck represents five, and each of four in the bottom represents one. The upper bead is distinguished from the bottom beads by colors.

Obviously, the early abacus calculation is evolved from the configurations of the counting rods. It is the same in numeration but different in calculation to present counting. Without accepted formulas and the threaded beads, the early abacus calculation might not be as convenient as counting rod arithmetic.

The academic circles still have different opinions on the interpretation of Zhen Luan's commentary. Then a controversy arises: whether the beads are threaded by rods and whether the upper and bottom decks are separated by the beam [1]. Xu Chunfang believes that the beads are arrayed in slots in the early abacus (see Fig. 27.1). But the Japanese scholar Totani Shinichi argues that beads are thread by rods (see Fig. 27.2). Anyway, most scholars agree that the early abacus recorded in Xu Yue's work *Shushujiyi* is the beginning of Chinese abacus.

After the middle Tang Dynasty, increasing demand for calculation in prosperous business requires urgent improvement in calculating methods. According to the mathematical booklist recorded in *Xin Tang Shu* (*New History of the Tang Dynasty*) and other historical literature, the reform of calculation in the Tang Dynasty aims to simplify multiplication and division into addition and subtraction by operating in 1 row instead of 3 rows. In the Song and Yuan Dynasties, pithy formulas specialized for quick operations and conversion regulations for non-decimal numbers were invented to facilitate the daily calculation. The pithy formulas designated for abacus today are basically borrowed from that in Yang Hui's works in the Southern Song Dynasty and Zhu Shijie's books in the Yuan Dynasty.

In coordination arises intolerably in the operations with the counting rods. Calculation with the help of pithy formulas for multiplying and dividing is so fast that people can't wait for the time-consuming layout of the counting rods. In the urgent

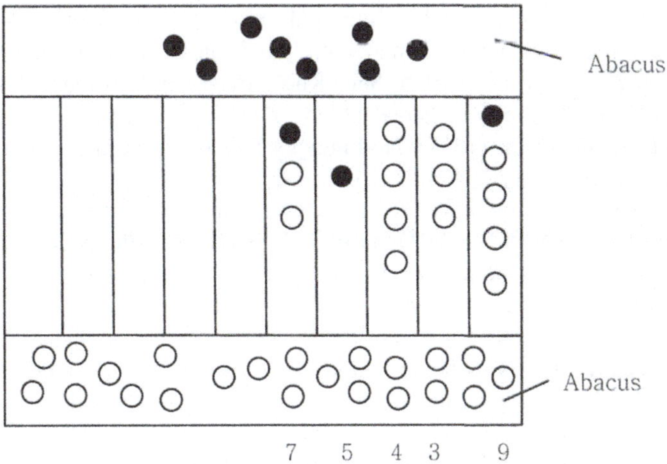

Fig. 27.1 Abacus in Xu Chunfang's *Zhongguo suanshu gushi* (*Stories of Chinese Mathematics*)

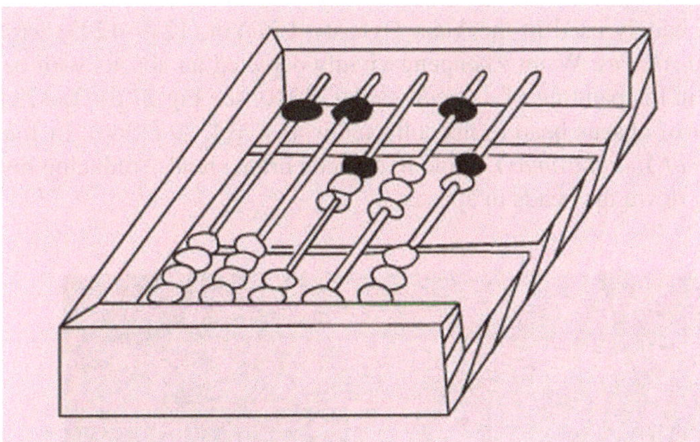

Fig. 27.2 Abacus in Totani Shinichi's Argument

need for better counting tool, the more powerful abacus with rods and abacus calculation methods are invented in this way: turning the abacus in Xu Yue's record into the abacus at present by reducing 3 decks to 2 decks and threading the beads with rods, and then redesigning the pithy formulas to be fit for abacus calculating instead of counting rod arithmetic.

The climax of reform in Chinese ancient calculating techniques came in Song and Yuan Dynasties, and multiplication and division were still the first to be dealt with. A large number of practical arithmetic bibliographies are collected in historical documents of this period, far more than that of the Tang Dynasty. Completely consistent with the current abacus calculation, new algorithms in Song and Yuan

Dynasties allow multiplying and dividing only in one row without needless operations. The practical examples of new algorithms can be found in many Chinese ancient mathematical works, such as Shen Kuo's *Mengxi bi tan* (about 1086–1093), Yang Hui's *Chengchu tongbian benmo* (1274), Zhu Shije's *Suanxue qimeng* (1299), Ding Ju's *Ding Ju suanfa* (1355), He Pingyu's *Xiangming suanfa* (1373), and Jia Heng's *Suanfa quanneng ji* (about 1373).

It is believed that Chinese traditional abacus with rods and threaded beads was invented no later than the Song Dynasty and abacus with threaded beads might be first used in the Northern Song Dynasty. In *Qingming shanghe tu* (*Along the River During the Qingming Festival*) painted by Zhang Zeduan in the Northern Song Dynasty, two square frames are clearly visible on the apothecary's counter and recognized as abacus by some researchers (see Fig. 27.3), even though other scholars argue that they are coin containers but not abacus. In addition, a wooden bead of 2.1 cm in diameter, unearthed in Julu, Hebei from the site of a city in the Northern Song Dynasty, looks exactly like an abacus bead in shape and size (see Fig. 27.4). More evidently, we can identify clearly the abacus beads and an abacus with rods in a gambling street painting by Liu Shengnian in the Southern Song Dynasty (see Fig. 27.5). All the evidences show that abacus with rods and threaded beads was widely used in the Yuan Dynasty. Liu Yin (1249–1293) wrote a short poem titled *Abacus*. Wang Zhenpeng vividly depicted an abacus with beam, rods, and beads in his painting of a street vendor (1310. see Fig 27.6). Tao Zongyi used a metaphor of abacus bead in his collected works, Vol. 29 (1366). In *Yuanqu xuan* (*Selections of Yuan Dramas*), a line in a poetic drama reads "reducing my life time like sliding down the beads in abacus."

Fig. 27.3 Image like abacus in *Qingming shanghe tu* (*Along the River during the Qingming Festival*)

Fig. 27.4 The wooden bead unearthed in Julu, Hebei

Fig. 27.5 Abacus in Liu Shengnian's painting of a gambling street

27.2 Popularization and Development of Abacus

From early to middle Ming Dynasty, abacus calculation had gained great development and popularization in line with the growth of commodity economy. The record of its usage is found in a picture book for children, *Kuiben duixiang siyan zazi* (1371). The illustration of abacus in this book clearly shows its frame, beam, rods, two beads in upper deck, and five beads in the bottom (see Fig. 27.7). Moreover, abacus, as one of the household necessities, is included in the list of the wooden furniture in *Luban mujing* (*Lu Ban's Classics*, sixteenth century).

All the above historical records prove that in the early fourteenth and early fifteenth centuries abacus was so widely used and truly common in daily life that it becomes one piece of goods supplied by street vendors; it serves as a well-known item of houseware written in the popular drama and it is selected in children's book as an illustration.

After its birth, abacus had been simultaneously used with the counting rods for a long time. At the beginning, abacus was adopted in daily practice while officials and scholars in the Song and Yuan Dynasties still applied the counting rods instead of abacus calculation in business and mathematical writings. The counting rod algorithm was the major method to elaborate the operations in most mathematical works before the early Ming Dynasty even though abacus calculation had prevailed for years in the Yuan Dynasty. Besides, it is difficult to tell if the counting method is used or not in mathematical works in the early Ming Dynasty. The simultaneous use and the interactions of abacus and counting rod last for a long time. The *Zhuanggui* algorithm of division was recorded early in *Suanxue qimeng* (*Introduction to Computational Studies*) and then became one of the key formulas in abacus calculation. In the turning of the Yuan and Ming Dynasties, multiplication table, *Guichu* formula, and *Zhuanggui* formula are all collected and then repeatedly performed with practical problems in mathematical works.

In the middle Ming Dynasty, most mathematicians still adopted expressions and norms of rod arithmetic in their writings. After the middle of the fifteenth century, discussions on abacus calculation appeared increasingly in mathematical works, such as Wu Jing's *Jiuzhang xiangzhu bilei suanfa daquan* (*The Annotations of Calculations in the Nine Chapters on the Mathematical Art*, 1450) and Wang Wensu's *Gujin suanxue baojian* (*Mirror of Mathematical Learning*, 1524). Wu Jing, Wang Wensu, Tang Shunzhi, Gu Yingxiang, and other Chinese ancient mathematicians contribute to the application and study of abacus. More importantly, Tang Shunzhi is very good at practical abacus operation. In historical records, a story about his contest on abacus calculation is very interesting:

> Tang Shunzhi went to Luzhou for business and tried to help counting the total amount of the grains recorded in the account book. He challenged a group of counting master-hands. When more than 10 opponents were working together for the answer, Tang worked out the exact number with an abacus. Even the professional accountant was greatly surprised at his quick abacus calculating [2].

Tang Shunzhi (1507–1560), known as Yingde (courtesy name) and Jingchuan (literary name), born in Wujin county, was a winner of metropolitan examination, a senior prosecutor and also a famous mathematician in the Ming Dynasty. The fact that Tang was so good at abacus calculation that the government applied it in its financial affairs indicates that abacus calculation substituted the rod arithmetic in algorithms and took a dominant position in counting practice.

Fig. 27.6 Abacus in Wang Zhenpeng's painting of a street vendor

Fig. 27.7 Abacus and counting rods in a children's book of the Ming Dynasty

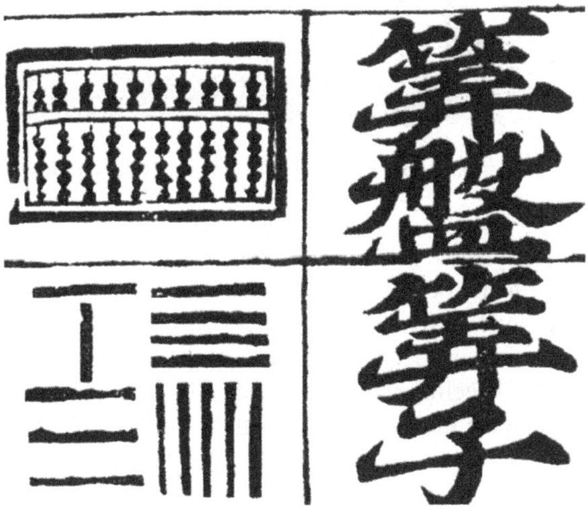

In 1987, a wooden abacus was unearthed in a Ming Dynasty tomb in Zhangpu County, Fujian. It is an abacus with 1 + 5 rhombus beads (see Fig. 27.8) and looks exactly the same as the Japanese abacus being widely used from seventeenth century till today. The tomb occupant Lu Weizhen, born in Zhangpu County, was once a top leader of Ministry of Labor and Ministry of Finance in the Ming Dynasty. In his official career, he was responsible for civil, financial, and tax affairs that needed a large number of calculations. The abacus in his tomb mirrors the wide applications of abacus in the entire China. In Fig. 27.9, we can see a masterpiece of abacus collected in the National Museum of China. It is an abacus with 11 rods and ivory beads,

Fig. 27.8 Abacus unearthed in Lu Weizhen's tomb

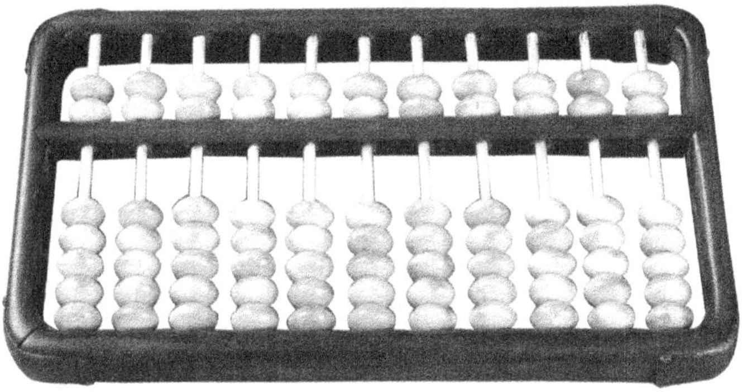

Fig. 27.9 Abacus of the Ming Dynasty in the National Museum of China

perfectly made in the Ming Dynasty. Obviously, the abacus is not a calculating tool for ordinary people, but a tool used by the upper class.

In the late sixteenth century, abacus calculation was widely used along with the emergence of a large number of special works for abacus calculation. Rod arithmetic began to decline and was overtaken by abacus calculation. Soon abacus calculation became the major counting method. In those special works for abacus calculation, Panzhu suanfa (Mathematical Methods for the Abacus, 1573) and Shuxue tonggui (Tracks to Mathematics, 1578) came out first, taking the lead in the popularization of abacus calculation.

Panzhu suanfa (*Mathematical Methods for the Abacus*, 1573) gives a detailed introduction to formulas, operations, and manipulations of abacus calculation. The formulas for abacus calculation in this book include three categories for addition, three for subtraction, three for division, two for multiplication, two for both division and multiplication, and an accumulating algorithm specially prepared for the beginners. In those formulas, there is something special about the multiplication table, which contains instructions for $10\times$ but not $\times 1$, such as $10 \times 2 = 20$, $10 \times 3 = 30$. This book, accompanied by 54 detailed abacus diagrams, shows that the practical

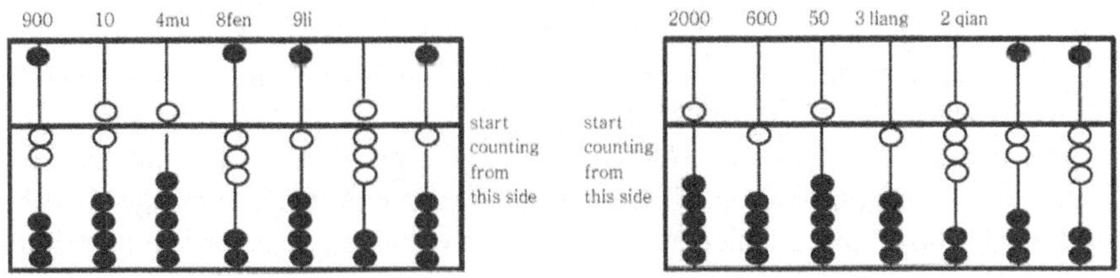

Fig. 27.10 Abacus diagrams in *Mathematical Methods for the Abacus*

application and operation of abacus formulas are used. And it is the first abacus calculation work with abacus diagrams attaching to formulas.

The following are two sample cases in the book:

Case 1: There is 914.89 *mu* (*mu*: a unit of area) of cropland and each *mu* produces 2.9 L of grain.

Question: What is the total amount of grain?

Answer: 2653.181 L.

Steps:

$9 \times 9 = 81, 2 \times 9 = 18$, then rid 2 and advance 1 in the left when adding 8;

$9 \times 8 = 72, 2 \times 8 = 16$, then rid 4 and advance 1 in the left when adding 6;

$9 \times 4 = 36$, then move up 1, rid 5 and advance 1 in the left when adding 6, $2 \times 4 = 8$, then move up 3, rid 5 and advance 1 in the left when adding 8, $9 \times 1 = 9$, rid 1 and advance 1 in the left when adding 9, go on the third digit;

$1 \times 2 = 2$, then move up 2, go on the second digit;

$9 \times 9 = 81$, then move down 5, rid 4 when adding 1, $9 \times 9 = 81$, then rid 2 and advance 1 in the left when adding 8. In the end, you can get the answer on abacus layout in Fig. 27.10(left).

Case 2: There is 2653.2 *liang* (*liang*: a unit of weight) of silver, and 515 people try to average them.

Question: How many *liang* of silver does each can get?

Answer: 5.151823 *liang*.

Steps:

$2 \div 5$ replaced by 4, $5 \div 5$ advance 1 in the left, rid 5 from 5, rid 25 from 55, rid 2 from 10 and pay back 8, rid 5 from 10 and pay back 5;

$5 \div 5$ advance 1 in the left, rid 1 from 1, rid 5 from 5, rid 5 from 10 and pay back 5;

$2 \div 5$ replaced by 4, $5 \div 5$ advance 1 in the left, rid 5 from 5, rid 25 from 5×5, rid 2 from 10 and pay back 8, rid 5 from 10 and pay back 5;

$5 \div 5$ advance 1 in the left, rid 1 from 1, move up 4, rid 5 when subtracting 1, rid 5 from 5, rid 5 from 10 and pay back 5;

4 ÷ 5 replaced by 8, rid 8 from 8, rid 8 from 10 and pay back 2, rid 40 from 5 × 8, move down 1, rid 5 when subtracting 4. In the end, you can get the answer on abacus layout in Fig. 27.10(right) [3].

Shuxue tonggui (*Tracks to Mathematics*, 1578) involves very comprehensive abacus techniques and almost all the abacus formulas used in later ages. In the late Ming Dynasty, abacus calculation was gradually finalized, and abacus algorithm became standardized and systematized step by step, which is particularly reflected in the emergence of the *Suanfa tongzong* (*General Source of Computational Methods*, 1592) by Cheng Dawei. In Cheng's work, formulas for addition and subtraction are left at the end, while multiplication is based on *liutou* (head-saved) method and division is based on *guichu* method. Cheng Dawei strongly advocated *liutou* multiplication, arguing that it is much better than the other multiplication algorithms. Following Cheng's opinion, subsequent works on abacus calculation mainly adopt *liutou* multiplication in the next 500 years. Inspired by Cheng's directions, the scattered formulas were further improved to be practically applied to the finger operations in abacus counting. For example, the formula "move up 4 and replace by 5 when adding 1" is changed into "move down 5 and rid 4 when adding 1." According to the latter formula, sliding down 1 bead for 5 in upper deck first and then ridding down 4 beads in bottom deck can be done in one direction and sequence without stopping. It is much more reasonable and efficient than the operation of the former formula with two antipodal actions.

The *liutou* multiplication is a traditional Chinese algorithm, which was named differently at different times. When counting with multiplier of more than three digits, the first digit of multiplier (head) is saved to be the last one multiplying the multiplicand. Hence, it is named *liutou* multiplication.

The multiplication starts with the last digit of multiplicand. Multiply the second digit of multiplier, go on to the third, the fourth till the last digit, and come back to the first digit of the multiplier. Keep multiplying in the head-saved way until the operations end with multiplication of the first digits of the multiplicand and multiplier. The number laid out on abacus is just the product. To show the product of multiplicand and a certain digit of multiplier on abacus, designate the units digit of product on the corresponding rod to the right side of multiplicand.

Let's illustrate the method in the following case:

Case 3: $753 \times 458 = 344874$

a. See Fig. 27.11a. Lay out multiplicand 753 on the left and multiplier on the right.
b. See Fig. 27.11b. Multiply the last digit 3 of multiplicand successively by the second digit 5, the last digit 8, and the first digit 4 of multiplier. The product is 1374.
c. See Fig. 27.11c. Multiply the next digit 5 of multiplicand successively by the second digit 5, the last digit 8, and the first digit 4 of multiplier. The product is added up to 24274.
d. See Fig. 27.11d. Multiply the first digit 7 of multiplicand successively by the second digit 5, the last digit 8, and the first digit 4 of multiplier. The product is added up to 344874.

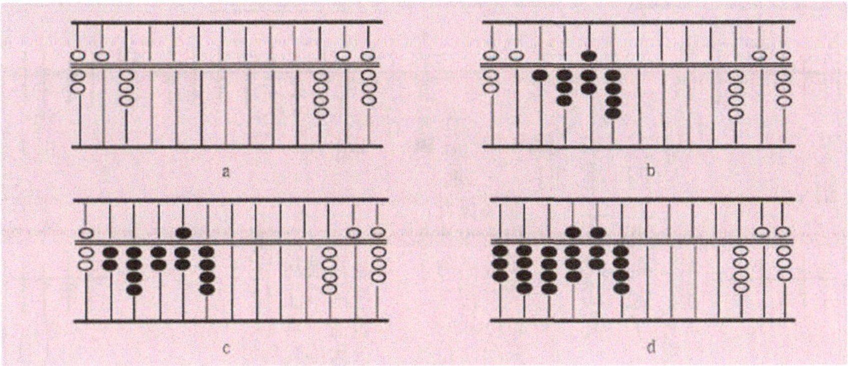

Fig. 27.11 Diagram of operations for *liutou* multiplication

In his book *Suanfa fameng* (*Elementary Algorithm*), Pan Fengxi, a mathematician in Qing Dynasty, gives detailed descriptions with diagrams of several multiplication methods. We can learn about the exact operations at that time. When introducing *liutou* multiplication, he explained: to count in the *liutou* method, multiply first by the second digit of the multiplier, go on multiplying to the last digit, and end up with the first digit. The following is a case described in the book.

Case 4: There are 538 workers and each is paid 246 *yuan*.

Question: What is the total expense? [4]

Answer: 132348 *yuan*.

Operation: Take the number of workers as the multiplicand and the number of average payment as the multiplier. Work out the result of 538 × 246 according to the instructions of the *liutou* multiplication in Fig. 27.12.

The detailed operations in Fig. 27.12, including diagrams and explanatory notes with marks for sequences, show us the explicit manipulation on abacus in ancient China. In Ming Dynasty, there are two methods of division on abacus, *shangchu* division (the way by a multiplication table) and *guichu* division (the way by a division table). The former originated from the traditional rod arithmetic and characterized by setting the quotient with mental arithmetic, which is quite similar to the manual division at present; the latter was invented for rod arithmetic in the Song Dynasty, setting the quotient with division formulas. In the Yuan Dynasty, the *guichu* division overtook the dominance of the *shangchu* division in calculation and became the key algorithm in abacus counting on account of its wide and practical use in abacus calculation. According to the *guichu* method, the division involving divisor of 1 digit is called *Jiugui* and the division involving divisor of more digits is called *guichu*. The *guichu* division is totally based on the division formulas compiling dividend, quotient, and remainder together in a pithy rhyme. The *gui* in *guichu* method means to divide the first digit of the dividend by the first digit of divisor and determine the quotient on abacus with the help of formulas. The *chu* in *guichu* method means to

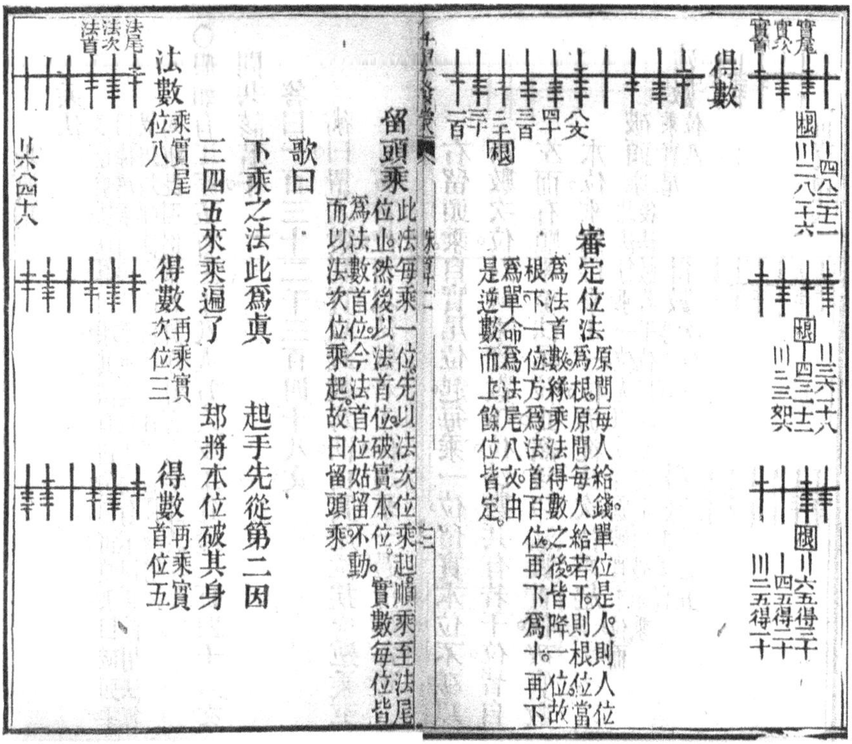

Fig. 27.12 Operations of the *liutou* multiplication in *Suanfa fameng* (*Elementary Algorithm*)

subtract the dividend, respectively, by the products of the other digits of the divisor and the quotient. For example, in *guichu* division of 35, the digit 3 is *gui* and the digit 5 is *chu*; in *guichu* division of 826, the digit 8 is *gui* and the digit 26 is *chu*.

The following two cases from *Suanfa zuanyao* (*General Survey of Algorithm*, 1598) by Cheng Dawei in the Ming Dynasty are a brief introduction to the calculating procedure of *jiugui* and *guichu* methods.

Case 5: There are 265.32 *liang* of silver, and it is shared by 2 people.

Question: How many *liang* can each get? [5]

Answer: 132.66 *liang*.

Operation: Take the amount of silver as the dividend on the left and the number of people as the divisor on the right. Work out the quotient of 265.32 ÷ 2 according to the instructions of the *jiugui* division.

The detailed operations on abacus in *suanfa zuanyao* can be interpreted in Arabic numerals by the following list:

Order	Dividend						Divisor		Formular
1		2	6	5	3	2		2	
2	One	○	6	5	3	2		2	Advance 1 when diving 2
3	One	Three	○	5	3	2		2	Advance 3 when diving 6
4	One	Three	Two	1	3	2		2	Advance 2 when diving 4
5	One	Three	Two	Five	3	2		2	Replace 1 ÷ 2 by 5
6	One	Three	Two	Six	1	2		2	Advance 1 when diving 2
7	One	Three	Two	Six	Five	2		2	Replace 1 ÷ 2 by 5
8	One	Three	Two	Six	Six			2	Advance 1 when diving 2

This division, with the divisor of only one digit 2, should be done with the *jiugui* method. The operations are based on the second one of the *jiugui* formulas: replace 1 ÷ 2 by 5, advance 1 when dividing 2, advance 3 when dividing 6, advance 4 when dividing 8. The quotients in the list above are expressed with Chinese numbers, such as 一 (one), 二 (two), 三 (three), and so on.

Case 6: There is 265.32 *liang* of silver, and it is shared by 11 people.

Question: How many *liang* can each get? [5]

Answer: 24.12 *liang*.

Operation: Take the amount of silver as dividend on the left and the number of people as the divisor on the right. Work out the quotient of 265.32 ÷ 11 according to the instructions of the *guichu* division.

This division, with the divisor of two digits 11, should be done with the *guichu* method. In operations, the first digit should be counted in *gui* method and the second digit in *chu* (subtract) method.

The detailed operations on abacus can be interpreted by the following list:

Order	Dividend						Divisor		Formular
1		2	6	5	3	2	1	1	
2	Two	○	6	5	3	2	1	1	Advance 2 when diving 2
3	Two	○	4	5	3	2	1	1	Rid 2 when 1 × 2
4	Two	Four	○	5	3	2	1	1	Advance 4 when diving 4
5	Two	Four	○	1	3	2	1	1	Rid 4 when 1 × 4
6	Two	Four	One	○	3	2	1	1	Advance 1 when diving 2
7	Two	Four	One	○	2	2	1	1	Rid 1 when 1 × 1

(continued)

(continued)

Order	Dividend						Divisor		Formular
8	Two	Four	One	Two	○	2	1	1	Advance 2 when diving 2
9	Two	Four	One	Two	○	○	1	1	Rid 2 when 1 × 2

During the reign of Emperor Wanli in the Ming Dynasty, the calculation of the root extraction was basically completed on abacus. The related records about extraction of square and cubic root can be found in *Suanxue xinshuo* (*Reflections on mathematics*, 1603) by Zhu Zaiyu and *Suanfa tongzong* (*General Source of Computational Methods*, 1592) by Cheng Dawei. Zhu Zaiyu, a famous scientist in the Ming Dynasty, wrote treatises on musical temperament, astronomy, physics, mathematics, and many other fields. Zhu was the sixth-generation descendant of the *Hongxi* Emperor and inherited the title the Prince of *Zheng* in 1593, but quickly resigned it, which was awarded to his cousin. In his book, Zhu elaborated his instructions on extracting 25 roots with an abacus of 567 beads to lay out more than 81 digits. He argued that the extraction of root could also be done with 4 or 5 abacuses arrayed together [6]. In Zhu's work, detailed operations on abacus to extract the square root of $\sqrt{200}$ are illustrated with the *guichu* method in the following diagram:

Step 1: Take 200 as the dividend and 10 as the first divisor.

Step 2: Subtract 1 from the first digit of the dividend to get 10 as the quotient. Do not subtract the result of the divisor and quotient in this step. Thus 100 is left in the dividend.

Step 3: Double the divisor to get 20 and make it the second divisor. Subtract the result of the divisor and quotient from now on.

Step 4: Subtract the first digit 1 in the dividend from $100 \div 20$ to get 4 as remainder in the first digit of the dividend. Put 4 to the divisor 20 to get another divisor 24.

Step 5: Subtract 42 from the second digit 20 to get 4 as the remainder in the dividend. Double the 4 in divisor to get another divisor 28.

Step 6: Subtract $28 \div 4$ from the second digit 8 in the divisor to get 0.1 as the remainder in the divisor. Put 0.1 to the divisor 28 to get another divisor 28.1.

Step 7: Subtract 1×8 from the third digit and then subtract 1×1 from the fourth digit to get 1.19 in the dividend [7]. The operations can be carried on to extraction of 25 root [8] (see Fig. 27.13).

Calculation performed with Abacus

200	
1100	20
1500	20
1420	24
1404	28
1412	281
141119	281
…	…

Fig. 27.13 Diagram of extraction of square root

While studying musical temperament, Zhu Zaiyu in the Ming Dynasty formulated the algorithm for converting novenary fraction into the decimal system with abacus. He skillfully organized the computing programs, so that the complex calculation of polynomial values could be simplified into the switched addition and subtraction on abacus. His method combines the simplicity of abacus calculation and the sequencing approach in ancient Chinese mathematics.

To convert the novenary 0.8376 *chi* (a unit of length) of *Dalvduchang* into the decimal 0.936442 *chi* of *Dalvduchang*, the operations on abacus and the modern interpretation goes in this way [8]:

Lay out the number on abacus 8376.

First division with *jiugui* formula 837666 divide the last digit 6 by 9, get 666…

Second division with *jiugui* formula 838518 divide the last 4 digits 7666 by 9, get 8517…, advance 1 to remainder.

Third division with *jiugui* formula 842798 divide the last 5 digits 38518 by 9, get 42797…, advance 1 to remainder.

Fourth division with *jiugui* formula 936442 divide the last 6 digits 842798 by 9, get 936442.

Zhu Zaiyu also gave an example of converting decimal into novenary system and summarized the general calculation program to invent the formula of temperament ratio for abacus operations. Zhu's algorithm for the interchange of the decimal and novenary fractions can be further applied to the interchange of the decimal and other fractions, which is an important achievement in mathematics [9].

Abacus calculation is believed to be originated from rod arithmetic, and they are the same in numeral system, in particular. Since the beads on the upper deck and the rods in upper row both have the value of 5, it is technically easy for people to advance from rod arithmetic to abacus calculation. The important difference between abacus calculation and rod arithmetic is its dependence and adaptability to formulas of algorithm. As a counting method, abacus calculation basically contains all functions of rod arithmetic and far overtakes the latter in speed. A skillful abacus user can work out the answer immediately when reading out formulas. The efficiency of abacus meets the needs of business community. Therefore, in the Ming Dynasty, abacus developed vigorously in the commercially prosperous society. In contrast, the Chinese traditional rod arithmetic became so rare that mathematicians in the Qing Dynasty didn't recognize it. At that time, people applied four methods in computation, including abacus calculation, written calculation, manual computation, and Napier's bones computation, but no Chinese traditional rod arithmetic. In 1713, He Guozhu, a famous calendarist in the Qing Dynasty, went to Korea for geodetic surveys. He came across Hong Jeong-Ha (1684–?), a Korean mathematician, who skillfully solved extraction of roots and simultaneous system of equations with rods. Hong was even faster than the Chinese peers proficient at western algorithms. He was so surprised that he asked Hong for 40 rods [10].

The popular abacus calculation in the Qing Dynasty was not much different from that in Ming Dynasty. The *liutou* and *guichu* methods in the Ming Dynasty continued to be used in the Qing Dynasty and *shangchu* method almost disappeared. In the Qing Dynasty, massive divisions with constant divisor were operated in *feigui* algorithm, for example, *feigui* algorithm for acreage of land. At late Qing Dynasty and early republic of China, popular abacus textbooks generally introduced the formulas of *feigui* algorithm, i.e., *guichu* formulas ranging from 1 to 9. In Fig. 27.14, we can see all the *guichu* formulas illustrated in *Zuixin jianming zhusuan keben* (*New Simplified Textbook of Abacus Calculation)* published in the first year of Xuantong in late Qing Dynasty. However, an excess of formulas impeded further popularization and promotion of abacus calculation. During 1930s and 1940s, some math educators advocated using *shangchu* method in abacus calculation instead of formulas of *guichu* division, so that *shangchu* method gradually retained great attention and revived in progress.

In Republic of China and early PRC, abacus gained better inheritance and development because all primary, commercial, and financial schools offered abacus courses and educators attached great importance to the teaching and research of abacus calculation.

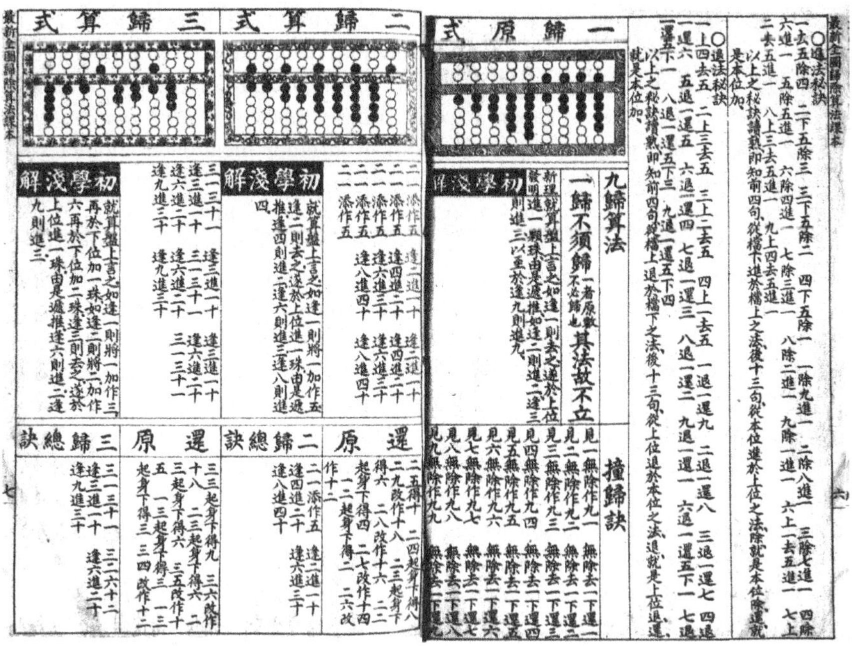

Fig. 27.14 Copy of *Zuixin jianming zhusuan keben* (*New Simplified Textbook of Abacus Calculation*)

27.3 The International Communication and Influence of Abacus

Rod arithmetic was introduced to Korea and Japan and then had an important impact on the development of mathematics in the two countries. In the same way, abacus calculation had played an important role in the development of mathematics in East Asia and gained popularity among the people.

Japan began to introduce Chinese mathematics and rod arithmetic in the Tang Dynasty. Rod arithmetic then became Japanese main computing tool. In sixteenth century, abacus calculation was introduced into Japan and soon became widely popular in their daily life.

Japanese already knew abacus long time ago. In the early Ming Dynasty (1336–1392), *Kuiben duixiang siyan zazi*, a children's book illustrated with abacus was brought to Japan and soon reprinted. The prosperous official and private trade between Japan and the Ming Dynasty inevitably urged Japanese businessmen and government to adopt more efficient computing tools. The earliest abacus in Japan in existence is believed to be the one owned and used by General Zenden Rika during 1592–1598, which now is collected in Zenden's private library (see Fig. 27.15).

The earliest record on Japanese abacus can be found in *Kulturgegensätze Europa-Japan* (1585) by Luis Frois (1532–1597), a Portuguese priest. Frois went to Japan in 1562 and discussed the difference between Japanese and European counting. He wrote in his book that Europeans counted with quill pens and rods but Japanese used the *jina* (abacus) for calculation [11]. In regard to *jina*, Joan Rodriguez (1561–1634), another Portuguese priest in Japan, explained in his book *Arte da ligoa de Japam* (1604): abacus, that is, *jina*, a counting tool. Rodriguez mentioned another word about abacus in his book *Vocabvlario da Lingoa de Iapam* (1603): Soroban, a small frame with bronze beads, used in Chinese and Japanese calculation [12].

After the introduction of abacus calculation into Japan, it gained popularity in a relatively short period of time and had a profound impact on Japanese mathematics and public education. From the perspective of mathematics history, abacus calculation inspires Japanese mathematicians, triggers the mathematic reform in Japan, and becomes the motive power of the birth of *Wasan* (Japanese mathematics).

The famous works on abacus calculation in the Ming Dynasty, *Panzhu suanfa* (*Mathematical Methods for the Abacus*, 1573), *Shuxue tonggui* (*Tracks to Mathematic,* 1578), and *Suanfa tongzong* (*General Source of Computational Methods*, 1592), were introduced to Japan soon after their publications. These works greatly promoted the development of abacus calculation and *Wasan* (Japanese mathematics). Referring to *Suanfa tongzong*, Japanese *Wasan* mathematician 嶋 den Tei wrote his book on algorithms in 1652 and published it next year. The original *Suanfa tongzong* issued in Japan was a private edition printed in 1593 and later annotated by Tousen Tokukore in 1675 (see Fig. 27.16). This annotated edition was widely distributed in Japan and became one of the most important reference books for Japanese *Wasan* mathematicians in Edo period.

Due to its commercial and practical values, the works on abacus calculation in the Ming Dynasty enjoy very high reputation in Japan. As a representative achievement of mathematics in the Ming Dynasty, *Suanfa tongzong* became the first choice for Japanese to study mathematics. Kichiden Kouyu, a mathematician from noble family, learned the algorithms from *Suanfa tongzong* and then wrote his book *Jinkouki*, a famous work on abacus calculation in Japan. With his book, mathematics from Ming Dynasty was widely spread. *Jinkouki* has great influence on Japanese mathematics. First published in 1627 and reprinted many times, *Jinkouki* shares huge circulation and more than 400 editions in later ages. That is why it has become a textbook of basic education in Japan and revised for many other purposes.

The abacus was introduced into Korea in the fifteenth and sixteenth centuries and was widely used among the people. In Fig. 27.17, it is an abacus in early Korea, now collected in Korean Folk Museum [13]. Interestingly, since abacus was just used by businessmen and common people and repelled by elites and mathematicians, abacus calculation had not gained a high position in computation for a long time. Korean mathematicians mainly used rod arithmetic at that time. Choi Seok-jeong, mathematician in the Joseon period of Korea, gave a brief introduction to abacus calculation in his book *Gusuryak* and some disguised criticism on abacus's replacement of rod arithmetic in China and Japan. Until the end of nineteenth century, abacus calculation had not been formally introduced in a mathematical work *Chouxue xinbian*, which unfortunately focused on the illustration of abacus's structure but not algorithm and formulas.

Fig. 27.15 Abacus used by General Zenden Rika

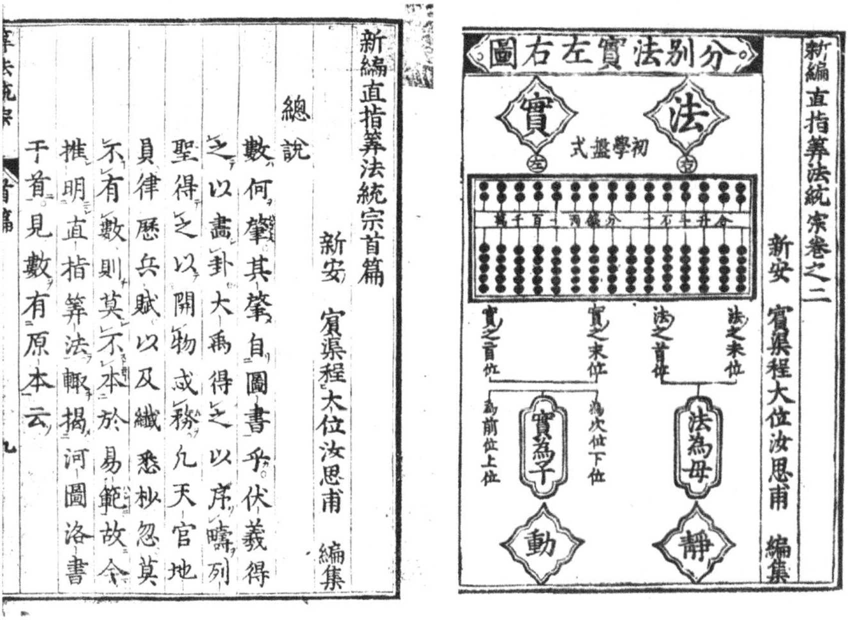

新編直指筭法統宗首篇

新安　賓渠程大位汝思甫　編集

總說

數何肇其肇自圖書乎伏羲得之以畫卦大禹得之以序疇列聖得之以開物成務凡天官地員律歷兵賦以及纖悉秘忽莫不有數則莫不本於易範故今推明直指筭法輒揭河圖洛書于首見數有原本云

新編直指筭法統宗卷之二

新安　賓渠程大位汝思甫　編集

分別法實左右圖

Fig. 27.16 Copy of annotated *Suanfa tongzong* in 1675

Fig. 27.17 Abacus collected in Korean Folk Museum

During the turn of the Ming and Qing Dynasties, abacus calculation also spread to Vietnam and Thailand (Siam). *Suanfa tongzong* (*General Source of Computational Methods*, 1592) by Cheng Dawei was widely used and promoted application in Vietnam. Some of Vietnamese extant algorithm books involve in abacus calculation. For instance, *Zhiming lichen suanfa*, an algorithm book published in 1820, was edited in direction of *Suanfa tongzong*. In author's preface, Pan hui hộp admitted that he got a lot of instructions and inspirations from *Suanfa tongzong*. There are abacus diagrams for beginners in this four-volume book, as practical as *Suanfa tongzong* [14]. In 1691 and 1693, Simon de la Loubere, envoy sent by Louis XIV to Siam, described the structure and usage of abacus and marveled at the efficiency of abacus counting in his book *Description du Royaume du Siam*.

It is quite appropriate to depict abacus calculation with expressions in today's computer science. Abacus can be regarded as hardware, counting methods as software while formulas as programming language. Thanks to its superiority, abacus calculation has greatly improved the computing power of Chinese people in the cultural circle of Chinese characters. The programmed algorithm of mathematics in East Asian cultural circle of Chinese characters should give credit to rod and abacus counting. In particular, abacus, as a practical tool, has played an important role in a long period of time. In terms of design philosophy, abacus is scientific and friendly in structure for user. In terms of operational principle, abacus calculation is fast and easy with the merchandized instructions of programmed formulas. In terms of functional practice, abacus calculation can be used in elementary arithmetic, extraction, and higher order equations. In 1960s, the extended abacuses were still applied to do high-order calculation in China's research project of "Two Bombs and One Star".

Abacus was listed the first one in "50 inventions that changed the world" released by *Times of India* on November 19, 2007. The newspaper praised, "Abacus is proved to be first used in China in about 190 A.D. Chinese abacus had been the fastest calculator for hundreds years, and is still faster than electronic calculator if in some professional contests [15]." In November 2013, The Atlantic announced *The 50 greatest technological breakthroughs since the wheel*. In those 50 great inventions selected by 12 scientists, historians and technologists, abacus, and the four great inventions of China are all included. Abacus is believed to be one of the first tools in favor of human intelligence [16] even though no evidence shows that abacus was invented in 3000 BC.

Abacus calculation is still used by many people in China, Japan, and some countries and regions in Southeast Asia. It is not only a skill to use tool, or a form of knowledge, scientific practice, but also a symbol of advanced mathematics in the cultural circle of Chinese characters and its neighbor, sharing multiple cultural functions and values. On December 4, 2013, abacus calculation was accepted in List of Intangible Cultural Heritage by the UNESCO Intergovernmental Committee for Safeguarding of the Intangible Cultural Heritage in Baku, capital of Azerbaijan.

The abacus calculation has been passed down from generation to generation and applied in so many practices of daily life. It is easy to use and offer the world an alternative algorithm and counting technology. We can always find the wisdom and intelligence of Chinese people in the long-lasting and precious abacus calculation.

References

1. Yinchun, Hua. 1987. *History of Chinese Zhusuan*, 9–12. Beijing: China Financial & Economic Publishing House.
2. Yao Zhiyin. *Notes on History of Yuan & Ming*, vol. 18.
3. Guo Shuchun. 2010. *History of Chinese Science and Technology—Mathematics* 578. Beijing: Science China Press.
4. Pan Fengxi. *Elementary Algorithm*. Zhusuan: Section 2.
5. Cheng Dawei. *General Survey of Algorithm*, vol. 2.
6. Zhu Zaiyu. 1931. Reflection on mathematics. In *Universal Libraries—Temperament*, vol. 6, 7. Shanghai: Commercial Press.
7. Zhu Zaiyu. 1931. Reflection on mathematics. In *Universal Libraries—Temperament*, vol. 6, 9–10. Shanghai: Commercial Press.
8. Guo Shuchun. 2010. *History of Chinese Science and Technology—Mathematics*, 598. Beijing: Science China Press.
9. Li Zhaohua. 2000. Commentary on Zhu Zaiyu's Mathematic works. In *Studies on Chinese Ancient Mathematics*, 283-297. Tianjin: Tianjin Science and Technology Press.
10. Hong Jeong-Ha. 1985. Collection of Counting Formulas. In *The History of Science and Technology in Korea—Mathematics*, vol. 2, 493. Lijiang Press.
11. Frois, Luis. 1992. *Kulturgegensätze Europa-Japan*, 112. Shanghai: Commercial Press.
12. Joan Rodriguez. 1960. *Vocabvlario da Lingoa de Iapam*. Iwanami Shoten.
13. Kim Yong Un. Kim Yong Guk. 1978. *History of Mathematics in Korea*, 284-285.
14. Han Qi. 1991. Sino-Vietnamese Exchanges of Astronomy & Mathematics. *The Chinese Journal for the History of Science and Technology* (2).
15. Inventions that changed the world. *Cankaoxiaoxi*, 2007.11.28.
16. The 50 greatest technological breakthroughs since the wheel. *Cankaoxiaoxi*, 2013.10.31.

Lisheng Feng Born in Wuhai, Inner Mongolia in 1962 graduated from Inner Mongolia Institute of Technology with a bachelors degree in engineering in 1983 then he got a Master of Science degree in 1987 from the Institute of History of Science. Inner Mongolia Normal University. In 1999 he got his doctor degree of science from the Department of Mathematics Northwestern University.

He was a professor at Inner Mongolia Normal University and currently the director of the Institute of Science and Technology History and Ancient Literature of Tsinghua University, and professor and doctoral supervisor of the Institute of Science and Technology History of Inner Mongolia Normal University. He has been a visiting scholar and visiting professor at Japan's Tohoku University, Tokyo University, and Central University. In 2004, he was selected into the "New Century Outstanding Talent Support Program" by the Ministry of Education. He also serves as a member of the National Intangible Cultural Heritage Protection Committee, executive director of the Chinese Academy of Science and Technology History, president of the Chinese Traditional Arts and Crafts Research Association, and vice chairman of the National Mathematical History Society.

Research interests: History of mathematics; History of Technology; Scientific and technological literature; Protection of Science and Technology Cultural Heritage.

Major works: *History of Ancient Chinese metrology, History of China–Japan Mathematical Relations.*

Awards: The First Prize of Scientific Research Achievements of the Inner Mongolia Normal University; *Co-editing Notes of Chou Ren Records* won the first prize of the National Outstanding Ancient Books Award; *The Encyclopedia of Color Map Science and Technology* that he participated in the compilation won the second prize of the 2009 National Science and Technology Progress Award; *History of Chinese Science and Technology: Mathematics* won the first prize of the Fourth Guo Moruo History Award.

Chapter 28
Ferment of Distiller's Yeast

Jiahua Zhou

Qu'nie fermentation or ferment of distiller's yeast is the core of traditional Chinese brewing technology. Qu is at the center of the Chinese wine-making techniques, which enriches China's alcoholic drinks and makes their unique taste and aroma.

Qu'nie fermentation is the core of the traditional Chinese brewing technology, especially wine-making techniques. Since ancient times, Chinese wine, vinegar, and sauce have formed their own characteristics and become unique in the world. Many wine lovers know that Chinese rice wine and liquor are different from the Western wine, whiskey, brandy, gin, vodka, and so on. Their differences lie in both the taste and the fragrance, which is mainly attributed to the radically different brewing techniques. The most distinctive feature of the traditional Chinese brewing technology is the use of Qu'nie, the soul of wine-making. In modern scientific terms, Qu is a multi-bacterial and multi-enzyme biological product. Actually, it is a special reagent for culturing mould, and optimizing and catalyzing fermentation reaction. Chinese ancestors invented Qu to screen out the beneficial microorganisms in brewing, domesticate, and cultivate them, so as to ensure that the brewing process is completed with the help of excellent strains, and high-quality wine is ultimately obtained. The second characteristic of the traditional Chinese brewing technology is the invention of a set of suitable techniques for temperature and humidity control in the making of Qu and wine. The purpose of applying these techniques is to create a normal breeding environment for the beneficial moulds in brewing, so that the fermentation process will be under control. In the whole process of Qu-making and wine brewing, Chinese ancestors dealt with moulds, a variety of microorganisms, and unconsciously accumulated much experience of harmonious coexistence and mutual utilization with them. Thanks to the use of Qu, kinds of beneficial moulds are involved in the fermentation

J. Zhou (✉)
The Institute for History of Natural Sciences, Chinese Academy of Sciences, Beijing, China
e-mail: zhoujiahua0129@sina.cn

J. Hua and L. Feng (eds.), *Thirty Great Inventions of China*,
https://doi.org/10.1007/978-981-15-6525-0_28

process, each, respectively, showing its ability to decompose starch into monosaccharides, protein into amino acids, or fat into polyols, and ultimately endowing Chinese wine with rich composition and unique flavor.

It is widely known that in China, mature vinegar, fumigated vinegar, aromatic vinegar, and rice vinegar are all made from cereal grains through fermentation, while in a Western family, vinegar made from grains is rarely seen. The common seasonings preferred by the Westerners mainly include tomato sauce, salad sauce, mustard sauce, and chili sauce, and the sour mainly comes from all kinds of verjuice, and even the fresh lemon juice. The fact proves that cereal vinegar is a specialty of the Orient, especially in China. The production of Chinese vinegar is based on the wine-making technology, that is to say, the cereals are first converted into wine, and then the wine is converted into vinegar under the catalysis of acetic acid bacteria in a specific environment. Therefore, it is beyond doubt that Chinese vinegar naturally inherits the merits and characteristics of Chinese wine, containing richer ingredients than fruit vinegar.

Owing to the lack of scientific preservation equipment and skills, foods were usually processed with sauces in ancient times for the convenience of transportation and long-term storage. Different sauces emerged in endless stream in ancient China, such as meat sauce, fish sauce, and vegetable sauce. Since the main products of planting industry are cereals and beans, the most common sauces are fermented flour sauce made from cereals and soybean sauce made of beans. Flour sauce, soybean sauce, and fermented soya beans can not only be eaten directly, but also be used to process other edible materials into tasty and durable food or condiments. Fermentation is also the key to sauce-making technology. Because of the different raw materials, especially the different strains during fermentation, the Qu used in sauce-making is also different from that in vinegar- and wine-making. Agriculture is also dominated by planting in other Eastern Asian nations, enabling them to inherit the Chinese sauce-making culture and skills. Whether in diet or cooking, sauce has been placed in a prominent position, forming a distinctive feature of East Asian diet.

Chinese ancestors created a unique fermentation technology system, which not only prevails in China, but is highly influential in East Asia and the whole world. As a common natural phenomenon, fermentation is ubiquitous on the earth and in practice almost all ethnic groups have their own ways, especially in food processing. However, in the long-term practice, by creatively utilizing microorganisms and domesticated moulds, Chinese ancestors scientifically developed a series of microbial engineering technologies of wine-making, vinegar-making, and sauce-making. These outstanding experiences and achievements are incomparable to those of other nations. The practical experience accumulated by Chinese ancestors in the above-mentioned microbial engineering not only provides a cognitive premise for the establishment of microbiology and microbial engineering science, but also lays a technological foundation for the development of contemporary biotechnology. In order to objectively understand this historical fact, it is necessary to review this history again.

Wine-making with fruits (including grapes) made no great achievement in ancient China. Dairy wine made from the animal milk was only popular among nomadic tribes in Northwest China. The large amount of wine could only be provided by brewing with cereals. Therefore, in ancient China, the wine-makers mainly brewed wine from cereal grains. Wine-making with fruit and milk only involves the process of alcoholization, but the production of cereal wine is much more complicated because of the ingredients contained in grains. The major components of grains are polysaccharide macromolecule compounds like starch, cellulose, and protein. They cannot be directly converted into ethanol by yeast. They have to be hydrolyzed and saccharified to monosaccharide before they can be fermented into wine. In other words, they must go through many steps before they can be converted into ethanol. To put it simply, the first step is to decompose starch into monosaccharides and disaccharides that can be used by yeasts, which is the saccharification process. The second step is to convert sugar into ethanol, which is named as the alcoholization process. If the two processes are carried out sequentially, the fermentation is named as single-mode fermentation; if both processes are carried out at the same time, this type of fermentation is called double-mode fermentation. Barley and wheat are abundant in the Mesopotamia of Western Asia, ancient Egypt, and ancient Greece. For a long time, people there have been using malt fermentation to make wine, that is, single-mode fermentation of malt saccharification before alcoholization. Malt wine, the original beer, is not only the earliest wine, but also has been handed down as the main wine. At that time, beer and bread, grape wine, and honey wine were probably the oldest varieties of fermented food in the Western nation. The initial production of the fermented food by ancient Egyptians and Babylonians can be traced back to 4000 BC.

Chinese ancestors made wine from grains by Qu. Qu contains moulds like Aspergillus, Rhizopus, and Mucor that can make starch saccharify (mould secretes Glucoamylase in reproduction, just as man secretes saliva which contains glucoamylase), and yeasts and bacteria that can make sugar alcoholize, so the fermentation of wine is double-mode fermentation. By comparison, double-mode fermentation is obviously more efficient than single-mode fermentation. More importantly, in the double-mode fermentation with Qu (mixed bacteria) as the starter, besides saccharification and alcoholization, complex biochemical reactions of organic substances such as protein, fat, and inorganic salts are also carried out. Therefore, the double-mode fermentation of grains is not only more complex than the single-mode fermentation, but also richer in the connotation of the produced wine which has both a mellow taste and an attractive aroma. A unique Chinese fermented raw juice wine (yellow wine) and distilled liquor (white wine) and their exquisite brewing technology have been created this way.

According to modern science, the saccharification and alcoholization processes of double-mode fermentation of grains are much more complicated than the chemical reactions mentioned above. Besides carbohydrates with starch as the main component, cereal grains contain nitrogen, fat, and pectin which will undergo a series of chemical changes during water infiltration and heating. For instance, starch granules in cereals are actually interwoven with cellulose, hemicellulose, protein, fat, inorganic salts, and other components. Starch granules themselves can resist external forces because they have an outer membrane. Starch granules are aggregated by many small needle-like crystals, and starch molecules are bundled by hydrogen bonds. Starch is a hydrophilic colloid, so when it meets water, water molecules will infiltrate into starch granules and increase their volume and mass. This phenomenon is called starch expansion. The expansion of starch increases accordingly with the rise of temperature. When the volume of the particles expands 50–100 times, the linkage between starch molecules will be weakened and they will eventually disintegrate into homogeneous viscous bodies. This phenomenon of infinite expansion of starch granules is called gelatinization. The temperature at the time is called gelatinization temperature. Different starches have different gelatinization temperatures. The initial expansion of starch releases a certain amount of heat, while the further gelatinization process is an endothermic process. After the starch has been gelatinized, if the product temperature continues to rise to about 130 °C, the hydrogen bond between macromolecules will be weakened; the original reticular structure will be destroyed; and the starch solution will become fluid with lower viscosity. This phenomenon is known as starch liquefaction or dissolution. It is in the process of starch gelatinization and liquefaction that enzymes in the system are activated (starting at 50–60 °C), which will decompose starch into dextrins and sugars. At the same time, further hexose changes, aminoglycan reactions, and caramel formation will occur. The sugar content of the original cereal raw materials was only about 4%. After gelatinization and liquefaction, all kinds of monosaccharides increase greatly.

Since starch is the main component of cereals, the most critical reaction in such a complex chemical reaction process is the decomposition of starch into monosaccharides, commonly known as saccharification. What really matters in saccharification is the catalysis of amylase. Amylases actually consist of α-amylase, glucoamylase, isoamylase, β-amylase, maltase, transferrin, and other enzymes. These enzymes act simultaneously in saccharification, producing fermentable sugars, dextrin, and oligosaccharides. Generally, monosaccharides (glucose, fructose, etc.) and disaccharides (sucrose, maltose, lactose, etc.) can be utilized by yeasts and are the most basic fermentative sugars.

In the saccharification stage, other substances contained in cereals are hydrolyzed by enzymes like redox enzymes to form medium and low molecular nitrogen compounds such as homoproteins and superproteins, providing timely nutrition for microorganisms such as yeasts. Lipid enzymatic hydrolysis also produces glycerol and fatty acids. Some types of glycerol are the nutritional sources of microorganisms. Some fatty acids produce a variety of lower fatty acids. Pectin is enzymatically hydrolyzed into pectic acid and methanol. The enzymatic hydrolysis of organic phosphoric compounds can release phosphoric acid, which provides phosphorus source for the growth of microorganisms such as yeasts. Enzymatic hydrolysis can hydrolyze some cellulose and hemicellulose to a small amount of glucose and xylose. In a word, the material changes in the whole saccharification process are complex, because they may be influenced by many factors, including the environment and the operation.

The following fermentation process is led by yeast, bacteria, and rhizopus, which converts sugar and amino acids into monohydric alcohols, polyols, and aromatic alcohols composed by ethanol. The reaction formula is as follows:

Glucose + ADP + Phosphoric Acid $\xrightarrow{\text{Alcohol Enzyme}}$ + Ethanol + Carbon Dioxide + ATP + 10.6 kJ.

ADP is the abbreviation of adenosine diphosphate, and ATP is short for adenosine triphosphate. They are the central substances for energy utilization and storage in organisms. They can be converted to each other under the catalysis of enzymes. ATP removes the phosphoryl group farthest from adenosine, which produces ADP and releases a large amount of energy that can help the chemical reaction. Conversely, inorganic phosphorus combines with ADP to form ATP, which can store the energy released by the oxidation of intracellular nutrients in the form of chemical energy in the high-energy phosphoric acid bonds of ATP molecules. Alcoholic enzymes are actually the general name of various enzymes and coenzymes in a series of biochemical reactions from glucose to alcohol, mainly including hexose phosphorylase, redox enzyme, enolase, decarboxylase, and phosphatase. These enzymes are intracellular enzymes of yeast. The reaction formula listed above reveals that no oxygen molecule is involved in the alcoholization reaction, so this process is an anaerobic fermentation process. It mainly includes two biochemical reactions: the fermentation of glucose and the anaerobic degradation of pyruvic acid. Four stages and twelve steps are included in the fermentation process of glucose before it is gradually transformed into ethanol by pyruvic acid–acetaldehyde. With yeasts converting glucose into ethanol, some other bacteria also do the same. When glucose is phosphorylated, some bacteria can turn it into ethanol through their own channels. However, the alcoholic fermentation ability of bacteria is not as good as that of yeast, and some fusel and polyols alcohols like butanol and organic acids like formic and acetic acid are produced simultaneously. In both the saccharification and the fermentation processes, biochemical reactions in fermented grains need to be carried out within a certain range of pH. Excessive or low pH value will be unfavorable to saccharification and fermentation.

From the above biochemical reactions, it can be clearly recognized that grain fermentation has undergone quite complex chemical reactions. Besides monosaccharides produced by starch hydrolysis and saccharification, and ethanol produced by monosaccharide alcoholization, amino acids, higher alcohols, fatty acids, various lipids, dextrins, organic acids, and other residual sugars that have not been alcoholized can also be produced. In short, the wine gained through fermentation is a kind of water solution of many nutrients and with alcohol as the main constituent. The superiority of Qu'nie fermentation over pure maltose saccharification or yeast alcoholization greatly lies in its full use of microorganisms to help complete this complex fermentation process.

In *Huainanzi*, written by Liu An in Han Dynasty, it is said that "the beauty of wine begins with ploughs and plowshares," which means the wine-making technology began from the rise of planting industry. And in *Shangshu*, it is mentioned that "when people make wine, Qu'nie must be used," which indicates that people have realized the decisive role of Qu'nie in wine-making in Shang Dynasty. From the Spring and Autumn Period and the Warring States Period to the Qin and Han Dynasties, the progress of wine-making technology was first manifested in the improvement of the technology of Qu-making. Long-term wine-making practice has made people realize that having a good Qu is the first step to make mellow wines, and that to enrich the wine, the variety of Qu should be increased. All the famous works about wine-making in ancient China spend a great length in introducing the Qu-making techniques first before switching to the detailed illustration of the wine-making technology. It is along this line that the development of the wine-making technology in China has been moving forward.

Viewed from modern science, Qu is mostly made of wheat (wheat and barley) and legumes such as peas and Adzuki beans that are crushed and watered into lumps or cakes in which natural microorganisms (moulds) are propagated and cultivated under certain temperature and humidity conditions. It contains dozens of microorganisms such as rhizopus, aspergillus, mucor, yeast, lactic acid bacteria, and acetic acid bacteria, offering a mixture of microorganisms needed for wine-making. Microorganisms grow and reproduce in this medium containing starch, protein, fat, and inorganic salts, producing a variety of enzymes. Enzyme is a kind of biocatalyst. Different enzymes have different catalytic decomposition abilities such as the saccharification ability to decompose starch into sugar, alcoholization ability to change sugar into ethanol, and protein into amino acids. Microorganisms can grow and reproduce only by obtaining nutrients from the enzymes they secrete. If starch is the main component of Qu, most of the microorganisms growing and proliferating in it must be strains with strong ability to decompose starch; if Qu contains more proteins, then more microorganisms with strong ability to decompose protein will be produced. Hence, one can see that the strains in Qu are cultivated through successive screening. The efficacy of Qu will be affected by the proportions of different raw materials, which will further exert an influence on the brewing process. Therefore, the quality of Qu determines the quality of wine (Figs. 28.1 and 28.2).

To further explain the characteristics of the traditional Chinese wine-making technology, comparisons are made between Chinese and Western technologies of brewed wine with raw juice and distilled liquor (See Fig. 28.3).

Wine-making technology in ancient Egypt can be seen from the murals of the tombs of Mori, one of the rulers of Egypt in the eighteenth Dynasty (about 1567 BC–1320 BC) (See Fig. 28.4). Figure 28.5 shows a tool for making kumiss. Two wooden models are exhibited in the British Museum, showing the production process of beer in the ancient West. Figure 28.6 shows a large workshop of bread and beer from the eleventh Dynasty (about 2050 BC) in Dier-el-Bahari, Egypt. Figure 28.7, from the Middle Dynasty of Egypt (about 1900 BC), shows the operation of a female worker in beer fermentation. Ancient Chinese rice wine production process at that time, as displayed in Figs. 28.8 and 28.9, is completely different. Figure 28.8 shows a stone sculpture unearthed in the Eastern Ear Room of No. 1 Han Tomb, Dahuting, Xinmi, Henan. It is divided into three columns: the upper column is a row of wine urns; the middle column is a row of wine into urns; and the lower column is a description of the brewing process.

As far as distilled wine production technology is concerned, the differences between China and the West are also obvious, which have been summarized in Table 28.1.

Fig. 28.1 Qu-making through tramping in Maotai liquor factory of Guizhou province

Fig. 28.2 Qu stacked in the storage room

Brandy, mainly produced in France, is traditionally regarded as noble and high-grade liquor because it is distilled from wine. As time develops, once the same production process is adopted, liquors made from other fruits are also named as brandy with the name of the fruit. Different distillation methods have great influence

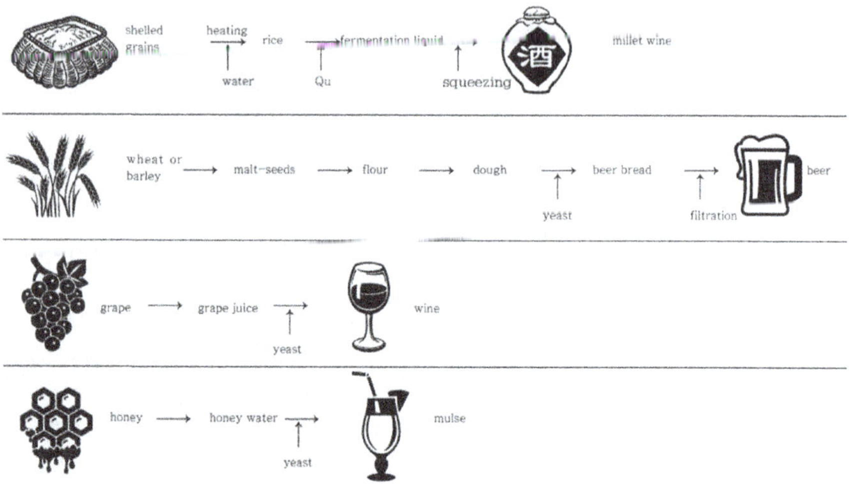

Fig. 28.3 Comparison of production techniques of fermented raw juice wine between China and Western countries

Fig.28.4 Ancient Egyptian murals: interpretation of wine technology 3,500 years ago

on the flavor of brandy products. The world-famous French Cognac uses the twice-distillation method: the grape juice wine is distilled twice in a pot-type steamer. However, Almanac brandy, another famous French liquor brand, prefers the continuous distillation in which the raw liquor fermented will be distilled once in the tower

Fig. 28.5 Leather bag and pounding mortar for brewing kumiss (Inner Mongolia Museum Collection)

Fig. 28.6 A workshop for making bread and beer

distillation equipment to meet the technological requirements. The quality of brandy increases with the storage time, especially when reserved in oak barrels.

Whisky is a distilled liquor made from germinated grains (in general, barley and wheat), whose varieties vary with the raw materials. After barley germinates, it will be dried and ground in a peat fire oven to produce fermentation mash. Hence, it has a flavor of peat smoke, forming the special flavor of Scottish whiskey. After two batches of distillation, monomer malt whiskey can be obtained from the fermented

Fig. 28.7 A woman stirring the fermented mash of raw dough with a wine fermentation tank in front of her

mash. A variety of monomer malt whiskey can be mixed together to produce "pure malt whiskey" or "mixed whiskey". Irish whiskey is made of barley malt plus wheat and rye. Barley malt is not baked in a peat pyrotechnic oven, so the finished wine has no aroma of sootiness.

Vodka means "lovely water" in Russian. Originating in Russia, Poland, Lithuania, and some Nordic countries, it is the national wine of these countries. It uses wheat, barley, potatoes, molasses (waste sugar beet molasses), and other starch-containing rhizome fruits as raw materials. These materials will produce edible alcohol through fermentation (before the nineteenth century, malt was used as saccharifying agent; in the twentieth century, artificial amylase has been gradually used as saccharifying agent and yeast as a starter) and distillation, which will go through the process of deodorization and impurities removal by birch charcoal whereby methanol, aldehydes, fusel oil, and higher fatty acids will be disposed of, making the wine refreshing, mellow, and mild. Vodka is a typical alcoholic beverage. The real vodka is mellow, tasteless, odorless, totally neutral, with only the aroma of pure alcohol.

Rum is a kind of distilled liquor mainly made from molasses in sugarcane or cane juice through fermentation, distillation, storage, and blending. The ordinary alcoholicity of rum is 40–43%. The major rum producers are the Caribbean countries that are rich in sugarcane, such as Jamaica, Cuba, Haiti, and Dominica. Sugarcane molasses or cane juice are fermented together with the specially selected aromatic yeast or ester-producing yeast, and then batch or continuous distillation is applied to

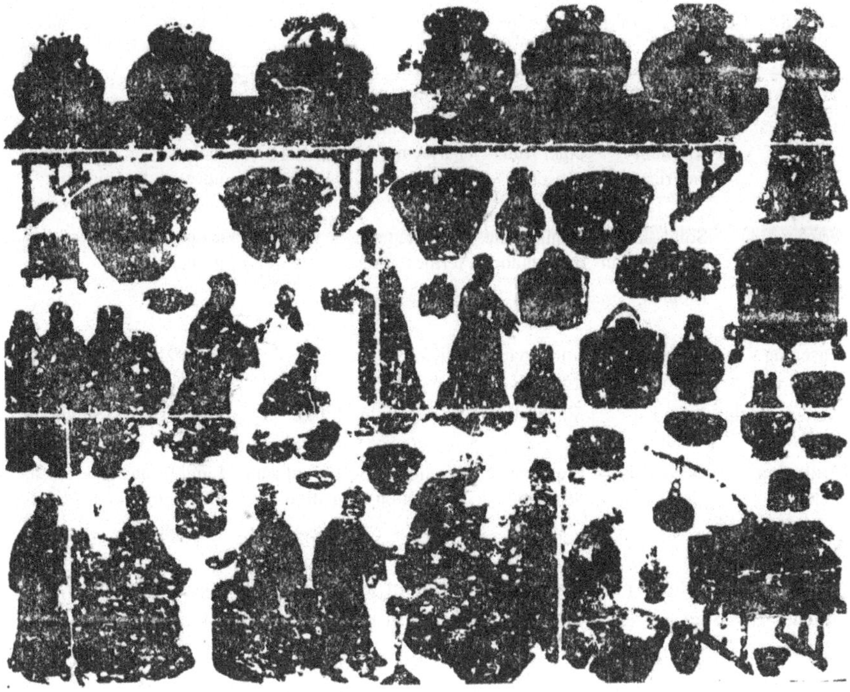

Fig. 28.8 Stone sculpture: wine-making

Fig. 28.9 Stone sculpture: A copy of the wine-making process

obtain liquor with alcoholicity up to 75%. These liquors should be preserved in oak barrels for several years before they are blended into liquors with alcoholicity from 40 to 43%.

Gin originates in the Netherlands and develops in Britain. It is distilled from edible alcohol by adding ginseng seeds and other spices (aromatic plants). Owing to the pleasant fragrance and diuretic function of ginseng, gin is actually a kind of alcoholic drink mixed with fruit juice.

Table 28.1 Comparison of distilled wine-making techniques between China and Western countries

Place of origin	China	Western countries				
Variety	White Spirit	Brandy	Whisky	Vodka	Rum	Gin
Material	Cereals like sorghum and rice	Grape or other fruits	Cereals and barley malt	Edible alcohol	Sugarcane molasses or juice	Edible alcohol, juniper berry, etc
Fermentation mode	Solid-state fermentation	Liquid-state fermentation	Liquid-state fermentation		Liquid-state fermentation	
Saccharifying agent	Mould		Amylase			
Leavening agent	Microzyme	Microzyme	Microzyme		Microzyme	
Microorganism	Mixed strains	Single strain	Single strain	Single strain	Single strain	Single strain
Distillation method	Solid-state distillation	Liquid-state distillation	Liquid-state distillation	Liquid-state distillation	Liquid-state distillation	Liquid-state distillation

In addition, there are Mexican tequila, brandy spirits brewed from wheat and potatoes in some Nordic countries, Polish Gibraltar (fermented with cow's favorite Gibraltar), Liqueur (aromatic spirits fermented with certain fruits, spices, and medicinal materials), etc.

The above introduction clearly shows that there are obvious differences in the fermentation technology of different kinds of wines in ancient East and West. Because of the distinct natural environments and cultural traditions, both the brewing technology and the products have their own typical features. The differences in natural environment are mainly reflected in two aspects. Firstly, the raw materials. Western wines usually take wheat and barley as the main material, both of which having a solid hull and are hard to be directly cooked and processed. In most cases, they are grinded into flour before being processed into pasta. Since the solid hull directly prevents fungi such as mould from growing and multiplying on its surface, yeast spores that roam in the air can only be accepted after the wheat and barley are processed and crushed. However, the situation is different for wine-making in the East. The main raw materials of Oriental wine-making are rice and millet that can be directly cooked after the soft hulls are removed. In hot and humid summer, cereal grains being processed, especially the ripe grains, naturally become the breeding ground for fungi. Thus, it will be a matter of pushing the boat along with the current to make wine through fermenting these grains. Secondly, the climate. It was hot and humid in summer in China, while it was hot and dry in ancient Sumeria and Egypt, which was not conducive to the reproduction of fungi. Additionally, there were few fungi in the local air, so grain fermentation caused by fungi rarely occurred. This may be one of the reasons why beer rather than rice wine can be produced in the West.

Before the Middle Ages, wine and beer were the norm in the West, while in the East, Chinese rice wine almost monopolized the market. Beer signifies the extension of bread making technology, while rice wine technology is the crystallization of natural fermentation of rice. In the West, as in beer production, starch saccharification and alcoholic fermentation are accomplished in two separate steps in sequence, and the microorganisms involved are basically yeasts. In China, starch saccharification and alcoholization proceed simultaneously shortly after the start of starch saccharification and alcoholization, in which both fungi and yeasts participate, and mixed strains are fermented in the form of Qu. By contrast, although the brewing technology in China is more complex, there are many advantages in combining two steps into one. Perhaps, the greatest advantage is that thanks to the participation of many microorganisms, the connotation of the wines produced will be more abundant and the taste will be better. In other words, mixed strains are introduced through Qu and each fights its own battle in fermentation, producing ethanol and many chemical compounds like other alcohols, organic acids, and esters, especially aromatic esters, and forming the unique characteristics of Chinese wine. An analysis will be made here from the perspective of ethanol concentration. In beer brewing, the saccharification process is mainly led by α-amylase and β-amylase in malt, both of which participate in starch hydrolysis. The process of beer alcoholization is completely operated by yeast, which hydrolyzes the maltose molecule produced by starch hydrolysis into two glucose molecules through its own maltase, and then ferments the glucose to ethanol. However, the presence of maltose is an inhibitor of β-amylase (because the formation of maltose from starch by β-amylase is a reversible reaction). Once the concentration of maltose reaches 7%, the β-amylase will stop working and the yield of maltose will decrease. Therefore, under normal circumstances, the ethanol concentration in beer is maintained at 3–4%. In Chinese rice wine brewing, people use Qu, and the fermentation process is quite a different situation. The introduction of fungi not only produces α-amylase and β-amylase, but also amyloglucosidase that can act on starch, releasing glucose molecules, which are directly fermented by yeast to produce ethanol. Nevertheless, since glucose does not inhibit the activity of amyloglucosidase, a higher concentration of glucose can be accumulated in the mash. Furthermore, the sensitivity of amyloglucosidase to ethanol is less than that of cereal β-amylase, so adding fresh culture medium, such as cooked rice or millet, to increase the concentration of starch–glucose in the mash can increase the concentration of ethanol. At that time, the only factor limiting ethanol concentration is the sensitivity of yeast to ethanol. Generally speaking, when the ethanol concentration in the mash reaches 11–12%, the fermentation of yeast slows down obviously. Therefore, the alcohol content of Chinese rice wine should have been able to reach about 10% in the Qin and Han Dynasties (in fact, due to technical reasons, only about 5%). However, the Chinese brewers have made many technological innovations in order to improve the alcoholicity of wine. For example, the Jiuyunchun wine-making method recommended by Cao Cao is to cultivate and exercise the working ability of mould and yeast by putting rice into mash in batches. In the Northern and Southern Dynasties especially, Qu already reserved for a long time was widely used to inoculate and cultivate new ones. Wine-makers at that time probably did not know that this technical measure was an important way to screen and domesticate high-quality fungi and yeasts. Long-term repeated operation of this technology has unconsciously cultivated strains and yeasts of strong vitality which can still maintain their activity when the alcohol concentration reaches 15–16%. Consequently, the alcohol content of Shaoxing rice wine and Fujian rice wine in modern times can reach 16.1–19.4%. Of course, there are more ways to improve the alcoholicity, such as the cultivation and utilization of distiller's yeast and acid slurry mentioned in Beishan Jiujing. In brief, the technical advantages of using Qu in Chinese wine-making are inexhaustible. No wonder the Western missionaries who came to China during the Ming and Qing Dynasties thought that Qu used in Chinese wine-making was very mysterious.

These missionaries felt that Chinese wine was different from what they used to drink. They were even more curious when they saw that Chinese brewers used the brick-shaped magic weapon—Qu. So, they tried to steal this mysterious brick and explore its secrets. When their knowledge could not help them to decrypt, they took great pains to smuggle these secrets back to Europe, asking their friends to help analyze and test. It was a French scholar who first demystified Qu.

In 1892, L. C. A. Calmette (1863–1933) not only noticed the unique method of Chinese wine-making, but also studied Qu brought back by the missionaries. With the help of his colleagues from Pasteur Institute, he isolated a class of microorganisms such as Mucor, Aspergillus oryzae, and Rhizopus from it. He found that these previously unnoticed fungi were the very microorganisms that played a key role in fermentation, thus initially revealing the unique functions of Chinese liquor starter. He was the first scientist to recognize the progressiveness of saccharification of moulds in liquor production in China. Carlmette was also a commercially minded scientist. In 1898, he applied for a patent in Europe for the use of Mucor in alcohol production and popularized this method in the alcohol industry. In the practice of popularization, they have developed several fungi that can be used as the carrier of enzymatic synthesis for fermentation production, which greatly promoted the development of world brewing industry and the improvement of brewing technology. For this amylase fermentation method, which imitated the Chinese brewing technology, changed unilateral fermentation in alcohol production into bilateral fermentation, it not only improved the production efficiency, but also reduced the cost and guaranteed the quality. Since then, scientists have further realized that the use of microorganisms such as fungi has great potential. They have successively developed fermentation technologies like Aspergillus oryzae, Aspergillus niger, and liquid Qu, and improved the technology of alcohol production with molasses, sugar beet molasses, and sugarcane molasses as raw materials. By using organic acid fermentation technology, they developed and produced citric acid and other products. And through utilizing amino acid fermentation technology, they found the way to produce kinds of amino acids like glutamic acid (sodium glutamate), L-lysine, and tryptophan. The most outstanding achievement is the invention and production of antibiotics. In September 1928, British scientist A. Fleming (1881–1955) discovered penicillin, the first antibiotic drug, and opened a new stage of the world's modern medical revolution. For Fleming, penicillin symbolizes significant progress in the treatment of

Fig. 28.10 The French scientist Pasteur who is making experiments

infectious diseases. At the same time, the value of fungal research, microbiology, and microbial engineering in the medical revolution was confirmed.

Brewing technology belongs to fermentation engineering, which is not only the oldest technology, but also a part of the most advanced bioengineering (modern biotechnology). It is old because the technology first appeared 8,000 years ago and people mastered it by imitating the phenomenon of natural fermentation. Despite the long history of fermentation, it was not until the second half of the nineteenth century that people gained a preliminary understanding of why cereals could be used to make wine. French scientist L. Pasteur (1822–1895, see Fig. 28.10) uncovered the mechanism of wine-making. In 1897, the German chemist E. Bucher (1860–1917, see Fig. 28.11) completed a well-designed and decisive scientific experiment, which proved the existence of acellular alcoholic fermentation and called this active fermentation agent "fermentation enzymes". The discovery of "fermentation enzymes" reveals that the essence of fermentation is an enzymatic chemical reaction process. Brewing enzymes are essentially proteins. We now know that saliva, gastric juice, pancreatic juice, and bile of animals can secrete a variety of enzymes: amylase for starch hydrolysis, protease for protein hydrolysis, lipase for lipid hydrolysis, and so on. It is these enzymes that help complete the digestion and metabolism of food. These enzymes are collectively called extracellular enzymes. Their discovery not only opens a new chapter in the study of biological metabolism, but also links the activity of yeast cells with enzymatic chemistry, promoting the development of microbiology, biochemistry, fermentation physiology, and enzymatic chemistry.

Fig. 28.11 The German scientist Bucher

From 1902 to 1909, Bucher studied other types of fermentation, such as the fermentation of lactic acid, acetic acid, citric acid, oil, and so on, and further confirmed that fermentation was a series of enzymatic reaction processes. The application of enzymatic chemistry constitutes the "enzymatic engineering" of biotechnology. The research into it is of great significance to the development of wine-making, vinegar-making, sauce-making, sugar-making, and the whole food industry, and also plays a special role in solving problems in the fields of medicine, hygiene, and biology. Because of his findings and achievements, Bucher won the Nobel Prize for Chemistry in 1907.

In the 1940s, the industrialization of penicillin production by submerged culture and fermentation was the highlight of the development of fermentation engineering. Since then, the knowledge of microbiology, biochemistry, and chemical engineering has been superimposed and progressed to promote the maturity of fermentation engineering. Since the 1960s, new technologies such as immobilized enzymes and immobilized cells have promoted the development of biochemical technology. With the progress of analysis, separation, and detection technology and the application of electronic computers, fermentation engineering has been innovated. In 1973, the emergence of the recombinant DNA technology made it possible to create new species by directionally altering the functions of the original species according to the blueprint of engineering design. This is what we call genetic engineering. The establishment of hybridoma through cell fusion to produce monoclonal antibodies has become a revolutionary progress in immunology. Through large-scale cultivation of animal and plant cells, we can produce a variety of substances needed by human beings in the same way as microbial fermentation. We can also cultivate new hybrid varieties that cannot be obtained by conventional methods, thus creating cell engineering.

Microbial fermentation engineering, enzyme engineering, cell engineering, and genetic engineering constitute together a comprehensive system called "bioengineering", which is also known as "modern biotechnology" with fermentation technology as the foundation and genetic engineering as the core. It is a newly developing engineering technology and represents the frontier of science and technology in the twenty-first century. Since its establishment in 1973, it has opened up new fields such as protein engineering and recombinant DNA engineering in the first 20 years, developed in experiments many new plant varieties with high yield and disease resistance, and reproduced animals such as cattle and sheep by using somatic cell cloning technology. Many of its research techniques and achievements are gradually applied in fields like medicine, agriculture, industry, and environmental protection, and the desired benefits are bound to be achieved.

China's achievements in the collection, breeding, domestication, preservation, and research of microbial strains, as well as the experience gained in the exploration, experiment, and innovation of fermentation and brewing technology, are precious treasures in the scientific journey of mankind. We are convinced that in the process of the development of modern biotechnology and the promotion of human civilization, Chinese people will also make their own new contributions like their predecessors.

Jiahua Zhou Born in Ruian, Zhejiang, in 1942, graduated from the Department of Chemistry Department of Guangxi University and is now a researcher of the Institute of Natural Science, Chinese Academy of Sciences.

Research interests: History of Chemical; Processing of traditional agricultural, livestock and mineral products; history of wine and so on.

Major works: *History of Chemical Thoughts, Biography of Chemists, History of Ancient Chinese Chemistry, History of Chinese Science and Technology: Chemistry, History of World Chemistry, Yongli and the Yellow Sea, Complete Works of Chinese Traditional Crafts: Brewing.*

Awards: *A brief history of science and technology in the 20th century (co-authored)* won the Second Prize of Natural Science in Chinese Academy of Sciences in 1989;

The Five Milestones of Human Understanding of the World (co-authored) won the Second Prize of the 2006 National Science and Technology Progress Award;

Principle—The Masterpiece of Times (co-authored) won 1988 National Outstanding Book Award;

Chinese Cultural Annals: Records of Science, Technology and Chemistry won the 2000 Book Honor Award;

History of Chinese Chemistry won the 14th China Book Award.

Chapter 29
Fire Arrow and Projection Firearm

Zhanhong You

Firearms refer to weapons that cause destruction by utilizing the chemical energy of the gunpowder. From physical killing (stabbing, chopping, splitting, and smashing) in the cold weapons era to chemical destruction (explosion and combustion) in the firearms times, weapons made a qualitative leap in their lethality. As chemical energy of gunpowder is the main source of the weapons' lethality, firepower becomes the first factor to measure their efficiency.

Black powder is characterized by its oxidizer (potassium nitrate) which can release oxygen, so when burning, it does not need to obtain oxygen from air. With such a self-oxygenated combustion system, black powder produces a large amount of heat based on a fierce reaction of oxidizer and combustible agents, such as carbon and sulfur.

Burning black powder in a sealed container can produce a large number of high-temperature and high-pressure gases in a limited space; so when the small solid powder suddenly expands its volume to thousands of times, the container explodes.

In hydromechanics, when a quantitative fluid is injected from high pressure to low pressure at a high speed, it produces a considerable reacting force. When the reacting force is applied to a moving object, it moves in the direction of the reacting force.

China had refined the original gunpowder containing nitrate, sulfur, and carbon in the third year of the reign of Emperor Xianzong of the Tang Dynasty (808) at the latest and had invented military gunpowder before the fourth year of the reign of Emperor Renzong in the Northern Song Dynasty (1044). After the invention of gunpowder, various firearms came into being. Firstly, gunpowder directly promoted the development of various military combustion and explosive materials and then

Z. You (✉)
Institute for History of Science and Technology & Ancient Texts, Tsinghua University, Beijing, China
e-mail: youzhh@lib.tsinghau.edu.cn

J. Hua and L. Feng (eds.), *Thirty Great Inventions of China*,
https://doi.org/10.1007/978-981-15-6525-0_29

spurred the development of metal tube firing weapons and fire arrows by combining with traditional casting technology.

"Flying Fire" is the earliest weapon manufactured with gunpowder. Its records can be found in *Jiu Guo Zhi* by Lu Zhen of the Song Dynasty [1].

From the Northern Song Dynasty to the mid-Ming Dynasty, a large number of firearms such as fire arrows, fireballs, guns, copper blunderbuss, and iron barrel guns were invented one after another, indicating that the importance of the firearms in war was increasing day by day. In the tenth century, China's military technology was ushered into the ear when cold weapons and firearms were combined in war.

29.1 Fire Arrows

Fire arrows in ancient times were invented based on the combustion and explosion of black powder. In other words, the thermal force generated by the combustion of black powder is transformed into the mechanical force of reaction to propel the arrow shaft to fly and shoot toward the predetermined target.

China is universally recognized as the earliest country in the world to invent fire arrows. The fire arrow in ancient China is tied with a black powder flask on its arrow shaft. After the gunpowder is ignited by the fuse, ejected gas from the tail produces force to push the arrow forward.

The gunpowder arrows were invented in the early Northern Song Dynasty when these primary firearms were not propelled by the gas jets as a result of gunpowder combustion. Tied with the gunpowder pack on the arrowheads (Fig. 29.1), they were launched with bow and crossbow to set fire to attack the enemy after the shell of the gunpowder pack was ignited. According to the *Army Log of the Song History*, in the third year of the reign of Emperor Song Taizu (970) of the Northern Song Dynasty, Feng Jisheng, the commander of the Military Ministry, proposed to test the method of launching fire arrows and then was granted clothes and silks as rewards by the emperor [3]. In the third year of the reign of Emperor Song Zhenzong (1000), Tang Fu, the captain of the navy, offered his self-made fire arrows, fireballs, and fire caltrops. Shipbuilders including Xiang Wan presented the design drawing of sea warships, so they all received rewards of money and silk from the emperor [3]. In September of the fifth year of the reign of Emperor Song Zhenzong (1002), Shi Pu, the military training officer of Jizhou Regiment, wrote to the emperor that he could make fireballs and fire arrows, and then he was summoned to the imperial palace to make demonstrations before the emperor and ministers [4].

Wu Jing Zong Yao (*General principles of martial arts*), a military book compiled during the reign of Song Renzong of the Northern Song Dynasty, specifically depicts the shape and design of the early gunpowder arrows [5].

Different from those burned with mugwort, grease, turpentine, and other burning materials in the past, these gunpowder arrows fired with the bow had not yet used fuse. When they are fired, the shell of the gunpowder pack needs to be ignited. After they hit the target, the gunpowder in the pack then causes intense combustion.

Fig. 29.1 Fire arrows [2]

Volume 12 (Fig. 29.2) of the *Wu Jing Zong Yao* also includes the image of the gunpowder arrows, showing that the gunpowder pack is bound near the head of the arrow shaft. Counter-thrust fire arrow was invented in the Southern Song Dynasty, and it was fired by the reaction force produced by the combustion gas of gunpowder to propel the arrow forward, with powder flask tied on the shaft. According to *Cheng Zhai Ji,* written by Yang Wanli (1127–1206) of the Southern Song Dynasty in the 31st year of Shaoxing period (1161), the army of the Jin Dynasty crossed the Yangtze River and fought with the naval forces of the Southern Song Dynasty in Cai Shi (the eastern coast of Maanshan City in Anhui today) when the army of Song fired "thunderbolts" to disturb the Jin soldiers. As recorded in the book, "A thunderbolt come from the boat, seemingly covered with paper but actually with lime and sulfur. The gunfire falls into the water from the air, while the burned sulfur jumps out and makes a thundering sound. Paper cracks and lime scatters into smoke, blinding the eyes of people and horses [8]." This gun is actually a kind of military rocket bomb, the impetus of which is just the counter-thrust of gunpowder.

In the Yuan Dynasty, inefficient gunpowder arrows were eliminated and directly replaced by counter-thrust fire arrows. According to *Long-range Firearms*, Volume 3 of *Ke Di Wu Lue Ying Huo Shen Ji*, the monograph of firearms at the end of the Yuan Dynasty in the Ming Dynasty manuscript, the single-shot arrows of the Yuan Dynasty were fired with a thick and long powder flask, carrying flying spears, knives, and swords to shoot the enemy [9]. The powder flask was 22 cm long and 6.2 cm in diameter. There were two layers of powder inside it, connected by a fuse, the lower

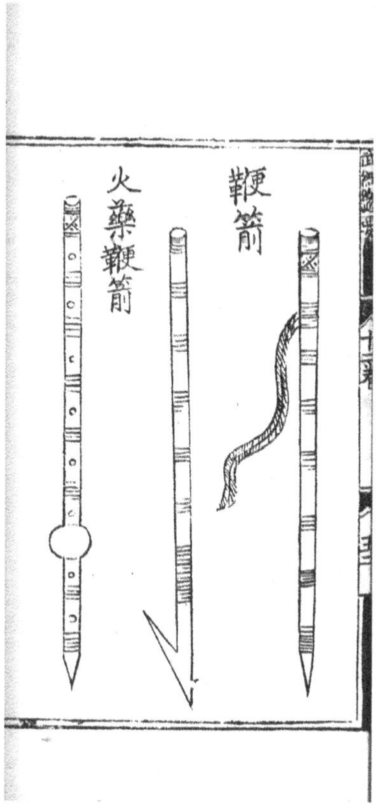

Fig. 29.2 The gunpowder arrow of the Northern Song Dynasty [5]

part of which kept exposed to the air, weighing 1.19 kg in total. *Wu Bei Zhi* of the Ming Dynasty provided the images of three kinds of fire arrows [9]: flying spear, flying knife, and flying sword (Fig. 29.3).

Obviously, the single-shot fire arrow of the Ming Dynasty was fired based on the counter-thrust produced by gunpowder combustion, and its making method is depicted in *Wu Bei Zhi* with pictures and texts (Figs. 29.4 and 29.5). The method uses the powder flask with a fuse as the power unit and the smearing poison on the arrowhead; its range can reach 500 steps (about 775 m; one step in the Ming Dynasty is equal to 1.55 m). *Wu Bei Zhi* clearly records that the key to making the fire arrows lies in the processing of the fuse hole of the gunpowder pack, which is better when made with iron rods than drilling. Whether the hole is straight or shallow, it directly affects the propellant thrust of gunpowder and the flight path of the arrow shaft, as written in the book: "If the hole is straight, the flight path will be straight; if the hole is slant, the flight path will be slant; if the hole is too deep, fire will be leaked from the bottom; if the hole is too shallow, the arrow will be powerless [6]."

In the Yuan Dynasty and the Ming Dynasty, the fire arrows with paralleled powder flasks, the winged fire arrows, the multi-stage fire arrows, and the multi-shot fire

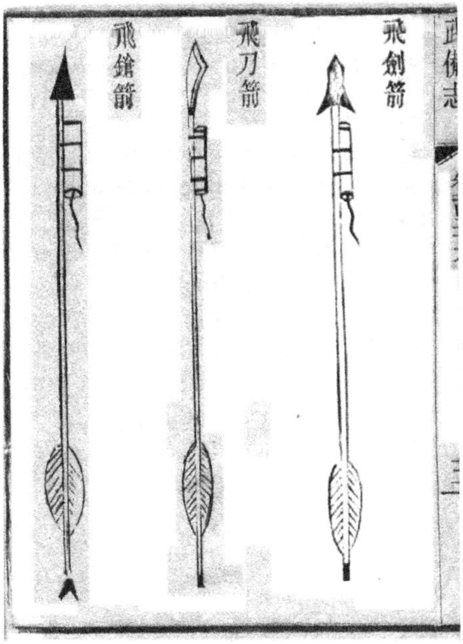

Fig. 29.3 The flying-spear arrow, flying knife arrow, flying-sword arrow recorded in *Wu Bei Zhi* [6]

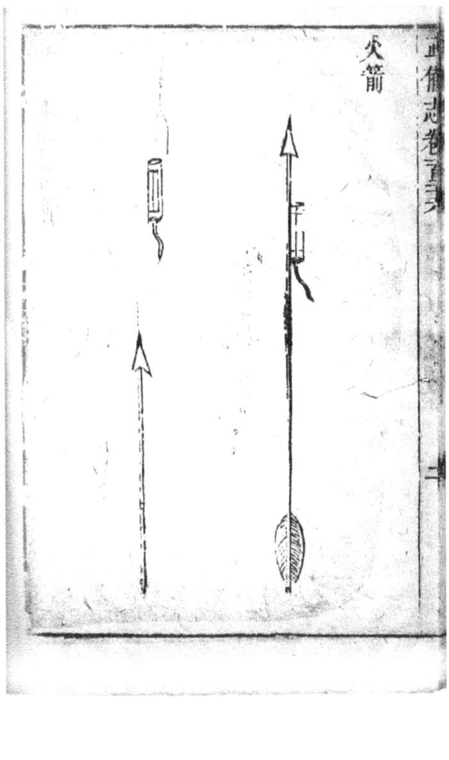

Fig. 29.4 The fire arrows of the Ming Dynasty [6]

Fig. 29.5 The restored fire arrows of the Ming Dynasty [7]

arrows were invented successively, and they were launched with special tube and cabinet, increasing their lethality and range.

For example, Fire Dragon out of Water is a multi-stage fire arrow invented in the Ming Dynasty (Figs. 29.6 and 29.7), which was used to fire enemy ships. The dragon body was made of bamboo tube, and dragon head and tail were wooden; a fire arrow was installed on its head and tail, respectively, and several fire arrows were contained in its belly. Upon being launched, the fire arrows on the head and tail were ignited to propel the dragon forward. When they were about to burn out, the arrows in the dragon's belly would continue the shooting toward the target. As recorded in *Huo Long Jing*, "in water warfare, the fire dragon is ignited three or four chi (a Chinese unit of length) above water surface, and then it flies two or three li (a Chinese unit of length) like a fire dragon out of water; when the gunpowder is burned out, the fire arrows inside will shoot to ruin the enemies and their ships [2]."

Fig. 29.6 Fire Dragon out of water of the Ming Dynasty [2]

Fig. 29.7 Restored model of Fire Dragon out of water of the Ming Dynasty [7]

"Swarm-of-Bees" is a kind of the multi-shot fire arrow invented in the Yuan Dynasty, and it was called "invincible Swarm-of-Bees Fire Cage" in *Long-range Firearms*, Volume 3 of *Ke Di Wu Lue Ying Huo Shen Ji*, the monograph of firearms at the end of the Yuan Dynasty in the Ming Dynasty manuscript (Fig. 29.8). It could launch 20 arrows at the same time to the targets on land and water, with a range of 4,500 steps (622–777.5 m), known as "the First Firearm" at that time [9]. According to *Wu Bei Zhi*, in the Ming Dynasty, the simultaneous discharge of the fire arrows by the "Swarm-of-Bees" increased to 32 (Figs. 29.9 and 29.10), tiger poison was smeared on the arrowhead, and the fuses of all the fire arrows were connected to a bus and installed in a wooden barrel to prevent from getting wet by rain. On the battlefield, "Swarm-of-Bees" is fixed on the chariot to shoot the fire arrows; just as described in *History of Chinese Gunpowder*, once the bus is ignited, the fire arrows will fly 300 steps (about 465 m) in a thunderous manner [10].

The Shenhuo Flying Raven is a multi-cartridge parallel propulsion fire arrow invented in the Yuan Dynasty (Fig. 29.11), which is used for arson attacks. It is roughly described in Volume 8 of *Ke Di Wu Lue Ying Huo Shen Ji*, the monograph of the firearms at the end of the Yuan Dynasty in the Ming Dynasty manuscript: "Made of bamboo strips, like birds, with wings on both sides, gunpowder suspended in the abdomen, fuse bound on the tail; in case of strong wind, it flies into the city, with fire covering the sky, creating chaos among the enemy [9]." In the fourth Yuan Dynasty, the Shenhuo Flying Raven was used to cover the troops to attack the city walls. Volume 131 of *Wu Bei Zhi* made a more specific description of the manufacturing method of Shenhuo Flying Raven, and the fire arrows bound on the Raven are clearly visible (Figs. 29.12 and 29.13). Made of bamboo strips, reed tubes, and cotton paper in the shape of a crow, it is filled with gunpowder in its abdomen, four fire arrows with gunpowder flask obliquely nailed underneath. The Shenhuo Flying Raven of the Ming Dynasty could also be used in water battle to burn the enemy's boats, according to *Wu Bei Zhi* [11].

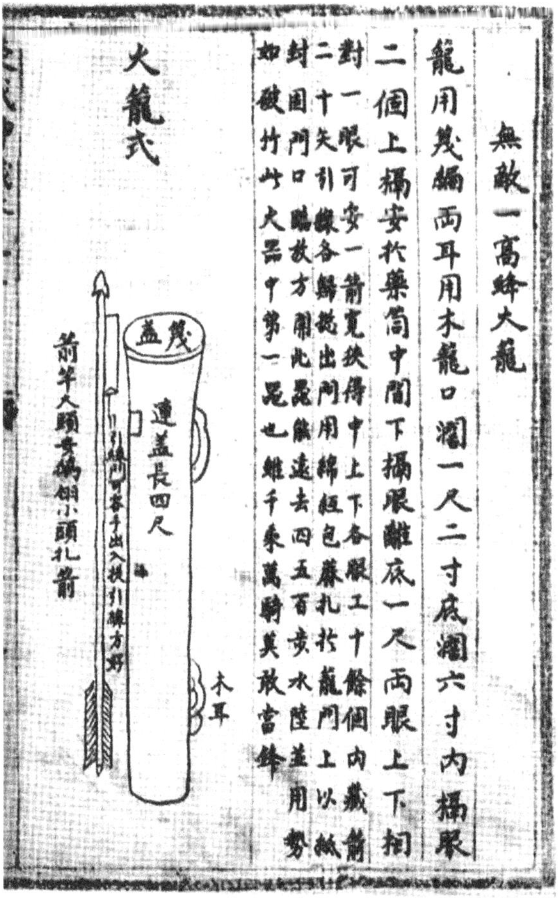

Fig. 29.8 Invincible Swarm-of-Bees Fire Cage of the Yuan Dynasty [9]

29.2 Explosive Firearms

The explosive firearms were originated from burning fireballs. In the early Northern Song Dynasty, fireball, a burning firearm with gunpowder, was invented to burn the enemy's battalions. Fireballs are usually centered on gunpowder, wrapped in paper, cloth, asphalt, turpentine, yellow wax, etc. After the outer layer of fireballs is penetrated by a red-hot iron cone, they are then be thrown by a trebuchet or by manpower, and then the gunpowder inside the fireballs starts to burn and produce flame or poisonous smoke. According to *Wu Bei Zhi*, such firearms include the ignition balls, Tribulus fireballs, thunderbolt fireballs, poison smoke balls, iron-mouth fire kite, bamboo fire kite, and so on.

The Tribulus fireball (Fig. 29.14) is a kind of fireball to spread iron Tribulus. There are three iron blades with six spikes in the center of the ball, wrapped in gunpowder. The ball wears a hemp rope about 12 *chi* (a Chinese unit of length) long in the

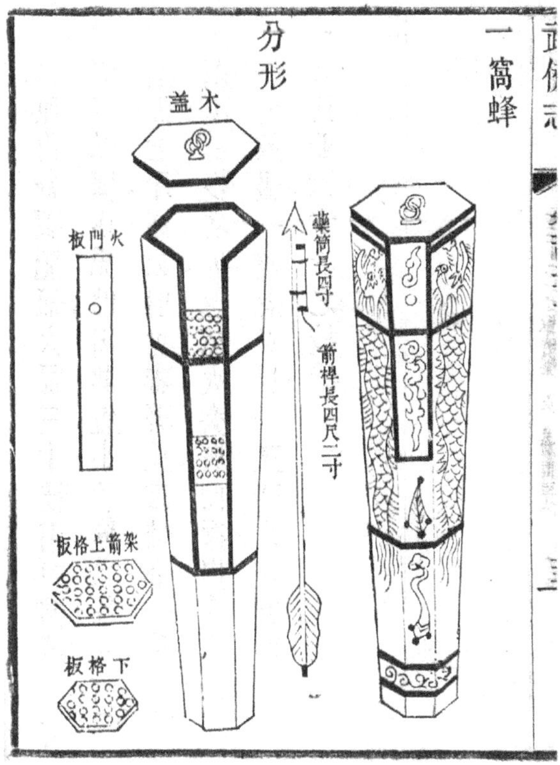

Fig. 29.9 Swarm-of-Bees of the Ming Dynasty [10]

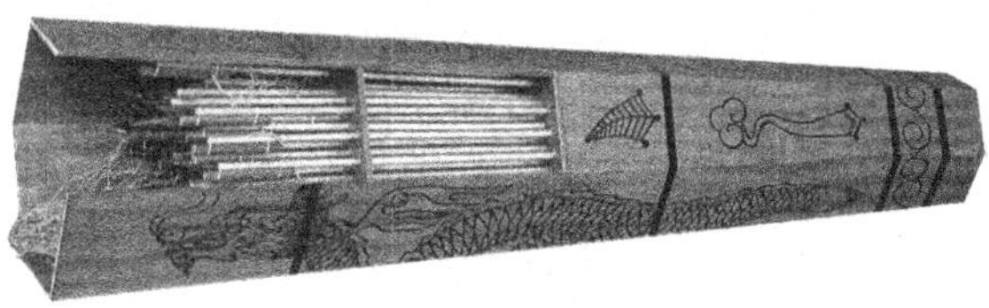

Fig. 29.10 The restored model of Swarm-of-Bees of the Ming Dynasty [7]

middle. Its outer shell is bound with paper and gunpowder. Eight iron Tribulus are placed on the ball. Upon being launched, firstly, iron corn red is burnt, followed by the spherical shell of the fireball thoroughly, and then it is thrown onto the enemy's position with a trebuchet. When the gunpowder starts to burn and produces a flame, the fireball cracks and the iron Tribulus scatters to block enemy's action.

Thunderbolt fireball (Fig. 29.15) is a kind of burning firearms for city defending. Its center is made of 30 pieces of thin porcelain tiles of the size of iron coin mixed with 34 kg of gunpowder, wrapped in an oval paper shell coated with flammable ingredients. When the enemy attacks the city by digging tunnels, the defending side would dig down a shaft underground, burn the thunderbolt fireballs with red-hot iron cones, throw them into the tunnels, and fan the smoke and flames resulting from the burning and bursting of gunpowder to the enemy.

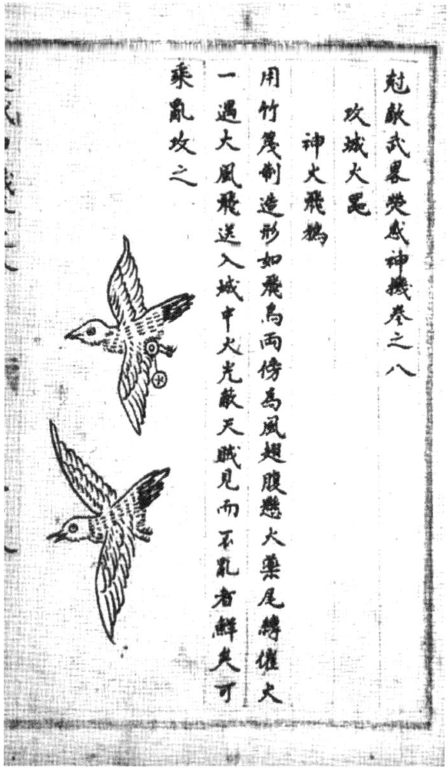

Fig. 29.11 The Shenhuo Flying Raven of the Yuan Dynasty [9]

Iron-mouth fire kite is a burning firearm used for defending the city. Its body is made of wood, mouth of iron, and the tail is bound with stalk grass and loaded with gunpowder. After its gunpowder is ignited, the irowen-mouth fire kite will be thrown by trebuchet to the crowds of enemy or the gathering places of provisions to cause burning. Bamboo fire kite (Fig. 29.16), also a kind of burning firearm, is woven with bamboo sheets into an oval bamboo cage. The cage contains 1 kg of gunpowder and is bound with 3–5 kg of straw at the tail. Iron bomb, a kind of iron-shelled bomb, was invented in the Southern Song Dynasty, which is filled with gunpowder with a shell of pig iron and a small hole on the shell for a fuse. After ignited, the high-pressure gas produced by gunpowder combustion will shatter the iron shell, causing many casualties among the enemies and their horses. In the Southern Song Dynasty, large quantities of iron bombs were produced.

According to the survey on weapons storage in Jingjiang (now Guilin City, Guangxi) by Li Zengbo, a minister of the Southern Song Dynasty, in 1257, there were more than 100,000 iron bombs in Jinghuai Area, while in Jiangling Prefecture (now Jiangling City, Hubei Province), 1,000–2,000 iron bombs could be produced per month [12]. As recorded in Volume 39 of *Jingding Jiankang Zhi*, Jiankang Prefecture (now south of Jiangning County, Jiangsu Province) could produce 22,044 1.5-kg

Fig. 29.12 The Shenhuo Flying Raven of the Ming Dynasty [2]

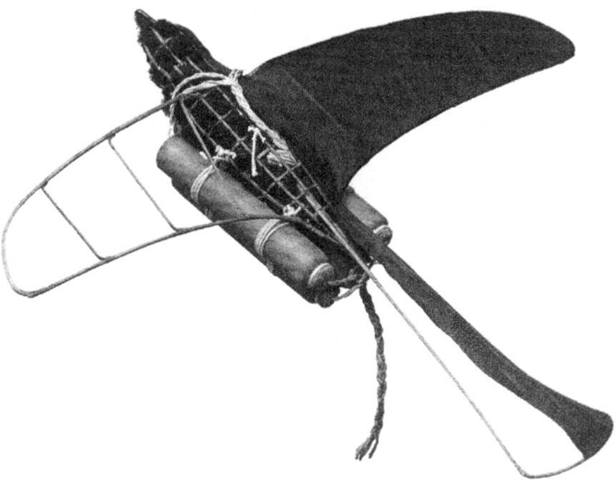

Fig. 29.13 Model of the Shenhuo Flying Raven of the Ming Dynasty [7]

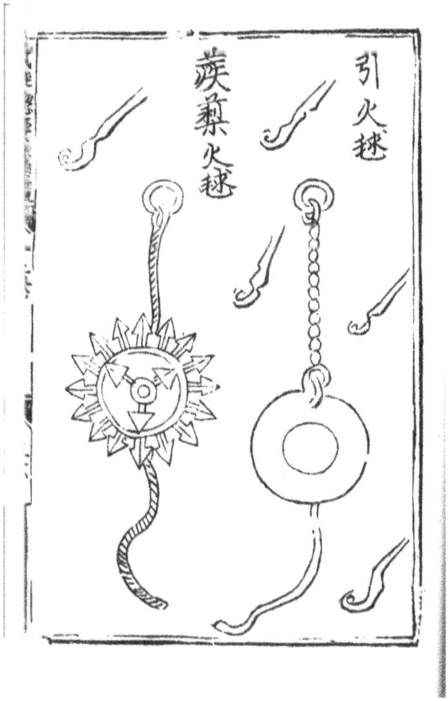

Fig. 29.14 The tribulus fireball of the Northern Song Dynasty [5]

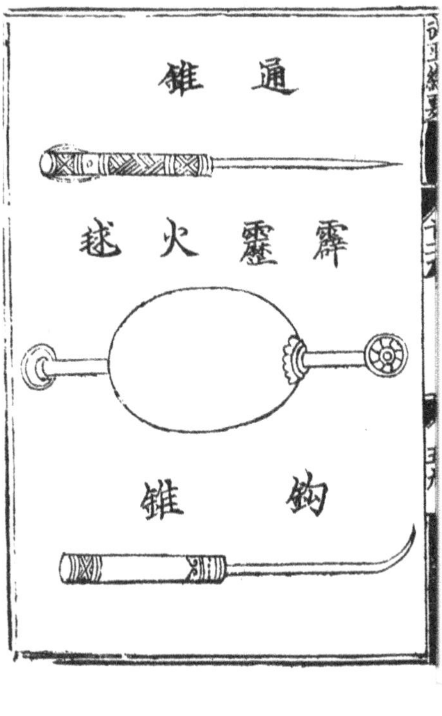

Fig. 29.15 The thunderbolt fireball of the Northern Song Dynasty [5]

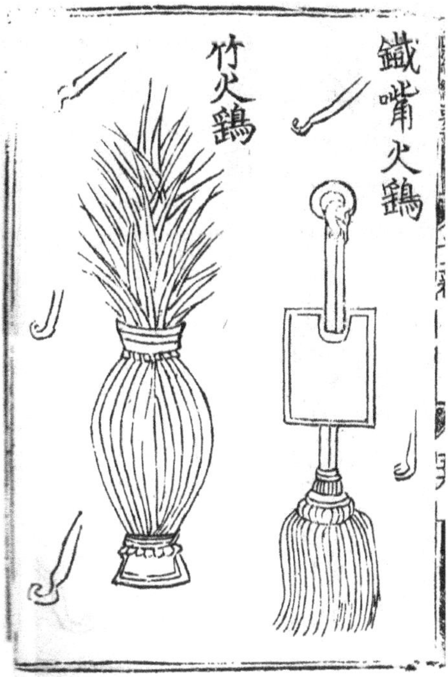

Fig. 29.16 The iron-mouth fire kite and bamboo fire kite of the Northern Song Dynasty [5]

heavy iron bombshells, 13,104 2.5-kg heavy iron bombshells, and 100 3-kg heavy iron shells within 2 years and 3 months [12]. The armies of the Southern Song, Jin, and Yuan dynasties often used iron bombs in the battle for fortress. For example, according to *Chizhanhexi Biography of the History of Jin Dynasty*, in 1232, the army of the Jin Dynasty employed iron bombs to destroy the siege weapons of Mongolian troops when the Mongolians were attacking Bianliang (now Kaifeng City, Henan Province) [13]. For another instance, according to *Maji Biography of the History of Song Dynasty*, in 1277, when the Yuan army broke through the city of Jingjiang (now Guilin, Guangxi), 250 soldiers of the Song Dynasty who were defending the city died at the same time when a big iron bomb exploded [14].

29.3 Tubular Firearms

The tubular firearms in ancient China were made of bamboo, paper, copper, or iron. Initially, a long-handled combat weapon, often spear, is tied with a gunpowder flask on the head (Fig. 29.17). In battle, first gunpowder was sprayed on the enemy and then fought with the spear. Later, bamboo, copper, and iron were gradually used to make barrels and fired projectiles with gunpowder to kill the enemy.

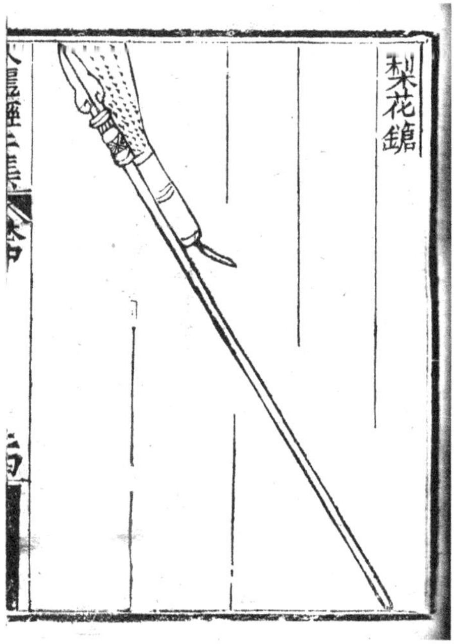

Fig. 29.17 The Pear Flower Spear of the Ming Dynasty

The earliest guns found in historical records are long bamboo pole guns and flying-fire guns invented in the Southern Song Dynasty. For example, in the second year of the Shaoxing period in the Southern Song Dynasty (1132), when Chen Gui was defending De'an Prefecture (now Anlu in Hubei), a long bamboo pole gun team was organized to burn the enemy's overpass to attack the city. As recorded in *Chen Gui Biography, the History of the Song Dynasty*, when the Huihao bridge was occupied by the enemy, Chen Gui led 60 people carrying guns came out from the west gate of the city to burn the enemy's overpass, which was destroyed in a moment with the help of fire cattle [15].

In the fifth year of the Shaoding period in the Southern Song Dynasty (1232), the Jin Army invented a flying-fire gun, a kind of flame-spraying weapon, which was to tie a powder flask at the head of a spear. When confronted with the enemy, they would first fire the gunpowder flask to spray flame onto the enemy and then use the spear to fight at close quarters. The making method of the flying-fire gun can be found in the records of *Puchaguannu Biography, the History of the Jin Dynasty* [16].

Sudden gun was invented in the Southern Song Dynasty, and it is the earliest used tubular firearm in the world. Manufactured in Shouchun Prefecture (Shouxian County, Anhui Province) in the first year of the Kaiqing period of the Song Dynasty (1259), sudden gun used giant bamboo as the launching tube and fired projectiles with gunpowder. As sudden gun has possessed three basic elements of tubular firearms—gun-barrel, gunpowder, and projectile, it can be regarded as the ancestor of guns in the world.

The copper blunderbuss (Fig. 29.18), a metal tubular firearm, was invented in the

Fig. 29.18 Copper bowl blunderbuss of the Yuan Dynasty [17]

Yuan Dynasty. A copper blunderbuss consists of front chamber, powder chamber, and tail. The powder chamber is filled with gunpowder, with the fire door on the top, and the front chamber is loaded with shots, which is fired when the gunpowder is ignited from the fire door with a fuse.

The Zhankou Blunderbuss in the second year of Dade period in the Yuan Dynasty (1298), collected in the Mongyuan Museum of Inner Mongolia, is the earliest existing blunderbuss in the world so far, 34 years earlier than that in the third year of Zhishun (1332) stored in the Chinese History Museum [18]. It is 34.7 cm long, 10.2 cm in outer caliber, 9.2 cm in inner caliber, and 27 cm in depth. Two lines of inscriptions of Phags-pa are inscribed on the bracket body, marking the time when it was manufactured [19].

The Zhishun Blunderbuss of the Yuan Dynasty also collected in the Chinese History Museum (Fig. 29.19) is 35.3 cm long, 10.5 cm in caliber, and 6.94 kg heavy, with the Chinese characters signifying its production date inscribed on the body [12].

Zhizheng Blunderbuss (Fig. 29.20) of the Yuan Dynasty collected in the Military Museum is 43.5 cm in length, 3 cm in caliber, weighing 4.75 kg, has six hoops from the front to the tail. Its bracket is also inscribed with the Chinese characters signifying its manufacturing time [12].

Besides, there are also unearthed non-inscribed blunderbusses of the Yuan Dynasty. For example, in 1970, a copper blunderbuss, named Acheng Blunderbuss, was unearthed in Acheng County in Heilongjiang; it is 34 cm long, 17.5 cm long in chamber, 2.6 cm in caliber, and weighs 3.55 kg, probably manufactured during

Fig. 29.19 Copper blunderbuss of the Yuan Dynasty [17]

Fig. 29.20 Copper blunderbuss of Zhizheng of the Yuan Dynasty [17]

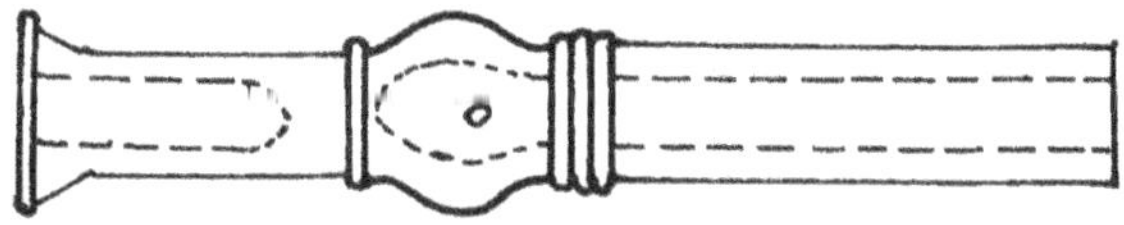

Fig. 29.21 Xi'an Copper Blunderbuss [12]

the period from the end of the thirteenth century to the beginning of the fourteenth century [12].

In 1971, a copper blunderbuss, named Heicheng Blunderbuss, was unearthed in Heicheng Commune, Togtoh County, Inner Mongolia; it is 29.5 cm long, 17.5 cm long in chamber, 2.5 cm in caliber, and weighs 2.3 kg, perhaps manufactured at the beginning of the fourteenth century [12].

In 1974, a copper blunderbuss, named Xi'an Blunderbuss (Fig. 29.21), was unearthed in Jinglongchi Alley, Xi'an City; it is 26.5 cm long, 14 cm long in chamber, 2.3 cm in caliber, and weighs 1.78 kg, also roughly manufactured at the beginning of the fourteenth century [12].

Compared with the bamboo firearms of the Southern Song Dynasty, the copper firearms of the Yuan Dynasty had the advantages of a long service life, fast firing speed, and great destructive power, and first used iron projectiles, which had better ballistic performance and greater lethality. The invention of the copper brass is an epoch-making technological change in the history of weapon development.

In the early Ming Dynasty, Baoyuan Bureau and Weapons Bureau were set up to make firearms. In the 28th year of Hongwu (1395), Arsenal Bureau was set up to specialize in the manufacture of various firearms. According to *Thirteen Firearms of Ministry of Industry*, Volume 193 of D*aming Huidian*, where firearms were made, blunderbusses were made of wrought copper or raw copper or both; The copper blunderbuss made in the ninth year of Hongzhi weighed from 5 or 6 to 10 *jin* (a Chinese unit of weight, half of a kilogram). Before the first year of Hongzhi (1488), the Weapons Bureau manufactured 3,000 bowl copper blunderbusses, 3,000 handled copper blunderbusses, 90,000 arrowheads, and 3,000 signal bullets every three years [20]; the Arsenal Bureau could produce several kinds of firearms like Major General, Second General, Third General, Lightning General, Divine Blunderbuss, Horse-killing Blunderbuss, handled copper blunderbuss, handled iron blunderbusses, bowl blunderbusses, and Zhankou blunderbuss.

In the seventh or eighth year of the Yongle period in the Ming Dynasty (1409–1410), Emperor Zhu Di ordered the formation of the world's first all-firearm army, Shenji Camp, which was composed of five troops, equipped with divine guns, fast guns, single-eyed blunderbusses, handled blunderbusses, bowl guns, general guns, single-shot fire arrow, Shenji arrows, and other firearms. He led many expeditions to the desert in Mongolia, each with Shenji Camp as the vanguard, killing a large number of enemy troops by the advantage of firepower.

Copper blunderbuss of the Ming Dynasty (Figs. 29.22 and 29.23) was well made, with a smaller caliber and longer barrel. In casting, the wall thickness from the gunpowder chamber to the mouth gradually decreased and the range was increased.

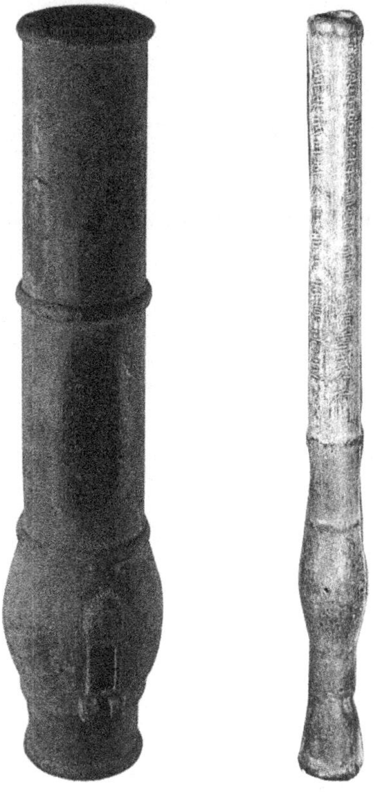

Fig. 29.22 The copper blunderbuss of "Qizi" (left) and "Shengzi" (right) of the Ming Dynasty [21]

The fire door has a rectangular groove with a protective cover, which can help trigger the shooting even in the wind and rain.

The multi-eyed blunderbuss is a multi-tube blunderbuss (Fig. 29.24) invented in the Jiajing period of the Ming Dynasty. Its tube is cast with copper or iron, with 3–10 holes on; several tubes were cast together, with a number of strong hoops on the outside and a wooden handle mounted at the tail. Each tube has a fire door, so multi-tube blunderbuss can be ignited continuously. The multi-eyed blunderbuss overcomes the defect of the single-tube blunderbuss and improves the shooting speed.

In the Hongwu period of the Ming Dynasty, a long-barrel gun with an iron tube was cast. For example, in the Shanxi Museum, three iron guns (Fig. 29.25) were cast in the tenth year of Hongwu period (1377), and each of them is 100 cm long, 21 cm in caliber, with a bigger caliber, a shorter barrel, and thicker tube wall [12]. It is the earliest large iron-cast gun in ancient China. Compared with the copper gun, the iron gun has longer barrel, larger caliber, and thicker tube wall, with more hoops cast on the barrel, which increases the strength of the gun; what's more, the ears on both sides can help adjust the shooting angle. The iron barrel cast during the Wanli period is slender, and the ratio of length to diameter increased significantly, which further increased the shooting range.

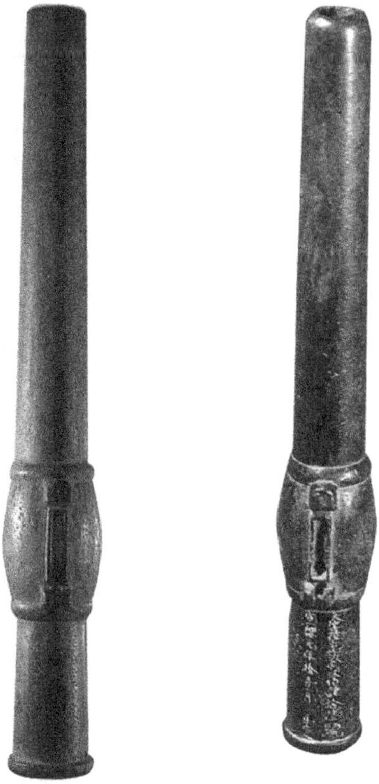

Fig. 29.23 "Tianzi" copper blunderbuss of the Ming Dynasty [21]

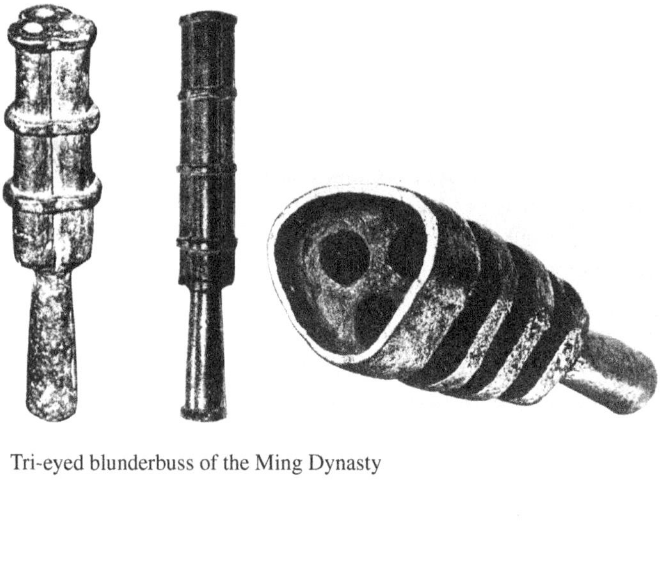

Fig. 29.24 Tri-eyed blunderbuss of the Ming Dynasty

Fig. 29.25 Iron gun of the Ming Dynasty [22]

29.4 The Western Transmission of Tubular Firearms and Fire Arrows

The Arabs played the role of bridge in the westward transmission of the gunpowder and firearm technology in ancient China.

During the Mamluk Dynasty of the Arab Empire (Cairo as the capital, 1250–1517), the technologies of manufacturing and using firearms in China began to spread to Arab countries. In 1260, the Mamluk army defeated the Mongolian army in Syria and captured hundreds of Mongolian soldiers. Thereafter, they acquired firearms and firearm craftsmen as more Mongolian troops surrendered to them. Two kinds of Chinese firearms were introduced into Mamluk at that time: one was called "Khitan Musket" and the other was "Khitan Fire Arrow" [23].

From the second half of the thirteenth century to the beginning of the fourteenth century, the Arabs imitated the firearms imported from China into a wooden tubular firearm, Mardafa (firearms in Arabic). Bombs, firearms, fire pot, and Mardafa are described in the manuscript of the Arabian military book *All the Technologies* compiled from 1300 to 1350. The book also provides graphics of Mardafa: wooden barrel, a chamber filled with gunpowder, spherical projectiles, or arrows loaded at the mouth of barrel, igniting from the door of fire to shoot, wooden handles of different length for easy holding [24].

In 1325, the Arabs brought Mardafa to Spain when they attacked this country. Then the Spaniard spread Mardafa to Western Europe. The Europeans, referring to Mardafa, made the earliest metal tubular firearm in Europe, called handgonne. The unearthed objects of handgonne and some murals handed down from generation to generation show that they are basically similar to the copper blunderbuss of the Yuan Dynasty and the early Ming Dynasty in terms of their basic structure and firing method [25].

In the fourteenth century, gunpowder and firearms made their debut in European wars. At the end of the fifteenth century, cannons that fired iron bullets with gunpowder effectively replaced trebuchet, enabling fortifications in medieval Europe to be collapsed easily. By the end of the seventeenth century, handgonne had completely taken the place of the cold weapons like spears, crossbows, and bows, which were once very lethal, putting an end to the era when cold weapons and firearms coexisted in Europe.

References

1. [Northern Song Dynasty] Zhen, Lu. 1985. *Jiu Guo Zhi*, vol. II, 29. Zhonghua Bookstore.
2. [Ming Dynasty] Yu, Jiao. *Huo Long Jing*, the Preface Version in the 10th Year of Yongle Period of the Ming Dynasty, vol. 2, 1412.
3. [Yuan Dynasty] Tuotuo et al. 1977. *Bing Zhi 11 of the History of Song Dynasty*. Zhonghua Bookstore.
4. [Northern Song Dynasty] Tao, Li. 1980. *Sequel of Zi Zhi Tong Jian*, vol. 52, 1153. Zhonghua Bookstore.
5. [Northern Song Dynasty] Gongliang, Zeng. 1959. *The Preface Collection of Wu Jing Ji Yao*. Photocopied ed. of Zhengde Period of the Ming Dynasty, vol. 12. Zhonghua Publishing House.
6. [Ming Dynasty] Yuanyi, Mao. *Firearms Picture Five. Wu Bei Zhi*, version of Lianxi Caotang of the Ming Dynasty, vol. 126.
7. The Museum of Chinese History and the State Administration of Cultural Relics. 1997. *Exhibition of Chinese Ancient Scientific and Technological Relics*. Chaohua Publishing House.
8. [Southern Song Dynasty] Wanli, Yang. 1936. *Hai Fu of Cheng Zhai Ji*, the first edition of four series, Book 64, vol. 44, 417–418. Shanghai Commercial Press.
9. Jixing, Pan. 2016. *History of Chinese Gunpowder (Collection of Illustrations)*, vol. 1. Shanghai Far East Publishing House.
10. [Ming Dynasty] Yuanyi, Mao. *Firearms Picture Six*, *Wu Bei Zhi*, version of Lianxi Caotang of the Ming Dynasty, vol. 127.
11. [Ming Dynasty] Yuanyi, Mao. *Firearms Picture Ten*, *Wu Bei Zhi*, version of Lianxi Caotang of the Ming Dynasty, vol. 131.
12. Zhaochun, Wang. 1991. *History of Firearms in China*. Military Science Press.
13. [Yuan Dynasty] Tuotuo et al. 1975. *Chizhanhexi Biography of the History of Jin Dynasty*, 2496. Zhonghua Bookstore.
14. [Yuan Dynasty] Tuotuo et al. 1975. *Maji Bilograph of the History of Song Dynasty*, 13270. Zhonghua Bookstore.

15. [Yuan Dynasty] Tuotuo et al. 1975. *Chen Gui Biography of the History of Song Dynasty*, 11643. Zhonghua Bookstore.
16. [Yuan Dynasty] Tuotuo et al. *Puchaguannu Biography, the History of Jin Dynasty*. Zhonghua Bookstore, 1975: p.2548.
17. Dong, Cheng, Zhong Shaoyi. 1990. *Atlas of Chinese Ancient Weapons*, Coloured ed. 27. PLA Press.
18. Dehe, Guo. 2006. *Ancient Weapons Bibliography, Military Encyclopedia of China*, 2nd ed., 137. China Encyclopedia Publishing House.
19. Shaoyi, Zhong, Qimude Dolgi, et al. 2004. The new discovery of copper blunderbuss of the Yuan Dynasty in inner mongolia and its significance. *Cult. Relics* 11: 65.
20. [Ming Dynasty] Dongyang, Li, et al. 1989. *Daming Huidian*, vol. 193, 2619. Jiangsu Guangling Ancient Books Engraving Agency.
21. Dong, Cheng, Zhong Shaoyi. 1990. *Atlas of Chinese Ancient Weapons*, Coloured ed. 29. PLA Press.
22. Dong, Cheng, Zhong Shaoyi. 1990. *Atlas of Chinese Ancient Weapons*, Coloured ed. 30. PLA Press.
23. Jiasheng, Feng. 1978. *The Invention and Westward Spread of Gunpowder*, 52. Shanghai People's Publishing House.
24. Zhaochun, Wang. 2007. *History of Firearms in China*, 56. Military Science Press.
25. Hong, Yang, Cheng Dong and Zhong Shaoyi. 1991. *Ancient Weapons Bibliography, Military Encyclopedia of China*, 16. Military Science Press.

Zhanhong You Born in Xinhua, Hunan in 1965, Doctorate of science at the Chinese Academy of Sciences, associate researcher at the Museum of Cartography at Tsinghua University, visiting researcher at the Chinese Character Museum.

Research interests:

History of Military science and technology

Major works:

Participating in the writing of the second part of *History of Chinese Mechanical Engineering Invention; Chinese Military Encyclopedia: History of World War.*

Papers: President Kennedy's Scientific consultant Wiesner and American Arms Control Policy; Technical Reflection on the Failure of Beiyang Navy; On the Historical Experience of Pugwash and Its Significance

Awards:

21 Lectures on Science and Technology History won the first prize of Tsinghua University's excellent textbooks in 2008;

In 2013, the research report *Thinking on the Development Trends of International Strategic Patterns and Countermeasures* that he co-authored won the first prize for outstanding achievements in military science of the Academy of Military Sciences.

Chapter 30
Variolation of Traditional Chinese Medicine Diagnosis and Treatment

Yahua Niu

Variolation, also known as vaccination, refers to the technique used in ancient China to prevent smallpox. The method involved was to produce vaccine from the smallpox scabs or liquid collected from the people who suffered from smallpox, and then inoculated healthy children so that they were mildly infected and produced antibodies. In this way, the invasion as well as the infection of variola virus was prevented.

Smallpox is a serious infectious disease caused by the variola virus. With a high death rate, the disease has taken millions of lives worldwide over thousands of years. According to relevant archaeological documents, the mummy of Egyptian pharaoh Ramses V who died in 1157 B.C. was left with smallpox scabs on the head. The plague that had been prevalent in Rome for 15 years from 251 A.D. to 266 A.D. was very likely smallpox. The serious round of smallpox spread in Rome in 312 A.D. caused heavy casualties, whose "destructive effect on the decline of the Roman Empire was much greater than that produced by wars and extravagant lifestyle" [1]. Smallpox had hit Europe many times, and only in the year 1719, 14,000 Parisians died. At the beginning of the sixteenth century, variola virus-carrying European colonists brought smallpox to the American continent. As a result, more than half of American Indians died from smallpox infection, some tribes were even on the verge of extinction. Smallpox became one of the main reasons causing death to Indians [2]. Throughout history worldwide, all countries had been ravaged by smallpox; countries like China, Japan, India, Australia, and so on set records of smallpox. In 1770, more than three million Indians died of smallpox [1]. It is estimated that smallpox had caused hundreds of millions of deaths; even if the infected person survived, he or she would be left with lifelong scabs on face after recovery. The severe cases would be disfigured or even go blind. Even up to now, no specific treatment can be adopted against smallpox. Fortunately, the immunization invented

Y. Niu (✉)
Institute of Information on Traditional Chinese Medicine, China Academy of Chinese Medical Sciences, Beijing, China
e-mail: niuyahua@hotmail.com

J. Hua and L. Feng (eds.), *Thirty Great Inventions of China*,
https://doi.org/10.1007/978-981-15-6525-0_30

can effectively prevent and eradicate this human killer. In May 1980, the World Health Organization announced that smallpox had been successfully annihilated. Smallpox thus became the first contagious disease that was completely eradicated by human beings, to which variolation certainly contributed a lot. The variolation was originated in China and was a significant invention made in the medicinal history of the world.

30.1 The Cognition of Smallpox in Ancient China

In Chinese history, smallpox was also known as "bean sore," "blister," "skin sore," "variola blotch," etc. Around the fifth century, smallpox was spread all over China through variola virus-carrying war prisoners. According to *Zhouhou Beiji Fang* (*Handbook of Prescriptions for Emergencies*) created by Ge Hong of the Jin Dynasty, "In recent years, an epidemic disease has prevailed, which causes spot sore on the skin… it is said that in the middle reign of Emperor Guangwu of the Han Dynasty the disease was infected from prisoners in Nanyang (now in Henan Province), thus called Luchuang (prisoner sore)." The book also gives a detailed description of the symptoms of smallpox: "The head and body of the infected are instantly covered by blister-like acne with white liquid in it. As soon as the old acne breaks, the new will emerge. Without prompt treatment, most severely infected people will die. Even if rescued from the jaws of death, the infected will be left dark-purple scabs on the face and body for a lifetime due to such vicious disease." [3]. *Zhu Bing Yuan Hou Lun* (*General Treatise on the Cause and Symptoms of Diseases*) written in the Sui Dynasty further explains the symptoms of smallpox: "If the root of the spots is red and the top white, the case is not severe; but if the spots are dark-purple in color and look like millets, the case is rather severe, thus also called miliary sore." [4]. In the Tang Dynasty, Wang Tao described in his *Wai Tai Mi Yao* (*Medical Secrets from the Royal Library*) the whole process of a smallpox attack: outbreak of rashes, formation of white liquid, festering, and scabbing. He advised that prognosis could be estimated according to the color, distribution, and the type of smallpox. Since the Song and Yuan Dynasties, smallpox swept across the whole country of China and also understood deeply. Qian Yi, an excellent pediatrician in the Song Dynasty, recorded the early symptoms of smallpox in his *Xiao Er Yao Zheng Zhi Jue* (*Key to Therapeutics of Children's Diseases*) and compared it with chickenpox and measles. Since the Ming Dynasty, smallpox turned to be more rampant and lethal, from which almost everyone at that time suffered. Wan Quan, a great medical scientist in the Ming Dynasty, recorded in his *Dou Zhen Shi Yi Xin Fa* (*Therapies for Acnes and Measles*): "In the spring of 1534, the outbreak of smallpox took the lives of nine-tenth of the infected." [5]. The severe situation at that time could be reflected. Wan Quan also recognized that "smallpox can only attack a person once in a lifetime. Once the infected recovers, he or she won't suffer from it for a second time." [6]. This recognition was crucial to the invention of variolation. With the further understanding of the symptoms, prognosis, and pathogenesis of smallpox, many methods for treating and

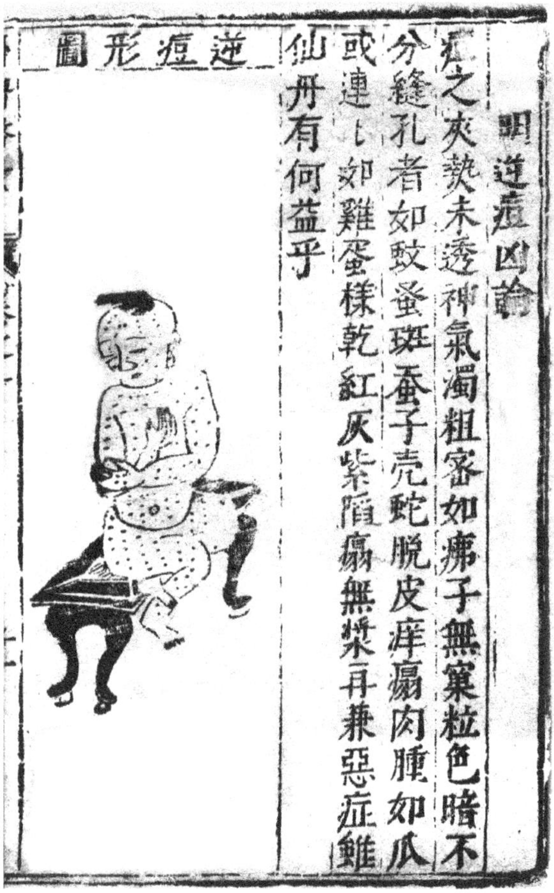

Fig. 30.1 *Wai Ke Qi Xuan* by Shen Douyuan of the Ming Dynasty

preventing smallpox had been summarized, and hundreds of medical books about it were published, among which the records of variolation were included (Figs. 30.1, 30.2 and 30.3).

30.2 The Invention of Variolation

As for the time when variolation was invented, there were different opinions. Dong Yushan of the Qing Dynasty records in his book *Niu Dou Xin Shu* (New Book on Vaccinia): "There had been no variolation until the Kaiyuan Period of the Tang Dynasty when a Zhao's family in the lower reaches of the Yangtze River began to carry out the variolation with lymph vaccine…" This means that the technique of variolation had been invented in the Kaiyuan Period of the Tang Dynasty. Zhu Chungu in his *Dou Zhen Ding Lun* (*Final Conclusions on Smallpox*) (1713, Fig. 30.4)

Fig. 30.2 "Picture of graine-like smallpox" in *Yi Zong Jin Jian* (*The Golden Mirror of Medicine*)

pointed out that "there is a highly skilled doctor living in the Emei Mountain who is good at conducting variolation and never misses," and the doctor was said to successfully vaccinate the son of Wang Dan, the Prime Minister of the Northern Song Dynasty. This proved that variolation had been a clinical treatment in the Northern Song Dynasty. The two opinions were recalled by the decedents, without adequate evidences to prove their authenticity. According to *Dou Ke Jin Jing Fu Ji Jie* (*The Solutions to Smallpox*) (1727) by Yu Maokun of the Qing Dynasty, "It is said that variolation originated from Taiping County of Ningguo Prefecture in the Longqing Period of the Ming Dynasty… It was created by a talented medicine family and then spread across the country. So far, most of the ones who are adept in carrying out variolation are from Ningguo." This statement turned out to be true. Lot of evidences support that variolation was invented around the Longqing Period of the Ming Dynasty or even earlier, at the beginning of the sixteenth century. It had been first spread in the folk for a while before it was documented.

The earliest record of the term "zhongdou (variolation)" discovered at present is given in *Quanya Wen Ji* (*Anthology of Quanya*) written by Shao Jingji of the Jiajing Period of the Ming Dynasty, in which an article entitled *Xiao Xiang Ji Wen*

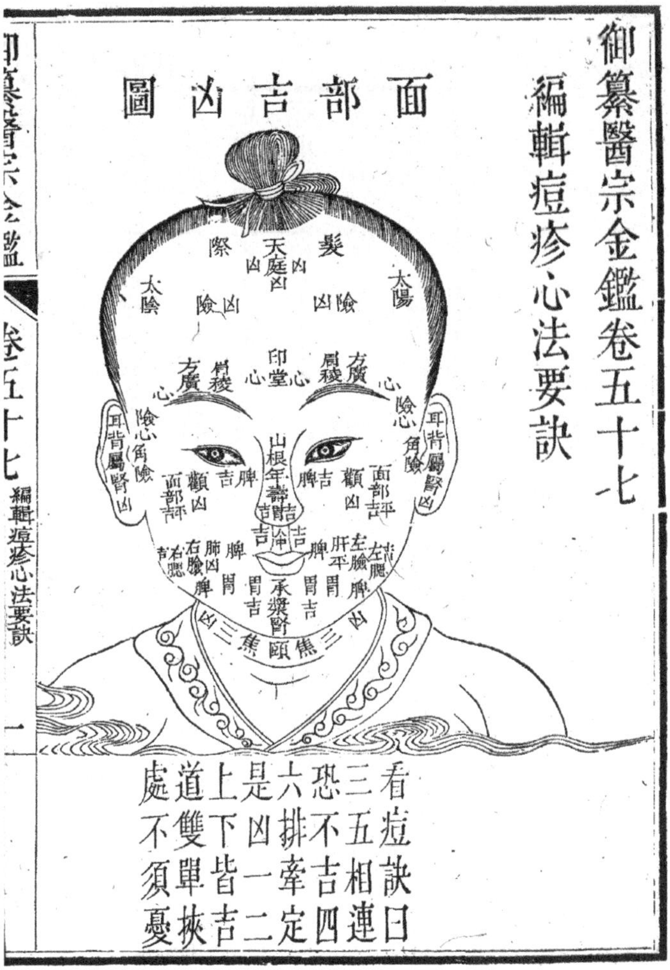

Fig. 30.3 The "picture of good or ill appearance" used to estimate prognosis in *Yi Zong Jin Jian* (*The Golden Mirror of Medicine*)

mentions that "your grandson has recently zhongdoued and recovered" [7]. These words, however, were not enough to determine whether the grandson was vaccinated or infected with smallpox. *Dou Zhen Shi Yi Xin Fa* by doctor Wan Quan of the Ming Dynasty also referred to "zhongdou," which was textually researched to be variolation [8]. But, the relevant original text was referred to be "a woman underwent zhongdou, so her menstruation came unexpectedly," which was not likely to refer to variolation, but to the infection of variola virus. According to *San Gang Shi Nue: Zhong Dou* by Dong Han in the early Qing Dynasty, "The Zhang's family in Anqing is said to carry out variolation for three generations. The method is: first collect smallpox liquid and store it in small porcelain bottle; then spread the liquid on the clothes of the child to

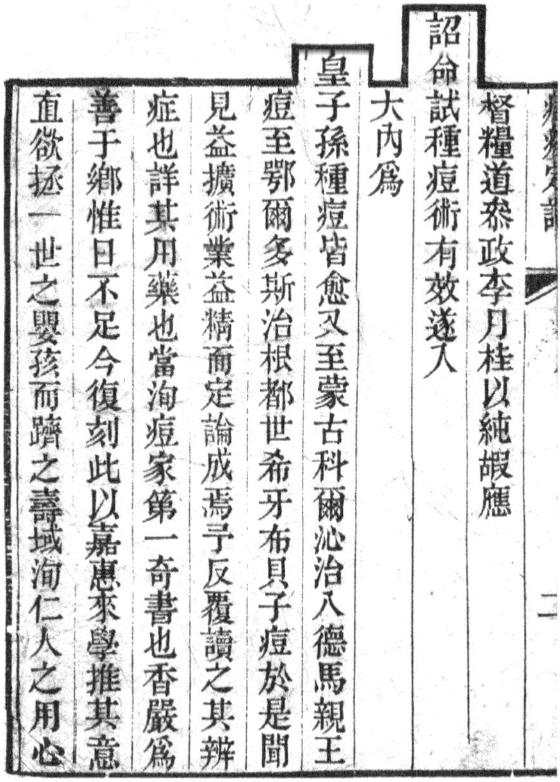
督糧道叅政李月桂以純嘏應
詔命試種痘術有效遂入
大內爲
皇子孫種痘皆愈又至蒙古科爾沁治八德馬親王
痘至鄂爾多斯治根都世希牙布貝子痘於是聞
見益擴術業益精而定論成焉予反覆讀之其辨
症也詳其用藥也當洵痘家第一奇書也香嚴爲
善于鄉惟日不足今復刻此以嘉惠來學推其意
直欲拯一世之嬰孩而躋之壽域洵仁人之用心

Fig. 30.4 Preface of the re-engraved version *Dou Zhen Ding Lun* by Zhu Chungu

be vaccinated." [9]. This was probably the earliest record about variolation, which had been handed down from the older generations of the family. It was said that Jiang Xishun (c. 1585–1668), a pediatrician in Jingde County of Anhui, wrote *Dou Zhen Yuan Zhu* and created variolation [10]. He would be the earliest one who carried out variolation in Jingde County. Huang Baijia (1643–1709), son of Huang Zongxi, once prefaced *Tian Hua Ren Shu* (*Treatments of Smallpox*) by Fu Shanglin in 1683. He referred that Fu's family was good at carrying out variolation, which had been handed down for three generations. They all belonged to the early practitioners of variolation.

In the early reign of Emperor Kangxi of the Qing Dynasty, variolation had been quite prevalent. According to *Wang Shan Tang Wen Ji* (*Anthology of Wangshantang*) by Zhang Fuyi, variolation had been popular in the area of Qianyang in Hunan in 1664. In the summer of that year, variolation technician Song Tai came here to carry out and spread the technique of variolation. In this single process, he effectively vaccinated more than 50 children. The method he adopted was the original variolation with clothes: let the healthy children sleep on the beddings of the children who suffered from smallpox; these healthy children will have a fever in 7–15 days and produce antibodies in 1 or 2 days. As a result, more than 40 children were vaccinated and then recovered, and only two or three in ten failed. According to this method, "to breed new smallpox virus, only one child or two are ordered to sleep on the virus-carrying bedding," [11] in case the virus fell short. Once the virus fell short, a special variolation technician was resorted to select and breed it. Variolation technician had become a profession in the local.

Although smallpox raged before the Manchu troops began to invade the Central Plains of China, the attack of smallpox was less severe than that in the Central Plains, since most people never got infected with variola virus. After establishing their regime in Beijing, the Manchu people were deeply threatened by smallpox, especially adults. Tan Qian, a historian in the early Qing Dynasty, recorded the then situation in his *Bei You Lu* (*Records of Traveling in the North*): "The Manchu people had not caught smallpox at all. After they had settled down in Chang'an (now Xi'an), many of them died of smallpox, thus addressing that they were infected by the Han people." [12]. Many royals died of smallpox, such as the wives of the Prince of Yu and the Prince of Ying. Even Emperor Shunzhi died of smallpox in the prime of his life [13]. So, in selecting the heir to the throne, German missionary Johann Adam Schall von Bell who served the Qing government then proposed that Xuanye, the third son of Emperor Shunzhi, be the proper heir as he had recovered from smallpox, which meant that he would have lifelong immunity against it [14]. Interestingly, such an important event as the alternation of imperial power was influenced by smallpox. Actually, Xuanye (later Emperor Kangxi) was the longest serving monarch in the Qing Dynasty who had achieved many political and military achievements. To some extent, smallpox had influenced the politics of the Qing Dynasty.

Variolation was quickly introduced into the Forbidden City. In 1678, Fu Weige, one of Fu Shanglin's relatives, successfully vaccinated the crown princes under Emperor Kangxi's request, and in 1680, he was called to do the same thing. In 1681, variolation technicians Zhu Chungu and Chen Tianxiang from Jiangxi entered the palace to vaccinate the princes and the Manchu nobles. Emperor Kangxi ordered all Manchu people and the Mongolian tribes to be vaccinated. According to *Ting Xun Ge Yan* (*Records of Emperor Kangxi's Words*), "At the beginning of the Qing Dynasty, our people were afraid of smallpox. Since I got the technique of variolation, the offspring of mine and my ministers were vaccinated, thus all free from smallpox. Nowadays, the people in the Inner Mongolia and the Mongolia Khalkha have been vaccinated under my commands. The ones who got vaccinated all recovered well. I still remember the time when we got vaccinated in the beginning, the elder doubted it. But I insisted on conducting it, so tens of millions of people survived from smallpox. Could it just be a coincidence?" [15]. It can be seen that Emperor Kangxi had promoted the technique of variolation in the Mongolian regions and received good results.

In the middle and late reign of Emperor Kangxi, variolation was spread rapidly in the folk. Zhang Lu, a famous doctor in the Qing Dynasty, described the promotion of variolation in his *The Zhang's Medical Book* (1695): "Recently variolation is prevailing, originating from the west of the Yangtze River, developing to Hebei and Shandong, and finally covering the whole country." He also recorded measures of variolation, such as the use of smallpox liquid, virus-carrying clothes, and so on:

"The vaccine used during variolation is nothing but the smallpox liquid extracted from smallpox-catching children. Dip a cotton ball into the smallpox liquid and then place it into the nostril of children, girls into the right one and boys into the left one. After seven days or so, children vaccinated will show the symptom of fever and spots occur, the number of spots varying from a dozen or so to hundreds. Some children's faces may slightly swell. Then the toxicity fades away. It is a truth that the variolation would be successfully completed if the variola virus is effective. If there is no smallpox liquid, the smallpox scabs may work as well. If smallpox scabs are also inaccessible, the underclothes worn by the children who had just suffered from smallpox can also be used to dress the children to be variolated." [16]. *Xiu Hu Xuan Chi Du* by Cao Yu records the story of Wu Dingchen, a famous variolation technician: "Dingchen can vaccinate with magical medicine, regarded as a magical variolation technician in the area of Loujiang. He has saved the lives of more than 1,200 infants in Loujiang, which is indeed magical." Two sons of Cao Yu were also vaccinated by Dingchen and both recovered well. Therefore, Cao invited Dingchen to vaccinate in the area of Taicang, advocating that "variolation can benefit much to children" [17]. It can be seen that experienced variolation technicians were very popular then. During the reign of Emperor Yongzheng, due to the promotion of variolation in the area of Jinhua of Zhejiang, the number of people who died of smallpox was greatly reduced. "The harm of smallpox had been rather heavy. Since the promotion of variolation, less people died of smallpox. Isn't the benefit of variolation remarkable?" [18]. At that time, people had figured out the influence of the quality of vaccines on the effect of variolation, so the safe and efficient vaccine was called "Dan (red) vaccine." According to *Dou Ke Jin Jing Fu Ji Jie*, Dan vaccine was very expensive, three *liang* (=50 g) of gold for one piece [19].

In the reign of Emperor Qianlong, the technique of variolation became more mature. *The New Book on Variolation* (in 1741, Fig. 30.5) written by Zhang Yan, son of Zhang Lu, is the earliest monograph on variolation. The book strongly advocates variolation: "I have engaged in variolation for decades. According to my observation, no one died of smallpox if vaccinated with high quality of vaccine, and eight tenth will survive when vaccinated with inferior vaccine. Variolation technicians always select the vaccine of high quality so that variolation will go well almost every time… I have visited many places and vaccinated more than 8,000 people, among whom less than 30 had been unsuccessfully saved. Under the rampant epidemic of smallpox, isn't it a kind of luck? Therefore, I advise everyone to have your children vaccinated. It is best to invite doctors to vaccinate children before smallpox spread to where you live. If smallpox has attacked, variolation technician will not dare to vaccinate hastily for fear that the child has already been infected. Even if the technician will, the chance of failure is high." [20]. Zhang Yan had obviously accumulated a lot of practical experience in variolation.

The Golden Mirror of Medicine: Concise Comprehension of Variolation clearly recognized the value of variolation. According to the book, "Smallpox is felt after one catches it, while variolation should be performed before one gets it; smallpox is cured at the time of its formation, while variolation should be inoculated when one is free of the disease. The vaccine is transmitted from the exterior to the interior,

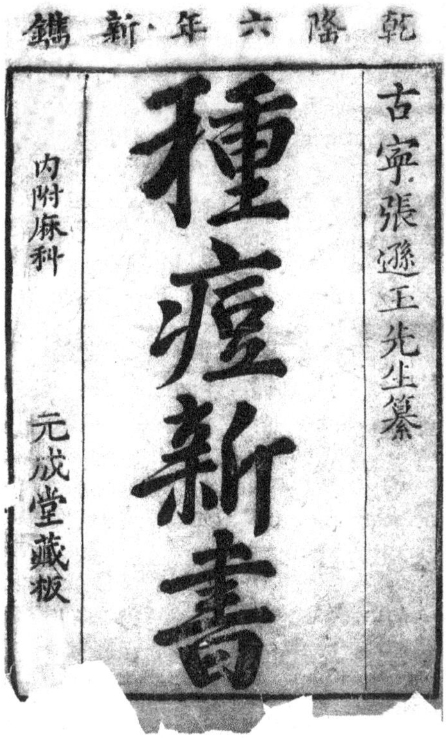

Fig. 30.5 The copy of *Zhong Dou Xin Shu* by Zhan Yan

and then from the interior to the exterior. If there is no other disease mingled with it and there are proper methods to guide, the exterior will gradually be influenced by fumigation and the fetus toxin will be expelled. In this case, we will not have to worry about the multifarious risks of variolation, whose unpredictable changes could make excellent physicians unknown of what to do. Variolation is sure to be a good way to reduce risk, avoid danger and achieve safety." [21].

The practice of variolation was opposed by some doctors due to its risk of smallpox infection and its certain proportion of failed cases caused by those unskilled physicians. Huang Tingjing (1704–?), an ophthalmologist of Zhang Yan's time, criticized variolation in his book *Mu Jing Da Cheng* (1741) [22]. However, still many physicians recognized the benefits of variolation. Xu Dachun pointed out in his *Lan Tan Gui Fan* (1764) that everyone would catch smallpox, and variolation could prevent children from infecting it. It was estimated that the death rate caused by smallpox was 80%–90% and that the failure rate of variolation was about 1%. The failure in variolation was completely due to the bad vaccine rather than the method itself [23]. According to *Zhong Dou Fang* (*The Method of Variolation*) by Zheng Wangyi, in the past, if the success rate of variolation of an attack of smallpox could reach 80–90%, smallpox would be called 'peaceful smallpox'; even if the death rate reached 50%, no physicians would be blamed. Nowadays, however, if four or five among the

100 children who were vaccinated died, the physician who practiced the variolation would be punished and prohibited from the practice of variolation. It was mentioned in many documents that there was even no single failure in a hundred practices of variolation. In the preface of *Zhong Dou Zhen Chuan* (Fig. 30.6) written in 1805, the author Fang Mingyi wrote that in his 30 years of learning variolation from Mr. Shi from Chongming and performing variolation in Fengtian (now Shenyang), "all practices of variolation I had carried out took effects." Taking his own vaccinated and non-vaccinated children as examples, he further demonstrated the necessity of variolation. Of all his children, seven who were vaccinated survived, and in the three who were not vaccinated, two suffered from smallpox, of which one died [24].

It can be seen that the practice of variolation had been greatly improved in safety with the advancement of technology (Fig. 30.7). Since cowpox was introduced into China, the variolation and the vaccination of cowpox had been adopted simultaneously for quite a long time. In 1985, some scholars surveyed a population of 653 who were over 65-year-old living in Shanghai, Zhejiang, and Jiangsu. Among the population, 511 had received cowpox vaccination, 115 human vaccination, and 27 no vaccination. Several years later, 16 cases (about 3.13%) received cowpox vaccination suffered from smallpox again and left scabs. Three cases (about 2.61%) failed in variolation, and two cases suffered from severe reactions and left scabs. The success rates of vaccination and variolation were 96.9% and 97.4%, respectively, and the difference between them remained non-significant. And in the people who were not vaccinated, there were 24 cases (89%) who suffered from smallpox and were left scabs [25].

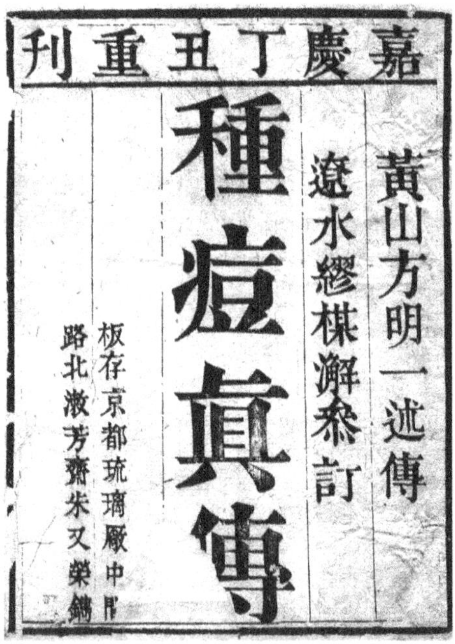

Fig. 30.6 The copy of *Zhong Dou Zhen Chuan* by Fang Mingyi

Fig. 30.7 The poster of Lay Buddhist Yunfeng encouraging variolation in the 8th year of Emperor Daoguang of the Qing Dynasty (Selected from *The Intellectual History of Chinese Preventive Medicine* by Fan Xingzhun)

30.3 Methods of Variolation

In Golden Mirror of Medicine: Concise Comprehension of Vaccination, variolation is divided into four types: (a) variolation by clothes: dress a healthy child for 2 to 3 days in the underclothes worn by the child who have caught smallpox, without taking off at night until the child has a fever with rash 9–11 days later, which indicates a success of the variolation; (b) wet variolation: dip a cotton ball into the smallpox liquid collected from the children who have caught smallpox and then put it into one nostril of a vaccinator who will accordingly develop smallpox; (c) dry variolation: grind smallpox scabs into powder, place it at one end of a curved silver tube and then blow it into one nostril of a vaccinator who will generally have a fever on the seventh day, which indicates a success of the variolation; and (d) water variolation: according to the age of vaccinators, grind 20–30 (or exactly 2–3) smallpox scabs into powder, mix it well with a few drops of clean water, then wrap it with a ultra-thin slice made by spreading a piece of new cotton, knead it into the shape of a date stone, and finally, put it into one nostril of a vaccinator pulled by a red line, which will be removed 12 h later; the vaccinator will thus get a fever on the seventh day, and 3 days later have smallpox, which is followed by complete rashes, the formation of vesicles, and finally scabs, which indicates a success of the variolation.

Among the four methods, variolation by clothes was gradually abandoned, as it was relatively primitive and much lower in the success rate. Wet variolation was much dangerous to adopt, so seldom used. Dry variolation is frequently adopted as it is simply and feasibly used, but if one fails to control the strength of blowing the smallpox vaccine, the nasal mucosa of the vaccinator will be irritated to produce much nasal discharge, which will wash away the smallpox vaccine and lead to a failure in the variolation. Water variolation is widely adopted due to its higher success rate and mild symptoms involved.

The variolation also concerns a series of technical approaches and processes, such as vaccine selection, vaccine storage, inoculation time, vaccinator selection, and recuperation. *Zhong Dou Xin Shu* introduces not only the specific procedures of variolation but also the methods of distinguishing the quality of vaccinia, vaccinia selection, vaccinia storage, distinguishing the quality of old and new vaccinia, vaccination methods, and avoidance after vaccination. *The Golden Mirror of Medicine: Concise Comprehension of Vaccination* also introduces a series of steps and points for attention, such as vaccinia selection, vaccinia storage, timing, auspicious days choosing, recuperation, warnings, availability and unavailability of vaccinia, inoculation method, and supplemental variolation.

Especially with the accumulating of practical experience, people get a deeper understanding of vaccinia selection. According to *Zhong Dou Xin Shu*, "If the vaccine is quite good, then no variolation would fail." [20]. And the method to select a vaccine is: "When extracting smallpox vaccine, one should investigate the source of the pox. If the smallpox is caught in villages, the pox will be mostly good, without abnormal development of smallpox. Such vaccine deserves to be extracted. Besides, the particles of the pox should be clearly demarcated and rare on the head, neck, face and chest, with the lymph adequate and the scabs thick and right, and the smallpox patient should be cured without any medicine. If so, the vaccine extracted will be full of Qi and blood, with its poison run out, really good vaccine." [20].

The vaccine made from the pus serum of natural smallpox is called "timely vaccine (vaccine at the beginning of smallpox)." The vaccine made from smallpox scabs or refined from the "timely vaccine" by the subculture is known as "wrought vaccine," which can be formed from preserved pus serum. *Zhong Dou Xin Shu* records a method of vaccine storage: "After vaccinia is extracted, wrap it with paper and put it into a small bamboo tube, whose ends will then be blocked to prevent leakage. Do not expose the tube to the sun, or roast it over a fire, but carry it with oneself, left to dry naturally. In addition, the date of vaccinia extraction should be written on vaccine bag." [20]. Joseph Needham, a well-known British historian of science, believed that "storing the vaccine at the body temperature (37 °C) or slightly below for over a month will certainly inactivate 80% of the live virus particles. However, due to the presence of these death proteins, when inoculated into human body, the vaccine will strongly stimulate the body to produce interferon, just like stimulating the production of antibodies." [26]. *The Golden Mirror of Medicine: Concise Comprehension of Vaccination* also recorded a method of vaccine storage: store the selected vaccinia in a clean porcelain bottle, and seal it in a clean and cool place for 1 month in spring or even 40–50 days in winter; the scabs preserved so long can still be available for variolation.

By continuous practice, physicians found that the variolation with "timely vaccine" was of great danger, while the variolation with cultivated and refined "wrought vaccine" was much safer. According to *Zhong Dou Xin Fa* (*Concise Comprehension of Variolation*) by Zhu Yiliang of the Qing Dynasty, "The more the vaccine is inoculated, the stronger the medicinal effect is; the more mature the selection and refining is, the more internal heat and poison will be expelled, thus with only vital essence left. Therefore, the method of variolation should be secure and harmless"; "if timely vaccine can be inoculated seven times in succession and be refined, it can develop into wrought vaccine." This method of selecting and breeding vaccine from human smallpox completely conformed to the scientific principle of modern vaccine preparation, and was completely consistent with the current principle of selecting attenuate "BCG vaccine" to expel the toxicity of bacterial strain and acquire antigenicity only [27].

Inoculation time and vaccine recipients should also be chosen carefully, usually in spring and winter. The child to be vaccinated should look high-colored, full of spirit and energy, free from any diseases; otherwise, he or she cannot be vaccinated. After vaccination, it is necessary to keep the room clean and the temperature indoors proper. The vaccinated child is required to moderately eat and to avoid strenuous activities and excitement. All these measures could reduce the possible risks caused by variolation. In historical records, variolation turned out to be high in success rate, and first rate in effect and safety.

30.4 The Spread of Variolation Abroad

During the late seventeenth and early eighteenth centuries, the variolation developed in China began to spread to such foreign countries as Russia, Turkey, European countries, the United States, Japan, Korea, etc.

Following the signing of *Treaty of Nerchinsk* in 1688, Russia began to send students to study in China. Because of the prevalence of smallpox in Russia at the time, there were Russian students who were specially dispatched to learn the treatment of smallpox. According to *Gui Si Cun Gao* by Yu Zhengxie, a famous scholar of the Qing Dynasty, "During the reign of Emperor Kangxi, Russia sent students to study variolation in Beijing of China…" From then on, variolation had been widely spread in Russia, via which variolation might then be introduced into Turkey.

It is in the early eighteenth century that variolation was introduced into Europe. In 1700, a business personnel from the British West India Company reported to the Royal Society the "method of being vaccinated against smallpox" he saw in China and gave a detailed description of the process of variolation: "Scratch one pustule on the patient who suffered from smallpox, dip some pus with a cotton ball and have it dried… then place it into one nostril of the person who may catch smallpox." After that, the vaccinator would become mildly ill with smallpox, and soon recover, thereby achieving a good effect of preventing smallpox. In a report given at the Royal Society, a doctor named Havers introduced the practice of variolation by Chinese.

Unfortunately, it did not attract the attention of the British medical community. 14 years later, in 1711, Dr. Tamune from the British Embassy in Turkey reported to the Royal Society the method of smallpox vaccine extraction and vaccination practiced in Constantinople (now Istanbul in Turkey), arguing that the variolation had been surely a secure and effective method. In 1718, Lady Mary Montagu, the British ambassador to Turkey, had her 6-year-old son vaccinated against smallpox. The following year, she returned to Britain, and personally participated in and actively promoted the practice of variolation, though which seemed to take little effect. In 1721, smallpox was prevalent in British. The Royal Society thus carried out a series of variolation tests, and finally confirmed the effect of variolation. Since then, the variolation began to spread in British to a certain extent [28].

Variolation was later introduced from England into France. However, it was not practiced immediately due to the opposition from French doctors, which aroused the dissatisfaction of French philosopher Voltaire. He said, "If the variolation is practiced in France, millions of lives will be saved," "and the 20,000 people who died of smallpox in Paris in 1723 might have been alive." He spoke highly of variolation: "I heard that Chinese people have developed this habit (variolation) for more than 100 years, which is a great precedent and model from a nation considered to be the smartest and politest in the world." [29].

In 1721, smallpox was prevalent in Boston of the United States. At that time, variolation had been practiced. For example, Dr. Zabdiel Boylston (1680–1766) vaccinated his two sons and a slave. The first American president Washington ordered all his family members and armies to be vaccinated against smallpox. Benjamin Franklin, a famous American scientist and statesman, exhorted all parents to vaccinate their children since his own son had died of smallpox. It can be seen that the practice of variolation has been promoted in the United States in the early eighteenth century. In the second half of the eighteenth century, variolation had become a medical treatment generally accepted in England, France, and the Netherlands [30].

In 1744, Li Renshan, a merchant from Fuzhou (now in Fujing Province), introduced Chinese variolation to Japan after he came to Nagasaki. Later, he vaccinated people there at the order of the chief General of Nagasaki and became the teacher of the physicians Liu Longyuan and Horie Dogen. Li Renshan also wrote *Zhong Dou Shuo* (*Theory of Variolation*), which was well received by Japanese physicians and entitled *Li Renshan Zhong Dou He Jie* (*Li Renshan's Therapy of Variolation*) after being revised in Japanese. In 1778, the volume of variolation in *The Golden Mirror of Medicine* was selected out and published with the title of *Zhong Dou Xin Fa*. The method of variolation thus began to be widely spread. In 1840, cowpox vaccination was introduced into Japan where the practice of variolation had been treated as the main way to prevent smallpox for about 100 years.

In 1763, Li Mu'an, a Korean, recorded the practice of Chinese variolation in a letter. In 1790, Korea sent envoys Piao Zhaijia and Piao Lingyang to the capital of China. When they returned, they brought with them the Chinese multi-volume medical series *The Golden Mirror of Medicine*, in which the chapter of *Concise*

Comprehension of Variolation in Pediatrics introduces variolation and its precautions. Later, a rural official was appointed by Piao Zhaijia to practice variolation according to the book, and succeeded.

Chinese variolation has effectively prevented the spread of smallpox, saved countless lives, and greatly impacted the invention of vaccinia. Edward Jenner (1749–1823), the inventor of vaccinia, used to be a village doctor. During the process he carried out variolation, he found that the milkmaids who had gotten cowpox always failed in variolation. In 1796, he tried to conduct a smallpox variolation test by replacing the human smallpox vaccine with cowpox vaccine, and finally succeeded. Some scholars suggested that the variolation had paved the way for both the invention of cowpox and the expectation of controlling other epidemics [31]. Louis Pasteur's work of developing vaccines could just be inspired by the variolation and the cowpox vaccination. In this sense, variolation can also be regarded as the pioneer of modern immunology.

References

1. [Italy] Castiglioni. 2003. *A History of Medicine*, 245. Cheng Zhifan [Trans.]. Guangxi Normal University Press.
2. Jian, Zhang. 2002. The origin, spreading, harm and prevention of smallpox. *Science Technology and Dialectics* 4: 54–57.
3. [Jin Dynasty] Hong, Ge. 1574. *Handbook of Prescriptions for Emergencies*, vol. 2. Engraved version by Li Shi of Jianjiang.
4. [Sui Dynasty] Yuanfang, Chao. 1997. *General Treatise on the Cause and Symptoms of Diseases*, 52. Liaoning Science and Technology Press.
5. [Ming Dynasty] Quan, Wan. *Therapies for Acnes and Measles*, vol. 11. Engraved version by Chen Yunsheng of the Ming Dynasty.
6. [Ming Dynasty] Quan, Wan. *Dou Zhen Ge Zhi Yao Lun*, vol. 1. Engraved version by Chen Yunsheng of the Ming Dynasty.
7. [Ming Dynasty] Jingji, Shao. *Quan Wen Ji*, vol. 6. Engraved version by Zhang Jingxian, Wang Xun, etc.
8. Boying, Ma. 2010. *The Cultural History of Chinese Medicine (Part I)*, 491. Shanghai People's Press.
9. [Qing Dynasty] Han, Dong. 2000. *San Gang Shi Lue*, vol. 2, 32. Zhi Zhi (Revised). Liaoning Education Press.
10. Xueliang, Tang. 1987. Research on Jiang Xishun. *Chinese Journal of Medical History* 2: 90.
11. Xuanying, Qu. 2009. (Compiled and edited). *Materials of Chinese Social History*, 686. Hunan Education Press.
12. [Ming Dynasty] Qian, Tan, Bei You Lu. 1971. *The Collection of the Historical Materials in the Ming and Qing Dynasties* (Book 7), vol. 11, ed. Shen Yunlong, 377. Wenhai Publishing House.
13. Jiaji, Du. 2004. The impact of smallpox on the administration of the early Qing Dynasty and the corresponding measures. *Journal of Seeking Truth* 6: 134–141.
14. [German] Alfons Vath, S. J. 1997. *Biography of Johann Adam Schall von Bell*. Yang Bingchen [Trans.]. Oriental Publishing House.
15. [Qing Dynasty] Emperor Yongzheng, et al. *Ting Xun Ge Yan*. Photocopied version of *Si Ku Quan Shu* by Wen Yuan Ge.
16. [Qing Dynasty] Lu, Zhang. *The Zhang's Medical Book*, vol. 13. Engraved version by Bao Han Lou.

17. [Qing Dynasty] Yu, Cao. *Xiu Hu Xuan Chi Du* (Book 2), vol.3. Engraved version by Chuan Wan Tang
18. [Qing Dynasty] Chongbing, Wang. Zhen Lie Zhuan. *Jinhua Zheng Xian Lue*, vol. 15. Engraved version in 1732.
19. [Qing Dynasty] Han, Dong. 2000. *San Gang Shi Nue*, vol. 2, 32. Zhi Zhi (Revised). Liaoning Education Press.
20. [Qing Dynasty] Yan, Zhang. *Zhong Dou Xin Shu*, vol. 3. Engraved version in the reign of Emperor Qianlong.
21. [Qing Dynasty] Qian, Wu, et al. 1995. *Yi Zhong Jin Jian*, 699. TCM Classics Publishing House.
22. [Qing Dynasty] Tingjing, Huang. *Mu Jing Da Cheng*, vol. 2. Engraved version by Da Dao Tang in the reign of Emperor Jiaqing.
23. [Qing Dynasty] Dachun, Xu. *Lan Tai Gui Fan*, vol. 8. The version of *Si Ku Quan Shu* by Wen Yuan Ge.
24. [Qing Dynasty] Mingyi, Fang. 1817. *Zhong Dou Zhen Chuan (Part I)*. Engraved version by Zhu Yourong from Shu Fang Zhai in Liulichang of Bejing in 1817.
25. Boying, Ma. 1991. Draw lessons from history: a retrospective survey of variolation in the late 19th and early 20th centuries. *Shanghai Journal of Traditional Chinese Medicine* 1: 43–46.
26. [British] Needham, Joseph. 1983. China and the origin of immunology. *Journal of Chinese Pharmacy* 5: 10.
27. Xijin, Liu. 1978. The ancient Chinese thought of immunity and the development of variolation. *Acta Microbiologica Sinica* 1: 3–7.
28. Shusheng, Xie, and Zhang Daqing. 2000. The spread of Chinese variolation to the western world and its influence. *Journal of Chinese Medical History* 3: 133–137.
29. [France] Voltaire. 2005. *Lettres Philosophiques*, 54. Gao Daguan, et al. [Trans.]. Shanghai People's Publishing House.
30. [USA] Magner, Lois N. 2009. *A History of Medicine*, 317–319. Liu Xueli [Trans.]. Shanghai People's Publishing House.
31. [USA] Magner, Lois N. 2009. *A History of Medicine*, 320. Liu Xueli [Trans.]. Shanghai People's Publishing House.

Yahua Niu Born in Hohhot, Inner Mongolia in 1961, graduated from Nanjing Pharmaceutical College in 1983 with a Bachelor degree of Science; in 1989 she graduated from Inner Mongolia Normal University with a master's degree in science; in 2005, she graduated from Northwestern University with a doctorate in science and technology history. He once served as director of the editorial department of the Journal of Inner Mongolia Medical College. She is currently a researcher and doctoral supervisor of the Institute of Traditional Chinese Medicine Information, Chinese Academy of Traditional Chinese Medicine. From 1994 to 1995, she has been studied at Japan Iwate Medical University.

Research interests: Medical history; Chinese medical literature.

Major works:

Papers: A Study on Two Earliest Translations of Western Anatomy; Study on Shen San Zi Fang.

Biography of Peng Sixun;

Notes on *Two kinds of Chinese and Western Medical Books; Western Medicine introduced in the Ming Dynasty;*

Chief editor of *The Collection of Traditional Chinese Medicine Classics in Qifen Shi.*

Chapter 31
Artemisinin (Qing Hao Su)

Yahua Niu

The history of applying Artemisia annua L. (*qinghao*) to cure malaria in China has been more than 1,700 years continuing from the *Elbow Backup Prescription for Emergency* (*Zhouhou Beiji Fang*) of Ge Hong in the Jin Dynasty to the 1950s. Based on the Traditional Chinese Medicine clinical experience Chinese scientists, represented by Tu Youyou have used the modern scientific methods to obtain artemisinin, a new antimalarial drug which is highly effective and low-toxic, and cured millions of malaria patients, making an important contribution to human health.

31.1 History of Malaria and Humans Against Malaria

Malaria is a disease caused by a type of protozoa called Plasmodium, which enters the human body with Anopheles as a medium. It invades red blood cells and reproduces inside, causing red blood cells to rupture, periodic or irregular high fever and chills of the body, as well as anemia and splenomegaly. Characterized by quick onset, acute malaria can cause symptoms such as coma and convulsions, and even be a threat to life. There are four types of parasitic Plasmodium of human malaria, namely Plasmodium vivax, P. malariae, P. falciparum, and P. ovale, which cause two-day, three-day, or irregular symptoms of chill and fever, respectively.

It is generally believed that the human malaria, whose history is as long as the human civilization, originated from Africa and then spread throughout almost the whole world. In ancient Greece, malaria is known as "swamp fever," and in the first century B.C., malaria was prevalent for a long time in Rome, resulting in physical

Y. Niu (✉)
Institute of Information on Traditional Chinese Medicine, Academy of Chinese Medical Sciences, Beijing, China
e-mail: niuyahua@hotmail.com

J. Hua and L. Feng (eds.), *Thirty Great Inventions of China*,
https://doi.org/10.1007/978-981-15-6525-0_31

weakness, land abandonment, national recession, and a decline of the Roman Empire [1].

There were also records of malaria epidemic in the Shang Dynasty in China. The inscriptions on the oracle have described diseases such as *nue* (malaria), *jie* (scabies), *gu* (venoms), *qu* (caries), *hui*(lavae), and *yi* (plague). Some scholars believe that rampant malaria was the direct cause for Pan Geng—the ancient emperor in the Shang Dynasty—to move the capital to Yin. In the chapter "Physician" of the ancient book *Rites of Zhou* (*Zhouli·Jiyi*), it is read that "there is malaria in autumn," which shows people at that time have realized that malaria is always prevalent in autumn. The damp and hot climate in southern China is much easier for the breeding of mosquitoes, and the incidence of malaria (called *zhang* or *zhangli*) is particularly high. "The Biography of Yan Zhu" in the *Book of Han* (*Han Shu·Yan Zhu Zhuan*) said: "The southern area is heat and wet, and the weather is hot in the summer. When living near the pool where it is easy for the snakes and noxious insects to breed, the diseases attack from time to time, and two or three people out of ten will die of the disease in time of peace." At the beginning of the Eastern Han Dynasty, the emperor sent armed forces to suppress the rebellion in Jiaozhi (now in Vietnam), and the "Biography of Ma Yuan" in the *Book of the Later Han* (*Hou Hanshu·Ma Yuan Zhuan*) said: "The troops suffered plague of the malignant malaria and almost half of the soldiers died of it. "Also in the "Treatise on the Southwestern Barbarians" of the *Book of the Later Han*, (*Hou Hanshu: Nanman Zhuan*) writes: "The climate in southern barbarian places is warm and humid, and there is poisonous gas which cost the lives of four or five of every ten people." Related studies show that in the Qin and Han Dynasty, for the malaria epidemics in the southern China are much more common than in the northern, the economy along the Yangtze River Basin and its south area is far behind the middle and lower reaches of Yellow River in the north [2].

Even in the nineteenth century, malaria was a common disease in continental Europe. Until now, there are still sporadic cases in developed countries. There are thousands of cases reported every year in China. For example, 2,718 malaria patients were reported in the year of 2012, mainly in five provinces of Yunnan, Guangxi, Jiangsu, Hunan, and Sichuan [3]. And people in Africa and Southeast Asia, where the economy is backward and the climate is hot and humid, are suffering much more from malaria. In 2007, the 60th World Health Organization Conference passed a resolution, that from 2008 onward, the date of April 25 was commemorated as the World Malaria Day every year, and all member states, relevant international organizations, and civil society should carry out the malaria prevention and control activities in an appropriate way. By 2010, there have been 106 malaria epidemic countries and regions worldwide with 216 million patients. About 65.5 million people died of malaria, of which 81% of the cases and 91% of deaths occur in the African region [4]. Scientists and historians believe that malaria can be taken as the most devastating disease in human history, in terms of the number of victims and the extent of injury as an index to assess its influence.

Humans have long been committed to fighting against malaria. In the history of fighting, both the East and West have made their own contribution.

China began very early to record how to treat malaria. The ancient Chinese medicine classics *The Inner Canon of Huangdi* (*Huangdi Neijing*) has two chapters on the disease, "On Malaria" (Nue Lun) and "Acupuncture Treatment to Malaria" (Ci Nue), describing in detail the reasons, classification, symptoms, treatment methods of the disease. Acupuncture treatment against malaria is the mainstay at that time. Zhang Zhongjing, a Chinese medical scientist in Eastern Han Dynasty wrote a book *Treatise on Cold Damages and Febrile Diseases* (*Shanghan Zabing Lun*), which had a prescription for malaria treatment: "If the symptom is like malaria and reoccurs in a day, the disease can be released by sweating. Patients should drink decoction of one dose of cassia twig (*guizhi*) and two of ephedra (*mahuang*) [5]; and if it is like malaria and the symptom attacks frequently, patients should drink the decoction of Common Goldenrop (*xiaochaihu*)" [6]. In the Eastern Jin Dynasty, Ge Hong's *Elbow Backup Prescription for Emergency (Zhouhou Beiji Fang)* has included "Treatment of Cold and Fever Malaria" containing more than 40 prescriptions, the second of which was Artemisia annua L. (*qinghao*), and the realgar (*xionghuang*) was also listed. The number of prescriptions including dichroa febrifuga (*changshan*) was as many as 14, demonstrating its importance to treat malaria. These records indicate that people at that time have accumulated considerable experience in the treatment of malaria. There are many discourses about malaria in the subsequent Chinese medical classics. In the ancient medical book *Prescriptions for Universal Relief* (*Puji Fang*) written in the Ming Dynasty, there were four volumes titled "Ways to Treat Malaria" concerning the disease and hundreds of prescriptions were included. Apparently, the prescriptions containing dichroa febrifuga and Artemisia annua L. are effective. In the influential classics *Compendium of Materia Medica* (*Bencao Gangmu*) compiled by Li Shizhen, the Chinese medical scientist in the Ming Dynasty, hundreds of drugs and prescriptions for the treatment of malaria were listed, which was a summary of the experience of treating malaria before the Ming Dynasty.

In Europe, laxatives, dieting, and bloodletting have been the main methods of treating malaria [7]. It was not until the seventeenth century that the finding of bark of the cinchona tree changed the situation. A widely circulated story told that in 1638, the Peruvian Governor's wife, Condessa de Chinchon, was infected with vivax malaria, which was cured with the powder from the bark of the cinchona tree. Later, the Governor's wife brought the medication back to her hometown—Spain—wherein the malaria was raging in the European continent and this bark powder quickly spread throughout Europe. However, according to scholars' research, this story was fictitious, and cinchona cream was brought back to Europe by the missionaries. Although quinine (commonly known as cinchona cream) may cause tinnitus, visual disturbances, headaches, etc., it was the first time that Europeans had obtained effective drugs for the treatment of malaria.

Soon the cinchona cream was brought to China. In 1693, the French missionary Jean de Fontaney used the cinchona cream to cure the malaria of Emperor Kangxi of China, and he was therefore treated with high courtesy by the royalty [8]. Another record was about Cao Xueqin's (famous Chinese writer in Qing Dynasty) grandfather Cao Yin, who was suffering from malaria and asked Emperor Kangxi for cinchona cream. Cao Yin's friend, Li Xu (supervisor for the Suzhou Weaving) submitted a memorial to Kangxi: "Cao Yin informed me that 'doctors have ended every medicine to treat my disease, yet without any effect. It must be only the holy medicine from Your Majesty that can save my life.'" Emperor Kangxi gave order to deliver the medicine to Cao through post horse day and night and repeated: "If it were not malaria, this medicine shall not be used. It must be taken seriously, note it!" [9]. Unfortunately, Cao Yin died before the drug arrived. The cinchona cream was named *jinjile* by Zhao Xuemin in *Supplement to the Compendium of Materia Medica* (*Bencao Gangmu Shiyi*) in which he wrote: "There is a kind of tree bark in the West, named *jinjile*, which can cure malaria as soon as it is been taken. In the fifth year of the reigns of the Emperor Jiaqing, the Emperor's brother Prince Jinzhai returned from the east Guangdong with this plant and showing it to people. The twigs are hollow, such like the core-removed polygala tenuifolia (*yuanzhi*) with a little pungent smell. It is said to have the effect to adjust the harmony of the inner body (*yingwei*). Generally as a hot-natured drug, it's good for gaining vital energy and improving the blood flow of the human body" [10].

In 1820, the French chemist Pelletier (1788–1842) and Caventou (1795–1877) separated two kinds of alkaloids quinine and cinchonine from the bark of the quia tree. In 1854, the botanist Hasskarl (1811–1894) transplanted Peruvian and Bolivian cinchona seeds to Indonesia and planted them on a large scale. Quinine began to be used in large quantities, and 97% of the quinine needed in the world is produced in Indonesia. Quinine has been the world's specific treatment for malaria until the application of the synthetic drugs such as chloroquine.

Although malaria was associated with swamps, the damp, and heat climate a long time ago, it was not until the end of the nineteenth century that medical scientists clarified the causes and route of transmission of malaria. French physician Charles Louis Alphonse Laveran (1845–1922), British tropical medicine pioneer Patrick Manson (1844–1922), and microbiologist Ronald Ross (1857–1932) all made a huge contribution to that.

Laveran is a military doctor who has been engaged in malaria research throughout his life. He suspected that the pathogen of malaria was a kind of parasite. In 1880, he observed the blood of malaria patients in a tent in the military camp of Algiers and found a small animal, the malaria parasite. How does it enter the human body? After repeated observations, he confirmed that it was not passed on from one person to another. As for the way of transmission, Laveran did not find the answer [11]. Later, Manson and Rose solved this mystery. Ross began research on malaria in 1892 and in 1894 he met the British tropical medicine pioneer Manson in London. Manson introduced to Ross the malaria specimens that Laveran discovered in 1880. In 1895, Ross studied the vector of malaria in India. He put the hatched Culex and Aedes mosquitoes on the malaria patients and then observed them under the microscope. Both ended with a negative result. On August 22, 1897, Ross used Anopheles to suck blood from the malaria patients, and after breeding and dissection, Plasmodium was finally found in the stomach cavity and stomach wall of Anopheles. In 1898, Ross found similar dyed cysts in the blood of birds suffering from malaria, and Plasmodium of Avian malaria was observed in the saliva of mosquitoes. Manson further confirmed that malaria is indeed transmitted by Anopheles [12]. Manson (whose Chinese name is Wan Bade) has long studied tropical diseases in Taiwan, Xiamen, and Hong Kong. He is the founder of Faculty of Medicine of the Chinese University of Hong Kong. Later, Grassi (1854–1925, Italy) conducted a more detailed study on the life history of the malaria parasite. Studies of malaria transmission pathways and life history of malaria parasites have created conditions for the prevention and treatment of malaria. Rose was awarded the Nobel Prize in Physiology or Medicine in 1902 for his contribution to the explanation of the etiology of malaria. In 1907, Laveran also won the Nobel Prize in Physiology or Medicine for discovering that the protozoa were also the culprit of the disease.

During the First World War, the source of quinine in Germany was cutoff, forcing the Germans to find alternatives to quinine, and Bayer Pharmaceuticals began chemical synthesis of antimalarial. Since 1932, a series of drugs such as atabrine have been synthesized, as well as the chloroquine synthesized in 1934. However, without careful experimentation, people assumed that chloroquine was too toxic and did not apply it into practice.

During the Second World War, Indonesia was occupied by the Japanese army. In many countries, the source of quinine was severed and there was an urgent need for alternative medicine. In 1944, two young chemists in the United States, Robert Burns Woodward (1917–1979) and William Doering, synthesized chloroquine, and later Berquela was also produced; In the United Kingdom, a series of drugs for the treatment and prevention of malaria are made, such as proguanil hydrochloride and pyrimethamine. Because of the high efficacy, less side effect, low prices, and convenience in use, these synthetic drugs soon played a role in the treatment of malaria, and have become the main antimalarial since World War II [13].

In China, after the outbreak of Antijapanese War, a large number of people in central China moved westward to malaria epidemic places such as Sichuan, Chongqing, Yunnan, Guizhou, and other areas. Though the malaria broke out constantly, the antimalarial drugs are deficient. Scientists have conducted research related to antimalarial drugs to seek self-sufficiency. The drug group of the Central Health Laboratory of Chongqing, under the leadership of Zhang Changsha, screened out antimalarial drugs from the Chinese herbal medicines. They clinically proved that the crude antifebrile dichroa extract has significant curative effect on malaria, and several kinds of dichroine were separated, three of which have a strong inhibitory effect to Plasmodium, and its effect is better than quinine. However, it failed to be used in clinical practice for it could cause severe vomiting after taking. In 1950, Wang Jinying reported that Brucea javanica had the antimalarial effect, and in 1952, it was put on trial in construction plant in Foziling and achieved good results [14].

Another important achievement in the antimalarial campaign during the Second World War is the discovery and widespread use of the chemical pesticide DDT. As the malaria vector and the life history of the malaria parasite became clear, people began looking for the pesticides and repellents to block the spread of malaria. As early as 1874, Otmar Hinderler obtained a compound dichlorodiphenyltrichloroethane (full chemical name of DDT), whose effect on pesticide was unknown at that time. In 1939, the Swiss chemist Paul Hermann Müller (1899–1965) discovered that it could kill a variety of pests. During the Second World War, DDT was used as a spray to combat insect-borne diseases such as malaria, yellow fever, typhus, and filariasis, which showed great power and saved many lives. At the same time, spraying DDT on livestock and grain also doubled their production. DDT, juxtaposed with penicillin and atomic bombs, was hailed as one of the three inventions of the Second World War. Muller was also awarded the Nobel Prize in Physiology or Medicine in 1948.

Through treating the source of infection with chloroquine, preventing with drugs such as primary primaquine, coupled with spraying DDT to kill mosquitoes, the global malaria incidence was once under control. In 1955, the campaign to eradicate malaria was launched in the world. By 1962, the incidence of malaria worldwide had dropped to a very low level.

By the end of the 1960s, malaria had staged a comeback. The main concern was that Plasmodium became resistant to quinines, especially in the hardest-hit areas such as Africa and Southeast Asia. Falciparum malaria against chloroquine was in the face of cure-free situation. In the 1960s, the US military on the battlefield in Vietnam was also suffering from malaria, and new drugs with special effects were urgently needed. The United States has invested a lot of manpower and resources to carry out research on antimalarial drugs and developed mefloquine. The drug is obtained by modifying quinine drugs, mainly for Falciparum malaria. Though a slow onset of action, it has a long duration of action and may have convulsions and liver and kidney toxicity. At the same time, at the request of the Vietnamese government, China mobilized the national forces to carry out research on antimalarial drugs, including insecticides and repellents, and developed a series of drugs with antimalarial effects, in which artemisinin is highly effective and low-toxic. Its remarkable efficacy in Plasmodium falciparum chloroquine gained the attention of the international community. Later artesunate, artemether, dihydroartemisinin, and other antimalarial drugs based on artemisinin were developed. Currently, artemisinin drugs are considered the safest rescue medication in the treatment of severe and multi-drug-resistant malaria, and it can even be used for pregnant women and children [15].

Fortunately, the World Malaria Report 2013 published by the World Health Organization estimated that since 2000, global efforts to control and eliminate malaria have saved the lives of 3.3 million people, and malaria mortality rates in the world and in Africa fell by 45% and 49%, respectively. The Artemisinin-based Combination Therapies (ACTs) recommended by WHO is becoming recognized in more and more places, the number of courses provided for all the countries is also increasing from 76 million times in 2006 to 331 million times in 2012 [16].

Artemisinin has helped millions of people with malaria and has made important contributions to human health. Tu Youyou won the 2015 Nobel Prize in Physiology or Medicine for the discovery of artemisini (Fig. 31.1). Artemisinin is derived from the clinical experience of TCM and is a shining example of modern research in traditional Chinese medicine.

Fig. 31.1 Tu Youyou accepts the 2015 Nobel Prize in Physiology or Medicine

31.2 Artemisia and Its Antimalarial Records in the Classics of Traditional Chinese Medicine (TCM)

The ancient Chinese people recognized artemisia as a kind of plant. *The Book of Songs*, the ancient Chinese literary work, recorded the name of artemisia. As quoted in *the Book of Songs · Xiaoya · Luming*, "the deer whined and ate the wild artemisia." The artemisia mentioned here is exactly the artemisia. The name of Tu Youyou, the discoverer of artemisinin, derives from this poem.

There is a long history of the artemisia as a medicine in China. Unearthed in Ma Wangdui No.3 Tomb of the Han Dynasty (around 168 B.C.), the silk manuscript called *Prescriptions for Fifty-two Diseases* recorded that the artemisia treated the hemorrhoid with particular illustration that the artemisia was called Qin in the land of Jin and Chu [17]. *Shennong's Classic of Materia Medica,* finished in about the Eastern Han Dynasty, also recorded the artemisia and listed it as a low-grade plant. At that time, it was called the feather cockscomb. In other words, the feather cockscomb (Fig. 31.2) was its formal name while the artemisia was its alternate name. "It is bitter in flavour, cold in property and is non-toxic. It grows in the river and can treat scabies, pruritus, crust and malignant sores, and it can kill pests. It can also keep the heat between the joints and improve the eyesight" [18]. Tao Hongjing (456–536) noted that the artemisia nowadays, grew everywhere and people also ate it with condiment. *The Handbook of Preions for Emergencies* written by Ge Hong in the Eastern Jin Dynasty recorded the treatment of artemisia for cold and heat resulted by the malarial, in other words, "taking a handful of the artemisia and soaking it with 2 l of water, then squeezing out the juice of the artemisia and drinking." It was the earliest record about the treatment of the artemisia against malarial, which inspired the artemisinin extraction at the low temperature.

Fig. 31.2 Artemisia (Artemisia annua L.)

New Materia Medica in Tang Dynasty followed *Shennong's Classic of Materia Medica,* which also regarded the feather cockscomb as the formal name of artemisia. *New Materia Medica* recorded that artemisia could be applied to cure serious wound, greatly stop bleeding, and help grow new flesh, and it is good at relieving the pain [19]. *A Supplement to Materia Medica* (739) written by Chen Zangqi noted, "the feather cockscomb is good at wiping out the pathogenic factors, restores women's vital energy as the blood tonic, relieving the abdominal distention, cold and heat, and the lingering dysentery. People tend to use its seeds in autumn and winter and its seedlings in spring and summer. They crush its seeds or seedlings to extract the juice and take it or expose it to the blazing sun and dry it, then take it. If people feel it is cold, they can boil it with the liquor [20].

The herbal classics of the Song Dynasty described the artemisia a lot, but more often cited it as the feather cockscomb. *Figure of the Medica* (1061) recorded that "the feather cockscomb, also known as the artemisia, shoots seedlings in spring and its leaves are very delicate and tender. People at that time also ate it with condiment. In summer, it grows to three or five feet; after autumn, its thin yellow flowers and bears seeds bloom as big as the millet under flowers. [19] *Classified Materia Medica* (1082) following recorded the feather cockscomb and collected the related previous records about it. It also wrote that Jiangdong people called it wormwood due to its similar smell while the Northern people called it the artemisia carvifolia [19].

The Dream Steam Essays (about 1086–1093) also talked about the artemisia and noted, "There are many kinds of the artemisia. For example, there are two kinds of artemisia: the yellow one and the blue one. Although *Chinese Materia Medica* called it artemisia, there might be some differences. The area between Sui and Yin in Shaanxi grew artemisia. Among the artemisia clusters, there sometimes were one or two strains which were very apparently blue. The natives called it the tarragon. Its stem and leaves were the same as the common artemisia. However, the common artemisia is green in color while the tarragon is verdant just like the pine; in the late autumn, other artemisia became yellow while the tarragon alone was blue and smelled fragrant. The tarragon was used by the ancient people and was regarded as the best [21].

It could be seen that the artemisia in the Song Dynasty refers to more than one plant. The artemisia is a general name, the so-called the artemisia class. Such discussion was also collected in *the Effective Prescriptions of Su and Shen.* It is written in *Materia Medica Yanyi* that "The feather cockscomb is the present artemisia. It grows everywhere and blooms in spring. People regard it as a kind of vegetable and its root is red, and the leaves are fragrant. The contemporary people call it the artemisia, but there are also some differences. In the artemisia class, people tend to take the blue one. The area between Sui and Yin in Shaanxi grew such artemisia. Among the artemisia clusters, there sometimes were one or two strains which were very apparently blue. The natives called it the tarragon. Its stem and leaves were the same as the common artemisia. However, the common artemisia was light blue in color while the tarragon was dark blue, and it smelt fragrant. It was likely that the ancient people used the dark one more frequently. Otherwise, why they did not like the other kind of Artemisia? [22].

Zongshi et al. most herbal works in the Song Dynasty didn't collect the relevant contents in *the Handbook of Preions for Emergencies* written by Ge Hong, and also didn't mention the antimalarial effect of artemisia. It could be seen that the artemisia was always known as the alternate name of the feather cockscomb in the Chinese materia medica until Song Dynasty while the medical formulary and other historical literatures used more frequently the artemisia as its name instead of the feather cockscomb.

According to volume 36, 71, and 168 of the book *Sheng Ji Zong Lu* (1111 AD–1118 AD), there were three artemisia soups with different recipes, treating spleen malaria, malaria, and pediatric hectic fever, respectively. In 1178, Zhou Qufei recorded a unique method of treating malaria in Lingnan area in his book, *Ling Wai Dai Da*, in which he said, "Malaria could not be treated only by medicine that treats typhoid fever in Zhongzhou (now Henan Province). If the fever becomes more serious, Chinese rhubarb and mirabilite should be added. If the weak patient remains critically ill, he or she can take some Artemisia powder, which is said to be given by a celestial being to the Tang's, a royal family in Jingjiang and which is also amazingly effective in treating malaria in the south. The formula of the artemisia powder constitutes artemisia, gypsum and herbal medicine. If a patient who takes the power has not been cured, it may be duo to the weak condition and seriousness of disease. However, if given aconite and cinnabar to eat, in most cases, the patient can be saved." There was another method, called pricking therapy, according to *Ling Nan Dai Da* which said, "Once a southern man who got fever malaria was cured by pricking therapy. One or two days after the southern man got fever malaria, prick his lips with acupuncture. The specific procedures are as follows: prick the inside lips of the patient, squeeze out the lip blood by hands and rub his tongue with folium broussonetiae. Then ask the patient to stand, with two feet together, and prick the Zhongqing Vein behind his two feet's joints causing lots of blood. In the end, let him take Artemisia annua and water, and he will recover soon." It can be seen that artemisia has long been used to treat malaria in places like Guangdong and Guangxi, which was very common. This treatment is also recorded in the book of *Ling Nan Wei Sheng Fang*.

In both *Jin Kui Gou Xuan* and *Dan Xi Xin Fa* written by Zhu Zhenheng (1281–1358) in the Yuan Dynasty, there are records of Artemisia pills for malaria. In his books, Zhu wrote, "One prescription includes 250 g of the Artemisia, some wax gourd leaves, Chinese cinnamon and vervain. Then bake and grind them into fine powder, and add some water to make pills as big as peppers. Divide every 50 g into 4 servings and finish taking them all 1 h before the onset of the illness. The other prescription consisted of 50 g of artemisia, 100 g of Chinese ilex leaves, vervain and Chinese cinnamon. Which one is the right formula is still unknown, so two are recorded for reference." In light of the Zhu's records of the two prescriptions, artemisia was widely used to treat malaria at that time. Until the Yuan Dynasty, most of the artemisia plants were not cooked at high temperatures so the efficacy of artemisia to treat malaria was certain.

The use of artemisia as an antimalarial drug increased in the Ming Dynasty. There were many records of artemisia treating of malaria in *Pu Ji Fang* in the early Ming Dynasty such as *the Magical Presciprtion, Hengshan Mountain Powder,* and *artemisia Powder* in Volume 197, *Killing Malarial Methods*. Also, according to articles like Fruit and Chinese Date Soup and Ginseng Powder with Strong Effect in Volume 199, artemisia was the basic remedy in these two medicines. In *the Magical Presciprtion*, "To treat malaria, prepare equal amount of the dried artemisia leaves picked on Dragon Boat Festival and Chinese Mosla Herb, and then grind them into fine powder. If feeling cold first, you should serve the powder with hot Chinese alcohol at 3 a.m to 5 a.m on the day when you are sick. If you get a fever first, then serve with the cool Chinese alcohol." "Ginseng Powder with Strong Effect" clearly said, "Ginseng Powder with Strong Effect could treat malaria caused by the irregular climate. No matter how long you have suffered malaria, be it chills or fever, or both, be it continuous or intermittent like every three or four days, the power can cure all of these symptoms. The prescription contains ginseng (with its rhizome cut out), antifebrile dichroa (filed), artemisia (with its root and stem cut out). All of them are equal amount. Then grind them into fine powder and make each serving 25 g." *Ying Tong Bai Wen* written by Lu Bosi in 1403 has records of Pear Syrup Yinzi (a Chinese traditional syrup without set drinking time and dosage), which is used to treat hectic fever, blood heat, internal heat, and malaria fever. They may happen every two or three days. Artemisia is also an important drug for accumulative heat, spleen heat, stomach heat, malaria fever, pathogenic heat, chills and fever, spleen malaria and night malaria, and so on. In 1551, *Yao Xing Cu Ping* (Fig. 31.3) made artemisia the formal name and made it clear that artemisia was a key treatment for chronic malaria fever [23].

According to the ancient books we have checked, in the medical books of past Dynasties, there are rare records of using artemisia as a herb. Perhaps *Compendium of*

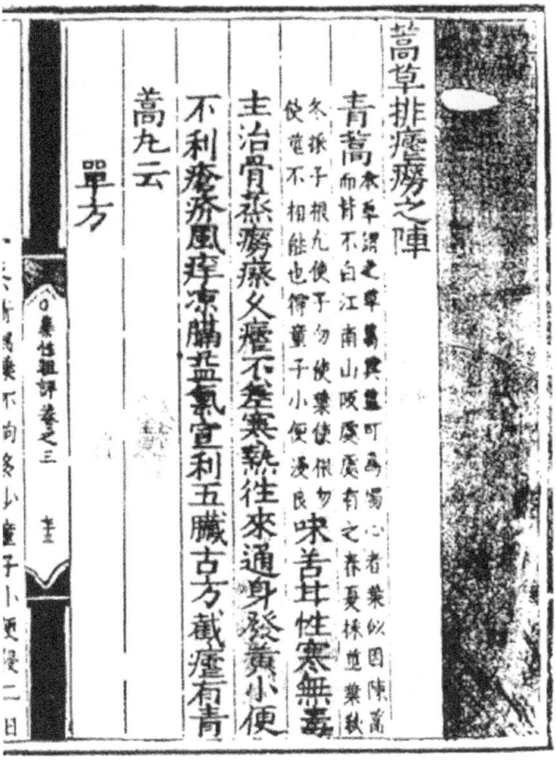
蒿草排瘧痨之陣

青蒿

味苦平性寒無毒

主治骨蒸痨瘵久瘧不差寒熱往來通身發黃小便

不利瘡疥風痒涼腸益氣宣利五臟古方截瘧有青

蒿丸云

單方

藥性粗評卷之三

Fig. 31.3 A picture of *Yao Xing Cu Ping*

material Medica is influenced by previous medical books, *Compendium of Materia Medica* changed the conventional method of previous herbal works, abandoned the name of feather cockscomb, and took the name of artemisia instead. The book comprehensively cites the previous expositions about artemisia, including Ge Hong's "*Handbook of Prescriptions for Emergency*" (Fig. 31.4), the content about using artemisia to treat malaria, and the description of the shape of artemisia: "Artemisia sprouts seedling in February, and its stem is thick as finger and the leaves' color is dark green. Its leaves are slightly like Yinchen, and its surface is green. The roots are white and hard. Its yellow flowers blooming in July and August are quite fragrant. The fruit with granules in it is as big as a pockmark." As for the main functions of artemisia, it has been clearly pointed out: "It treats malaria, coldness, and fever." In its volume III, *Principle Indications for Malaria*, more than 100 kinds of drugs for treating malaria are listed. As one of them, Artemisia can cure cold and fever malaria, and should be taken in the way of simmered juice or boiled with shaved cinnamon bark in liquor. If warm malaria is not cold, it should be served with yellow lead. To stop malaria, it is necessary to take artemisia with antifebrile dichroa and ginseng boiled with wine." In addition, it also lists the use of artemisia in combination with several other drugs to treat malaria, such as "Take citrus and the artemisia with wine." "For hot malaria, eat Hanshi noodles, and take artemisia water pills of 10 g."

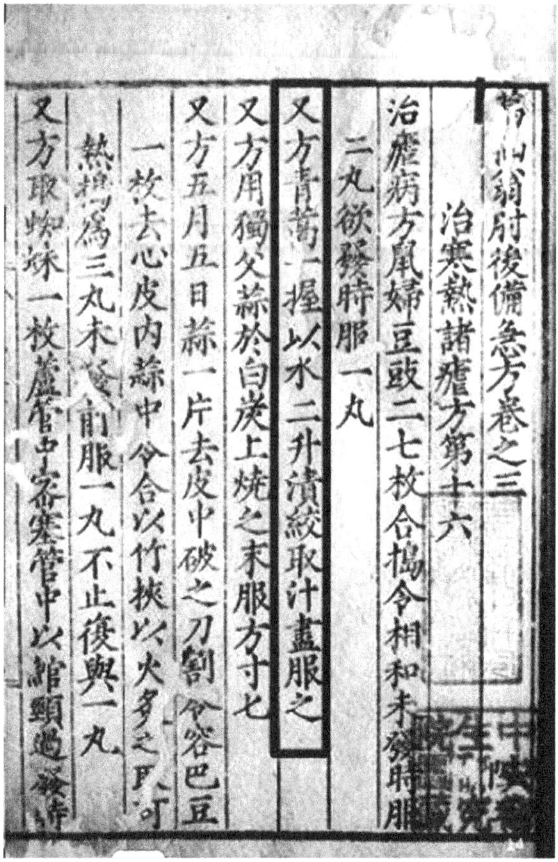
葛仙翁肘後備急方卷之三
治寒熱諸瘧方第十六
治瘧病方鼠婦豆豉二七枚合擣令相和未發時服
二丸欲發時服一丸
又方青蒿一握以水二升漬絞取汁盡服之
又方用獨父蒜於白炭上燒之末服方寸匕
又方五月五日蒜一片去皮中破之刀割令容巴豆
一枚去心皮內蒜中令合以竹挾以火炙之取可
熱擣爲三丸未發前服一丸不止復與一丸
又方取蜘蛛一枚蘆管中密塞管中以綰頸過發時

Fig. 31.4 A picture of *Zhou Hou Bei Ji Fang*

"To cure malaria, take pills made by white gourd leaves, artemisia, verbena, and Chinese cinnamon." "Jiziqing Pills should be cooked thoroughly to take. For malignant malaria, boil and drink the soup made of artemisia, rhizoma anemarrhenae, and peach seed." We can see that artemisia has been common in the treatment of malaria in the times of Li Shizhen.

The medical books of the late Ming Dynasty, such as *Zeng Ke Yi Bian*, *Wan's family Chao Fang*, *An Analysis of Diseases*, *Communication record of acne*, *Shi Cai San Shu*, *Fu XinYou Yi Fang*, *Quan Sheng Zhi Ming Fang*, and other books have cited "Anti-Malaria Artemisia Pills," or recorded the prescription of the artemisia to cure malaria, and noted "the wonderful effect on malaria." Some prescriptions adopt the method of boiling and frying or using the children's urine. Some use the dried artemisia leaves to make pills, and others advise to mash artemisia into water and then take it without cooking, as is recorded in *Ben Cao Yuan Shi* (1612) by Li Zhongli: "Artemisia should be mashed into water, drink it and then stick dregs on the body to treat the cold and fever caused by malaria."

In the Qing Dynasty's medical prescriptions and books of herbal medicines, the records of the treatment of malaria with artemisia have increased, as is quoted in *Za Bing Yuan Liu Xi Zhu* by Shen Jinxi, "Malaria has the syndrome of damp-heat. If hundreds of drugs don't work, detailed examination is especially needed. Or we can use artemisia, rhizoma atractylodis, and fruit of immature citron to cure this disease;" *Ai Lu Medical Case* recorded the cases of using artemisia to cure malaria; *Wen Bing Tiao Bian* has created the turtle shell artemisia soup; *Yi Jia Si Yao* recorded, "Artemisia can relieve heat and cure malaria and dysentery." In addition, *Jiang Xueyuan's Selected Notes on Ancient Prescriptions*, *Ben Cao Bei Yao*, and *Han's Family Medical Class* have records of malaria treatment with artemisia. Ye Tianshi, the famous doctor, often changed bupleurum into artemisia when treating malaria.

During the period from the late Qing Dynasty to the Republic of China, some pharmacies published items that were sold as medicines, such as *Catalogue of Suzhou Laosong Shoutang Pills Powder Ointment Dan Gum Dew*, *Collections of Peng Taihe Tang Pills Powder Ointment Dan Gum Dew*, *Collections of Shanghai Wang Hengchuntang Pills Powder Ointment Dan Gum Dew*, *Complete Works of Shanghai Leitong Juntang Pills Powder Ointment Dan*, *Collections of Jiang Yan Ze Tang Faji Pills Powder Ointment Dan*, *General Catalogue of Changsha Tongdetai Pills Powder Ointment Dan*, and *General Catalogue of Beiping Qingrentang Powder Ointment Dan*. They all contain sweet wormwood distillate or fresh sweet wormwood distillate, and *General Catalogue of Ye-Gang Detang Pills* contains Chen Qing wormwood distillate, which says "treatment of the bone injuries and heat because of labor." The sweet wormwood distillate has become a daily sales product of the drug stores, indicating that the treatment of malaria by Artemisia annua has been widely used in the late Qing Dynasty and the early Republic of China. In 1958, folks in Gaoyou, Jiangsu used the traditional method of treating malaria through Artemisia annua, and the effect was remarkable.

As can be seen from the above description, the classification of plants in ancient times is still not very detailed, such as Artemisia annua, which is called Artemisia scoparia in Herbal Works. Shen Kuo's *Meng Xi Bi Tan* also mentions that Artemisia annua has two different varieties, "Compendium of Materia Medica" abandoned the name of Artemisia, while Artemisia annua began to become its name, and at the same time, it added a drug of Artemisia annua L. to distinguish the two different plants contained in the Artemisia of the herbal books of the previous generation. The description of Artemisia annua L. in *Compendium of Materia Medica* is as follows: "Fragrant or smelly Artemisia is known as Artemisia. This Artemisia is similar to Artemisia annua, but this artemisia's color is green and kind of yellowish, and the smell is smelly and cannot be eaten. People sometimes pick it to make yellow wine." Its smell is bitter, cold, and non-toxic. Its main function is to attend children with fever and chills." Artemisia annua inherits the taste, efficacy, and the main function of the herb of the past of Artemisia annua L., increasing the function of treating malaria. *Compendium of Materia Medica* distinguishes between Artemisia annua and Artemisia annua L. in terms of odor and the difference of the main disease, as well as the morphological characteristics, distribution, and habits of "opening in a fine way like the yellow flowers" and "sturdy as pockmarks."

According to the research of Tu Youyou and others, in 1930, Japanese scholar しらいみつたろう, the author of *The First Annotation of Compendium of Materia Medica*, added the botanical name to each medicine in *Compendium of Materia Medica* and annotated "Artemisia apiacea hance" to "Qinghao" and "Artemisia annua L." to "Huanghuahao." Since then, such annotations have been quoted and followed by Chinese scholars [24]. In this case, artemisinin was thought to be extracted from Artemisia annua L., which is inconsistent with the records in *The Compendium of Materia Medica* and caused some disputes. Some people even think that the discovery of artemisinin has nothing to do with the experience of traditional Chinese medicine. In 1981, Tu Youyou discussed the authenticity of Artemisia apiacea hance through the research on ancient medicine literature, the distribution of original plants and resources, the comparison of chemical composition, pharmacological effects and curative effect, and concluded that the source of the plant should only be Artemisia A. annua [25]. Tu Youyou et al. investigated the commercially available Artemisia apiacea hance and confirmed that the medicinal material known as Artemisia apiacea hance on the market was actually Artemisia annua L. [26] Lin Yourun et al. believe that Qinhao and Huanghuahao recorded in *The Compendium of Materia Medica* were the same kind of plant, which is now known as "A. annua L." in botany. Hu Shilin believes that the description of Qinghao in *The Compendium of Materia Medica*, namely, "thin yellow flower" and "as big as a pockmark" is exactly the same as that of Artemisia annua L.. Since the characteristics of flowers and fruits are important evidences for classification and determination of species, we should correct the false statement of the Japanese scholar—Makino Tomitaro—that Qinghao in *Compendium of Materia Medica* is Artemisia apiacea hance, "a plant with inflorescence as big as a soybean, withering at the end of summer with no bitter taste, and not usable for malaria." From the above, we can see that the mismatch of name and object is caused by the mismatch of the Chinese medicine Qinghao and Huanghuahao to their Latin scientific names in modern times [27]. The role of Artemisia apiacea hance in the treatment of malaria has been recorded in ancient medical books. Li Shizhen has added the antimalarial function to the major functions of Artemisia apiacea hance on the basis of previous medical records or his own experience of drug use. The discovery of artemisinin further proves that the Japanese scholars mistakenly named Artemisia apiacea hance. Zhao Yuhuang, a famous Chinese pharmacognosist, investigated and identified the Compositae medical plants in Beijing in the 1930s. He believed that the original plant of Qinhao was Huanghuahao (Artemisia annua L.) and Qinghao is the "Huanghuahao" in *Compendium of Materia Medica*. He did not agree with the wrong conclusion the Japanese scholar Makino Tomitaro drew that "the description of Qinghao in *Compendium of Materia Medica* is quite close to the plant Artemisia apiacea hance (Seseli libanotis)" [28]. The 1977 edition of the Pharmacopoeia of the People's Republic of China defined the plant source of Qinghao as Artemisia annua L. and A. apiacea hance. The 1985 edition of the Pharmacopoeia of the People's Republic of China claimed that the Chinese herbal medicine Qinghao was the dry part of Artemisia annua L. above the ground. Given this, Hu Shilin appealed: "We have no reason to continue to mistakenly associate Qinghao with Artemisia apiacea."

From the above discussion, we can know that Qinghao, the traditional Chinese medicine used from ancient times up to now, is actually the plant known as Artemisia annua L. The practice of naming the new antimalarial drug as artemisinin has respected history and facts, just as the speech Tu Youyou made at Karolinska Institute "Artemisinin—A Gift from Traditional Chinese Medicine to the World."

31.3 Discovery of Artemisinin

31.3.1 "523" Project

In the 1960s, Plasmodium falciparum became more resistant to chloroquine and other quinolines, especially in Southeast Asia. A large number of the malaria patients were faced with a situation where there was no drug to treat. During the Vietnam War, malaria was prevalent in Vietnam, which not only threatened the health of the Vietnamese soldiers and civilians, but also caused non-combat attrition of the US troops. The search for new antimalarial drugs had become the focus of the pharmaceutical industry, and many countries have launched researches. Since 1960s, the United States had screened 300,000 kinds of compounds urgently needed for the war.

In 1964, at the invitation of the Vietnamese leaders, Chairman Mao Zedong and Premier Zhou Enlai instructed to set up a project to solve the problem of malaria of the troops in tropical areas as an important task of emergency aid and for combat readiness. The Chinese People's Liberation Army's Academy of Military Medical Sciences, the Second Military Medical University, and the military medical research institutes affiliated to the Guangzhou, Kunming, and Nanjing military regions carried out research on corresponding malaria prevention and the control-drug. In 1967, the Chinese People's Liberation Army Academy of Military Medical Sciences drafted a three-year research plan. After deliberation and discussion and examination and approval by the leading department, the State Science and Technology Commission and the General Logistics Department of the Chinese People's Liberation Army held a national coordination meeting in Beijing on May 23 with the participation of relevant ministries and commissions, leaders of military regions directly under the headquarters of the armed forces and relevant provinces, municipalities directly under the central government and autonomous regions, and their subordinate units. The three-year research plan drafted by the Academy of Military Sciences was discussed, revised, and finalized, and the "National Malaria Control Research Leading Group Office" (hereinafter referred to as "523 Office") was established. As this is an emergency military project involving the Vietnam War, for the concern of confidentiality, the date of the meeting was used as the code name, abbreviated as "523 Project." The leading group consists of one representative from each of the State Science and Technology Commission, the National Defense Science and Technology Commission, the General Logistics Department, the Ministry of Health, the Ministry of Chemical Industry and the Chinese Academy of Sciences, and it was directly under the leadership of the State Science and Technology Commission. The leading group consists of offices, mainly composed of 236 troops of the Chinese People's Liberation Army, and one member from each of the Chinese Academy of Sciences, the Chinese Academy of Medical Sciences, and the Chinese Pharmaceutical Industry Corporation. The office is located in Unit 236, which is responsible for daily research and the exchange of scientific research information. The goal of "523 Office" was to study and solve three problems in a short period of time: drug-resistant malaria control drugs, long-term drug-resistant malaria prevention drugs, and mosquito repellent. The policy of the research was to proceed from the actual operations of the troops in the tropical region, "combining far and near, combining traditional Chinese and western medicine, focusing on medicine, focusing on innovation, unified planning, and division of labor and cooperation."

The project had convened more than 500 people from more than 60 units in seven provinces and cities across the country to tackle the key problems from a multi-disciplinary perspective, attaching great importance to exploring new drugs from Chinese medicine. By 1969, more than 40,000 kinds of compounds and Chinese herbal medicines had been screened but without satisfactory results. In view of this, on January 21, 1969, "523 Office" invited the Institute of Traditional Chinese Medicine of the Institute of Traditional Chinese Medicine to participate in the work.

31.3.2 Discovery of Antimalaria Effect of Artesunate Ether Extract

Upon receiving the task, the Institute of Traditional Chinese Medicine set up a scientific research team and appointed Tu Youyou as its leader to take charge of the overall researching work [29].

Tu Youyou was born in Ningbo, Zhejiang in 1930. Her name comes from a verse in *Book of Songs Xiaoya Luming Literature.* Perhaps it is providence that Tu Youyou's life has become indissoluble with Artemisia. In 1951, she was admitted to the Department of Pharmacy, Medical College of Peking University, majoring in pharmacy. She graduated in 1955 and had been assigned to the newly established Institute of Traditional Chinese Medicine to work till now.

From 1959 to 1962, she took part in the course of western medicine and Chinese medicine for two and a half years. Upon graduation, she completed the article "Learning Experience of Chinese Medicine Identification Experience," published in the 6th issue of 1962 *Journal of Traditional Chinese Medicine.* She had received comprehensive modern pharmaceutical training with a deeper understanding of traditional Chinese medicine, which had laid a foundation for her future research work. In 1969, Tu Youyou joined the "523 Working Group" at the age of 39. At that time, she was young only with certain working experience. After receiving the task, she and her colleagues first systematically collected and sorted out the medical books and herbal literatures of the Chinese dynasties, searched the letters from the people since the establishment of the hospital, and also consulted the famous old Chinese medicine doctors in the hospital at that time. Pu Fuzhou once recommended "Leiji Powder" and "Sheng San Zi," and Yue Meizhong recommended ancient prescriptions such as "Mangzejian Decoction" and "Guizhi Baihu Decoction." In addition, she and her colleagues spent three months compiling more than 2,000 kinds of medicines for internal and external use, including plants, animals, minerals, and so on. On such basis, she compiled a *Malaria Traditional Prescriptions Collection*" with more than 640 kinds of medicines as the main ingredients, which was mimeographed into a book and sent to "523 Office" in April 1969. The "523 Office" transferred the task to the seven provinces and cities to jointly explore the prescriptions [30]. Among them, there was a prescription for Artemisia: Artemisia 0.5 kg; usage: take it by pounding juice or decocting in water or grinding into fine powder, and mix it with boiled water. Source: Fujian, Guizhou, Yunnan, Guangxi, Hunan, Jiangxi. There was also a remark: there were 13 prescriptions for malaria that use Artemisia in combination with other drugs. This shows that they had already paid attention to the problem of artemisia for treating malaria.

Soon, Tu Youyou and others took mice as models and carried out experimental research on the screening of traditional Chinese medicines (Fig. 31.5). By June 1969, researchers from the Institute of Traditional Chinese Medicine had screened the Chinese medicine such as Clematis chinensis, Portulaca oleracea, Gleditsia sinensis, Artemisia argyi leaf, asarum, capsicum, white pepper, pepper, Huang Dan, realgar, etc. They found that the inhibition rate of pepper extract on plasmodium was as high as 84%, but pharmacological experiments showed that the drug had unsatisfactory inhibitory effect on plasmodium. After screening more than 100 samples by experiments, they had to consider choosing new drugs.

In 1970, "523 Office" arranged the Institute of Traditional Chinese Medicine and the Military Medical Academy (hereinafter referred to as "Military Academy") to carry out cooperative research. The Military Academy sent Gu Guoming to work in the Institute of Traditional Chinese Medicine. Yu Yagang, referring to *The*

Fig. 31.5 Tu Youyou in the laboratory

Collection of Ancient and Modern Books, *Complete Records of Medical Department*, and *Taiping Huimin Heji Administration Prescription*, combed the antimalarial prescriptions included in *Malaria Album* compiled by Shanghai Traditional Chinese Medicine Literature Research Institute in 1965, excluded the prescriptions containing Changshan, and selected the drugs with both unilateral use experience and frequent occurrence in compound prescription, including Aconitum carmichaeli, Mume Fructus, Carapax Trionycis, Artemisia annua, etc., and he made them into corresponding preparations. Gu Guoming sent them to the Military Academy for screening of rat malaria models. Nearly a hundred prescriptions were screened, of which realgar had 90% inhibition of plasmodium and artemisia had 60–80% inhibition of plasmodium, but the effect was unstable. At that time, Yu Yagang paid more attention to realgar, but Gu Guoming thought realgar was toxic and difficult to be approved for use. They turned their attention to Artemisia. However, soon after Yu Yagang was transferred to the chronic bronchitis group, Gu Guoming returned to the Academy of Military Sciences, and they informed the director and team leader Tu Youyou of the results [31]. The Institute of Microbiology and Epidemiology of the Military Academy sent researcher Ning Dianxi to the Institute of Traditional Chinese Medicine to help establish an animal model of rat malaria. Since then, the research on Artemisia has been continued by the Institute of Traditional Chinese Medicine [32].

After that, Tu Youyou re-screened the medicine with higher efficacy that had been screened before. More than 100 samples were screened out this time, including Artemisia, but only with 40% or even 12% inhibition rate. Artemisia, therefore, was abandoned. She then returned to study the ancient literature. In the second half of 1971, she reread Ge Hong's "*Elbow Reserve Emergency Recipe*" of the Eastern Jin Dynasty. She thought why not using the traditional water to boil, realizing that high temperature or enzymolysis may destroy the effective component of Artemisia [30]. This inspiration knocked on the door of artemisinin discovery. She made water-decocted extract, 95% ethanol extract, from mature leaves of Artemisia collected in autumn in Beijing, and the volatile oil was ineffective against malaria in mice. The extract obtained by cold soaking with ethanol and controlling the temperature at 60 °C during concentration will increase the titer of mouse malaria, but will still be ineffective if the temperature is too high. The effect is greatly improved and stable by using ether reflux or cold leaching for extraction. They repeated the experiments and found that the key was to control the temperature below 60 °C. However, the antimalarial efficacy of the isolated artemisinin monomer is stable even though it is boiled in water for half an hour or refluxed in ethanol for 4 h. Therefore, it can be seen that only when the extract coexists with some substances in drugs, will the temperature rise destroy the antimalarial effect of Artemisia [33].

Although the extract of diethyl ether had good antimalarial effect, the problems of unstable potency and large dosage still existed, which also indicated toxicity of the extract. As for the unstable potency, they initially thought it was caused by the mixed variety of Artemisia annua. After analyzing the varieties of Artemisia annua, they learned that all the Artemisia annua they bought were Artemisia annua L produced in the suburbs of Beijing. They further searched for the reason and finally confirmed that the harvest season of Artemisia annua was different, which would affect the potency of Artemisia annua extract. After using the Artemisia annua harvested in the same season, the potency of the effective extract was more stable [34]. To solve the problems of large dosage and toxicity, it was necessary to furthermore extract and purified. They divided the extract into neutral and acidic parts, and found that the acidic part, which accounts for 2/3 of the total, was ineffective and toxic, while the neutral part was concentrated with little toxic and side effects. After repeated experiments, on October 4, 1971, the neutral extract sample of Artemisia annua No. 191 was isolated and obtained. After applying for internal use of the dose 1 g/kg for 3 days, and the inhibition rate on malaria plasmodium in mouse reached 100%, with low-toxic and side effects. Then more experiments were conducted on monkey, which had also achieved a 100% inhibition rate of malaria [35].

On March 28, 1972, the "523 Office" hosted a meeting in Nanjing. Tu Youyou, on behalf of the Chinese Medicine Institute, reported their above research results at the meeting of the "Chinese Medicine and Traditional Chinese Medicine Professional Group." Because it was during the "Cultural Revolution" and the topic was "Adopting Mao Zedong Thought to Guide the Exploration of Antimalaria Chinese Herbal Medicines," the participants were greatly concerned. Then the "523 Office" asked them to go to Hainan epidemic area for clinical observation that year.

After the meeting, the Institute of Traditional Chinese Medicine started the extraction work. Because the pharmaceutical companies were unwilling to undertake the extraction task, the researchers had to turn the laboratory into a workshop and fought day and night. Finally, on the eve of May 1 of that year, the medicine was extracted and the neutral part was made into capsules (Fig. 31.6). Before going to the clinic, they carried out safety experiments. Mice, cats, and dogs were used in the experiment

Fig. 31.6 Working scene of that tear

to observe acute toxicity and its effects on heart, liver, and kidney, respectively. No toxic lesions were found and only a few animals had slight effects on liver transaminase activity. In order to ensure the safety of medication, two more batches of human experiments were carried out. The first batch was taken by Tu Youyou, Yue Fengxian, and Lang Fulin on a trial basis. Upon approval, they were hospitalized in Dongzhimen Hospital in Beijing for observation. The dose was increased from 0.35 to 5 g daily, once a day for 7 days. The second batch consisted of five people, including Zhang Guozhen, Yan Shuchang, Pan Hengjie, Zhao Aihua, and Fang Wenxian. The dosage was increased to 3 g each time, twice a day for 3 days. Routine examinations of ECG, liver function, renal function, chest X-ray, blood, urine, and feces were carried out before, during and after taking the medicine, respectively, and the results showed no obvious influence on their bodies [36].

After human body test, Tu Youyou, Dai Shaode, Cao Qingshu, and other medical teams from the Institute of Traditional Chinese Medicine rushed to Changjiang, Hainan Island on August 24, 1972. At this time, it was the season of high incidence of malaria. They made clinical observation on 11 cases of vivax malaria, 9 cases of falciparum malaria, and 1 case of mixed infection among local and non-local population, while taking 4 cases of chloroquine as control group. At the same time, in Beijing 302 Hospital, 9 cases of vivax malaria were used as the validation group, accounting for 30 cases. Three dosage schemes, large, medium, and small, were effective. The high-dose group had the most obvious curative effect and low recurrence rate, higher than that of the chloroquine group. Apart from vomiting and diarrhea of individual patients, there were no other clinical side effects. After 2 months of clinical observation, it was completely consistent with the laboratory results, indicating that the ether extraction neutral part of Artemisia annua was effective for malaria treatment.

On November 17 of the same year, at the "523" plenary meeting held in Beijing, Tu Youyou reported the above 30 cases of clinical observation results for the medication efficiency, which once again attracted the attention of "523 Office." "523 Office" required them not only to expand clinical verification in 1973, but also to find effective ingredients as soon as possible. The release of the results of this clinical trial started the research on artemisinin. After that, Shandong Institute of Parasitic Diseases and Yunnan Institute of Medicine successively carried out the extraction and separation of artemisinin and clinical trials.

31.3.3 Extraction of Artemisinin

In the second half of 1972, the Institute of Traditional Chinese Medicine organized the extraction of artemisinin. At first, there were only five people participating in the work. The team leader was Tu Youyou. The team members included Ni Muyun, Zhong Yulong, Cui Shulian, and another technician. At that time, the extraction equipment of the Chinese medicine institute was insufficient, so several large cylinders were used as extraction containers (Fig. 31.7). In order to finish clinical verification in the autumn of 1973, Meng Huangrong, Tan Honggen, and others were included

Fig. 31.7 Container used for extracting artemisinin

to participate in the work, and several personnel were temporarily seconded from the research institute to support the extraction work. In the process of extracting, Ni Muyun mixed the neutral part of Artemisia annua ether extract with polyamide, percolated with 47% ethanol, concentrated the percolate, extracted with ether, and then eluted and separated on an alumina chromatographic column, but no monomer was separated. Zhong Yulong considered that neutral compounds should be separated by silica gel column, so she changed to silica gel column and eluted with petroleum ether and ethyl acetate-petroleum ether. On November 8, 1972, three kinds of crystals were eluted from the silica gel column. The first one was the square crystal with large content and the number was "Crystal I;" Then came needle-shaped crystals, numbered "Crystal II," with very little crystal content. Finally, another needle-shaped crystal was obtained with the number of "Crystal III." At the beginning of December, the malaria test on mouse proved that crystal III could turn all plasmodium test result into negative at a dose of 50–100 mg/kg, and was an effective antimalarial monomer in Artemisia annua. These three crystals were named artemisinin I, artemisinin II (Fig. 31.8), and artemisinin III [37].

The next step should have been to carry out systematic toxicity experiments on artemisinin II and then enter clinical trials. They stepped up their work, isolated more than 100 g of Artemisinin II, and conducted acute toxicity experiments in mice. In order to catch up with the onset season of malaria, they did not have time and conditions to conduct more toxicity experiments. The personnel in the institute were divided as to whether they could be used clinically without systematic toxicity test. Finally, the party committee of the institute decided to conduct human body test, and three people (two men and one woman) took the test [37]. No obvious problems were found. From September to October 1973, a medical team composed of Li Chuanjie, Liu Jufu, and others went to Hainan Changjiang Malaria Area again to carry out clinical verification. They first carried out clinical trial observation on

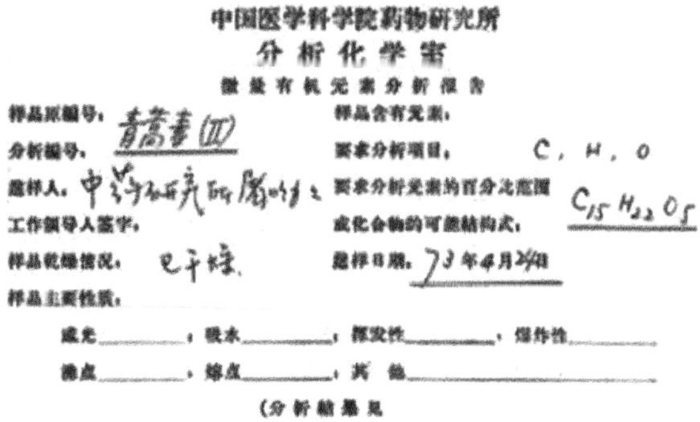

中国医学科学院药物研究所
分析化学室
微量有机元素分析报告

样品原编号：青高素(II)
分析编号：[illegible]
送样人：中药研究所[illegible]
工作领导人签字：
样品干燥情况：已干燥
样品主要性质：
旋光____，吸水____，挥发性____，爆炸性____
沸点____，熔点____，其他____
样品含有元素：
要求分析项目：C，H，O
要求分析元素的百分比范围或化合物的可能结构式：$C_{15}H_{22}O_5$
送样日期：73年4月2[illegible]日
(分析结果见

Fig. 31.8 Report on molecular weight determination results of artemisinin

five cases of falciparum malaria, and the results showed that one case was effective (protozoa over 70,000/mm^3, tablet dosage 4.5 g, defervescence 37 h, protozoa text result turning negative after 65 h, protozoa reappearing 6 days later); two patients stopped taking the medicine due to premature contraction of heart (one of them was first onset, protozoa was over 30,000/mm^3, fever was reduced 32 h after taking the medicine for 3 g, protozoa reappeared 1 day after stopping the medicine, and body temperature increased), and two patients were ineffective. The results were not satisfactory. The reason was immediately found to be the disintegration of the tablets. The artemisinin powder was immediately encapsulated and Zhang Guozhen, deputy director of the business, went to Hainan Changjiang for further experiments. Three cases of vivax malaria were observed among the non-local people. The total dosage of the capsule was 3–3.5 g. The average antipyretic time was 30 h. After 3 weeks of reexamination, 2 cases were cured and 1 case was effective (protozoa reappeared in 13 days). All of them were effective, which was confirmed that artemisinin II is indeed an effective antimalarial medication.

Shandong Institute of Parasitic Diseases drew lessons from the experience of the Institute of Traditional Chinese Medicine and extracted Artemisia annua (Artemisia annua) from Shandong with ether and alcohol. Animal tests showed that the antimalaria results on mouse were consistent with the report of the Institute of Traditional Chinese Medicine and made a written report to the “523 Office” on October 21, 1972. Later, they cooperated with Shandong Institute of Traditional Chinese Medicine and began to separate effective monomers. At that time, there were only two people doing related work. In November 1973, Shandong Institute of Traditional Chinese Medicine extracted 7 kinds of crystals from Artemisia annua L produced in Tai’an. Crystal No. 5 was an effective antimalarial crystal named “Artemisia annua L.”. In the first half of May 1974, the collaboration group of artemisinin in Shandong Province conducted clinical observation on 10 cases of vivax malaria patients in Zhu Zhuang Brigade of Chengguan East Commune in Juye County, Shandong. The dosage of artemisinin was 0.2 g × 3 d, 0.4 × 3 d, 5 cases each. The conclusion was that Artemisinin was a better quick-acting antimalarial drug, which seemed to be able to be used as an emergency drug. No obvious side effects were found in the treatment process, but the effect was not complete enough and the recurrence rate was high. It was considered that in order to effectively control the recurrence rate, it was difficult to increase the dosage of artemisinin alone and its compatibility with other antimalarial drugs should be considered.

At the end of 1972, Fu Liangshu, director of the "523 Office" in Kunming, went to Beijing to attend the annual meeting of the heads of the "523 Office" to learn about the researches on Artemisia annua by the Institute of Traditional Chinese Medicine. Upon his return, he called a meeting of relevant researchers from the Yunnan Institute of Medicine and conveyed the news, instructing the general screening of Artemisia plants of Compositae by taking advantage of the rich local plant resources. During the Spring Festival in 1973, Luo Zeyuan, from Yunnan Institute of Medicine, discovered a highly odorous Artemisia plant on the campus of Yunnan University. He collected plenty of them and took them back to the institute to dry and extract. At that time, he did not know this plant. Liu yuanfang, who studied plants, told him that it was "wormwood." Ethyl ether extract of Artemisia absinthium has antimalarial effect and high-repetition rate. In April 1973, they separated an effective antimalarial monomer and named it Artemisia dracunculus crystal III. After that, they sent the plant specimen of Artemisia dracunculus to Professor Wu Zhengyi, a botanist at Kunming Institute of Botany, Chinese Academy of Sciences, for his expert opinion. They determined that the name of this kind of Artemisia dracunculus was Artemisia annua L. F. Macrocephala Pamp, and renamed the crystal as artemisinin. Later, he bought Artemisia annua from Sichuan Chongqing Medicine Company, which originated from Youyang, Sichuan. The original plant was Artemisia annua L, from which higher content of artemisinin can be separated.

On September 8, 1974, Lu Weidong, Wang Xuezhong, and Huang Heng from Yunnan Clinical Cooperation Group took artemisinin to Yunxian and Chafang area for clinical observation. Since malaria was no longer common in these two areas, they had only treated one case with vivax malaria. Beijing Institute of Traditional Chinese Medicine failed to extract artemisinin for clinical use as required at the beginning of the year, so Liu Pu from the Institute was sent as an observer to join the Yunnan Clinical Collaboration Group. Liu Pu arrived in Yun County on October 6. They learned that there were falciparum malaria patients in Gengma County. Lu Weidong, Wang Xuezhong, and Liu Pu arrived in Gengma County on October 13. In Gengma County, they met a medical team led by Guangdong College of Traditional Chinese Medicine, which was carrying out treatment of cerebral malaria. Li Guoqiao had rich experience in malaria treatment. The three of them learned from Li Guoqiao and treated one vivax malaria and one falciparum malaria at the same time. After that, Li Guoqiao was entrusted to carry out clinical trials of artemisinin under the coordination of "523 Office." At the end of October and the beginning of November, Li Guoqiao led Guangdong's "523 team" to treat 3 patients with falciparum malaria and 2 patients with vivax malaria, and conducted medication trials, all of which were effective. Li Guoqiao decided to go to Nanla Hospital in Cangyuan County to look for cerebral malaria patients. This time, the effectiveness of medication of a total of 18 cases was verified, including 14 cases of falciparum malaria (1 case of cerebral malaria in pregnant women and 2 cases of icteric malaria), 4 cases of vivax malaria, and 3 cases of patients previously treated by Yunnan clinical cooperation group, all of which were successful treatment. Li Guoqiao believed that artemisinin is a quick-acting antimalarial medication and the first dose of 0.3–0.5 g can quickly control protozoa development. The rapid reproduction of protozoa and the rapid recurrence of symptoms may be due to the rapid excretion of the medication (or its rapid conversion into other substances in the body) and the short duration of the effective concentration in blood, which fails to completely kill protozoa. Li Guoqiao proved for the first time the curative effect of artemisinin on dangerous malaria, proposed that artemisinin has the characteristics of high efficiency, which can be used to treat malaria patients with severe symptoms, and suggested that artemisinin be made into injection as soon as possible [37].

The Institute of Traditional Chinese Medicine, Shandong Institute of Traditional Chinese Medicine, and Yunnan Institute of Medicine all obtained antimalarial effective monomers, named artemisinin, arteannuin (Shandong), and Artemisia annua L (Yunnan), respectively, which were identified as the same substance at that time. However, due to the unsatisfactory clinical effect of crystals separated from traditional Chinese medicine in academia sinica, and the indication of cardiotoxicity, it was being questioned. But according to their molecular weight and molecular formula, it should be artemisinin monomer. For a long time, there was no unified nomenclature for Artemisinin. The Chinese Pharmacopoeia Committee fully considered the customary usage of traditional Chinese medicines and included artemisinin in the second part of the 1995 edition of the Pharmacopoeia of the People's Republic of China. Although there is still controversy at present, it is believed that consensus will be reached with the in-depth study on the history of artemisinin.

31.3.4 Measurement of Artemisinin Chemical Structure

At the end of 1972, Tu Youyou and others began the structural determination of Artemisinin II. They were certained that Artemisinin II has a white needle crystal with melting point of 156–157 °C, optical rotation [¦Á] 17d = +66.3 (§^{3} = 1.64, chloroform). Chemical reaction confirmed that Artemisinin II has no nitrogen element and no bond. They entrusted the Institute of Medicine of the Chinese Academy of Medical Sciences to carry out elemental analysis. The April 1973 report showed that the possible molecular formula was $C_{15}H_{22}O_5$, the elemental analysis was 63.72% C, 7.86% H, and the relative molecular mass was 282. Later, under the guidance of Professor Lin Qishou of Beijing Medical College, it was inferred that Artemisinin II may be a sesquiterpene lactone, an antimalarial medication of a new structural type.

From May 28 to June 7, 1973, a symposium of leading comrades of malaria prevention and control research group was held in Shanghai, giving clear instructions to the determination of the chemical structure of effective antimalarial components of Artemisia annua: "While improving the promotion and use of dosage forms, cooperation in organizing efforts should be strengthened to work out the chemical structure and carry out the research on chemical synthesis in 1974."

Due to the shortage of chemical research strength and instruments and equipment in the Institute of Traditional Chinese Medicine, it was difficult to complete all structural identification work alone. Tu Youyou seeks cooperative units. She learned from the literature that Professor Liu Zhujin from the Shanghai Institute of Organic Chemistry of the Chinese Academy of Sciences (hereinafter referred to as "the Institute") was engaged in sesquiterpenes research, so she brought the relevant files to Shanghai to contact the Institute and was received by Chen Yuqun. In January 1974, Chen replied that the Institute of Traditional Chinese Medicine had agreed to send one person to work together. In February 1974, the Chinese Medicine Institute sent Ni Muyun with research materials and artemisinin to work in the organic institute. At that time, due to the change of research direction in Liu Zhujin, he moved to the office in charge of Zhou Weishan and was handed over to Wu Zhaohua and Ni Muyun for joint research. At first, he mainly repeated some experiments already done in Beijing and then did some research on chemical reactions and spectral data [36]. After that, Zhong Yurong, Fan Jufen, and Liu Jingming successively went to the organic institute to participate in the determination of Artemisinin II structure. They told Tu Youyou who stayed in Beijing about the progress of the structural measurement, Tu Youyou communicated with Lin Qishou or Professor Liang Xiaotian, etc. and consulted them, then fed back the results to Shanghai, providing reference opinions for the structural measurement in Shanghai. Wu Zhaohua will also inform Zhou Weishan of some experimental results, and Zhou Weishan would come to discuss during lunch break or after work in the evening.

The structural measurement of artemisinin had encountered many difficulties. The main problem was how to arrange 5 oxygen atoms in the skeleton of 15 carbon atoms. From the nuclear magnetic resonance spectrum, there is only one carbonyl carbon and one proton linked to the same carbon as the oxygen atom in the whole molecule. Peroxide has also been considered, but it is a very stable structure, which goes against the general idea that peroxide is easy to decompose.

When Li Ying of Shanghai Institute of Medicine attended the meeting in Chengdu in April 1975, he heard that Yu Dequan, Institute of Medicine, Chinese Academy of Medical Sciences, reported the determination of chemical structure of antimalarial monomer yingzhuajia A, which contained peroxygen groups, and told the news to Wu Yulin from the Institute of Organic Chemistry. Wu Yulin was inspired to speculate that artemisinin may also be peroxide. Through qualitative and quantitative analysis with Wu Zhaohua, it was proved that artemisinin does contain peroxide groups, suggesting that the peroxide groups are in possible structure of lactone ring [38].

While the Institute of Traditional Chinese Medicine and researchers from the Institute of Organic Medicine conducted chemical structure determination, in 1975, Tu Youyou contacted Liang Li and Allen Lee of the Institute of Biophysics of the Chinese Academy of Sciences (referred to as the "Institute of Biophysics") to determine the chemical structure of artemisinin by using the then advanced X-ray diffraction method in China. According to Tu Youyou, the chemical structure of artemisinin was confirmed on November 30, 1975. In order to ensure the accuracy, on January 14 and 18, 1976, the Institute of Traditional Chinese Medicine and the Institute of Biophysics invited Professor Liang Xiaotian to discuss and confirm. On January 26,

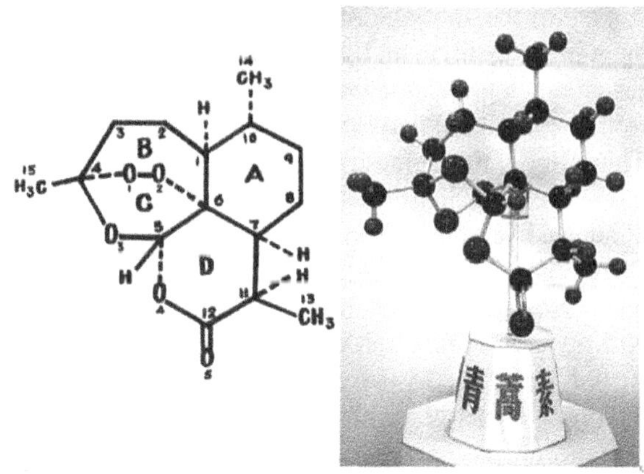

Fig. 31.9 Artemisinin structure

1976, Tu Youyou and Li Peng flew to the Shanghai Organic Institute to report the determination of artemisinin structure. Zhou Weishan, Wu Yulin, Wu Zhaohua, and others attended the meeting and confirmed [36]. In March 1977, the chemical structure of artemisinin was announced in the name of artemisinin collaboration group [39]. On the basis of fine measurement of anomalous scattering intensity data, Liang Li calculated with the computer of Beijing Computing Center and established its absolute configuration (Fig. 31.9) [40].

According to the *Late Report*, due to the fact that Artemisinin had not been extracted from Beijing Traditional Chinese Medicine Institute for a period of time, Shandong and Yunnan had provided some artemisinin crystals with higher purity for chemical structure determination through the arrangement of the national "523 Office [29]."

"Artemisinin 523 Secret News" said that Yunnan Institute of Medicine provided 100 mg of crystals to Institute of Traditional Chinese Medicine in autumn 1973 [41].

31.3.5 Synthesis of Artemisinin Derivatives

Artemisinin was discovered in 1972, and more than 6,000 clinical tests were carried out, which confirmed the efficacy of artemisinin. In 1986, artemisinin obtained a new drug certificate and successively developed artemisinin suppositories and tablets. Of course, artemisinin was not perfect. It was found in clinical application that the effectiveness by internal usage was low, solubility was small, and half-life period was short, which affected the curative effect. Therefore, researchers had carried out research on artemisinin derivatives.

As early as 1973, Tu Youyou and her colleagues conducted researches on artemisinin derivatives in order to determine the existence of functional group of artemisinin, namely carbonyl group. They found that the carbonyl peak disappeared after artemisinin was reduced by sodium borohydride, confirming the existence of carbonyl group, and obtained a compound with molecular formula $C_{15}H_{24}O_5$ and relative molecular mass 284, namely dihydroartemisinin. At the same time, they found that the peroxy group in artemisinin structure is the main active group for antimalaria. On the premise of retaining the peroxy group, modifying part of the structure can improve the biological activity. This result was publicly reported in the internal meeting of antimalaria research held in Henan Province [42], which provided ideas for the development of new drugs.

Since 1976, under the arrangement of "523 Office," the Shanghai Institute of Medicine of the Chinese Academy of Sciences had carried out the synthesis of artemisinin derivatives, and Guilin Pharmaceutical Factory had also carried out this work. In 1977, Liu Xu and others developed sodium artesunate. In 1978, Li Ying and others developed artemether [43]. In 1987, these two drugs were approved for production as a class of new drugs. Artemether injection and artesunate were listed in the World Health Organization's Essential Medicines Catalogue in 1997 and 2002, respectively [44]. In 1985, Tu Youyou and others resumed the previous dihydroartemisinin and further studied its structure-activity relationship.

In 1992, dihydroartemisinin and its tablets were developed and obtained a new drug certificate. Their trade names were Ketaixin and Ketaixin tablets. Artemisinin derivatives overcome the disadvantage of poor solubility of artemisinin and can be administered clinically by oral administration, injection, transdermal absorption, and other ways. When treating malaria in children, the drugs also show the advantages of low toxicity and wide antimalarial spectrum, and artesunate is especially effective in rescuing cerebral malaria and various critical malaria [45]. Since then, many scholars at home and abroad had carried out structural modification and total synthesis of artemisinin, which is still an active research field. In recent years, the biosynthetic pathway of artemisinin has attracted the attention of many researchers [46].

31.3.6 Pharmacological Research on Artemisinin

Since the advent of artemisinin drugs, researchers had conducted in-depth studies on their pharmacology and mechanism of action. The results show that artemisinin has a strong killing effect on all plasmodium, including chloroquine-resistant falciparum malaria and multi-drug-resistant malaria. The mechanism mainly acts on the membrane structure of plasmodium, causing its bubble membrane, nuclear membrane, and plasma membrane to be destroyed, mitochondria to swell and shrink, and inner and outer membranes to peel off. Artemisinin and its derivatives can block the supply of plasmodium nutrition by affecting the function of the membrane-mitochondria, thus achieving the purpose of antimalaria. Studies have shown that artemisinin drugs can rapidly kill early gametophytes of plasmodium, inhibit gametophytes of various stages, and interrupt the development of immature gametophytes. This inhibitory effect on gametophyte is not available in other antimalarial drugs, which is conducive to controlling malaria epidemic.

Artemisinin drugs not only have good killing effect on plasmodium, but also have certain inhibiting effect on other parasites. Studies have proved that artemisinin has antischistosomiasis effect and can kill schistosome in larval stage during the whole period of taking artemisinin drugs. In 1996, Artemether and artesunate were approved as schistosomiasis prevention drugs in China, and this achievement has attracted the attention of international peers. In addition, artemisinin drugs have effects on various parasitic diseases such as focusosis, paragonimiasis, toxoplasmosis, pneumocystis carinii, and leishmania infection.

Since the 1990s, the antitumor effect of artemisinin drugs had been reported at home and abroad. The antitumor effect in vitro is remarkable, and the antitumor spectrum is wide. In addition, there are reports on the antifungal, antiarrhythmic, and local anesthetic effects of artemisinin drugs [47].

In addition to the above work, Chinese researchers have also carried out research on the distribution of Artemisia annua resources, cultivation and breeding, chemical composition, analysis and determination, quality standards, clinical efficacy, and other aspects. Especially on the research of artemisinin compound preparation, Li Guoqiao made a compound preparation with artemisinin drugs and other antimalarial drugs such as benflumetol, which enhanced the curative effect, reduced drug resistance, and received good results. Artemisinin is commonly used as its derivative due to its low internal usage effectiveness. However, Li Guoqiao makes full use of the advantages of artemisinin such as quick effect, high efficiency, low toxicity, low price, long shelf life. When used in combination with other antimalarial drugs, it reduces the treatment cost and enables the poor to benefit from artemisinin [48] and obtain treatment.

After the chemical structure of artemisinin was published in 1977, it was included in American Chemical Abstracts (CA) and attracted the attention of international academic. In 1981, the World Health Organization sent a letter to the Chinese Ministry of Health requesting the first international conference on artemisinin to be held in China. During the meeting, participants believed that the discovery of artemisinin not only added new antimalarial drugs, but also had more important significance in finding the unique chemical structure of this compound, which pointed out the direction for further design and synthesis of new drugs. Artemisinin has been used clinically for more than 30 years.

Over the past years, thousands of lives have been saved and great contributions have been made to human health. The discoverers and researchers of artemisinin were respected all over the world. In January 2004, the discoverers of artemisinin won the Thai Mahilon Prize in Medicine. In 2011, Tu Youyou won the Lasker Prize for the discovery of artemisinin. In 2015, Tu Youyou became the first Chinese scientist to win the Nobel Prize in Science and the first Chinese scientist to win the Nobel Prize in Physiology or Medicine. It is also the highest award so far won by the Chinese medical community.

References

1. Frederick F. Cartwright, Michael Biddiss. 2004. *Disease and History*. Chen Zhongdan, Zhou Xiaozheng (Trans.), 11. Shandong Pictorial Publishing House.
2. Shengsheng, Gong. 1996. A Research on the Prevalence and Distribution of Malaria in China for the Period from Pre-Qin to Han. *Journal of Central China Normal University (Natural Sciences)* 4: 489–491.
3. Zhigui, Xia, Feng Jun, and Zhou Shuisen. 2013. Malaria Situation in the People's Republic of China in 2012. *Chinese Journal of Parasitology and Parasitic Diseases* 6: 413–418.
4. Jun, Ding. 2012. The Research Progress in the Status of Malaria Prevalence. *Chinese Journal of Public Health* 5: 717–718.
5. Zhang Zhongjing. 2000. *Treatise on Febrile and Other Diseases*. Liu Shien, Mao Shaofang (Reviewed), 61. Hualing Press.
6. Zhang Zhongjing. 2000. *Treatise on Febrile and Other Diseases*. Liu Shien, Mao Shaofang (Reviewed), 112. Hualing Press.
7. Frederick F. Cartwright, Michael Biddiss. 2004. *Disease and History*. Chen Zhongdan, Zhou Xiaozheng (Trans.), 4. Shandong Pictorial Publishing House.
8. Louis Pfister. 1997. *Notices biographiques et bibliographiques sur les jésuites de l'ancienne mission de Chine (1552–1773)*. Mei Chengqi, Mei Chengjun (Trans.), 494. Bureau Sinologique de Zi-ka-we.
9. *Archival Documents Concerning the Family of Cao, Administrator of Jiangning Textile Industry*. 1975. Zhonghua Book Company, 98–99.
10. Zhao Xuemin. 2007. *Supplement to Compendium of Materia Medica*, 194. China Traditional Chinese Medicine Press.
11. Ernst Bäumler. 2005. *Die großen Medikamente Rorscher und ihre Entdeckungen schenken uns Leben*. Zhang Rongchang (Trans.), 153. Guangxi Normal University Press.
12. Arturo Castiglionni. 2003. *History of Medicine*. Cheng Zhifan (Trans.), 747–748. Guangxi Normal University Press.
13. Kui, Zhang. 1953. *Chemical Treatment of Malaria*, 1–3. Shanghai: Shanghai Science and Technology Press.
14. Zhang Changsha. 1954. *Modern Chinese Medicine Research*. China Science Book and Instrument Company, 141–149.
15. Joel G. Hardman, Lee E. Limbird (Eds.). 2004. *Goodman & Gilman's The Pharmacological Basis ofTherapeutics (10th)*. Jin Youyu (Trans.), 828–829. People's Medical Publishing House.
16. WHO. 2013. Publishes the World Malaria Report 2013. *Disease Surveillance* (12), 1006.
17. The arrangement group of Ma Wangdui Han Tomb Silk Manuscript complied. Ma Wangdui Han Tomb Silk Manuscript: Prescriptions for Fifty-two Diseases. Cultural Relics Press, 1979: pp. 88–89; Shang Zhijun. *Drug Annotations for Prescriptions for Fifty-two Diseases*. Scientific Research Department of Wannan Medical College, 1985.
18. [Japan] The collection of Sen Lizhi: Shennong's Classic of Materia Medica. Wen Zhi Yao Bao Zang Zi, 1854.
19. [Song] Tang Shenwei compiled and Shang Zhijun and Zheng Jinsheng, etc. 1993. *Classified Materia Medica* (Revised), 288. Huaxia Publishing House.
20. [Tang] Chen Zangqi wrote and Shang Junzhi compiled and annotated. 2002. *A Supplement to Materia Medica*, 365. Anhui Science and Technology Press.
21. Shen Kuo, Hu Daojing (Collated). 1987. *The Collation of Meng Xi Bi Tan*, 873. Shanghai Chinese Classics Publishing House.
22. Kou Zongshi (in the Song Dynaty), Yan Zhenghua et. al. (collated). 1990. *Ben Cao Yan Yi*, 71. People's Medical Publishing House.
23. [Ming Dynasty] Xu Xizhou. Rough Comments on Medicinal Properties ·Artemisia Volume III. The 30th year in the reign of Jiajing.
24. Youyou, Tu. 2009. *Qinghao and Artemisinin-Based Drugs*, 6, 8. Beijing: Chemical Industry Press.

25. Youyou, Tu. 1987. Research on the Chinese Medicine Qinghao. *Traditional Chinese Medicine Journal* 1: 2–5.
26. Yourun, Lin. 1991. A Preliminary Study on Artemisia Argyi Plants in Ancient Chinese Pharmacopoeia. *Bulletin of Botanical Research* 1: 1–24.
27. Shilin, Hu, and Xv Youlin. 2005. Commemorate the 30th Anniversary of Artemisinin. *Modernization of Traditional Chinese Medicine and Materia Materia-World Science and Technology* 2: 1–2.
28. Yuhuang, Zhao. 2004. *Qi Zhou Yao Zhi*, 46. Fujian: Fujian Science and Technology Publishing House.
29. Zhang Fangjian. 2006. Editor-in-Chief: "Late Report-Project 523 and Artemisinin Research and Development Documentary" Preface. Yangcheng Evening News Publishing House.
30. Tu Youyou. 2009. *Artemisia Annua and Artemisinin Drugs*, 1. Chemical Industry Press.
31. Li Runhong. 2011. "Task 523" and rediscovery of artemisia annua's antimalarial effect. *Journal of Chinese History of Science and Technology*, vol. 32, 4, 488–500.
32. Zhang Fangjian. (Ed.) 2006. "Late Report-The 23rd Project and Artemisinin Research and Development Documentary" Preface. Yangcheng Evening News Publishing House.
33. Tu Youyou. 2009. *Artemisia Annua and Artemisinin Drugs*, 34. Chemical Industry Press.
34. Li Runhong. Rao Yi and Zhang Daqing. 2013. "523 Tasks" and Historical Exploration of Artemisinin Discovery. *Dialectics of Nature Newsletter* (1), 112.
35. Tu Youyou. 2009. *Artemisia Annua and Artemisinin Drugs*, 34–35. Chemical Industry Press.
36. Tu Youyou. 2009. *Artemisia Annua and Artemisinin Drugs*, 36–38. Chemical Industry Press.
37. Runhong, Li, Rao Yi, and Zhang Daqing. 2013. "523 Tasks" and Historical Exploration of Artemisinin Discovery. *Dialectics of Nature Newsletter* 1: 107–119.
38. Li Ying and Wu Yulin. 2007. Advances in Pharmaceutical Chemistry and Pharmacology of Artemisinin Compounds. See Li Ying's Editor-in-Chief. *Artemisinin Research*, 4. Shanghai Science and Technology Press.
39. Artemisinin structure research collaboration group. 1977. artemisinin, a new sesquiterpene lactone. *Scientific Bulletin* 3: 142.
40. Artemisinin Collaboration Group. 1979. Institute of Biophysics, Chinese Academy of Sciences. Crystal Structure and Absolute Configuration of Artemisinin. *Chinese Science* (11), 1114–1128.
41. Li. Artemisinin 523 Secret News. New West, 42, December 2011.
42. Tu Youyou. 2009. *Artemisia Annua and Artemisinin Drugs*, 187–194. Chemical Industry Press.
43. Yulin, Wu. 2009. Artemisinin-Revelation of History and Reality. *Chemical Progress* 11: 2365–2371.
44. Li Ying (Ed.) 2007. Artemisinin research Shanghai science and technology publishing house. *Artemisinin Research Memorabilia*, 1–6.
45. Luo Wei, Yang Liu, Conglin, Sun Li, Guo Chun. 2012. Research Progress of Artemisinin and Its Derivatives. *Chinese Journal of Medicinal Chemistry* (2), 155–163.
46. Jialian, Zhou, and Yang Henglin. 2008. Current Research Situation and Development Trend of Antimalarial Drugs. *Chinese Journal of Pathogenic Biology* 11: 865–866.
47. Luo Wei, Yang Liu, Conglin, Sun Li and Guo Chun. 2012. Research Progress of Artemisinin and Its Derivatives. *Chinese Journal of Medicinal Chemistry* (2), 155–163.
48. Shiying, Jin, and Liu Shuting. 2010. Professor Li Guoqiao Talks about Artemisinin and the Law of Eliminating Malaria. *Modern Hospital* 5: 83–84.

Yahua Niu Born in Hohhot, Inner Mongolia in 1961, graduated from Nanjing Pharmaceutical College in 1983 with a Bachelor degree of Science; in 1989 she graduated from Inner Mongolia Normal University with a master's degree in science; in 2005, she graduated from Northwestern University with a doctorate in science and technology history. He once served as director of the editorial department of the Journal of Inner Mongolia Medical College. She is currently a researcher and doctoral supervisor of the Institute of Traditional Chinese Medicine Information, Chinese Academy of Traditional Chinese Medicine. From 1994 to 1995, she has been studied at Japan Iwate Medical University.

Research interests: Medical history; Chinese medical literature.

Major works: Papers: A Study on Two Earliest Translations of Western Anatomy; Study on Shen San Zi Fang.

Biography of Peng Sixun;

Notes on *Two kinds of Chinese and Western Medical Books; Western Medicine introduced in the Ming Dynasty*;

Chief editor of *The Collection of Traditional Chinese Medicine Classics in Qifen Shi.*

Chapter 32
Hybrid Rice

Yeyun Xin

Produced by crossbreeding two rice varieties with certain genetic differences and good complementary traits, the hybrid rice is the first generation of crossbreeds with heterosis.

32.1 Introduction

As one of the world's major food crops, rice is the staple food of more than half of the world's population, which directly affects the world's food security.

Heterosis is utilized to improve crop yield and quality.

Heterosis is produced by crossing two parents with different heredity, and the hybrid generation is superior to their own parents in growth potential, viability, stress resistance, yield, and quality. Heterosis utilization is applying the first generation of hybrid to agricultural production in order to obtain the greatest economic benefits. Heterosis was discovered as early as 2,000 years ago when the cross of horses and donkeys produced mules in ancient China.

In the 1920s and 1930s, the United States adopted the suggestion of D. F. Jones, a maize geneticist and breeder, and carried out maize double crossing breeding. The popularization area of the hybrid maize reached 0.1% of the sown area of maize in the United States (about 3,800 hm^2). It pioneered the utilization of the plant heterosis. In 1954, J. C. Sterphens bred a sterile line 3197A of sorghum by crossing the West African sorghum with the South African sorghum, and bred restorer lines among Leitbayan 60 sorghum varieties. Using the "three-line method" to prepare the sorghum hybrids for production and application, he set a good example for heterosis utilization of cross-pollinated crops.

Y. Xin (✉)
China National Hybrid Rice R&D Center, Changsha, China
e-mail: xinyeyun@hhrrc.ac.cn

J. Hua and L. Feng (eds.), *Thirty Great Inventions of China*,
https://doi.org/10.1007/978-981-15-6525-0_32

The research on utilization of rice heterosis began in the early twentieth century. In 1926, J. W. Jones of the United States first proposed that rice had heterosis, which attracted the attention of breeders all over the world. Since then, studies on Rice Heterosis have been reported by India's Kadem (B. S. Kadem, 1937), Malaysia's Brown (F. B. Broun, 1953), Pakistan's Alim (A. Alim, 1957) and Japan's Okada Kuan (1958).

Scientists began their research on the utilization of Heterosis in Rice by breeding sterile lines. In 1958, Katsuo Kiyoshi of Tohoku University crossed the Chinese Mango wild rice with the Japanese japonica rice Fujisaka No. 5. After continuous backcross, the Fujisaka No. 5 male sterile line with the cytoplasm of Chinese Mango wild rice was bred. In 1966, Yoichi Shincheng of University of the Ryukyus crossed the Chinese Japonica Rice Taichung 65 with the Indian spring indica Chinsura Baoluo II. After continuous backcross, the Taichung 65 male sterile line with the cytoplasm of Chinsura Baoluo II was bred. In 1968, the Fujisaka 5 male sterile line with the cytoplasm of Myanmar Reed rice was bred by crossing the Myanmar indica reed rice with the Japanese japonica rice Fujisaka 5 by Watanabe of Japan Agricultural Technology Research Institute. However, none of these sterile lines could be used in production.

In 1964, Yuan Longping began to study the utilization of rice heterosis. In 1973, he successfully matched "three lines" and bred Nanyou 2 and other combinations, which were applied in production. Thus, it was clear that apart from the cross-pollinated crops and the often-cross-pollinated crops, the self-pollinating crops also had strong heterosis. Production practice proved that the hybrid rice had strong heterosis and great potential to increase yield, which was about 20% higher than the main conventional varieties. By 2012, hybrid rice had been popularized over 6 billion mu in China, and the total yield of the hybrid rice had increased by more than 60 billion kg.

In recent years, the annual planting area of the hybrid rice has exceeded 240 million mu, and the planting area of hybrid rice in China has exceeded 57%, accounting for 65% of the total rice production in China. The annual production of the hybrid rice is about 2.5 million tons, which can feed an additional 70 million people every year. China's hybrid rice is a great pioneering work, a primitive innovation in the history of rice breeding, which is in the leading position in the world. According to the data of FAO, the yield of the hybrid rice is 15–20% higher than that of the conventional varieties. In China, although the rice planting area dropped from 36.5 million hm^2 in 1975 to 30 million hm^2 in 2011, it can still meet the food needs of more than 1 billion people, thanks to the large-scale cultivation of the hybrid rice. At the same time, the hybrid rice has also been exported to more than 40 countries in Africa, America, and Southeast Asia, which has made outstanding contributions to ensuring the world food security.

32.2 Research and Development of the "Three-Line" Hybrid Rice

Produced by crossbreeding two rice varieties with certain genetic differences and good complementary traits, the hybrid rice is the first generation of crossbreeds with heterosis. Generally, hybrid rice refers to the first generation of crossbreeds which is formed by crossing two male sterile lines with different genetic backgrounds and restorer lines. At present, the hybrid rice widely promoted mainly uses the male sterile lines as genetic tools. China is the first country in the world to successfully develop and promote hybrid rice. The hybrid rice has a high degree of individual heterozygosity, and character segregation occurs in hybrid offspring, so annual seed production is required.

By chance, Yuan Longping began to study the hybrid rice in the early 1960s. He found a crop of rice which stood out in a large paddy field. As a result of segregation of character in the second year, he realized that it was "natural hybrid rice." Rice is a self-pollinating crop. The prevailing genetic viewpoint at that time was that the self-pollination of the cross-pollinating plants had degeneration phenomenon, and the hybridization had heterosis; the self-pollinating plants had no degeneration phenomenon and the hybridization had no heterosis. Enlightened by the "natural hybrid rice," Yuan Longping proved that rice did have heterosis through further experiments. He broke through the prevailing genetic viewpoint at that time and decided to utilize the heterosis of rice to raise the per unit yield.

Rice is monoecious with very small flower organs. The hybrid seeds produced by artificial hybridization cannot meet the needs of field production. Drawing on the experience of utilizing heterosis of maize and sorghum, Yuan Longping conceived the technical route of "three-line method," i.e., to breed a rice male sterile line and use the maintainer line to make the male sterile line propagate continuously; to breed a restorer line to restore the fertility of the male sterile line and generate heterosis. In this way, "three lines" matching can be realized to achieve the purpose of utilizing the rice heterosis in production.

On June 20, 1964, Yuan Longping started his research from searching for the male sterile plants of rice as he did on the hybrid maize and sorghum. At that time, the experimental field of the early rice in Anjiang Agricultural School of Hunan Province was heading and flowering. Yuan Longping carefully inspected it plant by plant for 15 days successively, but there was still no breakthrough. On July 5, after observing 140,000 panicles of rice, he finally found the first male sterile plant from the "Dongting Early Indica" variety. This meant that a key first step was taken to overcome the difficulties in hybrid rice breeding.

During the flowering period of the rice panicles (June to July) from 1964 to 1965, Yuan Longping inspected hundreds of thousands of the rice panicles. Six male sterile plants were found in four cultivars, i.e., Dongting early indica, Nante, Zaojing 4, and Shengli indica.

According to the situation of pollen abortion, there were three types: (1) non-pollen type (two plants, found in Shengli indica). The anthers were small, thin, white with non-cleavage, and contained no pollen or only a small number of very fine grains. It was at a complete male sterility, referred to as "indica non-cleavage." (2) Pollen abortion type (two plants, found from Nante No.): the anthers were small, yellow-white with non-cleavage. Pollens were few and incomplete, most of which were irregular and shrunken in shape and significantly smaller than the normal pollen. There was no blue-black reaction when exposed to potassium iodide solution, and they were completely male sterile. (3) Anther degeneration type: found in Zaojing 4 and Dongting early indica rice, the anthers were highly degenerated, the size of which was only 1/4 to 1/5 of the normal, and there was no pollen or few iodine pollens. They served as the starting point for the study of breeding "three lines." After two years of experiments, Yuan Longping had a better understanding of the rice male sterile materials and accumulated a lot of research data. In 1966, he wrote his first important paper, "The Male Sterility of Rice," published in the fourth issue of *Science Bulletin* in February 1966. In this paper, he elaborated on the male sterility of rice and put forward an assertion that "if we want to make use of the heterosis of rice, we should first make use of the male sterility." It was further predicted that the male sterile lines, maintainer lines, and restorer lines could be obtained through breeding, and the "three-line" matching could be realized, which made it possible to utilize the first-generation advantages of the hybrid rice and would bring about a significant increase in agricultural production.

During the Cultural Revolution (1966–1976), Yuan Longping's devoted research was severely impacted. More than 60 test bowls he used to cultivate the hybrid rice experimental seedlings were all smashed, and there were the "seedling destruction incidents" therein the male sterility experimental seedlings were pulled out and thrown into the wells, but these did not shake Yuan Longping's determination to study the hybrid rice at all.

In order to follow the instruction of the State Science and Technology Commission and help Yuan Longping to do a good job in the research of the rice male sterility, Hunan Science and Technology Commission selected Li Bihu and Yin Huaqi, the students as assistants of Yuan Longping, and organized a research group on the rice male sterility headed by Yuan. The establishment of the scientific research team which could ensure the experimental funds was a great help for Yuan to speed up the research. They used a wide range of cross-testing methods to select the male sterile lines and their maintainers from the existing rice male sterile plants. Since 1968, they have carried experimental materials, in order to speed up the breeding generation, to Guangxi, Yunnan, Guangdong, and Hainan Island every winter and spring to continue their research work. In this way, breeding from south to north increased from one generation per year in the past to three generations per year, and the research efficiency doubled. In 1969, they crossed more than 1000 varieties with the natural male sterile materials by extensive crossbreeding and "onion formula," and made more than 3800 combinations. They bred the pollen-free "Nanguang sticky" male sterile materials (referred to as "C system" for short), but the total sterility rate was less than 100%.

Because there was little progress in the artificial breeding of the sterile material (sterile material of C system), Yuan Longping analyzed and summarized several years' research. He believed that the problem was that most of the rice varieties tested and crossed were similar to "Nanguanglian." Old ancestors were "short-footed Nante" or "dwarf sticky." The genetic relationship between the original material (C system material) and the mating parent material was too close, so there was little genetic difference. Referring to the successful experience of breeding hybrid sorghum of which the male sterile line was using West African sorghum as female parent and South African sorghum as male parent, he believed that geographical distant hybridization could cause physiological disharmony and sterility. Therefore, he decided to search for distant parent materials for hybridization and distant hybridization.

So, Yuan Longping decided to create a new technology in sterile materials by realizing "distant hybridization between wild rice and cultivated rice," and decided to find a breakthrough from the wild rice. In the summer and autumn of 1970, Yuan Longping went to Sanya, Hainan with his assistants, to breed and collect the wild rice resources. On November 23, Feng Keshan, a technician of Hainan Red Farm, together with Li Bihu, inspected the heading and flowering of the wild rice in a common wild rice field in the pool marsh area of railway culvert between the farm and the Sanya Airport Highway. He found that the anthers of the three panicles were somewhat abnormal. The anthers were thin and arrow-shaped, light-colored and waterlogged, and did not split and disperse pollen. Therefore, it was preliminarily inferred that they might be a male sterile wild rice. After microscopic examination, it was confirmed that it was a wild rice with pollen abortion, and Yuan Longping immediately named it "wild abortion."

The discovery of the wild abortion added another sterile material to the study of the rice male sterility. The male sterile plants obtained by crossing wild abortion with several varieties could be inherited by 100%, and their descendants were male sterile plants in every generation. The discovery and transformation of wild abortion was a breakthrough for the matching of "three lines" and the successful research of the hybrid rice. Yuan Longping and his assistant took the lead in breeding the first practical rice male sterile line and its maintainer line 29Nan 1 A and B. In 1973, they realized the "three-line" matching, and bred the first strong and excellent combination "Nanyou 2." Following the Wild Abortive male sterile line, the breeding of the new cytoplasmic male sterile line also made gratifying progress. The Honglian male sterile line was bred in Hubei and the Ganggang male sterile line in Sichuan. However, statistical data showed that in the process of the hybrid rice breeding in China, the wild-rotten combinations accounted for about 90% of the total area of hybrid rice planted in the country because wild-rotten "three-line" combinations had strong advantages.

There are two experiences in successfully organizing research of the "three-line" hybrid rice. Firstly, great cooperation is the catalyst for the birth of hybrid rice and a powerful driving force.

In 1972, the study of the rice male sterility was listed as a major national research project in agriculture and forestry, co-chaired by the Chinese Academy of Agricultural and Forestry Sciences and Hunan Academy of Agricultural Sciences. With the increase of research institutes and personnel of rice male sterility, great collaboration of rice male sterility research in the whole province and the whole country became a trend.

In order to arouse the enthusiasm of scientists and technicians in many disciplines, such as plant genetics, plant physiology, ecology, breeding, cultivation, plant protection, soil, fertilizer, agrometeorology, and so on, agricultural units in Jiangxi, Guangdong, Guangxi, Fujian, Hubei, Anhui, Sichuan, Heilongjiang, Shanxi, and Shanghai and Xinjiang Production and Construction Corps sent researchers to Hunan successively. The Hunan Collaborative Group headed by Yuan Longping provided them with all the experimental materials unreservedly, and taught them the research methods of the rice "three lines" breeding. By 1972, when the study of rice male sterility was included in the major scientific research projects of agriculture and forestry in China, 19 provinces, municipalities directly under the Central Government and autonomous regions had carried out research work. On October 19 of the same year, the Chinese Academy of Agricultural and Forestry Sciences and the Hunan Academy of Agricultural Sciences co-chaired the first National Cooperative Conference on Hybrid Rice Research in Changsha, which marked the establishment of the National Cooperative Group.

From 1972 to 1982, the National Hybrid Rice Research Cooperative Group held nine collaborative meetings, showing the advantages of large-scale collaboration, which was conducive to overcoming difficult problems under the circumstances, clarifying the objectives of the next stage and the problems to be solved so as to speed up the research. The collaboration witnessed three difficult problems of Indica and Japonica hybrid rice, including "three lines" matching, strong superiority combination breeding, seed production technology, and the transformation of the "three lines" hybrid rice from scientific research to popularization, and the process of achieving brilliant achievements. The historical development showed that the reason why the hybrid rice develops so fast was that it not only organized the cooperation among units, but also developed the cooperation among disciplines and departments of scientific research, production, and promotion. It integrated various theories and viewpoints, opened up the train of thought of researchers, and promoted the research and development of the hybrid rice into a new stage of development in China. In 1976, the indica hybrid rice was widely popularized in China. The planting area of the hybrid rice in China was expanded to 207,000 mu. The yield of the hybrid rice increased by 20–30% compared with the conventional rice. In 1977, the area reached 31.95 million mu, 15 times that of 1976. After four years of development, the planting area of China reached 71.83 million mu in 1980, of which Hunan, Jiangxi, Sichuan, Jiangsu, and other provinces planted more than 10 million mu annually. Due to the improvement of the cultivation techniques and the continuous refreshment of high-yield records, the double-cropping rice area achieved "late rice super-early rice," and the yield of breeding and seed production also increased. A number of new sterile lines were selected and bred. Gratifying progress was made in basic theoretical research.

Secondly, Breeding in the South was not only a strategic decision, but also a major battle to promote the development of the hybrid rice. "Breeding in the South" was a new term appearing in the work of utilizing crop heterosis. It was another major measure to speed up agricultural scientific research and production under the tropical climate conditions in the southern part of the motherland based on the idea of "fighting for the day and night." Yuan Longping and his assistants once traveled north and south in Hunan, Yunnan, Hainan, Guangdong, or Guangxi, because these areas belong to the south subtropical zone, which provides superior natural conditions for rice breeding and accelerating the breeding process. Rice in these areas could be transferred to two or three generations in one year, greatly accelerating the breeding generation.

In order to strive for a nationwide promotion before 1976, the state leaders attached great importance to it in 1975. Hunan expanded breeding for four times, and formed a spectacular scene of large-scale seed production in Hainan. According to the limited conditions at that time, the Ministry of Agriculture and Forestry still provided 1.55 million yuan, of which 1.2 million yuan was the compensation for the transfer of seeds from Hunan, and 350,000 yuan was provided to Guangdong for the purchase of automobiles to transport the seeds to the south. In that year, more than 40,000 mu of the male sterile lines and more than 130,000 mu of seeds were produced in Hainan Island. The hybrid seeds produced by Southern propagation ensured the need of large-scale popularization of the hybrid rice in China.

The hybrid rice has been developed at an unprecedented rate in China in the history of world fine variety extension. By 2014, the hybrid rice had been popularized over 6 billion mu in China, and the total yield of the hybrid rice had increased by more than 60 billion kg. In recent years, the annual planting area of the hybrid rice has exceeded 240 million mu, and the annual output is about 2.5 million tons, which has made great contributions to solving the problem of the grain self-sufficiency in China.

32.3 Research and Development of the "Two-Line" Hybrid Rice

In early October 1973, Shi Mingsong, an agricultural technician in Mianyang County, Hubei, tried to find the male sterile plants in the field population of late Japonica rice. Three typical male sterile mutants were found in the field of Nongken 58, a late Japonica rice variety. This male sterile mutant was male sterile in summer and its pollen was abortive, but it became normal in autumn with its fertility recovered, becoming a photosensitive sterile type of rice. After six years of systematic experiments, he concluded that the late Japonica natural sterile plant bred from Nongken 58 had the fertility transformation characteristics under long light and short light.

Therefore, he published the paper "Preliminary Report on Breeding and Application of Late Japonica Natural Dual-purpose Lines" in the seventh issue of *Hubei Agricultural Sciences* in 1981, pointing out that the fertility-convertible rice could be used as a female parent in sterile period. Because of the dual-purpose line, self-bred male sterile line seeds could be used in Hybrid seed production in fertile period, and it was named dual-purpose line, namely Nongken 58S.

This new rice sterile resource provided a possibility for two-line method to utilize rice heterosis, opened up a new way for hybrid rice breeding and utilization, and became a new research topic for genetic breeding theory, which attracted attention both at home and abroad.

Since 1983, a collaborative group has been set up in Hubei, which includes Wuhan University, Huazhong Agricultural University, Hubei Academy of Agricultural Sciences, and Xiantao Photosensitive Genic Male Sterility Research Center. Through the collaborative group work, it was clear that the fertility of sterile materials discovered by Shi Mingsong was controlled by the length of sunshine, possibly by a pair of Recessive Genic Male Sterile main genes, and had nothing to do with the cytoplasm. In October 1985, at the identification meeting held in Wuyang, Hubei, the male sterile material was named "Hubei Photoperiod-sensitive Genic Male Sterile Rice" (shortened as HPGMR), which is of free allocation, wide recovery, and dual use.

The National Natural Science Foundation of China supported the discovery and research of the Photosensitive Genic Male Sterile Rice as a key and major project. The National Science and Technology Research Program and the 863 High-tech Development Program also funded this major research project. In 1986, according to the situation of the hybrid rice in the primary stage of development which has great potential for increasing yield and broad prospects for development, Yuan Longping put forward the strategy of hybrid rice breeding which was from the three-line method to the two-line method, and then through the two-line method to the one-line method, also known as from multiplication to simplicity but with higher efficiency in procedure. In terms of improving the level of heterosis, the strategy of hybrid rice breeding started from the utilization of inter-variety heterosis to the utilization of Inter-subspecific heterosis, and then to the utilization of distant heterosis between rice and other species, i.e., it was developing stronger heterosis. In 1987, Yuan Longping served as an expert in the national "863" project 1-01-101 on Two-line Hybrid Rice research, presiding over the cooperative tackling of key problems by 16 units throughout the country. The collaboration group organized experts to study the light–temperature conditions, genetic behavior of fertility, physiological and biochemical characteristics of pollen abortion, transfer effect of Photosensitive Genic Male Sterility and regional adaptability of Photosensitive Genic Male Sterility of Nongken 58S, a primitive male sterile line, and to breed a number of different types of Indica and japonica male sterile lines. For example, in August 1988, the first nuclear male sterile lines transformed from Nongken 58S and identified by the state were W6154S (Indica), N5047S (Japonica), 31111S (Japonica), and WD-IS (Japonica).

In the early stage of the study, many research institutes came to the conclusion that the fertility of Photoperiod-sensitive Male Sterile Rice was controlled by the length of sunshine, and the fertility transformation had nothing to do with temperature. After extensive and in-depth research by scientists and technicians, it was determined that this new type of nuclear sterility was controlled by recessive nuclear genes and had nothing to do with cytoplasm. Although this type of male sterility still belonged to the category of nuclear male sterility, it was different from the general type of nuclear male sterility, because its fertility expression was mainly regulated by light and temperature, i.e., in a certain period of development, under the conditions of high temperature and long sunshine, it showed male sterility; under the conditions of flat temperature and short sunshine, it returned to the characteristics of normal fertility transformation. It was a typical ecological genetic type called Photothermo-sensitive Genic Male Sterility (PTGMS) because it was controlled by both gene and phototemperature. During this period, more photothermo-sensitive genic male sterile materials were found, including the indica type: Annong S-1 discovered by Deng Huafeng, Anjiang Agricultural School of Hunan Province (1988); Hengnong S-1 discovered by Zhou Tingbo, Hengyang Institute of Agricultural Sciences of Hunan Province (1988); and 5460S discovered by Yang Rencui, Fujian Agricultural College (1989). Using this type of sterility to cultivate hybrid rice, we can use restorer lines to pollinate and produce the hybrid rice seeds when the sunshine time is long and the temperature is high in summer. In autumn or spring, when the temperature is low and the sunshine time is short, hybrid rice cultivation will be normal, and it can breed on its own in a sterile line irrespective of the maintainer line, thus eliminating the maintainer line, so it is called the "two-system method."

In 1989, the midsummer low temperature in China had a great impact on the fertility of the two-line male sterile lines. Many materials which had been identified as sterile became fertile, and the phenomenon of "swing" appeared, which resulted in serious setbacks in the study of the two-line method. The collaboration group once again made good use of the advantages of collaboration in tackling the key problems, using new materials of the photothermo-sensitive male sterile rice, and organized a number of national superior units to jointly tackle the key problems. The expert group led by Yuan Longping explored a set of operable implementation plans for the two-line breeding and guided breakthroughs in key technologies, including.

(1) The basic rules of fertility transformation between the photoperiod-thermo-sensitive male sterile lines and the relationship between light and temperature were revealed; (2) a set of technical schemes and systems for breeding practical photoperiod-thermo-sensitive male sterile lines were summarized; (3) a set of unique purification and seed production procedures for photoperiod-thermo-sensitive male sterile lines, which could keep the critical temperature relatively stable all the time, were designed; and (4) the technical strategies and techniques for selecting strong and excellent combinations among subspecies were put forward. At the same time, the strategy of combining photothermo-sensitive genic male sterile gene with the wide compatibility gene was adopted to further improve the yield of the Hybrid Rice by utilizing interspecific heterosis. Following the technical strategy of breeding the male sterile lines, many research institutes in China successively bred a number of

practical photothermo-sensitive male sterile lines and a large number of the high-yield and high-quality two-line hybrid rice combinations, which provided technical guarantee for the large-scale application of the Two-line Hybrid Rice and made it possible for the scientific research achievements to rapidly translate into practical productivity.

The differences between the three-line method and the two-line method are as follows:

(1) There are three parents involved in the three-line method (cytoplasmic-nuclear interaction male sterile line, the male sterile maintainer line, and the restorer line), and two parents in the two-line method (photothermo-sensitive male sterile line and restorer line of rice).
(2) The relationship among parents is different. There is a restoring relationship among the three parents in the three-line method, but there is no restoring relationship between the two parents in the two-line method, so they can be matched freely.
(3) The reproduction methods of CMS are different. The three-line method requires the corresponding maintainer line to propagate, and the two-line method requires self-fertilization under certain temperature and light conditions.
(4) The selection of the seed production base and season is different. The selection scope of the three-line method is wide as only one safe period (safe period of heading and flowering) is needed to be considered; and the selection scope of the two-line method is narrow as two safe periods (sensitive period of fertility and safe period of heading and flowering) are needed to be considered.
(5) The difference of purity preservation technology is that the two-line method is based on the three-line method, and has its own unique technology—fertility stabilization technology and fertility monitoring technology.

In 1995, the two-line hybrid rice was successful. The yield of the two-line hybrid rice was 5–10% higher than that of the three-line hybrid rice at the same maturity stage. In recent years, the planting area was about 60 million mu (Fig. 32.1). By making full use of all kinds of the high-quality rice germplasm resources, breeding a number of two-line high-yielding combinations that meet the national standard of high-quality rice, the rice quality is generally better, and we can better solve the difficult problem of coordination between the high-yielding and high-quality hybrid rice. Two-line hybrid rice is originally created in China. Its success is a major breakthrough in crop breeding. It is not only a world-leading scientific and technological achievement after the three-line hybrid rice in China, but also a new method to ensure food security in China and the world. It promotes the research and application of the two-line heterosis utilization of rape, sorghum, cotton, maize, wheat, and other crops in China, providing especially a new way for crops difficult to realize the use of three-line heterosis.

Fig. 32.1 Hybrid rice seed production in large scale

32.4 Research and Development of the Super Hybrid Rice

Breeding super-high yield rice is a key project of research institutes in many countries in recent 30 years. Japan took the lead in developing rice of Super-high Yield in 1981. It planned to increase the rice yield by 50% in 15 years, that is, from 410–520 kg per mu to 620–800 kg per mu at that time. During the eight years from 1981 to 1988, five varieties were bred, but most of them had problems in cold resistance, quality, and seed setting rate. By now, due to the inadequate technical route, research work in Japan has been in a dilemma. In 1989, the International Rice Research Institute proposed a super rice breeding plan: by 2000, the yield potential of the rice should be increased by 20–25%; the existing high-yield varieties with 10 tons potential per hectare should be increased to 12–12.5 tons per hectare, i.e., from 660 kg per hectare to 800–830 kg per hectare. In 1994, they announced that the breeding of new super rice varieties with new plant type and special germplasm resources had been successful. The yield of some strains in small area (300 m^2) test had exceeded 20–30% of the existing varieties. However, because of their resistance to brown planthopper, they could not be widely applied. The research of super rice breeding in IRRI was facing the situation of high index, difficulty, and confinement to morphological improvement. It was necessary to further improve its technical route and postpone the realization of its plan.

China is facing the severe pressure of population growth and the reduction of cultivated land. Its population is expected to increase to 1.6 billion by 2030 and the per capita arable land will decrease to 0.07 hectares. In order to ensure food security in China, in 1996, the Ministry of Agriculture of China and the Foundation for Agricultural Science and Education of China launched the project of "China Super Rice Breeding and Cultivation System," i.e., "China Super Rice Breeding Program."

The yield index of the hybrid rice was 700 kg/mu (10.5 t/hm^2) in the first stage (1996–2000) (demonstration of more than two 100 mu in the same ecological area). The second stage (2001–2005) was 800 kg/mu (12 t/hm^2). (In the same ecological area, two demonstration plots with more than 100 mu were produced for two consecutive years.

Yuan Longping believed that up to now, there have been only two effective ways to improve crop yield through breeding: morphological improvement and utilization of heterosis. Simple morphological improvement has limited potential; if heterosis is not combined with morphological improvement, the effect will be poor. Other approaches and techniques, including molecular breeding, must ultimately be applied to good morphology and strong heterosis in order to achieve good results. On the other hand, the higher-level development of crop breeding depends on the progress of modern biotechnology.

In 1997, Yuan Longping put forward the breeding program of Super Hybrid Rice in China, which should adopt a comprehensive technical route aiming at improving photosynthetic efficiency, combining morphological improvement with utilization of interspecific heterosis, supplemented by molecular means, and summarized the breeding theory and methods of the study of "Chinese super hybrid rice" from practice, including (1) In terms of morphological improvement, the super-high yield plant type model of "high canopy, low panicle, medium-large panicle and high lodging resistance," was proposed, requiring the upper three functional leaves to be long, straight, narrow, concave, and thick. The slender and straight leaves not only had larger leaf area, but also could be exposed to light on both sides without shading each other. Narrow and slightly concave leaves occupied smaller space area and had higher leaf area index. At the same time, concave shape could make the leaves stronger and more durable; thicker leaves had higher photosynthetic efficiency and were not easy to premature senescence. In short, the high canopy, i.e., the canopy height was more than 1.2 m; that is to say, the low spike layer, the spike tip was only 60–70 cm away from the ground at maturity, in order to reduce the center of gravity and high resistance to lodging. Only rice varieties with these morphological characteristics could have the maximum effective leaf area index and photosynthetic function, and provided sufficient photosynthate, namely organic source, for super-high yield. (2) Utilizing the interspecific Heterosis of Indica and japonica, the yield potential of Indica and japonica hybrid rice was more than 30% higher than that of the hybrid rice currently used in production. However, it was very difficult to utilize the Heterosis of Indica and japonica. The main reason was that indica and japonica were different subspecies, and their genetic relationship was farther away. There was incompatibility between them, resulting in abnormal fertilization and Fruiting of Indica and japonica hybrids. Generally, their fruits were only about 30%. According to the hypothesis of incompatibility between Indica and japonica revealed by H. Ikehashi, a Japanese scholar, and the "wide compatibility phenomenon" of low seed setting rate of hybrid rice, as well as the wide compatibility genes found in Japan rice varieties, the F1 generation was normally fruited by crossing some intermediate rice materials with the indica and japonica varieties. In this way, a breakthrough was made in overcoming the difficulty of the low seed setting rate of the Indica and japonica subspecies hybrid rice. At present, the Indica and Japonica hybrid materials have been used as parents. One of them has a wide compatibility gene and has been matched. Some of them have successfully solved the problem of the low seed setting rate of Indica and Japonica with strong heterosis.

The establishment of the theoretical and applied technical system of the Two-line Hybrid Rice promotes the use of the two-line method to breed super hybrid rice, which is not restricted by cytoplasm and restoring–preserving relationship, and has free mating, which can widely utilize the technical advantages of excellent traits of genetic resources aggregation. In order to establish the ideal plant type model of the super hybrid rice, morphological improvement and utilization of heterosis between the indica and japonica subspecies were adopted to breed a group of Two-line Super Hybrid rice. The first-class hybrid rice combinations provided technical support and solved the problem of long cycle and low efficiency in breeding super hybrid rice by the three-line method. At the same time, a number of super hybrid early rice combinations were bred by using various early-maturing and high-yielding rice germplasm resources, which provided key technical support for solving the long-standing problem of "early-maturing without high-yielding, high-yielding but not early-maturing" in yield and growth period of hybrid early rice in the middle and lower reaches of the Yangtze River.

Under the leadership of Yuan Longping, the scientific research team was unremittingly pursuing the super-high realm of hybrid rice. Luo Xiao and researcher of Hunan Hybrid Rice Research Center provided female parent and cooperated with Zou Jiangshi of Jiangsu Academy of Agricultural Sciences to develop pioneer combinations of the super hybrid rice such as Liangyoupeijiu and P64S/E32. In 1999, Liangyoupeijiu produced more than 700 kg per mu in Hunan alone, and in 2000, 16 100-mu and four 1000-mu demonstration patches reached the standard, and successfully achieved the target of Phase 1 Super Rice. In 1999, P64S/E32 set a record yield of 1,139 kg per mu in Yongsheng County, Yunnan. Because of its high yield and good quality, the planting area of Super Hybrid Rice expanded rapidly year by year. In recent years, the planting area of Super Hybrid Rice has approached 30 million mu. Among them, Liangyoupeijiu has grown about 20 million mu. It has become the largest hybrid rice in China. The average yield of Super Hybrid Rice has reached 550 kg per mu, which is 50 kg higher than that of the general high-yielding hybrid rice.

In 2001, Yuan Longping's scientific research team further carried out the second phase of super hybrid rice breeding (Fig. 32.2). Through the arduous efforts of the scientific researchers of the collaboration group, the new combination P88S/0293 of Hunan Hybrid Rice Research Center was demonstrated. The yield per mu of a 127-mu demonstration plot in Longshan County, Hunan, reached 817 kg, reaching the goal of 800 kg per mu in the second stage of the super rice for the first time. In 2003, five 100-mu demonstration sites in Hunan Xiangtan, Longhui, Guidong, Rucheng, and Zhongfang counties continued to demonstrate the new super rice combination 88S/0293 and Junliangyou 527, respectively. The acceptance yield reached 807.4 kg, 801.8 kg, 813.5 kg, 802 kg, and 808.9 kg, respectively. Among them, the 100-mu demonstration plot in Hunan Quantangzi Township was checked and accepted on site

Fig. 32.2 Yuan Longping observing super hybrid rice (provided by Xin Yeyun)

by the famous rice experts organized by the Ministry of Agriculture in our country. At the same time, a 100-mu demonstration plot in Fujian met the standard. In 2004, 30 100-mu and two 1000-mu patches were arranged in seven Southern provinces. Experience showed that the yield of 12 100-mu patches per mu was over 800 kg. Among them, three 100-mu plots were up to the standard for two consecutive years. The second phase of super rice in China was one year ahead of schedule, and was first achieved in Hunan in 2004. The yield per mu of super rice in China was about 50 kg higher than that in the first stage.

Based on the achievements and progress of the first and second phases of the Super Hybrid Rice research, in 2006, Yuan Longping proposed to the Ministry of Agriculture to carry out the third phase. In accordance with the principles of good varieties, good methods, good fields, and good ecological conditions put forward by Yuan Longping, the R&D team of the Super Hybrid Rice continuously explored the technology of new super hybrid rice varieties matching with the super-high yield cultivation and ecological environment. From 2011 to 2012, the key technologies of implementing good varieties and good methods were highlighted in order to select key experimental sites suitable for developing the super-high yield potential of varieties, as well as comprehensive land consolidation and soil plots. The comprehensive research and application of the Super-high Yield Cultivation techniques, such as force cultivation, suitable season planting, and application of special fertilizer for super hybrid rice, achieved remarkable results. In 2011, the average yield of "Y Liangyou 2" bred by researcher Deng Qiyun set a record high of 926.6 kg per mu in 108-mu demonstration area of Longhui County, Hunan. In 2012, overcoming adverse climatic factors, the yield of "Y Liangyou 08" bred by researcher Zhang Zhenhua reached 917.72 kg per mu in 103-mu demonstration area of Xupu County, Hunan. In the same year, Yongyou 12, developed by Ma Rongrong team of Ningbo Academy of Agricultural Sciences, also set a record of 963.65 kg per mu of demonstration plots, which marked the realization of the third phase of Super Hybrid Rice in China.

In 2013, the Ministry of Agriculture officially announced the establishment and launch of the fourth super hybrid rice breeding program after achieving the third phase super rice target. Yuan Longping, who was still not satisfied, boldly assumed that by increasing plant height and utilizing strong interspecific heterosis, a new type of Super-high Yield Combination with high lodging resistance could be developed. The first battle of the guiding key team was successful. On September 28, 2013, a 100-mu demonstration plot of YLiangyou 900, a pioneer combination of middle-season super hybrid rice, was selected by researcher Deng Qiyun in Niuxiang Village, Yangguao Township, Longhui County, Hunan. It was checked by a group of experts organized by the Ministry of Agriculture and its yield reached a new record of 988.1 kg per mu. In 2014, 28 counties and municipalities in 13 provinces and cities in the country demonstrated and tackled key problems at the same time, and implemented the supporting technology of "good varieties, good methods, good fields, and good state." On October 10, the Ministry of Agriculture organized experts to check the "Y Liangyou 900" 100-mu demonstration plot in Hongxing Village, Hengbanqiao Township, Xupu County, Hunan. As a result, the average yield per mu reached a new record of 1,026.7 kg, and a major breakthrough in tackling the fourth phase of the super rice was achieved.

Considering the solar energy utilization rate of rice, that is, the ratio of the energy of organic matter accumulated by photosynthesis to the solar energy irradiated on the ground, the maximum solar energy utilization rate of rice can reach about 5%, according to the estimates of Chinese scholars. If the calculation is based on 2.5% of the energy utilization rate, taking the solar radiation in Changsha area of Hunan as an example, the yield of early rice per mu can reach 1,000 kg, and the late rice yield can reach 1,000 kg. The yield of rice per mu can reach 1,100 kg, and that of medium-sized rice per mu can reach 1500 kg. In the future, the rice yield per mu still has the potential of 500 kg to be attained by scientific researchers. Just as the Chinese saying goes—an old steed in the stable still aspires to gallop a thousand Li-an old hero still cherishes high aspirations, Yuan Longping has launched the fifth phase of the super rice research plan with the target of 16 tons per hectare (1,067 kg/mu). He further suggested that new combinations with good morphology, moderate tightness, strong tillering capacity, and little difference between the main panicle and the tiller panicle should be developed from plant type improvement. In order to solve the high lodging resistance of Super-high-yield and high-stalk varieties, the use of strong interspecific combinations makes the root system very developed; Rice Panicles should droop so as to reduce the center of gravity, taking advantages of the rice seed resources with very hard stalks, short and thick basal internodes, and light foot. In this way, it is possible to select super-high yield combinations with high resistance to toppling. Current progress shows that the five 100-mu plots of the newly bred new super hybrid rice combination "Chaoyou 1000" planted in 35 key 100-mu plots in 2015, yielded more than 1,000 kg per mu. Among them, the average yield of the 100-mu plots planted in old ones in Yunnan reached 106.5 kg per mu, breaking the fifth yield target of 16 tons per hectare for the first time, and creating an average yield per hectare in large areas. In 2016, 81 key 100-mu plots were planted, of which Gejiu key 100-mu plots in Yunnan was commissioned by the Ministry of Science and Technology to organize production testing on September 14, with an average yield of 1,088 kg per mu, breaking through 16 tons per hectare again, and setting a new world record for the average yield per hectare of large-area planting.

Yuan Longping once said that he had two dreams: the first was to enjoy the cool under the standing grain; the second was to make the hybrid rice reach every corner of the world. In the first dream, he dreamed of the hybrid rice growing in experimental fields, taller than sorghum, spikes as long as brooms and grains as big as peanuts. So he was happy to sit in the cool with his colleagues and assistants under the waterfall-like rice ears. According to Yuan Longping's latest design of cultivating new super-high-yield and high-stalk varieties through improved plant type, the height of future super hybrid rice will be more than 2 m, so the dream of "enjoying the cool under the standing grain" will come true.

In the future, combining with the conventional breeding, molecular technology should be applied to tackle key problems and achieve higher goals of the super hybrid rice.

In order to put into use the achievements in hybrid rice research, since 2006, the "Sow Three, Yield Four" and "Three One" projects have been implemented successively, which have set an example for speeding up the application of new super hybrid rice achievements. The "Sow Three, Yield Four" is a project of using the technical achievements of the super hybrid rice to produce the total grain yield of four-mu land with three-mu land. The goal of Hunan Province was to strive to develop the planting area to 15 million mu by 2017, with an output equivalent to the total grain output of 20 million mu in 2005 and an increase of 5 million mu of farmland. The project has always been highly valued and supported by the Hunan Provincial Committee and the provincial government. It was first launched in 20 counties of Hunan Province in 2007 and had been implemented in 52 counties (cities and districts) by 2015. The demonstration area reached 1.196 million mu, and the total yield of rice increased to 1.386 billion kg. At present, Anhui, Henan, Guangdong, Guangxi, Yunnan, Guizhou, and other provinces have also actively followed up the implementation of the "Sow Three, Yield Four" project. It is expected that by 2020 the planting area in the whole country will have increased to 60 million mu and produced 80 million mu of grain, which is equal to increasing 20 million mu of farmland. Through the application of the Super Hybrid Rice technology, the "Three and One" project aims to produce 360 kg of grain per year in a planting plot of a third of mu, which is enough for a person's annual ration. The goal of Hunan Province is to develop the planting area to 5 million mu (about 9% of the cultivated land) by 2020 to produce food that can feed 24% of the population of Hunan Province. The "Three and One" project has been tested and demonstrated in Hunan, Guangxi, and Guangdong. In 2015, three annual production modes, i.e., double-cropping super hybrid rice, potato + one-season super hybrid rice, spring maize + super hybrid late rice, were adopted according to the local conditions. Fourteen of the sixteen counties (cities and districts) participating in the implementation of the project achieved the target yield with remarkable results, and will play a great exemplary role.

32.5 Hybrid Rice's Going Global

Yuan Longping always regards "developing hybrid rice for the benefit of the people of the world" as the greatest pursuit of his lifelong struggle. His wish is to realize the dream of hybrid rice reaching to every corner of the world.

In March 1980, Wilhelm, general manager of Yuanhuan Seed Company of the United States, came to China and initially signed a 20-year "Comprehensive Technology Transfer Contract for the Hybrid Rice" with China Seed Company in Beijing, which was formally approved by the State Import and Export Administration Commission, thus opening the prelude of China's hybrid rice to the world.

In order to make the hybrid rice benefit more countries in the world, Chinese agricultural scientists and technicians headed by Yuan Longping have carried out over the years, cooperative research and technical exchanges with the International Rice Research Institute, and taught hybrid rice technology to India, Vietnam, Myanmar, the Philippines, Bangladesh, and other countries. At the same time, more than 70 international training courses on hybrid rice have been held since the 1980s, amounting to 80. More than 3,000 government officials and agrotechnical experts have been trained in several countries. After returning home, these personnel acted as the backbone of local research and promotion of hybrid rice. In addition, in the 1990s, the Food and Agriculture Organization of the United Nations (FAO) adopted the promotion of the hybrid rice as the preferred measure to solve the food shortage problem in developing countries. Yuan Longping was appointed as the chief international consultant. This measure promoted the introduction and research of the hybrid rice in more than 40 countries and regions in Southeast Asia, South Asia, South America, Africa, and North America, and achieved remarkable yield-increasing benefits.

The hybrid rice has attracted more and more attention in the world, especially in Asia, Africa, and other countries. Under Yuan Longping's advocacy and suggestion, "Hybrid Rice Diplomacy" is becoming an important part of China's "Going Global" strategy, which has been highly valued by relevant departments. At the same time, it has become an important symbol of China's scientific development, peaceful rise, and the display of sense of responsibility and harmony of a great power to the world.

As the main crop, rice is widely cultivated in 120 countries and regions in the world. More than half of the world's population lives on rice, but the average yield of rice per mu is only about 200 kg. The hybrid rice technology enjoys the reputation of "Oriental Magic Rice." By 2013, the hybrid rice had been introduced and popularized in more than 40 countries around the world, which opened up an effective way for rice-producing countries to increase their grain production. India, Vietnam, Bangladesh, Indonesia, the Philippines, the United States, and Brazil had become countries that grow hybrid rice on a large scale.

In 2015, their planting area totaled 6 million hectares, and the average yield per hectare was about 2 tons higher than that of the local fine varieties. Take the Philippines as an example; if it imports hybrid rice, whose yield is 2 tons per hectare more than that of the regular rice, and plans to develop the planting area to 1 million hectares, it will basically solve the long-standing problem of food shortage. In March 2016, six countries in the Lancang-Mekong Region held a cooperation meeting in Sanya, China, and took the development of the hybrid rice as the preferred cooperation project. As Robert Ziegler, the then director of the International Rice Research Institute, said, "We need technologies to improve rice production, and hybrid rice is one of the most needed."

Facing the fact that 852 million people in the world still suffer from frequent starvation and 50,000 children die of hunger and malnutrition every year, the hybrid rice still has a long way to go in order to play an active role in ensuring world food security and maintaining the world peace. The rice planting area in the world is 2.25 billion mu. According to the statistics of FAO, if 10% or about 200 million mu of hybrid rice is planted, the increased grain yield will account for 20% of the world's total rice production. If the planting area of the hybrid rice accounts for about 50% of the world's total rice area (150 million hectares), the world's annual grain yield increase can feed 400 million to 500 million people. In this way, there is great hope of eliminating hunger, which is Yuan Longping's greatest wish.

Yuan Longping devoted his whole life to the scientific research of hybrid rice. He persisted in pursuing and climbing the peak for more than 50 years. The scientific research achievements made by Yuan Longping have made China a world leader in hybrid rice and super hybrid rice breeding. The promotion of the hybrid rice has not only solved the problem of food self-sufficiency in China, but also made outstanding contributions to the world food security. He has won more than 20 national and international awards, including the National Prize for Invention, the first National Highest Science and Technology Awards, the Special Award for National Science and Technology Progress Awards, the UNESCO Prize for Science, and the United States Prize for World Food. But he did not stop his work because of these honors; he is still innovating and working hard for the further development of the hybrid rice. As Don Palberg, a famous American agricultural economist, said, Yuan (Longping) is leading us to a well-fed and well-clothed world.

Yeyun Xin Born in Linli, Hunan in 1966, and is now the researcher of National Hybrid Rice Engineering and Technology Research Center and Director of Academician's Office.

Research interests: Study on the superiority mechanism and utilization of rice hybridization; Research and development of hybrid rice technology.

Major works: *Oral Autobiography of Yuan Longping*;

Hybrid Rice Technology Development—Ensuring China's Food Security;

Development of Hybrid Rice Technology—Guarantee the Food Safety in China;

Heterosis markers and QTL analysis for yield of super hybrid rice liangyoupeijiu.

Awards: "The Breeding and Application of Super Hybrid Rice Liangyupeijiu Yield by Using Molecular" won the First Prize for Science and Technology Progress in Jiangsu Province 2003;

Oral Autobiography of Yuan Longping was selected into the third "Three One Hundred" original book publishing project of the General Administration of Press and Publication.

The manufacturer's authorised representative in the EU is Springer Nature Customer Service Centre GmbH, Europaplatz 3, 69115 Heidelberg, Germany. If you have any concerns regarding our products, please contact ProductSafety@springernature.com

Printed and bound by CPI Group (UK) Ltd, Croydon, CR0 4YY
30/09/2026
02225288-0020